Algorithms and Computation in Mathematics • Volume 18

Editors

Arjeh M. Cohen   Henri Cohen
David Eisenbud   Michael F. Singer
Bernd Sturmfels

Anton Betten · Michael Braun
Harald Fripertinger · Adalbert Kerber
Axel Kohnert · Alfred Wassermann

Error-Correcting Linear Codes

Classification by Isometry and Applications

With 51 Figures and 102 Tables

 Springer

Anton Betten
Department of Mathematics
Colorado State University
Fort Collins, CO 80523
USA
betten@math.colostate.edu

Michael Braun
Siemens
CT IC 3
D-81739 München
Germany
mic.braun@siemens.com

Harald Fripertinger
Institut für Mathematik
Karl-Franzens-Universität Graz
A-8010 Graz
Austria
harald.fripertinger@uni-graz.at

Adalbert Kerber
Axel Kohnert
Alfred Wassermann
Mathematisches Institut
Universität Bayreuth
D-95440 Bayreuth
Germany
kerber@uni-bayreuth.de
axel.kohnert@uni-bayreuth.de
alfred.wassermann@uni-bayreuth.de

Library of Congress Control Number: 2006929536

Mathematics Subject Classification (2000): 05E, 51E, 94B, 05B25, 05E20, 11D04

ISSN 1431-1550

ISBN-10 3-540-28371-4 Springer Berlin Heidelberg New York
ISBN-13 978-3-540-28371-3 Springer Berlin Heidelberg New York

Springer is a part of Springer Science+Business Media

springer.com

© Springer-Verlag Berlin Heidelberg 2006

Typeset by the authors using a Springer LaTeX macro package
Production: LE-TeX Jelonek, Schmidt & Vöckler GbR, Leipzig
Cover design: design & production GmbH, Heidelberg

Printed on acid-free paper   46/3100YL - 5 4 3 2 1 0

Preface

The fascinating theory of error-correcting codes is a rather new addition to the list of mathematical disciplines. It grew out of the need to communicate information electronically, and is currently no more than 60 years old. Being an applied discipline by definition, a surprisingly large number of pure mathematical areas tie into Coding Theory. If one were to name just the most important connections, one would start of course with Linear Algebra, then list Algebra and Combinatorics, and further mention Number Theory and Geometry as well as Algebraic Geometry.

Being a thorough introduction to the field, this book starts from the very beginning, which is the channel model of communication in the presence of noise. From there, we develop the fundamental concepts of error-correcting codes, like the Hamming metric and the maximum likelihood decoding principle. After discussing dual codes and simple decoding procedures, this book takes an unusual turn. The standard approach would be to move on from there and introduce either more theory or present standard constructions of codes. The approach taken here is different.

After raising the question of what it means for codes to be "essentially different" we consider the metric Hamming space together with its isometries, which are the maps preserving the metric structure. This in turn will lead to a rigorous definition of equivalence of codes. In fact, we will call codes isometric if they are equivalent as subspaces of the Hamming space. After that, the discussion shifts to a more abstract analysis of the different kinds of isometries. This is laying out the more general theme behind this book. In essence, this book serves two purposes. On the one hand, the book introduces the fundamentals of the theory of error correcting codes like parameters, bounds as well as known classes of codes including the important class of cyclic codes (Chapters 1–4). Also included is a decent introduction to the theory of finite fields (Chapter 3). Moreover the application of coding theory to CD-players is discussed in detail. On the other hand, the second part of the book covers more advanced and specialized topics which so far have not yet made it into the standard textbooks in the area.

Chapters 1–4 are the core of everyone's understanding of the theory of error correcting codes. Chapter 1 discusses basic concepts, including isometry, weight enumerators, systematic encoding, and a section on the explicit computation of the minimum distance of a code. Chapter 2 is also classic. We discuss bounds on the parameters of codes. This involves both direct combinatorial bounds and also bounds which are obtained from modifications of codes. Particular series of codes are introduced, like the general form of the

Hamming-codes, the extended binary Golay-code of length 24, the class of Reed–Muller-codes as well as a general discussion of MDS-codes.

Chapter 3 is devoted to the theory of finite fields. No book on linear codes can do without such a chapter. Usually, this discussion is placed somewhat after the general introduction of codes and before things get more involved. We have tried to keep the theory of finite fields together in one single chapter, at perhaps the expense of keeping the reader waiting longer than usual. We hope that this decision pays off in that the body of the theory of finite fields can be presented in a single entity. We start with algebraic elements, discuss minimal polynomials, finite field extensions and their automorphisms, i.e. the Galois group. The discussion moves on to finite group actions, and their applications to field theory. Using the notion of Lyndon words, this allows the construction of irreducible polynomials and hence the construction of finite fields of any given order. We discuss two distinct ways of representing field elements on a computer, and the final section is devoted to a brief introduction of the projective geometry over a finite field.

Chapter 4 is dedicated to cyclic codes. The reader will learn about the very important classes of Reed–Solomon and BCH-codes. The close relationship between cyclic codes and ideals in the group algebra is discussed. Furthermore, we discuss quadratic-residue-codes, Golay-codes, idempotent generators and the Fourier Transform, Alternant-codes and Goppa-codes, the structure of cyclic codes as modules, general Reed–Muller-codes, and encoding and decoding issues. It should be noted that Alternant-codes and Goppa-codes are not in general cyclic. The reason for discussing them in the context of cyclic codes is that the algebraic tool of the Chinese Remainder Theorem makes it easy to treat these codes in that context.

Chapter 5 introduces the reader to the application of Coding Theory in connection with the compact disc, a technique which was developed in the early 1980's by the Philips and Sony companies. The reader will see decoding algorithms for BCH-codes, interleaving methods, product codes and an introduction to Fourier analysis and Shannon's Sampling Theorem.

In Chapters 6–9, we start the second part of the above-mentioned division. Here, we come back to the fundamental question "To what extent do good codes exist and how can we find them?" This is of course the fundamental problem of Coding Theory, and in a sense mostly anything studied so far is concerned with certain aspects of this problem. Nevertheless, to answer this question qualitatively, we must go beyond the scope of standard texts. The first step to tackle this problem was made by David Slepian in the 1960's, when he pioneered the application of techniques from Combinatorics to this problem. In fact, he used a technique which is known as Pólya's Theory of Enumeration to the problem of determining the number of isometry classes of codes. In this way, he was able to determine how many classes there are for any given set of parameters. The method involves a fairly detailed study of the way isometries act on the Hamming space. His delicate and powerful computations are brought up again here, and they are refined and adapted to match all different types of isometries we consider. The presentation in Chapter 6 will introduce the reader to this very versatile topic of Combinatorics. At the end, numbers of isometry classes of codes will be presented. This chapter also features sections on random generation of codes, the notion of critical indecomposable codes as introduced by E. Assmus, and a section on the explicit construction of normal bases of finite fields.

After the enumeration of codes, we move on to the construction of representatives of the isometry classes. As a matter of fact, enumeration theory does not tell us about the minimum distance of the codes. For this, we have to construct codes explicitly. There are essentially two different strategies. Both rely on the close connection between codes and geometry, more precisely configurations in finite projective spaces. Each of the two methods allows the restriction to "good" codes, i.e. codes with high minimum distance. This is facilitated by specifying a lower bound on the minimum distance and then constructing only those codes whose minimum distance is bounded below by the given value. In the extreme case, the algorithm would prove that there are no codes with the desired minimum distance.

The first approach (Chapter 8) makes yet another assumption, namely on the presence of symmetries, or – as we shall call them – automorphisms. This is a method which has had its successes recently in other areas, like the theory of combinatorial designs, and it proves to be powerful in that objects can be constructed which would otherwise be out of reach. In fact, the method of lattice actions combined with a construction of configurations called minihypers allows the construction of good codes with a preassigned group of automorphisms. This chapter is based on results of Chapter 7, which discusses lattice methods. A lattice is a set of vectors which are integer linear combinations of a given set of linearly independent vectors in a finite dimensional vector space. These structures are studied in Number Theory. Here, we use lattices to solve integer equations, also known as Diophantine linear equations. Finding the integral solutions of such systems of equations is known to be a very hard problem, since there is no discrete analogue of Gaussian elimination. To solve these systems, combinatorial techniques like lattice basis reduction are applied, combined with enumeration techniques to search through lattices for "short" vectors. As it turns out, the construction problem of codes with preassigned group of automorphisms can be reduced to solving such a system of integer equations, so Chapters 7 and 8 may be considered as a sequence. Several hundred new optimal codes could be constructed with this method which in essence relies on an enormous data reduction because of the group action. That is, the assumption on the existence of nontrivial automorphisms is essential for reducing the size of the problem. On the other hand, the general construction problem (i.e. without making any assumption on the presence of automorphisms) is another topic.

This is where the second method comes in. This time, there will be no further assumption other than the lower bound on the minimum distance. To tackle the "general case" (Chapter 9), a full search on all codes is facilitated, using isomorph rejection in order to construct each isometry class of codes exactly once. This technique searches through the set of all possible codes according to the lexicographical ordering, and is therefore known as orderly generation. In fact, the technique was developed in the 1970's for the construction of graphs, and has since been refined and applied to a plethora of different problems. It was only a matter of time that this technique would find its way to the construction problem of codes. This book will finish with a brief account on the orderly generation of linear codes with a prescribed minimum distance. This involves a fair amount of algorithmic background for dealing with permutation groups. We will present the reader with essential concepts of how to work with permutation groups on a computer and how to solve orbit type problems. We describe in detail the theoretical aspects of dealing with the projective linear and semilinear groups. In the end, we give tables classifying the optimal linear codes for small or moderate parameters over various finite fields.

Chapter A, the appendix, contains an introduction to the attached compact disc. It describes the installation of the software in both a Windows and a Linux environment. It also gives a survey on the accompanying data. The most recently updated version of the programs should be found at the website

http://linearcodes.uni-bayreuth.de

The included software allows one to compute the minimum distance and the weight distribution of given codes, construct codes with a given minimum distance and randomly generate linear codes which are uniformly distributed over the isometry classes of codes with given parameters. The dynamic tables describe the isometry classes of linear codes. In the precomputed tables the reader will find enumerative results on numbers of semilinear isometry classes. Moreover, there are tables containing information on optimal linear codes. The largest possible minimum distance is given, together with the number of semilinear isometry classes of such optimal codes. In addition, corresponding generator matrices can be found. Altogether, around 2 million isometry classes of codes have been computed, of which more than 800 000 are optimal codes. Nearly 200 000 generator matrices can be found on the attached compact disc, of which more than 70 000 generate optimal codes. The complete set of computed generator matrices can be downloaded from the web-site mentioned above. These codes are all pairwise inequivalent. More precisely, they are representatives of different semilinear isometry classes.

On the side of the reader we assume only a basic knowledge of Linear Algebra and Algebra. Many fundamental notions are reviewed in the text. Readers with a background in field theory may skip Chapter 3. We should also mention what this book for one reason or the other does not cover. For instance, we do not discuss algebraic geometric codes, in particular the generalized version of Goppa-codes. Also not included are convolutional codes, Turbo codes, LDPC codes, codes over rings, e.g. $\mathbb{Z}_4$, and decoding methods using Tanner-graphs. All this, as well as the connection between codes and designs and a deeper account on the theory of self-dual codes had to be left out. To this end, we refer the interested reader to the excellent literature on these topics, for instance the recent books by Pless and Huffman [94], Moon [153], Nebe, Rains and Sloane [157]. The "classic" for nearly 30 years, the book by MacWilliams and Sloane [139] from 1977 is still astonishingly comprehensive. The connections between codes and designs are described in the book by Assmus and Key [5]. The Handbook of Coding Theory [163], edited by Pless and Huffman, has articles on many of these topics written by experts in the field.

The authors wish to express their sincerest thanks in particular to Karl-Heinz Zimmermann, coauthor of the German precursor book "Codierungs-theorie". We also thank our colleague Reinhard Laue, to whom we owe great thanks for many stimulating discussions on the constructive theory of finite structures using group actions. Thanks are also due, of course, to our students and to the scientific community, in particular to Andries Brouwer, who maintains a very helpful table on parameters of optimal linear codes, as well as to the editors of the important two volumes of the Handbook of Coding Theory, Vera S. Pless and W. Cary Huffman. Furthermore, we greatly appreciate receiving helpful comments and suggestions from Georg Wolfgang Desch, Evi Haberberger, Oscar Jenkins, Steve Linton, Rebecca Lynn, Eric Moorhouse, Eammon O'Brien, Tim Penttila, Jens Schwaiger, Ákos Seress, Justyna Sikorska and Johannes Zwanzger.

We acknowledge financial support by the *Deutsche Forschungsgemeinschaft* and the *Österreichischer Fonds zur Förderung der wissenschaftlichen Forschung* for helpful financial support of several research projects on these topics. These projects contributed very much to the development of the theory and to the implementation of corresponding software, as well as to the collection of data which are now available for the interested reader via email, Internet and the attached compact disc.

Last but not least, we should like to thank Ruth Allewelt, Martin Peters and Thomas Wurm from the Springer publishing company for their patient and diligent handling of this book project.

Fort Collins, Munich, Graz, Bayreuth
July 4, 2006

Anton Betten
Michael Braun
Harald Fripertinger
Adalbert Kerber
Axel Kohnert
Alfred Wassermann

Table of Contents

List of Tables

List of Figures

List of Symbols

I_k	the unit matrix, 27
$[x]_R$	the equivalence class of $x \in X$ with respect to R, 29
S_X	the symmetric group on the set X, 30
$e^{(i)}$	the i-th unit vector, 31
$\mathbb{F}_q^*$	$\mathbb{F}_q \setminus \{0\}$, the multiplicative group of $\mathbb{F}_q$, 31
$M_{(\varphi;\pi)}$	matrix representation of a linear isometry given by $\varphi \in (\mathbb{F}_q^*)^n$ and $\pi \in S_n$, 31
$M_n(q)$	the group of linear isometries, 32
$_GX$	an action of the group G on X from the left, 33
δ	a permutation representation, 33
$\sim_G$	the equivalence relation defined by $_GX$, 33
$G(x)$	the orbit of x under the action of G, 33
$G \backslash\backslash X$	the set of all G-orbits on X, 33
Y^X	the set of all mappings from X to Y, 34
$H \wr_X G$	the wreath product of H and G w.r.t. $_GX$, 34
$GL_k(q)$	the set of all regular $k \times k$-matrices over $\mathbb{F}_q$, 36
$\mathcal{U}(n,k,q)$	the set of all k-dimensional subspaces of $\mathbb{F}_q^n$, 37
$U \trianglelefteq G$	U is a normal subgroup of G, 38
$_{G/U}(U \backslash\backslash X)$	the factor action of G with respect to U, 39
$\mathbb{F}_q^{k \times n}$	the set of all $k \times n$-matrices over $\mathbb{F}_q$, 40
$\mathbb{F}_q^{k \times n,r}$	the set of all $k \times n$-matrices of rank r over $\mathbb{F}_q$, 40
$S_{\mathbb{F}_q^*}$	the symmetric group on $\mathbb{F}_q^*$, 44
$W_C(x,y)$	the weight enumerator of C, homogeneous form, 52
$w_C(x)$	the weight enumerator, inhomogeneous form, 52
$GL(V)$	the general linear group of V, 53
D	a linear representation of a group, 53
χ^D	the character of the representation D, 54
$a_1(\overline{g})$	the number of fixed points of $\overline{g}$, 54
$\mathbb{F}^*$	the multiplicative group of $\mathbb{F}$, 54
X_g	the set of points fixed under g, 59
X_G	the set of invariants of G on X, 59
σ_m	an elementary symmetric polynomial, 60
$\text{supp}(v)$	the support of the vector v, 61
$X \Delta Y$	the symmetric difference of sets X and Y, 61
$\exp(G)$	the exponent of the group G, 63
χ_g	the character corresponding to $g \in G$, 63

$\hat{f}$	the Discrete Fourier Transform of $f \in \mathbb{C}^G$, 63
$K_i^{n,q}(x)$	a Krawtchouk polynomial, 64
$(I_k \mid A)$	a systematic generator matrix, 65
$D_n(q)$	the regular $n \times n$ diagonal matrices over $\mathbb{F}_q$, 66
$n_{\min}(k,d,q)$	the least length n of a linear code over $\mathbb{F}_q$ of dimension k and with minimum distance d, 82
$d_{\max}(n,k,q)$	the maximal minimum distance of an (n,k)-code over $\mathbb{F}_q$, 82
$k_{\max}(n,d,q)$	the maximal dimension k of a linear code over $\mathbb{F}_q$ of length n and minimum distance d, 82
$\lceil r \rceil$	the least integer $\geq r$, 88
$\rho(C)$	the covering radius of C, 90
$P(C)$	the parity extension of C, 95
$Pu(C)$	the punctured code of C, 96
$C_0 + C_1$	$\{(c \mid c') \mid c \in C_0,\ c' \in C_1\}$, 97
(C_0, C_1)	the (u,v)-construction of C_0 and C_1, 98
$C_0 \mid C_1$	$\{(c \mid c + c') \mid c \in C_0,\ c' \in C_1\}$, the $(u \mid u+v)$-construction of C_0 and C_1, 100
$S(C)$	a shortening of C, 100
G_{24}	the binary Golay-code of length 24, 106
$C \downarrow \mathbb{F}$	the intersection $C \cap \mathbb{F}^n$, 108
$\mathrm{Bl}_B(C)$	the blow up of C with respect to B, 108
(Lb, n, k, d, q)	there exists an (n,k,d,q)-code, 109
(Ub, n, k, d, q)	there does not exist an (n,k,d,q)-code, 109
$Q_1 \leq Q_2$	the bound Q_2 is at least as strong as Q_1, 112
$M_1 \leq M_2$	the operation M_2 modifies code bounds at least as good as M_1, 112
$U \otimes V$	the tensor product of the vector spaces U and V, 114
$\mathcal{B}_m^q$	the $\mathbb{F}_q$-algebra of Boolean functions, 119
$\mathrm{RM}_{m,t}^q$	the t-th order Reed–Muller-code over $\mathbb{F}_q$, 120
$N_q(k)$	the maximal length of a k-dimensional MDS-code over $\mathbb{F}_q$, 132
$\mathbb{Z}_p$	the residue class ring consisting of the integers taken modulo p, 139
$\mathbb{P}$	the prime subfield of $\mathbb{F}$, 140
$\mathbb{Z}_p[x]$	the ring of polynomials over $\mathbb{Z}_p$, 140
$\deg f$	the degree of the polynomial f, 140

$I(f)$	the ideal generated by the polynomial f, 141
$\mathbb{Z}_p[x]/I(f)$	the factor ring of $\mathbb{Z}_p[x]$ modulo the ideal $I(f)$, 142
$\gcd(a,b)$	the greatest common divisor of a and b, 145
$\mathrm{ord}(g)$	the order of the group element g, 149
$\langle g \rangle$	the subgroup generated by $g \in G$, 149
$|G|$	the order of the group G, 149
$\simeq$	isomorphy of groups, rings or fields, 153
$\mathbb{F}(\alpha)$	the smallest field extension of $\mathbb{F}$ containing α, 156
$\mathbb{F}(\alpha_0,\ldots,\alpha_{n-1})$	the smallest field extension of $\mathbb{F}$ which contains $\alpha_0,\ldots,\alpha_{n-1}$, 156
$\overline{\mathbb{F}_q}$	the algebraic closure of $\mathbb{F}_q$, 160
$m_p(d)$	the number of monic, irreducible polynomials in $\mathbb{F}_p[x]$ which are of degree d, 162
ζ	the (number theoretic) zeta function, 162
μ	the (number theoretic) Möbius function, 162
$\ker(\phi)$	the kernel of the homomorphism ϕ, 163
$\mathrm{Gal}\,[\,\mathbb{F}_{p^n} : \mathbb{F}_{p^m}\,]$	the Galois group of $\mathbb{F}_{p^n}$ over $\mathbb{F}_{p^m}$, 167
G_x	the stabilizer of x in G, 170
$L(G)$	the lattice of subgroups of G, 172
$U \wedge V$	the intersection $U \cap V$ of the subgroups U, V, 172
$U \vee V$	the group generated by the union $U \cup V$, 172
$G\backslash\!\!\backslash_{\widetilde{U}}X$	the set of orbits of type $\widetilde{U}$, 172
$Z(L(G),\leq)$	the zeta matrix of the subgroup lattice, 173
$M(L(G),\leq)$	the Möbius matrix of the subgroup lattice, 173
l_{mn}	$|C_n\backslash\!\!\backslash_{\widetilde{1}}m^n|$, a Dedekind number, 178
$L_n(m)$	the Lyndon words of length n over m, 179
$L(m)$	the Lyndon set over the alphabet m, 179
$l_0(f)\ldots l_{\lambda(f)-1}(f)$	a decomposition of f into Lyndon words, 180
$\widetilde{U}$	the conjugacy class of $U \leq G$, 181
$\mathrm{PG}_{k-1}(q)$	the $(k-1)$-dimensional projective space, 205
$\mathrm{PG}_{k-1}^*(q)$	the set of punctured one-dimensional subspaces of $\mathbb{F}_q^k$ without the zero vector, 205
$\mathrm{PG}(\mathbb{F}_q^k)$	the projective geometry, 205
$\theta_{k-1}(q)$	the cardinality of $\mathrm{PG}_{k-1}(q)$, 206
$\mathcal{Z}_k$	the center of the linear group $\mathrm{GL}_k(q)$, 206
$\mathrm{PGL}_k(q)$	the projective linear group, 207

$_R M$	M is an R-left-module, 292
M/U	the factor module where U is a submodule of M, 293
$\langle T \rangle$	the linear closure of T, 293
$M \simeq_R N$	two isomorphic R-modules M and N, 294
$M \simeq N$	two isomorphic modules M and N, 294
$\ell_R(M)$	the length of the R-module M, 299
$C_{i,j}$	the cyclic code $\mathrm{I}(g_{i,j})/\mathrm{I}(x^n - 1)$, 305
$\mathrm{rem}_g(f)$	remainder of f modulo division by g, 335
$\mathrm{DR}(g)$	the division shift register with recoupling polynomial g, 336
$\mathrm{AG}(V)$	the affine geometry of V, 355
$\mathrm{AG}_n(q)$	the n-dimensional affine geometry over $\mathbb{F}_q$, 356
$L^p(I)$	the set of all measurable functions $f : I \to \mathbb{C}$ for which $|f(x)|^p$ is integrable on I, 370
$\hat{f}$	the Fourier Transform of $f \in L^1(\mathbb{R})$, 375
$C^{(\lambda)}$	the λ-way interleave of C, 405
$\mathcal{U}(n,q)$	the set of all subspaces of $\mathbb{F}_q^n$, 444
$\mathcal{V}(n,k,q)$	the set of $U \in \mathcal{U}(n,k,q)$ without zero columns, 452
$\overline{\mathcal{V}}(n,k,q)$	the set of all projective $U \in \mathcal{V}(n,k,q)$, 452
$\overline{\mathcal{U}}(n,k,q)$	the set of all injective $U \in \mathcal{U}(n,k,q)$, 452
$\mathcal{U}_{n,k,q}$	the set of linear isometry classes in $\mathcal{U}(n,k,q)$, 452
$\mathcal{V}_{n,k,q}$	the set of linear isometry classes in $\mathcal{V}(n,k,q)$, 452
$\overline{\mathcal{U}}_{n,k,q}$	the set of linear isometry classes in $\overline{\mathcal{U}}(n,k,q)$, 452
$\overline{\mathcal{V}}_{n,k,q}$	the set of linear isometry classes in $\overline{\mathcal{V}}(n,k,q)$, 452
$\mathcal{T}_{n,k,q}$	the set of linear isometry classes of linear codes of dimension $\leq k$ in $\mathcal{U}(n,q)$, 453
$\overline{\mathcal{T}}_{n,k,q}$	the set of linear isometry classes of linear codes of dimension $\leq k$ in $\overline{\mathcal{U}}(n,k,q)$, 453
T_{nkq}	the number of $\mathrm{GL}_k(q) \times S_n$-orbits on $\mathrm{PG}_{k-1}^*(q)^n$, 453
$\overline{T}_{nkq}$	the number of $\mathrm{GL}_k(q) \times S_n$-orbits on the set of injective functions in $\mathrm{PG}_{k-1}^*(q)^n$, 453
V_{nkq}	the number of linear isometry classes of nonredundant (n,k)-codes over $\mathbb{F}_q$, 453
$\overline{V}_{nkq}$	the number of linear isometry classes of projective (n,k)-codes over $\mathbb{F}_q$, 453
U_{nkq}	the number of linear isometry classes of (n,k)-codes over $\mathbb{F}_q$ which may contain columns of zeros, 453

$\overline{U}_{nkq}$	the number of linear isometry classes of injective (n,k)-codes over $\mathbb{F}_q$, 453
$(a_1(\overline{g}),\dots,a_{|X|}(\overline{g}))$	the cycle type of the permutation $\overline{g}$, 454
$a_i(\overline{g})$	the number of cyclic factors of length i of the permutation $\overline{g}$, 455
$C(G,X)$	the cycle index of the action $_G X$, 455
$\mathrm{ord}(f)$	the order of the formal series f, 456
$\mathbb{Q}\,[\![x]\!]$	the ring of formal power series in the indeterminate x over $\mathbb{Q}$, 457
Y^X_{inj}	the set of injective $f \in Y^X$, 460
$[n]$	$1 + x + \dots + x^{n-1} \in \mathbb{Q}\,[x]$, 460
$\left[{n \atop k}\right]$	a Gauss-polynomial, 460
$c(\overline{g})$	the number of cycles in the decomposition of $\overline{g}$, 461
$A_0 \dotplus \dots \dotplus A_{n-1}$	the outer sum of the matrices $A_0, \dots, A_{n-1}$, 468
$\mathcal{R}_{nkq}$	the set of linear isometry classes of nonredundant, indecomposable (n,k)-codes over $\mathbb{F}_q$, 472
R_{nkq}	the number of linear isometry classes of nonredundant indecomposable (n,k)-codes over $\mathbb{F}_q$, 472
$\overline{\mathcal{R}}_{nkq}$	the set of linear isometry classes of (nonredundant), indecomposable, projective (n,k)-codes over $\mathbb{F}_q$, 472
$\overline{R}_{nkq}$	the number of linear isometry classes of (nonredundant), indecomposable, projective (n,k)-codes over $\mathbb{F}_q$, 472
$[q]_k$	the order of the group $GL_k(q)$, 477
$a \vdash n$	$a = (a_1, \dots, a_n)$ is a cycle type of n, 477
M_A	the minimal polynomial of the endomorphism A, 478
$C(f)$	the companion matrix of f, 480
$H(f^n)$	the hyper companion matrix of f^n, 480
$D(f,a)$	a block-diagonal matrix, consisting of companion and hyper companion matrices of f, 480
χ_A	the characteristic polynomial of A, 481
$b(d,a)$	the order of the centralizer of $D(f,a)$, 482
$sc(A,v)$	a subcycle expression, 486
$SC(\mathrm{GL}_k(q),\mathbb{F}_q^k \setminus \{0\})$	the subcycle index of $\mathrm{GL}_k(q)$ on $\mathbb{F}_q^k \setminus \{0\}$, 486
$\mathrm{Exp}(f)$	the exponent of the polynomial f, 486
$E(d,q)$	the set of all exponents of monic, irreducible polynomials of degree d over $\mathbb{F}_q$, 487

γ_n	the Hermite constant, 585
$\gamma_{*,j}^\top$	the j-th column of the matrix (γ_{ij}), 616
$\gamma_{i,*}$	the i-th row of the matrix (γ_{ij}), 616
$H(v)$	the dual space of $\langle v \rangle$, 616
$\{\!\!\{\ldots\}\!\!\}$	denotes a multiset, 618
$\widetilde{\widetilde{\Gamma}}$	the multiset defined by the columns of Γ, 618
$\widetilde{\widetilde{\Gamma}} \downarrow H$	the restriction of $\widetilde{\widetilde{\Gamma}}$ to the hyperplane H, 618
$(X, \le)$	a poset, 625
$_G(X, \le)$	a poset action, 625
$(X, \wedge, \vee)$	a lattice, 625
$_G(X, \wedge, \vee)$	a lattice action, 625
$A^\wedge$	a Plesken matrix, 630
$A^\vee$	a Plesken matrix, 630
G^*	the dual group of G, 634
$\Gamma(S)$	a generator matrix from a set of projective points, 666
$(G(x), \mathcal{E})$	the Schreier-tree, 672
$\mathrm{orbit}(G, X)$	the orbit data structure, 675
$\mathrm{rk}_{k,q}$	the rank function for $\mathbb{F}_q^k$, 677
$\mathrm{rk}_{d;q}$	the rank function for $\mathrm{PG}_d(q)$, 679
$\mathcal{P}(X)$	the power set of the set X, 682
$\mathcal{P}_k(X)$	the set of subsets of size k of the set X, 682
rk_X	the rank function for $\mathcal{P}(X)$, the power set of X, 685
$\mathrm{rk}_{X,k}$	the rank function for the set of k-subsets of X, 686
G_R	the setwise stabilizer of R in G, 688
$G_{r_0,\ldots,r_{s-1}}$	the pointwise stabilizer of $r_0, \ldots, r_{s-1}$ in G, 688
$G^{(i)}$	the i-th term in a stabilizer chain for G, 717
$\mathcal{O}^{(i)}$	the i-th basic orbit, 718
ℓ_i	the length of the i-th basic orbit, 718
$\sigma_{i,j}$	a coset representative in a stabilizer chain, 718
$S^{(i)}$	a subset of the strong generating set S, 718
$\overset{\leftarrow}{L}$	a reversed sequence, 723
$\mathrm{PG}_{d\backslash s}(q)$	the points of $\mathrm{PG}_d(q)$ which do not lie in a $\mathrm{PG}_s(q)$, 727
$\theta_{d\backslash s}(q)$	the cardinality of $\mathrm{PG}_{d\backslash s}(q)$, 727
$\mathrm{rk}_{d\backslash s;q}$	the rank function for $\mathrm{PG}_s(q)$, 728
$F_{n,i}$	a distorted identity matrix, 728
P	a 2×2 permutation matrix, 730

Chapter 1
Linear Codes

1

1

1 Linear Codes

In the first chapter, we introduce the basic definitions, methods and results from the theory of error-correcting linear codes and its applications.

1.1 Introduction

As Claude Shannon, the founding father of modern Information Theory puts it in [178],

> *"The fundamental problem of communication is that of reproducing at one point either exactly or approximately a message selected at another point."*

Error-correcting codes are used to improve the reliability of such communication systems. We may think of communication across large distances such as spacecraft communication, or basically any form of information transmission, including playing back a piece of music which was recorded previously and stored on some media, for instance. In any case, the goal is to transmit and reproduce the information as accurately as possible, even under unfavorable circumstances, like in an error-prone environment.

In order to make a mathematical approach possible, we introduce the following communication model. It has a sender and a receiver and they are supposed to communicate in one direction, so that information passes from the sender to the receiver. Thus, we suppose that between the sender and the receiver there is a *communication system*, a *channel*, and all information passes through this directed channel:

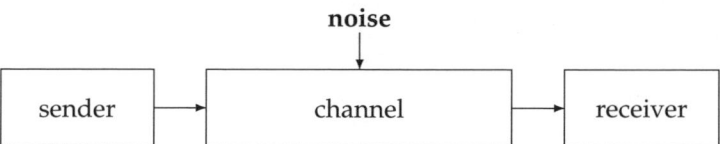

This channel may be unreliable, e.g. we are expecting that information may be altered as it is passing through. Often, this is called a *noisy channel*, appealing to the common experience that in a noisy room full of people it is usually impossible to understand a word someone has said at the other end of the room. Of course, we will make assumptions about the behavior of the channel, and it is clear that the channel should not be too bad, i.e. we require that a certain amount of information passes through correctly. Hence, we assume that the output of the channel is a more or less damaged version of the original input.

We suppose, for example, that the length of the signals, or *codewords* as we call them, that are sent through the channel is never changed by noise, and that, if noise inflicts a codeword, a letter is changed into another letter with the same probability for each letter. Such channels are called *symmetric*. On the receiving end, a process which is called *error-correction* takes place. Given the altered or damaged codeword, one tries to recover the original one by correcting errors. Of course, this is a difficult task as it is not clear where the error may have occurred (or if an error has occurred at all).

On the other end of the channel, the sender is trying to help by manipulating the messages *before they are transmitted*. This can be done by adding redundancy, for example, by repeating the message. The purpose of this is to protect the message, so that the influence of noise can later be corrected up to a certain degree.

A *message* is defined to be a finite sequence of elements of a given alphabet. Subsequently, such a sequence is also referred to as a *word*. There is no restriction in assuming that all messages are of a fixed length, say k. If the message is very long, we may break it up into pieces, and each such piece may be considered a message by itself. Hence without loss of generality, we assume that the messages are of size k. To enable error-correction, a process called *encoding* takes place. Here, we replace the message words by possibly longer sequences over the same alphabet, the codewords. The added redundancy will later enable the receiver to correct errors which may have occurred during transmission. The only requirement at this point is that the encoding map shall be injective, for otherwise the receiver would not be able to decide which message was sent, even under the most favorable circumstances when no error has occurred during transmission. It is customary to denote the length of codewords by n. The process of correcting errors and obtaining back the message is called *decoding*. This refined communication model is depicted in Fig. 1.1.

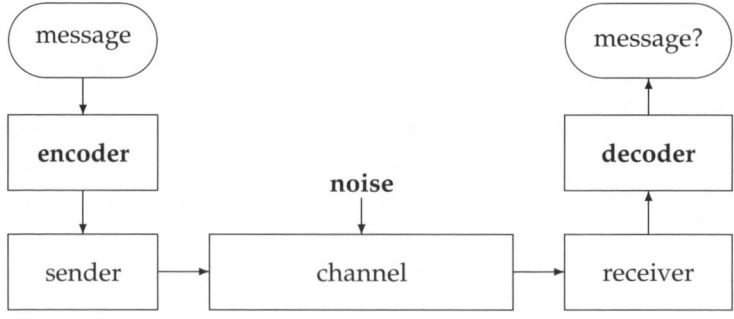

Fig. 1.1 The refined communication system

The arrows indicate that the communication is one-way. In particular, the possibility of asking for retransmission shall be excluded. Hence, error *correction* will be the main topic while error *detection* will be less important. As an aside, it should be mentioned that there are indeed codes which are based on the idea that bidirectional communication is possible. An important example is the well-known *ISBN-code,* which assigns to each book a unique number which may be used to identify the book. This number is composed of a certain number of digits. One digit plays the special role of a check digit. This check digit allows the detection of single errors and of interchanges of adjacent digits – a very frequent mistake. Of course, the idea is that the message can be repeated once the receiver has flagged the first transmission as erroneous. See Exercises 1.1.1–1.1.3 for details on this and similar codes. For further reading, see [98], [99], and [47].

Here is a very easy example that demonstrates the *metric principle* which is used for error detection and error correction. Suppose that we want to transmit over a noisy channel a message which is just one of the two answers "yes" or "no". These two messages can be identified with the one-element sequences 1 for "yes" and 0 for "no". So, $k = 1$ in this case, and the alphabet consists of the two symbols 0 and 1. A particularly simple way of adding redundancy is to use the well-known pedagogical principle of repetition of the message in question, or, in other words, to use the *repetition code.* Assume we have agreed to use three-fold repetition, i.e. $n = 3 \cdot k = 3 \cdot 1 = 3$. Then we intend to send as codewords sequences which are either 111 (for "yes") or 000 (for "no"). Thus, the *code* in question is the set $\{000, 111\}$. Moreover, we assume that this fact is known to the receiver. It motivates the following strategy:

— If neither 111 nor 000 but another sequence of length 3 is received, then the transmission has been distorted and there must be at least one error in the received sequence.

Hence, to begin with, we see that this repetition code is able to *detect* a certain amount of errors, in our particular example one or two. This means that in these cases we are certain that the transmission of the message is erroneous. But we should be aware of the fact that in the case of three errors the receiver cannot detect them. Moreover, the receiver can *correct* the received word into the original one, provided that not too may errors have occurred.

— If the receiver obtains one of the sequences 110, 101, 011, 100, 010, or 001, and if in addition we *assume that there was only one symbol changed by noise,* then we can simply decode the received word into the codeword which is *most similar,* i.e. into 111 in the first three cases, and into 000 in the latter three cases.

Hence, using the three-fold repetition code we are in a position to *correct a single error.* Our strategy is based on the fact that the messages that have to be sent, namely the sequences 111 or 000, differ in *three* places. The other six possibly received sequences differ from the original sequences in either one or two places. In fact, for each of the $2^3 = 8$ sequences which can be received, there is *exactly one* of the two vectors 111 and 000 which is *most similar.*

In order to demonstrate this principle in more detail, let us see what happens if we use four-fold repetition. Besides the correct sequences 1111 and 0000 there are fourteen other ones that can be received when errors have occurred. In this case we are in a slightly better position than with three-fold repetition: We can detect up to three errors (to be exact, we realize that errors have occurred), but we are not able to correct more than one error. The error correcting property has *not* improved. The reason is that in the case of two errors the received sequence consists of two letters 1 and two letters 0. In this situation we are unable to tell which codeword has been sent. In the case of five-fold repetition we can recognize up to four errors, and we can correct at most *two* errors, and so on.

The *metric principle* used is based on the fact that all the sequences that can occur at the receiver side differ from the correct sequences $1 \ldots 1$ and $0 \ldots 0$, of length n, in at most $\lfloor n/2 \rfloor$ many places. The number of places (or coordinates) in which two codewords differ is called the *Hamming distance* of the two codewords in question. The least Hamming distance between any two distinct codewords is called the *minimum distance* of the code. If d is the minimum distance of a linear code – not necessarily a repetition code – then up to $t := \lfloor (d-1)/2 \rfloor$ errors can be corrected, while up to $d-1$ errors are recognized. That is, if no more than $d-1$ errors occur, we detect that something is wrong. Later on, we will make this more precise.

The quality of a code with messages of length k and codewords of length n is indicated by

— the quotient k/n, the *information rate* of the code, which measures the effort needed for the transmission of an encoded message,

— the *relative minimum distance* d/n which gives roughly twice the proportion of errors that can be corrected in each encoded message (it is also called the *error-correction rate*),

— the *complexity of the encoding and of the decoding procedure.*

Summarizing, the main goal of Coding Theory is to provide codes with high information rate, high error correction rate and with a low complexity of encoding and decoding.

An important and intensely studied class of codes are the *linear codes*. These are just the k-dimensional subspaces of an n-dimensional vector space over a finite field. They form a subclass of the more general class of *block codes*, which are merely subsets of an n-dimensional space. The structure of linear codes can be analyzed using methods of Linear Algebra and Algebra as well as Combinatorics and Geometry.

In this introductory chapter, our goal is to discuss the fundamentals of the theory of linear codes. We also classify linear codes according to their error-correcting qualities. Codes with similar metric structure are collected into *isometry classes* of codes. Finally, we present an algorithm to determine the minimum distance of a given linear code. In later chapters we will deepen the theory, the construction, and the generation of linear codes and their application, and we will describe some important families of codes.

Exercises

Exercise The ISBN-code ("International Standard Book Number") is a sequence of ten elements $x_{10}, \ldots, x_1$ taken from the set $\{0, 1, \ldots, 9, X\}$. This sequence is divided into four parts of variable length, which must be separated by hyphens or spaces. The hyphens or spaces increase the readability and indicate the borders between the four different parts. These characters, however, do not influence the ability of the code to detect and correct errors.

E.1.1.1

1. The first subsequence $x_{10}, \ldots$ (mostly of length 1) encodes the language, or, rather, the language region in which the book was printed. 0 stands for English speaking countries, 3 for the German speaking ones.

2. The next subsequence encodes the publishing company. It consists of at least two entries.

3. The sequence of the following entries, $\ldots, x_2$, is a number chosen by the publisher to identify the book.

The entries $x_{10}, \ldots, x_2$ are taken from the set $\{0, 1, \ldots, 9\}$.

4. The final entry, x_1, is the residue modulo 11 of $-\sum_{i=2}^{10} x_i \cdot i$. If this residue happens to be equal to 10, one puts $x_1 := X$.

Show that this code recognizes a single error as well as an interchange of two neighboring entries, and that it allows the reconstruction of an unreadable entry.

E.1.1.2   **Exercise** The ISSN-code ("International Standard Serial Number") is a sequence of eight elements $x_8, \ldots, x_1$ taken from the set $\{0, 1, \ldots, 9, X\}$. This sequence is divided into two parts, each consisting of four digits, which must be separated by a hyphen. Similar to the ISBN-code, the entries $x_8, \ldots, x_2$ are taken from the set $\{0, 1, \ldots, 9\}$, and the final entry x_1 is determined such that $\sum_{i=1}^{8} x_i \cdot i \equiv 0 \bmod 11$ is satisfied. If $x_1 = 10$, then x_1 is represented as X. This code has exactly the same properties as the ISBN-code.

1. Determine the ISSN-number of the Bayreuther Mathematische Schriften from the sequence ISSN 0172-?062 where the digit x_4 is not readable.

2. The number ISSN 0174-1062 is not a valid ISSN-number. Assuming that exactly one error occurred, determine all valid ISSN-numbers which could be represented by the given one.

E.1.1.3   **Exercise** The EAN-code ("European Article Number") has two basic formats, the 8 and 13 digit variants. The 13 digit code is more common, so we discuss it here. The 8 digit code is generally used where space is restricted. The EAN code is intended as a world wide standard (although some countries use other systems), therefore, no two products may have the same EAN number. To ease the administration of number allocation, each country using EAN has a country identifier at the start of the code. For instance the digits 00 to 13 identify the USA and Canada, 40 to 44 Germany, and 90 to 91 Austria. Other countries have 2 or 3 digit prefixes. The rest of the EAN code is divided into the manufacturer number which can be of variable length, the item reference number, assigned by the manufacturer, and the check digit. In general, both the manufacturer number and the item reference number consist of 5 digits. This means that in this case a manufacturer can have up to 10^5 products. For that reason, those manufacturers which produce a smaller number of products get longer manufacturer codes. The check digit is the last number. All 13 digits $x_{13}, \ldots, x_1$ are taken from the set $\{0, 1, \ldots, 9\}$. The check digit x_1 is determined by the other digits such that

$$\sum_{i \equiv 1 \bmod 2} x_i + 3 \cdot \sum_{i \equiv 0 \bmod 2} x_i \equiv 0 \bmod 10$$

is satisfied.

1. Show that the EAN-code recognizes a single error and allows the reconstruction of an unreadable entry, but in general it does not detect a swap of two neighboring entries.

2. The EAN of books can easily be obtained from their ISBN-number. As prefix, the three digits 978 are used, regardless of the country in which the

book was published. Then the ISBN-number with the check digit stripped is appended. Finally, the EAN check digit is computed from these 12 digits as described above. Compute the EAN-code of the present book!

The EAN is also coded in a machine-readable version as a barcode. For this purpose, the EAN is encoded as a binary sequence of bars and spaces. A 1 in the code is represented by a *bar section* and a 0 by a *space section*. Consecutive 1's or 0's are combined to form wider bars or spaces. The EAN barcode consists of the following parts:

— The left-hand starting section of the form 101,

— binary encodings of the digits $x_{12}, \ldots, x_7$,

— the center pattern of the form 01010,

— binary encodings of the digits $x_6, \ldots, x_1$,

— the right-hand closing section of the form 101.

For the encoding of $x_{12}, \ldots, x_1$ three different codes are used, codes A, B, and C. (See also [104, 1.2.5 Beispiel].) The codes A and B are applied for encoding $x_{12}, \ldots, x_7$, and code C is used for encoding $x_6, \ldots, x_1$. So far the digit x_{13} has not been encoded. Depending on the value of x_{13}, different sequences of the codes A and B are applied for the encoding of $x_{12}, \ldots, x_7$. They are given in the table below:

x_{13}	x_{12}	x_{11}	x_{10}	x_9	x_8	x_7		digit	code A	code B	code C
0	A	A	A	A	A	A		0	0001101	0100111	1110010
1	A	A	B	A	B	B		1	0011001	0110011	1100110
2	A	A	B	B	A	B		2	0010011	0011011	1101100
3	A	A	B	B	B	A		3	0111101	0100001	1000010
4	A	B	A	A	B	B		4	0100011	0011101	1011100
5	A	B	B	A	A	B		5	0110001	0111001	1001110
6	A	B	B	B	A	A		6	0101111	0000101	1010000
7	A	B	A	B	A	B		7	0111011	0010001	1000100
8	A	B	A	B	B	A		8	0110111	0001001	1001000
9	A	B	B	A	B	A		9	0001011	0010111	1110100

We realize that for encoding x_{12}, code A is always used. If $x_{13} = 0$ then all digits $x_{12}, \ldots, x_7$ are encoded with code A. In all other cases, codes A and B are each used to encode three digits.

The three codes A, B, and C encode each digit as a binary word of length 7. Each codeword consists of two bar and two space sections. No bar or space is longer than four elements. All codewords of codes A and B start with 0 and end with 1. All codewords of code C start with 1 and end with 0. Actually, it would be enough to describe the codewords of code A. In order to obtain the codewords of code C from code A, exchange the 0's and 1's in the codewords of A. In order to obtain the codewords of code B from code C, reverse the order of each codeword of C.

Moreover, we realize that no codeword occurs in two different codes, and no codeword of A is the reverse of a codeword in C. This fact, together with the rule that x_{12} is always encoded with code A allows the determination of the direction (from left to right or from right to left) in which a barcode is read. When reading from left to right, after the left-hand starting section 101, the reader comes across an element of code A. When reading from right to left, after the reverse of the right-hand closing section, which is again 101, the reader comes across the reverse of an element of code C. Consequently, after reading the first digit it is possible to determine the direction of reading.

Finally, let us consider the following example. The book *Codierungstheorie, Springer, Berlin, 1998,* by some of the present authors and K.-H. Zimmermann, has the ISBN 3-540-64502-0. First, this number is encoded as an EAN of the form 9783540645023 where the last 3 is the EAN check digit. Since $x_{13} = 9$, we have to apply the codes $ABBABA$ for the encoding of $x_{12}, \ldots, x_7$. This way we obtain the following binary representation of the bar code of 9783540645023.

101	left-hand starting
0111011 0001001 0100001 0110001 0011101 0001101	$x_{12} \ldots x_7$
01010	center pattern
1010000 1011100 1001110 1110010 1101100 1000010	$x_6 \ldots x_1$
101	right-hand closing

This gives a barcode of the form:

9 783540 645023

1.2 Linear Codes, Encoding and Decoding

As we have seen, the goal of Coding Theory is to provide methods which improve the reliability of communication via a noisy channel. For example, we may think of the transmission of satellite photos taken in space and sent back to earth. For this purpose, one decomposes the picture into a large number of pixels (which stands for "picture elements"), each pixel having a certain grey value, for example. The grey value is then mapped to a number, which in turn is converted to a binary sequence by means of the binary representation of an integer (i.e. one of 0, 1, 10, 11, 100, 101, 110, 111 etc.). Hence, the messages, i.e. the grey values of the pixels, can be considered as *words* over the *alphabet* $\{0,1\}$.

For example, in the case of six values of grey, we have the messages

$$0, 1, 10, 11, 100, 101.$$

By padding with zeros up-front, we can make the sequences all have length 3. We can also add the remaining sequences of that length over the given alphabet, which in our case gives

$$000, 001, 010, 011, 100, 101, 110, 111.$$

The reader certainly knows that the two elements 0 and 1 are the elements of a field $\mathbb{F}$, the binary field. The above sequences can be considered as the vectors of $\mathbb{F}^3$, the three-dimensional vector space over $\mathbb{F}$,

$$\mathbb{F}^3 = \{000, 001, 010, 011, 100, 101, 110, 111\}.$$

This vector space is called the *message space*.

In more general situations it will be a k-dimensional vector space $\mathbb{F}^k$ over a finite field $\mathbb{F}$, which may be different from the field of two elements, of course. Later on we will see that the order defines a field up to isomorphism. Therefore, a field consisting of q elements is indicated by $\mathbb{F}_q$. Moreover, it will turn out that the orders q of finite fields are exactly the prime powers $q = p^m$, p a prime and $m \in \mathbb{N}^* := \mathbb{N} \setminus \{0\}$. For example, for each prime number p, the integers $0, 1, 2, \ldots, p-1$ form the field $\mathbb{F}_p$ with respect to addition and multiplication modulo p. In the case when the original finite alphabet A does not consist of elements of a finite field, then we simply take a suitable finite field $\mathbb{F}$ with at least $|A|$ elements and rename the letters of A by elements of $\mathbb{F}$.

As we have seen, the encoding map should be injective. This means that we are looking for an *embedding* of $\mathbb{F}^k$ into some space $\mathbb{F}^n$, where $n \geq k$. In order to add redundancy, we usually let n be strictly larger than k. The encoding of messages is done using an *encoder*

$$\gamma \colon \mathbb{F}^k \to \mathbb{F}^n,$$

an injective linear mapping of the message space $\mathbb{F}^k$ into $\mathbb{F}^n$. For example, we may simply repeat the messages twice, which yields the following embedding of $\mathbb{F}^3$ into $\mathbb{F}^6$:

$$\gamma(\mathbb{F}^3) = \{000000, 001001, \ldots, 110110, 111111\} \subset \mathbb{F}^6.$$

1.2.1   **Definition (linear codes, generator matrices)** The image

$$C = \gamma(\mathbb{F}^k)$$

of the encoder γ is a subspace of $\mathbb{F}^n$ which is isomorphic to the message space $\mathbb{F}^k$. We call C a *linear (n,k)-code* or briefly an *(n,k)-code* over $\mathbb{F}$. The number k is its *dimension* and the number n will be called the *block length* or simply the *length* of the code C. The vectors in C are the *codewords* or *codevectors*.

The encoder can be expressed as multiplication by a matrix Γ of rank k. Using the *row convention*, i.e. by writing vectors as row-vectors, Γ turns out to be a $k \times n$-matrix. The embedding is then given by the map

$$\gamma : \mathbb{F}^k \to \mathbb{F}^n \ : \ v \mapsto v \cdot \Gamma,$$

and we obtain that

$$C = \gamma(\mathbb{F}^k) = \{v \cdot \Gamma \mid v \in \mathbb{F}^k\}.$$

For this reason, the matrix Γ, which is in general not uniquely determined, is called a *generator matrix* of C. Its rows form a *basis* of C.   ◇

1.2.2   **Example** If $k = 1$ and $\mathbb{F}^1 = \{0, 1\}$ is the message space, then the *three-fold repetition code* $C = \{000, 111\}$, which is an embedding of $\mathbb{F}^1$ into $\mathbb{F}^3$, has the generator matrix

$$\Gamma = (1 \quad 1 \quad 1).$$

In this particular case, the generator matrix is uniquely determined, but this is exceptional. For example, in the case of $k = n = 3$, each regular 3×3-matrix over $\mathbb{F}$ is a generator matrix.   ◇

Hence, there are usually plenty of generator matrices of a given code C, and it is clear from Linear Algebra that they can be obtained from Γ by applications of invertible matrices:

1.2.3   **Theorem** *The set of all generator matrices of a linear code with generator matrix Γ is*

$$\{B \cdot \Gamma \mid B \in \mathrm{GL}_k(\mathbb{F})\},$$

where $\mathrm{GL}_k(\mathbb{F})$ is the set of all regular $k \times k$-matrices over $\mathbb{F}$.   □

Now we describe a way of considering $\mathbb{F}^n$ as a *metric space* in order to justify the metric principle used in the decoding process as described in the introduction. Usually we indicate vectors $v \in \mathbb{F}^n$ in the form

$$v = (v_0, v_1, \ldots, v_{n-1}).$$

Throughout the book we consistently use the recursive definition of natural numbers as *sets*,

$$n := \{0, \ldots, n-1\}, \text{ if } n > 0, \ 0 := \emptyset.$$

Thus, a vector v can be considered as a mapping from this *set* n to $\mathbb{F}$, with v_i the image of $i \in n$ under v,

$$v : n \to \mathbb{F} : i \mapsto v_i.$$

In this sense, the vector space can be identified with a set of mappings:

$$\mathbb{F}^n = \{v \mid v : n \to \mathbb{F}\}.$$

Of course, we also use the natural number n as the *cardinality* of sets of this order, but it should be always clear from the context if n means a set or a cardinality of a set.

The metric principle which we are going to describe is based on the following fact:

Definition and Theorem (Hamming metric) *The function*   1.2.4

$$d : \mathbb{F}^n \times \mathbb{F}^n \to \mathbb{N} : (u, v) \mapsto |\{i \mid i \in n, \ u_i \neq v_i\}|$$

is a metric on the vector space $\mathbb{F}^n$, *the* Hamming *metric. This means that the function* d *satisfies*

$$
\begin{aligned}
d(u, v) &= 0 \iff u = v, \\
d(u, v) &= d(v, u), \\
d(u, v) &\leq d(u, w) + d(w, v),
\end{aligned}
$$

for all $u, v, w \in \mathbb{F}^n$. *The nonnegative integer* $d(u, v)$ *is called the* Hamming *distance between the vectors* $u, v \in \mathbb{F}^n$. *Hence, the pair* $(\mathbb{F}^n, d)$ *is a metric space, the* Hamming space *of dimension n over $\mathbb{F}$. It will also be denoted by*

$$H(n, \mathbb{F}) \quad \text{or by} \quad H(n, q),$$

if $\mathbb{F} = \mathbb{F}_q$. *The Hamming distance is invariant under translation and multiplication by nonzero scalars: For* $u, v, w \in H(n, \mathbb{F})$ *and* $\lambda \in \mathbb{F}, \lambda \neq 0$,

$$d(u, v) = d(u + w, v + w), \quad \text{and} \quad d(u, v) = d(\lambda u, \lambda v).$$

Proof: The equations $d(u,v) = 0$ and $u = v$ are obviously equivalent, and the symmetry of d is trivial. To show the triangle inequality

$$d(u,v) \leq d(u,w) + d(w,v),$$

we only note that the i-th component contributes to the left hand side if and only if $u_i \neq v_i$, in which case it also contributes to the sum on the right hand side, since then $u_i \neq w_i$ or $v_i \neq w_i$. The invariance under translation and scalar multiplication follows from $u_i = v_i \iff u_i + w_i = v_i + w_i$ and from $u_i = v_i \iff \lambda u_i = \lambda v_i$. This completes the proof. □

A measure for the error correction capabilities of a linear code C is the least Hamming distance between two distinct codewords. The reason is that this value determines the *packing radius* of C, which is the largest integer t such that the balls of radius t centered at codewords are all disjoint (intuitively, we can "pack" the balls).

1.2.5   **Definition (minimum distance)** If C denotes a linear code, then its *minimum distance* is defined as

$$\mathrm{dist}(C) := \min\{d(c,c') \mid c,c' \in C,\ c \neq c'\}.$$ ◇

A glance at Fig. 1.2 shows that the packing radius is the greatest integer which is strictly less than half the value of the minimum distance.

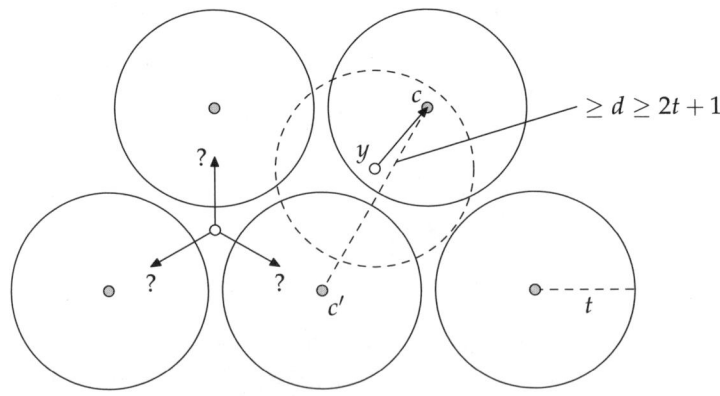

Fig. 1.2 The maximum-likelihood-decoding method

1.2.6   **Corollary** *The packing radius of a linear code C is* $\lfloor (\mathrm{dist}(C) - 1)/2 \rfloor$. □

The crucial point is the following decoding method, which is based on the concept of the packing radius:

Maximum-likelihood-decoding *It is possible to correct up to*   **1.2.7**

$$t := \lfloor (\operatorname{dist}(C) - 1)/2 \rfloor$$

errors in the following way (cf. Fig. 1.2):

— *Using* maximum-likelihood-decoding, *a vector $y \in \mathbb{F}^n$ is decoded into a codeword $c \in C$ which is closest to y with respect to the Hamming metric. In formal terms: y is decoded into a codeword $c \in C$, such that*

$$d(c,y) \leq d(c',y), \text{ for all } c' \in C.$$

If there are several $c \in C$ with this property, one of them is chosen at random.

— *If the codeword $c \in C$ was sent and no more than t errors have occurred during transmission, the received vector is*

$$y = c + e \in \mathbb{F}^n,$$

where e denotes the error vector. *It satisfies*

$$d(c,y) = d(e,0) \leq t,$$

and hence c is the unique *element of C which lies in a ball of radius t around y. A maximum likelihood decoder yields this element c, and so we obtain the correct codeword.*   □

We mention that codes of dimension $0 < k = n$ obviously have minimum distance $d = 1$, and soon we will see that in the case $k = n - 1$ we have $d \leq 2$.

If we want to *evaluate* the minimum distance of a code, we can, of course, evaluate the distances $d(c,c')$ of all

$$\binom{|C|}{2} = \binom{|\mathbb{F}|^k}{2}$$

unordered pairs of different codewords. But this is a very inefficient way to do so. A better approach is the following. For a vector v, we denote by

$$\operatorname{wt}(v) := d(v,0),$$

the *Hamming weight* of v. It is just the number of components of v which are nonzero. For a code C, the *minimum weight* of C is defined as

$$\min\{\operatorname{wt}(c) \mid c \in C, \ c \neq 0\},$$

and it is not difficult to show (Exercise 1.2.8) that, because of linearity, the following is valid:

1.2.8   **Corollary** *The minimum distance of linear codes is the minimum weight:*

$$\mathrm{dist}(C) = \min \left\{ \mathrm{wt}(c) \mid c \in C \setminus \{0\} \right\}. \qquad \square$$

An (n,k)-code C with minimum distance d is called an (n,k,d)-*code* or a linear code of *type* (n,k,d). If C is an (n,k,d)-code over a field with q elements, we also say that it is an (n,k,d,q)-code.

1.2.9   **Corollary** *Using an (n,k,d)-code in connection with maximum-likelihood-decoding, we can correct up to*

$$t := \lfloor (d-1)/2 \rfloor$$

errors. For this reason, (n,k,d)-codes are sometimes called t-error-correcting linear codes, for $t := \lfloor (d-1)/2 \rfloor$. Moreover, such a code is $(d-1)$-error-detecting since a word which was received with at least one and at most $d-1$ errors can never be another codeword. $\qquad \square$

This is of course the reason why the minimum distance of a code is of such importance.

We are now able to refine our communication model. Denoting by m a message, and by M the message space $\mathbb{F}^k$, we are faced with the situation of Fig. 1.3.

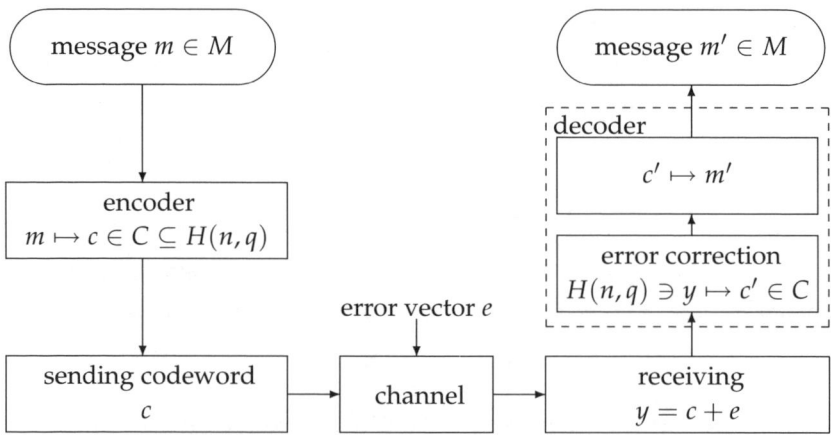

Fig. 1.3 The refined communication system once again

Of course, it is not always true that the message m' after decoding is equal to the message m which was sent originally. The main point is that maximum likelihood decoding ensures that $m = m'$, provided that $\mathrm{wt}(e) \leq \lfloor (d-1)/2 \rfloor$.

Exercises

Exercise Assume that V and W are two finite dimensional vector spaces over   **E.1.2.1**
$\mathbb{F}$ with bases

$$\mathcal{B} = (b^{(0)}, \ldots, b^{(k-1)}) \quad \text{and} \quad \mathcal{C} = (c^{(0)}, \ldots, c^{(n-1)}).$$

Prove the following: Every vector space homomorphism $\varphi : V \to W$ is uniquely determined by its values on a basis of V. Assume that

$$\varphi(b^{(i)}) = \sum_{j \in n} \kappa_{ij} c^{(j)}, \quad \kappa_{ij} \in \mathbb{F},$$

for $i \in k$. If we collect the elements κ_{ij} in form of a matrix we obtain the *matrix
representation*

$$M_{\mathcal{B},\mathcal{C}}(\varphi) := (\kappa_{ij}) = \begin{pmatrix} \kappa_{00} & \kappa_{01} & \cdots & \kappa_{0,n-1} \\ \kappa_{10} & \kappa_{11} & \cdots & \kappa_{1,n-1} \\ \vdots & \vdots & \ddots & \vdots \\ \kappa_{k-1,0} & \kappa_{k-1,1} & \cdots & \kappa_{k-1,n-1} \end{pmatrix}$$

of φ with respect to the bases $\mathcal{B}$ and $\mathcal{C}$. Conversely, prove that any $k \times n$-matrix
(κ_{ij}) over $\mathbb{F}$ determines a vector space homomorphism $\varphi : V \to W$ such that
$M_{\mathcal{B},\mathcal{C}}(\varphi) = (\kappa_{ij})$. The homomorphism φ is given by

$$\varphi\left(\sum_{i \in k} v_i b^{(i)}\right) = \sum_{j \in n} w_j c^{(j)}$$

with

$$(w_0, \ldots, w_{n-1}) = (v_0, \ldots, v_{k-1}) \cdot (\kappa_{ij}).$$

This shows that a generator matrix Γ of an (n, k)-code over $\mathbb{F}$ describes a vector
space homomorphism $\varphi : \mathbb{F}^k \to \mathbb{F}^n$.

In particular, if $\mathcal{B} = (b^{(0)}, \ldots, b^{(k-1)})$ is a basis of V, every endomorphism
of V can be represented as a $k \times k$-matrix over $\mathbb{F}$ with respect to this basis.

Exercise Let V and W be two finite dimensional vector spaces over $\mathbb{F}$ of di-   **E.1.2.2**
mension k and n respectively. Show that a homomorphism $\varphi : V \to W$ is injective if and only if $\dim(\varphi(V)) = \dim(V)$. Hence, the rows of any matrix
representation (κ_{ij}) of φ are linearly independent, and the rank of (κ_{ij}) equals
k, the number of its rows. Moreover, $\varphi : V \to W$ is an isomorphism if and only
if φ is injective and $n = k$.

E.1.2.3   **Exercise** Assume that V is a k-dimensional vector space over $\mathbb{F}$ with basis $\mathcal{B} = (b^{(0)}, \ldots, b^{(k-1)})$. The matrix representation $M_{\mathcal{B},\mathcal{B}}(\varphi)$ of any automorphism φ of V is a $k \times k$-matrix.

1. Show that the rank of this matrix equals k, which means that it is a *regular* matrix.

2. Conversely, prove that any regular $k \times k$-matrix over $\mathbb{F}$ determines an automorphism of V.

3. Prove that $M_{\mathcal{B},\mathcal{B}}(\varphi_1 \varphi_2)$ equals the product $M_{\mathcal{B},\mathcal{B}}(\varphi_2) \cdot M_{\mathcal{B},\mathcal{B}}(\varphi_1)$ for all automorphisms $\varphi_1, \varphi_2 \in \mathrm{Aut}(V)$.

4. Deduce from the previous result that the set $\mathrm{GL}_k(\mathbb{F})$ of all regular $k \times k$-matrices over $\mathbb{F}$ forms a group with respect to matrix multiplication, the *general linear group*.

5. Show that the mapping $\theta \colon \mathrm{Aut}(V) \to \mathrm{GL}_k(\mathbb{F})$ which maps a vector space automorphism φ of V to $M_{\mathcal{B},\mathcal{B}}^{\top}(\varphi)$, the transpose of its matrix representation, is a group isomorphism.

6. Changing the basis of V from $\mathcal{B} = (b^{(0)}, \ldots, b^{(k-1)})$ to $\mathcal{C} = (c^{(0)}, \ldots, c^{(k-1)})$ is described by the matrix representation $M_{\mathcal{B},\mathcal{C}}(\mathrm{id})$ of the identity mapping. This is also a regular quadratic matrix. Express $M_{\mathcal{C},\mathcal{C}}(\varphi)$ in terms of these matrices.

E.1.2.4   **Exercise** Prove 1.2.3. Hint: Use the fact that any encoding γ of a linear code C is a vector space isomorphism from $\mathbb{F}^k$ to C.

E.1.2.5   **Exercise** Which code has an invertible generator matrix?

E.1.2.6   **Exercise** Show that in a linear code over $\mathbb{F} = \{0,1\}$ either all codewords begin with 0, or exactly half of them begin with 0 and half of them begin with 1.

E.1.2.7   **Exercise** Give a formal proof of 1.2.6.

E.1.2.8   **Exercise** Give a formal proof of 1.2.8.

E.1.2.9   **Exercise** Assume that G is an abelian group which contains a subset A with the following three properties:

1. If $a_1, a_2 \in A$, then $a_1 - a_2 \in A$,

2. if $b_1, b_2 \in G \setminus A$, then $b_1 - b_2 \in A$,

3. if $a \in A$ and $b \in G \setminus A$, then $a + b \notin A$.

Show that $A = G$ or A is a proper subset of G and there is an element b_0 of $G \setminus A$ such that $G = A \cup (b_0 + A)$ where $b_0 + A = \{b_0 + a \mid a \in A\}$. (This exercise generalizes the last two exercises.)

Exercise   E.1.2.10

1. Show that there are $\binom{n}{m}$ binary vectors of length n and weight m.

2. Show that there are $(q-1)^m \binom{n}{m}$ vectors in $\mathbb{F}^n$ of weight m, provided that $\mathbb{F}$ consists of q elements.

3. For $x \in H(n,q)$, how many vectors $y \in H(n,q)$ satisfy $d(x,y) \leq m$?

Exercise Let u and v be binary vectors of length n with $d(u,v) = d$. For $r, s \in \mathbb{N}$ E.1.2.11
determine z by

$$z := \left| \{ w \in \{0,1\}^n \mid d(u,w) = r \text{ and } d(v,w) = s \} \right|.$$

Prove the following statements: If $d + r - s \geq 0$ and $d + r - s$ is even then

$$z = \binom{d}{i} \binom{n-d}{r-i},$$

where $i = (d + r - s)/2$. If $d + r - s$ is odd or $d + r - s < 0$, then $z = 0$. If $r + s = d$, then $z = \binom{d}{r}$.

Exercise Let u, v and w be binary vectors which are pairwise at distance d.   E.1.2.12
Show that d is even and that there exists exactly one vector which is at distance $d/2$ from u, v, w. If u, v, w and x are binary vectors which are pairwise at distance d, show that there exists at most one vector at distance $d/2$ from u, v, w and x.

Exercise Show that if C is a binary linear code, and $a \in \{0,1\}^n \setminus C$, then   E.1.2.13
$C \cup (a + C)$ is also a linear code.

Exercise Define the "intersection" of two binary vectors u and v to be the   E.1.2.14
vector

$$u \wedge v := (u_0 v_0, \ldots, u_{n-1} v_{n-1})$$

which has ones only where both u and v have ones. Also, let

$$u \vee v := (1 - (1 - u_0)(1 - v_0), \ldots, 1 - (1 - u_{n-1})(1 - v_{n-1}))$$

be the "union" of u and v, i.e. the vector which is one if at least one of u or v is one. Show that

$$\text{wt}(u+v) = \text{wt}(u) + \text{wt}(v) - 2\,\text{wt}(u \wedge v) = \text{wt}(u \vee v) - \text{wt}(u \wedge v).$$

1.3 Check Matrices and the Dual Code

Let us now address the important issue of decoding. It turns out that Linear Algebra helps to understand the problem quite a bit. We will discuss a decoding method using what is called the *coset leader algorithm*. Nevertheless, this problem is computationally hard and may only be practical for small parameters. However, it illustrates some very important concepts of Coding Theory which are also useful for other purposes, too.

An (n,k)-code $C \subseteq H(n, \mathbb{F})$ can be considered both as the *image of an injective linear mapping* $\gamma \colon \mathbb{F}^k \to \mathbb{F}^n$, and as the *kernel of a surjective linear mapping* $\delta \colon \mathbb{F}^n \to \mathbb{F}^{n-k}$ (Exercise 1.3.1). This leads to the following

1.3.1 **Definition (check matrices)** Let C be an (n,k)-code over $\mathbb{F}$. There exists an $(n-k) \times n$-matrix Δ over $\mathbb{F}$ which is of rank $n-k$ and satisfies

$$C = \ker(\delta) = \{w \in \mathbb{F}^n \mid w \cdot \Delta^\top = 0\},$$

where $\Delta^\top$ denotes the transpose of the matrix Δ. Any such matrix is called *check matrix* of C.   ◇

Codes over the field $\mathbb{F}_2 := \{0,1\}$ of two elements are called *binary codes*. Codes over the field $\mathbb{F}_3 := \{0,1,2\}$ of three elements are called *ternary codes*, whereas codes over a four-element field $\mathbb{F}_4$ are called *quaternary*.

1.3.2 **Example** Consider the following check matrix over the field $\mathbb{F}_2 = \{0,1\}$ of two elements, consisting of a single row of length $n \geq 2$,

$$\Delta := \left(\begin{array}{cccc} 1 & 1 & \ldots & 1 \end{array}\right).$$

It is a check matrix of a binary $(n, n-1)$-code C. Each codeword

$$c = (c_0, \ldots, c_{n-1}) \in C$$

is of even weight, since

$$0 = c \cdot \Delta^\top = c_0 + \ldots + c_{n-1} \equiv \text{wt}(c) \bmod 2.$$

Conversely, $w \cdot \Delta^\top = 0$ for each vector $w \in \mathbb{F}_2^n$ of even weight, i.e., C consists of the vectors of even weight in $\mathbb{F}_2^n$, and so C has minimum distance $d = 2$. This shows that C can detect one error. It is called a *parity check code*, since C can be obtained in the following way: Take $C' := \mathbb{F}_2^{n-1}$ as the message space and add to each of its elements $c' = (c_0, \ldots, c_{n-2})$ a further coordinate c_{n-1}, a single bit called a *parity check bit*, given by

$$c_{n-1} := \begin{cases} 1 & \text{if } \mathrm{wt}(c') \text{ is odd,} \\ 0 & \text{otherwise.} \end{cases}$$

The purpose of the parity check bit is to ensure that each codeword of the extended code C has even weight. A generator matrix of C is

$$\Gamma = \begin{pmatrix} 1 & & & 1 \\ & 1 & 0 & 1 \\ & 0 & \ddots & \vdots \\ & & 1 & 1 \end{pmatrix}.$$

$\diamond$

Let us abbreviate the all-one vector $(1, \ldots, 1)$ as $\mathbf{1}$ and the vector whose entries are all zero by $\mathbf{0}$. We also write $\mathbf{1}_n$ or $\mathbf{0}_n$ for such vectors of length n. For instance, the check matrix and the generator matrix of the above example can be written as

$$\Delta = (\,\mathbf{1}_n\,) \quad \text{and} \quad \Gamma = (\,I_{n-1} \mid \mathbf{1}_{n-1}^\top\,),$$

respectively, where I_{n-1} indicates the identity matrix of rank $n-1$.

Now we introduce for each linear code C another code which is closely related to C via its check and its generator matrices. Using the *standard bilinear form*

$$\langle w, w' \rangle := \sum_{i \in n} w_i w_i' \in \mathbb{F},$$

we associate with C the following subspace:

Definition (the dual code, self-orthogonal and self-dual codes) The *dual code* to $C \subseteq H(n, \mathbb{F})$ is defined to be the space of vectors that are orthogonal to C with respect to the standard bilinear form:

$$C^\perp := \{ w \in \mathbb{F}^n \mid \forall\, c \in C : \langle c, w \rangle = 0 \}.$$

1.3.3

A code C is called *self-orthogonal* if $C \subseteq C^\perp$ and we say that it is *self-dual* if $C = C^\perp$.

$\diamond$

The standard bilinear form has the following property:

$$\langle w, w' \rangle = 0, \text{ for all } w \in \mathbb{F}^n \iff w' = 0 \in \mathbb{F}^n.$$

For $v \in \mathbb{F}^k$, $w \in \mathbb{F}^n$ and a generator matrix Γ of C it follows from

$$\langle v \cdot \Gamma, w \rangle = \langle v, w \cdot \Gamma^\top \rangle$$

that

$$C^\perp = \{ w \in \mathbb{F}^n \mid w \cdot \Gamma^\top = 0 \}.$$

This shows that the generator matrix Γ of C is a check matrix of $C^\perp$. Consequently, $C^\perp$ is a linear $(n, n-k)$-code. Since $(C^\perp)^\perp = C$ (cf. Exercise 1.3.12), the converse is true as well, and we obtain

1.3.4   **Corollary** *The check matrices of a code C are the generator matrices of the dual code $C^\perp$ and vice versa. Dually, the check matrices of the dual code are the generator matrices of the code.*   □

It is now time to present an example of a linear code which can correct one error. This is everyone's first code which is not a repetition code or any of the other trivial examples. It was introduced by Hamming. Before we define this code, let us make one more definition.

1.3.5   **Definition** Let $b \geq 2$ be an integer. Every nonnegative integer $m \leq b^k - 1$ can be expressed in the form

$$m = \sum_{i \in k} a_i b^i, \quad \text{where} \quad 0 \leq a_i < b, \ \text{for } i = 0, 1, \ldots, k-1.$$

We call this the base b representation of m. The a_i are called the *digits* in the representation and we write

$$m = (a_{k-1}, \ldots, a_1, a_0)_b.$$

The integer b is called *base*. The expression is unique up to the number of leading zeros (we do not distinguish between two such representations which only differ in the number of leading zeros). The case $b = 10$ is of course the usual representation of integers in decimal, whereas $b = 2$ gives us the binary numbers. Notice the ubiquitous reverse ordering of the digits with respect to the index set k.   ◇

The announced code is described in the following

1.3.6   **Example** Consider the binary representations of the numbers from 1 to 7, $(0,0,1)_2$, $(0,1,0)_2$, $(0,1,1)_2$, $(1,0,0)_2$, $(1,0,1)_2$, $(1,1,0)_2$, and $(1,1,1)_2$, respectively. We may form the binary matrix

$$A = \begin{pmatrix} 1 & 0 & 0 & 1 & 1 & 0 & 1 \\ 0 & 1 & 0 & 1 & 0 & 1 & 1 \\ 0 & 0 & 1 & 0 & 1 & 1 & 1 \end{pmatrix},$$

whose columns are exactly these binary representations (slightly "mixed up" however). We may take Δ to be the check matrix of a binary code of length 7. The rowspace of Δ is the dual space of the code, and hence the code is the set of vectors c with $c \cdot \Delta^\top = 0$. Using Linear Algebra, we can find a basis for this 4-dimensional space, and writing the basis vectors in the rows of a matrix we find that the code is generated by

$$\Gamma = \begin{pmatrix} 1 & 1 & 0 & 1 & 0 & 0 & 0 \\ 1 & 0 & 1 & 0 & 1 & 0 & 0 \\ 0 & 1 & 1 & 0 & 0 & 1 & 0 \\ 1 & 1 & 1 & 0 & 0 & 0 & 1 \end{pmatrix}.$$

This is the $(7,4)$-Hamming-code. Actually, it is a member of a whole class of codes which are all called *Hamming-codes*. The more general definition of a Hamming-code will follow at the beginning of Chapter 2. By enumerating the 16 codewords and counting weights, one can easily determine that the minimum distance of this code is 3. Note that it cannot be larger than that since we see vectors of weight 3 in the rows of the generator matrix Γ. However, we need to convince ourselves that there is no word of *lower* weight. Hence this code has type $(n,k,d,q) = (7,4,3,2)$. By 1.2.7, it is a 1-*error correcting code*. Its information rate is $k/n = 4/7 \approx 0{,}57$. By comparison, the 1-error correcting repetition code of length 3 has information rate $1/3 \approx 0{,}33$. This is already a good improvement.   ◇

Using check matrices, we can easily formulate an important decoding procedure which will turn out to agree with maximum-likelihood-decoding. For this purpose, we recall the definition of a *coset* of C, which is a subset of $\mathbb{F}^n$ of the form

$$a + C := \{a + c \mid c \in C\},$$

where a is an element of $\mathbb{F}^n$. It is possible to decompose $\mathbb{F}^n$ into pairwise disjoint cosets of C (cf. Exercise 1.3.4),

$$\mathbb{F}^n = \bigcup_i (a^{(i)} + C).$$

As coset representatives we use *coset leaders* $a^{(i)}$, which are elements of smallest weight in their coset,

$$\mathrm{wt}(a^{(i)}) \leq \mathrm{wt}(a^{(i)} + c), \quad \text{for all } c \in C.$$

The decoding algorithm itself can be described as follows:

1.3.7   **Syndrome decoding** Let Δ be a check matrix of C and suppose that the Hamming space $H(n, \mathbb{F}) \supseteq C$ is decomposed into cosets $a^{(i)} + C$ such that the chosen representatives $a^{(i)}$ are coset leaders. For each vector $w \in \mathbb{F}^n$ we call the vector

$$w \cdot \Delta^\top$$

its *syndrome*. Assume that the vector y has been received. To determine the coset $a^{(i)} + C$ containing y we proceed as follows:

— If $y \in a^{(i)} + C$, say $y = a^{(i)} + c$, then

$$y \cdot \Delta^\top = (a^{(i)} + c) \cdot \Delta^\top = a^{(i)} \cdot \Delta^\top + c \cdot \Delta^\top = a^{(i)} \cdot \Delta^\top,$$

i.e. the received vector y has the same syndrome as its coset leader.

— Syndromes of different $a^{(i)}$ are distinct, since

$$a^{(i)} \cdot \Delta^\top = a^{(j)} \cdot \Delta^\top \Rightarrow (a^{(i)} - a^{(j)}) \cdot \Delta^\top = 0 \Rightarrow a^{(i)} - a^{(j)} \in C \Rightarrow i = j.$$

Consequently, we can deduce the coset number i from the syndrome of y by comparing it to the (pairwise distinct!) syndromes of the coset leaders.

— Having the coset number i of y and its coset leader $a^{(i)}$ at hand, we simply subtract $a^{(i)}$ from y in order to obtain a codeword c. This is called the *syndrome decoding method*:

$$y \mapsto c := y - a^{(i)}.$$

For short: *Subtract from the received vector its coset leader!*

In fact, this is the maximum-likelihood-decoding method, since $y = a^{(i)} + c$ implies that

$$d(y, c) = \text{wt}(y - c) = \text{wt}(a^{(i)}),$$

as we have seen already. Therefore, since $a^{(i)}$ is a leader, c is one of the codewords next to y. $\diamond$

1.3.8   **Example** Consider the check matrix

$$\Delta = \begin{pmatrix} 1 & 1 & 0 & 1 & 0 & 0 \\ 1 & 1 & 1 & 0 & 1 & 0 \\ 0 & 1 & 1 & 0 & 0 & 1 \end{pmatrix}$$

of a binary $(6,3)$-code. The following table presents the coset leaders and the corresponding syndromes:

$a^{(i)}$	$a^{(i)} \cdot \Delta^\top$
000000	000
000001	001
000010	010
000100	100
100000	110
000101	101
001000	011
010000	111

The reader should carefully note that coset leaders are usually not uniquely determined. In our example there are *several coset leaders* admissible for the syndrome 101. They are 000101, 101000 and 010010.   ◇

The following theorem provides an important characterization of the minimum distance in terms of check matrices. Remember that the check matrix is not unique. The statement holds true for any check matrix.

Theorem *The check matrix Δ of an (n,k,d)-code over $\mathbb{F}$ with $0 < k < n$ has the following properties:*   **1.3.9**

1. *Δ is an $(n-k) \times n$-matrix over $\mathbb{F}$ of rank $n-k$,*

2. *any $d-1$ columns are linearly independent, and*

3. *there exist d columns that are linearly dependent.*

Conversely, any matrix Δ satisfying these properties is a check matrix of an (n,k,d)-code over $\mathbb{F}$.

Proof: To begin with, we assume that Δ is a check matrix of such a code. By 1.3.4, Δ is a generator matrix of the dual code, i.e. an $(n-k) \times n$-matrix over $\mathbb{F}$ of rank $n-k$. Now let c be a word in C of minimum weight d. Then $c \cdot \Delta^\top = 0$, since the rows of Δ are a basis for $C^\perp$. But this means that there is a nontrivial linear combination of d columns of Δ that gives the zero vector (namely, the columns corresponding to the nonzero entries of c). Moreover, since there is *no* codeword $c \neq 0$ with Hamming weight strictly less than d, any $d-1$ columns of Δ are linearly independent.

Conversely, if we are given such a matrix Δ, the rank condition implies that the set
$$\{w \in \mathbb{F}^n \mid w \cdot \Delta^\top = 0\}$$
is a subspace of dimension k. Moreover, as before, we find that it is a code of minimum distance d.   □

From this result we deduce the following criterion which can be used in many cases:

1.3.10   **Corollary** *Each $(n - k) \times n$-matrix over $\mathbb{F}$ of rank $n - k$ with the property that any $d - 1$ of its columns are linearly independent is a check matrix of a linear (n, k)-code C over $\mathbb{F}$ with minimum distance $\mathrm{dist}(C) \geq d$, for short: of an $(n, k, \geq d)$-code.*   □

The excluded case when $k = n$ is obviously trivial, since in this case $d = 1$.

Exercises

E.1.3.1   **Exercise** Check that in fact any (n, k)-code C can be described as the kernel of a surjective linear mapping from $\mathbb{F}^n$ to $\mathbb{F}^{n-k}$. Hint: Assume that Γ is a generator matrix of C. Let $\{b^{(0)}, \ldots, b^{(n-k-1)}\}$ be a basis of the solution space of the homogeneous linear system $\Gamma \cdot x^\top = 0$, where $x \in \mathbb{F}^n$. Then C is the kernel of the mapping $\mathbb{F}^n \to \mathbb{F}^{n-k} : w \mapsto w \cdot \Delta^\top$ with

$$
\Delta = \left(\begin{array}{c} \overline{b^{(0)}} \\ \vdots \\ \overline{b^{(n-k-1)}} \end{array} \right),
$$

an $(n - k) \times n$-matrix over $\mathbb{F}$. Thus, Δ is a check matrix of C.

E.1.3.2   **Exercise** List all codewords of the binary codes C_0 and C_1 with the check matrices

$$
\Delta_0 = \left(\begin{array}{cccc} 1 & 1 & 0 & 0 \\ 1 & 0 & 1 & 1 \end{array} \right) \quad \text{and} \quad \Delta_1 = \left(\begin{array}{cccc} 0 & 1 & 1 & 1 \\ 1 & 0 & 1 & 1 \end{array} \right).
$$

How are these two codes related?

E.1.3.3   **Exercise** Assume that Δ is the check matrix of a linear code C. Describe the set of all check matrices of C.

E.1.3.4   **Exercise** Verify that $\mathbb{F}^n$ is the union of pairwise disjoint cosets of C.

E.1.3.5   **Exercise**

1. Check that the rowspace of the matrix Γ in 1.3.6 is indeed the dual space of the rowspace of Δ.
2. Verify the claim about the minimum distance of the $(7, 4)$ Hamming-code made in 1.3.6.

Exercise Prove that for a binary code with check matrix Δ, the syndrome is   E.1.3.6
the transpose of the sum of the columns of Δ where the errors have occurred.

Exercise Compute coset leaders for the binary code generated by   E.1.3.7

$$\Gamma = \begin{pmatrix} 1 & 0 & 1 & 1 & 0 & 1 \\ 0 & 1 & 1 & 0 & 1 & 1 \\ 0 & 0 & 0 & 1 & 1 & 1 \end{pmatrix}.$$

Decode the vectors $(1,1,0,1,0,0)$ and $(1,1,1,1,1,1)$ using the method of 1.3.7.

Exercise Evaluate the minimum distances of the binary codes which are gen-   E.1.3.8
erated by

$$\begin{pmatrix} 1 & 0 & 1 & 0 & 1 & 0 & 1 \\ 0 & 1 & 1 & 0 & 0 & 1 & 1 \\ 0 & 0 & 0 & 1 & 1 & 1 & 1 \end{pmatrix} \text{ and } \begin{pmatrix} 1 & 1 & 1 & 1 & 1 & 1 & 1 & 1 & 0 & 0 \\ 1 & 1 & 1 & 1 & 0 & 0 & 0 & 0 & 1 & 0 \\ 1 & 1 & 1 & 0 & 1 & 0 & 0 & 0 & 0 & 1 \end{pmatrix}.$$

Exercise Let C be an (n,k)-code. Consider the block matrix $\Gamma = (I_k \mid A)$,   E.1.3.9
where A is a $k \times (n-k)$-matrix and I_k denotes the unit matrix. Show that Γ is
a generator matrix of C if and only if $\Delta = (-A^\top \mid I_{n-k})$ is a check matrix of C.

Exercise Prove that for each generator matrix Γ and every check matrix Δ of   E.1.3.10
C the products $\Gamma \cdot \Delta^\top$ and $\Delta \cdot \Gamma^\top$ are zero matrices.

Exercise Over any finite field $\mathbb{F}$, the $(n, n-1)$ parity check code C is obtained   E.1.3.11
from the message space $\mathbb{F}^{n-1}$ by adding a parity check bit $c_n := -\sum_{i=0}^{n-1} c_i$.
Find a generator matrix for this code and determine the minimum distance.

Exercise Verify that $(C^\perp)^\perp = C$.   E.1.3.12

Exercise Assume that C and C' are linear codes of length n and let $C + C' :=$ E.1.3.13
$\{c + c' \mid c \in C,\ c' \in C'\}$. Show that $(C + C')^\perp = C^\perp \cap C'^\perp$.

Exercise A linear code C is self-orthogonal if and only if $\langle c, c' \rangle = 0$ for all   E.1.3.14
$c, c' \in C$. Show that C is self-dual if and only if C is self-orthogonal and C is of
dimension $k = n/2$ (and hence n is even).

Exercise Construct binary self-dual codes of lengths 4 and 8.   E.1.3.15

E.1.3.16   **Exercise** Let C be a binary, self-orthogonal code.

1. Show that each word of C is even and that $C^\perp$ contains the all-one vector **1**.

2. Assume in addition that the length n of C is odd and that the dimension of C is $(n-1)/2$. Show that

$$C^\perp = C \cup (\mathbf{1} + C).$$

E.1.3.17   **Exercise** Show that a code with check matrix $\Delta = (I_k \mid A)$ is self-dual if and only if A is a square matrix with $A \cdot A^\top = -I_k$.

E.1.3.18   **Exercise** Show the following:

1. If $u, v \in \mathbb{F}_2^n$, then $\langle u, v \rangle \equiv \mathrm{wt}(u \wedge v) \bmod 2$ (where $u \wedge v$ is as in Exercise 1.2.14).

2. If $u \in \mathbb{F}_2^n$, then $\langle u, u \rangle \equiv \mathrm{wt}(u) \bmod 2$.

3. If $u \in \mathbb{F}_3^n$, then $\langle u, u \rangle \equiv \mathrm{wt}(u) \bmod 3$.

E.1.3.19   **Exercise** If C is a binary, self-orthogonal code, show that every codeword has even weight. Furthermore, if each row of the generator matrix Γ of C has weight divisible by 4, then so does every codeword.

E.1.3.20   **Exercise** Let C be a ternary, self-orthogonal code. Show that $\mathrm{wt}(c) \equiv 0 \bmod 3$ for every codeword $c \in C$.

E.1.3.21   **Exercise** Let C be a code whose generator matrix Γ has the property that no column of Γ is zero and no two columns of Γ are linearly dependent. Show that $\mathrm{dist}(C^\perp) \geq 3$. (Such codes will be called *projective* in 6.1.14.)

1.4 Classification by Isometry

As we have seen, the coding theoretic properties of a code depend primarily on the Hamming distances between different codewords and between codewords and non-codewords. For example, the closest pair of codewords determines the error-correction rate of a code. Moreover, it may be that one code can be mapped onto another by means of a map which preserves the Hamming distances. Clearly, in any practical application, one code would be as good as the other, as far as error-correction is concerned. It seems natural to call such codes equivalent. In this section we study a corresponding notion of equivalence, by means of which codes can be classified.

Of course, only the types of essentially distinct – i.e. nonequivalent – codes are of interest. In fact, there are various ways in which such an equivalence relation can be defined. We discuss three such relations. These relations are indeed only refinements of each other, meaning that there is one relation which is strongest. The other relations are "weaker" in the sense that codes which are equivalent under the strongest relation may be inequivalent under the other two relations. The three relations are motivated by concepts from Projective Geometry, see Section 3.7 for more on that.

Recall that an *equivalence relation* R on a set X is a subset of $X \times X$ such that for all $x, y, z \in X$ we have

— $(x, x) \in R$ (reflexivity),

— $(x, y) \in R$ if and only if $(y, x) \in R$ (symmetry),

— $(x, y), (y, z) \in R$ implies that also $(x, z) \in R$ (transitivity).

The equivalence class of $x \in X$ with respect to R is the set

$$[x]_R := \{y \in X \mid (x, y) \in R\},$$

and the set of all equivalence classes with respect to R is indicated as X/R. It forms a decomposition of X into pairwise disjoint and nonempty subsets. Instead of $(x, y) \in R$ we usually write $x \sim y$ where $\sim$ denotes the equivalence relation.

Two (n, k)-codes $C, C' \subseteq H(n, q)$ are of the same quality if there exists a mapping

$$\iota \colon H(n, q) \to H(n, q)$$

with $\iota(C) = C'$ which preserves the Hamming distance, i.e.

$$d(w, w') = d(\iota(w), \iota(w')), \quad \text{for all } w, w' \in H(n, q).$$

Mappings with the latter property are called *isometries*. Using this notion we introduce the following concept which is in fact *the central concept of the present book:*

Definition (isometric codes) Two linear codes $C, C' \subseteq H(n, q)$ are called *isometric* if there exists an isometry of $H(n, q)$ that maps C onto C'. ◇   **1.4.1**

Obvious isometries are the permutations of the coordinates. These isometries will be called *permutational isometries*. Recall that the set of bijections from a set X to itself forms a group, the *symmetric group*

$$S_X := \{\pi \mid \pi\colon X \to X,\ \pi \text{ is bijective}\}.$$

The multiplication in this group is the composition of mappings,

$$(\pi \circ \rho)(x) := \pi(\rho(x)).$$

We write S_n for the symmetric group on the set $X = n = \{0,\ldots,n-1\}$.

1.4.2   **Definition (permutationally isometric codes)** Two linear codes $C, C' \subseteq H(n,q)$ are *permutationally isometric* if there exists a permutational isometry of $H(n,q)$ that maps C onto C'. This means that there is a permutation π in the symmetric group S_n such that

$$C' = \pi(C) = \{\pi(c) \mid c \in C\}, \ \text{ and } \ d(c,\tilde{c}) = d(\pi(c), \pi(\tilde{c})),$$

for all $c, \tilde{c} \in C$, where

$$\pi(c) = \pi(c_0,\ldots,c_{n-1}) := (c_{\pi^{-1}(0)},\ldots,c_{\pi^{-1}(n-1)}). \qquad \diamond$$

Isometries which are also linear mappings are called *linear isometries* (with respect to the Hamming metric). Linear isometries leave the Hamming weight invariant, since by linearity we have $\iota(0) = 0$, and therefore also

$$\mathrm{wt}(v) = d(v,0) = d(\iota(v),\iota(0)) = d(\iota(v),0) = \mathrm{wt}(\iota(v)).$$

1.4.3   **Definition (linearly isometric codes)** Two linear codes $C, C' \subseteq H(n,q)$ are *linearly isometric* if there exists a linear isometry of $H(n,q)$ that maps C onto C'. $\qquad \diamond$

We remark that what we call linearly isometric is often called isometric (unqualified) or monomially isometric in the literature. Our reason for calling it linearly isometric is two-fold. First, we will see shortly that this, together with the special case of permutational isometry, is not the only way in which codes can be isometric. Secondly, concerning the notion of equivalence, we felt that the concept of monomial mapping is not that well-known. Hence we chose to make reference to the fact that these isometries are induced by linear mappings.

We might have imposed a seemingly weaker condition by asking for the existence of a *local* linear isometry between C and C' only, i.e. an isometry of C and not necessarily of $H(n,q)$, that maps C onto C'. It can be shown, see

6.8.4, that each such local linear isometry can be extended to a linear isometry of $H(n,q)$. Later on we will see that the isometry relation is an equivalence relation on the set of codes with block length n over $\mathbb{F}_q$, and in later chapters we will consider the corresponding isometry classes in detail.

In order to characterize linear isometries, we have to study linear maps of the vector space $\mathbb{F}_q^n$ and investigate their effect on the Hamming distance. That is, we study linear maps of $H(n,q)$. Recall that any linear map is defined by the images of the unit vectors. Since linear isometries preserve the Hamming weight, a unit vector $e^{(i)}$ is mapped to a nonzero multiple of a unit vector, i.e.

$$\iota(e^{(i)}) = \kappa_j e^{(j)}, \text{ for suitable } j \in n, \; \kappa_j \in \mathbb{F}_q^* := \mathbb{F}_q \setminus \{0\} \, .$$

Moreover, the sum of two *different* unit vectors is of weight 2, and so different unit vectors are mapped under ι to nonzero multiples of different unit vectors. Hence, there exists a unique permutation π in the symmetric group S_n and a unique mapping φ from $n = \{0, \ldots, n-1\}$ to $\mathbb{F}_q^*$ such that

$$\iota(e^{(i)}) = \varphi(\pi(i))e^{(\pi(i))}.$$

Therefore, we may record ι as a pair of mappings,

$$\iota = (\varphi; \pi).$$

In terms of these mappings, applying ι to $v := \sum_{i \in n} v_i e^{(i)}$ gives

$$\iota(v) = (\varphi; \pi)(v) = \sum_{i \in n} v_i \varphi(\pi(i)) e^{(\pi(i))} = \sum_{i \in n} \varphi(i) v_{\pi^{-1}(i)} e^{(i)},$$

i.e.

$$(\varphi; \pi)((v_0, \ldots, v_{n-1})) = (\varphi(0)v_{\pi^{-1}(0)}, \ldots, \varphi(n-1)v_{\pi^{-1}(n-1)}).$$

Using matrix multiplication, we could also write

$$(\varphi; \pi)((v_0, \ldots, v_{n-1})) = (v_0, \ldots, v_{n-1}) \cdot M_{(\varphi;\pi)}^\top,$$

where $M_{(\varphi;\pi)}$ is the matrix whose k-th column is zero except for the (i,k)-entry which is $\varphi(i)$. Here $i = \pi(k)$, so that

$$M_{(\varphi;\pi)} := \begin{pmatrix} & & & 0 & & & \\ & & & \vdots & & & \\ & & & 0 & & & \\ 0 & \cdots & 0 & \varphi(i) & 0 & \cdots & 0 \\ & & & 0 & & & \\ & & & \vdots & & & \\ & & & 0 & & & \end{pmatrix} \quad i = \pi(k).$$

Conversely, any linear mapping with

$$e^{(i)} \longmapsto \varphi(\pi(i))e^{(\pi(i))},$$

for $\varphi \colon n \to \mathbb{F}_q^*$ and $\pi \in S_n$, is a linear isometry. Moreover, linear isometries are invertible, and the composition of two of them is again a linear isometry. A straightforward calculation shows (Exercise 1.4.1) that

$$(\psi;\rho)((\varphi;\pi)(v)) = (\psi\varphi_\rho;\rho\pi)(v), \qquad v \in H(n,q),$$

where $\psi\varphi_\rho(i) := \psi(i)\varphi(\rho^{-1}(i))$. Summarizing we obtain

---

1.4.4   **Corollary** *The linear isometries form the group*

$$\left\{ (\varphi;\pi) \ \middle| \ \varphi \colon n \to \mathbb{F}_q^*, \ \pi \in S_n \right\},$$

called the group of linear isometries *of the Hamming space. Multiplication in this group is given by the formula*

$$(\psi;\rho)(\varphi;\pi) = (\psi\varphi_\rho;\rho\pi).$$

The matrices representing the elements of this group form

$$M_n(q) := \left\{ M_{(\varphi;\pi)} \ \middle| \ \varphi \colon n \to \mathbb{F}_q^*, \ \pi \in S_n \right\},$$

and they multiply according to the rule

$$M_{(\psi;\rho)} \cdot M_{(\varphi;\pi)} = M_{(\psi\varphi_\rho;\rho\pi)}.$$

The correspondence between a linear map and the associated matrix with respect to a fixed basis constitutes the isomorphism

$$(\varphi;\pi) \longmapsto M_{(\varphi;\pi)}$$

between these two groups.   □

The application of the linear isometry group to the Hamming space is our central concept, and it is a special case of the general notion of group action which we will use in other situations, too. Hence, we carefully introduce the basic definitions and results on group actions at this point.

Actions of groups on sets play an important role in Algebra, in Combinatorics, in Topology, but also in the sciences (Chemistry, Computer Science and Physics, in particular). For more details on group actions we refer the reader to [108].

An *action* of a group G (which we assume to be written multiplicatively) from the left on a nonempty set X is defined by a mapping

$$G \times X \to X \ : \ (g, x) \mapsto gx$$

with the properties

$$(gg')x = g(g'x) \text{ and } 1x = x,$$

for $x \in X$, $g, g' \in G$ and the identity element 1 of G. We abbreviate such an action of G on X from the left by

$$_G X.$$

An equivalent characterization of a group action is as follows (Exercise 1.4.4).

Lemma *Let $_G X$ be a group action. Then the mapping*   1.4.5

$$\delta : G \to S_X \ : \ g \mapsto \overline{g}, \text{ where } \overline{g} : x \mapsto gx,$$

from G to the symmetric group S_X is a homomorphism. The **kernel** *of the action is by definition the kernel of this homomorphism, i.e. the set of group elements that fix each $x \in X$.*   □

We call δ the *permutation representation* induced by the action of $_G X$, $\overline{g} = \delta(g)$ is the permutation *induced by g on X* and $\overline{G} := \delta(G)$ the permutation group *induced by G on X*. Actions from the right are defined similarly. In the following, we define the basic notions for actions from the left. It is clear that corresponding notions can be introduced for actions from the right as well.

The crucial point is that $_G X$ induces the following relation $\sim_G$ on X:

$$x \sim_G y \ :\Longleftrightarrow \ \exists g \in G \ : \ gx = y.$$

It is easy to prove that $\sim_G$ is indeed an *equivalence relation* on X (Exercise 1.4.4). The proof is based on the following fact which is immediate from the definition of group actions and which is of fundamental importance:

$$gx = x' \ \Longleftrightarrow \ x = g^{-1}x'.$$

The equivalence class

$$G(x) = \{gx \mid g \in G\}$$

of $x \in X$ is called the *G-orbit* or, briefly, the *orbit* of x. We use the notation

$$G \backslash\backslash X := \{G(x) \mid x \in X\}$$

to denote the set of orbits of G on X. A minimal but complete set T of orbit representatives is called a *transversal* of the orbits. Since $\sim_G$ is an equivalence relation on X, $G\backslash\backslash X$ is a *set partition* of X, i.e. a complete dissection of X into pairwise disjoint and nonempty subsets $G(t)$, for $t \in T$:

1.4.6
$$X = \overset{\cdot}{\underset{\omega \in G\backslash\backslash X}{\bigcup}} \omega = \bigcup_{t \in T} G(t).$$

Several basic examples of group actions appearing in Group Theory and Combinatorics are described in Exercises 1.4.5 to 1.4.7. It is easy to check that for any group action $_G X$ the orbits $G(x)$ and $\overline{G}(x)$, $x \in X$, coincide, whence, $G\backslash\backslash X = \overline{G}\backslash\backslash X$. A group action is called *finite* if both G and X are finite. If X is finite, then the action $_{\overline{G}} X$ is always finite.

We are now going to introduce an important action of a group on a set of mappings. This action will be the prototype action for the enumeration of isometry classes of codes later on. For nonempty sets X and Y, the set of mappings from X to Y is denoted as

$$Y^X := \{f \mid f : X \to Y\}.$$

If G acts on X, then we can define an action of G on Y^X as follows:

1.4.7
$$G \times Y^X \to Y^X : (g, f) \mapsto f \circ \overline{g}^{-1}.$$

Here $\overline{g}$ is the permutation induced by g on X as introduced in 1.4.5. Thus, under this action, we associate to the pair (g, f) the composition $f \circ \overline{g}^{-1}$, i.e. the mapping $\widetilde{f} \in Y^X$ with $\widetilde{f}(x) = f(g^{-1}x)$, for all $x \in X$.

Let us now introduce the wreath product of two groups. As it turns out, the linear isometry group of the Hamming space will be such a product.

1.4.8   **Definition (wreath product)** Consider an action $_G X$ and a group H. The *wreath product* of H with G, with respect to $_G X$, consists of the set

$$H \wr_X G := H^X \times G = \{(\varphi; g) \mid \varphi : X \to H, \, g \in G\},$$

with multiplication defined by

$$(\varphi; g)(\varphi'; g') := (\varphi \varphi'_g; gg'),$$

where $(\varphi \varphi'_g)(x) := \varphi(x) \cdot \varphi'_g(x)$ and $\varphi'_g(x) := \varphi'(g^{-1}x)$, for $x \in X$. The identity element is

$$1_{H \wr_X G} = (\epsilon; 1_G),$$

where $\epsilon \in H^X$ is the constant mapping $\epsilon : x \mapsto 1_H$, and $1_G, 1_H$ denote the identity elements of G and H, respectively. The inverse of $(\varphi; g) \in H \wr_X G$ is

$$(\varphi; g)^{-1} = (\varphi_{g^{-1}}^{-1}; g^{-1}),$$

where
$$\varphi^{-1}(x) := \varphi(x)^{-1} \quad \text{and} \quad \varphi_{g^{-1}}^{-1} := (\varphi_{g^{-1}})^{-1} = (\varphi^{-1})_{g^{-1}}. \qquad \diamond$$

So, the wreath product $H \wr_X G$ comes together with an action of G on X. It may happen that the group H acts on another set Y, say. In this case, we can define an action of $H \wr_X G$ on the set of mappings Y^X in the following way.

$$H \wr_X G \times Y^X \to Y^X \; : \; ((\varphi; g), f) \mapsto \widetilde{f}, \text{ where } \widetilde{f}(x) := \varphi(x) f(g^{-1}x). \qquad \textbf{1.4.9}$$

This action is a host of further actions, some of which will be described next. These further actions are in fact actions of various subgroups of $H \wr_X G$ (cf. Exercise 1.4.5). The first case is when the group G is trivial and all mappings $\varphi : X \to H$ are constant. In this situation, only the group H acts on the set Y, such that the corresponding action on the set of functions Y^X is

$$H \times Y^X \to Y^X \; : \; (h, f) \mapsto \overline{h} \circ f. \qquad \textbf{1.4.10}$$

Another action is given by the direct product $H \times G$ of the groups H and G, which acts as follows:

$$(H \times G) \times Y^X \to Y^X \; : \; ((h, g), f) \mapsto \overline{h} \circ f \circ \overline{g}^{-1}. \qquad \textbf{1.4.11}$$

The purpose of Exercise 1.4.11 is to show that these definitions yield group actions. The action of the wreath product 1.4.9 is a generalization of 1.4.7, 1.4.10, and 1.4.11.

Example (the linear isometry group) Our paradigmatic example of an action   **1.4.12**
as in 1.4.9 is the following one. Take as H the multiplicative group $\mathbb{F}_q^*$ of the field $\mathbb{F}_q$. Let G be the symmetric group S_n acting on the set $n = \{0, \ldots, n-1\}$. Thus
$$H \wr_X G := \mathbb{F}_q^* \wr_n S_n = \left\{ (\varphi; \pi) \; \middle| \; \varphi : n \to \mathbb{F}_q^*, \; \pi \in S_n \right\}.$$
The action on $Y^X := \mathbb{F}_q^n$ is given in the following way:

$$\mathbb{F}_q^* \wr_n S_n \times \mathbb{F}_q^n \to \mathbb{F}_q^n \; : \; ((\varphi; \pi), v) \mapsto \left(\varphi(0) v_{\pi^{-1}(0)}, \ldots, \varphi(n-1) v_{\pi^{-1}(n-1)} \right).$$

Equivalently, in terms of Linear Algebra, we could also write

$$M_n(q) \times H(n, q) \to H(n, q) \; : \; (M_{(\varphi; \pi)}, v) \mapsto v \cdot M_{(\varphi; \pi)}^\top.$$

Since $M_n(q) \simeq H \wr_n S_n$ is called the *full monomial group of degree n over H*, the group of linear isometries of the Hamming space is the full monomial group of degree n over the multiplicative group of the field. $\diamond$

We are now in a position to formulate linear isometry in terms of group actions.

1.4.13   **Remarks** Let us apply what we know about linear isometry groups, their actions on vector spaces and the general theory of group actions on sets of mappings Y^X. We iterate this process of constructing actions in the following way:

— We start from the action of the linear isometry group of $H(n,q)$,

$$\mathbb{F}_q^* \wr_n S_n \left(\mathbb{F}_q^n \right) = M_n(q) \left(H(n,q) \right).$$

— Then we use the fact that the set of mappings

$$2^{H(n,q)} = \{ F \colon H(n,q) \to \{0,1\} \}$$

can be identified with the power set of $H(n,q)$ by identifying F with the inverse image $F^{-1}(\{1\})$ of 1, which is a subset of $H(n,q)$.

— The given action of the linear isometry group of $H(n,q)$ induces the action

$$_G \left(Y^X \right) := {}_{\mathbb{F}_q^* \wr_n S_n} \left(2^{\mathbb{F}_q^n} \right) = {}_{M_n(q)} \left(2^{H(n,q)} \right).$$

— Correspondingly, the orbits in

$$M_n(q) \,\backslash\backslash\, 2^{H(n,q)}$$

are the linear isometry classes of sub*sets* of $H(n,q)$ or *block codes*.

— Linear subspaces of $H(n,q)$ are of course also subsets of $H(n,q)$, and the previous remarks apply to them as well. It turns out that each element in the orbit of a linear subspace under the isometry group is again a linear subspace (this follows since the isometry group $M_n(q)$ is linear). Thus, these are the orbits we are interested in most. They are *the linear isometry classes of linear codes*.

In later chapters we will enumerate these classes, construct representatives and provide a method for randomly generating subsets of $\mathbb{F}_q^n$ that are uniformly distributed over these classes.   ◇

Next, we describe linear codes and their isometry classes as orbits under certain group actions by using results from the Exercises 1.4.14, 1.4.15, and 1.4.16, replacing the subspaces by generator matrices, i.e. by bases, so that they can be handled by a computer as well:

1.4.14   **Theorem**

1. *Assume that $\mathbb{F}_q^{k \times n,k}$ denotes the set of all $k \times n$ matrices of rank k over $\mathbb{F}_q$, $k \geq 1$, and $GL_k(q)$ the set of all regular $k \times k$-matrices over $\mathbb{F}_q$. The set of all generator matrices of the linear (n,k)-code C with generator matrix $\Gamma \in \mathbb{F}_q^{k \times n,k}$ is the orbit*

$\mathrm{GL}_k(q)(\Gamma) = \{B \cdot \Gamma \mid B \in \mathrm{GL}_k(q)\}$. *Whence the set of all linear (n,k)-codes over $\mathbb{F}_q$, we indicate it as $\mathcal{U}(n,k,q)$, can be identified with*

$$\mathrm{GL}_k(q) \backslash\backslash \mathbb{F}_q^{k \times n, k}.$$

2. *The linear isometry group $M_n(q)$ acts on $\mathcal{U}(n,k,q), k \geq 1$, according to*

$$M_n(q) \times \mathcal{U}(n,k,q) \to \mathcal{U}(n,k,q) : (M_{(\varphi;\pi)}, C) \mapsto \left\{ c \cdot M_{(\varphi;\pi)}^\top \ \middle|\ c \in C \right\}.$$

The linear isometry class of the linear (n,k)-code C is the orbit

$$M_n(q)(C).$$

Hence, the set of linear isometry classes of linear (n,k)-codes is

$$M_n(q) \backslash\backslash \mathcal{U}(n,k,q).$$

3. *The direct product $\mathrm{GL}_k(q) \times M_n(q), k \geq 1$, acts on $\mathbb{F}_q^{k \times n, k}$ by*

$$\left(\mathrm{GL}_k(q) \times M_n(q)\right) \times \mathbb{F}_q^{k \times n, k} \to \mathbb{F}_q^{k \times n, k} : \left((B, M_{(\varphi;\pi)}), \Gamma\right) \mapsto B \cdot \Gamma \cdot M_{(\varphi;\pi)}^\top$$

and so the set of linear isometry classes of linear (n,k)-codes corresponds to the set of orbits

$$\left(\mathrm{GL}_k(q) \times M_n(q)\right) \backslash\backslash \mathbb{F}_q^{k \times n, k}. \qquad \square$$

Exercises

Exercise Show that   E.1.4.1

— linear isometries are invertible,

— the composition of two of them is again a linear isometry,

— the composition satisfies

$$(\psi;\rho)((\varphi;\pi)(v)) = (\psi\varphi_\rho;\rho\pi)(v), \qquad v \in H(n,q),$$

where $\psi\varphi_\rho(i) := \psi(i)\varphi(\rho^{-1}(i))$, and

— the representing matrices satisfy

$$M_{(\psi;\rho)} \cdot M_{(\varphi;\pi)} = M_{(\psi\varphi_\rho;\rho\pi)}.$$

Exercise Let U be a nonempty subset of a finite group G (written multiplica-   E.1.4.2
tively). Show that U is a subgroup if and only if U is closed under multiplication, i.e.

$$u, u' \in U \Longrightarrow u \cdot u' \in U.$$

E.1.4.3   **Exercise** Verify 1.4.5.

E.1.4.4   **Exercise** Check that $\overline{g}$ is in fact a permutation and $\sim_G$ an equivalence relation.

E.1.4.5   **Exercise** If $_GX$ is a group action and U is a subgroup of G, prove that

$$U \times X \to X \ : \ (u,x) \mapsto ux$$

is a group action of U on X, the *restriction of $_GX$ to U*. Prove that each orbit $G(x)$ is a union of U-orbits.

E.1.4.6   **Exercise** If G is a group, prove that both

$$G \times G \to G \ : \ (g,x) \mapsto gx$$

and

$$G \times G \to G \ : \ (g,x) \mapsto xg^{-1}$$

are group actions of G on G. They are called the *left regular* or *right regular representation* of G, respectively. Prove that $G(x) = G$ for any $x \in G$. A group action with just one orbit is called *transitive*. Hence, the left regular and the right regular representation are transitive group actions.

Let U be a subgroup of G. Determine the orbits of the restricted action $_UG$. In the first case they are called *right cosets*, in the second case *left cosets* of U. Prove that all orbits $U(x)$ for $x \in G$ are of the same size. If G is a finite group, deduce that the order of U divides the order of G. This is *Lagrange's Theorem*.

E.1.4.7   **Exercise** Show that an action of G on a set X induces natural actions of G on $\binom{X}{k}$, the set of all k-subsets of X, for $0 \le k \le |X|$, and on 2^X, the power set of X, which is the set of all subsets of X. This natural action of g on the subset A of X is given by $(g, A) \mapsto \{gx \mid x \in A\}$.

E.1.4.8   **Exercise** Consider a group action $_GX$, a normal subgroup $U \trianglelefteq G$ and the restricted action $_UX$. Prove the following facts:

— For each orbit $U(x)$ and any $g \in G$, the set $gU(x)$ is again an orbit of U on X. Indeed $gU(x) = U(gx)$.

— The group G acts on the set $U\backslash\backslash X$ of the U-orbits by

$$G \times U\backslash\backslash X \to U\backslash\backslash X \ : \ (g, U(x)) \mapsto U(gx).$$

— The factor group G/U acts on the set $U\backslash\backslash X$ via

$$G/U \times U\backslash\backslash X \to U\backslash\backslash X : (gU, U(x)) \mapsto U(gx).$$

We call this action a *factor action* of G with respect to U and denote it by

$$_{G/U}(U\backslash\backslash X).$$

— Up to identification of the U-orbits with the set of their elements, the following equations hold:

$$G\backslash\backslash X = G\backslash\backslash(U\backslash\backslash X) \text{ and } G\backslash\backslash X = (G/U)\backslash\backslash(U\backslash\backslash X).$$

Exercise Use Exercise 1.4.8 in order to prove: An action of the direct product $H \times G$ on X induces both a natural action of H on the set of orbits of the restricted action $_G X$:   **E.1.4.9**

$$H \times (G\backslash\backslash X) \to G\backslash\backslash X : (h, G(x)) \mapsto G(hx),$$

and a natural action of G on the orbits of the restricted action $_H X$:

$$G \times (H\backslash\backslash X) \to H\backslash\backslash X : (g, H(x)) \mapsto H(gx).$$

Show that the orbit of $G(x) \in G\backslash\backslash X$ under H is the set of orbits of G on X that form $(H \times G)(x)$, while the orbit of $H(x) \in H\backslash\backslash X$ under G consists of the orbits of H on X, that form $(H \times G)(x)$. Hence

$$(H \times G)(x) = \bigcup_{h \in H} G(hx) = \bigcup_{g \in G} H(gx).$$

Prove the following identity for a finite set X:

$$|H\backslash\backslash(G\backslash\backslash X)| = |G\backslash\backslash(H\backslash\backslash X)| = |(H \times G)\backslash\backslash X|.$$

Exercise Assume that both $_G X$ and $_H X$ are group actions with $g(hx) = h(gx)$ **E.1.4.10**
for all $g \in G, h \in H$, and $x \in X$. Prove that

$$(H \times G) \times X \to X : ((h, g), x) \mapsto h(gx)$$

describes an action of the direct product $H \times G$ on X.

Exercise Assume that X and Y are sets and H is a group which acts on Y.   **E.1.4.11**
Prove that 1.4.10 describes an action of H on Y^X.

If, in addition, $_G X$ is another group action, then use Exercise 1.4.10 to show that 1.4.11 defines an action of $H \times G$ both on the domain and the range of these mappings. Note that $\bar{g}$ stands for the permutation representation of g acting on X, whereas $\bar{h}$ denotes the permutation representation of h acting on Y.

E.1.4.12 **Exercise** Let V be a vector space over $\mathbb{F}$. Show that the multiplicative group $\mathbb{F}^*$ acts on V by

$$\mathbb{F}^* \times V \to V \ : \ (\lambda, v) \mapsto \lambda v.$$

Prove that the orbit of 0 is of size one, and all the other orbits are of the same length. For $v \neq 0$ the orbit $\mathbb{F}^*(v)$ describes a punctured one-dimensional subspace of V, i.e. the subspace without the zero vector. If $\mathbb{F} = \mathbb{F}_q$, then the orbit of $v \neq 0$ is of size $q - 1$.

E.1.4.13 **Exercise** Show that the group of regular $k \times k$-matrices over $\mathbb{F}$ acts on $\mathbb{F}^k$ by

$$\mathrm{GL}_k(\mathbb{F}) \times \mathbb{F}^k \to \mathbb{F}^k \ : \ (B, v) \mapsto \left(B \cdot v^\top \right)^\top = v \cdot B^\top.$$

Prove that the orbit of 0 is of size one. Moreover, show that this action commutes with the action of $\mathbb{F}^*$ described in Exercise 1.4.12, and deduce from Exercise 1.4.10 that the direct product $\mathrm{GL}_k(\mathbb{F}) \times \mathbb{F}^*$ acts on $\mathbb{F}^k$. Describe the orbits $(\mathrm{GL}_k(\mathbb{F}) \times \mathbb{F}^*)(v)$ with the methods of Exercise 1.4.9.

E.1.4.14 **Exercise** Let the set of $k \times n$-matrices over $\mathbb{F}_q$ be denoted by $\mathbb{F}_q^{k \times n}$, and the set of $k \times n$-matrices of rank r by $\mathbb{F}_q^{k \times n, r}$. Show that $\mathrm{GL}_k(q) := \mathrm{GL}_k(\mathbb{F}_q), k \geq 1$, acts both on $\mathbb{F}_q^{k \times n}$ and $\mathbb{F}_q^{k \times n, r}$ by

$$(B, \Gamma) \mapsto B \cdot \Gamma$$

where $B \in \mathrm{GL}_k(q)$ is a regular matrix, and Γ is a $k \times n$-matrix.

From 1.2.3 deduce that the orbit $\mathrm{GL}_k(q)(\Gamma)$ of $\Gamma \in \mathbb{F}_q^{k \times n, k}$ determines the set of all generator matrices of the code C with Γ. Thus the set of all linear (n, k)-codes over $\mathbb{F}_q$ can be identified with the set of orbits $\mathrm{GL}_k(q) \backslash\backslash \mathbb{F}_q^{k \times n, k}$.

E.1.4.15 **Exercise** Show that the full monomial group $M_n(q)$ acts on $\mathcal{U}(n, k, q)$ by

$$M_n(q) \times \mathcal{U}(n, k, q) \to \mathcal{U}(n, k, q) \ : \ (M_{(\varphi; \pi)}, C) \mapsto \left\{ c \cdot M_{(\varphi; \pi)}^\top \ \middle| \ c \in C \right\}.$$

E.1.4.16 **Exercise** Show that $M_n(q)$ acts both on $\mathbb{F}_q^{k \times n}$ and $\mathbb{F}_q^{k \times n, r}$ by

$$(M_{(\varphi; \pi)}, \Gamma) \mapsto \Gamma \cdot M_{(\varphi; \pi)}^\top$$

where $M_{(\varphi; \pi)} \in M_n(q)$ is a monomial matrix, and Γ is a $k \times n$-matrix. Moreover, show that this action commutes with the action of $\mathrm{GL}_k(q)$ described in Exercise 1.4.14 and thus deduce from Exercise 1.4.10 that the direct product $\mathrm{GL}_k(q) \times M_n(q)$ acts on $\mathbb{F}_q^{k \times n}$ and $\mathbb{F}_q^{k \times n, r}$. Describe $(\mathrm{GL}_k(q) \times M_n(q))(\Gamma)$ with the methods of Exercise 1.4.9.

From Exercise 1.4.14 deduce that for $\Gamma \in \mathbb{F}_q^{k \times n, k}$, a generator matrix of the (n, k)-code C, the orbit $(GL_k(q) \times M_n(q))(\Gamma)$ consists of all generator matrices of codes which are linearly isometric to C. Therefore, the set of orbits $(GL_k(q) \times M_n(q)) \backslash\backslash \mathbb{F}_q^{k \times n, k}$ is in bijection to the linear isometry classes of linear (n, k)-codes over $\mathbb{F}_q$.

1.5 Semilinear Isometry Classes of Linear Codes

It is, of course, a legitimate question to ask for *generalizations* of the concept of linear isometry *by relaxing* the condition of *linearity*. The only requirement in addition to isometry will be that the admissible isometries map subspaces onto subspaces. To be more precise the image of a subspace under an isometry is again a subspace of $\mathbb{F}_q^n$. Under these assumptions we derive for $n \geq 3$ that these mappings preserve the dimension, i.e. they map (n, k)-codes to (n, k)-codes, and that they are the *semilinear* isometries of $\mathbb{F}_q^n$ (cf. 1.5.7). In order to prove this we need a more detailed analysis of isometries. At first we prove that it suffices to investigate isometries ι of $\mathbb{F}_q^n$ with $\iota(0) = 0$.

Lemma *If $\iota \colon \mathbb{F}_q^n \to \mathbb{F}_q^n$ is an isometry, then*

$$\iota' \colon \mathbb{F}_q^n \to \mathbb{F}_q^n \; : \; \iota'(v) := \iota(v) - \iota(0), \qquad v \in \mathbb{F}_q^n,$$

is again an isometry of $\mathbb{F}_q^n$ and $\iota'(0) = 0$.

Conversely, if $\iota' \colon \mathbb{F}_q^n \to \mathbb{F}_q^n$ is an isometry with $\iota'(0) = 0$, then for any $w \in \mathbb{F}_q^n$ the mapping

$$\iota \colon \mathbb{F}_q^n \to \mathbb{F}_q^n \; : \; \iota(v) := \iota'(v) + w, \qquad v \in \mathbb{F}_q^n,$$

is an isometry with $\iota(0) = w$. □

This result, the proof of which is left to the reader as Exercise 1.5.1, shows that it suffices to consider only isometries ι with $\iota(0) = 0$. For example, if ι maps subspaces onto subspaces, then this condition always holds, since the null space $\{0\}$ is mapped onto $\{0\}$. If $\iota(0) = 0$, then ι also preserves the weight, since

$$\mathrm{wt}(\iota(v)) = d(\iota(v), 0) = d(\iota(v), \iota(0)) = d(v, 0) = \mathrm{wt}(v), \qquad v \in \mathbb{F}_q^n.$$

Lemma *Each isometry ι on a finite vector space $\mathbb{F}_q^n$ is bijective. If it satisfies $\iota(0) = 0$, then it permutes the orbits*

$$\mathbb{F}_q^*(e^{(i)}) = \{\kappa e^{(i)} \mid \kappa \in \mathbb{F}_q^*\}$$

of the unit vectors with respect to the action of $\mathbb{F}_q^*$ by left multiplication. In formal terms:

$$\exists\, \pi \in S_n\ \forall\, i \in n:\ \iota\,(\mathbb{F}_q^*(e^{(i)})) = \mathbb{F}_q^*(e^{(\pi(i))}).$$

Proof: 1. It is easy to see that ι is injective: $\iota(u) = \iota(v)$ implies

$$0 = d(\iota(u), \iota(v)) = d(u, v),$$

and so $u = v$. Since ι is a map from the *finite* set $\mathbb{F}_q^n$ to itself, it is also one-to-one.

2. Now we note that, for each $i \in n$ and $\lambda \in \mathbb{F}_q^*$, there exists $k \in n$ and $\mu \in \mathbb{F}_q^*$ such that

$$\iota(\lambda e^{(i)}) = \mu e^{(k)}.$$

This follows from $1 = \mathrm{wt}(\lambda e^{(i)}) = \mathrm{wt}(\iota(\lambda e^{(i)}))$.

3. Moreover, this index k does not depend on λ: Suppose that for $\lambda = 1$ we have $\iota(e^{(i)}) = v e^{(j)}$. Then, for $\lambda \neq 1$ we get

$$1 = d(\lambda e^{(i)}, e^{(i)}) = d(\iota(\lambda e^{(i)}), \iota(e^{(i)})) = d(\mu e^{(k)}, v e^{(j)}),$$

and this implies $j = k$.

4. Thus we obtain, for the index j defined by $\iota(e^{(i)}) = v e^{(j)}$,

$$\iota(\mathbb{F}_q^*(e^{(i)})) \subseteq \mathbb{F}_q^*(e^{(j)}).$$

The bijectivity of ι implies that $\iota(\mathbb{F}_q^*(e^{(i)}))$ is in fact *equal* to $\mathbb{F}_q^*(e^{(j)})$, and it assures the existence of some $\pi \in S_n$ which satisfies

$$\iota\,(\mathbb{F}_q^*(e^{(i)})) = \mathbb{F}_q^*(e^{(\pi(i))}),$$

for all $i \in n$.   □

1.5.3   **Lemma** *Let ι be an isometry of $\mathbb{F}_q^n$ with $\iota(0) = 0$. For $i \neq k$ and $\lambda, \mu \in \mathbb{F}_q^*$ we have,*

$$\iota(\lambda e^{(i)} + \mu e^{(k)}) = \iota(\lambda e^{(i)}) + \iota(\mu e^{(k)}).$$

Proof: 1. The assumption implies that

$$2 = \mathrm{wt}(\lambda e^{(i)} + \mu e^{(k)}) = \mathrm{wt}(\iota(\lambda e^{(i)} + \mu e^{(k)})),$$

and so

$$\iota(\lambda e^{(i)} + \mu e^{(k)}) = v e^{(j_i)} + \rho e^{(j_k)},$$

for suitable $v, \rho \in \mathbb{F}_q^*$ and $j_i \neq j_k$.

2. Using

$$1 = d(\lambda e^{(i)}, \lambda e^{(i)} + \mu e^{(k)}) = d(\mu e^{(k)}, \lambda e^{(i)} + \mu e^{(k)})$$
$$= d(\iota(\lambda e^{(i)}), \iota(\lambda e^{(i)} + \mu e^{(k)})) = d(\iota(\mu e^{(k)}), \iota(\lambda e^{(i)} + \mu e^{(k)}))$$

we can deduce from 1. that

$$1 = d(\iota(\lambda e^{(i)}), v e^{(j_i)} + \rho e^{(j_k)}) = d(\iota(\mu e^{(k)}), v e^{(j_i)} + \rho e^{(j_k)}).$$

Thus, by $j_i \neq j_k$, either $\iota(\lambda e^{(i)}) = v e^{(j_i)}$ or $\iota(\lambda e^{(i)}) = \rho e^{(j_k)}$, and similarly either $\iota(\mu e^{(k)}) = \rho e^{(j_k)}$ or $\iota(\mu e^{(k)}) = v e^{(j_i)}$.

3. Since ι permutes the orbits of the unit vectors, by 1.5.2, we get from 2. that

$$\iota(\lambda e^{(i)}) + \iota(\mu e^{(k)}) = v e^{(j_i)} + \rho e^{(j_k)} = \iota(\lambda e^{(i)} + \mu e^{(k)}),$$

as stated.   □

Generalizing this approach we prove

Corollary *Let ι be an isometry of $\mathbb{F}_q^n$ with $\iota(0) = 0$, then, for $v \in \mathbb{F}_q^n$,*   **1.5.4**

$$\iota(v) = \iota\left(\sum_{i \in n} v_i e^{(i)}\right) = \sum_{i \in n} \iota(v_i e^{(i)}).$$

Proof: Let k be the number of nonzero components of v. For $0 \leq k \leq 2$ the assertion is true by assumption, by 1.5.2 and 1.5.3. Now we consider $2 < k \leq n$ and assume that the assertion is valid for all vectors with at most $k - 1$ nonzero components. We prove that it holds true for the vector $v = \sum_{r \in k} v_{i_r} e^{(i_r)}$ with k nonzero components. Thus we assume that $i_r \in n$ for $r \in k$, $i_r \neq i_s$ for $r, s \in k$, $r \neq s$, and $v_{i_r} \neq 0$ for $r \in k$. Then

$$d\left(\sum_{r=1}^{k-1} v_{i_r} e^{(i_r)}, \sum_{r=0}^{k-1} v_{i_r} e^{(i_r)}\right) = 1 = d\left(\sum_{r=0}^{k-2} v_{i_r} e^{(i_r)}, \sum_{r=0}^{k-1} v_{i_r} e^{(i_r)}\right)$$

whence

$$d\left(\iota\left(\sum_{r=1}^{k-1} v_{i_r} e^{(i_r)}\right), \iota\left(\sum_{r=0}^{k-1} v_{i_r} e^{(i_r)}\right)\right) = 1 = d\left(\iota\left(\sum_{r=0}^{k-2} v_{i_r} e^{(i_r)}\right), \iota\left(\sum_{r=0}^{k-1} v_{i_r} e^{(i_r)}\right)\right)$$

and by the induction hypothesis

$$d\left(\sum_{r=1}^{k-1} \iota(v_{i_r} e^{(i_r)}), \iota\left(\sum_{r=0}^{k-1} v_{i_r} e^{(i_r)}\right)\right) = 1 = d\left(\sum_{r=0}^{k-2} \iota(v_{i_r} e^{(i_r)}), \iota\left(\sum_{r=0}^{k-1} v_{i_r} e^{(i_r)}\right)\right).$$

According to 1.5.2 there exists some $\pi \in S_n$ and $\tilde{v}_{i_r} \in \mathbb{F}_q^*$, $r \in k$, so that

$$\sum_{r=1}^{k-1} \iota(v_{i_r} e^{(i_r)}) = \sum_{r=1}^{k-1} \tilde{v}_{i_r} e^{(\pi(i_r))} \quad \text{and} \quad \sum_{r=0}^{k-2} \iota(v_{i_r} e^{(i_r)}) = \sum_{r=0}^{k-2} \tilde{v}_{i_r} e^{(\pi(i_r))}.$$

Therefore, necessarily we have

$$\iota\left(\sum_{r=0}^{k-1} v_{i_r} e^{(i_r)}\right) = \sum_{r=0}^{k-1} \tilde{v}_{i_r} e^{(\pi(i_r))} = \sum_{r=0}^{k-1} \iota\left(v_{i_r} e^{(i_r)}\right). \qquad \square$$

We are now in a position to describe the group of isometries ι which satisfy $\iota(0) = 0$ as a wreath product. Since

$$\iota(v_i e^{(i)}) \in \iota(\mathbb{F}_q^*(e^{(i)})) = \mathbb{F}_q^*(e^{(\pi(i))}),$$

we can obtain the scalar factor of $e^{(\pi(i))}$ in $\iota(v_i e^{(i)})$ (if $v_i \neq 0$, otherwise we can simply neglect this summand since $\iota(0) = 0$) by the application of a suitable permutation $\varphi(\pi(i))$ of the scalars that keeps 0 fixed,

$$\iota(v_i e^{(i)}) = \varphi(\pi(i))(v_i) e^{(\pi(i))}.$$

Or, in formal terms and since we have to take all the indices into account, there exists a mapping

$$\varphi \colon n \longrightarrow S_{\mathbb{F}_q^*},$$

from n to the symmetric group

$$S_{\mathbb{F}_q^*} := \{\rho \mid \rho \colon \mathbb{F}_q \to \mathbb{F}_q, \ \rho \text{ is bijective and } \rho(0) = 0\}$$

on $\mathbb{F}_q^*$ (considered as the subgroup of the symmetric group $S_{\mathbb{F}_q}$ on $\mathbb{F}_q$ consisting the permutations ρ of $\mathbb{F}_q$ that keep the zero element fixed: $\rho(0) = 0$), which satisfies

$$\iota(v_0, \ldots, v_{n-1}) = (\varphi(0)(v_{\pi^{-1}(0)}), \ldots, \varphi(n-1)(v_{\pi^{-1}(n-1)})).$$

This proves the following useful description of the group of isometries:

1.5.5   **Theorem** *The group of isometries ι, with $\iota(0) = 0$, of the finite vector space $\mathbb{F}_q^n$, is the wreath product*

$$S_{\mathbb{F}_q^*} \wr_n S_n$$

of the symmetric group $S_{\mathbb{F}_q^*}$ on $\mathbb{F}_q$ and the symmetric group S_n on n. The action is the following one:

$$S_{\mathbb{F}_q^*} \wr_n S_n \times \mathbb{F}_q^n \to \mathbb{F}_q^n \ : \ ((\varphi; \pi), v) \mapsto (\varphi(0)(v_{\pi^{-1}(0)}), \ldots, \varphi(n-1)(v_{\pi^{-1}(n-1)})).$$

$\square$

It is easy to check that all these $(\varphi; \pi) \in S_{\mathbb{F}_q^*} \wr_n S_n$ are isometries which map 0 onto 0. Together with 1.5.1 we obtain

Theorem *The group of all isometries ι on the finite vector space $\mathbb{F}_q^n$ is the wreath*   **1.5.6**
product

$$S_{\mathbb{F}_q} \wr_n S_n$$

of the symmetric group $S_{\mathbb{F}_q}$ on $\mathbb{F}_q$ and the symmetric group S_n on n. The action is the following one:

$$S_{\mathbb{F}_q} \wr_n S_n \times \mathbb{F}_q^n \to \mathbb{F}_q^n \; : \; ((\varphi; \pi), v) \mapsto (\varphi(0)(v_{\pi^{-1}(0)}), \ldots, \varphi(n-1)(v_{\pi^{-1}(n-1)})).$$

$\square$

It is easy to check that all these $(\varphi; \pi) \in S_{\mathbb{F}_q} \wr_n S_n$ are isometries.

There exist isometries of $\mathbb{F}_q^n$ such that the image of a subspace of $\mathbb{F}_q^n$ is not a subspace. For instance, if $\iota(0) \neq 0$, then the null space $\{0\}$ is not mapped onto a subspace of $\mathbb{F}_q^n$. If $\iota(0) = 0$ consider, for example, the linear $(2,1)$-code C over $\mathbb{F}_5 = \mathbb{Z}/5\mathbb{Z}$ with generator matrix $\Gamma = (1\ 1)$. It contains the five codewords $(0,0)$, $(1,1)$, $(2,2)$, $(3,3)$, and $(4,4)$. The image of C under the isometry $\iota = (\varphi; \mathrm{id}) \in S_{\mathbb{F}_5^*} \wr_n S_n$ with $\varphi(0) = \mathrm{id}_{\mathbb{F}_q}$ and $\varphi(1) = \begin{pmatrix} 0 & 1 & 2 & 3 & 4 \\ 0 & 3 & 2 & 1 & 4 \end{pmatrix}$ is $\{(0,0), (1,3), (2,2), (3,1), (4,4)\}$, which is not a subspace of $\mathbb{F}_5^2$.

Now we want to show that isometries which *map subspaces onto subspaces* belong to the following class of mappings, if $n \geq 3$:

Definition (semilinear mappings) The mapping $\sigma \colon \mathbb{F}_q^n \to \mathbb{F}_q^n$ is called *semilinear*   **1.5.7**
if there exists an automorphism α of $\mathbb{F}_q$ such that, for all $u, v \in \mathbb{F}_q^n$ and all $\kappa \in \mathbb{F}_q$ we have

$$\sigma(u+v) = \sigma(u) + \sigma(v), \qquad \sigma(\kappa u) = \alpha(\kappa)\sigma(u).$$

An isometry which is also a semilinear mapping is called *semilinear isometry* (with respect to the Hamming metric). $\diamond$

Lemma *If the isometry $\iota \colon \mathbb{F}_q^n \to \mathbb{F}_q^n$, $n \geq 3$, maps subspaces onto subspaces, then*   **1.5.8**
for each $u \in \mathbb{F}_q^n$ we have

$$\iota(\mathbb{F}_q^*(u)) = \mathbb{F}_q^*(\iota(u)).$$

Moreover, there exists an automorphism α of $\mathbb{F}_q$ such that, for each $\kappa \in \mathbb{F}_q$,

$$\iota(\kappa u) = \alpha(\kappa)\iota(u).$$

Proof: 1. Since ι maps subspaces onto subspaces, the space $\{0\}$ must be mapped onto itself, whence $\iota(0) = 0$. Therefore, the assertion is obviously true for $u = 0$.

2. Assume that $u \neq 0$. Since ι is bijective and since it maps subspaces to subspaces, $\iota(\langle u \rangle)$ is a one-dimensional subspace, and so, using $\iota(u) \neq 0$, we obtain

$$\iota(\langle u \rangle) = \langle \iota(u) \rangle.$$

Moreover, as $\iota(0) = 0$,

$$\iota(\mathbb{F}_q^*(u)) = \mathbb{F}_q^*(\iota(u)).$$

Hence, there is a permutation of the scalars

$$\Phi_u \in S_{\mathbb{F}_q^*} \leq S_{\mathbb{F}_q},$$

depending possibly on the vector u, which satisfies

$$\iota(\kappa u) = \Phi_u(\kappa)\iota(u).$$

We have to show that Φ_u is independent of u and that it is a field automorphism.

3. For the special case $e := \sum_{i \in n} e^{(i)}$ we have

$$\iota(\kappa e) = \Phi_e(\kappa)\iota(e) = \Phi_e(\kappa) \sum_{i \in n} \varphi(\pi(i))(1)e^{(\pi(i))}, \qquad \kappa \in \mathbb{F}_q^*,$$

as well as

$$\iota(\kappa e) = \sum_{i \in n} \varphi(\pi(i))(\kappa)e^{(\pi(i))}, \qquad \kappa \in \mathbb{F}_q^*,$$

so that we obtain

1.5.9
$$\forall\, i \in n\, :\, \Phi_e(\kappa) = \frac{\varphi(\pi(i))(\kappa)}{\varphi(\pi(i))(1)}, \qquad \kappa \in \mathbb{F}_q^*.$$

4. Now we prove that $\Phi_e(\kappa\mu) = \Phi_e(\kappa)\Phi_e(\mu)$, for $\kappa, \mu \in \mathbb{F}_q$. The assertion is trivial for $\kappa = 0$ or $\mu = 0$. So it is possible to restrict attention to $\kappa, \mu \in \mathbb{F}_q^*$. To begin with, we consider another special case (recalling that $n > 2$, by assumption): Let

$$w := e^{(0)} + \mu e^{(i)},$$

for $i \neq 0$ and $\mu \in \mathbb{F}_q^*$. The corresponding equation

$$\iota(\kappa w) = \Phi_w(\kappa)\iota(w), \qquad \kappa \in \mathbb{F}_q^*,$$

implies that

$$\varphi(\pi(0))(\kappa)e^{(\pi(0))} + \varphi(\pi(i))(\kappa\mu)e^{(\pi(i))}$$
$$= \Phi_w(\kappa)\big(\varphi(\pi(0))(1)e^{(\pi(0))} + \varphi(\pi(i))(\mu)e^{(\pi(i))}\big).$$

Comparing the coefficients of the basis vectors on both sides we obtain two useful identities. The coefficients of $e^{(\pi(0))}$ give

$$\varphi(\pi(0))(\kappa) = \Phi_w(\kappa)\varphi(\pi(0))(1),$$

so that we can deduce

$$\Phi_w(\kappa) = \frac{\varphi(\pi(0))(\kappa)}{\varphi(\pi(0))(1)} = \Phi_e(\kappa), \qquad \kappa \in \mathbb{F}_q^*,$$

and hence $\Phi_w = \Phi_e$ in this particular situation. The second identity, obtained by comparing the coefficients of $e^{(\pi(i))}$, is

$$\varphi(\pi(i))(\kappa\mu) = \Phi_w(\kappa)\varphi(\pi(i))(\mu).$$

Using $\Phi_w = \Phi_e$ and dividing both sides by $\varphi(\pi(i))(1)$ we derive that

$$\Phi_e(\kappa\mu) = \Phi_e(\kappa)\Phi_e(\mu), \qquad \kappa, \mu \in \mathbb{F}_q^*,$$

i.e. Φ_e is multiplicative.

5. We want to show that $\Phi_u = \Phi_e$, for all $u \neq 0$. According to 1.5.4 and 1.5.9, for $u = \sum_{i \in n} u_i e^{(i)}$ we get

$$
\begin{aligned}
\iota(u) &= \sum_{i \in n} \iota(u_i e^{(i)}) = \sum_{i \in n} \varphi(\pi(i))(u_i) e^{(\pi(i))} \\
&= \sum_{i \in n} \Phi_e(u_i)\varphi(\pi(i))(1) e^{(\pi(i))}.
\end{aligned}
$$

Since Φ_e is multiplicative, we derive for $\kappa \in \mathbb{F}_q^*$ that

$$
\begin{aligned}
\iota(\kappa u) &= \sum_{i \in n} \Phi_e(\kappa u_i)\varphi(\pi(i))(1) e^{(\pi(i))} \\
&= \Phi_e(\kappa)\sum_{i \in n} \Phi_e(u_i)\varphi(\pi(i))(1) e^{(\pi(i))} \\
&= \Phi_e(\kappa)\iota(u),
\end{aligned}
$$

which can be compared with the identity

$$\iota(\kappa u) = \Phi_u(\kappa)\iota(u),$$

obtaining $\Phi_e(\kappa) = \Phi_u(\kappa)$ for all $\kappa \in \mathbb{F}_q^*$. Hence we have proved that in fact $\Phi_u = \Phi_e$, as stated.

6. It remains to show that Φ_e is additive, i.e.

$$\Phi_e(\lambda + \mu) = \Phi_e(\lambda) + \Phi_e(\mu), \qquad \lambda, \mu \in \mathbb{F}_q.$$

Since $\Phi_e(0) = 0$, this formula is true for $\lambda = 0$ or $\mu = 0$. By assumption $n \geq 3$, and so we can consider

$$u := e^{(0)} + e^{(1)}, \quad w := e^{(1)} + e^{(2)}$$

and the subspace $U := \langle\{u, w\}\rangle$ generated by these two vectors. For $\lambda, \mu \in \mathbb{F}_q^*$, the vectors $\iota(\lambda u)$, $\iota(\mu w)$ and $\iota(\lambda u) + \iota(\mu w)$ are contained in the subspace $\iota(U)$. Hence, there exists some $z \in U$, for which $\iota(z) = \iota(\lambda u) + \iota(\mu w)$. Then

$$
\begin{aligned}
\iota(z) = \; & \Phi_e(\lambda)\varphi(\pi(0))(1)e^{(\pi(0))} + \Phi_e(\lambda)\varphi(\pi(1))(1)e^{(\pi(1))} \\
& + \Phi_e(\mu)\varphi(\pi(1))(1)e^{(\pi(1))} + \Phi_e(\mu)\varphi(\pi(2))(1)e^{(\pi(2))}.
\end{aligned}
$$

On the other hand, since

$$
z = z_0 e^{(0)} + (z_0 + z_2)e^{(1)} + z_2 e^{(2)},
$$

we have

$$
\begin{aligned}
\iota(z) = \; & \Phi_e(z_0)\varphi(\pi(0))(1)e^{(\pi(0))} + \Phi_e(z_0 + z_2)\varphi(\pi(1))(1)e^{(\pi(1))} \\
& + \Phi_e(z_2)\varphi(\pi(2))(1)e^{(\pi(2))}.
\end{aligned}
$$

Since $\varphi(\pi(i))(1) \neq 0$, we derive from these two representations of $\iota(z)$ that $\Phi_e(z_0) = \Phi_e(\lambda)$ and $\Phi_e(z_2) = \Phi_e(\mu)$. Since Φ_e is a bijection on $\mathbb{F}_q$, we obtain $z_0 = \lambda$, $z_2 = \mu$ and

$$
\Phi_e(\lambda) + \Phi_e(\mu) = \Phi_e(z_0 + z_2) = \Phi_e(\lambda + \mu),
$$

which completes the proof of the additivity.

7. Hence, $\alpha := \Phi_e$ is in fact an automorphism of $\mathbb{F}_q$ which satisfies

$$
\iota(\kappa u) = \alpha(\kappa)\iota(u), \qquad \kappa \in \mathbb{F}_q, \; u \in \mathbb{F}_q^n.
$$

Finally

$$
\begin{aligned}
\iota(u + v) &= \iota\left(\sum_{i \in n}(u_i + v_i)e^{(i)}\right) \\
&= \sum_{i \in n}\iota\big((u_i + v_i)e^{(i)}\big) \\
&= \sum_{i \in n}\alpha(u_i + v_i)\varphi(\pi(i))(1)e^{(\pi(i))} \\
&= \sum_{i \in n}\alpha(u_i)\varphi(\pi(i))(1)e^{(\pi(i))} + \sum_{i \in n}\alpha(v_i)\varphi(\pi(i))(1)e^{(\pi(i))} \\
&= \iota(u) + \iota(v),
\end{aligned}
$$

which completes the proof. $\qquad\square$

Summarizing, an isometry of $\mathbb{F}_q^n$, $n \geq 3$, which maps subspaces onto subspaces is semilinear and is described by three mappings

$$
\varphi \colon n \to S_{\mathbb{F}_q^*}, \quad \alpha \in \mathrm{Aut}(\mathbb{F}_q), \quad \pi \in S_n.
$$

It acts on a vector $v \in \mathbb{F}_q^n$ by

$$\iota(v_0, \ldots, v_{n-1}) = (\alpha(v_{\pi^{-1}(0)})\varphi(0)(1), \ldots, \alpha(v_{\pi^{-1}(n-1)})\varphi(n-1)(1)).$$

The permutations $\varphi(i)$ are contained in $S_{\mathbb{F}_q^*}$, and so each factor $\varphi(i)(1)$ is contained in $\mathbb{F}_q^*$. Since we only need to know the values $\varphi(i)(1)$, $i \in n$, we can replace the mapping φ by

$$\psi : n \to \mathbb{F}_q^* \; : \; \psi(i) := \varphi(i)(1), \qquad i \in n.$$

Therefore, we can write ι as the triple $(\psi; (\alpha, \pi))$, where $(\psi; \pi)$ is a linear isometry. In other words $(\psi; \pi)$ belongs to the wreath product $\mathbb{F}_q^* \wr_n S_n$. This allows the slightly simpler expression for $\iota(v)$ given by

$$(\psi; (\alpha, \pi))(v_0, \ldots, v_{n-1}) = (\alpha(v_{\pi^{-1}(0)})\psi(0), \ldots, \alpha(v_{\pi^{-1}(n-1)})\psi(n-1)).$$

We collect these results in the following

Theorem *For $n \geq 3$, the isometries of $\mathbb{F}_q^n$ which map subspaces onto subspaces are exactly the semilinear mappings of the form $(\psi; (\alpha, \pi))$, where $(\psi; \pi)$ is a linear isometry and α is a field automorphism. These mappings form a group, the* group of semilinear isometries. □

1.5.10

In Section 6.7, we will describe this group as a generalized wreath product.

Definition (semilinearly isometric codes) Two (n, k)-codes C and C' over $\mathbb{F}_q$ are called *semilinearly isometric* if and only if there exists an automorphism α in $\mathrm{Aut}(\mathbb{F}_q)$ and a linear isometry $(\psi; \pi)$ in $\mathbb{F}_q^* \wr_n S_n$, such that the mapping

1.5.11

$$(c_0, \ldots, c_{n-1}) \mapsto (\psi(0)\alpha(c_{\pi^{-1}(0)}), \ldots, \psi(n-1)\alpha(c_{\pi^{-1}(n-1)}))$$

maps C onto C'. The orbits of the group of semilinear isometries on the set of subspaces of $H(n, q)$ are the *semilinear isometry classes* of linear codes of length n over $\mathbb{F}_q$. ◇

In addition, we mention the following facts (the first one is obvious, the second one will become clear in the chapter on finite fields):

1. The group of linear isometries of $H(n, 2)$ is isomorphic to the symmetric group S_n, since $\mathbb{F}_2^* = \{1\}$.

2. The group of semilinear isometries of $H(n, q)$ is the same as the group of linear isometries if and only if q is a prime p. The reason is that the field $\mathbb{F}_q$ has only the trivial automorphism if and only if $q = p$.

Hence, if the linear and semilinear isometry groups differ, we expect to see different numbers of orbits. This is indeed the case. The smallest examples are for $q = 4$, $n = 8$ and $k \geq 3$ (see Tables 6.9 and 6.31 in Chapter 6). What happens is that two linear isometry classes form a single semilinear isometry class. For example, we consider two codes over a field consisting of four elements. We take the field $\mathbb{F}_4 = \{0, 1, \alpha, \alpha + 1\}$ subject to the relation $\alpha^2 = \alpha + 1$ (see Chapter 3 for more details on finite fields). The code C_1 generated by

$$
\Gamma_1 = \begin{pmatrix}
1 & 1 & 1 & 1 & 1 & 0 & 0 & 0 \\
\alpha & 1 & 1 & 0 & 0 & 1 & 0 & 0 \\
\alpha & 1 & 0 & 1 & 0 & 0 & 1 & 0 \\
\alpha + 1 & 0 & 1 & 1 & 0 & 0 & 0 & 1
\end{pmatrix}
$$

is semilinearly equivalent to C_2 generated by

$$
\Gamma_2 = \begin{pmatrix}
1 & 1 & 1 & 1 & 1 & 0 & 0 & 0 \\
\alpha & 1 & 1 & 0 & 0 & 1 & 0 & 0 \\
\alpha & 1 & 0 & 1 & 0 & 0 & 1 & 0 \\
\alpha & 0 & 1 & 1 & 0 & 0 & 0 & 1
\end{pmatrix}.
$$

To see that the codes are semilinearly equivalent, add the first row of Γ_1 to the second and third row. This gives

$$
\begin{pmatrix}
1 & 1 & 1 & 1 & 1 & 0 & 0 & 0 \\
\alpha + 1 & 0 & 0 & 1 & 1 & 1 & 0 & 0 \\
\alpha + 1 & 0 & 1 & 0 & 1 & 0 & 1 & 0 \\
\alpha + 1 & 0 & 1 & 1 & 0 & 0 & 0 & 1
\end{pmatrix}.
$$

Now swap pairwise the second and the fifth and the third and the fourth column to get

$$
\begin{pmatrix}
1 & 1 & 1 & 1 & 1 & 0 & 0 & 0 \\
\alpha + 1 & 1 & 1 & 0 & 0 & 1 & 0 & 0 \\
\alpha + 1 & 1 & 0 & 1 & 0 & 0 & 1 & 0 \\
\alpha + 1 & 0 & 1 & 1 & 0 & 0 & 0 & 1
\end{pmatrix}.
$$

Application of the field automorphism $x \mapsto x^2$ takes the resulting matrix to Γ_2. It can be proved that the codes C_1 and C_2 are linearly inequivalent. This shows that two linear isometry classes may be combined under the semilinear group. Indeed, the number of linear isometry classes of codes which may join is bounded from above by the number of field automorphisms, which is of course two in this case.

For $n = 1$ or $n = 2$ the groups of isometries that map subspaces onto subspaces are described in

1.5.12   **Theorem** *For $n = 1$, the isometries of $\mathbb{F}_q$ which map subspaces onto subspaces are exactly the isometries of $\mathbb{F}_q$ which map 0 onto 0. According to 1.5.5 these are the elements of $S_{\mathbb{F}_q^*}$.*

For $n = 2$, the isometries of $\mathbb{F}_q^2$ which map subspaces onto subspaces are exactly the mappings of the form $(\psi; (\alpha, \pi))$, where $(\psi; \pi)$ is a linear isometry and α is a group automorphism of the multiplicative group $\mathbb{F}_q^*$. □

One can check that there exist group automorphisms of $\mathbb{F}_q^*$ which cannot be extended to field automorphisms of $\mathbb{F}_q$.

In conclusion, in large parts of the present book we will be concerned with orbits of the linear or semilinear isometry group on the set of subspaces of $\mathbb{F}_q^n = H(n, q)$.

Exercises

Exercise Prove 1.5.1.	E.1.5.1

Exercise Complete the proofs of 1.5.5 and 1.5.6 by showing that all elements of the corresponding wreath products are isometries.	E.1.5.2

Exercise In order to complete the proof of 1.5.10, show that any semilinear mapping of the given form is an isometry which maps subspaces onto subspaces. Moreover, prove 1.5.12.	E.1.5.3

1.6 The Weight Enumerator

1.6

An important issue is to find out when two $k \times n$ generator matrices over $\mathbb{F}_q$ define isometric codes. In general, this is not an easy task since normal forms of generator matrices are expensive to find (cf. Chapter 9). But there are *invariants* of linear codes which may help to distinguish between different codes. An invariant is simply a quantity (or a property) which we can associate to a code, and which is equal for codes of the same equivalence class (i.e. a "fingerprint"). One of these invariants will be introduced next, it is the weight distribution of a code. Essentially, this distribution records how many words of a code have a given Hamming weight. It is usually recorded as the coefficients of a polynomial, the weight enumerator. The permutational, linear or semilinear isometries of 1.4 and 1.5 preserve Hamming distances and Hamming weights. Codes with different weight enumerators are definitely not permutationally, linearly or semilinearly isometric. In Chapter 8, we will introduce a method for the evaluation of generator matrices that automatically provides the weight distribution as well.

We display the weight distribution of a linear code C of length n in terms of a *generating polynomial*. For this purpose we use commuting indeterminates

x and y over $\mathbb{C}$, and we indicate by $A_i = A_i(C)$ the number of codewords of weight i in C. For example, if $\text{dist}(C) = d > 1$, then $A_0 = 1$, $A_1 = \ldots = A_{d-1} = 0$ and $A_d \neq 0$.

1.6.1   **Definition (weight enumerator)** The homogeneous *weight enumerator* of a linear code C of length n is defined as

$$W_C(x,y) := \sum_{c \in C} x^{\text{wt}(c)} y^{n-\text{wt}(c)} = \sum_{i=0}^{n} A_i x^i y^{n-i} \in \mathbb{C}[x,y].$$

Notice that this is a homogeneous polynomial of degree n. Setting $y = 1$ yields the inhomogeneous weight enumerator

$$w_C(x) := \sum_{c \in C} x^{\text{wt}(c)} = \sum_{i=0}^{n} A_i x^i \in \mathbb{C}[x]. \qquad \diamond$$

For example, the 4-fold binary repetition code

$$C = \{0_4, 1_4\}$$

has weight enumerator

$$W_C(x,y) = x^4 + y^4 \text{ and } w_C(x) = x^4 + 1.$$

The following result is often useful. It follows from Exercise 1.2.14.

1.6.2   **Lemma** *For any two vectors* $u, v \in \mathbb{F}_2^n$ *we have the equivalence*

$$\text{wt}(u+v) \equiv \text{wt}(v) \bmod 2 \iff \text{wt}(u) \equiv 0 \bmod 2. \qquad \square$$

This means that adding a vector $u \in \mathbb{F}_2^n$ to a vector $v \in \mathbb{F}_2^n$ keeps the congruence class modulo two of $\text{wt}(v)$ fix if and only if the weight of u is even. Hence, adding a vector of odd weight in a binary code C to vectors of even weight gives vectors of odd weight and vice versa. Since the set of vectors of a linear code is closed under addition, this leaves only two possible cases. Either there is no vector of odd weight in a binary code C, or the set of vectors of C falls into two categories of equal size, one consisting of the vectors of even weight and the other containing all vectors whose weight is odd.

1.6.3   **Corollary** *For binary codes* C *the following holds true:*

— *The codewords of even weight form the subspace*

$$C_e := \{c \in C \mid \text{wt}(c) \text{ is even}\}.$$

— If there exists a codeword c of odd weight, then the complement $C \setminus C_e$ of C_e is equal to $c + C_e$.

— Hence either there is no element of odd weight or exactly half of the codewords in C have odd weight. We can express this fact as follows: If

$$R := \sum_{i=0}^{\lfloor n-1/2 \rfloor} A_{2i+1} \text{ and } S := \sum_{i=0}^{\lfloor n/2 \rfloor} A_{2i},$$

then either $R = 0$ or $R = S$.   □

In 1.6.9 we will derive an identity which is due to MacWilliams. It shows that the weight enumerator of a code and that of its dual code mutually determine each other. In order to prepare for a proof of this identity, we introduce the notion of a *linear* representation of a group. This notion generalizes the concept of a *permutation* representation or action of a group which has already been used on several occasions.

According to 1.4.5, a finite action $_G X$ is essentially the same as a *permutation representation* of G on X. This is a homomorphism

$$\delta : G \to S_X : g \mapsto \delta(g),$$

from G into S_X, where $g \in G$ is mapped onto $\delta(g) = \overline{g}$, the permutation $x \mapsto gx$ of X, an element of the symmetric group S_X. A *linear representation* D of G over a field $\mathbb{F}$ is defined to be a homomorphism

$$D : G \to \mathrm{GL}(V) : g \mapsto D(g),$$

from G into the group $\mathrm{GL}(V)$ of invertible linear mappings on a finite dimensional vector space V over $\mathbb{F}$. The vector space V is called the *representation space* and its dimension f^D is called the *dimension* of D. $\mathbb{F}$ is said to be the *groundfield* of D.

Two representations $D : G \to \mathrm{GL}(V)$ and $D' : G \to \mathrm{GL}(V')$ of G over $\mathbb{F}$ are considered *equivalent* if there exists an invertible linear mapping $T : V \to V'$ such that

$$\forall g \in G : TD(g) = D'(g)T.$$

Every choice of a basis $\{b^{(0)}, \ldots, b^{(f^D-1)}\}$ of V yields invertible matrices $\mathbf{D}(g)$ which describe $D(g)$ with respect to the given basis. Therefore, a *matrix representation* $\mathbf{D}$ of G over $\mathbb{F}$ is a homomorphism

$$\mathbf{D} : G \to \mathrm{GL}_{f^D}(\mathbb{F}) : g \mapsto \mathbf{D}(g)$$

from G to the *general linear group* $\mathrm{GL}_{f^D}(\mathbb{F})$, the group consisting of all invertible matrices over $\mathbb{F}$ with f^D rows and columns. Conversely, it is clear that each matrix representation $\mathbf{D} \colon G \to \mathrm{GL}_{f^D}(\mathbb{F})$ yields a representation $D \colon G \to \mathrm{GL}(V)$ where V is an f^D-dimensional vector space over $\mathbb{F}$. Equivalence of matrix representations is defined correspondingly. Hence we are free to consider either representations or matrix representations. Which concept we choose will depend on the situation in question. In the present section we are mainly concerned with matrix representations and their characters.

Let D be a representation of G. Consider the map

$$\chi^D \colon G \to \mathbb{F} \ : \ g \mapsto \sum_{i \in f^D} d_{ii}(g) = \mathrm{trace}(\mathbf{D}(g)),$$

which takes $g \in G$ to $\chi^D(g)$, the trace of $\mathbf{D}(g) = (d_{ij}(g))$. From Linear Algebra it is clear that the trace of a matrix $\mathbf{D}(g)$ corresponding to the linear mapping $D(g)$ is independent of the choice of a basis. The map χ^D is called the *character* of D. Representations and characters over the field $\mathbb{C}$ of complex numbers are called *ordinary*. In the case when the groundfield is finite, they are called *modular*.

1.6.4   Examples

- Every finite action $_GX$ yields a representation on the space $\mathbb{F}^X$, the vector space over $\mathbb{F}$ which has a basis whose elements are indexed by the elements of X. Thus we already have a wealth of examples at hand.

- The trivial representation of G arising from the *trivial action* $_G\{x\}$ of G on a set of cardinality one, where $gx := x$, is called the *identity representation* or the *trivial representation* and it is indicated as

$$I \colon G \to \mathrm{GL}(\mathbb{F}) \ : \ g \mapsto id_{\mathbb{F}},$$

where $\mathbb{F}$ is the 1-dimensional vector space over $\mathbb{F}$. Its character χ^I has the value $\chi^I(g) = 1_{\mathbb{F}}$, for each $g \in G$.

- In general, any finite action $_GX$ gives rise to a linear representation of G on $\mathbb{F}^X$. This representation associates to g the permutation $\overline{g}$ of the basis elements X (recall 1.4.5). Its character is

$$\chi(g) = a_1(\overline{g}) := |\{x \in X \mid gx = x\}|, \qquad g \in G,$$

which counts the number of fixed points of g. More precisely, $\chi(g) = a_1(\overline{g}) \cdot 1_{\mathbb{F}}$. In the *ordinary* case, i.e. if $\mathbb{F} = \mathbb{C}$, this character is the character of the action of the group.

- A *one-dimensional character* of G is the character of a one-dimensional linear representation, whence a homomorphism from G into the multiplicative group $\mathbb{F}^*$ of the groundfield. Therefore, for each such character χ we have

$$\chi(g \cdot h) = \chi(g)\chi(h) \ \text{ and } \ \chi(1_G) = 1_{\mathbb{F}},$$

provided that G is written multiplicatively. If G is written additively, we have correspondingly

$$\chi(g + h) = \chi(g)\chi(h) \text{ and } \chi(0_G) = 1_{\mathbb{F}}.$$

— For our purposes, the one-dimensional characters are of particular interest. A simple example is a one-dimensional character of the additive group of the field $\mathbb{F}_q$. Consider the group

$$G := \mathbb{Z}_p := \{\bar{z} \mid z \in p\}$$

of residue classes $\bar{z}$ of integers $z \in \mathbb{Z}$ modulo the prime number p. For more details on residue classes see Exercise 3.1.3. Addition in the group is done modulo the prime p. If

$$\zeta := e^{\frac{2\pi i}{p}} \in \mathbb{C}$$

denotes a primitive p-th root of unity, and if $j \in p$, the mapping

$$\chi^{(j)} : \mathbb{Z}_p \to \mathbb{C}^* : \bar{z} \mapsto \zeta^{j \cdot z}$$

is a one-dimensional character of G. It is not difficult to see that these characters are in fact *all* one-dimensional characters over $\mathbb{C}$ of this group, but we do not need this fact. We just remark that the character $\chi^{(j)}$ is nontrivial for $j \neq 0$.

— We can easily generalize this to a direct product

$$G := \mathbb{Z}_p \times \cdots \times \mathbb{Z}_p$$

of $m \geq 2$ factors of such groups. If $(\bar{z}_0, \ldots, \bar{z}_{m-1}) \in G$, and $j_i \in p$, then

$$\chi^{(j_0, \ldots, j_{m-1})} : (\bar{z}_0, \ldots, \bar{z}_{m-1}) \mapsto \zeta^{\sum_i j_i z_i}$$

is a one-dimensional character of $G = \mathbb{Z}_p \times \cdots \times \mathbb{Z}_p$. Moreover, this character is nontrivial if and only if $j_i \neq 0$ for at least one i.

— Later on in 3.1.6 we will see that for $q = p^m$ with p prime, the additive group of $\mathbb{F}_q$ is isomorphic to $G := \mathbb{Z}_p \times \cdots \times \mathbb{Z}_p$ (with m factors $\mathbb{Z}_p$). Hence, we have established the existence of nontrivial one-dimensional characters of the additive group of any finite field. This fact is all we need in the present section. $\diamond$

In particular we use the following result on the sum of character values:

Lemma *Let χ be a nontrivial character of a finite group G over a field $\mathbb{F}$. Then*   1.6.5

$$\sum_{g \in G} \chi(g) = 0.$$

Proof: Since χ is nontrivial, there exists an element $h \in G$ such that $\chi(h) \neq 1$. From

$$\chi(h) \sum_{g \in G} \chi(g) = \sum_{g \in G} \chi(h \cdot g) = \sum_{g \in G} \chi(g),$$

we obtain that

$$(\chi(h) - 1) \sum_{g \in G} \chi(g) = 0,$$

and this implies the statement since $\chi(h) \neq 1$. □

Suppose that $\chi \colon \mathbb{F}_q \to \mathbb{C}^*$ is a nontrivial one-dimensional ordinary character of $\mathbb{F}_q$, whose existence was established in 1.6.4. Fix an element $0 \neq v \in \mathbb{F}_q^n$. Using the standard bilinear form on $\mathbb{F}_q^n$, we introduce a character of the additive group $G := \mathbb{F}_q^n$ as follows:

1.6.6
$$\chi_{(v)} \colon \mathbb{F}_q^n \to \mathbb{C}^* \; : \; w \mapsto \chi(\langle v, w \rangle).$$

It is not difficult to see that this is a nontrivial one-dimensional character of $\mathbb{F}_q^n$.

Let us return to the weight enumerator W_C and consider the *weight function* in its homogeneous form,

$$f \colon \mathbb{F}_q^n \to \mathbb{C}[x, y] \; : \; v \mapsto x^{\mathrm{wt}(v)} y^{n - \mathrm{wt}(v)}.$$

Together with the weight function we examine a second function, a *Discrete Fourier Transform* of f (see also Exercise 1.6.9). It is defined by

$$\hat{f} := \sum_{v \in \mathbb{F}_q^n} f(v) \cdot \chi_{(v)},$$

where $\chi_{(v)}$ is the character defined by 1.6.6. To begin with, we prove

1.6.7   **Lemma** *For* $w \in \mathbb{F}_q^n$ *we have*

$$\hat{f}(w) = (y - x)^{\mathrm{wt}(w)} \big(y + (q - 1)x\big)^{n - \mathrm{wt}(w)}.$$

Proof: Let χ denote a nontrivial one-dimensional ordinary character of the additive group $G := \mathbb{F}_q$. For $\alpha \in \mathbb{F}_q$ we define

$$|\alpha| := \begin{cases} 1 & \text{if } \alpha \neq 0, \\ 0 & \text{otherwise.} \end{cases}$$

For each $w \in \mathbb{F}_q^n$ we compute

$$\begin{aligned} \hat{f}(w) &= \sum_{v \in \mathbb{F}_q^n} \chi(\langle v, w \rangle) f(v) \\ &= \sum_{v \in \mathbb{F}_q^n} \chi(\langle v, w \rangle) x^{\mathrm{wt}(v)} y^{n - \mathrm{wt}(v)} \end{aligned}$$

$$= \sum_{v_0 \in \mathbb{F}_q} \cdots \sum_{v_{n-1} \in \mathbb{F}_q} \chi \left(\sum_{i \in n} v_i w_i \right) x^{|v_0| + \ldots + |v_{n-1}|} y^{(1-|v_0|) + \ldots + (1-|v_{n-1}|)}$$

$$= \sum_{v_0 \in \mathbb{F}_q} \cdots \sum_{v_{n-1} \in \mathbb{F}_q} \prod_{i \in n} \chi(v_i w_i) x^{|v_i|} y^{1-|v_i|}$$

$$= \prod_{i \in n} \sum_{g \in G} \chi(g w_i) x^{|g|} y^{1-|g|}.$$

For the fourth equation we used that χ is a homomorphism.

— If $w_i = 0$ then $\chi(g w_i) = \chi(0) = 1$, and so

$$\sum_{g \in G} \chi(g w_i) x^{|g|} y^{1-|g|} = y + (q-1)x.$$

— On the other hand, if $w_i \neq 0$, we obtain

$$\sum_{g \in G} \chi(g w_i) x^{|g|} y^{1-|g|} = y + \sum_{g \in G \setminus \{0\}} \chi(g w_i) x$$

$$= y + \sum_{g \in G \setminus \{0\}} \chi(g) x$$

which, by 1.6.5, equals $y - \chi(0)x = y - x$.   □

Lemma *If C is an (n,k)-code over $\mathbb{F}_q$, then*   **1.6.8**

$$\sum_{c \in C} \hat{f}(c) = q^k \sum_{v \in C^\perp} f(v).$$

Proof: We know that

$$\sum_{c \in C} \hat{f}(c) = \sum_{c \in C} \sum_{v \in \mathbb{F}_q^n} \chi_{(v)}(c) f(v)$$

$$= \sum_{v \in \mathbb{F}_q^n} \sum_{c \in C} \chi(\langle v, c \rangle) f(v)$$

$$= \sum_{v \in C^\perp} \sum_{c \in C} \chi(\langle v, c \rangle) f(v) + \sum_{v \in \mathbb{F}_q^n \setminus C^\perp} \sum_{c \in C} \chi(\langle v, c \rangle) f(v).$$

In the first sum we have $\chi(\langle v, c \rangle) = \chi(0) = 1$ for all $v \in C^\perp$ and all $c \in C$. In order to simplify the second sum we recall that the map $c \mapsto \langle v, c \rangle$ is a linear form $C \to \mathbb{F}_q$. Since v belongs to $\mathbb{F}_q^n \setminus C^\perp$, this linear form is surjective, whence its kernel has dimension $k - 1$. Therefore, for each $g \in \mathbb{F}_q$, there are q^{k-1} vectors $c \in C$ such that $\langle v, c \rangle = g$. For this reason we can continue as follows:

$$\sum_{c \in C} \hat{f}(c) = q^k \sum_{v \in C^\perp} f(v) + q^{k-1} \sum_{v \in \mathbb{F}_q^n \setminus C^\perp} f(v) \sum_{g \in G} \chi(g) = q^k \sum_{v \in C^\perp} f(v),$$

by 1.6.5.   □

We are now in a position to prove the announced identity of MacWilliams [137] for the weight distribution of the dual code:

1.6.9   **The MacWilliams-identity** *The weight enumerator of an (n,k)-code C over $\mathbb{F}_q$ is related to the weight enumerator of its dual code in the following way:*

$$W_{C^\perp}(x,y) = q^{-k}W_C(y-x,y+(q-1)x).$$

Proof:

$$W_{C^\perp}(x,y) = \sum_{c \in C^\perp} f(c) \overset{1.6.8}{=} q^{-k} \sum_{c \in C} \hat{f}(c) \overset{1.6.7}{=} q^{-k}W_C(y-x,y+(q-1)x). \quad \square$$

1.6.10   **Example** Recall from 1.6.1 that the 4-fold binary repetition code $C = \{0_4, 1_4\}$ has weight enumerator $W_C(x,y) = x^4 + y^4$. By the MacWilliams-identity, the weight enumerator of its dual code is

$$W_{C^\perp}(x,y) = \frac{1}{2}((y-x)^4 + (y+x)^4) = y^4 + 6x^2y^2 + x^4. \qquad \diamond$$

It is sometimes useful to apply the MacWilliams-identity with exchanged roles of C and $C^\perp$.

1.6.11   **Example** Consider the $(7,4)$-Hamming-code of 1.3.6. The dual code $C^\perp$, generated by Δ, has 8 codewords. The 7 nonzero words are all of weight 4. Hence $C^\perp$ has weight enumerator

$$W_{C^\perp}(x,y) = y^7 + 7x^4y^3.$$

By the MacWilliams-identity 1.6.9, the weight enumerator of the $(7,4)$ Hamming-code $C^{\perp\perp} = C$ is determined as

$$\begin{aligned}
W_C(x,y) &= \frac{1}{2^3}W_{C^\perp}(y-x,y+x) \\
&= \frac{1}{8}((y+x)^7 + 7(y-x)^4(y+x)^3) \\
&= y^7 + 7x^3y^4 + 7x^4y^3 + y^7.
\end{aligned}$$

This shows that C has 7 words of weight 3 and 4 each. Together with the zero and the all-one-vector, this amounts to all 16 words in the code. $\diamond$

Particular cases of interest are the *self-dual codes* which we have introduced in 1.3.3. These are the linear codes C satisfying $C = C^\perp$. For these codes we have $k = n - k$, n is therefore even and $k = n/2$. Since the weight enumerator is a homogeneous polynomial of degree n, this implies the following:

Corollary *If C is self-dual, then*   1.6.12

$$W_C(x,y) = W_C(-x,-y) = W_C\left(\frac{y-x}{\sqrt{q}}, \frac{y+(q-1)x}{\sqrt{q}}\right).$$ □

This corollary shows that the weight enumerator of a self-dual code is a fixed point, i.e. an invariant of a group acting on a polynomial ring, in the following sense:

Definition (fixed point, invariant) Let $_G X$ be an action of a group G on a set   1.6.13
X. An element $x \in X$ is called a *fixed point* of an element $g \in G$ if $gx = x$. The set of *all fixed points of g* is denoted by

$$X_g := \{x \in X \mid gx = x\},$$

and we let

$$X_G := \{x \in X \mid \forall g \in G : gx = x\} = \bigcap_{g \in G} X_g$$

be the set of common fixed points of all elements $g \in G$. The elements in X_G are also called the *invariants* of G on X.   ◇

We note that the linear group $\mathrm{GL}_n(\mathbb{C})$ of invertible matrices of rank n over the complex field acts on the polynomial ring $\mathbb{C}[x_0, \ldots, x_{n-1}]$ in the following way:

$$\mathrm{GL}_n(\mathbb{C}) \times \mathbb{C}[x_0, \ldots, x_{n-1}] \to \mathbb{C}[x_0, \ldots, x_{n-1}],$$

$$(B, f(x_0, \ldots, x_{n-1})) \mapsto (Bf)(x_0, \ldots, x_{n-1}) := f((x_0, \ldots, x_{n-1}) \cdot B^\top).$$

For example, if $B := -I_2 := \begin{pmatrix} -1 & 0 \\ 0 & -1 \end{pmatrix}$ then

$$((-I_2)W_C)(x,y) = W_C\left((x,y) \cdot \begin{pmatrix} -1 & 0 \\ 0 & -1 \end{pmatrix}\right) = W_C(-x,-y),$$

which shows that the weight enumerator W_C of a self-dual code is a fixed point of $-I_2 \in \mathrm{GL}_2(\mathbb{C})$. We may also express this by saying that W_C is an invariant of the group $G := \{I_2, -I_2\}$ of order two.

Any subgroup $G \leq \mathrm{GL}_n(\mathbb{C})$ induces a subaction, and the set of common fixed points

$$\mathbb{C}[x_0, \ldots, x_{n-1}]_G = \{f \in \mathbb{C}[x_0, \ldots, x_{n-1}] \mid \forall B \in G : Bf = f\}$$

is the set of *invariants* of G on $\mathbb{C}[x_0, \ldots, x_{n-1}]$. The standard example is

$$\mathbb{C}[x_0, \ldots, x_{n-1}]_{S_n},$$

the set of invariants of the symmetric group S_n. A polynomial in this set is invariant under all possible permutations of its variables. Hence the invariants of the symmetric group consist of the *symmetric polynomials*.

An important series of symmetric polynomials is the series of *elementary symmetric* polynomials

$$\sigma_m := \sum_{0 \le i_1 < ... < i_m \le n-1} x_{i_1} \cdots x_{i_m}, \quad 1 \le m \le n, \ \sigma_0 := 1.$$

They generate the ring of symmetric polynomials, i.e. any symmetric polynomial can be written in a unique way as a polynomial in the elementary symmetric polynomials. Moreover, the coefficients of polynomials can be expressed in terms of their roots, using elementary symmetric polynomials. For example

$$\prod_{i \in n} (x - \kappa_i) = \sigma_0 \cdot x^n + ... + (-1)^n \sigma_n(\kappa_0, ..., \kappa_{n-1}).$$

For the other coefficients see Exercise 1.6.13.

From 1.6.12 we derive that the weight enumerator W_C of a self-dual linear code C is an invariant of the group

$$G := \langle -I_2, B \rangle \quad \text{where } B = \frac{1}{\sqrt{q}} \begin{pmatrix} -1 & 1 \\ q-1 & 1 \end{pmatrix}.$$

It is easy to check that this group has four elements. It is isomorphic to the Klein four-group V_4, and so we have obtained:

1.6.14   **Corollary** *The weight enumerator of a self-dual linear code is an invariant of the Klein four-group:*

$$W_C(x,y) \in \mathbb{C}[x,y]_{V_4}.$$ □

Binary self-dual codes have an additional property:

1.6.15   **Definition (divisible codes)** A linear code C is called *r-divisible* if each codeword has a weight which is divisible by r.   ◇

A 2-divisible code is called *even*, a 4-divisible code is called *doubly even*. A code which is 2-divisible but not 4-divisible is called *singly even*. Notice that a binary self-dual code is even, since each word is orthogonal to itself, which means that its Hamming weight is even.

1.6.16   **Lemma** *If C is self-dual and r-divisible, then its weight distribution W_C is an invariant of the group*

$$G := \langle -I_2, B, D \rangle, \quad \text{where } B = \frac{1}{\sqrt{q}} \begin{pmatrix} -1 & 1 \\ q-1 & 1 \end{pmatrix}, \ D = \begin{pmatrix} \epsilon & 0 \\ 0 & 1 \end{pmatrix}$$

and where $\epsilon \in \mathbb{C}$ denotes a primitive r-th root of unity. Formally,

$$W_C \in \mathbb{C}[x, y]_G.$$

Proof:

$$DW_C(x, y) = \sum_{i=0}^{n} A_i x^i \epsilon^i y^{n-i} = \sum_{i=0}^{n} A_i x^i y^{n-i} = W_C(x, y),$$

since $\epsilon^i = 1$ for i divisible by r, and $A_i = 0$ in the other cases.   □

Finally, we show the following result for ternary codes, which is the converse of the assertion made in Exercise 1.3.20. We will use the notion of the *support* of a vector, which is just the set of coordinates where the vector is nonzero. We denote it as

$$\text{supp}(v) = \{i \in n \mid v_i \neq 0\}, \qquad v \in \mathbb{F}^n.$$

In particular, $|\text{supp}(u)| = \text{wt}(u)$.

Lemma *Let C be a ternary linear code. Then C is 3-divisible if and only if C is self-orthogonal.*   1.6.17

Proof: If C is self-orthogonal, then, according to Exercise 1.3.20, it is 3-divisible. Conversely, assume that C is 3-divisible. Consider two codewords u and v in C. Let $X = \text{supp}(u)$ and $Y = \text{supp}(v)$. Furthermore, we introduce the sets $E = \{i \mid u_i = v_i \neq 0\}$ (E for "equal") and $N = \{i \mid 0 \neq u_i \neq v_i \neq 0\}$ (N for "non-equal") and $Z = \{i \mid u_i = v_i = 0\}$ (Z for "zero"). Then E and N partition $X \cap Y$, and the sets

$$X \setminus Y, \quad Y \setminus X, \quad E, \quad N, \quad \text{and} \quad Z$$

partition the set of all coordinates. Furthermore, using the notion of the *symmetric difference* of two sets, which is defined as

$$X \Delta Y = (X \setminus Y) \cup (Y \setminus X),$$

we have for any i

$$u_i + v_i \begin{cases} \neq 0 & \text{if } i \in X \Delta Y, \\ \neq 0 & \text{if } i \in E, \\ = 0 & \text{if } i \in N, \\ = 0 & \text{if } i \in Z, \end{cases} \quad \text{and} \quad u_i - v_i \begin{cases} \neq 0 & \text{if } i \in X \Delta Y, \\ = 0 & \text{if } i \in E, \\ \neq 0 & \text{if } i \in N, \\ = 0 & \text{if } i \in Z. \end{cases}$$

Therefore, $\text{wt}(u + v) = |X \Delta Y| + |E| \equiv 0 \mod 3$ and $\text{wt}(u - v) = |X \Delta Y| + |N| \equiv 0 \mod 3$, so that $|E| \equiv |N| \mod 3$. We conclude that

$$\langle u, v \rangle = \sum_{i} u_i v_i = \sum_{i \in E} u_i v_i + \sum_{i \in N} u_i v_i = |E| - |N| \equiv 0 \mod 3.$$

This shows that $C \subseteq C^{\perp}$.   □

Exercises

E.1.6.1   **Exercise** Let C be a linear code over $\mathbb{F}_q$. Show that each $c \in C$ satisfies

$$\left|\{c' \in C \mid d(c, c') = i\}\right| = \left|\{c' \in C \mid \operatorname{wt}(c') = i\}\right| = A_i$$

for $0 \leq i \leq n$.

E.1.6.2   **Exercise** By Exercise 1.3.16, a binary self-orthogonal code is even. What about the converse?

E.1.6.3   **Exercise** Let C be a binary linear code of length n containing the all-one vector. Show that $A_i = A_{n-i}$ for $0 \leq i \leq n$.

E.1.6.4   **Exercise** Consider vectors $u, v, w \in \mathbb{F}_2^n$ satisfying $d(u, v) \equiv d(v, w) \bmod 2$. Then $d(u, w) \equiv 0 \bmod 2$. Show that this is in fact equivalent to 1.6.2.

E.1.6.5   **Exercise** Prove the following properties of one-dimensional characters of a *finite* and multiplicative group G:

- $\chi(g)$ is a $|G|$-th root of unity, i.e. $\chi(g)^{|G|} = 1_{\mathbb{F}}$.
- $\chi(g^{-1}) = \chi(g)^{-1}$. In particular, if $\mathbb{F}$ is the field $\mathbb{C}$ of complex numbers then $\chi(g^{-1}) = \overline{\chi(g)}$, where $\overline{\chi(g)}$ denotes the complex conjugate of $\chi(g)$.
- Show that the one-dimensional ordinary characters of G form a group $\hat{G}$ with respect to pointwise multiplication.

E.1.6.6   **Exercise** Let $(G, +)$ be a group. For $n \in \mathbb{Z}$ and $g \in G$ the *n-fold sum* of g is defined by

$$n \cdot g := \begin{cases} 0 & \text{if } n = 0, \\ (n - 1) \cdot g + g & \text{if } n > 0, \\ (-n) \cdot (-g) & \text{if } n < 0, \end{cases}$$

where $(-g)$ is the additive inverse of g. Prove the following:

- $(n + m) \cdot g = n \cdot g + m \cdot g$ and $(nm) \cdot g = n \cdot (m \cdot g)$ for all $n, m \in \mathbb{Z}$ and $g \in G$.
- For an abelian group $(G, +)$, $n \cdot (g_1 + g_2) = n \cdot g_1 + n \cdot g_2$ for all $n \in \mathbb{Z}$, and $g_1, g_2 \in G$.
- If G is a ring then $n \cdot (g_1 g_2) = (n \cdot g_1) g_2$ is satisfied for all $n \in \mathbb{Z}$ and $g_1, g_2 \in G$.

If $(G, \cdot)$ is a multiplicative group, then, correspondingly, for $n \in \mathbb{Z}$ we use powers in order to indicate the *n-fold product* of $g \in G$ defined by

$$g^n := \begin{cases} 1 & \text{if } n = 0, \\ g^{n-1} \cdot g & \text{if } n > 0, \\ (\tilde{g})^{-n} & \text{if } n < 0, \end{cases}$$

where $\tilde{g}$ is the multiplicative inverse of g. Analogously to the *n*-fold sum, formulate the corresponding assertions for the *n*-fold product.

Exercise Recall (see e.g. [101]), that each finite abelian group G is isomorphic to a direct product of suitable cyclic groups:

$$G \simeq \mathbb{Z}_{n_0} \times \ldots \times \mathbb{Z}_{n_{r-1}},$$

E.1.6.7

where $\prod_{i \in r} n_i = |G|$. This decomposition is unique provided that $n_i \mid n_{i+1}$ for $0 \leq i < r - 1$. In this case the number n_{r-1} is the *exponent* $\exp(G)$ of G. It is the smallest positive integer m such that the m-fold sum satisfies

$$m \cdot g = 0 \qquad \forall\, g \in G.$$

Hence, any element $g \in G$ can be written as a tuple $(g_0, \ldots, g_{r-1})$ with $g_i \in \mathbb{Z}_{n_i}$. Consider a primitive n_r-th root of unity $\xi \in \mathbb{C}$ and prove that the mapping

$$\phi : G \to \hat{G} \; : \; g \mapsto \left(h \mapsto \prod_{i \in r} \xi^{\frac{n_{r-1}}{n_i} g_i h_i} \right)$$

into the group of one-dimensional characters (cf. Exercise 1.6.5) is a group isomorphism, where $g = (g_0, \ldots, g_{r-1})$ and $h = (h_0, \ldots, h_{r-1})$. If we indicate the character $\phi(g)$ by χ_g, then

$$\chi_g(h) = \prod_{i \in r} \xi^{\frac{n_{r-1}}{n_i} g_i h_i} \text{ for each } h \in G.$$

Exercise Verify the following *orthogonality relation* for the characters of a finite abelian group G over $\mathbb{C}$: For each $g, g' \in G$ we have

E.1.6.8

$$\frac{1}{|G|} \sum_{h \in G} \chi_{-g}(h) \chi_{g'}(h) = \begin{cases} 1 & \text{if } g = g', \\ 0 & \text{otherwise.} \end{cases}$$

Exercise Let G be a finite abelian group. We associate to each $f : G \to \mathbb{C}$ its *Discrete Fourier Transform*

E.1.6.9

$$\hat{f}(h) := \sum_{g \in G} f(g) \chi_g(h), \qquad h \in G.$$

Show that

$$f(h) = \frac{1}{|G|} \sum_{g \in G} \hat{f}(g)\chi_{-g}(h), \qquad h \in G.$$

Therefore, the set $\hat{G}$ of one-dimensional characters of G forms a generating system of the vector space $\mathbb{C}^G$, whence (compare the dimension) even a basis.

E.1.6.10   **Exercise** Prove the *Lemma of Cauchy–Frobenius for Representations*: Let D denote a representation of a finite group G on a vector space over a field $\mathbb{F}$ of characteristic prime to $|G|$. Then the space

$$V_G = \{v \in V \mid \forall\, g \in G\colon\ D(g)v = v\}$$

of invariants of the group $D(G)$ is of dimension

$$\dim(V_G) = \frac{1}{|G|} \sum_{g \in G} \chi^D(g).$$

Hint: The linear mapping

$$\varphi := \frac{1}{|G|} \sum_{g \in G} D(g)$$

is a *projection*, i.e. $\varphi^2 = \varphi$.

E.1.6.11   **Exercise** Use the MacWilliams-identity in order to express $A_i^\perp := A_i(C^\perp)$ in terms of the $A_i = A_i(C)$. Rephrase your result in terms of the *Krawtchouk polynomial*

$$K_i^{n,q}(x) = \sum_{j=0}^{i}(-1)^j(q-1)^{i-j}\binom{x}{j}\binom{n-x}{i-j},$$

where

$$\binom{x}{j} := \frac{x\cdots(x-j+1)}{j!}, \qquad \binom{n-x}{i-j} := \frac{(n-x)\cdots(n-x-(i-j)+1)}{(i-j)!}.$$

E.1.6.12   **Exercise** Show that the parity extension of the $(7,4)$ binary Hamming-code (cf. Example 1.3.6) is self-orthogonal and hence self-dual, with minimum distance 4. Write down the MacWilliams-identity for its weight enumerator.

E.1.6.13   **Exercise** Express the coefficients of $\prod_{i \in n}(x - \kappa_i)$ in terms of elementary symmetric polynomials and the roots κ_i.

1.7 Systematic Encoding, Information Sets

It may happen that a set of k coordinates of the codewords of a fixed code always determines the remaining coordinate values. This means that if we are given the values of a codeword on those k coordinates, then the remaining $n - k$ coordinates are determined *uniquely*. We say that such a set of k coordinates forms an *information set*. The elements of an information set, i.e. the coordinates which are part of it, are called *information places*. If a k-set of coordinates is an information set, then we say that the remaining $n - k$ coordinates form a *redundancy set*. Its elements are of course called *redundancy places*. They are also called *check bits,* since they may be used for error detection and error correction.

Any code has at least one information set. It corresponds to a maximal set of columns of a generator matrix which are linearly independent. Recall that a generator matrix Γ is a $k \times n$-matrix of rank k. Such a matrix always has a set of k columns which are linearly independent. Gaussian elimination for example will reveal such a set of columns. The columns holding the pivot elements have the property that they are linearly independent. If necessary, we permute these columns up-front, for example by means of a linear isometry. This means that we may have to change to a code C' which is isometric to the original code C, which is of course no real restriction. This code C' then has the following nice property:

Corollary *Each (n, k)-code C with generator matrix Γ is linearly isometric to a code C' with generator matrix of the form*

$$\Gamma' = (I_k \mid A),$$

where I_k denotes the $k \times k$-unit matrix.   □

We say that a generator matrix of the form $(I_k \mid A)$ is *systematic*. The corresponding encoding map $v \mapsto v \cdot \Gamma'$ is called *systematic*. We have seen that up to linear (or semilinear) isometry, any code can be generated systematically.

When using systematic encoding $v \mapsto v \cdot \Gamma' = w$, the first k coordinate places of w simply repeat the k components of the message v. The remaining $n - k$ coordinates of w can then be used for error correction (note however, that errors may also have occurred in the first k coordinates, so decoding by simply reading out the first k coordinate values does not work). Here is an example of a generator matrix Γ and a linear isometry which determines a systematic generator matrix Γ' of a ternary code. The code generated by

$$\Gamma = \begin{pmatrix} 1 & 2 & 1 & 2 \\ 2 & 1 & 1 & 0 \end{pmatrix}$$

is linearly isometric to the code generated by

$$
\Gamma' := \begin{pmatrix} 0 & 2 \\ 1 & 1 \end{pmatrix} \cdot \Gamma \cdot \begin{pmatrix} 0 & 0 & 2 & 0 \\ 2 & 0 & 0 & 0 \\ 0 & 0 & 0 & 2 \\ 0 & 2 & 0 & 0 \end{pmatrix} = \begin{pmatrix} 1 & 0 & 2 & 1 \\ 0 & 1 & 0 & 1 \end{pmatrix}.
$$

Both are $(4,2)$-codes over $\mathbb{F}_3$.

Now we examine the effect of the linear isometries on $H(n,q) = \mathbb{F}_q^n$ that correspond to multiplication of columns by nonzero elements of the field. We want to show that such isometries map a systematic code onto a systematic one.

Isometries obtained by multiplications are described by regular $n \times n$ diagonal matrices D, and these matrices form a normal subgroup $D_n(q)$ of $M_n(q)$ (see Exercise 1.7.4). An (n,k)-code C with generator matrix $(I_k \mid A)$, where A is the $k \times (n-k)$-matrix

$$
A = \begin{pmatrix} a_{0,k} & \cdots & a_{0,n-1} \\ \vdots & & \vdots \\ a_{k-1,k} & \cdots & a_{k-1,n-1} \end{pmatrix},
$$

is mapped under that kind of isometries onto an isometric code C' with generator matrix

$$
(I_k \mid A) \cdot D.
$$

Any matrix which can be obtained from the above generator matrix via left multiplication by a regular $k \times k$-matrix is a generator matrix of C' as well. Suppose we choose for the left multiplication the upper left part of the multiplicative inverse of $D = (d_{ij}) \in D_n(q)$, i.e. the matrix

$$
D' := \begin{pmatrix} d_{0,0}^{-1} & & 0 \\ & \ddots & \\ 0 & & d_{k-1,k-1}^{-1} \end{pmatrix},
$$

then we obtain the systematic generator matrix

$$
D' \cdot (I_k \mid A) \cdot D = (I_k \mid D * A)
$$

of C', where

1.7.2
$$
D * A := \left(d_{ii}^{-1} a_{ij} d_{jj} \right)_{0 \le i < k, k \le j < n}.
$$

This proves the following

Lemma *For each $D \in D_n(q)$, the systematic matrices $(I_k \mid A)$ and $(I_k \mid D * A)$* $\qquad$ **1.7.3**
generate linearly isometric codes. $\qquad\qquad\qquad\qquad\qquad\qquad\qquad\qquad\qquad\qquad$ $\square$

Another characterization of information sets is the following. Let J be a set of column indices, i.e. $J \subseteq \{0, \dots, n-1\} = n$. Denote the complement of J as

$$\bar{J} := n \setminus J.$$

Then

$$\mathbb{F}_q^{(J)} := \left\{ (w_0, \dots, w_{n-1}) \in \mathbb{F}_q^n \ \middle|\ w_j = 0 \text{ for all } j \in \bar{J} \right\} \qquad \textbf{1.7.4}$$

is a subspace of $\mathbb{F}_q^n$ of dimension $|J|$. In particular,

$$\mathbb{F}_q^{(J)} \oplus \mathbb{F}_q^{(\bar{J})} = \mathbb{F}_q^n. \qquad\qquad \textbf{1.7.5}$$

Theorem *An (n,k)-code C possesses a k-subset $J \subseteq n$ as an information set if and* $\qquad$ **1.7.6**
only if $C \oplus \mathbb{F}_q^{(\bar{J})} = \mathbb{F}_q^n$ holds true. $\qquad\qquad\qquad\qquad\qquad\qquad\qquad\qquad\qquad\quad$ $\square$

The proof of this theorem is Exercise 1.7.7. Since each element of $\mathbb{F}_q^{(J)}$ is of Hamming weight at most $|J|$, we obtain the following

Theorem *Consider $d \in \mathbb{N}^*$. For each linear code C of length n over $\mathbb{F}_q$, the following* $\qquad$ **1.7.7**
conditions are equivalent:

— *C has minimum weight at least d.*

— *For each $J \subseteq n$, where $|J| < d$, we have*

$$C \cap \mathbb{F}_q^{(J)} = \{0\}. \qquad\qquad\qquad\qquad \square$$

Now we want to describe the close connection between systematic generator matrices and systematically encoded linear codes.

Theorem *Assume that $1 \leq k \leq n-1$. The mapping* $\qquad\qquad\qquad\qquad\qquad$ **1.7.8**

$$A \longmapsto \left\{ v \cdot (I_k \mid A) \ \middle|\ v \in \mathbb{F}_q^k \right\}$$

is a bijection between the set of $k \times (n-k)$-matrices A over $\mathbb{F}_q$ and the set of systematically encoded (n,k)-codes over $\mathbb{F}_q$.

Proof: The given mapping is obviously surjective. In order to prove injectivity, we consider two $k \times (n-k)$-matrices A and B over $\mathbb{F}_q$ which differ in their i-th row for some i. If $e^{(i)}$ denotes the i-th unit vector, then the codewords $e^{(i)} \cdot (I_k \mid A)$ and $e^{(i)} \cdot (I_k \mid B)$ are distinct. However, the two codewords agree in all of the first k coordinates, whose values by Theorem 1.7.6 determine the codeword uniquely. The only possibility for this is that the codes generated by $(I_k \mid A)$ and $(I_k \mid B)$ are distinct. This proves the statement. $\qquad$ $\square$

When classifying linear codes we want to obtain complete lists of representatives of the isometry classes for given parameters n, k and q. It is most convenient to describe the representatives by systematic generator matrices. Therefore, it is of interest to determine all systematic generator matrices of codes belonging to a single isometry class.

1.7.9   **Remark** [184, 2.10] Assume that $\Gamma = (I_k \mid A)$ is a systematic generator matrix of an (n,k)-code over $\mathbb{F}_q$ with $k < n$. The systematic generator matrices of codes (semi)linearly isometric to C can be obtained as follows. Apply a (semi)linear isometry so that the first k columns of the resulting matrix Γ' are linearly independent. Then pre-multiply Γ' by a suitable matrix from $\mathrm{GL}_k(q)$. There are several types of isometry operations which guarantee that the first k columns of Γ' are linearly independent. They can be generated by repeated application of the following isometry operations:

— Considering permutational isometries we obtain: The permutations of columns that replace the first k columns of Γ by linearly independent columns can be generated by repeated application of three types of permutations:
  1. Interchange the columns with index i and j, where $i, j < k$. After interchanging the i-th and j-th row of Γ', the resulting matrix is again systematic.
  2. Interchange the columns with index i and j, where $i, j \geq k$, then the resulting matrix is systematic.
  3. Interchange the columns with index i and j, where $i < k \leq j$. This is only possible in case $a_{ij} \neq 0$, for otherwise the first k columns of Γ' would no longer be linearly independent. In order to obtain a systematic matrix, multiply the i-th row of Γ' by a_{ij}^{-1}, and for $\ell \neq i$ subtract this new row multiplied by $a_{\ell j}$ from the ℓ-th row of Γ'.

— Furthermore, using linear isometries it is possible to multiply columns of Γ by nonzero field elements. If we multiply the i-th column of Γ by $\kappa \in \mathbb{F}_q^*$, say, then the resulting matrix either is already systematic (namely if $i \geq k$), or can be brought into a systematic form (namely by multiplying the i-th row by κ^{-1}).

— When considering also semilinear isometries, apply an automorphism $\alpha \in \mathrm{Aut}(\mathbb{F}_q)$ to each entry of Γ. The resulting matrix is again systematic. ◇

Exercises

E.1.7.1   **Exercise** Use the existence of systematic generator matrices in order to show that each linear code with $k = n - 1$ has a minimum distance at most 2.

Exercise Let C be an (n,k)-code with minimum distance d. Show that d is   E.1.7.2
the largest integer with the property that any $n - d + 1$ coordinate positions
contain an information set.

Exercise There are three types of elementary $k \times k$-matrices over $\mathbb{F}$. For $\lambda \in \mathbb{F}^*$ E.1.7.3
and for $i_0, j_0 \in k$ with $i_0 \neq j_0$ they are given by:

— $B_{i_0,\lambda}^{(1)}$ is the unit matrix I_k in which the entry 1 occurring in position (i_0, i_0)
  is replaced by λ, thus it is a diagonal matrix $(b_{ij})_{i,j \in k}$ with

$$b_{ij} = \begin{cases} \lambda & \text{if } i = j = i_0, \\ 1 & \text{if } i = j \neq i_0, \\ 0 & \text{else.} \end{cases}$$

— $B_{i_0,j_0,\lambda}^{(2)}$ is the unit matrix I_k with an additional entry λ in position (i_0, j_0),
  thus it is the matrix $(b_{ij})_{i,j \in k}$ with

$$b_{ij} = \begin{cases} 1 & \text{if } i = j, \\ \lambda & \text{if } i = i_0 \text{ and } j = j_0, \\ 0 & \text{else.} \end{cases}$$

— $B_{i_0,j_0}^{(3)}$ is the unit matrix I_k in which the rows (or columns) of index i_0 and j_0
  are exchanged, thus it is the matrix $(b_{ij})_{i,j \in k}$ with

$$b_{ij} = \begin{cases} 1 & \text{if } i = j \text{ and } (i \neq i_0 \text{ or } j \neq j_0), \\ 1 & \text{if } (i, j) = (i_0, j_0) \text{ or } (i, j) = (j_0, i_0), \\ 0 & \text{else.} \end{cases}$$

Prove that all these matrices are regular, and that the inverse of an elementary
matrix is again elementary. Deduce then that every matrix of $\mathrm{GL}_k(q)$ can be
written as a product of elementary matrices.

Show that the following holds true: Multiplying a $k \times n$-matrix Γ from
the left with an elementary matrix B yields an elementary row operation on
Γ. Hence, $B \cdot \Gamma$ is a composition of elementary row operations on Γ for all
$B \in \mathrm{GL}_k(q)$. Multiplying an $n \times k$-matrix Γ from the right with an elementary
matrix yields an elementary column operation on Γ. Hence, $\Gamma \cdot B$ is a composition of elementary column operations on Γ for all $B \in \mathrm{GL}_k(q)$.

Exercise Check that the regular $n \times n$ diagonal matrices over $\mathbb{F}_q$ form a *normal*   E.1.7.4
subgroup $D_n(q)$ of $M_n(q)$, which means that $D_n(q)$ is a subgroup of $M_n(q)$ and
that $M^{-1} \cdot D \cdot M \in D_n(q)$ for each $D \in D_n(q)$ and $M \in M_n(q)$.

E.1.7.5   **Exercise** Verify that the composition $*$ of 1.7.2 satisfies

$$D_1 * (D_2 * A) = (D_1 \cdot D_2) * A \ \text{ and } \ I_n * A = A.$$

Here, D_1 and D_2 are elements of $D_n(q)$ and A is any $k \times k$-matrix. In particular, this operation is a group action.

E.1.7.6   **Exercise** Prove that 1.7.5 holds for all subsets $J \subseteq n$.

E.1.7.7   **Exercise** Prove 1.7.6.

E.1.7.8   **Exercise** Assume that W and W' are subspaces of the vector space V. Prove that the following two statements are equivalent:

1. $V = W \oplus W'$ (which means $V = W + W'$ and $W \cap W' = \{0\}$).

2. For each $v \in V$ there exist uniquely determined $w \in W$ and $w' \in W'$ such that $v = w + w'$.

E.1.7.9   **Exercise** Assume that W and W' are subspaces of the finite dimensional vector space V. Prove that the following two statements are equivalent:

1. $V = W \oplus W'$.

2. $V = W + W'$ and $\dim(V) = \dim(W) + \dim(W')$.

1.8   1.8 A Minimum Distance Algorithm

As we have seen, the minimum distance is a very important parameter of a linear code. Nevertheless, *evaluating* this parameter for a given code may turn out to be surprisingly hard. As example 1.3.8 shows, the minimum distance of a code can be less than the minimum weight of the rows of a particular generator matrix. Here we present an algorithm, which is a variation of an idea of A. Brouwer and due to K.-H. Zimmermann. Information sets play an important role in this algorithm. It uses an iteration of Gaussian elimination and it works efficiently if the code under consideration has many information sets which are pairwise disjoint.

Algorithm (MinDist) To compute the minimum distance of a given linear (n, k)-code C. The *input* is a generator matrix Γ of C and the *output* is the minimum distance $d = \text{dist}(C)$ of C.

— Recall that the code in question is linearly isometric to a systematic one. This means that there exist matrices $M \in M_n(q)$ and $B \in \text{GL}_k(q)$ such that $B \cdot \Gamma \cdot M^\top$ is a systematic generator matrix. In fact, recalling the Gaussian Algorithm, we can obtain a systematic generator matrix by elementary row operations, i.e. by multiplying from the left with a matrix $B_1 \in \text{GL}_k(q)$, and a suitable column permutation, i.e. a multiplication from the right by the transpose of a permutation matrix $M_{\pi_1} := M_{(\epsilon;\pi_1)}$ (cf. 1.4.8),

$$\Gamma_1 := B_1 \cdot \Gamma \cdot M_{\pi_1}^\top = (\, I_{k_1} \mid A_1 \,),$$

where $k_1 = k$.

— If A_1 is neither empty nor a zero matrix, its rank is k_2 with $0 < k_2 \leq k_1$. Applying Gaussian elimination, we can obtain k_2 different unit vectors in the remaining $n - k_1$ columns. Of course, this process may distort the original unit matrix I_{k_1} in the leftmost k_1 columns. In other words, we can multiply Γ_1 from the left by an element B_2 of $\text{GL}_k(q)$ and from the right by the transpose of a permutation matrix M_{π_2} with $\pi_2(j) = j$ for $0 \leq j < k_1$, obtaining a generator matrix of a linearly isometric code,

$$\Gamma_2 := B_2 \cdot \Gamma_1 \cdot M_{\pi_2}^\top = \left(\begin{array}{c|c|c} A_2' & I_{k_2} & A_2 \\ \hline & 0 & 0 \end{array} \right).$$

The matrix A_2' is a $k \times k_1$-matrix and A_2 is a $k_2 \times (n - k_1 - k_2)$-matrix. The zeros indicate zero matrices.

— Assume that for $i \geq 2$ the matrix A_i which has just been computed is neither empty nor a zero matrix. Then its rank is k_{i+1} with $0 < k_{i+1} \leq k_i$. We continue this way, obtaining regular matrices $B_{i+1} \in \text{GL}_k(q)$, permutation matrices $M_{\pi_{i+1}} \in M_n(q)$, with $\pi_{i+1}(j) = j$ for $0 \leq j < k_1 + \ldots + k_i$, and generator matrices

$$\Gamma_{i+1} := B_{i+1} \cdot \Gamma_i \cdot M_{\pi_{i+1}}^\top = \left(\begin{array}{c|c|c} A_{i+1}' & I_{k_{i+1}} & A_{i+1} \\ \hline & 0 & 0 \end{array} \right),$$

A_{i+1}' a $k \times (k_1 + \ldots + k_i)$-matrix and A_{i+1} a $k_{i+1} \times (n - k_1 - \ldots - k_{i+1})$-matrix. We repeat this procedure. Eventually, we will obtain a generator matrix Γ_m, say, such that

$$\Gamma_m := B_m \cdot \Gamma_{m-1} \cdot M_{\pi_m}^\top = \left(\begin{array}{c|c|c} A_m' & I_{k_m} & A_m \\ \hline & 0 & 0 \end{array} \right),$$

where A_m is either empty (which means it has no columns) or it is a zero matrix. Then, $k_1 + \ldots + k_m \leq n$ and A_m has $n - k_1 - \ldots - k_m$ columns. Consequently, the generator matrix Γ_m has $n - k_1 - \ldots - k_m$ zero columns, whence all elements of the code generated by Γ_m have weight at most $k_1 + \ldots + k_m$.

— Let $\tilde{C}$ be the code generated by Γ_m. We note that C is linearly isometric to this code, whereas the matrices

$$\Gamma_1, \ldots, \Gamma_m$$

generate codes which are *linearly isometric* to $\tilde{C}$ but not necessarily *equal* to $\tilde{C}$ (except for Γ_m, of course). For this reason we put

$$\tilde{\Gamma}_i := \Gamma_i \cdot M_{\pi_{i+1}}^\top \cdots M_{\pi_m}^\top = B_i \cdots B_1 \cdot \Gamma \cdot M_{\pi_1}^\top \cdots M_{\pi_m}^\top = \tilde{B}_i \Gamma_m,$$

$\tilde{B}_i \in GL_k(q)$, so that the matrices

$$\tilde{\Gamma}_1, \ldots, \tilde{\Gamma}_m$$

generate the *same* code $\tilde{C}$. Moreover, the leftmost $k_1 + \ldots + k_i$ columns of Γ_i and $\tilde{\Gamma}_i$ are the same, whence $\tilde{\Gamma}_i$ has the unit matrix I_{k_i} in the same position as Γ_i.

— Using these matrices, we define for $1 \leq i \leq k$ the following subsets $\tilde{C}_i$ of C:

$$\tilde{C}_i := \bigcup_{j=1}^{m} \left\{ v \cdot \tilde{\Gamma}_j \mid v \in \mathbb{F}_q^k, \ \mathrm{wt}(v) \leq i \right\}.$$

Clearly, these sets form the ascending chain

$$\tilde{C}_1 \subseteq \tilde{C}_2 \subseteq \ldots \subseteq \tilde{C}_k = \tilde{C}$$

of subsets of $\tilde{C}$, and hence the minimum weights

$$\bar{d}_i := \min \left\{ \mathrm{wt}(c) \mid c \in \tilde{C}_i, \ c \neq 0 \right\},$$

form the decreasing sequence

$$\bar{d}_1 \geq \bar{d}_2 \geq \ldots \geq \bar{d}_k = \mathrm{dist}(\tilde{C}) = \mathrm{dist}(C).$$

In most cases, we do not need to compute all of these values. In fact, the computation of $\bar{d}_k$ is just the evaluation of $\mathrm{dist}(\tilde{C})$ as the least weight of all codewords c in $\tilde{C} \setminus \{0\}$, which we want to avoid, if possible. As a matter of fact, in the first step we just compute $\bar{d}_1$. Later, if $\bar{d}_i$ has been computed for some $i \geq 1$, we will compare it with $\underline{d}_i$, which is a lower bound for the weight of the elements in $\tilde{C} \setminus \tilde{C}_i$. If $\bar{d}_i \leq \underline{d}_i$ we are finished. Otherwise, if $\bar{d}_i > \underline{d}_i$, we proceed to compute the exact value of $\bar{d}_{i+1}$.

— Hence, we try to find lower bounds for the weights in the complements $\widetilde{C} \setminus \widetilde{C}_i$. For this purpose we pick an element $c \in \widetilde{C} \setminus \widetilde{C}_i$. Since $c \notin \widetilde{C}_i$, there exists, for each j, a vector $v^{(j)} \in \mathbb{F}_q^k$ such that

$$c = v^{(j)} \cdot \widetilde{\Gamma}_j, \quad 1 \le j \le m, \text{ and } \mathrm{wt}(v^{(j)}) \ge i+1.$$

In order to estimate the weight of c, we consider each of these representations of c by using the various information places in $\widetilde{\Gamma}_j$, the columns of which contain the unit matrix I_{k_j}. These are the columns of index r for $k_1 + \ldots + k_{j-1} \le r < k_1 + \ldots + k_j$. We are especially interested in the k_j coordinates c_r of $c = v^{(j)} \cdot \widetilde{\Gamma}_j$ corresponding to these k_j columns. Since $v^{(j)}$ is of length k, these entries of c contribute at least the value $i + 1 - (k - k_j)$ to the weight of c. Since these sets of places are disjoint, for different j, we obtain

$$\mathrm{wt}(c) \ge \sum_{j=1}^{m} (i + 1 - (k - k_j)).$$

We can restrict our attention to positive summands, which gives the lower bound

$$\mathrm{wt}(c) \ge \sum_{j:k-k_j \le i} (i + 1 - (k - k_j)) =: \underline{d}_i.$$

Obviously, since the first summand is $i + 1$, the sequence of these bounds is increasing:

$$2 \le \underline{d}_1 < \underline{d}_2 < \underline{d}_3 < \ldots < \underline{d}_k.$$

In addition,

$$\underline{d}_k = m + \sum_{j=1}^{m} k_j.$$

— Since $\mathrm{wt}(c) \le k_1 + \ldots + k_m$ for all $c \in \widetilde{C}$, there exists a smallest index i_0 such that

$$\overline{d}_{i_0} \le \underline{d}_{i_0}.$$

For this i_0 we have that

$$\overline{d}_{i_0} := \min \left\{ \mathrm{wt}(c) \mid c \in \widetilde{C}_{i_0},\ c \ne 0 \right\},$$

and the inequality

$$\underline{d}_{i_0} \le \min \left\{ \mathrm{wt}(c) \mid c \in \widetilde{C} \setminus \widetilde{C}_{i_0} \right\}$$

holds true. Hence,

$$\overline{d}_{i_0} = \mathrm{dist}(\widetilde{C}),$$

and a codeword of weight $\mathrm{dist}(\widetilde{C})$ is contained in $\widetilde{C}_{i_0}$.

— To simplify the algorithm, it is possible to do all these computations with Γ_i instead of $\widetilde{\Gamma}_i$. Notice that the values of $\overline{d}_j$ and d_j do not change when we use Γ_i instead of $\widetilde{\Gamma}_i$ since

$$\mathrm{wt}\left(v \cdot \widetilde{\Gamma}_i\right) = \mathrm{wt}\left(v \cdot \Gamma_i \cdot M_{\pi_{i+1}}^\top \cdots M_{\pi_m}^\top\right) = \mathrm{wt}(v \cdot \Gamma_i).$$

Moreover, we can replace the sets $\widetilde{C}_i$ by the *isometric sets*

$$C_i := \bigcup_{j=1}^{m} \left\{v \cdot \Gamma_j \mid v \in \mathbb{F}_q^k,\ \mathrm{wt}(v) \leq i\right\}. \qquad \square$$

Here is a summary of the algorithm **MinDist**.

1.8.2　**Algorithm** Compute the minimum distance of a given linear (n,k)-code C.
　　　　Input:　　A systematic generator matrix $\Gamma_1 = (I_k \mid A_1)$ of C.
　　　　Output:　The minimum distance $\mathrm{dist}(C)$.

(1)　$m := 2$
(2)　$k_1 := k$
(3)　**repeat**
(4)　　Apply Gaussian elimination and possibly permutations of the
　　　　columns to the matrix A_{m-1} from $\Gamma_{m-1} = \left(A'_{m-1} \begin{array}{c|c} I_{k_{m-1}} & A_{m-1} \\ \hline 0 & 0 \end{array} \right)$
　　　　to obtain a generator matrix $\Gamma_m = \left(A'_m \begin{array}{c|c} I_{k_m} & A_m \\ \hline 0 & 0 \end{array} \right)$
(5)　**until** $\mathrm{rank}(A_m) = 0$
(6)　$C_0 := \{0\}$
(7)　$i := 0$
(8)　**repeat**
(9)　　$i := i+1$
(10)　$C_i := C_{i-1} \cup \bigcup_{j=1}^{m} \{v \cdot \Gamma_j \mid v \in \mathbb{F}(q)^k, \mathrm{wt}(v) = i\}$
(11)　$\overline{d}_i := \min\{\mathrm{wt}(c) \mid c \in C_i, c \neq 0\}$
(12)　$\underline{d}_i := \sum_{\substack{j=1 \\ k-k_j \leq i}}^{m} (i+1) - (k - k_j)$
(13)　**until** $\overline{d}_i \leq \underline{d}_i$
(14)　**return** $\overline{d}_i$　　　　　　　　　　　　　　　　　$\square$

Example We apply the algorithm **MinDist** to the binary $(7,3)$-code C with generator matrix   1.8.3

$$\Gamma_1 = \begin{pmatrix} 1 & 0 & 0 & 1 & 0 & 1 & 1 \\ 0 & 1 & 0 & 1 & 1 & 0 & 1 \\ 0 & 0 & 1 & 0 & 1 & 1 & 1 \end{pmatrix}.$$

This matrix has the information set $\{0,1,2\}$. The algorithm successively computes the generator matrices

$$\Gamma_2 = \begin{pmatrix} 0 & 1 & 1 & 1 & 0 & 0 & 1 \\ 1 & 1 & 0 & 0 & 1 & 0 & 1 \\ 1 & 1 & 1 & 0 & 0 & 1 & 0 \end{pmatrix}$$

with information set $\{3,4,5\}$ and the generator matrix

$$\Gamma_3 = \begin{pmatrix} 0 & 1 & 1 & 1 & 0 & 0 & 1 \\ 1 & 0 & 1 & 1 & 1 & 0 & 0 \\ 1 & 1 & 1 & 0 & 0 & 1 & 0 \end{pmatrix}$$

with information set $\{6\}$. The set C_1 consists of all rows of the three generator matrices Γ_1, Γ_2 and Γ_3. Each of them is of weight 4, whence $\bar{d}_1 = 4$. The lower bound for the minimum weight of the vectors outside of C_1 is $\underline{d}_1 = 4$. Hence, $d = \bar{d}_1 = 4$ is the minimum distance of C. ◇

Example We apply the algorithm **MinDist** to the binary $(15,5)$-code C with generator matrix   1.8.4

$$\Gamma = \begin{pmatrix} 1 & 0 & 1 & 0 & 0 & 1 & 1 & 0 & 1 & 1 & 1 & 0 & 0 & 0 & 0 \\ 0 & 1 & 0 & 1 & 0 & 0 & 1 & 1 & 0 & 1 & 1 & 1 & 0 & 0 & 0 \\ 0 & 0 & 1 & 0 & 1 & 0 & 0 & 1 & 1 & 0 & 1 & 1 & 1 & 0 & 0 \\ 0 & 0 & 0 & 1 & 0 & 1 & 0 & 0 & 1 & 1 & 0 & 1 & 1 & 1 & 0 \\ 0 & 0 & 0 & 0 & 1 & 0 & 1 & 0 & 0 & 1 & 1 & 0 & 1 & 1 & 1 \end{pmatrix}.$$

This code will be constructed in 4.3.5, it is a BCH-code. The systematic matrices are

$$\Gamma_1 = \begin{pmatrix} 1 & 0 & 0 & 0 & 0 & 1 & 0 & 1 & 0 & 0 & 1 & 1 & 0 & 1 & 1 \\ 0 & 1 & 0 & 0 & 0 & 1 & 1 & 1 & 1 & 0 & 1 & 0 & 1 & 1 & 0 \\ 0 & 0 & 1 & 0 & 0 & 0 & 1 & 1 & 1 & 1 & 0 & 1 & 0 & 1 & 1 \\ 0 & 0 & 0 & 1 & 0 & 1 & 0 & 0 & 1 & 1 & 0 & 1 & 1 & 1 & 0 \\ 0 & 0 & 0 & 0 & 1 & 0 & 1 & 0 & 0 & 1 & 1 & 0 & 1 & 1 & 1 \end{pmatrix},$$

$$\Gamma_2 = \begin{pmatrix} 1 & 1 & 0 & 1 & 1 & 1 & 0 & 0 & 0 & 0 & 1 & 0 & 1 & 0 & 0 \\ 1 & 0 & 1 & 1 & 0 & 0 & 1 & 0 & 0 & 0 & 1 & 1 & 1 & 1 & 0 \\ 0 & 1 & 0 & 1 & 1 & 0 & 0 & 1 & 0 & 0 & 0 & 1 & 1 & 1 & 1 \\ 0 & 1 & 1 & 1 & 0 & 0 & 0 & 0 & 1 & 0 & 1 & 0 & 0 & 1 & 1 \\ 1 & 0 & 1 & 1 & 1 & 0 & 0 & 0 & 0 & 1 & 0 & 1 & 0 & 0 & 1 \end{pmatrix},$$

and

$$\Gamma_3 = \begin{pmatrix} 1 & 0 & 1 & 0 & 0 & 1 & 1 & 0 & 1 & 1 & 1 & 0 & 0 & 0 & 0 \\ 1 & 1 & 1 & 1 & 0 & 1 & 0 & 1 & 1 & 0 & 0 & 1 & 0 & 0 & 0 \\ 0 & 1 & 1 & 1 & 1 & 0 & 1 & 0 & 1 & 1 & 0 & 0 & 1 & 0 & 0 \\ 1 & 0 & 0 & 1 & 1 & 0 & 1 & 1 & 1 & 0 & 0 & 0 & 0 & 1 & 0 \\ 0 & 1 & 0 & 0 & 1 & 1 & 0 & 1 & 1 & 1 & 0 & 0 & 0 & 0 & 1 \end{pmatrix}.$$

The information sets are

$$\{0,1,2,3,4\}, \quad \{5,6,7,8,9\}, \quad \text{and} \quad \{10,11,12,13,14\}.$$

The minimum weight of the rows in these matrices is $\overline{d}_1 = 7$, whereas $\underline{d}_1 = 2 - (5-5) + 2 - (5-5) + 2 - (5-5) = 6$. Since $7 > 6$, we continue by considering linear combinations of any two rows of the Γ_i. For example, if $i = 1$ we look at vectors v and codewords $v \cdot \Gamma_1$ where

v	$v \cdot \Gamma_1$	$\mathrm{wt}(v \cdot \Gamma_1)$
$(1,1,0,0,0)$	$(1,1,0,0,0,0,1,0,1,0,0,1,1,0,1)$	7
$(1,0,1,0,0)$	$(1,0,1,0,0,1,1,0,1,1,1,0,0,0,0)$	7
$(1,0,0,1,0)$	$(1,0,0,1,0,0,0,1,1,1,1,0,1,0,1)$	8
$(1,0,0,0,1)$	$(1,0,0,0,1,1,1,1,0,1,0,1,1,0,0)$	8
$(0,1,1,0,0)$	$(0,1,1,0,0,1,0,0,0,1,1,1,1,0,1)$	8
$(0,1,0,1,0)$	$(0,1,0,1,0,0,1,1,0,1,1,1,0,0,0)$	7
$(0,1,0,0,1)$	$(0,1,0,0,1,1,0,1,1,1,0,0,0,0,1)$	7
$(0,0,1,1,0)$	$(0,0,1,1,0,1,1,1,0,0,0,0,1,0,1)$	7
$(0,0,1,0,1)$	$(0,0,1,0,1,0,0,1,1,0,1,1,1,0,0)$	7
$(0,0,0,1,1)$	$(0,0,0,1,1,1,1,1,0,1,0,1,1,0,0,1)$	8

If $i = 2$, we have

v	$v \cdot \Gamma_2$	$\mathrm{wt}(v \cdot \Gamma_2)$
$(1,1,0,0,0)$	$(0,1,1,0,1,1,1,0,0,0,0,1,0,1,0)$	7
$(1,0,1,0,0)$	$(1,0,0,0,0,1,0,1,0,0,1,1,0,1,1)$	7
$(1,0,0,1,0)$	$(1,0,1,0,1,1,0,0,1,0,0,0,1,1,1)$	8
$(1,0,0,0,1)$	$(0,1,1,0,0,1,0,0,0,1,1,1,1,0,1)$	8
$(0,1,1,0,0)$	$(1,1,1,0,1,0,1,1,0,0,1,0,0,0,1)$	8
$(0,1,0,1,0)$	$(1,1,0,0,0,0,1,0,1,0,0,1,1,0,1)$	7
$(0,1,0,0,1)$	$(0,0,0,0,1,0,1,0,0,1,1,0,1,1,1)$	7
$(0,0,1,1,0)$	$(0,0,1,0,1,0,0,1,1,0,1,1,1,0,0)$	7
$(0,0,1,0,1)$	$(1,1,1,0,0,0,0,1,0,1,0,0,1,1,0)$	7
$(0,0,0,1,1)$	$(1,1,0,0,1,0,0,0,1,1,1,1,0,1,0)$	8

and for $i = 3$ we obtain

v	$v \cdot \Gamma_3$	$\mathrm{wt}(v \cdot \Gamma_3)$
$(1,1,0,0,0)$	$(0,1,0,1,0,0,1,1,0,1,1,1,0,0,0)$	7
$(1,0,1,0,0)$	$(1,1,0,1,1,1,0,0,0,1,0,1,0,0)$	7
$(1,0,0,1,0)$	$(0,0,1,1,1,1,0,1,0,1,1,0,0,1,0)$	8
$(1,0,0,0,1)$	$(1,1,1,0,1,0,1,1,0,0,1,0,0,0,1)$	8
$(0,1,1,0,0)$	$(1,0,0,0,1,1,1,1,0,1,0,1,1,0,0)$	8
$(0,1,0,1,0)$	$(0,1,1,0,1,1,1,0,0,0,1,0,1,0)$	7
$(0,1,0,0,1)$	$(1,0,1,1,1,0,0,0,0,1,0,1,0,0,1)$	7
$(0,0,1,1,0)$	$(1,1,1,0,0,0,0,1,0,1,0,0,1,1,0)$	7
$(0,0,1,0,1)$	$(0,0,1,1,0,1,1,1,0,0,0,0,1,0,1)$	7
$(0,0,0,1,1)$	$(1,1,0,1,0,1,1,0,0,1,0,0,0,1,1)$	8

This shows that $\bar{d}_2 = 7$. On the other hand, $\underline{d}_2 = 3 - (5 - 5) + 3 - (5 - 5) + 3 - (5 - 5) = 9$ which is greater than 7, i.e. the minimum distance has been determined to be 7. In this example, we have looked at $15 + 3 \cdot 10 = 45$ codewords, which is actually worse than the original problem.   ◇

We see that the algorithm may actually be worse than the original problem. But in many cases, in particular when the codes get bigger, there is a benefit. For example, the minimum distance of the binary extended Golay code of length 24 and dimension 12 (presented in 2.3.12) is computed by looking at 596 rather than $2^{12} = 4096$ codewords.

Exercises

Exercise Prove the remaining statements about $\underline{d}_i$ for $1 \le i \le k$ in the description of 1.8.1.   **E.1.8.1**

Exercise Use the algorithm **MinDist** in order to evaluate the minimum distance of the binary $(7,4)$-code with generator matrix   **E.1.8.2**

$$\begin{pmatrix} 1 & 1 & 0 & 1 & 0 & 0 & 0 \\ 0 & 1 & 1 & 0 & 1 & 0 & 0 \\ 0 & 0 & 1 & 1 & 0 & 1 & 0 \\ 0 & 0 & 0 & 1 & 1 & 0 & 1 \end{pmatrix}.$$

Check your result using the attached software.

Chapter 2

Bounds and Modifications

2

2

2 Bounds and Modifications

The fundamental parameters of a linear code are the length n, the dimension k, the minimum distance d and the size q of the finite field over which it is defined. For applications, we are interested in the information rate k/n and the relative minimum distance d/n both being large. We may think of this as a typical *packing problem of combinatorics*. Is it possible to pack a large number of vectors (codewords) into the Hamming space $H(n,q)$ such that no two words are close? Of course, these are contradicting aims. To see this, we think of the balls of radius $\lfloor (d-1)/2 \rfloor$ which are centered around codewords, since, for unique decoding using the maximum likelihood principle, we require that these balls should never overlap. It is intuitively clear that a large number of codewords can only be achieved if the balls are small. Conversely, if the balls are large then this tends to limit the number of codewords (or balls) which can be packed. This is the fundamental dilemma of Coding Theory (cf. Fig. 2.1).

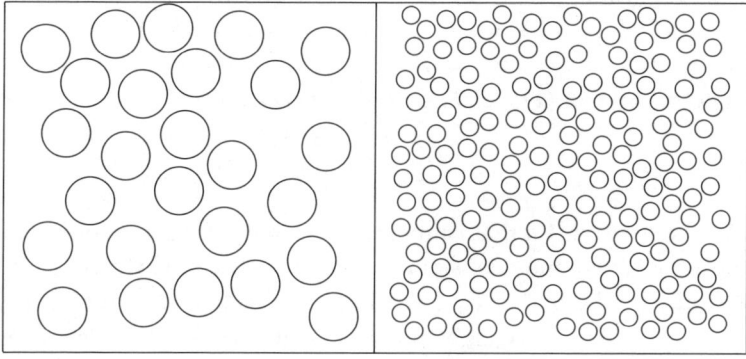

Fig. 2.1 The fundamental dilemma

In order to understand the situation better, we are going to study various bounds for the parameters of codes. We consider (n,k,d)-codes *optimal* if they optimize one parameter given the other two (the parameter q is kept fixed). Furthermore, we will discuss various constructions of new codes from old. These constructions in turn lead to bounds. Interesting classes of codes are obtained by analyzing whether bounds can be met with equality. Also, we will meet further series of codes which are connected to the above-mentioned constructions, or which are of interest because they meet one of the bounds which will be presented. Examples are the Hamming- and simplex-codes, perfect codes, Reed–Muller- and MDS-codes.

2.1 Combinatorial Bounds for the Parameters

For the purpose of applications we certainly prefer linear codes with *optimal* properties. The search for optimal codes can be described in three ways:

1. For given parameters k, d, q find a linear code of *least length*
$$n_{\min}(k,d,q) := \min \{ \, n \mid \text{there exists an } (n,k,d,q)\text{-code} \, \}.$$

2. For given parameters n, k, q find a linear code of *maximal minimum distance*
$$d_{\max}(n,k,q) := \max \{ \, d \mid \text{there exists an } (n,k,d,q)\text{-code} \, \}.$$

3. For given parameters n, d, q find a linear code of *maximal dimension*
$$k_{\max}(n,d,q) := \max \{ \, k \mid \text{there exists an } (n,k,d,q)\text{-code} \, \}.$$

To begin with, we derive a few direct combinatorial bounds for the parameters of a code. Each such result in turn yields a bound for $n_{\min}(k,d,q)$, $d_{\max}(n,k,q)$ and $k_{\max}(n,d,q)$. After that, we will tabulate the best bounds we have obtained at that point. In the following section we will investigate the two functions $n_{\min}(k,d,q)$ and $d_{\max}(n,k,q)$ more thoroughly.

2.1.1   **The Singleton-bound** *For each linear (n,k,d)-code C over $\mathbb{F}_q$ we have the inequality*
$$d \le n - k + 1.$$

Proof: We know from 1.4 that isometric codes have the same coding theoretic properties. By 1.7 we may consider a code isometric to C which is generated systematically by the matrix $(I_k \mid A)$. Then, for each unit vector $e^{(i)} \in \mathbb{F}_q^k$, the vector $e^{(i)} \cdot (I_k \mid A)$ is of weight not greater than $n - k + 1$. This proves the statement, since by 1.2.8 the minimum distance is the minimum weight of a nonzero codeword.   □

2.1.2   **Definition (MDS-codes)** Codes with minimum distance $d = n - k + 1$ are called *MDS-codes* (an abbreviation of *maximum distance separable*).   ◇

Trivial MDS-codes are the $(n,1)$-repetition-codes, the $(n, n-1)$-parity check codes (cf. Exercise 1.3.11), and the (n,n)-codes. We will discuss MDS-codes in Section 2.5.

2.1.3   **The Hamming-bound** *The parameters of each (n,k,d,q)-code satisfy the inequality*
$$\sum_{i=0}^{\lfloor (d-1)/2 \rfloor} \binom{n}{i} (q-1)^i \le q^{n-k}.$$

Equality holds if and only if the closed balls of radius $\lfloor (d-1)/2 \rfloor$ around codewords cover the whole Hamming space $H(n,q)$.

Proof: The number of vectors of Hamming distance i from a given vector is $\binom{n}{i}(q-1)^i$, since the first factor counts the number of ways of choosing i coordinates out of the n coordinate positions and the second term is the number of possibilities to change such an i-tuple in each place. We may think of these vectors as forming a ball of radius i around a given codeword. Summing over $i = 0,\ldots,r$ yields the number of vectors in a ball of radius r. Since the balls of radius $r = \lfloor (d-1)/2 \rfloor$ around codewords are all disjoint, the left hand side of the inequality multiplied by q^k is less than or equal to $|H(n,q)| = q^n$. Dividing by q^k yields the statement. $\square$

Definition (perfect codes) Codes whose parameters attain the Hamming-bound are called *perfect*. $\diamond$

2.1.4

Important examples of perfect codes are the Hamming-codes, which we will introduce next. Further perfect codes are the Golay-codes G_{23} and G_{11}; they will be presented in Section 4.4 (cf. Exercise 2.1.2). Trivial perfect codes are described in Exercise 2.1.1. A. Tietäväinen [191] and, independently, V.A. Zinovjev and V.K. Leontjev [207] have shown that there are no further perfect linear codes. However, there exist other perfect codes which are not linear.

The general form of the *Hamming-codes* was introduced first by M.J.E. Golay [70] and R.W. Hamming [80]. The binary $(7,4)$-Hamming-code is indeed older. It is mentioned in the seminal paper of C.E. Shannon [178]. The Hamming-codes form an infinite family of perfect, 1-error-correcting linear codes. The following definition specifies this class of codes up to isometry.

Definition (Hamming-codes, simplex-codes) Let Δ be any matrix whose columns form a system of nonzero representatives of the one-dimensional subspaces of $\mathbb{F}_q^m$. A linear code C which has Δ as its check matrix is called an *m-th order q-ary Hamming-code*. The dual code of a Hamming-code, i.e. the code which is *generated* by the matrix Δ (cf. 1.3.4) is called an *m-th order q-ary simplex-code*. Of course, both the Hamming- and the simplex-code are only defined up to isometry. $\diamond$

2.1.5

In 1.3.6, we have already met the third order binary Hamming-code.

Theorem *For $m \geq 2$ the m-th order q-ary Hamming-code is a perfect code with parameters*

2.1.6

$$(n,k,d,q) = \left(\frac{q^m - 1}{q-1}, \frac{q^m - 1}{q-1} - m, 3, q \right).$$

Proof: The statement about the length is clear from the definition, since the number of one-dimensional subspaces of $\mathbb{F}_q^m$ is $(q^m - 1)/(q - 1)$. A check matrix Δ of a Hamming-code contains in particular the m unit vectors (or nonzero scalar multiples thereof). Hence Δ is of rank m and the dimension of the code is $\dim(C) = (q^m - 1)/(q - 1) - m$, by 1.3.1. It remains to determine the minimum distance. For this, we note that any two columns of Δ are by definition linearly independent. Furthermore, since $m \geq 2$ there exist three columns which are dependent. By 1.3.9 this implies that the minimum distance is $d = 3$. Finally, we see that this code is perfect, since

$$\sum_{i=0}^{\lfloor (d-1)/2 \rfloor} \binom{n}{i} (q-1)^i \;=\; \sum_{i=0}^{1} \binom{(q^m - 1)/(q - 1)}{i} (q-1)^i$$

$$= \; 1 + \frac{q^m - 1}{q - 1}(q - 1) = q^m. \qquad \square$$

2.1.7   **Theorem** *The m-th order q-ary simplex-code C has parameters*

$$(n, k, d, q) = \left(\frac{q^m - 1}{q - 1}, m, q^{m-1}, q \right).$$

All nonzero codewords have weight q^{m-1}, i.e.

$$w_C(x) = 1 + (q^m - 1)x^{q^{m-1}}, \quad \text{and} \quad W_C(x,y) = y^{\frac{q^m - 1}{q - 1}} + (q^m - 1)x^{q^{m-1}} y^{\frac{q^{m-1} - 1}{q - 1}}.$$

Proof: Consider the matrix Δ from the proof of 2.1.6. This time, regard Δ as a generator matrix. The statement about the length is clear, and the value for the dimension follows again from 1.3.1 together with 1.3.4. It remains to show that each nonzero codeword has weight q^{m-1}. For this, we consider the encoding map $v \mapsto v \cdot \Delta$. Write

$$\Delta = \left(u^{(0)\top} \;\middle|\; \dots \;\middle|\; u^{(n-1)\top} \right)$$

with $u^{(i)} \in \mathbb{F}_q^m$. Then, using the standard bilinear form, we have for $v \in \mathbb{F}_q^m$

$$v \cdot \Delta = \left(\langle v, u^{(0)} \rangle, \dots, \langle v, u^{(n-1)} \rangle \right).$$

Fix an element $v \in \mathbb{F}_q^m \setminus \{0\}$. The mapping $u \mapsto \langle v, u \rangle$ for $u \in \mathbb{F}_q^m$ is a surjective linear form, as already pointed out in the proof of 1.6.8. It takes on each value of $\mathbb{F}_q$ exactly q^{m-1} times. Thus, for exactly $q^{m-1}(q - 1)$ vectors $u \in \mathbb{F}_q^m$ the value of $\langle v, u \rangle$ is nonzero. By linearity, we have $\langle v, \lambda u \rangle = \lambda \langle v, u \rangle$ for all $\lambda \in \mathbb{F}_q$. In particular, the value of $\langle v, w \rangle$ is either always zero or always nonzero for elements w of the form $w = \lambda u$, where $\lambda \in \mathbb{F}_q^*$. This means that the fact that $\langle v, u \rangle$ is zero or nonzero only depends on the one-dimensional subspace containing $u \neq 0$. Now recall that the $u^{(i)}$ form a transversal of the one-dimensional subspaces (disregarding the zero vector, which is in every subspace). This means that the products $\lambda \cdot u^{(i)}$ where $\lambda \in \mathbb{F}_q^*$ and $0 \leq i < (q^m - 1)/(q - 1)$ take on every nonzero vector $u \in \mathbb{F}_q^m$ exactly once. The previous remark implies that the $q^{m-1}(q - 1)$ vectors $u \in \mathbb{F}_q^m$ with $\langle v, u \rangle \neq 0$ ($u = 0$ is not one of them!) are contained in exactly q^{m-1} one-dimensional subspaces. Thus

$$\mathrm{wt}(v \cdot \Delta) = \mathrm{wt}(\langle v, u^{(0)} \rangle, \dots, \langle v, u^{(n-1)} \rangle) = q^{m-1}$$

for any $v \neq 0$. The statement about the weight enumerator $w_C(x)$ is clear. Using the identity $(q^m - 1)/(q - 1) = 1 + q + q^2 + \dots q^{m-1}$ we obtain the homogeneous version $W_C(x, y)$. This finishes the proof. □

Example The third order ternary Hamming-code is a $(13, 10, 3, 3)$-code. It has a check matrix of the form

$$\begin{pmatrix} 1 & 0 & 1 & 2 & 0 & 1 & 2 & 0 & 1 & 2 & 0 & 1 & 2 \\ 0 & 1 & 1 & 1 & 0 & 0 & 0 & 1 & 1 & 1 & 2 & 2 & 2 \\ 0 & 0 & 0 & 0 & 1 & 1 & 1 & 1 & 1 & 1 & 1 & 1 & 1 \end{pmatrix},$$

and its dual code is a ternary simplex-code of type $(13, 3, 9, 3)$. ◇

2.1.8

The next bound is an *explicit* bound for the minimum distance:

The Plotkin-bound *For each linear (n, k, d, q)-code C the following holds:*

$$d \leq \frac{nq^{k-1}(q - 1)}{q^k - 1}.$$

2.1.9

Proof: Consider the double sum of distances

$$D := \sum_{c \in C} \sum_{c' \in C} d(c, c').$$

It is bounded from below, since for each $c \neq c'$ we have $d(c, c') \geq d$, and this implies $D \geq q^k(q^k - 1)d$.

We may evaluate D in a different way. For this purpose we label the elements of $\mathbb{F}_q$ by $\{\kappa_1, \dots, \kappa_q\}$. For $0 \leq j < n$ and $1 \leq i \leq q$ let D_{ij} denote the number of codewords which have as their j-th component the element κ_i. In terms of this notation we obtain

$$D = \sum_{j \in n} \sum_{i=1}^{q} D_{ij}(q^k - D_{ij}).$$

Since $\sum_{i=1}^{q} D_{ij} = q^k$, we get

$$D = nq^{2k} - \sum_{j\in n}\sum_{i=1}^{q} D_{ij}^2.$$

For $j \in n$ the following is true:

$$0 \leq \sum_{i=1}^{q}\sum_{t=i+1}^{q} (D_{ij} - D_{tj})^2$$

$$= \underbrace{\sum_{i=1}^{q}\sum_{t=i+1}^{q} D_{ij}^2}_{=\sum_i(q-i)D_{ij}^2} + \underbrace{\sum_{i=1}^{q}\sum_{t=i+1}^{q} D_{tj}^2}_{=\sum_i(i-1)D_{ij}^2} - \sum_{i=1}^{q}\sum_{t=i+1}^{q} 2D_{ij}D_{tj}.$$

This yields the estimate

$$q\sum_{i=1}^{q} D_{ij}^2 \geq \sum_{i=1}^{q} D_{ij}^2 + \sum_{i=1}^{q}\sum_{t=i+1}^{q} 2D_{ij}D_{tj} = \left(\sum_{i=1}^{q} D_{ij}\right)^2 = q^{2k}.$$

Thus

$$\sum_{i=1}^{q} D_{ij}^2 \geq q^{2k-1}, \qquad j \in n,$$

from which we obtain

$$D = nq^{2k} - \sum_{j\in n}\sum_{i=1}^{q} D_{ij}^2 \leq nq^{2k-1}(q-1).$$

Combining these two bounds for D we conclude that

$$q^k(q^k - 1)d \leq D \leq nq^{2k-1}(q-1),$$

and the statement now follows by comparing the left hand side and the right hand side. $\square$

A few remarks concerning this bound are in order.

2.1.10 Remarks

1. Since the term on the right hand side of the bound may evaluate to a fraction, the bound can actually be read as

$$d \leq \left\lfloor \frac{nq^{k-1}(q-1)}{q^k - 1} \right\rfloor.$$

However, since we want to investigate what happens if the bound is met with equality, let us consider the inequality as stated in 2.1.9.

2. Equality holds in 2.1.9 under the following two conditions:
   (a) The distance between any two distinct codewords is equal to a constant (such a code is called *equidistant*.)
   (b) At any coordinate position, each field element appears equally often.
   An example for such a code is the simplex-code (cf. Exercise 2.1.5).

3. We may reformulate the Plotkin-bound as a bound for the number of codewords or, equivalently, for the dimension of a linear code of length n and minimum distance d over $\mathbb{F}_q$ as

$$q^k \leq \frac{d}{d - n(q-1)/q} ,$$ 2.1.11

provided that $d > n(q-1)/q$.   ◇

Next we collect some facts about $n_{\min}(k, d, q)$.

Lemma *If there exists an (n, k, d, q)-code with $d > 1$, then for each $1 \leq d' < d$ there*   2.1.12
exist (n, k, d', q)-codes.

Proof: Since $d > 1$, we may assume without loss of generality that the (n, k, d)-code C whose existence we assume has a systematic generator matrix

$$\Gamma = (I_k \mid A) = \left(e^{(0)^\top} \mid \cdots \mid e^{(k-1)^\top} \mid u^{(k)^\top} \mid \cdots \mid u^{(n-1)^\top} \right),$$

where A is a $k \times (n-k)$-matrix with $n - k \geq 1$. Replacing in Γ a column $u^{(j)^\top}$, $k \leq j < n$, by a column of zeros, we obtain a code C' with parameters (n, k, d) or $(n, k, d-1)$. The minimum distance of C' equals $d - 1$ if and only if there exists a codeword $c \in C$ of weight d such that $c_j \neq 0$. Summarizing, the replacement of a column of A by a column of zeros either leaves the minimum distance of C unchanged or decreases it by 1.

We start with the code C of minimum distance $d > 1$. Replacing one by one all the columns of A by columns of zeros, we eventually obtain a matrix of the form $(I_k \mid 0)$ which is the generator matrix of an $(n, k, 1)$-code. In each step, the minimum distance either stays put or decreases by 1. Whence by this procedure we construct (n, k, d')-codes for all $1 \leq d' < d$.   □

Lemma *Let q be a power of a prime and let k and d be positive integers. Then:*   2.1.13

1. $n_{\min}(k, 1, q) = k$, *for $k \geq 1$.*

2. $n_{\min}(1, d, q) = d$, *for $d \geq 1$.*

3. *If $d' \leq d$, then $n_{\min}(k, d', q) \leq n_{\min}(k, d, q)$ for $k \geq 1$.*

4. *If $k \geq 2$, then $n_{\min}(k, d, q) \geq d + n_{\min}(k - 1, \lceil d/q \rceil, q)$ for $d \geq 1$, where $\lceil r \rceil$ denotes as usual the least integer greater than or equal to r.*

Proof: The first two assertions are clear. In order to prove the third, we consider an $(n_{\min}(k, d, q), k, d, q)$-code C. According to 2.1.12, there also exists an $(n_{\min}(k, d, q), k, d', q)$-code for any $d' \leq d$. Whence $n_{\min}(k, d', q) \leq n_{\min}(k, d, q)$. In other words, the function $n_{\min}(k, d, q)$ is monotone increasing in the second argument. Lastly, in order to prove the final assertion, we consider again a code C of type $(n_{\min}(k, d, q), k, d, q)$. Let w be a vector of weight d in C, and assume that this vector is an element of a basis of C. Permuting columns and multiplying them with suitable constants, if necessary, we can assume that $w = (1_d, 0)$ and we see that C is linearly isometric to a code with generator matrix

$$\Gamma := \left(\frac{w}{*} \right) := \left(\begin{array}{c|c} 1_d & 0 \\ \hline \Gamma_1 & \Gamma_2 \end{array} \right).$$

Here, the top row w of Γ is of Hamming weight d, and Γ_1 and Γ_2 are matrices of size $(k - 1) \times d$ and $(k - 1) \times (n_{\min}(k, d, q) - d)$, respectively.

We claim that Γ_2 is of rank $k - 1$. Assume not. Then the rank of Γ is at most $k - 2$, and we can assume that the first row of Γ_2 contains only zeros. By the condition on the minimum distance, all elements in the corresponding row of Γ_1 are nonzero. Using an elementary row transformation, we can transform at least one further element of the top row of Γ into zero, which of course contradicts the fact that w was a word of minimum weight d. This proves the claim.

At this point, we know that the code C_2 generated by Γ_2 has the parameters $(n_{\min}(k, d, q) - d, k - 1, d_2)$, for some d_2 which is not yet known. Therefore

2.1.14 $$n_{\min}(k - 1, d_2, q) \leq n_{\min}(k, d, q) - d.$$

Now we consider $c = (c^{(1)}, c^{(2)}) \in C$ where $c^{(2)} \in C_2$ and $\text{wt}(c^{(2)}) = d_2$. By the pigeon-hole principle, there exists an element $\alpha \in \mathbb{F}_q$ which occurs at least $\lceil d/q \rceil$ many times in $c^{(1)}$. Without loss of generality, we can assume that $\alpha \in \mathbb{F}_q^*$. (If $\alpha = 0$, then choose any $\alpha_0 \in \mathbb{F}_q^*$ and replace c by the codeword $c + \alpha_0 w$ in which the element α_0 occurs at least $\lceil d/q \rceil$-many times among the first d components. Moreover, the last $n_{\min}(k, d, q) - d$ components of $c + \alpha_0 w$ are the same as in c.) Hence, subtracting the α-fold of the top row w of Γ from c yields the estimate

$$d \leq \text{wt}(c - \alpha w) \leq (d - \lceil d/q \rceil) + d_2,$$

and so $d_2 \geq \lceil d/q \rceil$. Furthermore, since the function $n_{\min}(k,d,q)$ is monotone in the second argument (this was proved in 3), we get

$$n_{\min}(k-1,d_2,q) \geq n_{\min}(k-1,\lceil d/q \rceil,q).$$

This together with 2.1.14 implies the desired inequality.   □

We remark that the fourth result of the previous Lemma is also known under the name "one-step Griesmer-bound". We will see that this result is essential for the Griesmer-bound, which we will present next. Recall that the Singleton-bound implies a bound for the length of (n,k,d)-codes,

$$n_{\min}(k,d,q) \geq k+d-1.$$ **2.1.15**

A better estimate is obtained from the following bound, whose binary version was discovered by Griesmer [76]. We present the form for general q which is due to Solomon and Stiffler [186].

The Griesmer-bound *Each linear (n,k,d,q)-code satisfies*   **2.1.16**

$$n \geq \sum_{i \in k} \lceil d/q^i \rceil.$$

Proof: The case $k = 1$ is trivial, so we may assume that $k \geq 2$. Applying the inequality of the fourth item of 2.1.13 iteratively we obtain the statement (see also Exercise 2.1.7):

$$
\begin{aligned}
n &\geq n_{\min}(k,d,q) \\
&\geq d + n_{\min}(k-1,\lceil d/q \rceil,q) \\
&\geq d + \lceil d/q \rceil + n_{\min}\left(k-2,\underbrace{\left\lceil \frac{\lceil d/q \rceil}{q} \right\rceil}_{=\lceil d/q^2 \rceil},q\right) \\
&\geq \ldots \\
&\geq \sum_{i \in k-1} \lceil d/q^i \rceil + \underbrace{n_{\min}(1,\lceil d/q^{k-1} \rceil,q)}_{=\lceil d/q^{k-1} \rceil} \\
&= \sum_{i \in k} \lceil d/q^i \rceil.
\end{aligned}
$$

□

Example We claim that there is no binary $(31,10,13)$-code. To see this, we   **2.1.17**
apply the Griesmer-bound, which gives us

$$n \geq 13 + \left\lceil \frac{13}{2} \right\rceil + \left\lceil \frac{13}{4} \right\rceil + \left\lceil \frac{13}{8} \right\rceil + \left\lceil \frac{13}{16} \right\rceil + \ldots + \left\lceil \frac{13}{512} \right\rceil$$
$$= 13 + 7 + 4 + 2 + 1 + 1 + 1 + 1 + 1 + 1 = 32.$$

But our code has length 31, which is a contradiction.   ◇

For each (n,k,d,q)-code C, the nonnegative integer $n - (k+d-1)$ has been called the *defect* of C (see 2.1.15). It can be estimated by an application of the Griesmer-bound:

2.1.18   **Theorem** *For each (n,k,d,q)-code with defect s we have:*

1. *If $k \geq 2$, then $d \leq q(s+1)$.*

2. *If $k \geq 3$ and $d = q(s+1)$, then $s+1 \leq q$.*

Proof: Both statements follow from the Griesmer-bound: As $\lceil d/q^i \rceil \geq 1$ we have

$$n \geq \sum_{i \in k} \lceil d/q^i \rceil \geq d + \lceil d/q \rceil + (k-2).$$

Assume indirectly that $d > q(s+1)$, and hence $\lceil d/q \rceil \geq s+2$. Then the right hand side is $\geq n+1$, which is clearly a contradiction.

Similarly, from $k \geq 3$ and $d = q(s+1)$ we obtain

$$d + k - 1 + s = n \geq \sum_{i \in k} \lceil d/q^i \rceil \geq d + (s+1) + \left\lceil \frac{s+1}{q} \right\rceil + (k-3),$$

thus $s+1 \geq s + \lceil (s+1)/q \rceil$, which implies that $s+1 \leq q$.   □

After these upper bounds we now derive two important *lower bounds*. Such bounds are essentially existence results: they state the existence of good codes. There is a catch, however. It may not always be easy to explicitly *find* the code whose existence is predicted by the lower bound. The first bound, due to Gilbert [67] resembles the Hamming-bound quite astonishingly. The second bound, due to Varshamov [194], turns out to be stronger than the Gilbert-bound (cf. Exercise 2.1.8). Nevertheless, asymptotically the two bounds agree.

2.1.19   **The Gilbert-bound** *Let q be a power of a prime and $n,k,d \in \mathbb{N}^*$ with $n \geq k,d$. The inequality*

$$\sum_{i \in d} \binom{n}{i} (q-1)^i < q^{n-k+1}$$

implies the existence of a linear (n,k)-code over $\mathbb{F}_q$ with minimum distance at least d.

Proof: Let C be a linear (n,k',d,q)-code with k' maximal. This means that there is no (n,k'',d)-code with $k'' > k'$. Let $\rho(C) = \max_{x \in H(n,q)} \min_{c \in C} d(x,c)$ be the *covering radius* of C, which measures how far away a word in the Hamming space can be from the given code. We claim that $\rho(C) \leq d-1$. Assume otherwise. Let $x \in H(n,q)$ be a vector with

$$d(x,C) := \min_{c \in C} d(x,c) \geq d.$$

Then, for $\lambda, \mu \in \mathbb{F}_q^*$ and $c, c' \in C$, we have

$$d(c + \lambda x, c' + \mu x) = d((\lambda - \mu)x, c' - c) \geq d,$$

unless $\lambda = \mu$ and $c = c'$. This is clear when $\lambda - \mu = 0$, as C has distance d. Otherwise, it follows from the fact that x is at distance $\geq d$ from $c' - c \in C$. The inequality just proved implies that the span of C and x, i.e. $C \oplus \langle x \rangle$, has minimum distance at least d. But $C \oplus \langle x \rangle$ is a linear code of dimension $k' + 1$, which contradicts the fact the k' was the largest possible dimension of such a code. Thus we have proved that $\rho(C) \leq d - 1$.

Now consider an $(n, k, \geq d, q)$-code. If the inequality $q^k \sum_{i \in d} \binom{n}{i} (q - 1)^i < q^n$ is satisfied, then the balls of radius $d - 1$ around codewords do not cover $H(n, q)$, i.e. there is a word $x \in H(n, q)$ with $d(x, C) \geq d$, i.e. $\rho(C) \geq d$. But this means that k is not maximal, i.e. there is a bigger code. The code whose existence is claimed can now be constructed directly. Start with the zero-code C, with $k = 0$. As long as the inequality $\sum_{i \in d} \binom{n}{i} (q - 1)^i < q^{n-k}$ is satisfied, there is a vector $x \in H(n, q)$ with $d(x, C) > d$. Replace C by $C \oplus \langle x \rangle$, a code of dimension $k + 1$ and repeat the procedure. We stop the procedure if $\sum_{i \in d} \binom{n}{i} (q - 1)^i \geq q^{n-k}$ and $q^{k-1} \sum_{i \in d} \binom{n}{i} (q - 1)^i < q^{n-(k-1)}$. Thus we end up with a linear $(n, k, \geq d, q)$-code as claimed. $\square$

The Varshamov-bound *Let q be a power of a prime and $n, k, d \in \mathbb{N}^*$ with $n \geq k, d$.*   **2.1.20**
The inequality

$$\sum_{i \in d-1} \binom{n-1}{i} (q - 1)^i < q^{n-k}$$

implies the existence of a linear (n, k)-code over $\mathbb{F}_q$ with minimum distance at least d.

Proof: If $n = k$ the inequality is satisfied only for $d = 1$. In this case there exists the trivial $(n, n, 1)$-code. Now we assume that $n - k \geq 1$. First we prove that $d - 1 \leq n - k$. Assume on the contrary that $d - 1 > n - k$. We obtain, since $d - 2 \geq n - k$ and $n - 1 \geq n - k$,

$$\sum_{i \in d-1} \binom{n-1}{i} (q - 1)^i \geq \sum_{i \in n-k+1} \binom{n-k}{i} (q - 1)^i = q^{n-k},$$

which contradicts our assumption.

Inductively, we will now construct an $(n - k) \times n$-matrix Δ of rank $n - k$, any $d - 1$ columns of which are linearly independent. Then, according to 1.3.10, Δ is a check matrix of an (n, k, d')-code C with $d' \geq d$. We start with the matrix $\Delta_{n-k} = I_{n-k}$ which consists of the $n - k$ unit vectors of length $n - k$. It is of rank $n - k$ and any $d - 1$ columns are linearly independent. If Δ_i with $n - k \leq i < n$ is an $(n - k) \times i$-matrix with the desired properties, we try to find a vector $u \in \mathbb{F}_q^{n-k}$ such that $\Delta_{i+1} = (\Delta_i \mid u^\top)$ also satisfies these properties. This vector u must be chosen from the set of elements of $\mathbb{F}_q^{n-k}$ which cannot be expressed as a linear combination of at most $d - 2$ columns of Δ_i. Of course, any linear combination of at most $d - 2$ columns of Δ_i is uniquely defined by its nonzero coefficients. Hence at most

$$\sum_{j \in d-1} \binom{i}{j} (q-1)^j$$

vectors can be written as linear combinations of at most $d - 2$ columns of Δ_i. Since

$$\sum_{j \in d-1} \binom{i}{j} (q-1)^j \le \sum_{j \in d-1} \binom{n-1}{j} (q-1)^j < q^{n-k},$$

there exists a vector u in $\mathbb{F}_q^{n-k}$ such that the system consisting of u and any $d - 2$ columns of Δ_i is linearly independent. Therefore Δ_{i+1} is of rank $n - k$ and any $d - 1$ columns of Δ_{i+1} are linearly independent. Finally, Δ can be chosen as the matrix Δ_n.   □

2.1.21   **Example** In the following table, we display upper and lower bounds for the optimal minimum distance $d_{\max}(n, k, 2)$ of binary codes with a given length n and dimension $k \le n$. For a given pair (n, k), the table shows either the exact value of $d_{\max}(n, k, 2)$, or an interval consisting of a lower bound and an upper bound. Subscripts are used to indicate which rule led to the bound. The subscripts V, S, H, G, or P stand for the Varshamov, Singleton, Hamming, Griesmer, or Plotkin-bound, respectively. For example, the table entry for $n = 8$ and $k = 2$ reads $4_V 5_P$ which stands for the two bounds $4 \le d_{\max}(8, 2, 2)$ by Varshamov and $d_{\max}(8, 2, 2) \le 5$ due to Plotkin.

$n\backslash k$	1	2	3	4	5	6	7	8
1	$1_{V,S}$							
2	$2_{V,S}$	$1_{V,S}$						
3	$3_{V,S}$	$2_{V,S}$	$1_{V,S}$					
4	$4_{V,S}$	$2_{V,H}$	$2_{V,S}$	$1_{V,S}$				
5	$5_{V,S}$	$3_{V,P}$	$2_{V,H}$	$2_{V,S}$	$1_{V,S}$			
6	$6_{V,S}$	$3_V 4_H$	$3_{V,P}$	$2_{V,H}$	$2_{V,S}$	$1_{V,S}$		
7	$7_{V,S}$	$4_{V,P}$	$3_V 4_H$	$3_{V,P}$	$2_{V,H}$	$2_{V,S}$	$1_{V,S}$	
8	$8_{V,S}$	$4_V 5_P$	$4_{V,H}$	$3_V 4_H$	$2_{V,H}$	$2_{V,H}$	$2_{V,S}$	$1_{V,S}$
9	$9_{V,S}$	$5_V 6_H$	$4_{V,G}$	$3_V 4_H$	$3_V 4_H$	$2_{V,H}$	$2_{V,H}$	$2_{V,S}$
10	$10_{V,S}$	$5_V 6_P$	$4_V 5_P$	$4_{V,G}$	$3_V 4_H$	$3_V 4_H$	$2_{V,H}$	$2_{V,H}$

This table will be improved in the next section, and the intervals will be replaced by exact values.   ◇

Exercises

E.2.1.1   **Exercise** Show that the following codes are perfect:

1. the (n, n)-code over $\mathbb{F}_q$ for any $n \geq 1$,

2. the n-fold repetition code over $\mathbb{F}_2$ for n odd.

Exercise Verify that the following parameter sets attain the Hamming-bound: $\quad$ E.2.1.2
$(23, 12, 7, 2)$, $(11, 6, 5, 3)$, $(90, 78, 5, 2)$. (Note that there exist perfect codes only
for the first two parameters.)

Exercise Prove that a linear code C is perfect if and only if $\rho(C) = \mathrm{dist}(C)$. $\quad$ E.2.1.3

Exercise Prove that 2.1.11 is equivalent to the Plotkin-bound. $\quad$ E.2.1.4

Exercise Check that the m-th order q-ary simplex-code meets the Griesmer- $\quad$ E.2.1.5
bound and the Plotkin-bound.

Exercise Let C be the m-th order binary Hamming-code of length $n = 2^m - 1$. $\quad$ E.2.1.6

1. Show that the homogeneous weight enumerator is

$$W_C(x, y) = \frac{1}{n+1} \left((x+y)^n + n(y-x)^{\frac{n+1}{2}} (x+y)^{\frac{n-1}{2}} \right).$$

2. Show that the coefficients A_i in $W_C(x, y) = \sum_{i=0}^{n} A_i x^i y^{n-i}$ satisfy the following recursion:

$$i A_i = \binom{n}{i-1} - A_{i-1} + (i - 2 - n) A_{i-2}$$

for $i \geq 3$ with initial conditions $A_0 = 1$, $A_1 = A_2 = 0$. Hint: Compute the
formal derivative w'_C of $w_C(x) = W_C(x, 1)$ and verify that

$$(1 - x^2) w'_C(x) + (1 + nx) w_C(x) = (1 + x)^n.$$

After that, compare coefficients.

Exercise Prove the following formula for positive integers r, s, t: $\quad$ E.2.1.7

$$\left\lceil \frac{\lceil r/s \rceil}{t} \right\rceil = \left\lceil \frac{r}{st} \right\rceil.$$

E.2.1.8   Exercise

1. Verify that the Varshamov-bound 2.1.20 is sometimes stronger than the Gilbert-bound 2.1.19. For example, the Varshamov-bound guarantees the existence of a $(7,4,3,2)$-code, whereas the Gilbert-bound only predicts the existence of a $(7,3,3,2)$-code.

2. Prove that the Varshamov-bound is always at least as strong as the Gilbert-bound. Do this by showing that the validity of the inequality in 2.1.19 implies that the inequality in 2.1.20 holds as well. Hint: put $f(x) = \sum_{i \in d-1} \binom{n-1}{i} x^i$ and $g(x) = \sum_{i \in d} \binom{n}{i} x^i$ and verify that $g(x) = (1+x)f(x) + x^{d-1}\binom{n-1}{d-1}$. Then put $x = q - 1$.

2.2

2.2 New Codes from Old and the Minimum Distance

Now we describe modifications of codes that permit the construction of new codes from given ones. An interesting application is, for example, that step by step we are able to improve our knowledge on the maximal minimum distance of (n,k)-codes over $\mathbb{F}_q$.

Recall the table obtained in 2.1.21. It contains bounds for maximal minimum distances $d_{\max}(n,k,2)$ of binary codes for $n \leq 10$ and $k \leq 8$. In several places it contains the *exact value* of $d_{\max}(n,k,2)$ while, in a few other places, it gives an *interval* containing the desired value $d_{\max}(n,k,2)$:

2.2.1

$n\backslash k$	1	2	3	4	5	6	7	8
1	1							
2	2	1						
3	3	2	1					
4	4	2	2	1				
5	5	3	2	2	1			
6	6	3 – 4	3	2	2	1		
7	7	4	3 – 4	3	2	2	1	
8	8	4 – 5	4	3 – 4	2	2	2	1
9	9	5 – 6	4	3 – 4	3 – 4	2	2	2
10	10	5 – 6	4 – 5	4	3 – 4	3 – 4	2	2

H.J. Helgert and R.D. Stinaff [85] gave such a table in 1973, containing lower and upper bounds for $d_{\max}(n,k,2)$, where $k \leq n \leq 127$. T. Verhoeff [195] improved it in 1987 by taking into account certain modifications. This work has been continued by Brouwer, who maintains an Internet database [32] with information on the best linear codes. A description of his methods and results can be found in [33]. Further tables can be found at [13], [27] as well as on the attached compact disc. In this section we introduce elementary modifications, which produce new codes from given ones, and discuss their influence on the table of lower and upper bounds for $d_{\max}(n, k, q)$.

Clearly, the entries in the leftmost column and the elements of the main diagonal are

$$d_{\max}(n, 1, q) = n \quad \text{and} \quad d_{\max}(n, n, q) = 1, \qquad n \geq 1.$$

Also, from 2.1.12 it follows that each value $0 < d \leq d_{\max}(n, k, q)$ occurs as a minimum distance of a suitable (n, k)-code over $\mathbb{F}_q$.

Parity extension Let C be an (n, k, d, q)-code with generator matrix $\hspace{2cm}$ 2.2.2

$$\Gamma = (\gamma_0 \mid \ldots \mid \gamma_{n-1}),$$

where γ_i denotes the i-th column vector of the matrix. Then the *parity extension* of C is the code $P(C)$ with generator matrix

$$\Gamma' := (\gamma_0 \mid \gamma_1 \mid \ldots \mid \gamma_{n-1} \mid -\sum_{i \in n} \gamma_i),$$

the additional last column of which contains the negative sum of the columns of Γ. The code $P(C)$ is an $(n+1, k)$-code with minimum distance at least d. ◇

Example In the binary case, we obtain $P(C)$ by simply adding an entry 0 to all $\hspace{1cm}$ 2.2.3 even codewords, and an entry 1 to all codewords of odd weight. In any case, the resulting codewords of $P(C)$ will have even Hamming weight. $\hspace{2cm}$ ◇

Corollary *If C denotes an $(n, k, d, 2)$-code with* odd *minimum distance d, then $P(C)$* $\hspace{1cm}$ 2.2.4 *is an $(n+1, k, d+1, 2)$-code.* $\hspace{5cm}$ □

Example For the binary $(7, 4)$-Hamming-code, the parity extension yields $\hspace{1cm}$ 2.2.5

$$\Gamma = \begin{pmatrix} 1 & 1 & 0 & 1 & 0 & 0 & 0 \\ 1 & 0 & 1 & 0 & 1 & 0 & 0 \\ 0 & 1 & 1 & 0 & 0 & 1 & 0 \\ 1 & 1 & 1 & 0 & 0 & 0 & 1 \end{pmatrix} \rightarrow \Gamma' = \begin{pmatrix} 1 & 1 & 0 & 1 & 0 & 0 & 0 & 1 \\ 1 & 0 & 1 & 0 & 1 & 0 & 0 & 1 \\ 0 & 1 & 1 & 0 & 0 & 1 & 0 & 1 \\ 1 & 1 & 1 & 0 & 0 & 0 & 1 & 0 \end{pmatrix}.$$

More generally, by 2.1.6 and 2.2.4 the *extended m-th order binary Hamming-code* is a $(2^m, 2^m - m - 1, 4)$-code. Furthermore, the parity extension of $H(n, q)$ is an $(n+1, n, 2)$-code. $\hspace{5cm}$ ◇

Let us see what the parity extension gives for the bounds for $d_{\max}(n, k, 2)$ of 2.2.1. In three places, we have codes of length n and dimension k whose minimum distance is odd. These are the $(5,2,3)$, $(6,3,3)$ and $(7,4,3)$-codes. We deduce that there exist $(6,2,4)$, $(7,3,4)$ and $(8,4,4)$-codes. In the table, we replace the intervals $3-4$ by an exact bound, which is 4, indicated by the boxed entries in 2.2.6. A further consequence is the existence of a $(9,4,4)$-code which results from the $(8,4,4)$-code by attaching a zero coordinate to every codeword. This improves the bound for $d_{\max}(9,4,2)$ to 4, which is shown underlined in the table.

2.2.6

$n\backslash k$	1	2	3	4	5	6	7	8
1	1							
2	2	1						
3	3	2	1					
4	4	2	2	1				
5	5	3	2	2	1			
6	6	$\boxed{4}$	3	2	2	1		
7	7	4	$\boxed{4}$	3	2	2	1	
8	8	$4-5$	4	$\boxed{4}$	2	2	2	1
9	9	$5-6$	4	$\underline{4}$	$3-4$	2	2	2
10	10	$5-6$	$4-5$	4	$3-4$	$3-4$	2	2

The last operation can be formulated as follows:

2.2.7

Corollary *For given q and k, the entries of the table $(d_{\max}(n,k,q))_{n,k}$, are weakly increasing downwards in each column, i.e.,*

$$d_{\max}(n+1,k,q) \geq d_{\max}(n,k,q), \qquad n \geq 1.$$ □

The next modification shows that the entries in these columns increase by at most 1:

2.2.8

Puncturing a code Assume that C is an (n,k)-code with $k < n$ and generator matrix

$$\Gamma = (\gamma_0 \mid \ldots \mid \gamma_{n-1}).$$

Then, without loss of generality (recall the definition of linear isometry of codes), we assume that there exists an information set *to which the last coordinate does not belong*. When canceling this component in all codewords, the resulting code $Pu(C)$, which is called *punctured code of C*, has the generator matrix

$$\Gamma' = (\gamma_0 \mid \gamma_1 \mid \ldots \mid \gamma_{n-2}).$$

According to our choice of the information set of C and of the canceled coordinate, the dimension k of the code is not changed and, therefore, $Pu(C)$ is an $(n-1,k)$-code. Its minimum distance is at least $d-1$. ◇

Using 2.1.12, we obtain

Corollary *For $k < n$, the existence of an (n,k,d,q)-code implies that there is also*   2.2.9
an $(n-1,k,d-1,q)$-code. In particular, the entries in a column of the matrix
$(d_{\max}(n,k,q))_{n,k}$ increase by at most 1 at a time, i.e.,

$$d_{\max}(n+1,k,q) - d_{\max}(n,k,q) \leq 1, \qquad n \geq 1. \qquad \square$$

Puncturing improves the preceding table in the following two boxed entries, whereas the underlined value follows from 2.2.4:

$n\backslash k$	1	2	3	4	5	6	7	8
1	1							
2	2	1						
3	3	2	1					
4	4	2	2	1				
5	5	3	2	2	1			
6	6	4	3	2	2	1		
7	7	4	4	3	2	2	1	
8	8	4 – 5	4	4	2	2	2	1
9	9	5 – 6	4	4	$\boxed{3}$	2	2	2
10	10	5 – 6	4 – 5	4	$\underline{4}$	$\boxed{3}$	2	2

2.2.10

Another way of combining codes is the concatenation, and there are essentially *two* different ways of doing this:

The concatenation (outer direct sum) Let C_i be an (n_i, k_i, d_i, q)-code with   2.2.11
generator matrix Γ_i for $i = 0,1$. The *outer direct sum* of C_0 and C_1 is defined as

$$C_0 \dotplus C_1 := \{(c \mid c') \mid c \in C_0, c' \in C_1\}.$$

It is clear that $C_0 \dotplus C_1$ is an $(n_0 + n_1, k_0 + k_1, \min\{d_0, d_1\}, q)$-code with generator matrix

$$\left(\begin{array}{c|c} \Gamma_0 & 0 \\ \hline 0 & \Gamma_1 \end{array} \right).$$

The outer direct sum can be expressed as

$$C_0 \dotplus C_1 = \left\{ (u \cdot \Gamma_0 \mid v \cdot \Gamma_1) \mid u \in \mathbb{F}_q^{k_0}, v \in \mathbb{F}_q^{k_1} \right\}.$$

in terms of the generator matrices. $\diamond$

Since the minimum distance of the outer direct sum is the minimum of the
minimum distances of the summands, this construction is not very exciting as
far as $d_{\max}$ is concerned. But it leads to another concatenation. In the particular
case $k_0 = k_1$ we can consider a subset of the outer sum which is, in a certain
sense, a diagonal:

2.2.12   **The diagonal concatenation ((u,v)-construction)** Let C_i be an (n_i, k, d_i, q)-code with generator matrix Γ_i for $i = 0, 1$. Then there exists an $(n_0 + n_1, k, d, q)$-code $C := (C_0, C_1)$, with $d \geq d_0 + d_1$, called the *diagonally concatenated code* or the (u,v)-*construction* applied to C_0 and C_1. It is generated by $\Gamma := (\Gamma_0 \mid \Gamma_1)$,

$$C := \left\{ (w \cdot \Gamma_0 \mid w \cdot \Gamma_1) \mid w \in \mathbb{F}_q^k \right\}. \qquad \diamond$$

For example, we know from 2.2.10 that there exist both a $(5,2,3,2)$-code and a $(3,2,2,2)$-code, and so we obtain via diagonal concatenation of these codes an $(8,2,5,2)$-code: Since $d_{\max}(8,2,2) \in \{4,5\}$, we get $d_{\max}(8,2,2) = 5$. In the same way we deduce from $d_{\max}(6,3,2) = 3$ and $d_{\max}(4,3,2) = 2$ that $d_{\max}(10,3,2) = 5$. Moreover, using 2.2.4 we obtain that $d_{\max}(9,2,2) = 6$, whereas $d_{\max}(10,2,2) = 6$ follows from the fact that the values in each column are increasing, as shown in 2.2.7. This way we improve the preceding table, obtaining

$n \backslash k$	1	2	3	4	5	6	7	8
1	1							
2	2	1						
3	3	2	1					
4	4	2	2	1				
5	5	3	2	2	1			
6	6	4	3	2	2	1		
7	7	4	4	3	2	2	1	
8	8	5	4	4	2	2	2	1
9	9	6	4	4	3	2	2	2
10	10	6	5	4	4	3	2	2

as the upper left hand corner of the table $(d_{\max}(n,k,2))_{n,k}$.

Hence, the upper left hand part of the desired table of maximal minimum distances of binary codes looks as follows:

$$
(d_{\max}(n,k,2))_{n,k} =
\begin{pmatrix}
1 & & & & & & & & & \\
2 & 1 & & & & & & & & \\
3 & 2 & 1 & & & & & & & \\
4 & 2 & 2 & 1 & & & & & & \\
5 & 3 & 2 & 2 & 1 & & & & & \\
6 & 4 & 3 & 2 & 2 & 1 & & & & \\
7 & 4 & 4 & 3 & 2 & 2 & 1 & & & \\
8 & 5 & 4 & 4 & 2 & 2 & 2 & 1 & & \\
9 & 6 & 4 & 4 & 3 & 2 & 2 & 2 & 1 & \\
10 & 6 & 5 & 4 & 4 & 3 & 2 & 2 & 2 & 1 \\
\vdots & & & & & & & & & \ddots
\end{pmatrix}.
$$

From this table we can directly deduce that the upper left hand corner of the matrix of $n_{\min}(k,d,2)$ is given by

$$(n_{\min}(k,d,2))_{k,d} = \begin{pmatrix} 1 & 2 & 3 & 4 & 5 & 6 & 7 & \cdots \\ 2 & 3 & 5 & 6 & 8 & 9 & \cdots \\ 3 & 4 & 6 & 7 & 10 & \cdots \\ 4 & 5 & 7 & 8 & \cdots \\ 5 & 6 & 9 & 10 & \cdots \\ 6 & 7 & 10 & \cdots \\ 7 & 8 & \cdots \\ 8 & 9 & \cdots \\ 9 & 10 & \cdots \\ \vdots \end{pmatrix}.$$

Moreover, for $k_{\max}(n,d,q)$ we obtain

$$(k_{\max}(n,d,2))_{n,d} = \begin{pmatrix} 1 & & & & & & & \\ 2 & 1 & & & & & & \\ 3 & 2 & 1 & & & & & \\ 4 & 3 & 1 & 1 & & & & \\ 5 & 4 & 2 & 1 & 1 & & & \\ 6 & 5 & 3 & 2 & 1 & 1 & & \\ 7 & 6 & 4 & 3 & 1 & 1 & 1 & \\ 8 & 7 & 4 & 4 & 2 & 1 & 1 & 1 \\ \vdots & & & & & & & & \ddots \end{pmatrix}.$$

We have seen that the entries in each column of the matrix $(d_{\max}(n,k,q))_{n,k}$ weakly increase and that the difference between two neighbors in the same column is at most 1. Now we note that the diagonal concatenation

$$\Gamma' := (\Gamma \mid I_k)$$

of a generator matrix Γ of an (n,k,d,q)-code and the identity matrix I_k generates an $(n+k,k,d',q)$-code with $d' \geq d+1$.

Corollary *For $k < n$ the existence of an (n,k,d,q)-code implies the existence of an $(n+k,k,d',q)$-code with $d' > d$. In particular, this shows that the entries in a column of the matrix $(d_{\max}(n,k,q))_{n,k}$ increase by at least 1 within an interval of k values for the length, and so each column of this matrix contains every positive integer at least once.*   2.2.14
□

A slight modification of the outer direct sum construction is

2.2.15   **The $(u \mid u+v)$-construction** For $i = 0,1$ let C_i be an (n, k_i, d_i, q)-code with generator matrix Γ_i. We define a linear code $C_0 \mid C_1$ by putting

$$C_0 \mid C_1 := \{(c, c+c') \mid c \in C_0, c' \in C_1\}.$$

This code is called the $(u \mid u+v)$-*construction* of C_0 and C_1. It is also known as the *semidirect sum* or *Plotkin construction* of C_0 and C_1. A generator matrix of it is

$$\left(\begin{array}{c|c} \Gamma_0 & \Gamma_0 \\ \hline 0 & \Gamma_1 \end{array}\right).$$

The $(u \mid u+v)$-construction $C_0 \mid C_1$ has the parameters

$$(2n, k_0 + k_1, \min\{2d_0, d_1\}, q).$$

Proof: The statements on the generator matrix, the length, and the dimension of $C_0 \mid C_1$ are clearly true. For the Hamming distance of two different codewords $(c, c+c')$ and $(w, w+w')$ of $C_0 \mid C_1$ the following holds:

$$d(c, w) + d(c+c', w+w') = \mathrm{wt}(c-w) + \mathrm{wt}(c-w+c'-w').$$

In the case when $c' = w'$ this sum is $2d(c, w) \geq 2d_0$, while otherwise we obtain a lower bound:

$$\mathrm{wt}(c-w) + \mathrm{wt}(c-w+c'-w') \geq$$
$$\mathrm{wt}(c-w) + \mathrm{wt}(c'-w') - \mathrm{wt}(c-w) = \mathrm{wt}(c'-w') \geq d_1. \qquad \square$$

2.2.16   **Example** The binary code C_0 with check matrix $\Delta = \mathbf{1}_4$ is a $(4,3)$-code. Each $c \in C_0$ has even parity because of $c \cdot \Delta^\top = c_0 + c_1 + c_2 + c_3 = 0$. Hence, C_0 consists of all vectors of even weight in $\mathbb{F}_2^4$. We deduce that C_0 is a $(4,3,2)$-parity check code. If C_1 denotes the $(4,1,4)$-repetition code, then $C_0 \mid C_1$ is an $(8,4,4)$-code. $\diamond$

The next construction allows us to deduce properties of the entries in the *subdiagonals* of $(d_{\max}(n, k, q))_{n,k}$, the entries $d_{\max}(n, n-i, q)$, for $i \in \mathbb{N}$ fixed.

2.2.17   **Shortening a code** Assume that the generator matrix $\Gamma = (\gamma_{ij})$ of C with $k > 1$ does not contain a column of zeros and that it is (after a permutation of rows) of the form

$$\left(\begin{array}{c|c} * & \gamma_{0,n-1} \\ \hline \Gamma' & \mathbf{0}^\top \end{array}\right), \quad \text{where } \gamma_{0,n-1} \neq 0.$$

We indicate the code generated by the submatrix Γ' by $S(C)$,

$$S(C) := \{(c_0, \ldots, c_{n-2}) \mid (c_0, \ldots, c_{n-2}, 0) \in C\}.$$

It is an $(n-1, k-1, d')$-code with $d' \geq d$ and it is called a *shortening* of C (in its last coordinate). $\diamond$

Let $k > 1$. If there is a codeword of C of weight d the last coordinate of which is zero, then the shortening $S(C)$ has minimum distance $d' = d$. This implies

Corollary *If $n \geq k > 1$, then we obtain from the existence of (n, k, d, q)-codes the existence of $(n-1, k-1, d, q)$-codes. This means for the table $(d_{max}(n, k, q))_{n,k}$, for fixed q, that its entries are weakly decreasing down each subdiagonal:* **2.2.18**

$$d_{max}(n-1, k-1, q) \geq d_{max}(n, k, q). \qquad \square$$

This corollary, together with 2.2.4, 2.2.7, 2.2.9, and 2.2.14, yields

Theorem *The matrix $(d_{max}(n, k, q))_{n,k}$ of maximal minimum distances of has the following properties:* **2.2.19**

1. *It is a lower triangular matrix.*

2. *Its main diagonal consists of 1's.*

3. *The entries in each column are weakly increasing from top to bottom.*

4. *Each column contains every positive integer at least once.*

5. *The entries in each subdiagonal are weakly decreasing from top left to bottom right.*

6. *In the binary case each odd positive integer occurs in each column* exactly once.
$\qquad \square$

Moreover, we obtain via shortening several inequalities for $n_{min}(k, d, q)$:

Lemma *The least length $n_{min}(k, d, q)$ satisfies:* **2.2.20**

1. *If $k \geq 2$, then $n_{min}(k, d, q) \geq n_{min}(k-1, d, q) + 1$.*

2. *If $d \geq 2$, then $n_{min}(k, d, q) > k$.*

3. *If $d \geq 2$, then $n_{min}(k, d, q) \geq n_{min}(k, d-1, q) + 1$.*

Proof: 1. Assume that C is an $(n_{min}(k, d, q), k, d)$-code, $k \geq 2$. Shortening C yields the $(n_{min}(k, d, q) - 1, k-1, d')$-code $S(C)$ with $d' \geq d$. Consequently

$$n_{min}(k-1, d, q) \leq n_{min}(k-1, d', q) \leq n_{min}(k, d, q) - 1.$$

2. The second statement can be proved by induction on k, using the second assertion of 2.1.13.

3. We again assume that C is an $(n_{\min}(k,d,q),k,d)$-code. Since $d \geq 2$, we obtain from the second assertion that $k < n_{\min}(k,d,q)$. The punctured code $Pu(C)$ is an $(n_{\min}(k,d,q) - 1, k, d')$-code with $d' \geq d - 1$. Consequently

$$n_{\min}(k,d-1,q) \leq n_{\min}(k,d',q) \leq n_{\min}(k,d,q) - 1,$$

which completes the proof.   □

Exercises

E.2.2.1   **Exercise** Prove that the weight enumerator of the outer direct sum $C_0 \dotplus C_1$ is $W_{C_0 \dotplus C_1}(x,y) = W_{C_0}(x,y) \cdot W_{C_1}(x,y)$.

E.2.2.2   **Exercise** Let C be a linear code over $\mathbb{F}_q$. For $\alpha \in \mathbb{F}_q$ let $\sigma(\alpha)$ be the number of codewords $c \in C$ whose parity sum $\sum_{i=0}^{n} c_i$ equals α. Prove that either $\sigma(0) = q^k$ and $\sigma(\alpha) = 0$ for $\alpha \in \mathbb{F}_q^*$, or $\sigma(\alpha) = q^{k-1}$ for all $\alpha \in \mathbb{F}_q$. Hint: The parity sum is a vector space homomorphism $C \to \mathbb{F}_q : c \mapsto \sum_{i \in n} c_i$.

2.3 2.3 Further Modifications and Constructions

We continue the description of modifications and constructions.

2.3.1   **Prolongation** A *prolongation* of an (n,k)-code C is an $(n+1,k+1)$-code obtained by adding an information place to C.   ◇

2.3.2   **Binary Augmentation** If Γ is the generator matrix of a binary (n,k,d)-code C which does not contain the all-one vector, then the code generated by

$$\begin{pmatrix} \mathbf{1}_n \\ \Gamma \end{pmatrix}$$

is called the *(binary) augmentation* of C. It contains all codewords of C and also the complement of each codeword. (The complement of a binary vector is obtained by replacing each 0 by 1 and vice versa.) The augmentation of C is an $(n,k+1)$-code with minimum distance equal to $\min\{d, n-d'\}$, where $d' := \max\{\mathrm{wt}(c) \mid c \in C\}$ is the *maximum weight* of C.   ◇

In the proof of the Griesmer-bound we encountered another modification called

The A-construction Any binary (n, k, d)-code C is linearly isometric to a code   **2.3.3**
with generator matrix

$$\left(\begin{array}{c|c} \mathbf{1}_d & \mathbf{0}_{n-d} \\ \hline * & \Gamma' \end{array} \right) ,$$

whose first row contains a codeword of minimum weight whose entries 1 are
left-aligned. As shown in 2.1.13, the matrix Γ' generates an $(n - d, k - 1)$-
code, $A(C)$, called the *A-construction*. The minimum distance of $A(C)$ is at
least $\lceil d/q \rceil$.   ◇

Example The A-construction enables us to prove that there cannot be a bi-   **2.3.4**
nary $(16, 6, 7)$-code. Assume on the contrary that there is a $(16, 6, 7)$-code
C. Using the A-construction we obtain a binary $(9, 5, 4)$-code $A(C)$ so that
$9 \geq n_{\min}(5, 4, 2)$, which contradicts our previous result that $n_{\min}(5, 4, 2) = 10$.
                                                                                   ◇

Corollary *The existence of an (n, k, d, q)-code implies the existence of a linear code of*   **2.3.5**
type

$$(n - d, k - 1, \geq \lceil d/q \rceil, q).$$ □

The next modification uses the check matrix of a code.

The Y1-construction Without loss of generality, we assume that the check ma-   **2.3.6**
trix Δ of an (n, k, d)-code with $n - 1 > k$ is of the form

$$\Delta = \left(\begin{array}{c|c} \mathbf{1}_{d^\perp} & \mathbf{0}_{n-d^\perp} \\ \hline * & \Delta' \end{array} \right) ,$$

where the first row is an element of minimum weight $d^\perp$ belonging to $C^\perp$. If
$d^\perp \leq k$, then the submatrix Δ' is the check matrix of an $(n - d^\perp, k - d^\perp + 1)$-
code, whose minimum distance is at least d by 1.3.10. This construction is
called the *Y1-construction*.   ◇

A generalization of the Y1-construction is

The B-construction Assume the existence of an (n, k, d, q)-code C with $n - 1 >$ **2.3.7**
k and $d_{\max}(n, n - k, q) \leq k$, which guarantees that $d^\perp \leq k$. From the (n, k, d, q)-
code C we obtain by Y1-construction an $(n - d^\perp, k - d^\perp + 1, d', q)$-code C' with
$d' \geq d$. Hence, for all s with $d^\perp \leq s \leq k$, the *B-construction* yields, by successive
shortening, $(n - s, k - s + 1)$-codes $B_s(C)$ with minimum distance at least d. ◇

2.3.8   **Example** Using the B-construction, one can give another proof that there is no binary $(16,6,7)$-code. Assume on the contrary that there is such a code. In order to apply the B-construction, we need an upper bound on the minimum distance of the dual code, which is a $(16,10)$-code. The Hamming-bound shows that there is no $(16,10,5)$-code, since

$$\binom{16}{0} + \binom{16}{1} + \binom{16}{2} = 1 + 16 + 8 \cdot 15 = 137 \not\leq 2^{16-10} = 64.$$

Thus $d_{\max}(16,10,2) \leq 4 = s \leq 6 = k$. The assumptions for the B-construction are satisfied, and we can produce from the $(16,6,7)$-code a $(16 - 4, 6 - 4 + 1, 7) = (12,3,7)$-code. But such a code does not exist because the parameters do not satisfy the Plotkin-bound:

$$d = 7 \not\leq \left\lfloor \frac{12 \cdot 2^2 \cdot 1}{2^3 - 1} \right\rfloor = \left\lfloor \frac{48}{7} \right\rfloor = \lfloor 6 + 6/7 \rfloor = 6.$$

This shows that the assumption was invalid, i.e. there does not exist a $(16,6,7)$-code, i.e. $d_{\max}(16,6,2) \leq 6$.   ◇

Another interesting combination of codes is

2.3.9   **The X-construction** It applies to chains of codes

$$C_1 \subset C_0 \subseteq \mathbb{F}_q^n,$$

which means that C_1 is a proper *subcode* of the code C_0. We can assume that C_1 is generated by the k_1-rowed submatrix Γ_1 of the generator matrix

$$\Gamma_0 = \left(\frac{\Gamma'}{\Gamma_1} \right)$$

of C_0 with $1 \leq k_1 < k_0$. If C_2 denotes an $(n_2, k_0 - k_1, d_2)$-code with generator matrix Γ_2, then

$$\Gamma = \left(\begin{array}{c|c} \Gamma' & \Gamma_2 \\ \hline \Gamma_1 & 0 \end{array} \right)$$

generates a code C called the *X-construction*, which is of type $(n_0 + n_2, k_0, d, q)$ with $d \geq \min\{d_1, d_0 + d_2\}$.

Proof: The statements on the length and on the dimension are obviously true. The surjective linear mapping

$$\phi: C_0 \to C_2 : v \cdot \Gamma_0 \mapsto v \cdot \left(\frac{\Gamma_2}{0} \right), \qquad v \in \mathbb{F}_q^{k_1},$$

is well-defined and has kernel C_1. Therefore, the code C has the form

$$C = \{(c, \phi(c)) \mid c \in C_0\}.$$

For each nonzero $c \in C_0$ the following holds:

$$\text{wt}(c, \phi(c)) = \text{wt}(c) + \text{wt}(\phi(c)) \geq \begin{cases} d_1 & \text{if } c \in C_1, \\ d_0 + d_2 & \text{else.} \end{cases}$$

$\square$

Example The binary $(5,3,1)$-code C_0 generated by   2.3.10

$$\Gamma_0 = \begin{pmatrix} 0 & 0 & 1 & 1 & 0 \\ 0 & 0 & 0 & 0 & 1 \\ 1 & 1 & 1 & 0 & 1 \end{pmatrix}$$

contains a $(5,1,4)$-subcode C_1 with generator matrix $\Gamma_1 = (1\ \ 1\ \ 1\ \ 0\ \ 1)$. Together with the binary $(3,2,2)$-code C_2, generated by

$$\Gamma_2 = \begin{pmatrix} 1 & 1 & 0 \\ 0 & 1 & 1 \end{pmatrix}$$

we obtain via X-construction an $(8,3,3)$-code with generator matrix

$$\Gamma = \left(\begin{array}{ccccc|ccc} 0 & 0 & 1 & 1 & 0 & 1 & 1 & 0 \\ 0 & 0 & 0 & 0 & 1 & 0 & 1 & 1 \\ 1 & 1 & 1 & 0 & 1 & 0 & 0 & 0 \end{array} \right).$$

$\diamond$

Now we introduce a construction that gives, for example, one of the most famous codes, the binary Golay-code G_{24}.

The $(u+w \mid v+w \mid u+v+w)$-**construction** For $i = 0,1$ let C_i be an   2.3.11
(n, k_i, d_i, q)-code, generated by Γ_i. The $(u+w \mid v+w \mid u+v+w)$-*construction*, applied to C_0 and C_1, is the linear code with generator matrix

$$\left(\begin{array}{c|c|c} \Gamma_0 & 0 & \Gamma_0 \\ \hline \Gamma_1 & \Gamma_1 & \Gamma_1 \\ \hline 0 & \Gamma_0 & \Gamma_0 \end{array} \right).$$

It is, therefore, the following set:

$$\{(u+w \mid v+w \mid u+v+w) \mid u,v \in C_0, w \in C_1\}.$$

It is a $(3n, 2k_0 + k_1)$-code. $\diamond$

Here is the announced prominent example:

2.3.12   **Example** Let C_0 be the extended third-order binary Hamming-code with generator matrix Γ_0 as in 2.2.5. Now reverse the columns of the (unextended) Hamming-code, and let C_1 be the parity extension of this code, i.e. C_1 is generated by

$$
\Gamma_1 := \begin{pmatrix}
0 & 0 & 0 & 1 & 0 & 1 & 1 & 1 \\
0 & 0 & 1 & 0 & 1 & 0 & 1 & 1 \\
0 & 1 & 0 & 0 & 1 & 1 & 0 & 1 \\
1 & 0 & 0 & 0 & 1 & 1 & 1 & 0
\end{pmatrix}.
$$

We know that C_0 and C_1 are both $(8,4,4,2)$-codes. From the $(u+w \mid v+w \mid u+v+w)$ construction, we obtain the following generator matrix $\Gamma = \Gamma_{24}$ of a binary $(24,12)$-code.

2.3.13

$$
\left(
\begin{array}{cccccccc|cccccccc|cccccccc}
1 & 1 & 0 & 1 & 0 & 0 & 0 & 1 & 0 & 0 & 0 & 0 & 0 & 0 & 0 & 0 & 1 & 1 & 0 & 1 & 0 & 0 & 0 & 1 \\
1 & 0 & 1 & 0 & 1 & 0 & 0 & 1 & 0 & 0 & 0 & 0 & 0 & 0 & 0 & 0 & 1 & 0 & 1 & 0 & 1 & 0 & 0 & 1 \\
0 & 1 & 1 & 0 & 0 & 1 & 0 & 1 & 0 & 0 & 0 & 0 & 0 & 0 & 0 & 0 & 0 & 1 & 1 & 0 & 0 & 1 & 0 & 1 \\
1 & 1 & 1 & 0 & 0 & 0 & 1 & 0 & 0 & 0 & 0 & 0 & 0 & 0 & 0 & 0 & 1 & 1 & 1 & 0 & 0 & 0 & 1 & 0 \\
0 & 0 & 0 & 1 & 0 & 1 & 1 & 1 & 0 & 0 & 0 & 1 & 0 & 1 & 1 & 1 & 0 & 0 & 0 & 1 & 0 & 1 & 1 & 1 \\
0 & 0 & 1 & 0 & 1 & 0 & 1 & 1 & 0 & 0 & 1 & 0 & 1 & 0 & 1 & 1 & 0 & 0 & 1 & 0 & 1 & 0 & 1 & 1 \\
0 & 1 & 0 & 0 & 1 & 1 & 0 & 1 & 0 & 1 & 0 & 0 & 1 & 1 & 0 & 1 & 0 & 1 & 0 & 0 & 1 & 1 & 0 & 1 \\
1 & 0 & 0 & 0 & 1 & 1 & 1 & 0 & 1 & 0 & 0 & 0 & 1 & 1 & 1 & 0 & 1 & 0 & 0 & 0 & 1 & 1 & 1 & 0 \\
0 & 0 & 0 & 0 & 0 & 0 & 0 & 0 & 1 & 1 & 0 & 1 & 0 & 0 & 0 & 1 & 1 & 1 & 0 & 1 & 0 & 0 & 0 & 1 \\
0 & 0 & 0 & 0 & 0 & 0 & 0 & 0 & 1 & 0 & 1 & 0 & 1 & 0 & 0 & 1 & 1 & 0 & 1 & 0 & 1 & 0 & 0 & 1 \\
0 & 0 & 0 & 0 & 0 & 0 & 0 & 0 & 0 & 1 & 1 & 0 & 0 & 1 & 0 & 1 & 0 & 1 & 1 & 0 & 0 & 1 & 0 & 1 \\
0 & 0 & 0 & 0 & 0 & 0 & 0 & 0 & 1 & 1 & 1 & 0 & 0 & 0 & 1 & 0 & 1 & 1 & 1 & 0 & 0 & 0 & 1 & 0
\end{array}
\right).
$$

One can show (see below) that its minimum distance is 8. This code is the binary *Golay-code* G_{24}, one of the most prominent linear codes. In fact, it can be shown that this code is the unique (up to linear isometry) code with parameters $(24,12,8,2)$. It played an important role during the Voyager 1 and 2 missions to Jupiter and Saturn in the late 1970s. A reason for its importance is that it carries many interesting combinatorial structures (like Steiner systems, etc.), and it was used even in the classification of finite simple groups (cf. [40]).

2.3.14   **Theorem** *The binary code C generated by the matrix Γ_{24} of 2.3.13 is a self-dual $(24,12,8)$-code.*

Proof: The codes C_0 and C_1 consist of 16 words of length 8 each, as shown in Table 2.1.

Table 2.1 The words of C_0 and C_1

message v	$v \cdot \Gamma_0$	$v \cdot \Gamma_1$
$(0,0,0,0)$	$(0,0,0,0,0,0,0,0)$	$(0,0,0,0,0,0,0,0)$
$(1,0,0,0)$	$(1,1,0,1,0,0,0,1)$	$(0,0,0,1,0,1,1,1)$
$(0,1,0,0)$	$(1,0,1,0,1,0,0,1)$	$(0,0,1,0,1,0,1,1)$
$(1,1,0,0)$	$(0,1,1,1,1,0,0,0)$	$(0,0,1,1,1,1,0,0)$
$(0,0,1,0)$	$(0,1,1,0,0,1,0,1)$	$(0,1,0,0,1,1,0,1)$
$(1,0,1,0)$	$(1,0,1,1,0,1,0,0)$	$(0,1,0,1,1,0,1,0)$
$(0,1,1,0)$	$(1,1,0,0,1,1,0,0)$	$(0,1,1,0,0,1,1,0)$
$(1,1,1,0)$	$(0,0,0,1,1,1,0,1)$	$(0,1,1,1,0,0,0,1)$
$(0,0,0,1)$	$(1,1,1,0,0,0,1,0)$	$(1,0,0,0,1,1,1,0)$
$(1,0,0,1)$	$(0,0,1,1,0,0,1,1)$	$(1,0,0,1,1,0,0,1)$
$(0,1,0,1)$	$(0,1,0,0,1,0,1,1)$	$(1,0,1,0,0,1,0,1)$
$(1,1,0,1)$	$(1,0,0,1,1,0,1,0)$	$(1,0,1,1,0,0,1,0)$
$(0,0,1,1)$	$(1,0,0,0,0,1,1,1)$	$(1,1,0,0,0,0,1,1)$
$(1,0,1,1)$	$(0,1,0,1,0,1,1,0)$	$(1,1,0,1,0,1,0,0)$
$(0,1,1,1)$	$(0,0,1,0,1,1,1,0)$	$(1,1,1,0,1,0,0,0)$
$(1,1,1,1)$	$(1,1,1,1,1,1,1,1)$	$(1,1,1,1,1,1,1,1)$

By inspection, we see that $C_0 \cap C_1 = \{0_8, 1_8\}$. Also, we know that C_0 and C_1 are both self-dual with weight enumerator $1 + 14x^4 + x^8$. The statement about the dimension of C is clear, since the 12 vectors of the form $(u, 0, u)$, $(0, v, v)$ and (w, w, w) are linearly independent, provided that u and v run through a basis of C_0 and w is taken from a basis for C_1. It is easy to check that C is self-orthogonal, and hence self-dual (Exercise 2.3.2). By Exercise 1.3.19, in a self-orthogonal code, the sum of 4-divisible codewords is 4-divisible. In C, any word can be written as a sum of vectors of the form $(u, 0, u)$, $(0, v, v)$ and (w, w, w), with $u, v \in C_0$ and $w \in C_1$. Since u, v and w are all 4-divisible, so are the three vectors and hence any vector in C. To show that the minimum distance of C is 8, we need to exclude the existence of words of weight 4. For this, let us assume that $c \in C$ is a word of weight less than 8. By Exercise 1.2.14, the sum of even vectors is even. Hence each of the three components of $c = (u + w, v + w, u + v + w)$ is even. In order to have weight either 4 or 0, at least one of the components must be zero. But $u, v \in C_0$ and $w \in C_1$, and we have seen that $C_0 \cap C_1$ consists of 0_8 and 1_8. Consider the case $w = 0_8$. Then $c = (u, v, u + v)$. Since C_0 only has words of weight $0, 4$ and 8, we have $c = 0$. Otherwise, if $w = 1_8$, then $c = (u + 1, v + 1, u + v + 1)$. Again it follows that $c = 0$. This proves the assertion. $\square$

A further important way of combining two codes is the following product:

2.3.15   **The tensor product** We recall from multilinear algebra that the *tensor product* $C_0 \otimes C_1$ of two linear codes C_0, C_1 can be defined as follows: It consists of the elements $c \otimes c'$, where $c \in C_0$ and $c' \in C_1$, and

$$c \otimes c' := \left(c_0 c_0', \ldots, c_0 c_{n_1-1}', \ldots, c_{n_0-1} c_0', \ldots, c_{n_0-1} c_{n_1-1}' \right).$$

In other words, the generator matrix is the Kronecker product

$$\Gamma := \Gamma_0 \otimes \Gamma_1 := \left(\begin{array}{c|c|c} \gamma_{00} \Gamma_1 & \cdots & \gamma_{0,n_0-1} \Gamma_1 \\ \hline \cdots & \cdots & \cdots \\ \hline \gamma_{k_0-1,0} \Gamma_1 & \cdots & \gamma_{k_0-1,n_0-1} \Gamma_1 \end{array} \right).$$

If C_i is a (n_i, k_i, d_i)-code, then by Exercise 2.3.5 the parameters of $C := C_1 \oplus C_2$ are

$$(n, k, d, q) = (n_0 n_1, k_0 k_1, d_0 d_1, q). \qquad \diamond$$

2.3.16   **Examples** The product $C_0 \otimes C_0$ of the binary $(7,4)$-Hamming-code C_0 with itself is a binary $(49, 16, 9)$-code. If we denote by C_1 the binary $(7,1)$-repetition code, then each word of the product code $C_0 \otimes C_1$ can be obtained as a 7-fold repetition of a codeword in C_0. $\qquad \diamond$

The next two constructions modify the field over which the codes are considered. For the reader not familiar with the theory of finite fields, the missing details will be presented in Chapter 3.

2.3.17   **Restriction** The *restriction* of a code C over $\mathbb{F}_q$ of length n to a subfield $\mathbb{F}$ of $\mathbb{F}_q$ is the code

$$C \downarrow \mathbb{F} := C \cap \mathbb{F}^n,$$

when considered as a linear code over $\mathbb{F}$. $\qquad \diamond$

A different way of constructing from a code over $\mathbb{F}_q$ a code over a subfield $\mathbb{F}$ uses the fact that $\mathbb{F}_q$ is a vector space over $\mathbb{F}$.

2.3.18   **Blowing up** If m is the $\mathbb{F}$-dimension of $\mathbb{F}_q$ and

$$B = \{\beta_0, \ldots, \beta_{m-1}\}$$

is an $\mathbb{F}$-basis of $\mathbb{F}_q$, then we obtain from the (n, k)-code C over $\mathbb{F}_q$ a linear code of length mn over $\mathbb{F}$ by replacing the components of the codewords in C by the m-tuples with respect to the basis B. This new code is called the *blow up* of C with respect to B. We denote it by $\mathrm{Bl}_B(C)$. Formally speaking, we obtain $\mathrm{Bl}_B(C)$ as the image $\psi_B(C)$ of C under the linear map

$$\psi_B : \mathbb{F}_q^n \to \mathbb{F}^{mn} : (c_0, \ldots, c_{n-1}) \mapsto (\phi_B(c_0) \mid \ldots \mid \phi_B(c_{n-1})),$$

which is the n-fold extension of the coordinate map

$$\phi_B : \mathbb{F}_q \to \mathbb{F}^m : \sum_{i \in m} \kappa_i \beta_i \mapsto (\kappa_0, \ldots, \kappa_{m-1}),$$

with respect to the basis B. This shows that each $\mathbb{F}_q$-basis $\{b^{(0)}, \ldots, b^{(k-1)}\}$ of C yields an $\mathbb{F}$-basis $\{\psi_B(\beta_i b^{(j)}) \mid i \in m,\ j \in k\}$ of $\mathrm{Bl}_B(C)$. The code $\mathrm{Bl}_B(C)$ is therefore an (mn, mk)-code over $\mathbb{F}$. Its minimum distance d' satisfies $d' \geq d$, since from $c_i \neq 0$ we obtain $\phi_B(c_i) \neq 0$, and so $\mathrm{wt}(\phi_B(c_0), \ldots, \phi_B(c_{n-1})) \geq \mathrm{wt}(c)$ holds true for each $c = (c_0, \ldots, c_{n-1}) \in C$.   ◇

Example The field $\mathbb{F}_4$ consists of the elements   2.3.19

$$0,\ 1,\ \alpha,\ \alpha^2,$$

where α is a root of the polynomial $x^2 + x + 1$ and, therefore, $\alpha^2 = \alpha + 1$. We consider the $(3, 2)$-code C over $\mathbb{F}_4$ with generator matrix

$$\Gamma = \begin{pmatrix} 1 & \alpha^2 & 0 \\ 0 & 1 & \alpha^2 \end{pmatrix}.$$

It consists of the following codewords:

000	$01\alpha^2$	$1\alpha^2 0$	$\alpha^2 01$
$0\alpha 1$	$\alpha 10$	10α	$0\alpha^2\alpha$
$\alpha^2\alpha 0$	$\alpha 0\alpha^2$	111	$\alpha\alpha\alpha$
$\alpha^2\alpha^2\alpha^2$	$1\alpha\alpha^2$	$\alpha\alpha^2 1$	$\alpha^2 1\alpha$.

This shows that C has minimum distance 2. Its blow up $\mathrm{Bl}_B(C)$ with respect to the $\mathbb{F}_2$-basis $B = \{\alpha, \alpha^2\}$ of $\mathbb{F}_4$ is a binary code, consisting of the words

000000	001101	110100	010011
001011	101100	110010	000110
011000	100001	111111	101010
010101	111001	100111	011110.

Hence, $\mathrm{Bl}_B(C)$ is a $(6, 4, 2)$-code. The restriction of C to $\mathbb{F}_2$ is the repetition code $\{000, 111\}$.   ◇

Let us summarize the results on lower and upper bounds for $d_{\max}(n, k, q)$. Following the ideas of T. Verhoeff [195], we may express the bounds in terms of two predicates,

$$(Lb, n, k, d, q) :\Longleftrightarrow \text{ there exists an } (n, k, d, q)\text{-code}$$

and

$$(Ub, n, k, d, q) :\Longleftrightarrow \text{ there does not exist an } (n, k, d, q)\text{-code},$$

so that

$$(Lb, n, k, d_1, q) \wedge (Ub, n, k, d_2, q) \Longrightarrow d_1 \leq d_{\max}(n, k, q) < d_2.$$

For example, the predicates

$$(Lb, n, n, 1, q), \ (Ub, n, n, 2, q), \ (Lb, n, 1, n, q), \ \text{and} \ (Ub, n, 1, n+1, q)$$

hold true, since over any field $\mathbb{F}_q$ and for any length n there is the $(n, n, 1)$-code $H(n, q)$ and the $(n, 1, n)$-repetition code.

If M denotes one of the modifications of codes described above, then we may deduce further predicates, which we shall denote as $M(b, n, k, d, q)$. Here, b stands for either Lb or Ub and (b, n, k, d, q) denotes a previously known predicate. Thus, we can consider the modifications as operators on the set of predicates. The goal is to tabulate the best known lower and upper bounds for the minimum distance of a linear code with a given length n and dimension k. This can be done in a systematic way by applying all modifications to an initial set of predicates. If this process is repeated sufficiently often, the resulting table will eventually be invariant under these modifications. Let

$$LB(n, k, q) := \max \{d \mid Lb(n, k, d, q)\}, \ UB(n, k, q) := \min \{d \mid Ub(n, k, d, q)\}.$$

In the following, we will restrict our attention to binary codes and therefore we will omit the parameter $q = 2$ from the list of arguments. For the nonbinary case, see Exercise 2.3.11.

2.3.20   **Theorem** *For binary codes the following is true:*

1. *Parity extension:*

$$P(Lb, n, k, d) = \begin{cases} (Lb, n+1, k, d+1) & \text{if } d \text{ is odd,} \\ (Lb, n+1, k, d) & \text{otherwise,} \end{cases} \quad n \geq 1,$$

$$P(Ub, n, k, d) = \begin{cases} (Ub, n-1, k, d-1) & \text{if } d \geq 2 \text{ is even,} \\ (Ub, n-1, k, d) & \text{otherwise,} \end{cases} \quad n > k \geq 1.$$

2. *Puncturing:*

$$Pu(Lb, n, k, d) = (Lb, n-1, k, d-1) \ \text{for } n > k \text{ and } d > 1,$$
$$Pu(Ub, n, k, d) = (Ub, n+1, k, d+1).$$

3. *Shortening:*

$$S(Lb, n, k, d) = (Lb, n-1, k-1, d) \ \text{for } k > 1,$$
$$S(Ub, n, k, d) = (Ub, n+1, k+1, d).$$

4. *A-construction:*

$$A(Lb, n, k, d) = (Lb, n - d, k - 1, \lceil d/2 \rceil) \text{ for } k > 1,$$
$$A(Ub, n, k, d) = (Ub, n + 2d, k + 1, 2d).$$

5. *B-construction:*

$$B_1(Lb, n, k, d) = (Lb, n - s, k - s + 1, d)$$
$$\text{for } UB(n, n - k) - 1 \leq s \leq k.$$
$$B_2(Ub, n, \ell, s + 1) = (Lb, n - s, n - \ell - s + 1, LB(n, n - \ell))$$
$$\text{for } UB(n, \ell) - 1 \leq s \leq n - \ell.$$
$$B_3(Ub, n, k, d) = (Ub, n + s, k + s - 1, d),$$
$$\text{for } UB(n + s, n - k + 1) - 1 \leq s.$$
$$B_4(Ub, n, \ell, s + 1) = (Ub, n, n - \ell, UB(n - s, n - \ell - s + 1))$$
$$\text{for } UB(n, \ell) - 1 \leq s \leq n - \ell.$$

Proof: The statements concerning parity extension, puncturing, shortening, and the A-construction are obvious. The B-construction gives, for $k \geq s \geq UB(n, n - k) - 1$,

$$(Lb, n, k, d) \wedge (Ub, n, n - k, s + 1) \implies (Lb, n - s, k - s + 1, d) \qquad \textbf{2.3.21}$$

and

$$(Ub, n, k, d) \wedge (Ub, n + s, n - k + 1, s + 1) \implies (Ub, n + s, k + s - 1, d). \qquad \textbf{2.3.22}$$

B_1 and B_2 come from 2.3.21, by keeping the first, respectively the second member of the conjunction fixed. Analogously we obtain B_3 and B_4 from 2.3.22. The details are left to the reader (Exercise 2.3.9). ☐

An invariant table of bounds can be improved by externally obtained bounds or by applications of non-primitive operations. Good lower bounds can be obtained from cyclic codes, from generalized Reed–Solomon-codes, from Alternant-, or Goppa-codes. We introduce these codes later in the Sections 4.5 and 4.6. Typical nonprimitive operations that can be used for such improvements of parameter tables are code combinations like the outer direct sum, (u, v)-construction, $(u \mid u + v)$construction, or the tensor product. In the case when we use prolongation methods, then we obtain infinitely many entries, in which case we must restrict attention to a maximal block length n_{max}.

In case a new predicate $Q = (b, n, k, d, q)$ has been found, the invariance of the parameter table can be restored by the following recursive algorithm:

2.3.23   **Algorithm** To enter a bound in a table of parameter bounds:
     Input: A predicate Q, a table of parameters.
     Output: The invariant table of parameters that takes Q into account.

Update(Q)
(1) **if** Q improves the table **then**
(2)   insert Q into the table;
(3)   **for each** primitive modification M **do**
(4)   **Update**($M(Q)$)
(5)   **end do**
(6) **end if**   □

An application of this algorithm to a table of code parameters usually produces many primitive operations that do not improve the table. If we are given two lower bounds for the minimum distance of (n, k)-codes over $\mathbb{F}_q$, then the larger one is considered better. Similarly, the smaller upper bound is preferred. In terms of predicates, with $Q_1 = (b, n, k, d_1, q)$ and $Q_2 = (b, n, k, d_2, q)$ we put

$$(b, n, k, d_1, q) \leq (b, n, k, d_2, q) :\Longleftrightarrow \begin{cases} d_1 \leq d_2 & \text{if } b = Lb, \\ d_1 \geq d_2 & \text{if } b = Ub. \end{cases}$$

Therefore, $Q_1 \leq Q_2$ means that the predicate Q_2 is an estimate which is at least as sharp for $d_{\max}(n, k, q)$ as Q_1. This notion can also be used to compare primitive modifications M_1 and M_2. We write $M_1 \leq M_2$ in order to indicate that for each predicate Q contained in the range of both M_1 and M_2, the inequality $M_1(Q) \leq M_2(Q)$ holds true. We can use that in order to define

$$M_1 = M_2 :\Longleftrightarrow M_1 \leq M_2 \wedge M_2 \leq M_1.$$

For example, the operations A and S commute in the binary case, since

$$\begin{aligned} (S \circ A)(Lb, n, k, d, 2) &= S(Lb, n - d, k - 1, \lceil d/2 \rceil, 2) \\ &= (Lb, n - d - 1, k - 2, \lceil d/2 \rceil, 2) \\ &= A(Lb, n - 1, k - 1, d, 2) \\ &= (A \circ S)(Lb, n, k, d, 2) \end{aligned}$$

and

$$\begin{aligned} (S \circ A)(Ub, n, k, d, 2) &= S(Ub, n + 2d, k + 1, 2d, 2) \\ &= (Ub, n + 2d + 1, k + 2, 2d, 2) \\ &= A(Ub, n + 1, k + 1, d, 2) \\ &= (A \circ S)(Ub, n, k, d, 2). \end{aligned}$$

A detailed analysis of the primitive modifications allows the reduction of the number of recursive calls of functions in **Update** (see Exercise 2.3.11).

Besides the primitive operations we have also discussed some methods for the combination of linear codes. Now we describe how they can be used to improve a table of bounds for $d_{\max}(n, k, d, q)$. Among others we have obtained the following rules:

Corollary   2.3.24

1. *Outer direct sum:*

$$(Lb, n_1, k_1, d_1, q) \wedge (Lb, n_2, k_2, d_2, q) \Rightarrow (Lb, n_1 + n_2, k_1 + k_2, \min\{d_1, d_2\}, q).$$

2. $(u \mid u + v)$-*construction:*

$$(Lb, n, k_1, d_1, q) \wedge (Lb, n, k_2, d_2, q) \Rightarrow (Lb, 2n, k_1 + k_2, \min\{2d_1, d_2\}, q).$$

3. *Tensor product:*

$$(Lb, n_1, k_1, d_1, q) \wedge (Lb, n_2, k_2, d_2, q) \Rightarrow (Lb, n_1 n_2, k_1 k_2, d_1 d_2, q). \qquad \square$$

We refrain from giving the corresponding upper bounds since their influence on the quality of a parameter table has shown to be rather small [203]. Further details on the construction of an invariant table of parameters and its improvement by using code combinations can be found in Exercise 2.3.11.

Exercises

Exercise For binary codes, prove the following expression for the weight of   E.2.3.1
the elements in a $(u + w \mid v + w \mid u + v + w)$-construction:

$$\mathrm{wt}(u + w \mid v + w \mid u + v + w) = 2 \cdot \mathrm{wt}(u \vee v) - \mathrm{wt}(w) + 4 \cdot s,$$

where $s := |\{i \mid u_i = v_i = 0, w_i = 1\}|$ and $u \vee v$ is as defined in Exercise 1.2.14. Derive from this equation that the minimum distance of G_{24} is 8.

Exercise Verify that the code generated by Γ_{24} in 2.3.12 is self-orthogonal.   E.2.3.2

Exercise Confirm the parameters of the augmentation of a linear code given   E.2.3.3
in 2.3.2.

E.2.3.4   **Exercise** In Multilinear Algebra the *tensor product* $U \otimes V$ of the $\mathbb{F}_q$-vector spaces U and V *of finite dimension* is defined to be the factor group

$$U \otimes V := \mathbb{Z}^{U \times V}/T.$$

Here $\mathbb{Z}^{U \times V}$ means the free abelian group over the cartesian product $U \times V$, the set of mappings f from $U \times V$ to $\mathbb{Z}$ with pointwise addition. The set T indicates the subgroup of $\mathbb{Z}^{U \times V}$ generated by the elements of the following forms

$$(u + u', v) - (u, v) - (u', v),$$
$$(u, v + v') - (u, v) - (u, v'),$$
$$(u, \alpha v) - (\alpha u, v),$$

with $u, u' \in U$, $v, v' \in V$, and $\alpha \in \mathbb{F}_q$. The pair $(u, v) \in U \times V$ stands for the element $f_{(u,v)} \in \mathbb{Z}^{U \times V}$, defined by

$$f_{(u,v)}(x, y) = \begin{cases} 1 & \text{if } (u, v) = (x, y), \\ 0 & \text{else.} \end{cases}$$

The elements in $U \otimes V$ are called *tensors*.

1. Prove that the canonical mapping from $\mathbb{Z}^{U \times V}$ onto the factor group, i.e.

$$\otimes : \mathbb{Z}^{U \times V} \to \mathbb{Z}^{U \times V}/T \; : \; (u, v) \mapsto u \otimes v := (u, v) + T,$$

    satisfies the rules

$$(u + u') \otimes v = u \otimes v + u' \otimes v,$$
$$u \otimes (v + v') = u \otimes v + u \otimes v',$$
$$u \otimes (\alpha v) = (\alpha u) \otimes v.$$

2. Verify that $U \otimes V$ turns into an $\mathbb{F}_q$-vector space via

$$\alpha \sum_i \left(u^{(i)} \otimes v^{(i)} \right) := \sum_i \left((\alpha u^{(i)}) \otimes v^{(i)} \right), \qquad \alpha \in \mathbb{F}_q.$$

    The elements of $U \otimes V$ are finite sums $\sum_i \left(u^{(i)} \otimes v^{(i)} \right)$ with $u^{(i)} \in U$ and $v^{(i)} \in V$.

3. Show that, if B is a basis of U and B' a basis of V, then

$$\{ b \otimes b' \mid b \in B, \; b' \in B' \}$$

    is a basis of $U \otimes V$.

4. Check that each element of $U \otimes V$ can uniquely be expressed in the form

$$\sum_{b \in B, \; b' \in B'} \alpha_{bb'} \left(b \otimes b' \right), \qquad \alpha_{bb'} \in \mathbb{F}_q,$$

and we have

$$\dim(U \otimes V) = \dim(U) \cdot \dim(V).$$

5. Show that the mapping

$$\Phi_{m,n} \colon \mathbb{F}_q^m \otimes \mathbb{F}_q^n \to \mathbb{F}_q^{m \times n} \; : \; \sum_{i \in m} \sum_{j \in n} \left(\alpha_{ij} e^{(i)} \otimes f^{(j)} \right) \mapsto (\alpha_{ij})_{i,j},$$

(where $e^{(i)}$ and $f^{(j)}$ denote the respective unit vectors in $\mathbb{F}_q^m$ and $\mathbb{F}_q^n$) is an $\mathbb{F}_q$-isomorphism. So the elements of $\mathbb{F}_q^m \otimes \mathbb{F}_q^n$ can be written as $m \times n$-matrices, and we can speak of rows and columns of a tensor.

Exercise Assume that C_i is a linear (n_i, k_i, d_i, q)-code for $i = 0, 1$. Show that $C_0 \otimes C_1$ is an $(n_0 n_1, k_0 k_1, d_0 d_1, q)$-code.   E.2.3.5

Exercise Let C be a binary $(3,2)$-parity check code. Evaluate the elements of the product code $C \otimes C$.   E.2.3.6

Exercise Evaluate a generator matrix of the binary code obtained in 2.3.18 by blowing up.   E.2.3.7

Exercise Suppose that C is an (n, k, d)-code with $n > k > 1$ and $c' \in C^\perp$ has $\mathrm{wt}(c') = d'$. Show that an $(n - 1, k - 1, d)$-code exists, the dual code of which contains a codeword of weight $d' - 1$.   E.2.3.8

Exercise Prove 2.3.20 and rephrase it for nonbinary codes.   E.2.3.9

Exercise Assume that M is a primitive modification on codes. Iterating the operation M until it does not change the parameters any more is denoted by M^*. Prove that the following assertions (cf. [203]) are true: E.2.3.10

$$P \circ Pu = id \;\; \text{for even } d$$
$$P \circ Pu \leq id \;\; \text{for each } d$$
$$Pu \circ P = id \;\; \text{for odd } d$$
$$Pu \circ P \leq id \;\; \text{for each } d$$
$$P \circ S = S \circ P$$
$$Pu \circ S = S \circ Pu,$$

where *id* denotes the identity mapping on the predicates.

Show that for lower bounds we have:

$$A \circ P \le P^* \circ A$$
$$A \circ Pu \le A$$
$$A \circ S = S \circ A,$$

while for upper bounds

$$A \circ P = P \circ A \text{ for even } d$$
$$(Pu)^3 \circ A = A \circ Pu$$
$$A \circ S = S \circ A.$$

E.2.3.11   **Exercise** Implement a database for the lower and upper bounds of linear binary codes, i.e. of 5-tuples of the form (b, n, k, d, q) where $b = Lb$ or $b = Ub$ and $q = 2$.

1. Implement each of the primitive modifications of 2.3.20.

2. Write a procedure that initializes the database with the "trivial" bounds

$$(Lb, n, n, 1, q), \quad (Lb, n, 1, n, q), \quad (Ub, n, n, 2, q), \quad (Ub, n, 1, n+1, q)$$

   for all nonnegative n up to a user defined maximal length n_{max}.

3. Allow for input of external lower and upper bounds to the parameter table. Note that this addition should be combined with an application of the procedure **Update**.

4. Develop a procedure which applies, for fixed block length $n \le n_{max}$, the following rules (see 2.3.24) to the entries (lower bounds) of the tables and which inserts newly found lower bounds for codes of length n:
   — Outer direct sum:
$$(Lb, n_0, k_0, d_0, q) \wedge (Lb, n - n_0, k_1, d_1, q) \Rightarrow (Lb, n, k_0 + k_1, \min\{d_0, d_1\}, q).$$
   — $(u \mid u + v)$-construction:
$$(Lb, \frac{n}{2}, k_0, d_0, q) \wedge (Lb, \frac{n}{2}, k_1, d_1, q) \Rightarrow (Lb, n, k_0 + k_1, \min\{2d_0, d_1\}, q).$$
   — Tensor product:
$$(Lb, n_0, k_0, d_0, q) \wedge (Lb, \frac{n}{n_0}, k_1, d_1, q) \Rightarrow (Lb, n, k_0 k_1, d_0 d_1, q).$$

5. Use the program in order to search for good codes. After the initialization of the table, add lower bounds from the existence results of Chapter 9. Also, use parameters of Reed–Muller-codes (cf. Section 2.4), BCH-codes (Section 4.3) as lower bounds. Then apply the combination methods described above. Compare the results with the list of best known binary linear codes [32].

Exercise Assume that C_i is a linear code with check matrix Δ_i for $i = 0, 1$.   **E.2.3.12**
Show that

$$\left(\begin{array}{c|c} \Delta_0 & 0 \\ \hline 0 & \Delta_1 \end{array} \right)$$

is a check matrix of $C_0 \dotplus C_1$.

Exercise Let C_0, C_1 and C_2 be linear codes. Prove the following properties of   **E.2.3.13**
the outer direct sum:

- If C_0 is linearly isometric to C_0' and C_1 linearly isometric to C_1', then $C_0 \dotplus C_1$
  is linearly isometric to $C_0' \dotplus C_1'$.
- $C_0 \dotplus C_1$ is linearly isometric to $C_1 \dotplus C_0$.
- $C_0 \dotplus (C_1 \dotplus C_2) = (C_0 \dotplus C_1) \dotplus C_2$.
- $(C_0 \dotplus C_1)^\perp = C_0^\perp \dotplus C_1^\perp$.

Exercise Let A, B, C and D be matrices over a field $\mathbb{F}$. Prove the following   **E.2.3.14**
properties of the Kronecker product:

- $A \otimes (B \otimes C) = (A \otimes B) \otimes C$.
- $(A \otimes B)^\top = B^\top \otimes A^\top$.
- If the number of columns of A respectively B coincides with the number of
  rows of C respectively D, then $(A \otimes B) \cdot (C \otimes D) = (A \cdot C) \otimes (B \cdot D)$.
- If A is an $r \times s$-matrix and B a $t \times u$-matrix, then there exist permutations
  $\pi \in S_{rt}$ and $\sigma \in S_{su}$, so that $A \otimes B = M_\pi \cdot (B \otimes A) \cdot M_\sigma$, where M_π and
  M_σ are the permutation matrices corresponding to π and σ. Determine the
  two permutations π and σ which depend only on the numbers r, s, t, and u
  but not on the particular values of the matrices A and B.

Exercise Let C_0, C_1 and C_2 be linear codes. Prove the following properties of   **E.2.3.15**
the tensor product:

- If C_0 is linearly isometric to C_0' and C_1 linearly isometric to C_1', then $C_0 \otimes C_1$
  is linearly isometric to $C_0' \otimes C_1'$.
- $C_0 \otimes C_1$ is linearly isometric to $C_1 \otimes C_0$.
- $C_0 \otimes (C_1 \otimes C_2)$ is linearly isometric to $(C_0 \otimes C_1) \otimes C_2$.
- $C_0 \otimes (C_1 \dotplus C_2)$ is linearly isometric to $(C_0 \otimes C_1) \dotplus (C_0 \otimes C_2)$.
- In general, $(C_0 \otimes C_1)^\perp$ is not linearly isometric to $C_0^\perp \otimes C_1^\perp$.

E.2.3.16   **Exercise** Assume that C_i is a linear (n_i, k_i)-code with a systematic generator matrix $(I_{k_i} \mid A_i)$ for $i = 0, 1$. Show that $C_0 \otimes C_1$ is linearly isometric to the code generated by

$$(I_{k_0} \otimes I_{k_1} \mid I_{k_0} \otimes A_1 \mid A_0 \otimes I_{k_1} \mid A_0 \otimes A_1).$$

If we denote the last $n_0 n_1 - k_0 k_1$ columns of this matrix by B, prove that $(I_{n_0 n_1 - k_0 k_1} \mid -B^\top)$ is a check matrix of a code linearly isometric to $C_0 \otimes C_1$.

E.2.3.17   **Exercise** Let C_0, C_1 and C_2 be linear codes and denote the linear isometry of a linear code C by $\hat{C}$. Deduce from Exercise 2.3.13 and Exercise 2.3.15 that the following sum and product of linear isometry classes

$$\hat{C}_0 \dotplus \hat{C}_1 := \widehat{C_0 \dotplus C_1}, \qquad \hat{C}_0 \otimes \hat{C}_1 := \widehat{C_0 \otimes C_1}$$

are well-defined. Moreover, prove the following assertions:

- $\hat{C}_0 \dotplus \hat{C}_1 = \hat{C}_1 \dotplus \hat{C}_0.$
- $\hat{C}_0 \dotplus (\hat{C}_1 \dotplus \hat{C}_2) = (\hat{C}_0 \dotplus \hat{C}_1) \dotplus \hat{C}_2.$
- $\hat{C}_0 \otimes \hat{C}_1 = \hat{C}_1 \otimes \hat{C}_0.$
- $\hat{C}_0 \otimes (\hat{C}_1 \otimes \hat{C}_2) = (\hat{C}_0 \otimes \hat{C}_1) \otimes \hat{C}_2.$
- $\hat{C}_0 \otimes (\hat{C}_1 \dotplus \hat{C}_2) = (\hat{C}_0 \otimes \hat{C}_1) \dotplus (\hat{C}_0 \otimes \hat{C}_2).$
- The linear $(1,1)$-code D with generator matrix $\Gamma = (1)$ satisfies $\hat{C} \otimes \hat{D} = \hat{D} \otimes \hat{C} = \hat{C}$ for all linear isometry classes $\hat{C}$.

2.4 Reed–Muller-Codes

From 1969 until 1977, spacecrafts of NASA were equipped with a 7-error-correcting binary $(32, 6)$-code, a Reed–Muller-code. This is a low rate code with good error correction capabilities. A very prominent mission was Mariner 9, which was devoted to the photographic observation of the surface of Mars. Mariner 9 actually entered a Martian orbit in 1971 and became a satellite. The mission was complicated by a heavy dust storm which engulfed the whole Martian surface. It was not until 1972 that the storm subsided and the first clear photos arrived and changed our view of that planet so profoundly. We introduce the Reed–Muller-codes following the original ideas of D.E. Muller [154], who discovered their binary version. However, we will present the more general version of these codes which works for all finite fields $\mathbb{F}_q$. Later on, we will specialize to the binary case.

Reed–Muller-codes are subspaces of the vector space of all mappings

$$f : \mathbb{F}_q^m \to \mathbb{F}_q \; : \; (u_0, \dots, u_{m-1}) \mapsto f(u_0, \dots, u_{m-1})$$

with pointwise addition $(f + g)(u) := f(u) + g(u)$ and scalar multiplication $(\alpha f)(u) := \alpha \cdot f(u)$ for $u \in \mathbb{F}_q^m$ and $\alpha \in \mathbb{F}_q$. Together with pointwise multiplication $(fg)(u) := f(u)g(u)$, this set of mappings forms the $\mathbb{F}_q$-algebra (Exercise 2.4.1)

$$\mathcal{B}_m^q.$$

In the case $q = 2$ these are the well-known Boolean functions or switching functions of degree m. It is helpful to note that these functions f are polynomial, i.e. for each $f \in \mathcal{B}_m^q$ there exists a polynomial $\tilde{f} \in \mathbb{F}_q[x_0, \dots, x_{m-1}]$ such that $f(u) = \tilde{f}(u_0, \dots, u_{m-1})$ for all $u \in \mathbb{F}_q^m$. For this purpose, we consider both $\mathcal{B}_m^q$ and the space of polynomial functions as vector spaces. Our first goal is to exhibit a basis for this space.

The "unit vectors" of $\mathcal{B}_m^q$ are the functions f_u for $u = (u_0, \dots, u_{m-1}) \in \mathbb{F}_q^m$ with

$$f_u(v) = \begin{cases} 1 & \text{if } v = u, \\ 0 & \text{else.} \end{cases}$$

A function from $\mathbb{F}_q^m$ to $\mathbb{F}_q$ that takes exactly the same values as f_u is obtained from the polynomial

$$\tilde{f}_u(x_0, \dots, x_{m-1}) := \prod_{i \in m} \left(1 - (x_i - u_i)^{q-1} \right) \in \mathbb{F}_q[x_0, \dots, x_{m-1}]. \qquad \textbf{2.4.1}$$

Since $u^{q-1} = 1$, for each element $u \in \mathbb{F}_q^*$ (see 3.2.2), $(x_i - u_i)^{q-1} = 1$ if $x_i \neq u_i$, and so it is clear that this polynomial takes the value 1 exactly at

$$(x_0, \dots, x_{m-1}) = (u_0, \dots, u_{m-1}) \in \mathbb{F}_q^m$$

and 0 elsewhere. Any f in $\mathcal{B}_m^q$ is a linear combination

$$f = \sum_{u \in \mathbb{F}_q^m} f(u) f_u \qquad \textbf{2.4.2}$$

of unit vectors f_u, i.e. every element of $\mathcal{B}_m^q$ is a polynomial function. Hence the f_u generate $\mathcal{B}_m^q$ as a vector space. However, the representation is not unique. The non-uniqueness lies in the fact that $x^q - x$ is identically zero on $\mathbb{F}_q$. Thus two polynomials f and g in $\mathbb{F}_q[x_0, \dots, x_{m-1}]$ induce the same function if and only if f and g are congruent modulo $x_0^q - x_0, \dots, x_{m-1}^q - x_{m-1}$. This means that f and g cannot be distinguished from their functions if and only if their difference $f - g$ is a polynomial in the terms $x_i^q - x_i$ for $i = 0, \dots, m - 1$. Let us see what this condition means in terms of monomials. We use multi-index notation and let x^b denote the monomial $x_0^{b_0} \cdots x_{m-1}^{b_{m-1}}$ for $b = (b_0, \dots, b_{m-1})$.

Applying the relation $x_i^q - x_i$ means reducing the exponent b_i modulo $q - 1$ in the following sense: If b_i is either zero or not divisible by $q - 1$ then $x_i^{b_i}$ may be replaced by $x_i^{a_i}$ where a_i is the remainder after dividing b_i by $q - 1$, i.e. a_i is the unique integer in $b_i = c(q - 1) + a_i$ with $0 \leq a_i < q - 1$ (where c is another suitable integer). If $q - 1$ divides $b_i \neq 0$ then $x_i^{b_i}$ may be replaced by x_i^{q-1}. It is clear that any polynomial $f \in \mathbb{F}_q[x_0, \ldots, x_{m-1}]$ may be *reduced* to one whose monomials x^a satisfy $0 \leq a_i \leq q - 1$ for $i = 0, \ldots, m - 1$. The main point is that if we restrict to polynomials in $\mathbb{F}_q[x_0, \ldots, x_{m-1}]$ which are reduced in this sense then any function in $\mathcal{B}_m^q$ can be expressed uniquely as a reduced polynomial. We summarize this as

2.4.3   **Theorem** *The $\mathbb{F}_q$-algebra $\mathcal{B}_m^q$ is isomorphic to the ring of polynomials*

$$\mathbb{F}_q[x_0, \ldots, x_{m-1}]$$

modulo $x_0^q - x_0, \ldots, x_{m-1}^q - x_{m-1}$. An $\mathbb{F}_q$-basis is given by the reduced polynomials

$$\left\{ x_0^{b_0} \ldots x_{m-1}^{b_{m-1}} \,\middle|\, b_i \in q \right\}. \qquad \square$$

Because of this result, we will identify the elements f of $\mathcal{B}_m^q$ with polynomial functions in the following.

In the theory of switching functions, the multinomials $x^b = x_0^{b_0} \ldots x_{m-1}^{b_{m-1}}$ are called *minterms*. The *degree* of x^b is the sum of its exponents $\sum_i b_i$, and the degree of $f \in \mathcal{B}_m^q$ is defined to be the largest degree of a multinomial x^b which occurs in a reduced expression of f with a nonzero coefficient (which is at most $m(q - 1)$ by the preceding discussion).

Bounding the degree of the polynomials to any number $t \leq m(q - 1)$ results in a vector subspace of $\mathcal{B}_m^q$ (but not a sub-algebra). This enables us to define the Reed–Muller-codes in the following way:

2.4.4   **Definition (Reed–Muller-code)** Assume that $0 \leq t \leq m(q - 1)$. The *t-th order Reed–Muller-code* of degree m over $\mathbb{F}_q$ is defined to be

$$\mathrm{RM}_{m,t}^q := \left\{ f \in \mathcal{B}_m^q \,\middle|\, \deg f \leq t \text{ or } f = 0 \right\}. \qquad \diamond$$

The considerations above show that the elements of this code can be described in two ways, either as mappings or as polynomials. If we think of them as mappings, we may display the images of all vectors. We may do so by defining another vector of length q^m whose i-th entry is the value of the i-th vector of $\mathbb{F}_q^m$. Of course, one needs to fix an ordering on the elements of $\mathbb{F}_q^m$ for this. Here are a few examples:

Examples 2.4.5

— The 0-th order binary Reed–Muller-code of length $n = 2^m$ consists of the two constant functions 0 and 1. Hence $\mathrm{RM}^2_{m,0}$ is the n-th order binary repetition code.

— The m-th order binary Reed–Muller-code of length 2^m consists of all vectors in $\mathbb{F}_2^{2^m}$.

— The first order binary Reed–Muller-code $\mathrm{RM}^2_{2,1}$ of degree 2 is of length 4 and consists of the vectors in the following table. (In the left column we list the polynomial f and in the right column the values of the corresponding polynomial function.)

f	$f(00)$	$f(10)$	$f(01)$	$f(11)$
0	0	0	0	0
1	1	1	1	1
x_0	0	1	0	1
x_1	0	0	1	1
$x_0 + x_1$	0	1	1	0
$1 + x_0$	1	0	1	0
$1 + x_1$	1	1	0	0
$1 + x_0 + x_1$	1	0	0	1

— The second order binary Reed–Muller-code $\mathrm{RM}^2_{2,2}$ of degree 2 is of length 4 and contains the elements of $\mathrm{RM}^2_{2,1}$ *together* with the codewords shown in the following table:

f	$f(00)$	$f(10)$	$f(01)$	$f(11)$
$x_0 x_1$	0	0	0	1
$1 + x_0 x_1$	1	1	1	0
$x_0 + x_0 x_1$	0	1	0	0
$x_1 + x_0 x_1$	0	0	1	0
$x_0 + x_1 + x_0 x_1$	0	1	1	1
$1 + x_0 + x_0 x_1$	1	0	1	1
$1 + x_1 + x_0 x_1$	1	1	0	1
$1 + x_0 + x_1 + x_0 x_1$	1	0	0	0

— A closer examination of $\mathrm{RM}^2_{2,2}$ shows its recursive structure: Each of the 16 polynomials f in

$$\{0, 1, x_0, x_1, \ldots, 1 + x_1 + x_0 x_1, 1 + x_0 + x_1 + x_0 x_1\}$$

can be written as $f = h + x_1 g$, where both h and g are polynomials in the single indeterminate x_0, and therefore uniquely determined. For example,

$$1 + x_1 + x_0 x_1 = 1 + x_1(1 + x_0) = h + x_1 g.$$

The mappings $h \, (= 1)$ and $g \, (= 1 + x_0)$ take $\mathbb{F}_2^1 = \{(0), (1)\}$ to $\mathbb{F}_2$, and so

$$f \, : \, \{(00), (10), (01), (11)\} \rightarrow \mathbb{F}_2 \, : \, (x_0, x_1) \mapsto h(x_0) + x_1 g(x_0),$$

is of the form

$$f = (h(0), h(1), h(0) + 1 \cdot g(0), h(1) + 1 \cdot g(1)).$$

In terms of code constructions (recall 2.2.15),

$$f = (h \mid h + g),$$

i.e. we obtain

$$\mathrm{RM}_{2,2}^2 = \underbrace{\mathrm{RM}_{1,2}^2}_{= \mathrm{RM}_{1,1}^2} \mid \mathrm{RM}_{1,1}^2,$$

an $(u \mid u + v)$-construction!   ◇

More generally, any polynomial f in $\mathrm{RM}_{m,t}^2$ can be expressed (uniquely) in the form

$$f(x_0, \ldots, x_{m-1}) = h(x_0, \ldots, x_{m-2}) + x_{m-1} g(x_0, \ldots, x_{m-2}),$$

where $\deg h \leq t$ and $\deg g \leq t - 1$ (Exercise 2.4.2), and we obtain

---

2.4.6   **Corollary**   *The Reed–Muller-code $\mathrm{RM}_{m,t}^2$ is the $(u \mid u + v)$-construction of two Reed–Muller-codes, namely*

$$\mathrm{RM}_{m,t}^2 = \mathrm{RM}_{m-1,t}^2 \mid \mathrm{RM}_{m-1,t-1}^2, \qquad 1 \leq t \leq m.$$

(Note that $\mathrm{RM}_{m,t}^2 = \mathrm{RM}_{m,m}^2$, if $t > m$.) Hence, if $\Gamma_{m,t}$ generates $\mathrm{RM}_{m,t}^2$, then

$$\Gamma_{m,t} = \left(\begin{array}{c|c} \Gamma_{m-1,t} & \Gamma_{m-1,t} \\ \hline 0 & \Gamma_{m-1,t-1} \end{array} \right).$$ □

Its parameters are as follows:

2.4.7   **Theorem**   *The binary Reed–Muller-code $\mathrm{RM}_{m,t}^2$ is of type*

$$\left(2^m, \sum_{i=0}^{t} \binom{m}{i}, 2^{m-t}, 2 \right).$$

Proof: $\text{RM}^2_{m,t}$ is a linear $(2^m, k)$-code with

$$k = \binom{m}{0} + \binom{m}{1} + \ldots + \binom{m}{t},$$

since it has a basis consisting of the multinomials x^b, $0 \le b_i \le 1$, $\sum_i b_i \le t$.

In order to evaluate its minimum distance, we use induction both on m and t. For $m = 1$ and $t = 0, 1$ the statement is clearly true. Now assume that $m > 1$. As we have seen already, the code $\text{RM}^2_{m,0}$ consists of only the two vectors $\mathbf{0}_{2^m}$ and $\mathbf{1}_{2^m}$. Thus $\text{RM}^2_{m,0}$ is the repetition code of length 2^m with minimum distance 2^m, and so the statement is true in this case. Therefore, we can assume that $t \ge 1$. By the induction hypothesis, the Reed–Muller-code $\text{RM}^2_{m-1,s}$ has minimum distance 2^{m-1-s}. From 2.4.6 and 2.2.15 we deduce that $\text{RM}^2_{m,t}$ has minimum distance

$$\min\left\{ 2 \cdot 2^{m-1-t}, 2^{m-1-(t-1)} \right\} = 2^{m-t}. \qquad \square$$

For example, the above-mentioned code $\text{RM}^2_{5,1}$ used during Mariner missions is of type $(32, 6, 16)$. Therefore, this code can indeed correct 7 errors.

Finally, we also consider the codes which are dual to Reed–Muller-codes:

Theorem *For $0 \le t < m$, the code dual to $\text{RM}^2_{m,t}$ is $\text{RM}^2_{m,m-t-1}$.*   2.4.8

Proof: Consider $f \in \text{RM}^2_{m,t}$ and $g \in \text{RM}^2_{m,m-t-1}$. Their product $h = fg$ is of degree not greater than $m - 1$. Hence, h is in $\text{RM}^2_{m,m-1}$ and one can show (Exercise 2.4.3) that h has even weight. Now identify $\mathbb{F}_2^m$ with the set $\{0, \ldots, 2^m - 1\} = 2^m$ via the bijection $(a_0, \ldots, a_{m-1}) \mapsto \sum_{i \in m} a_i 2^i$. Represent f, g, and h as $\mathbb{F}_2^{2^m}$-vectors $(f(i))_{i \in 2^m}$, $(g(i))_{i \in 2^m}$ and $(h(i))_{i \in 2^m}$, respectively. The inner product of f and g is

$$\langle f, g \rangle = \sum_{i \in 2^m} f(i)g(i) = \sum_{i \in 2^m} (fg)(i) = \sum_{i \in 2^m} h(i) = 0,$$

since h has even weight. For this reason, $\text{RM}^2_{m,m-t-1}$ is contained in the dual of $\text{RM}^2_{m,t}$. Moreover, the dimension of $\text{RM}^2_{m,m-t-1}$ is

$$\sum_{i \in m-t} \binom{m}{i} = \sum_{i=t+1}^{m} \binom{m}{i} = 2^m - \dim(\text{RM}^2_{m,t}) = n - k,$$

whence $\text{RM}^2_{m,m-t-1}$ is the dual of $\text{RM}^2_{m,t}$, as stated. $\square$

Another more algebraic description of Reed–Muller-codes will be presented in Section 4.10.

The binary Reed–Muller-code $\mathrm{RM}^2_{m,m}$, which is $\mathbb{F}_2^{2^m}$, has a generator matrix with a highly recursive structure. (See also [84, first edition, Section 8.11.2].) Clearly, for $m = 0$

$$\Gamma_0 := (\,1\,)$$

is a generator matrix of $\mathrm{RM}^2_{0,0}$. According to 2.4.6, for $m > 0$ the Reed–Muller-code $\mathrm{RM}^2_{m,m}$ is the $(u \mid u + v)$-construction

$$\mathrm{RM}^2_{m,m} = \mathrm{RM}^2_{m-1,m-1} \mid \mathrm{RM}^2_{m-1,m-1},$$

since obviously $\mathrm{RM}^2_{m-1,m} = \mathrm{RM}^2_{m-1,m-1}$. Therefore, it has a generator matrix of the form

$$\Gamma_m := \left(\begin{array}{c|c} \Gamma_{m-1} & \Gamma_{m-1} \\ \hline 0 & \Gamma_{m-1} \end{array} \right)$$

where Γ_{m-1} is a generator matrix of $\mathrm{RM}^2_{m-1,m-1}$. The matrix Γ_m is an upper triangular matrix. In order to describe it in more detail and to show further properties of Reed–Muller-codes, in particular relations to Hamming- and simplex-codes, we label its rows (respectively columns) from top to bottom (respectively from left to right) with values from 0 to $2^m - 1$. We express the row number i in binary form, $i = \sum_{j \in m} b_j 2^j$ and identify i with the characteristic set $B_i := \{j \in m \mid b_j \neq 0\}$. Finally, we associate B_i with the monomial $\prod_{j \in B_i} x_j$.

We also express the column index i in binary form as $i = \sum_{j \in m} t_j 2^j$. This means that t_j takes the value 0 in all columns with index

$$i \in \bigcup_{r \in 2^{m-j-1}} \left\{ s \in \mathbb{N} \mid 2r2^j \leq s < (2r+1)2^j \right\}.$$

In all other columns t_j takes the value 1. The (i,j)-th entry of Γ_m is the monomial associated with the characteristic set B_i evaluated at $(x_0, \ldots, x_{m-1}) = (t_0, \ldots, t_{m-1}) \in \mathbb{F}_2^m$ where $(t_0, \ldots, t_{m-1})$ is determined by j.

From this description it is easy to compute directly (i.e. without recursion) the entries of the i-th row $(y_0, \ldots, y_{2^m-1})$ of Γ_m. Let B_i be the characteristic set of i, then for

$$t \in \bigcup_{j \in B_i} \left(\bigcup_{r \in 2^{m-j-1}} \left\{ s \in \mathbb{N} \mid 2r2^j \leq s < (2r+1)2^j \right\} \right)$$

we have $y_t = 0$. Otherwise $y_t = 1$.

If we have two characteristic sets B_i and B_j, then $B_i \cup B_j$ is also a characteristic set, of row ℓ say. There occurs the entry 1 in the t-th position of the ℓ-th row if and only if both in the i-th row and in the j-th row there is the

#	bin	set																															
0	00000	∅	1	1	1	1	1	1	1	1	1	1	1	1	1	1	1	1	1	1	1	1	1	1	1	1	1	1	1	1	1	1	1	1
1	00001	{0}		1		1		1		1		1		1		1		1		1		1		1		1		1		1		1		1
2	00010	{1}			1	1			1	1			1	1			1	1			1	1			1	1			1	1			1	1
3	00011	{1,0}				1				1				1				1				1				1				1				1
4	00100	{2}					1	1	1	1					1	1	1	1					1	1	1	1					1	1	1	1
5	00101	{2,0}						1		1						1		1						1		1						1		1
6	00110	{2,1}							1	1							1	1							1	1							1	1
7	00111	{2,1,0}								1								1								1								1
8	01000	{3}									1	1	1	1	1	1	1	1									1	1	1	1	1	1	1	1
9	01001	{3,0}										1		1		1		1										1		1		1		1
10	01010	{3,1}											1	1			1	1											1	1			1	1
11	01011	{3,1,0}												1				1												1				1
12	01100	{3,2}													1	1	1	1													1	1	1	1
13	01101	{3,2,0}														1		1														1		1
14	01110	{3,2,1}															1	1															1	1
15	01111	{3,2,1,0}																1																1
16	10000	{4}																	1	1	1	1	1	1	1	1	1	1	1	1	1	1	1	1
17	10001	{4,0}																		1		1		1		1		1		1		1		1
18	10010	{4,1}																			1	1			1	1			1	1			1	1
19	10011	{4,1,0}																				1				1				1				1
20	10100	{4,2}																					1	1	1	1					1	1	1	1
21	10101	{4,2,0}																						1		1						1		1
22	10110	{4,2,1}																							1	1							1	1
23	10111	{4,2,1,0}																								1								1
24	11000	{4,3}																									1	1	1	1	1	1	1	1
25	11001	{4,3,0}																										1		1		1		1
26	11010	{4,3,1}																											1	1			1	1
27	11011	{4,3,1,0}																												1				1
28	11100	{4,3,2}																													1	1	1	1
29	11101	{4,3,2,0}																														1		1
30	11110	{4,3,2,1}																															1	1
31	11111	{4,3,2,1,0}																																1

$$t_0 = 0\;1\;0\;1\;0\;1\;0\;1\;0\;1\;0\;1\;0\;1\;0\;1\;0\;1\;0\;1\;0\;1\;0\;1\;0\;1\;0\;1\;0\;1\;0\;1$$
$$t_1 = 0\;0\;1\;1\;0\;0\;1\;1\;0\;0\;1\;1\;0\;0\;1\;1\;0\;0\;1\;1\;0\;0\;1\;1\;0\;0\;1\;1\;0\;0\;1\;1$$
$$t_2 = 0\;0\;0\;0\;1\;1\;1\;1\;0\;0\;0\;0\;1\;1\;1\;1\;0\;0\;0\;0\;1\;1\;1\;1\;0\;0\;0\;0\;1\;1\;1\;1$$
$$t_3 = 0\;0\;0\;0\;0\;0\;0\;0\;1\;1\;1\;1\;1\;1\;1\;1\;0\;0\;0\;0\;0\;0\;0\;0\;1\;1\;1\;1\;1\;1\;1\;1$$
$$t_4 = 0\;0\;0\;0\;0\;0\;0\;0\;0\;0\;0\;0\;0\;0\;0\;0\;1\;1\;1\;1\;1\;1\;1\;1\;1\;1\;1\;1\;1\;1\;1\;1$$

Fig. 2.2 Recursive structure of a generator matrix of $\mathrm{RM}^2_{5,5}$

entry 1 in the t-th coordinate. Hence, knowing the rows corresponding to all characteristic sets of cardinality 1, it is easy to write down any other row of Γ_m.

We recall the recursive structure of the generator matrix in the general case of a binary Reed–Muller-code. For $0 \le t \le m$,

$$\Gamma_{m,t} = \left(\begin{array}{c|c} \Gamma_{m-1,t} & \Gamma_{m-1,t} \\ \hline 0 & \Gamma_{m-1,t-1} \end{array} \right)$$

Of course $\Gamma_{m,m} = \Gamma_m$ is a generator matrix of $\mathrm{RM}^2_{m,m}$. The matrix $\Gamma_{m,0}$ contains just one vector, the all-one vector, which is the top row of Γ_m (cf. Exercise 2.4.4). It is a generator matrix of the repetition code $\mathrm{RM}^2_{m,0}$. The identification of characteristic sets and monomials shows that $\Gamma_{m,t}$ is a generator matrix of $\mathrm{RM}^2_{m,t}$. We call it the *canonical generator matrix* of $\mathrm{RM}^2_{m,t}$.

2.4.9   **Example** Figure 2.2 shows a generator matrix of $\mathrm{RM}_{5,5}^2$. The rows are labeled by integers, the corresponding binary numbers and characteristic sets. The columns are labeled by the $(t_0, \ldots, t_4) \in \mathbb{F}_2^5$.   ◇

2.4.10   **Theorem** *For $0 \le t < m$, the Reed–Muller-code $\mathrm{RM}_{m,t}^2$ is even.*

Proof: For $t = 0$, the Reed–Muller-code is the binary repetition code of length 2^m. From the recursive construction of Γ_m it is clear that each row with exception of the last one has even weight. If $t > 0$, by puncturing $\mathrm{RM}_{m,t}^2$ in the last component, we obtain the code $Pu(\mathrm{RM}_{m,t}^2)$. All rows in its generator matrix have odd weight. Hence by 1.6.3, exactly half of its codewords have odd weight. Thus, the Reed–Muller-code $\mathrm{RM}_{m,t}^2$ is the parity extension of $Pu(\mathrm{RM}_{m,t}^2)$ which contains codewords of even weight only.   □

Adding the first row to all remaining rows of $\Gamma_{m,s}$ we obtain another generator matrix $\tilde{\Gamma}_{m,s}$ of $\mathrm{RM}_{m,s}^2$. Apart from the first row, any row of $\tilde{\Gamma}_{m,s}$ is the complement of the corresponding row of $\Gamma_{m,s}$. The last column of $\tilde{\Gamma}_{m,s}$ is $(1, 0, \ldots, 0)^\top$. Due to 2.4.8, for $0 \le s \le m - 1$ the matrix $\tilde{\Gamma}_{m,m-s-1}$ is a check matrix of $\mathrm{RM}_{m,s}^2$. Hence, from Exercise 1.3.10 we derive that $\Gamma_{m,s} \cdot \tilde{\Gamma}_{m,m-s-1}^\top = 0$. Moreover, if $s < m - 1$ then we may write

$$\Gamma_{m,s} = (\Gamma \mid \mathbf{1}^\top) \quad \text{and} \quad \tilde{\Gamma}_{m,m-s-1} = \begin{pmatrix} 1 & 1 \\ \hline \tilde{\Gamma} & \mathbf{0}^\top \end{pmatrix}$$

where Γ is a generator matrix of $Pu(\mathrm{RM}_{m,s}^2)$ and $\tilde{\Gamma}$ is a matrix with $\sum_{i=1}^{m-s-1} \binom{m}{i}$ rows and $2^m - 1$ columns. The rows of $\tilde{\Gamma}$ are orthogonal to the rows of Γ. Therefore, $\tilde{\Gamma}$ is the generator matrix of the dual code $Pu(\mathrm{RM}_{m,s}^2)^\perp$.

If $s = m - 2$, the columns of the $m \times (2^m - 1)$-matrix $\tilde{\Gamma}$ are exactly all nonzero vectors in $\mathbb{F}_2^m$. Thus $\tilde{\Gamma}$ is a generator matrix both of the m-th order binary simplex-code and of $Pu(\mathrm{RM}_{m,m-2}^2)^\perp$. In other words, $\tilde{\Gamma}$ is a check matrix of the m-th order binary Hamming-code and a check matrix of $Pu(\mathrm{RM}_{m,m-2}^2)$. Conversely, the matrix

$$\begin{pmatrix} 1 \\ \tilde{\Gamma} \end{pmatrix}$$

is a generator matrix of $Pu(\mathrm{RM}_{m,1}^2)$, whence $Pu(\mathrm{RM}_{m,1}^2)$ is the augmentation of the m-th order binary simplex-code (cf. 2.3.2). We collect these results in the following

2.4.11   **Theorem**

1. $\mathrm{RM}_{m,m-2}^2$ is the parity extension of the m-th order binary Hamming-code.

2. $Pu(\mathrm{RM}^2_{m,1})$ is the augmentation of the m-th order binary simplex-code.

3. The weight distributions of $Pu(\mathrm{RM}^2_{m,1})$ and $\mathrm{RM}^2_{m,1}$ are given by

$$w_{Pu(\mathrm{RM}^2_{m,1})}(x) = 1 + (2^m - 1)x^{2^{m-1}-1} + (2^m - 1)x^{2^{m-1}} + x^{2^m-1}$$

and

$$w_{\mathrm{RM}^2_{m,1}}(x) = 1 + 2(2^m - 1)x^{2^{m-1}} + x^{2^m},$$

respectively.

Proof: The first two assertions follow directly from the considerations above. The weight distribution of the simplex-code was determined in 2.1.7. The final statement follows from the definitions of augmentation and puncturing. □

Example From the matrix

$$\Gamma_{5,1} = \begin{pmatrix} 1111111111111111111111111111111 \\ 0101010101010101010101010101010101 \\ 0011001100110011001100110011 \\ 0000111100001111000011110000 \\ 0000000011111111000000001111 \\ 0000000000000000111111111111 \end{pmatrix}$$

we obtain the matrix

$$\tilde{\Gamma}_{5,1} = \begin{pmatrix} 1111111111111111111111111111111 \\ 1010101010101010101010101010 \\ 1100110011001100110011001100 \\ 1111000011110000111100001111 0000 \\ 1111111100000000111111110000 0000 \\ 1111111111111110000000000000000 \end{pmatrix}$$

and

$$\tilde{\Gamma} = \begin{pmatrix} 1010101010101010101010101010101 \\ 1100110011001100110011001100110 \\ 1111000011110000111100001111000 \\ 1111111100000000111111110000000 \\ 1111111111111110000000000000000 \end{pmatrix},$$

a generator matrix of the 5-th order binary simplex-code. ◇

2.4.12

Exercises

Exercise Prove that $\mathcal{B}^q_m$ is in fact an $\mathbb{F}_q$-algebra, which means that it is both a vector space over $\mathbb{F}_q$ and a ring so that

$$\alpha(f \cdot g) = (\alpha f) \cdot g = f \cdot (\alpha g)$$

holds true, for all $\alpha \in \mathbb{F}_q$ and $f, g \in \mathcal{B}^q_m$.

E.2.4.1

E.2.4.2   **Exercise** Show that each polynomial f in $\mathrm{RM}^2_{m,t}$ can be uniquely expressed in the form

$$f(x_0, \ldots, x_{m-1}) = h(x_0, \ldots, x_{m-2}) + x_{m-1}g(x_0, \ldots, x_{m-2}),$$

where $\deg h \le t$ and $\deg g \le t - 1$.

E.2.4.3   **Exercise** Verify that the $(m-1)$-th order binary Reed–Muller-code of length 2^m consists of all vectors of even weight in $\mathbb{F}_2^{2^m}$.

E.2.4.4   **Exercise** Check that the top row and the rightmost column in the canonical form of $\Gamma_{m,t}$ consist of all-one vectors.

E.2.4.5   **Exercise** Write down Pascal's triangle, reduce the entries modulo 2 and compare with the generator matrix of the Reed–Muller-code in Fig. 2.2. Do you see a connection?

2.5 MDS-Codes

As MacWilliams and Sloane [139] put it, *we come now to one of the most fascinating chapters in all of coding theory: MDS-codes.* As we have seen in 2.1.1, for any (n, k, d)-code over any field we have $d \le n - k + 1$. Codes with $d = n - k + 1$ have been called *maximum distance separable*, or MDS for short. These codes have a wide range of applications, and they tie in well with structures in projective geometry. The compact disc stores music using linear $(32, 28, 5)$ and $(28, 24, 5)$-MDS-codes over $\mathbb{F}_{2^8}$ (for details see Chapter 5). *Trivial* MDS-codes are the codes of types $(n, 1, n)$, $(n, n-1, 2)$, and $(n, n, 1)$, which exist over any field $\mathbb{F}_q$ (cf. Exercise 2.5.1). Here we collect some properties characterizing MDS-codes:

2.5.1   **Theorem** *For linear (n, k)-codes C the following properties are equivalent:*

1. *C is an MDS-code.*

2. *In each check matrix of C any $n - k$ columns are linearly independent.*

3. *In each generator matrix of C any k columns are linearly independent.*

4. *$C^\perp$ is an MDS-code.*

Proof: The equivalence of the first two statements follows from 1.3.9 and the Singleton-bound 2.1.1. Together with 1.3.4 we obtain the equivalence of the third and the fourth property.

Now assume that C is an MDS-code, i.e. that $d = n - k + 1$. To show that $C^\perp$ is also MDS, we prove that its minimum distance $d^\perp$ equals $n - (n - k) + 1 = k + 1$. Assume, indirectly, that $C^\perp$ contains an element $c \neq 0$ of Hamming weight at most k. Each nonzero element of the dual code can occur as a row in a check matrix of C. Let Δ be a check matrix of C containing c as its top row. Now consider the columns of Δ which are zero in their top component. By assumption, there are at least $n - k$ of them. Since Δ has $n - k$ rows, these columns are dependent. According to 1.3.9, these columns give rise to a word in C of weight at most $n - k$, which contradicts the assumption. This shows that $d^\perp \geq k + 1$. From the Singleton-bound we obtain $d^\perp \leq n - (n - k) + 1 = k + 1$ so finally $d^\perp = k + 1$, which means that $C^\perp$ is MDS.

A symmetric argument shows that C is MDS provided that $C^\perp$ has this property.   □

The third item of 2.5.1 yields

Corollary *An (n, k)-code is MDS if and only if any k coordinates form an information set.*   **2.5.2**
                                                                          □

Recall that $\mathbb{F}_q^{(J)}$ has been defined in 1.7.4 as the set of vectors which are zero on all of $\bar{J}$. It is a subspace of dimension $|J|$.

Theorem *For each (n, k, d, q)-code C the following statements are equivalent:*   **2.5.3**

1. *C is an MDS-code.*

2. *For each $J \subseteq n$ with $|J| = d - 1$ we have $C \oplus \mathbb{F}_q^{(J)} = \mathbb{F}_q^n$.*

3. *For each $J \subseteq n$ with $|J| = k$ we have $C \oplus \mathbb{F}_q^{(\bar{J})} = \mathbb{F}_q^n$.*

Proof: Let C denote an MDS-code and consider a set $J \subseteq n$ with $|J| = d - 1$. By 1.7.7

$$C \oplus \mathbb{F}_q^{(J)} \subseteq \mathbb{F}_q^n.$$

Counting dimensions we see that the space on the left hand side is of dimension $k + (d - 1) = n$, since C is MDS. Thus $C \oplus \mathbb{F}_q^{(J)} = \mathbb{F}_q^n$.

Conversely, assume that $C \oplus \mathbb{F}_q^{(J)} = \mathbb{F}_q^n$ for some $J \subseteq n$ with $|J| = d - 1$. Then $k = \dim(C) = n - d + 1$, i.e. C is MDS. The equivalence between the first and the third statement can be derived from 2.5.2 together with 1.7.6.   □

2.5.4   **Theorem** *Suppose that C is an (n,k)-code with systematic generator matrix $\Gamma = (I_k \mid A)$. Then C is MDS if and only if, for each $i = 1, \ldots, \min\{k, n-k\}$, all $i \times i$-submatrices of A are regular.*

Proof: We assume that C is an MDS-code. We introduce some notation. For a matrix M, and for X and Y subsets of the sets of row and column indices, let $M_{X,Y}$ be the submatrix containing the elements of A which are at the intersection of rows indexed by elements of X and columns indexed by elements of Y. Moreover, $\overline{X}$ denotes the complement of X in the set of row indices. Now consider $\Gamma = (I_k \mid A)$ and let $k = \{0, \ldots, k-1\}$ and $\{k, \ldots, n-1\} = n \setminus k$ be index sets for the matrix $A = (a_{ij})_{i \in k, \, j \in n \setminus k}$. Assume that $A' = A_{X,Y}$ is a submatrix of A consisting of $i \leq \min\{k, n-k\}$ rows and columns, i.e. with $X \subseteq k$, and $Y \subseteq \{k, \ldots, n-1\}$ and $|X| = |Y| = i$. Consider the matrix of k columns of Γ

$$A'' = \left(\begin{array}{c|c} I_{X,\overline{X}} & A_{X,Y} \\ \hline I_{\overline{X},\overline{X}} & A_{\overline{X},Y} \end{array} \right) = \left(\begin{array}{c|c} 0 & A' \\ \hline I_{k-i} & * \end{array} \right),$$

where 0 denotes the $i \times (k-i)$ zero matrix and where $*$ denotes a $(k-i) \times i$-matrix. According to 2.5.1, A'' is regular and hence $\det A'' = \pm \det A' \neq 0$.

The converse of this statement follows directly from 2.5.1.   □

2.5.5   **Example** We consider the $(4,2)$-code C over the field with four elements $\mathbb{F}_4 = \{0, 1, \alpha, \alpha+1\}$ subject to the relation $\alpha^2 = 1 + \alpha$ with $\Gamma = (I \mid A)$, where

$$A = \left(\begin{array}{cc} \alpha & \alpha^2 \\ \alpha^2 & \alpha \end{array} \right).$$

Since each $i \times i$-submatrix of A is regular ($i = 1, 2$), C is MDS.   ◇

2.5.6   **Theorem** *Up to linear isometry, each MDS-code is generated systematically by a matrix $\Gamma = (I \mid B)$, where B is of the form*

$$B = \left(\begin{array}{c|c} 1 & 1 \\ \hline 1^\top & * \end{array} \right).$$

Proof: Up to isometry we may assume that the code is generated systematically by the matrix $\Gamma = (I \mid A)$ where $A = (a_{ij})_{i \in k, j \in n \setminus k}$. By 2.5.4, all entries of A are nonzero, so that

$$D = \operatorname{diag}\left(1, \frac{a_{1,k}}{a_{0,k}}, \ldots, \frac{a_{k-1,k}}{a_{0,k}}, a_{0,k}^{-1}, \ldots, a_{0,n-1}^{-1} \right)$$

is a regular diagonal matrix. From 1.7.3 we know that $(I \mid A)$ and $(I \mid D * A) = (I \mid B)$ generate linearly isometric codes. Moreover, we know from 1.7.2 that

$$b_{ij} = d_{ii}^{-1} a_{ij} d_{jj}.$$

Hence, an easy check shows that the leftmost column and the top row consist of all-one vectors, as stated.   □

Let us now discuss the question of when MDS-codes exist.

Theorem *For each (n,k)-MDS-code over $\mathbb{F}_q$ with $2 \le k \le n - 2$ the inequality*

$$q \ge \max\{k, n - k\} + 1$$

holds true.

Proof: By 2.5.6, each MDS-code is linearly isometric to an MDS-code with a systematic generator matrix $\Gamma = (I \mid B)$, where

$$B = \left(\begin{array}{c|c} 1 & 1 \\ \hline 1^{\top} & B' \end{array} \right).$$

By 2.5.4, each $i \times i$-submatrix of B is regular. In particular, the 2×2-submatrices containing two elements of the highest row or of the left column have determinants

$$\det \left(\begin{array}{cc} 1 & 1 \\ \alpha & \alpha' \end{array} \right) = \alpha' - \alpha \quad \text{and} \quad \det \left(\begin{array}{cc} 1 & \alpha \\ 1 & \alpha' \end{array} \right) = \alpha' - \alpha$$

distinct from zero. Consequently, the elements in the rows and in the columns of B' are pairwise distinct and also distinct from 0 and 1. For this reason, the alphabet $\mathbb{F}_q$ contains at least $\max\{k - 1, n - k - 1\} + 2$ elements.   □

Example By the previous theorem, there are no nontrivial binary MDS-codes. For $n \ge 4$, $(n, 2)$-MDS-codes over $\mathbb{F}_q$ with $q \ge n - 1$ exist, they are generated by $(I \mid B)$ where B is any matrix of the form

$$B = \left(\begin{array}{cccc} 1 & 1 & \cdots & 1 \\ 1 & b_{1,3} & \cdots & b_{1,n-1} \end{array} \right),$$

where $b_{1,3}, \ldots, b_{1,n-1}$ are pairwise different field elements, which are all distinct from both 0 and 1.   ◇

Theorem *For any $n \ge 6$ there exists an $(n, 3)$-MDS-code over $\mathbb{F}_q$ with $q = 2^m$ and $q \ge n - 2$.*

Proof: Assume that $0, b_3 := 1, b_4, \ldots, b_{n-1}$ are pairwise distinct elements in $\mathbb{F}_q$ with $q = 2^m$. From $q \ge n - 2$ it follows that $m \ge 2$. We form the matrix

$$B = \left(\begin{array}{cccc} 1 & 1 & \cdots & 1 \\ b_3 & b_4 & \cdots & b_{n-1} \\ b_3^2 & b_4^2 & \cdots & b_{n-1}^2 \end{array} \right),$$

and consider the $(n,3)$-code C over $\mathbb{F}_q$ generated by $(I \mid B)$. From 2.5.4 it follows that C is MDS: To begin with, each 3×3-submatrix of B is a Vandermonde matrix and hence regular (cf. Exercise 2.5.2). Furthermore, each 2×2-submatrix

$$
\begin{pmatrix} b_i & b_j \\ b_i^2 & b_j^2 \end{pmatrix} \quad \text{and} \quad \begin{pmatrix} 1 & 1 \\ b_i & b_j \end{pmatrix}, \qquad 3 \le i < j \le n-1,
$$

is non-singular. Finally, the elements $0, 1, b_4^2, \ldots, b_{n-1}^2$ are pairwise distinct, since by 3.2.13 the Frobenius mapping $\mathbb{F}_{2^m} \ni \alpha \mapsto \alpha^2 \in \mathbb{F}_{2^m}$ is an automorphism. Hence, the submatrices

$$
\begin{pmatrix} 1 & 1 \\ b_i^2 & b_j^2 \end{pmatrix}, \qquad 3 \le i < j \le n-1,
$$

are also regular. □

The condition $q = 2^m$ turns out to be necessary (see Exercise 2.5.5).

2.5.10   **Example** By 2.5.9, the $(6,3)$-code over $\mathbb{F}_4 = \{0, 1, \alpha, \alpha+1\}$ $(\alpha^2 = \alpha + 1)$ with generator matrix

$$
\Gamma = \begin{pmatrix} 1 & 0 & 0 & 1 & 1 & 1 \\ 0 & 1 & 0 & 1 & \alpha & \alpha^2 \\ 0 & 0 & 1 & 1 & \alpha^2 & \alpha \end{pmatrix}
$$

is MDS. This code is known as the *hexacode*. ◇

Let $N_q(k)$ be the maximal length of an MDS-code of dimension k over $\mathbb{F}_q$. From 2.5.7 and 2.5.9 we obtain the important

2.5.11   **Corollary** *For $q = 2^m$ we have $N_q(3) = q + 2$.* □

For all other cases we have the following conjecture ([139], p. 328):

$$
N_q(k) = \begin{cases} q+1 & \text{if } 2 \le k \le q, \\ k+1 & \text{if } q < k. \end{cases}
$$

Finally, let us consider the weight distribution of MDS-codes.

2.5.12   **Theorem** *Suppose that C is an (n,k)-MDS-code over $\mathbb{F}_q$. We denote by A_i the number of codewords in C of weight i. Then the following holds:*

— $A_0 = 1, A_1 = A_2 = \ldots = A_{n-k} = 0.$

— *For each $i \in \{0, 1, \ldots, k-1\}$:*

$$
A_{n-i} = \sum_{m=i}^{k-1} (-1)^{m-i} \binom{m}{i} \binom{n}{m} (q^{k-m} - 1).
$$

Proof: For each subset J of $n = \{0, \ldots, n-1\}$, let $C(\bar{J})$ denote the set of codewords in C whose components c_j for $j \in J$ are all zero. i.e.

$$C(\bar{J}) := \{c \in C \mid \forall j \in J : c_j = 0\}.$$

Since each nonzero codeword in C has at most $k-1$ zero entries, $C(\bar{J}) = \emptyset$ for each J with $|J| \geq k$.

For each m with $0 \leq m \leq k-1$ we determine the cardinality of the set

$$\mathcal{S} = \left\{ (J,c) \;\middle|\; J \in \binom{n}{m}, \, 0 \neq c \in C(\bar{J}) \right\}$$ 2.5.13

in two different ways. (Here $\binom{n}{m}$ indicates the set of all m-subsets of the set n.) On the one hand, each k-subset of n is an information set of C, and hence by 1.7.6 for each $J \subseteq n$ with $|J| \leq k-1$ we have

$$|C(\bar{J})| = q^{k-|J|}.$$

Thus the set $\mathcal{S}$ is of cardinality

$$\binom{n}{m} \cdot (q^{k-m} - 1).$$

On the other hand, we may decompose the set of codewords of C into sets

$$C_i := \{c \in C \mid \mathrm{wt}(c) = n - i\} \quad \text{for } 0 \leq i \leq n.$$

Thus, the coefficients of the weight distribution are $A_{n-i} = |C_i|$. If $i \geq m$, for each $c \in C_i$ there are exactly $\binom{i}{m}$ subsets J of n of cardinality m such that $c \in C(\bar{J})$. Hence, there exist exactly $\binom{i}{m} \cdot A_{n-i}$ pairs of the form (J,c) with $c \in C_i \cap C(\bar{J})$. Thus, the set $\mathcal{S}$ of 2.5.13 is of cardinality

$$\sum_{i=m}^{k-1} \binom{i}{m} \cdot A_{n-i}.$$

This way we obtain the following system of k equations in the k indeterminates $A_{n-k+1}, \ldots, A_n$:

$$\sum_{i=m}^{k-1} \binom{i}{m} \cdot A_{n-i} = \binom{n}{m} \cdot (q^{k-m} - 1), \qquad 0 \leq m \leq k-1.$$

Rearranging the indeterminates as $A_n, \ldots, A_{n-k+1}$, the coefficient matrix of this system of equations turns out to be upper triangular with ones along its main diagonal, i.e. with determinant 1. Therefore this system has a unique solution, which is given by (Exercise 2.5.6)

$$A_{n-i} = \sum_{m=i}^{k-1} (-1)^{m-i} \binom{m}{i} \binom{n}{m} (q^{k-m} - 1), \qquad 0 \leq i \leq k-1. \qquad \square$$

The weight distribution gives an upper bound for $N_q(k)$.

2.5.14   **Lemma** *For each $k \geq 2$ we have $N_q(k) \leq q + k - 1$.*

Proof: Every (n,k)-MDS-code over $\mathbb{F}_q$ satisfies

$$
\begin{aligned}
A_{n-k+2} &= \binom{n}{k-2}(q^2 - 1) - (k-1)\binom{n}{k-1}(q-1) \\
&= \binom{n}{k-2}(q-1)(q-1-(n-k)).
\end{aligned}
$$

As A_{n-k+2} cannot be negative, the factor $q - 1 - n + k$ is ≥ 0. $\square$

Exercises

E.2.5.1   **Exercise** Prove that trivial MDS-codes of length n exist over every finite field.

E.2.5.2   **Exercise** Consider field elements $\alpha_0, \ldots, \alpha_{n-1}$. Show that the *Vandermonde matrix*

$$
\begin{pmatrix}
1 & \alpha_0 & \alpha_0^2 & \cdots & \alpha_0^{n-1} \\
1 & \alpha_1 & \alpha_1^2 & \cdots & \alpha_1^{n-1} \\
\vdots & & & & \vdots \\
1 & \alpha_{n-1} & \alpha_{n-1}^2 & \cdots & \alpha_{n-1}^{n-1}
\end{pmatrix}
$$

has as determinant the expression

$$
\prod_{i<j}(\alpha_j - \alpha_i) = \prod_{i=0}^{n-2}\prod_{j=i+1}^{n-1}(\alpha_j - \alpha_i).
$$

In particular, this determinant is nonzero provided that $\alpha_i \neq \alpha_j$ for $i \neq j$.

E.2.5.3   **Exercise** Show that every shortened MDS-code is again MDS.

E.2.5.4   **Exercise** Construct a $(5,2)$-MDS-code over the field $\mathbb{F}_5$.

E.2.5.5   **Exercise** Prove that there is no $(7,4)$-MDS-code over $\mathbb{F}_5$. Hint: use 2.5.4.

E.2.5.6   **Exercise** Prove

$$
\binom{n}{m}\binom{m}{p} = \binom{n}{p}\binom{n-p}{m-p}, \qquad p \leq m \leq n.
$$

Verify that for $0 \leq m \leq n$ the identity

$$\sum_{k=m}^{n} (-1)^{n-k} \binom{n}{k} \binom{k}{m} = \begin{cases} 0 & \text{if } n \neq m, \\ (-1)^{n-m} & \text{if } n = m \end{cases}$$

holds true. Fill in the missing details of the proof of 2.5.12.

Exercise Show that the nonzero coefficients in the weight enumerator of the hexacode of Example 2.5.10 are $A_0 = 1$, $A_4 = 45$ and $A_6 = 18$.

E.2.5.7

Exercise Determine the weight enumerator of the $(6,3)$-MDS-code over $\mathbb{F}_5$ which is generated by

E.2.5.8

$$\begin{pmatrix} 1 & 1 & 1 & 4 & 0 & 0 \\ 3 & 2 & 1 & 0 & 4 & 0 \\ 4 & 3 & 1 & 0 & 0 & 4 \end{pmatrix}.$$

Exercise Prove that for $n - k + 1 \leq r \leq n$ the coefficients A_r in the weight distribution of an (n,k)-MDS-code over $\mathbb{F}_q$ are given by

E.2.5.9

$$A_r = \binom{n}{r} \sum_{i \in r - d + 1} (-1)^i \binom{r}{i} (q^{r-i-d+1} - 1).$$

Exercise Show that the two $(10,5)$-codes over $\mathbb{F}_9$ generated by

E.2.5.10

$$\begin{pmatrix} 1 & 1 & 1 & 1 & 1 & 2 & 0 & 0 & 0 & 0 \\ 2+\eta & 1+\eta & \eta & 2 & 1 & 0 & 2 & 0 & 0 & 0 \\ 1+\eta & 1+2\eta & 2\eta & \eta & 1 & 0 & 0 & 2 & 0 & 0 \\ 2\eta & 2+2\eta & 1+\eta & 2+\eta & 1 & 0 & 0 & 0 & 2 & 0 \\ 1+2\eta & 2 & 2+2\eta & 2\eta & 1 & 0 & 0 & 0 & 0 & 2 \end{pmatrix}$$

and

$$\begin{pmatrix} 1 & 1 & 1 & 1 & 1 & 2 & 0 & 0 & 0 & 0 \\ 2+\eta & 1+\eta & \eta & 2 & 1 & 0 & 2 & 0 & 0 & 0 \\ 2+2\eta & 2+\eta & 2\eta & 1+\eta & 1 & 0 & 0 & 2 & 0 & 0 \\ 2\eta & 2+2\eta & 2 & 1+2\eta & 1 & 0 & 0 & 0 & 2 & 0 \\ 1+2\eta & \eta & 2+\eta & 2+2\eta & 1 & 0 & 0 & 0 & 0 & 2 \end{pmatrix}$$

are semilinearly inequivalent and MDS. Here, we have $\mathbb{F}_9 = \{a + b\eta \mid a, b \in \mathbb{F}_3\}$ subject to the relation $\eta^2 = 2\eta + 1$. (The first code is obtained from a rational normal curve, with automorphism group $\mathrm{P\Gamma L}_2(9)$ of order 1440. The second code is obtained from the Glynn-arc [69] in $\mathrm{PG}_4(9)$, with automorphism group $\mathrm{PGL}_2(9)$ of order 720. Both automorphism groups act transitively on the 10 coordinates. It is known that there is no $(11,6)$-MDS-code over $\mathbb{F}_9$.)

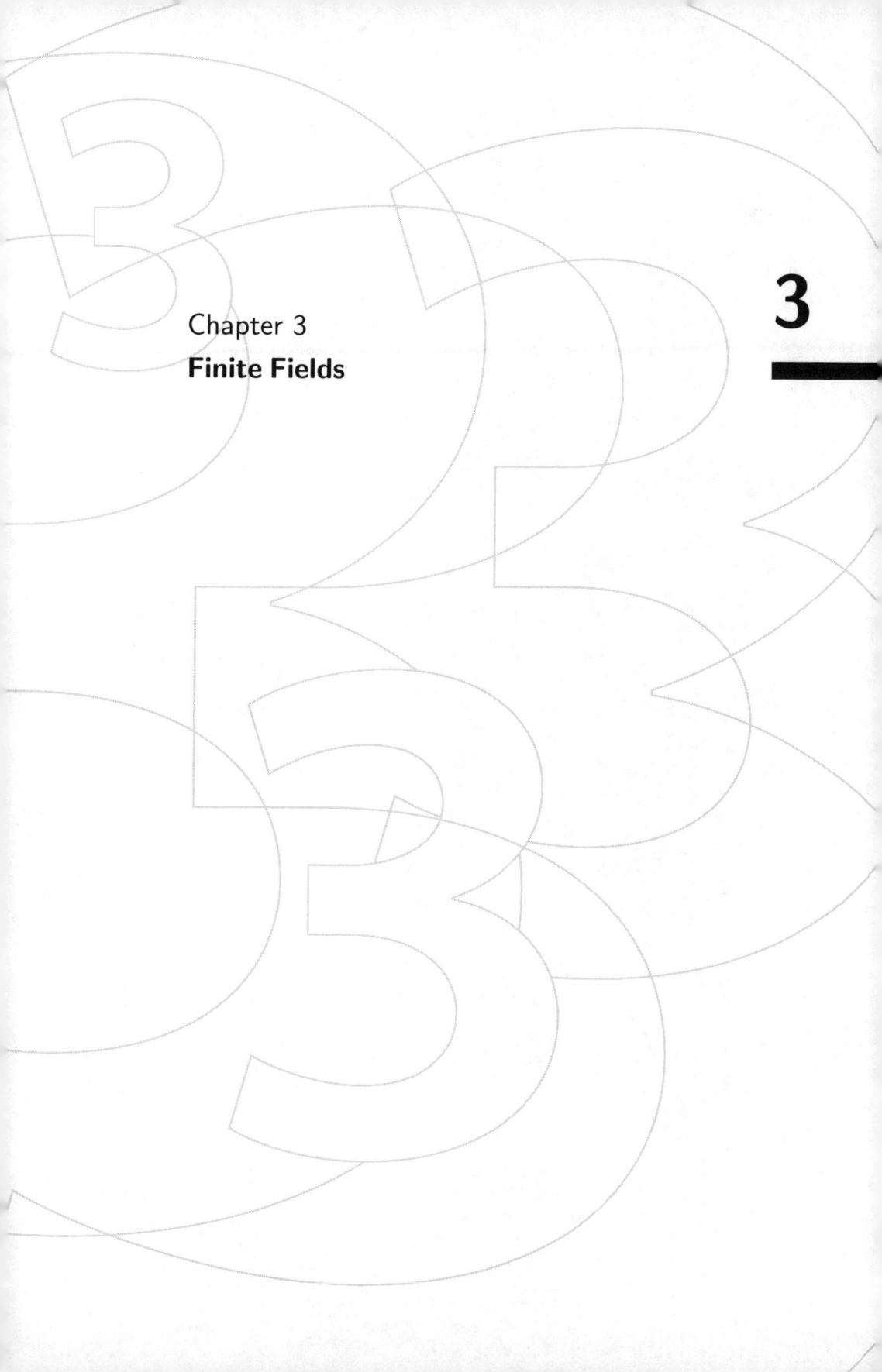

Chapter 3
Finite Fields

3

3

3   Finite Fields

A linear (n,k)-code is a subset of the n-dimensional vector space $\mathbb{F}^n$ over a finite field $\mathbb{F}$. This vector space consists of all words of length n over the alphabet $\mathbb{F}$. In the examples we have seen so far, we have always considered the most simple situations, the smallest field $\mathbb{F} = \mathbb{F}_2 = \{0,1\}$, the binary field of two elements, or the ternary field $\mathbb{F} = \mathbb{F}_3 = \{0,1,2\}$, or the quaternary field $\mathbb{F} = \mathbb{F}_4$. As we also intend to consider more complicated cases in full detail, we need a summary of the most important facts on finite fields in general, their properties and how they can be constructed. Afterwards, the reader should be able to construct, at least in principle, each finite field and implement it on a computer.

3.1 Finite Fields – An Introduction

Let $\mathbb{F}$ denote a finite field and $(\mathbb{F},+)$ its *additive* group. To begin with, we study the subgroup
$$\langle\, 1\, \rangle$$
of $(\mathbb{F},+)$ which is generated by the (multiplicative) identity element 1 of $\mathbb{F}$. Since $\mathbb{F}$ is finite, there is a least integer $p > 0$ so that the p-fold sum (cf. Exercise 1.6.6) of the identity element 1 of $\mathbb{F}$ gives the zero element
$$p \cdot 1 = \underbrace{1 + \ldots + 1}_{p \text{ times}} = 0.$$

This positive integer p is the cardinality of $\langle\, 1\, \rangle$. Since $1 \neq 0$, we have $p \geq 2$. Also, p must be prime since a field has no zero divisors (see Exercise 3.1.1).

Definition (characteristic of a field) The smallest number $p \in \mathbb{N}^*$ such that $p \cdot 1 = 0$ is called the *characteristic* of $\mathbb{F}$:
$$\operatorname{char}(\mathbb{F}) := p := |\langle\, 1\, \rangle|. \qquad \diamond$$

For each element κ of a field of characteristic p the following holds:
$$p \cdot \kappa = p \cdot (1\kappa) = (p \cdot 1)\kappa = 0.$$

A prominent field is $\mathbb{Z}_p$, the ring of residue classes of integers modulo p, equipped with addition and multiplication modulo p (cf. Exercise 3.1.3). It consists of p elements and its characteristic is p. It is not difficult to check that the subgroup $\langle\, 1\, \rangle$ of the additive group of an arbitrary field $\mathbb{F}$ of characteristic p is isomorphic to $\mathbb{Z}_p$. Moreover, $\langle\, 1\, \rangle$ is the smallest subfield of $\mathbb{F}$ with respect to inclusion.

3.1.3   **Definition (prime field)** Each finite field $\mathbb{F}$ of characteristic p contains a sub-field $\mathbb{P}$ isomorphic to $\mathbb{Z}_p$. It consists of the multiples $n \cdot 1$ of the identity element:

$$\mathbb{P} := \{n \cdot 1 \mid 0 \leq n \leq p - 1\} = \langle 1 \rangle \simeq \mathbb{Z}_p.$$

For this reason, $\mathbb{P}$ is called the *prime field of* $\mathbb{F}$, whereas $\mathbb{Z}_p$ is called *the* prime field of characteristic p.   ◇

Apart from being a field, $\mathbb{F}$ is also a vector space over the subfield $\mathbb{P}$. This follows from the fact that addition in $\mathbb{F}$ and multiplication by "scalars" from $\mathbb{P}$ satisfy the usual vector space axioms. Since $\mathbb{F}$ is finite, the dimension of $\mathbb{F}$ as a vector space over $\mathbb{P}$ is finite, too. Let n denote this dimension. Then $\mathbb{F}$ contains exactly $|\mathbb{P}|^n = p^n$ elements.

3.1.4   **Corollary** *The order of a finite field is p^n for some prime p and some positive integer n.*   □

If $\mathbb{K}$ is a field containing the field $\mathbb{F}$ as a subfield, then we say that $\mathbb{K}$ is an *extension field* of $\mathbb{F}$. We call it a *finite* extension if the dimension of $\mathbb{K}$ as a vector space over $\mathbb{F}$ is finite.

In coding theory, we take finite fields as the alphabets over which messages and codewords are defined. In order to handle codewords we should be able to write down the elements of a finite field. This raises the question how to *construct*, for a given prime number p, integer $n \geq 1$, and starting from the well known prime field $\mathbb{Z}_p$, all the other finite fields of order p^n. It will turn out that in fact there is a field of order p^n for any prime p and any integer $n \geq 1$. We will see in 3.2.10 that there is up to isomorphism a unique finite field of each such order. Thus our problem reduces to the problem of constructing, for each prime p and for each positive integer n, *one* field of order p^n.

The classical *construction* of a field of order p^n is by means of *residue class rings* of the polynomial ring

$$\mathbb{Z}_p[x] := \left\{ f = \sum_{i=0}^{n} \kappa_i x^i \ \middle| \ n \in \mathbb{N}, \ \kappa_i \in \mathbb{Z}_p \right\}.$$

3.1.5   **The ring of polynomials over a field** Let $\mathbb{F}$ be a field and let x be an indeterminate over $\mathbb{F}$. Then $\mathbb{F}[x]$ is an *integral domain*, i.e. it is a commutative ring with 1 which is free from zero divisors (cf. Exercise 3.1.1 and Exercise 3.1.4).

If f is a nonzero polynomial of the form $f = \sum_{i=0}^{n} \kappa_i x^i \in \mathbb{F}[x]$ with $\kappa_n \neq 0$, then n is called the *degree* of f, abbreviated as

$$\deg f = n.$$

A polynomial of degree 1 is called *linear*. The coefficient κ_n, i.e. the coefficient of the highest power of x appearing in f, is called the *leading coefficient* of f. A polynomial is *monic* if its leading coefficient is equal to 1. For technical reasons, we define the degree of the zero polynomial to be $-\infty$.

We call $g \in \mathbb{F}[x]$ a *divisor* of $f \in \mathbb{F}[x]$ if there exists a polynomial $h \in \mathbb{F}[x]$ such that $f = gh$. A polynomial $f \in \mathbb{F}[x]$ of degree at least 1 is called *irreducible* if it cannot be written as a product of two polynomials in $\mathbb{F}[x]$ which are both not constant. Thus, if $f = gh$ is irreducible with $g, h \in \mathbb{F}[x]$, then at least one of these factors is a constant polynomial, whence a nonzero element of the multiplicative group of the underlying field. Thus $g \in \mathbb{F}^*$ or $h \in \mathbb{F}^*$. If f is not irreducible, it is called *reducible*.

Instead of writing "$f \in \mathbb{F}[x]$ is irreducible", we also say that "f is irreducible over $\mathbb{F}$". As a matter of fact, an irreducible polynomial over $\mathbb{F}$ can be reducible over a field extension $\mathbb{L}$ of $\mathbb{F}$.

We use *ideals* in the polynomial ring $\mathbb{F}[x]$, i.e. nonempty subsets I which contain, together with two elements f_1 and f_2, also their difference $f_1 - f_2$ as well as any multiple $f \cdot f_1$ for $f \in \mathbb{F}[x]$. It is not difficult to see (Exercise 3.1.11) that each ideal in $\mathbb{F}[x]$ is generated by a single polynomial which is uniquely determined up to a nonzero factor $\kappa \in \mathbb{F}^*$. This means that each ideal I consists of all the multiples of a single generator, thus $\mathbb{F}[x]$ is a *principal ideal domain*. The polynomial f is a generator of I if

$$I = \mathrm{I}(f) := \{fg \mid g \in \mathbb{F}[x]\}.$$

If α denotes an element of $\mathbb{F}$, then we can evaluate any polynomial in $\mathbb{F}[x]$ at α, simply by substituting α for x. The *evaluation map*,

$$\mathbb{F}[x] \to \mathbb{F} \ : \ f \mapsto f(\alpha),$$

maps the polynomial $f = \sum_{i=0}^{n} \kappa_i x^i$ to $f(\alpha) := \sum_{i=0}^{n} \kappa_i \alpha^i$. This is a ring epimorphism (cf. Exercise 3.1.7).

An element $\alpha \in \mathbb{F}$ is called a *root* of the polynomial $f \in \mathbb{F}[x]$ if $f(\alpha) = 0$. If α is a root of f, then the linear polynomial $x - \alpha$ divides f. The largest positive integer k such that $(x - \alpha)^k$ is a divisor of f (cf. Exercise 3.1.8) is called the *multiplicity* of the root α. A *simple root* is a root with multiplicity 1. All other roots are called *multiple*.

Let $f \in \mathbb{F}[x]$ be a polynomial of degree $n \geq 0$ and let $\alpha_0, \dots, \alpha_{r-1}$ be distinct roots of f with multiplicities $k_0, \dots, k_{r-1}$. Then $(x - \alpha_0)^{k_0} \cdots (x - \alpha_{r-1})^{k_{r-1}}$ divides f, and comparison of degrees shows that $k_0 + \dots + k_{r-1} \leq n$. Consequently, f can have at most n distinct roots.

Let f be a monic, irreducible polynomial in $\mathbb{F}[x]$, and let $\mathbb{L}$ be an extension of $\mathbb{F}$. If $\alpha \in \mathbb{L}$ is a root of f, then f is the monic polynomial of least degree in $\mathbb{F}[x]$ which has α as a root. Being also monic, f is uniquely determined by these two properties. The polynomial f is called the *minimal polynomial* of α over $\mathbb{F}$ (see Exercise 3.1.9).

For the announced construction of a finite field of cardinality p^n we choose a *monic, irreducible* polynomial $f \in \mathbb{Z}_p[x]$ of degree n. Further down in the proof of 3.2.25 we will show that for any given n there is at least one irreducible polynomial of degree n. We form the ideal $I(f)$ which is generated by f. Then we consider the residue classes of polynomials in $\mathbb{Z}_p[x]$ modulo $I(f)$, i.e. the sets of the form $g + I(f)$, with $g \in \mathbb{Z}_p[x]$. Two polynomials are in the same residue class if and only if their difference is in $I(f)$. The residue classes are the elements of the factor ring

$$\mathbb{Z}_p[x]/I(f) := \{g + I(f) \mid g \in \mathbb{Z}_p[x]\}.$$

Using the Division Theorem and comparing degrees (Exercises 3.1.5 and 3.1.6) we see that each residue class in $\mathbb{Z}_p[x]/I(f)$ contains exactly one polynomial g with $\deg g < n$. This proves that

$$\mathbb{Z}_p[x]/I(f) = \{g + I(f) \mid g \in \mathbb{Z}_p[x], \ \deg g < n\}.$$

The set $\mathbb{Z}_p[x]/I(f)$ together with addition

$$(g + I(f)) + (h + I(f)) := (g + h) + I(f)$$

and multiplication

$$(g + I(f)) \cdot (h + I(f)) := gh + I(f)$$

forms a commutative ring with zero element $0 + I(f) = I(f)$ and with identity element $1 + I(f)$. The multiplicative inverse of a nonzero element $g + I(f)$ can be obtained by an application of Bézout's Identity (Exercise 3.1.6): since f is irreducible and $\deg g < \deg f$, it follows that $\gcd(f, g) = 1$. Thus there exist $r, s \in \mathbb{Z}_p[x]$ such that

$$rf + sg = 1.$$

Since $rf \in I(f)$ and therefore $rf + I(f) = I(f)$, we have

$$1 + I(f) = rf + sg + I(f) = sg + I(f) = (s + I(f))(g + I(f)),$$

which means that $s + I(f)$ is the multiplicative inverse of $g + I(f)$.

3.1.6 **Corollary** *If f is a monic, irreducible polynomial over $\mathbb{Z}_p$ of degree $n \geq 1$, then the residue class ring*

$$\mathbb{F} := \mathbb{Z}_p[x]/I(f) = \{g + I(f) \mid g \in \mathbb{Z}_p[x], \ \deg g < n\}$$

is a field with p^n elements. $\square$

In order to implement such a finite field, we should, of course, *not* implement it as a residue class ring and carry $I(f)$ all the time with us. As a matter of fact we can express each of its elements as a $\mathbb{Z}_p$-linear combination of powers of certain elements α. From the last corollary we deduce that the field $\mathbb{F} = \mathbb{Z}_p[x]/I(f)$ consists of the residue classes of the polynomials of degree less than n, where n is the degree of the monic irreducible polynomial f,

$$\mathbb{F} = \left\{ \left(\sum_{i \in n} \kappa_i x^i \right) + I(f) \;\middle|\; \kappa_i \in \mathbb{Z}_p,\ i \in n \right\}.$$

According to the definition of addition and multiplication in $\mathbb{Z}_p[x]/I(f)$ the following is true:

$$\left(\sum_{i \in n} \kappa_i x^i \right) + I(f) = \sum_{i \in n} (\kappa_i x^i + I(f)) = \sum_{i \in n} \kappa_i (x^i + I(f)) = \sum_{i \in n} \kappa_i (x + I(f))^i.$$

Corollary *Let f be a monic, irreducible polynomial over $\mathbb{Z}_p$ of degree $n \geq 1$, and let* **3.1.7**
$\mathbb{F} = \mathbb{Z}_p[x]/I(f)$. Putting $\alpha := x + I(f) \in \mathbb{F}$, we see that $\mathbb{F}$ is generated, as a vector space over $\mathbb{Z}_p$, by the set $\{\alpha^0 = 1, \alpha^1, \ldots, \alpha^{n-1}\}$. □

Moreover, we note that the residue class $\alpha := x + I(f)$ is a root of f: If the monic, irreducible polynomial f is given by $f = \sum_{i=0}^{n} \kappa_i x^i$ with $\kappa_i \in \mathbb{Z}_p$ for $0 \leq i \leq n$ and $\kappa_n = 1$, then

$$f(\alpha) = \sum_{i=0}^{n} \kappa_i \alpha^i = \sum_{i=0}^{n} \kappa_i x^i + I(f) = f + I(f) = 0 + I(f) = 0 \in \mathbb{F}.$$

Since α is a root of f and f is irreducible, f is a polynomial of least degree in $\mathbb{Z}_p[x] \setminus \{0\}$ such that $f(\alpha) = 0$.

Corollary *Let f be a monic, irreducible polynomial over $\mathbb{Z}_p$ of degree $n \geq 1$, and let* **3.1.8**
$\alpha := x + I(f) \in \mathbb{F}$. Then f is the minimal polynomial of α over $\mathbb{Z}_p$. □

This helps to show that the set $B := \{1, \alpha, \ldots, \alpha^{n-1}\}$ is a generating set of $\mathbb{F}$ which is linearly independent over $\mathbb{Z}_p$, i.e. it is a basis for $\mathbb{F}$ considered as a vector space over $\mathbb{Z}_p$. If $\sum_{i \in n} \mu_i \alpha^i = 0$ for some $\mu_0, \ldots, \mu_{n-1} \in \mathbb{Z}_p$, then $g = \sum_{i \in n} \mu_i x^i$ is a polynomial in $\mathbb{Z}_p[x]$ of degree less than n which has α as a root. By Exercise 3.1.9, the minimal polynomial f of α divides g. From $\deg f > \deg g$ we obtain that g is the zero polynomial, i.e. $\mu_0 = \ldots = \mu_{n-1} = 0$.

Corollary *Let f be a monic, irreducible polynomial over $\mathbb{Z}_p$ of degree n, and let* **3.1.9**
$\mathbb{F} = \mathbb{Z}_p[x]/I(f)$. Then the set $\{1, \alpha, \ldots, \alpha^{n-1}\}$ with $\alpha := x + I(f) \in \mathbb{F}$ is a $\mathbb{Z}_p$-basis of $\mathbb{F}$. Hence, each element in $\mathbb{F}$ can be expressed in a unique way as a linear combination of α^i, $i \in n$, with coefficients in $\mathbb{Z}_p$. □

3.1.10   **Example** Let us construct a field $\mathbb{F}$ of four elements. For this purpose we start with the field $\mathbb{Z}_2 = \{0,1\}$ of integers modulo 2. For sake of simplicity, we denote the elements as 0 and 1 rather than $\bar{0}$ and $\bar{1}$, as in Exercise 3.1.3. To construct $\mathbb{F}$, we need an irreducible polynomial of degree 2. It is easy to check that $f := 1 + x + x^2$ is irreducible over $\mathbb{Z}_2$.

The residue class ring $\mathbb{F} := \mathbb{Z}_2[x]/I(f)$ consists of the residue classes of the polynomials 0, 1, x and $1 + x$. Writing α instead of the residue class $x + I(f)$, we get

$$\mathbb{F} = \{0, 1, \alpha, 1 + \alpha\}.$$

Being a root of $f = 1 + x + x^2$, the element α satisfies the relation $\alpha^2 = 1 + \alpha$. Multiplication in $\mathbb{F}$ is done modulo this relation, which leads to the following addition and multiplication tables:

$+$	0	1	α	$1+\alpha$
0	0	1	α	$1+\alpha$
1	1	0	$1+\alpha$	α
α	α	$1+\alpha$	0	1
$1+\alpha$	$1+\alpha$	α	1	0

$\cdot$	0	1	α	$1+\alpha$
0	0	0	0	0
1	0	1	α	$1+\alpha$
α	0	α	$1+\alpha$	1
$1+\alpha$	0	$1+\alpha$	1	α

$\diamond$

It is worth noting that the construction of extension fields using irreducible polynomials can be applied to any field, not just to fields of prime order. The resulting residue class ring will always be a field. Conversely, it is true that any finite extension of a field can be obtained this way. Hence, for example, fields of order $p^{m \cdot n}$ can be constructed in different ways, using irreducible polynomials of degree $m \cdot n$ over $\mathbb{Z}_p$, or using irreducible polynomials of degree m over fields of order p^n, or using irreducible polynomials of degree n over fields of order p^m. Later on, however, we will see that the result will always be the same. Namely, we will show that there is up to isomorphism only one finite field of any given order p^n.

Exercises

E.3.1.1   **Exercise** A ring R is said to be free of *zero divisors* if the product of two nonzero elements of R is always nonzero. That is, if $r, s \in R \setminus \{0\}$ then $rs \neq 0$. Show that any field $\mathbb{F}$ and any polynomial ring $\mathbb{F}[x]$ with coefficients from a field $\mathbb{F}$ is free of zero divisors.

Exercise

— *Division Theorem* for integers: Prove that for integers a and $b \neq 0$ there exist uniquely determined integers q, r such that

$$a = qb + r \text{ with } 0 \leq r < |b|.$$

— *Euclidean Algorithm* for integers: Consider two integers a_1 and $a_2 \neq 0$ with $|a_1| \geq |a_2|$. Prove that the following sequence of calculations

$$
\begin{aligned}
a_1 &= q_1 a_2 + a_3 \text{ with } 0 < a_3 < |a_2|, \\
a_2 &= q_2 a_3 + a_4 \text{ with } 0 < a_4 < a_3,
\end{aligned}
$$

$$\cdots$$

$$
\begin{aligned}
a_{l-1} &= q_{l-1} a_l + a_{l+1} \text{ with } 0 < a_{l+1} < a_l, \\
a_l &= q_l a_{l+1}
\end{aligned}
$$

exists. The integer $d := |a_{l+1}|$ is the *greatest common divisor* of a_1 and a_2, for short $d = \gcd(a_1, a_2)$. This means that d is a divisor of both a_1 and a_2, every common divisor of a_1 and a_2 is a divisor of d, and d is positive. (The last mentioned property assures that d is uniquely determined.)

— *Bézout's Identity* for integers: Prove that $d = \gcd(a, b)$ is a linear combination of a and b, which means that there exist $r, s \in \mathbb{Z}$ such that

$$d = ra + sb.$$

Exercise For $n \geq 2$ we define a relation $\equiv_n$ on $\mathbb{Z}$ by

$$a \equiv b \bmod n :\Longleftrightarrow a \equiv_n b :\Longleftrightarrow n \mid a - b.$$

Prove the following statements:

— The relation $\equiv_n$ is an equivalence relation on $\mathbb{Z}$.

— The equivalence class of a with respect to $\equiv_n$, i.e. the set

$$\{b \in \mathbb{Z} \mid a \equiv b \bmod n\},$$

will be indicated by $\bar{a}$ and is called the *residue class* of a modulo n. Show that the set $\{0, 1, \ldots, n-1\}$ is a complete set of representatives with respect to $\equiv_n$. The elements $0, 1, \ldots, n-1$ of $\mathbb{Z}$ are usually called the *canonical representatives* of the classes $\bar{0}, \bar{1}, \ldots, \overline{n-1}$, respectively. The set

$$\mathbb{Z}_n := \{\bar{0}, \bar{1}, \ldots, \overline{n-1}\}$$

is called the *set of residue classes modulo n*.

— The relation $\equiv_n$ respects addition and multiplication, i.e. $\bar{a} = \bar{b}$ together with $\bar{c} = \bar{d}$ imply both $\overline{a+c} = \overline{b+d}$ and $\overline{a \cdot c} = \overline{b \cdot d}$.

— The set $\mathbb{Z}_n$ together with addition

$$\bar{a} + \bar{b} := \overline{a+b}, \qquad \bar{a}, \bar{b} \in \mathbb{Z}_n,$$

and multiplication

$$\bar{a} \cdot \bar{b} := \overline{a \cdot b}, \qquad \bar{a}, \bar{b} \in \mathbb{Z}_n,$$

is a commutative ring with identity element $\bar{1}$. This ring is called the *residue class ring modulo n*.

— The group of the (multiplicative) units of $\mathbb{Z}_n$ is

$$\mathbb{Z}_n^* = \{\bar{r} \in \mathbb{Z}_n \mid \gcd(n, r) = 1\}.$$

— The ring $\mathbb{Z}_n$ is a field if and only if n is a prime number.

E.3.1.4   **Exercise** Let R be a ring. Consider the set P of all sequences $(r_n)_{n \geq 0}$ with $r_n \in R$ for all $n \geq 0$, where only finitely many r_n are different from 0. Prove that this set forms a ring with addition and multiplication defined by

$$(r_n)_{n \geq 0} + (s_n)_{n \geq 0} = (r_n + s_n)_{n \geq 0}, \qquad (r_n)_{n \geq 0}, (s_n)_{n \geq 0} \in P,$$

$$(r_n)_{n \geq 0} \cdot (s_n)_{n \geq 0} = (t_n)_{n \geq 0}, \quad t_n = \sum_{i=0}^{n} r_i s_{n-i}, \ n \geq 0, \quad (r_n)_{n \geq 0}, (s_n)_{n \geq 0} \in P.$$

Furthermore show that

— P is commutative if and only if R is commutative;

— P is a ring with 1 if and only if R is a ring with 1;

— P is an integral domain if and only if R is an integral domain.

We identify the elements r of R with the sequences $(r, 0, 0, \ldots)$ in P. Consider the particular element $x = (0, 1, 0 \ldots) \in P$. If $r_n = 0$ for all $n \geq n_0$, then $(r_n)_{n \geq 0}$ describes the polynomial

$$\sum_{n=0}^{n_0} r_n x^n,$$

where x^n is the n-fold product of x introduced in Exercise 1.6.6. Show that $P = R[x]$, which means that P is the smallest ring containing R and x.

Exercise Prove the *degree formulae*: For any polynomials $f, g \in \mathbb{F}[x]$ we have   E.3.1.5

$$\deg(f + g) \leq \max\{\deg f, \deg g\}$$

and

$$\deg(fg) = \deg f + \deg g.$$

We use the convention $-\infty < n$ for all $n \in \mathbb{N}$, and

$$(-\infty) + n = n + (-\infty) = (-\infty) + (-\infty) = -\infty.$$

Exercise   E.3.1.6

— *Division Theorem* for polynomials: Prove that for any two polynomials $f, g \in \mathbb{F}[x]$ with $g \neq 0$ there exist (unique) polynomials $q, r \in \mathbb{F}[x]$ such that

$$f = qg + r \text{ with } \deg r < \deg g.$$

— *Euclidean Algorithm* for polynomials: Consider two polynomials $f_1, f_2 \in \mathbb{F}[x]$, $f_2 \neq 0$ with the property $\deg f_2 \leq \deg f_1$. Prove that there is a sequence of equations

$$
\begin{aligned}
f_1 &= q_1 f_2 + f_3 \text{ with } 0 \leq \deg f_3 < \deg f_2, \\
f_2 &= q_2 f_3 + f_4 \text{ with } 0 \leq \deg f_4 < \deg f_3,
\end{aligned}
$$

$$\cdots$$

$$
\begin{aligned}
f_{l-1} &= q_{l-1} f_l + f_{l+1} \text{ with } 0 \leq \deg f_{l+1} < \deg f_l, \\
f_l &= q_l f_{l+1}
\end{aligned}
$$

and some $\kappa \in \mathbb{F}$ so that the monic polynomial $d = \kappa \cdot f_{l+1}$ is the *greatest common divisor* of f_1 and f_2, for short $d = \gcd(f_1, f_2)$. This means that d is a divisor of both f_1 and f_2, each common divisor of f_1 and f_2 is a divisor of d, and d is monic. (The last mentioned property assures that d is uniquely determined.)

— *Bézout's Identity* for polynomials: Verify that the greatest common divisor d of two polynomials $f, g \in \mathbb{F}[x]$ can be written as a linear combination

$$d = rf + sg \text{ with } r, s \in \mathbb{F}[x].$$

Exercise Prove that for $\alpha \in \mathbb{F}$ the mapping   E.3.1.7

$$\mathbb{F}[x] \to \mathbb{F} : f \mapsto f(\alpha),$$

which takes the polynomial $f = \sum_{i=0}^n \kappa_i x^i$ to $f(\alpha) := \sum_{i=0}^n \kappa_i \alpha^i$ is a ring epimorphism.

E.3.1.8   **Exercise** Show that $\alpha \in \mathbb{F}$ is a root of $f \in \mathbb{F}[x]$ if and only if $(x - \alpha)$ is a divisor of f.

E.3.1.9   **Exercise** Assume that f is a monic, irreducible polynomial in $\mathbb{F}[x]$, and $\mathbb{L}$ is an extension of $\mathbb{F}$. Show that if $\alpha \in \mathbb{L}$ is a root of f, then f is a polynomial of smallest degree in $\mathbb{F}[x]$ such that $f(\alpha) = 0$. This polynomial is the minimal polynomial of α over $\mathbb{F}$.

Let g be any polynomial in $\mathbb{F}[x]$. Prove that $g(\alpha) = 0$ if and only if the minimal polynomial f of α over $\mathbb{F}$ is a divisor of g.

Assume that $f \in \mathbb{F}[x]$ is the minimal polynomial of $\alpha \in \mathbb{L}$. Show that if f has roots different from α in $\mathbb{L}$, then f is also the minimal polynomial of these roots over $\mathbb{F}$.

E.3.1.10   **Exercise** For an arbitrary monic polynomial $f \in \mathbb{F}[x]$ of degree $n \geq 0$, define a relation $\equiv_f$ on $\mathbb{F}[x]$ by

$$g \equiv h \bmod I(f) : \Longleftrightarrow g \equiv_f h : \Longleftrightarrow f \mid g - h \Longleftrightarrow g - h \in I(f).$$

Prove the following:

- The relation $\equiv_f$ is an equivalence relation on $\mathbb{F}[x]$.

- Let
$$\overline{g} := \{h \in \mathbb{F}[x] \mid g \equiv h \bmod I(f)\}$$
be the equivalence class of g with respect to $\equiv_f$. Let $\mathbb{F}[x]_f$ be the set of all equivalence classes $\overline{g}$ for $g \in \mathbb{F}[x]$. Show that
$$\mathbb{F}[x]_f = \{\overline{g} \mid g \in \mathbb{F}[x], \ \deg g < n\}.$$
The unique element g of $\overline{g}$ with $\deg g < n$ is called the *canonical representative* of $\overline{g} \in \mathbb{F}[x]_f$.

- The relation $\equiv_f$ respects addition and multiplication. That is, if $\overline{g} = \overline{h}$ and $\overline{u} = \overline{v}$ then we can deduce that $\overline{g + u} = \overline{h + v}$ and $\overline{g \cdot u} = \overline{h \cdot v}$.

- The set $\mathbb{F}[x]_f$ together with addition
$$\overline{g} + \overline{h} := \overline{g + h}, \qquad \overline{g}, \overline{h} \in \mathbb{F}[x]_f,$$
and multiplication
$$\overline{g} \cdot \overline{h} := \overline{g \cdot h}, \qquad \overline{g}, \overline{h} \in \mathbb{F}[x]_f,$$
forms a commutative ring with identity element $\overline{1}$.

— The group of (multiplicative) units of $\mathbb{F}[x]_f$ is

$$(\mathbb{F}[x]_f)^* = \{\bar{g} \in \mathbb{F}[x]_f \mid \gcd(f,g) = 1\}.$$

— The map

$$\mathbb{F}[x]/\mathrm{I}(f) \to \mathbb{F}[x]_f \ : \ g + \mathrm{I}(f) \mapsto \bar{g}$$

is a ring isomorphism.

— The ring $\mathbb{F}[x]_f$ is a field if and only if f is irreducible over $\mathbb{F}$.

Exercise Show that each ideal I of $\mathbb{F}[x]$ is a *principal ideal*, which means that it is generated by a single element,

$$I = \mathrm{I}(f) = \{fg \mid g \in \mathbb{F}[x]\}$$

for some $f \in \mathbb{F}[x]$.

E.3.1.11

Exercise Show that a polynomial $f \in \mathbb{F}[x]$ with $2 \le \deg f \le 3$ is irreducible over $\mathbb{F}$ if and only if f does not have a root in $\mathbb{F}$. Why is this characterization wrong for polynomials of degree ≥ 4? What about polynomials of degree 1?

E.3.1.12

3.2 Existence and Uniqueness of Finite Fields

3.2

Let us now consider the *multiplicative group* $\mathbb{F}^* := \mathbb{F} \setminus \{0\}$ of a finite field $\mathbb{F}$. We want to prove that this group is cyclic, i.e. generated by an element which is called a *primitive element* of $\mathbb{F}^*$. For example, the field $\mathbb{F}$ with four elements from 3.1.10 has the following nonzero elements:

$$1 = \alpha^0, \ \alpha, \ \alpha^2 = 1 + \alpha.$$

Thus, α is a primitive element for this field (as is $\alpha + 1$). We begin by introducing some terminology from Group Theory.

Let G be a multiplicatively written, finite group and let g be an element of G. Then there exists a positive integer k such that $g^k = 1$, since otherwise g^n with $n \in \mathbb{N}$ is an infinite number of powers of g. The smallest positive integer k with $g^k = 1$ is called the *order* of g in G, and is denoted as $\mathrm{ord}(g) := k$. Every $g \in G$ generates a subgroup of G which is denoted as $\langle g \rangle := \{1, g, \dots, g^{\mathrm{ord}(g)-1}\}$. Its order $|\langle g \rangle|$ is equal to $\mathrm{ord}(g)$. Hence, according to Lagrange's Theorem (cf. Exercise 1.4.6), the order of any element of G is a divisor of the order of G, i.e.

$$\mathrm{ord}(g) \mid |G|.$$

3.2.1

If we write $|G| = m \cdot \mathrm{ord}(g)$, for some integer m which depends on g, then this means that $g^{|G|} = (g^{\mathrm{ord}(g)})^m = 1$ for all $g \in G$. In particular, since the multiplicative group of a finite field $\mathbb{F}$ has order $q - 1$, we get that $\kappa^{q-1} = 1$ for all $\kappa \in \mathbb{F}^*$. Equivalently, we have

3.2.2
$$\kappa^q = \kappa \quad \text{for all } \kappa \in \mathbb{F}.$$

This means that each element $\kappa \in \mathbb{F}$ is a root of the polynomial $x^q - x \in \mathbb{F}[x]$. Therefore, $x - \kappa$ divides $x^q - x$ for every $\kappa \in \mathbb{F}$, and thus

3.2.3
$$x^q - x = \prod_{\kappa \in \mathbb{F}}(x - \kappa),$$

since both polynomials are monic of the same degree and since $\prod_{\kappa \in \mathbb{F}}(x - \kappa)$ divides the polynomial on the left hand side by the above argument.

Let us discuss some further properties of the polynomial ring $\mathbb{F}[x]$.

3.2.4   **Lemma** *If an irreducible polynomial $f \in \mathbb{F}[x]$ divides a product $f_0 \cdots f_{m-1}$ of polynomials in $\mathbb{F}[x]$, then f is a divisor of at least one f_j.*

Proof: Applying the mapping

$$\pi: \mathbb{F}[x] \to \mathbb{F}[x]/\mathrm{I}(f) \; : \; \pi(h) = h + \mathrm{I}(f), \quad h \in \mathbb{F}[x],$$

which is a ring homomorphism, to the product $f_0 \cdots f_{m-1}$ we get

$$(f_0 + \mathrm{I}(f)) \cdots (f_{m-1} + \mathrm{I}(f)) = 0 + \mathrm{I}(f),$$

since f is a divisor of $f_0 \cdots f_{m-1}$. According to Exercise 3.1.10 the factor ring $\mathbb{F}[x]/\mathrm{I}(f)$ is a field, whence it does not have any zero divisors. Consequently, there exists at least one $j \in m$ such that $f_j + \mathrm{I}(f) = 0 + \mathrm{I}(f)$, which means that f is a divisor of f_j. $\square$

3.2.5   **Unique factorization in $\mathbb{F}[x]$** *Any polynomial $f \in \mathbb{F}[x]$ of positive degree can be written as*

$$f = \kappa f_0^{n_0} \cdots f_{k-1}^{n_{k-1}},$$

where $\kappa \in \mathbb{F} \setminus \{0\}$, $k \in \mathbb{N}^*$, $f_0, \ldots, f_{k-1}$ are distinct monic, irreducible polynomials in $\mathbb{F}[x]$, and $n_0, \ldots, n_{k-1}$ are positive integers. This factorization is unique apart from the order in which the factors occur.

Proof: The first assertion of this theorem is proved by induction on the degree of f. If $\deg f = 1$, then f is irreducible and its leading coefficient κ is different from 0. So $f = \kappa(\kappa^{-1}f)$, where $\kappa^{-1}f$ is a monic, irreducible polynomial. Now let $n > 1$ and assume that the desired factorization into irreducible polynomials is possible for all nonconstant polynomials of degree strictly less than n. If $\deg f = n$ and f is irreducible we can proceed as in the case of a polynomial of degree 1. Then $f = \kappa(\kappa^{-1}f)$ is the factorization of f. Otherwise, f is reducible, thus there exist $g, h \in \mathbb{F}[x]$, both of degree at least 1 and less than n, such that $f = gh$. By induction hypothesis, g and h can be factored into $g = \kappa_g g_0 \cdots g_{k_g-1}$ and $h = \kappa_h h_0 \cdots h_{k_h-1}$ with monic, irreducible polynomials $g_0, \ldots, g_{k_g-1}, h_0, \ldots, h_{k_h-1}$ and with $\kappa_g, \kappa_h \neq 0$. Thus, $f = (\kappa_g \kappa_h) g_0 \cdots g_{k_g-1} h_0 \cdots h_{k_h-1}$ is the desired factorization of f.

In order to prove uniqueness, assume that f has two factorizations of the form

$$f = \kappa_f f_0^{n_0} \cdots f_{k-1}^{n_{k-1}} = \kappa_g g_0^{m_0} \cdots g_{\ell-1}^{m_{\ell-1}}$$

into monic, irreducible polynomials $f_0, \ldots, f_{k-1}$ and $g_0, \ldots, g_{\ell-1}$, elements $\kappa_f, \kappa_g \in \mathbb{F} \setminus \{0\}$ and positive integers $n_0, \ldots, n_{k-1}$ and $m_0, \ldots, m_{\ell-1}$. Moreover, we may assume that $f_i \neq f_j$ for $i \neq j$ and $g_i \neq g_j$ for $i \neq j$. Comparing the leading coefficients of the two factorizations yields $\kappa_f = \kappa_g$. The irreducible polynomial f_0 is a divisor of $g_0^{m_0} \cdots g_{\ell-1}^{m_{\ell-1}}$. Hence by 3.2.4, there exists $j \in \ell$ such that f_0 divides g_j. Since g_j is irreducible, $g_j = \kappa f_0$ for some $\kappa \in \mathbb{F}$. Moreover, f_0 and g_j are monic, whence $\kappa = 1$ and $f_0 = g_j$. Thus, one occurrence of f_0 can be canceled with one occurrence of g_j. Proceeding in this way, we can cancel each term f_i on the left hand side with a corresponding factor g_j of equal degree on the right hand side. After $n_0 + \ldots + n_{k-1}$ such steps we arrive at a constant on the left hand side. Therefore, the right hand side must also be a constant. Thus we see that both sides of the equation show the same irreducible factors, possibly in different orders.   □

Now we take into account that the left hand side of 3.2.3 is a polynomial over $\mathbb{Z}_p$. According to 3.2.5, we decompose it into its monic, irreducible factors over $\mathbb{Z}_p$. Then 3.2.3 shows that each κ is a root of *exactly one* of these irreducible and monic factors.

Now we prove the main theorem on the multiplicative group of a finite field.

Theorem *The multiplicative group $\mathbb{F}^*$ of a finite field $\mathbb{F}$ is cyclic.*   3.2.6

Proof: We assume that $\mathbb{F}$ is a finite field with $q = p^n$ elements for some prime p and some positive integer n. Among the nonzero elements of $\mathbb{F}$, let β be an element of maximal order. We claim that β generates the multiplicative group $\mathbb{F}^*$ or, equivalently, that the order of β is $q - 1$.

1. We prove that

$$\mathrm{ord}(\kappa) \mid \mathrm{ord}(\beta)$$ 3.2.7

for each $\kappa \in \mathbb{F}^*$. Suppose on the contrary that there exists some $\kappa \in \mathbb{F}^*$ such that $\mathrm{ord}(\kappa)$ does *not* divide $\mathrm{ord}(\beta)$. Then there exists a prime u and an integer

$i \geq 0$ as well as further positive integers r, s, for which

$$\text{ord}(\beta) = u^i s \text{ and } \text{ord}(\kappa) = u^{i+1} r,$$

where u does not divide s. In particular,

$$(\beta^{u^i})^s = \beta^{u^i s} = 1 \text{ and } (\kappa^r)^{u^{i+1}} = \kappa^{u^{i+1} r} = 1.$$

According to Exercise 3.2.1, this implies that

$$\text{ord}(\beta^{u^i}) = s \text{ and } \text{ord}(\kappa^r) = u^{i+1}.$$

Consequently, $\text{ord}(\beta^{u^i})$ and $\text{ord}(\kappa^r)$ are relatively prime. Since $\mathbb{F}^*$ is abelian, we have $\text{ord}(\beta^{u^i}\kappa^r) = \text{ord}(\beta^{u^i})\text{ord}(\kappa^r) = su^{i+1} > su^i = \text{ord}(\beta)$ (see Exercise 3.2.2). This contradicts the maximality of $\text{ord}(\beta)$.

2. Summarizing, for $l = \text{ord}(\beta)$, we have shown that $\kappa^l = 1$ for each $\kappa \in \mathbb{F}^*$. This shows that every $\kappa \in \mathbb{F}^*$ is a root of the polynomial $x^l - 1 \in \mathbb{Z}_p[x]$, i.e., $x - \kappa$ divides $x^l - 1$. Thus

$$\prod_{\kappa \in \mathbb{F}^*} (x - \kappa) \,\Big|\, x^l - 1.$$

Comparing degrees leads to $l \geq |\mathbb{F}^*| = q - 1$. On the other hand, the order of any group element divides the group order (see 3.2.1) which gives $l \mid q - 1$. This shows that $l = \text{ord}(\beta) = q - 1$. As remarked above, this means that $\mathbb{F}^*$ is cyclic. Namely, it is generated by β.   □

3.2.8   **Definition (primitive element)** Generators of the multiplicative group of the finite field $\mathbb{F}$ are called *primitive elements* of $\mathbb{F}$. The minimal polynomials of primitive elements are called *primitive polynomials*.   ◇

3.2.9   **Example** Let us construct a field with 16 elements. For this purpose, we choose the irreducible polynomial $f = x^4 + x + 1 \in \mathbb{Z}_2[x]$. Let α denote the root $x + I(f)$ of f. Then $\alpha^4 = \alpha + 1$, and α turns out to be a primitive element of the field $\mathbb{F} := \mathbb{Z}_2[x]/I(f)$ of 16 elements. This follows from Table 3.1, since the powers of α run through the set of all nonzero elements of $\mathbb{F}$. This table also shows the vectors of coefficients of the nonzero elements with respect to the basis $(1, \alpha, \alpha^2, \alpha^3)$.

  If we construct a field of 16 elements by using the irreducible polynomial $g = x^4 + x^3 + x^2 + x + 1 \in \mathbb{Z}_2[x]$, and if β denotes one of the roots of g, then we have $\beta^4 + \beta^3 + \beta^2 + \beta + 1 = 0$. This gives $\beta^5 = 1$, and so β is *not* a primitive element of $\mathbb{L} := \mathbb{Z}_2[x]/I(g)$ and, hence, g is *not* a primitive polynomial.   ◇

Table 3.1 The 15 elements $\neq 0$ of $\mathbb{Z}_2[x]/\mathrm{I}(1+x+x^4)$

$\alpha^0 = 1$	$(1,0,0,0)$
$\alpha^1 = \alpha$	$(0,1,0,0)$
α^2	$(0,0,1,0)$
α^3	$(0,0,0,1)$
$\alpha^4 = \alpha + 1$	$(1,1,0,0)$
$\alpha^5 = \alpha^2 + \alpha$	$(0,1,1,0)$
$\alpha^6 = \alpha^3 + \alpha^2$	$(0,0,1,1)$
$\alpha^7 = \alpha^3 + \alpha + 1$	$(1,1,0,1)$
$\alpha^8 = \alpha^2 + 1$	$(1,0,1,0)$
$\alpha^9 = \alpha^3 + \alpha$	$(0,1,0,1)$
$\alpha^{10} = \alpha^2 + \alpha + 1$	$(1,1,1,0)$
$\alpha^{11} = \alpha^3 + \alpha^2 + \alpha$	$(0,1,1,1)$
$\alpha^{12} = \alpha^3 + \alpha^2 + \alpha + 1$	$(1,1,1,1)$
$\alpha^{13} = \alpha^3 + \alpha^2 + 1$	$(1,0,1,1)$
$\alpha^{14} = \alpha^3 + 1$	$(1,0,0,1)$

Theorem *Any two finite fields of the same order are isomorphic.* **3.2.10**

Proof: Let $\mathbb{L}$ and $\mathbb{F}$ denote two finite fields with p^n elements. The prime fields of $\mathbb{F}$ and $\mathbb{L}$ are isomorphic to $\mathbb{Z}_p$, so we identify them with $\mathbb{Z}_p$. Consider $\mathbb{F}$. If α denotes a primitive element of $\mathbb{F}$, and f_α is its minimal polynomial over $\mathbb{Z}_p$, then f_α is a divisor of $x^{p^n} - x$ of degree n (cf. Exercise 3.1.9). Now we apply to the polynomial ring $\mathbb{Z}_p[x]$ the evaluation at α (cf. Exercise 3.1.7). Because of Exercise 3.1.9, the kernel of this ring epimorphism is the ideal $\mathrm{I}(f_\alpha)$. By the Homomorphism Theorem (Exercise 3.2.3), we obtain the *isomorphism*

$$\mathbb{Z}_p[x]/\mathrm{I}(f_\alpha) \simeq \mathbb{F} \ : \ g + \mathrm{I}(f_\alpha) \mapsto g(\alpha).$$

We now turn to $\mathbb{L}$. According to 3.2.2, the primitive element $\alpha \in \mathbb{F}$ is a root of $x^{p^n} - x \in \mathbb{Z}_p[x]$. On the other hand, from 3.2.3 we deduce that

$$x^{p^n} - x = \prod_{\lambda \in \mathbb{L}} (x - \lambda).$$

Recall from 3.2.5 that the decomposition of a monic polynomial into the product of monic, irreducible factors is unique up to ordering of the factors. Since f_α divides the left hand side of this equation, it has an element β of $\mathbb{L}$ as one of its roots. The fact that f_α is monic and irreducible then implies that f_α is the minimal polynomial of $\beta \in \mathbb{L}$ over $\mathbb{Z}_p$ (Exercise 3.1.9). Using the same arguments as before, we obtain $\mathbb{Z}_p[x]/\mathrm{I}(f_\alpha) \simeq \mathbb{L}$, which shows that the fields $\mathbb{F}$ and $\mathbb{L}$ are indeed isomorphic. □

Since finite fields are uniquely determined by their order up to an isomorphism, we can denote the finite field containing p^n elements by $\mathbb{F}_{p^n}$. Often such fields are also indicated as $\mathrm{GF}(p^n)$, where GF stands for *Galois field*.

3.2.11   **Theorem** *The field $\mathbb{F}_{p^m}$ is a subfield of $\mathbb{F}_{p^n}$ if and only if m is a divisor of n.*

Proof: Let m be a divisor of n, say $n = md$, for a suitable $d \in \mathbb{N}^*$. Then, in each ring R we have for every $r \in R$ the identity

$$r^n - 1 = (r^m)^d - 1 = (r^m - 1)(r^{(d-1)m} + \ldots + r^m + 1).$$

If we choose $R = \mathbb{Z}$, then from $m \mid n$ we obtain $p^m - 1 \mid p^n - 1$. In the case $R = \mathbb{F}_p[x]$, we get from $m \mid n$ directly that $x^m - 1 \mid x^n - 1$; in particular, $p^m - 1 \mid p^n - 1$ implies that $x^{p^m-1} - 1$ divides $x^{p^n-1} - 1$. Hence, it follows from 3.2.3 that $\mathbb{F}_{p^m}$ is a subfield of $\mathbb{F}_{p^n}$.

Conversely, assume that $\mathbb{F}_{p^m}$ is a subfield of $\mathbb{F}_{p^n}$. Then $\mathbb{F}_{p^n}$ is a finite dimensional vector space over $\mathbb{F}_{p^m}$, say of dimension d. Then $p^n = (p^m)^d = p^{md}$, and so $n = md$. □

For example, $\mathbb{F}_{16} = \mathbb{F}_{2^4}$, constructed in 3.2.9, contains the subfield $\mathbb{F}_4 = \mathbb{F}_{2^2} = \{0, 1, \alpha^5, \alpha^{10}\}$, but *not* the field $\mathbb{F}_8 = \mathbb{F}_{2^3}$, since 3 does not divide 4.

For $\alpha, \beta \in \mathbb{F}_{p^n}$ and $t \in \mathbb{N}$, the *binomial formula*, is

$$(\alpha + \beta)^t = \sum_{i=0}^{t} \binom{t}{i} \alpha^{t-i} \beta^i,$$

where the binomial coefficients mean positive integers, while the powers of α and β are elements of the finite field. The i-th summand is the $\binom{t}{i}$-fold sum (cf. Exercise 1.6.6) of $\alpha^{t-i}\beta^i$,

$$\binom{t}{i} \alpha^{t-i} \beta^i := \underbrace{\alpha^{t-i}\beta^i + \ldots + \alpha^{t-i}\beta^i}_{\binom{t}{i} \text{ times}}.$$

If $t = p$, the characteristic of $\mathbb{F}_{p^n}$, then, since p divides the binomial coefficient $\binom{p}{i}$ for $1 \leq i \leq p - 1$ (Exercise 3.2.4), we deduce by using 3.1.2 that the middle terms vanish, so that for fields of characteristic p we have

$$(\alpha + \beta)^p = \alpha^p + \beta^p.$$

This is the binomial theorem for fields of characteristic p. By induction on m, this implies that

3.2.12
$$(\alpha + \beta)^{p^m} = \alpha^{p^m} + \beta^{p^m}.$$

Furthermore, $(\alpha\beta)^p = \alpha^p \beta^p$. Consequently, the *Frobenius mapping*

$$\sigma : \mathbb{F}_{p^n} \to \mathbb{F}_{p^n} \ : \ \alpha \mapsto \alpha^p$$

is a homomorphism. Its kernel is an ideal in $\mathbb{F}_{p^n}$. Each ideal in a field is trivial, and so the Frobenius mapping is either the zero map, or its kernel is trivial, i.e. consists only of the zero element. The first case cannot happen since fields have no zero divisors. Hence the Frobenius mapping is injective, and therefore also surjective, i.e. bijective. Thus we have proved:

Theorem *The Frobenius mapping* $\sigma \colon \mathbb{F}_{p^n} \to \mathbb{F}_{p^n} \ \colon \ \alpha \mapsto \alpha^p$ *is an automorphism,* 3.2.13
the Frobenius automorphism. $\qquad\qquad\qquad\qquad\qquad\qquad\qquad\qquad\qquad\qquad\qquad\qquad$ $\square$

Since $\sigma(\alpha) = \alpha^p = \alpha$ for all α in the prime field $\mathbb{F}_p$, the Frobenius automorphism fixes each element of $\mathbb{F}_p$. We call this kind of automorphisms an *automorphism of* $\mathbb{F}_{p^n}$ *over* $\mathbb{F}_p$. In order to stress that the Frobenius automorphism σ fixes the elements of $\mathbb{F}_p$, it is also called the *Frobenius automorphism over* $\mathbb{F}_p$.

Example The Frobenius automorphism 3.2.14

$$\sigma \colon \mathbb{F}_{2^4} \to \mathbb{F}_{2^4} \ \colon \ \kappa \mapsto \kappa^2$$

maps 0 onto itself, and the nonzero elements of $\mathbb{F}_{2^4}$, represented as powers of the primitive element α (see, e.g. 3.2.9), to powers of α:

$$
\begin{aligned}
&0 \mapsto 0, \\
&1 \mapsto 1, \\
&\alpha \mapsto \alpha^2 \mapsto \alpha^4 \mapsto \alpha^8, \\
&\alpha^3 \mapsto \alpha^6 \mapsto \alpha^{12} \mapsto \alpha^9, \\
&\alpha^5 \mapsto \alpha^{10}, \\
&\alpha^7 \mapsto \alpha^{14} \mapsto \alpha^{13} \mapsto \alpha^{11}.
\end{aligned}
$$

The last element in each of these rows is mapped by the Frobenius automorphism onto the first element in that row: $0 \mapsto 0$, $1 \mapsto 1$, $\alpha^8 \mapsto \alpha$, $\alpha^9 \mapsto \alpha^3$, and so on. The elements which are kept fix by the automorphism form the prime field $\mathbb{F}_2 = \{0,1\}$.

This shows that the Frobenius automorphism σ gives rise to a *permutation* $\bar{\sigma}$ of the elements in $\mathbb{F}_{2^4}$. It *induces*, as we say, the following permutation whose *cyclic factors* contain the elements of the above rows in the same order as they occur in these rows, i.e.

$$\bar{\sigma} = (0)(1)(\alpha, \alpha^2, \alpha^4, \alpha^8)(\alpha^3, \alpha^6, \alpha^{12}, \alpha^9)(\alpha^5, \alpha^{10})(\alpha^7, \alpha^{14}, \alpha^{13}, \alpha^{11}). \qquad \diamond$$

Assume that $\mathbb{F}_{q^n}$ is an extension field of $\mathbb{F}_q$. In a similar way the Frobenius automorphism of $\mathbb{F}_{q^n}$ over $\mathbb{F}_q$ can be introduced. If $q = p^r$, then the Frobenius automorphism σ over $\mathbb{F}_p$ must be replaced by its r-th power $\tau := \sigma^r$, $\kappa \mapsto \kappa^q = \kappa^{(p^r)} = \kappa^{p^r}$ in order to obtain the Frobenius automorphism of $\mathbb{F}_{q^n}$ over $\mathbb{F}_q$.

The main property of the Frobenius automorphism τ is that it maps roots of polynomials with coefficients in $\mathbb{F}_q$ onto roots of the same polynomial: Assume that $\beta \in \mathbb{F}_{q^n}$ is a root of the polynomial $f = \sum_{i=0}^{m} \kappa_i x^i \in \mathbb{F}_q[x]$. Using the fact that $\tau(0) = 0$ we deduce

3.2.15
$$0 = \tau\left(\sum_{i=0}^{m} \kappa_i \beta^i\right) = \sum_{i=0}^{m} \tau(\kappa_i \beta^i) = \sum_{i=0}^{m} \tau(\kappa_i)\tau(\beta^i) = \sum_{i=0}^{m} \kappa_i (\tau(\beta))^i.$$

From 3.2.3 and the fact that a polynomial of degree n has at most n roots we immediately obtain

3.2.16 **Corollary** *Let τ be the Frobenius automorphism of $\mathbb{F}_{q^n}$ over $\mathbb{F}_q$ and $\kappa \in \mathbb{F}_{q^n}$. Then $\tau(\kappa) = \kappa$ if and only if $\kappa \in \mathbb{F}_q$.* $\square$

If $\mathbb{F}$ is a field and α belongs to an extension field $\mathbb{L}$, then $\mathbb{F}(\alpha)$ indicates the smallest subfield $\mathbb{K}$ of $\mathbb{L}$ which contains both $\mathbb{F}$ and α. For $\alpha_0, \ldots, \alpha_{n-1} \in \mathbb{L}$, $n > 1$, we define recursively $\mathbb{F}(\alpha_0, \ldots, \alpha_{n-1}) := \mathbb{F}(\alpha_0, \ldots, \alpha_{n-2})(\alpha_{n-1})$.

For arbitrary fields $\mathbb{F}$ and $\mathbb{L}$ we can prove the

3.2.17 **Embedding Theorem** *Let $\psi \colon \mathbb{F} \to \mathbb{L}$ be an embedding, i.e. an injective homomorphism. Let x be an indeterminate over $\mathbb{F}$ and $\mathbb{L}$. Consider an irreducible polynomial $f = \sum_{i=0}^{n} f_i x^i \in \mathbb{F}[x]$ and a field extension $\mathbb{K}$ over $\mathbb{F}$. Assume that $\alpha \in \mathbb{K}$ is a root of f, and that $\beta \in \mathbb{L}$ is a root of $\psi(f) := \sum_{i=0}^{n} \psi(f_i) x^i \in \mathbb{L}[x]$. Then there exists an embedding $\varphi \colon \mathbb{F}(\alpha) \to \mathbb{L}$ such that $\varphi(\alpha) = \beta$, and φ restricted to $\mathbb{F}$ is equal to ψ. Moreover, φ is an isomorphism between $\mathbb{F}(\alpha)$ and $\psi(\mathbb{F})(\beta)$.*

Proof: According to Exercise 3.2.7, the polynomial $\psi(f)$ is irreducible over $\psi(\mathbb{F})$. The elements of $\mathbb{F}(\alpha)$ are of the form $g(\alpha)$ for some $g \in \mathbb{F}[x]$. Therefore, we define $\varphi(g(\alpha)) := \psi(g)(\beta)$. In order to prove that φ is well-defined and injective we assume that $g_1(\alpha) = g_2(\alpha)$ for $g_1, g_2 \mathbb{F}[x]$. Then equivalently we have:

- $g_1 - g_2 = fh$ for some $h \in \mathbb{F}[x]$.
- $\psi(g_1) - \psi(g_2) = \psi(f)\psi(h)$ for some $h \in \mathbb{F}[x]$.
- $\psi(g_2)(\beta) = \psi(g_1)(\beta) - (\psi(g_1)(\beta) - \psi(g_2)(\beta))$
   $= \psi(g_1)(\beta) - \psi(f)(\beta)\psi(h)(\beta) = \psi(g_1)(\beta).$

Reading these implications from the top to the bottom, we deduce that φ is well-defined. Reading them the other way round we find that φ is injective.

It is clear by definition that $\varphi(\kappa) = \psi(\kappa)$ for $\kappa \in \mathbb{F}$. Moreover, it is easy to verify that φ is a field homomorphism.

Finally, φ is an isomorphism between $\mathbb{F}(\alpha)$ and $\psi(\mathbb{F})(\beta)$, since each element γ of $\psi(\mathbb{F})(\beta)$ is of the form $h(\beta)$ for some $h \in \psi(\mathbb{F})[x]$ of degree less than $\deg f = \deg \psi(f)$. Let $g = \psi^{-1}(h)$, then $\deg g < \deg f$ and γ is equal to $\varphi(g(\alpha))$, whence it belongs to $\varphi(\mathbb{F}(\alpha))$. $\square$

In 3.2.5 we have described the decomposition of a nonconstant polynomial in irreducible factors.

Definition (splitting field) A field extension $\mathbb{L}$ of $\mathbb{F}$ is called a *splitting field* of   3.2.18
a nonconstant polynomial $f \in \mathbb{F}[x]$ if all factors in the decomposition of f into irreducible factors over $\mathbb{L}$ are of degree one, and if f does not factor completely into linear factors over any proper subfield of $\mathbb{L}$. We also say that f splits over $\mathbb{L}$ into linear factors. $\diamond$

It will be shown in 3.2.22 that the splitting field of f over $\mathbb{F}$ is unique up to isomorphism. We collect the most important results about minimal polynomials over finite fields, their roots, and their splitting fields in the next

---

Theorem *Assume that $f \in \mathbb{F}_q[x]$ is a monic, irreducible polynomial of degree $n > 0$,*   3.2.19
and let τ be the Frobenius automorphism $\tau \colon \mathbb{F}_{q^n} \to \mathbb{F}_{q^n}$ given by $\tau(\kappa) = \kappa^q$ for all
$\kappa \in \mathbb{F}_{q^n}$. Then

1. *f has a root in $\mathbb{F}_{q^n}$.*

2. *All roots of f are simple and belong to $\mathbb{F}_{q^n}$. If α denotes one of the roots of f, then the n distinct elements $\alpha, \alpha^q = \tau(\alpha), \ldots, \alpha^{q^{n-1}} = \tau^{n-1}(\alpha)$ comprise the set of all roots of f.*

3. *$\mathbb{F}_{q^n}$ is a splitting field of f. Any two splitting fields of f are isomorphic.*

Proof: In 3.1.8 we have proved that $\alpha = x + I(f)$ is a root of f. According to 3.1.6 and 3.1.7, α is contained in the field $\mathbb{F}_q[x]/I(f)$ which is of cardinality q^n. This field is isomorphic to $\mathbb{F}_{q^n}$ by 3.2.10. Moreover, $\mathbb{F}_{q^n}$ is the smallest extension of $\mathbb{F}_q$ which contains α because of 3.1.7.

Since α belongs to $\mathbb{F}_{q^n}$, it is a root of $g(x) = x^{q^n} - x$, so f, being the minimal polynomial of α, is a divisor of g. From 3.2.3 it follows that $\mathbb{F}_{q^n}$ is a splitting field of g, whence all roots of f belong to $\mathbb{F}_{q^n}$, and $\mathbb{F}_{q^n}$ is a splitting field of f. Since g is of degree $q^n = |\mathbb{F}_{q^n}|$, all roots of g and therefore also all roots of f are simple.

According to 3.2.15, $\tau(\alpha)$ is a root of f. It suffices to prove that $\tau^j(\alpha)$ for $j \in n$ are n pairwise distinct elements. Now assume, on the contrary, that $\tau^j(\alpha) = \tau^k(\alpha)$ for $0 \le j < k < n$. Then $\tau^{n-k}(\tau^j(\alpha)) = \tau^{n-k}(\tau^k(\alpha))$, thus $\alpha^{q^{n-k+j}} = \alpha^{q^n} = \alpha$. Consequently, the minimal polynomial f of α is a divisor of $x^{q^{n-k+j}} - x$ and $\alpha \in \mathbb{F}_{q^{n-k+j}}$. Since $\mathbb{F}_{q^n}$ is the smallest field extension of $\mathbb{F}_q$ containing α, it must be included in $\mathbb{F}_{q^{n-k+j}}$, whence $q^n \le q^{n-k+j}$ and $k \le j$

which is a contradiction. Thus, for $j \in n$ the powers $\tau^j(\alpha)$ yield all the n different roots of f.

Assume that $\mathbb{L}$ is a field extension of $\mathbb{F}_q$ and let $\beta_0, \ldots, \beta_{n-1} \in \mathbb{L}$ be the roots of f. We want to prove that there exists a field isomorphism φ from $\mathbb{F}_q(\alpha)$ to $\mathbb{F}_q(\beta_0, \ldots, \beta_{n-1})$ with $\varphi(\alpha) = \beta_0$ such that the restriction of φ to $\mathbb{F}_q$ is the identity. By 3.2.17, there exists an isomorphism $\varphi \colon \mathbb{F}_q(\alpha) \to \mathbb{F}_q(\beta_0)$. Since $\varphi(\tau^j(\alpha)) = \tau^j(\beta_0)$, $j \in n$, are the n distinct roots of f, we obtain that $\mathbb{F}_q(\beta_0, \ldots, \beta_{n-1}) = \mathbb{F}_q(\beta_0)$. This shows that any two splitting fields of the irreducible polynomial f are isomorphic. □

3.2.20   **Theorem** *For every prime power p^n, the polynomial $x^{p^n} - x$ is the product of all monic, irreducible polynomials over $\mathbb{F}_p$ whose degree divides n.*

Proof: We first show that only polynomials whose degree divides n can arise as factors. Let $f \in \mathbb{F}_p[x]$ be a monic, irreducible polynomial of degree m dividing $x^{p^n} - x$. By 3.2.3, f has a root α in $\mathbb{F}_{p^n}$, and f is the minimal polynomial of α over $\mathbb{F}_p$. We know that the field $\mathbb{F}_{p^m}$ consists of the p^m linear combinations $a_0 + a_1\alpha + \ldots + a_{m-1}\alpha^{m-1}$ over $\mathbb{F}_p$. Since α is contained in $\mathbb{F}_{p^n}$, each of these linear combinations is also contained in $\mathbb{F}_{p^n}$. Hence, $\mathbb{F}_{p^m}$ is a subfield of $\mathbb{F}_{p^n}$. From 3.2.11 we obtain that m divides n.

Now we are going to verify that each of these polynomials in fact occurs as a factor of $x^{p^n} - x$. Let $f \in \mathbb{F}_p[x]$ be a monic, irreducible polynomial of degree m, where m divides n. Then, by 3.2.19, each root of f is an element of $\mathbb{F}_{p^m}$, and so it is also a root of $x^{p^m} - x$ according to 3.2.3. This implies that f is a divisor of $x^{p^m} - x$ (Exercise 3.1.9). Since n is a multiple of m, $x^{p^m} - x$ is a divisor of $x^{p^n} - x$ (cf. the proof of 3.2.11). Thus, f is also a divisor of $x^{p^n} - x$. Moreover, f is the minimal polynomial of each of its roots.

As each element of $\mathbb{F}_{p^n}$ has a minimal polynomial over $\mathbb{F}_p$, the product of all these monic, irreducible polynomials is $x^{p^n} - x$. □

3.2.21   **Example** The polynomial $x^{16} - x = x^{16} + x \in \mathbb{F}_2[x]$ has the following decomposition over $\mathbb{F}_2[x]$:

$$x^{16} + x = x(x+1)(x^2 + x + 1)(x^4 + x + 1)(x^4 + x^3 + 1)(x^4 + x^3 + x^2 + x + 1).$$

◇

We now prove the announced result that the splitting field of a nonconstant polynomial is essentially unique.

3.2.22   **Theorem** *Any nonconstant polynomial $f \in \mathbb{F}_q[x]$ has a splitting field. Any two splitting fields of f are isomorphic.*

Proof: 1. According to 3.2.5 there exists a unique factorization of f into irreducible polynomials $f_i \in \mathbb{F}_q[x]$ of the form

$$f = \kappa \prod_{i \in k} f_i^{r_i}$$

with $\kappa \in \mathbb{F}_q$, $r_i \geq 1$ and $\deg f_i = n_i$, $i \in k$. A splitting field of the irreducible polynomial f_i is $\mathbb{F}_{q^{n_i}}$ by 3.2.19. Therefore, a splitting field of f is the smallest field which contains $\bigcup_{i \in k} \mathbb{F}_{q^{n_i}}$. By 3.2.11, this is the field $\mathbb{F}_{q^N}$ for $N = \mathrm{lcm}(n_0, \ldots, n_{k-1})$.

2. If $\mathbb{L}$ is a splitting field of f, then it is a finite field isomorphic to $\mathbb{F}_{q^M}$ for some $M \in \mathbb{N}$. It contains the splitting fields of each irreducible divisor f_i. According to 3.2.19, they are isomorphic to $\mathbb{F}_{q^{n_i}}$, whence $\mathbb{L}$ contains $\mathbb{F}_{q^N}$ and $N := \mathrm{lcm}(n_0, \ldots, n_{k-1})$ is a divisor of M. Since $\mathbb{L}$ is the smallest field extension where f splits into linear factors, $\mathbb{L}$ is isomorphic to $\mathbb{F}_{q^N}$, and any two splitting fields of f are isomorphic. □

Let $\mathbb{L}$ be an extension field of $\mathbb{K}$. The element $\alpha \in \mathbb{L}$ is *algebraic* over $\mathbb{K}$ if there exists a nonzero polynomial $f \in \mathbb{K}[x]$ so that $f(\alpha) = 0$. The field $\mathbb{L}$ is *algebraic* over $\mathbb{K}$ if each element of $\mathbb{L}$ is algebraic over $\mathbb{K}$. For instance each finite extension is algebraic (Exercise 3.2.8). Thus, the splitting field of $f \in \mathbb{K}[x] \setminus \mathbb{K}$ is algebraic over $\mathbb{K}$. Moreover, if β is contained in an algebraic extension field of $\mathbb{L}$ and if $\mathbb{L}$ is algebraic over $\mathbb{K}$, then β is algebraic over $\mathbb{K}$ (cf. Exercise 3.2.11).

A field $\mathbb{F}$ is *algebraically closed* if every nonconstant polynomial over $\mathbb{F}$ has a root in $\mathbb{F}$. It then follows easily that every nonconstant polynomial $f \in \mathbb{F}[x]$ splits over $\mathbb{F}$ in linear factors.

Let $\mathbb{K}$ be a field. Any field $\mathbb{L}$ containing $\mathbb{K}$ which is algebraically closed and algebraic over $\mathbb{K}$ is called *algebraic closure* of $\mathbb{K}$. We will show that any finite field $\mathbb{F}_q$ possesses an algebraic closure and that this algebraic closure is essentially unique.

Theorem *There exists an algebraic closure of* $\mathbb{F}_q$. *If* $\mathbb{K}$ *and* $\mathbb{L}$ *are two algebraic closures of* $\mathbb{F}_q$, *then there exists an isomorphism* $\varphi \colon \mathbb{K} \to \mathbb{L}$ *so that* φ *restricted to* $\mathbb{F}_q$ *is the identity.*　　　　3.2.23

Proof: 1. Since $\mathbb{F}_q$ is a finite field, the set of nonconstant polynomials over $\mathbb{F}_q$

$$\mathbb{F}_q[x] \setminus \mathbb{F}_q = \bigcup_{r \geq 1} \{f \in \mathbb{F}_q[x] \mid \deg f = r\}$$

is a countable union of finite sets, whence it is countable. This means that there exists a bijection $\mathbb{N} \to \mathbb{F}_q[x] \setminus \mathbb{F}_q$ mapping $i \mapsto f_i$. This is a labeling of all nonconstant polynomials over $\mathbb{F}_q$. By 3.2.22 we find a splitting field $\mathbb{K}_0$ of f_0

over $\mathbb{F}_q$. Then for each $i \in \mathbb{N}^*$ we find a splitting field $\mathbb{K}_i$ of $f_0 \cdots f_i$ containing $\mathbb{K}_{i-1}$ as a subfield. We claim that the union

$$\mathbb{K} := \bigcup_{i \geq 0} \mathbb{K}_i$$

is an algebraic closure of $\mathbb{F}_q$. The set $\mathbb{K}$ is a field, since two elements of $\mathbb{K}$ lie in some $\mathbb{K}_n$ so their sum, product and quotient (if the denominator is not zero) are defined in $\mathbb{K}_n$. Since $\mathbb{K}_n$ is a subfield of $\mathbb{K}_m$ for $n < m$, this sum, product and quotient do not depend on the choice of n. Furthermore, $\mathbb{K}$ is algebraic over $\mathbb{F}_q$, since each element of $\mathbb{K}$ lies in some $\mathbb{K}_n$ which is algebraic over $\mathbb{F}_q$. Moreover, $\mathbb{K}$ is algebraically closed. By construction, any $f \in \mathbb{F}_q[x]$ splits into linear factors over $\mathbb{K}$. Suppose that $f \in \mathbb{K}[x]$ is irreducible and let $\mathbb{L}$ be an extension of $\mathbb{K}$ containing a root α of f. Then α is algebraic over $\mathbb{K}$ and $\mathbb{K}$ is algebraic over $\mathbb{F}_q$. So, by Exercise 3.2.11, α is algebraic over $\mathbb{F}_q$. Let $M_\alpha \in \mathbb{F}_q[x]$ be the minimal polynomial of α over $\mathbb{F}_q$. Then there exists an integer n so that M_α splits into linear factors over $\mathbb{K}_n$, whence $\alpha \in \mathbb{K}_n \subset \mathbb{K}$. Since α is a root of the irreducible polynomial $f \in \mathbb{K}[x]$, it follows that f is linear. Thus every irreducible polynomial in $\mathbb{K}[x]$ is linear and, consequently, $\mathbb{K}$ is algebraically closed.

2. Assume that $\mathbb{K} = \bigcup_{i \geq 0} \mathbb{K}_i$ and $\mathbb{L}$ are two algebraic closures of $\mathbb{F}_q$. Let $\psi = \mathrm{id}$ be the trivial embedding of $\mathbb{F}_q$ into $\mathbb{L}$. By the Embedding Theorem 3.2.17, there exists an embedding $\varphi_0 : \mathbb{K}_0 \to \mathbb{L}$ so that φ_0 restricted to $\mathbb{F}_q$ is the identity. Then for each $i \in \mathbb{N}^*$ there exists an embedding $\varphi_i : \mathbb{K}_i \to \mathbb{L}$ so that the restriction of φ_i to $\mathbb{K}_{i-1}$ is φ_{i-1}. Now we define $\varphi : \mathbb{K} \to \mathbb{L}$ as follows. Consider some $\alpha \in \mathbb{K}$, then there exists some $n \in \mathbb{N}$ so that $\alpha \in \mathbb{K}_m$ for $m \geq n$ and $\alpha \notin \mathbb{K}_m$ for $m < n$. We define $\varphi(\alpha) := \varphi_n(\alpha)$. (By construction, $\varphi(\alpha) = \varphi_m(\alpha)$ for all $m \geq n$.) Then φ is an embedding of $\mathbb{K}$ into $\mathbb{L}$, and the restriction of φ to $\mathbb{K}_n$ is φ_n.

Finally we claim that φ is an isomorphism between $\mathbb{K}$ and $\mathbb{L}$. Since $\mathbb{K}$ is algebraically closed also $\varphi(\mathbb{K})$ is algebraically closed. By construction $\mathbb{F}_q \subset \varphi(\mathbb{K}) \subseteq \mathbb{L}$. Since $\mathbb{L}$ is algebraic over $\mathbb{F}_q$ it is also algebraic over $\varphi(\mathbb{K})$. Consider some $\beta \in \mathbb{L}$, then β is algebraic over $\varphi(\mathbb{K})$, thus it belongs to $\varphi(\mathbb{K})$, whence $\mathbb{L} \subseteq \varphi(\mathbb{K})$. $\qquad\square$

The algebraic closure of $\mathbb{F}_q$ is often denoted by $\overline{\mathbb{F}_q}$.

3.2.24   **Partially ordered sets and lattices** Let X be a set. A binary relation $\leq$ on X which is

— reflexive, i.e. $x \leq x$ for all $x \in X$,

— transitive, i.e. $x \leq y$ and $y \leq z$ imply $x \leq z$ for all $x, y, z \in X$,

— antisymmetric, i.e. $x \leq y$ and $x \neq y$ imply $y \not\leq x$ for all $x, y \in X$,

is called a *partial order* on X. The pair $(X, \leq)$ is called a *partially ordered set* or a *poset*.

Two elements x, y of a poset $(X, \leq)$ are called *comparable* if $x \leq y$ or $y \leq x$ is satisfied. Otherwise, they are called *incomparable*. For $x, y \in X$ we write $x < y$ to express that $x \leq y$ and $x \neq y$. Moreover, $x \geq y$ or $x > y$ are used as synonyms for $y \leq x$ or $y < x$, respectively. If any two elements of $(X, \leq)$ are comparable, then X is a *totally ordered set* and $\leq$ is a *total order* on X.

Let $(X, \leq)$ be a poset. For $x, y \in X$, the *interval* $[x, y]$ is the set

$$[x, y] := \{z \in X \mid x \leq z \leq y\}.$$

A poset is called *locally finite* if every interval in X is finite.

Let $(X, \leq)$ be a poset and assume that $Y \subseteq X$. The element $x_0 \in X$ is an *upper bound* of Y if $y \leq x_0$ for all $y \in Y$. The element $x_0 \in X$ is called *supremum* (or *least upper bound*) of Y if x_0 is an upper bound of Y and $x_0 \leq z$ for any upper bound z of Y.

Correspondingly, the *lower bound* and the *infimum* (or *greatest lower bound*) of Y are defined. It is customary to denote the supremum (and infimum, respectively) of a finite set $S \subseteq X$ as $\vee S$ (and $\wedge S$). If $S = \{x, y\}$ contains only two elements, the notation $x \vee y$ and $x \wedge y$ is common.

A partially ordered set $(X, \leq)$ is called a *lattice* $(X, \wedge, \vee)$ if for any two elements $x, y \in X$ both the supremum $x \vee y$ and infimum $x \wedge y$ exist.

A bijection $f : X \to Y$ between two partially ordered sets $(X, \leq)$ and $(Y, \preceq)$ is called an *order isomorphism* if

$$x_1 \leq x_2 \iff f(x_1) \preceq f(x_2), \qquad x_1, x_2 \in X,$$

and is called an *order anti-isomorphism* if

$$x_1 \leq x_2 \iff f(x_2) \preceq f(x_1), \qquad x_1, x_2 \in X.$$

A bijection $f : X \to Y$ between two lattices $(X, \wedge, \vee)$ and $(Y, \sqcap, \sqcup)$ is called a *lattice isomorphism* if

$$f(x_1 \wedge x_2) = f(x_1) \sqcap f(x_2) \text{ and } f(x_1 \vee x_2) = f(x_1) \sqcup f(x_2), \qquad x_1, x_2 \in X,$$

and is called a *lattice anti-isomorphism* if

$$f(x_1 \wedge x_2) = f(x_1) \sqcup f(x_2) \text{ and } f(x_1 \vee x_2) = f(x_1) \sqcap f(x_2), \qquad x_1, x_2 \in X.$$

At the end of this section we prove the *existence* of finite fields for every prime power order.

3.2.25   **Theorem** *For every power $q = p^n$ of a prime number p there exists a finite field with q elements.*

Proof: We denote by $m_p(d)$ the number of monic, irreducible polynomials in $\mathbb{F}_p[x]$ which are of degree d. From 3.2.20 we derive by comparing degrees that

$$p^n = \sum_{d|n} d \cdot m_p(d) = \sum_{d=1}^{n} \zeta(d,n) \cdot d \cdot m_p(d),$$

if we put

$$\zeta(d,n) := \begin{cases} 1 & \text{if } d \text{ divides } n, \\ 0 & \text{otherwise.} \end{cases}$$

This means that ζ indicates the *zeta function* of the partially ordered set $\langle \mathbb{N}^*, | \rangle$, i.e. the set $\mathbb{N}^*$ of positive integers, equipped with the partial order

$$m \leq n :\Longleftrightarrow m \mid n \Longleftrightarrow m \text{ divides } n.$$

The matrix

$$Z(\mathbb{N}^*, |) := (\zeta(m,n))_{m,n \in \mathbb{N}^*} = \begin{pmatrix} 1 & 1 & 1 & 1 & 1 & 1 & 1 & \cdots \\ & 1 & 0 & 1 & 0 & 1 & 0 & \cdots \\ & & 1 & 0 & 0 & 1 & 0 & \cdots \\ & & & 1 & 0 & 0 & 0 & \cdots \\ & & & & 1 & 0 & 0 & \cdots \\ & & & & & 1 & 0 & \cdots \\ & & & & & & 1 & \cdots \\ & 0 & & & & & & \ddots \end{pmatrix}$$

with the values of the zeta function as entries, is an upper triangular matrix with ones along its main diagonal. It is therefore invertible over $\mathbb{Z}$, and the elements of its inverse

$$M(\mathbb{N}^*, |) := Z(\mathbb{N}^*, |)^{-1} =: (\mu(m,n))_{m,n \in \mathbb{N}^*}$$

$$= \begin{pmatrix} 1 & -1 & -1 & 0 & -1 & 1 & -1 & \cdots \\ & 1 & 0 & -1 & 0 & -1 & 0 & \cdots \\ & & 1 & 0 & 0 & -1 & 0 & \cdots \\ & & & 1 & 0 & 0 & 0 & \cdots \\ & & & & 1 & 0 & 0 & \cdots \\ & & & & & 1 & 0 & \cdots \\ & & & & & & 1 & \cdots \\ & 0 & & & & & & \ddots \end{pmatrix}$$

are the values of the *Möbius function* on the poset $\langle \mathbb{N}^*, | \rangle$. This function is used very often in number theory and hence it is called the *number theoretic*

Möbius function. Since the value $\mu(d,n)$ depends only on the quotient n/d (intervals $[d,n]$ and $[d',n']$ in the partial order $(\mathbb{N}^*, |\,)$ are order isomorphic if $n/d = n'/d'$), one can replace the notation $\mu(d,n)$ by $\mu(n/d)$. Later on we will encounter zeta and Möbius functions of other posets. The definition of the Möbius function as the inverse of the zeta function yields

$$m_p(n) = \frac{1}{n} \sum_{d|n} \mu(n/d) p^d.$$

3.2.26

On the right hand side we see the summand p^n (for $d = n$, since $\mu(n,n) = 1$). Moreover, the values of the Möbius function belong to the set $\{0, 1, -1\}$ (for the exact values of the number theoretic Möbius function and for the general Möbius inversion see Exercises 3.2.15 and 3.2.16). This gives the following lower bound for $m_p(n)$:

$$m_p(n) \geq \frac{1}{n} \left(p^n - (p^{n-1} + \ldots + p + 1) \right) = \frac{1}{n} \left(p^n - \frac{p^n - 1}{p - 1} \right).$$

This number is greater than 0 for every prime number p. It implies the existence of at least one irreducible polynomial of degree n in $\mathbb{F}_p[x]$, and so, according to the construction described above, the existence of at least one field with p^n elements.   □

Exercises

Exercise Let $(G, \cdot)$ be a finite group. Assume that $g \in G$ is of order r. Prove that for $n \in \mathbb{N}$ the order of g^n is equal to $r/\gcd(r,n)$.

E.3.2.1

Exercise Let G denote a finite abelian group. Prove that the order of the product ab of two elements $a, b \in G$ with relatively prime orders is

$$\mathrm{ord}(ab) = \mathrm{ord}(a) \cdot \mathrm{ord}(b).$$

E.3.2.2

Exercise The *Homomorphism Theorem* (for rings): Let R, S denote rings and $\phi: R \to S$ be a ring epimorphism. Check that the kernel of ϕ, which is defined as $\ker(\phi) := \{r \in R \mid \phi(r) = 0\}$, is an ideal in R and that the mapping

$$R/\ker(\phi) \to S \;:\; r + \ker(\phi) \mapsto \phi(r)$$

is an isomorphism of rings.

E.3.2.3

Exercise Let p denote a prime number. Show that p divides the binomial coefficients $\binom{p}{i}$ for $1 \leq i \leq p - 1$.

E.3.2.4

E.3.2.5   **Exercise** Use the primitive polynomial $x^3 + 2x + 1$ of $\mathbb{F}_3[x]$ in order to describe the field $\mathbb{F}_{27}$ as an $\mathbb{F}_3$-vector space.

E.3.2.6   **Exercise** Show that a field has only the trivial ideals $I(0)$ and $I(1)$.

E.3.2.7   **Exercise** Let $\psi \colon \mathbb{F} \to \mathbb{L}$ be an embedding and $f \in \mathbb{F}[x]$ an irreducible polynomial. Show that $\psi(f)$ is irreducible over $\psi(\mathbb{F})$.

E.3.2.8   **Exercise** Prove that every finite field extension is algebraic.

E.3.2.9   **Exercise** Assume that $\mathbb{L}$ over $\mathbb{K}$ and $\mathbb{K}$ over $\mathbb{F}$ are finite field extensions. Show that $\mathbb{L}$ over $\mathbb{F}$ is also a finite extension.

E.3.2.10   **Exercise** Consider a field extension $\mathbb{L}$ over $\mathbb{K}$ and $\alpha \in \mathbb{L}$. Prove that α is algebraic over $\mathbb{K}$ if and only if $\mathbb{K}(\alpha)$ is a finite field extension over $\mathbb{K}$.

E.3.2.11   **Exercise** Assume that $\mathbb{L}$ is algebraic over $\mathbb{K}$. If β in an extension field of $\mathbb{L}$ is algebraic over $\mathbb{L}$, show that β is algebraic over $\mathbb{K}$. Hint: If β is algebraic over $\mathbb{L}$ prove that $\mathbb{K}(\beta)$ is a finite extension. By assumption there exists $g(x) = \sum_{i=0}^{n} \alpha_i x^i \in \mathbb{L}[x]$ so that $g(\beta) = 0$. Then β is algebraic over $\mathbb{K}(\alpha_0, \ldots, \alpha_n)$. Deduce that $\mathbb{K}(\beta, \alpha_0, \ldots, \alpha_n)$ is a finite extension of $\mathbb{K}(\alpha_0, \ldots, \alpha_n)$. And show that $\mathbb{K}(\alpha_0, \ldots, \alpha_n)$ is a finite extension of $\mathbb{K}$.

E.3.2.12   **Exercise** Prove that $(\mathbb{N}^*, |)$ is a poset.

E.3.2.13   **Exercise** Prove that a binary relation $\prec$ on a set X is antisymmetric if and only if

$$x \prec y \text{ and } y \prec x \text{ imply } x = y \text{ for all } x, y \in X.$$

E.3.2.14   **Exercise** Assume that $(X, \leq)$ is a poset with two operators $\wedge, \vee \colon X \times X \to X$. Prove that $(X, \wedge, \vee)$ is a lattice, where $\vee$ is the supremum and $\wedge$ the infimum function, if and only if

1. $\vee$ and $\wedge$ are associative, commutative, and $x \vee x = x \wedge x = x$ for all $x \in X$,

2. $x \wedge (x \vee y) = x = x \vee (x \wedge y)$ for all $x, y \in X$,

3. $x \wedge y = x \iff x \vee y = y \iff x \leq y$ for all $x, y \in X$.

Exercise Show that the function $\mu\colon \mathbb{N}^* \to \mathbb{Z}$, defined by   E.3.2.15

$$\mu(d) = \begin{cases} 1 & \text{if } d = 1, \\ (-1)^r & \text{if } d \text{ is a product of } r \text{ pairwise distinct primes,} \\ 0 & \text{otherwise,} \end{cases}$$

has the following properties:

$$\sum_{d|n} \mu(d) = \begin{cases} 1 & \text{if } n = 1, \\ 0 & \text{otherwise.} \end{cases}$$

In fact, this is the *number theoretic Möbius function*. Prove that, for any functions $f, g\colon \mathbb{N}^* \to G$ and for every abelian group G, the expression

$$\forall\, n \in \mathbb{N}^* \;:\; g(n) = \sum_{d|n} f(d)$$

is equivalent to

$$\forall\, n \in \mathbb{N}^* \;:\; f(n) = \sum_{d|n} \mu(d) g(n/d).$$

The equivalence of these two expressions permits us to replace one by the other. This replacement is called the *number theoretic Möbius inversion*.

Exercise Let $(P, \le)$ be a locally finite poset, and let $\mathbb{K}$ be a field of characteristic   E.3.2.16
zero. The *incidence algebra* over P is the set

$$\mathrm{IA}(P) := \{f\colon P \times P \to \mathbb{K} \mid x \not\le y \Longrightarrow f(x,y) = 0\},$$

together with addition

$$(f + g)(x,y) := f(x,y) + g(x,y), \qquad f, g \in \mathrm{IA}(P),\ x,y \in P,$$

scalar multiplication

$$(\lambda \cdot f)(x,y) := \lambda f(x,y), \qquad f \in \mathrm{IA}(P),\ \lambda \in \mathbb{K},\ x,y \in P,$$

and convolution

$$(f * g)(x,y) := \begin{cases} \displaystyle\sum_{\substack{z \in P \\ x \le z \le y}} f(x,z) g(z,y) & \text{if } x \le y, \\ 0 & \text{else,} \end{cases} \qquad f, g \in \mathrm{IA}(P),\ x,y \in P.$$

Two particular elements of $\mathrm{IA}(P)$ are

— the *zeta function* defined by

$$\zeta(x,y) := \begin{cases} 1 & \text{if } x \le y, \\ 0 & \text{else,} \end{cases}$$

— the *delta function* defined by

$$\delta(x,y) := \begin{cases} 1 & \text{if } x = y, \\ 0 & \text{else.} \end{cases}$$

Prove:

1. The incidence algebra is an associative $\mathbb{K}$-algebra, where δ is a neutral element with respect to the convolution.

2. There exists an inverse element of $f \in IA(P)$ with respect to $*$ if and only if

$$f(x,x) \neq 0 \quad \text{for all } x \in P.$$

3. The inverse of the zeta function exists. We call it the *Möbius function* and denote it by μ. It has the following properties:

$$\mu(x,x) = 1, \qquad x \in P,$$

$$\mu(x,y) = - \sum_{x < z \leq y} \mu(z,y) = - \sum_{x \leq z < y} \mu(x,z), \qquad x < y.$$

4. *General inversion formula:* Assume that $f, g, h \in IA(P)$ and h^{-1} is the inverse of h with respect to $*$. Then

$$f = g * h \Longleftrightarrow g = f * h^{-1}.$$

5. *Möbius inversion:*
   — Assume that for $f \in \mathbb{K}^P$ there exists an element $p \in P$ such that $x \not\geq p$ implies that $f(x) = 0$, and let

$$g(x) := \sum_{y \leq x} f(y).$$

Then

$$f(x) = \sum_{y \leq x} g(y)\mu(y,x).$$

   — Assume that for $f \in \mathbb{K}^P$ there exists an element $q \in P$ such that $x \not\leq q$ implies that $f(x) = 0$, and let

$$g(x) := \sum_{y \geq x} f(y).$$

Then

$$f(x) = \sum_{y \geq x} \mu(x,y)g(y).$$

3.3 The Galois Group and Normal Bases   3.3

Let us consider again the number $m_p(n)$ of irreducible polynomials of degree n over $\mathbb{F}_p$. We will show that the expression for $m_p(n)$ in 3.2.26 can be deduced directly from results on finite group actions. In fact, it is the number of orbits of the automorphism group of $\mathbb{F}_{p^n}$ which are of maximal length. This way, the number of such polynomials can not only be evaluated, but it can be *understood* as well. Before we do that, let us study the automorphism group of a finite field. We will come back to the numbers $m_p(n)$ in the next section. Here we introduce the Galois group.

Definition (Galois group) Consider positive integers m and n such that m di-   3.3.1
vides n, and recall that this implies the inclusion $\mathbb{F}_{p^m} \leq \mathbb{F}_{p^n}$ by 3.2.11. The *Galois group*

$$\mathrm{Gal}\,[\,\mathbb{F}_{p^n} : \mathbb{F}_{p^m}\,]$$

of $\mathbb{F}_{p^n}$ over $\mathbb{F}_{p^m}$ consists of *all* automorphisms of $\mathbb{F}_{p^n}$ which fix the subfield $\mathbb{F}_{p^m}$ elementwise. The field $\mathbb{F}_{p^m}$ is called the *fixed field* of the Galois group $\mathrm{Gal}\,[\,\mathbb{F}_{p^n} : \mathbb{F}_{p^m}\,]$.   ◇

A case of particular interest is the Galois group $\mathrm{Gal}\,[\,\mathbb{F}_{p^n} : \mathbb{F}_p\,]$. Since every field automorphism fixes the identity element, it also fixes the whole prime field $\mathbb{F}_p$. This shows that the Galois group of a finite field $\mathbb{F}_{p^n}$ over its prime field is the automorphism group $\mathrm{Aut}(\mathbb{F}_{p^n})$. By the same argument,

$$\mathrm{Gal}\,[\,\mathbb{F} : \mathbb{P}\,] = \mathrm{Aut}(\mathbb{F}).$$ 3.3.2

Let σ be the Frobenius automorphism $\sigma(\kappa) = \kappa^p$, $\kappa \in \mathbb{F}_{p^n}$. It is clear that the cyclic subgroup $\langle \sigma \rangle$ is contained in $\mathrm{Gal}\,[\,\mathbb{F}_{p^n} : \mathbb{F}_p\,]$. Now we prove that these groups coincide.

Theorem *The Galois group is the cyclic group generated by the Frobenius automor-   3.3.3
phism. For any prime number p we have*

$$\mathrm{Gal}\,[\,\mathbb{F}_{p^n} : \mathbb{F}_p\,] = \langle\, \sigma : \kappa \mapsto \kappa^p \,\rangle,$$

and for $q = p^r$ with $r \in \mathbb{N}^*$ we have

$$\mathrm{Gal}\,[\,\mathbb{F}_{q^n} : \mathbb{F}_q\,] = \langle\, \tau : \kappa \mapsto \kappa^q \,\rangle.$$

Proof: It suffices to show that every automorphism $\theta \in \mathrm{Gal}\,[\,\mathbb{F}_{q^n} : \mathbb{F}_q\,]$ belongs to $\langle \tau \rangle$. Let α denote a primitive element of $\mathbb{F}_{q^n}$, and let f be the minimal polynomial of α over $\mathbb{F}_q$. Then f is a monic, irreducible polynomial of degree n (see Exercise 3.3.2). By 3.2.19, all its roots can be obtained in the form $\tau^j(\alpha)$

for $j \in n$. Since θ belongs to the Galois group, a similar computation as in 3.2.15 shows that θ maps α again to a root of f, whence $\theta(\alpha) = \tau^j(\alpha)$ for a suitable j. Because α is primitive, the two automorphisms θ and τ^j coincide, whence $\theta \in \langle \tau \rangle$.   □

If $\alpha \in \mathbb{F}_{q^n}$, then the elements $\alpha, \alpha^q, \ldots, \alpha^{q^{n-1}}$ are called *conjugates* of α with respect to $\mathbb{F}_q$. As we have seen already, they are the roots of the same minimal polynomial. Moreover, they have the same order in the multiplicative group $\mathbb{F}_{q^n}^*$ (cf. Exercise 3.2.1).

The action of the Galois group on the elements of $\mathbb{F}_q$ is an example of a group action. We have met this important concept already in Section 1.4, when we studied isometry classes of codes.

The Galois group

$$\mathrm{Gal}\,[\,\mathbb{F}_{q^n} : \mathbb{F}_q\,] = \langle\, \tau : \kappa \mapsto \kappa^q \,\rangle$$

acts on $\mathbb{F}_{q^n}$ in the following way:

$$\langle\, \tau \,\rangle \times \mathbb{F}_{q^n} \to \mathbb{F}_{q^n} \; : \; (\tau^i, \kappa) \mapsto \tau^i(\kappa) = \kappa^{q^i}, \qquad i \in n.$$

This will be our main example in the present section. We recall that an action $_G X$ of G on X induces the equivalence relation $\sim_G$ on X. The equivalence classes $G(x) := \{gx \mid g \in G\}$ are called orbits, and for the set of *all* orbits of G on the set X we have introduced the symbol $G \backslash\!\backslash X$. In 3.2.14 we have seen that the set of orbits

$$\mathrm{Gal}\,[\,\mathbb{F}_{2^4} : \mathbb{F}_2\,]\backslash\!\backslash\mathbb{F}_{2^4}$$

of the Galois group is equal to

$$\{\{0\}, \{1\}, \{\alpha, \alpha^2, \alpha^4, \alpha^8\}, \{\alpha^3, \alpha^6, \alpha^{12}, \alpha^9\}, \{\alpha^5, \alpha^{10}\}, \{\alpha^7, \alpha^{14}, \alpha^{13}, \alpha^{11}\}\}.$$

These orbits are simply the elements in the cyclic factors of the permutation $\bar{\sigma}$ induced on $\mathbb{F}_{2^4}$ by the generating element σ of the Galois group.

3.3.4   **Theorem** *Let α be an element of $\mathbb{F}_{q^n}$. Let $\tau : \kappa \mapsto \kappa^q$ be the Frobenius automorphism of $\mathbb{F}_{q^n}$. Assume that $\mathrm{Gal}(\alpha)$ is the orbit of α under the Galois group. Then*

$$M_\alpha := \prod_{\kappa \in \mathrm{Gal}(\alpha)} (x - \kappa)$$

is the minimal polynomial of α over $\mathbb{F}_q$.

Proof: In order to show that M_α is a polynomial over $\mathbb{F}_q$, we note that the coefficients of it are elementary symmetric functions of its roots. Hence, these coefficients are invariant under the action of the Galois group, which implies that they are elements of the subfield $\mathbb{F}_q$ which is fixed elementwise by the

Galois group. Moreover M_α is monic. Since α is a root of $M_\alpha \in \mathbb{F}_q[x]$, all conjugates of α, i.e. all elements of $\mathrm{Gal}(\alpha)$, are also roots of M_α. This shows that M_α is the monic polynomial of least degree in $\mathbb{F}_q[x]$ which has α as a root. Therefore, it is the minimal polynomial of α over $\mathbb{F}_q$.   □

Now we introduce a special kind of basis of $\mathbb{F}_{q^n}$ over its subfield $\mathbb{F}_q$.

Definition (normal basis) A basis B of $\mathbb{F}_{q^n}$ over $\mathbb{F}_q$ is called *normal basis* if there exists some $\kappa \in \mathbb{F}_{q^n}$ such that $B = \{\kappa, \tau(\kappa), \ldots, \tau^{n-1}(\kappa)\}$ where τ is the Frobenius automorphism of $\mathbb{F}_{q^n}$ over $\mathbb{F}_q$.   ◇

  3.3.5

It is clear that if $\{\kappa, \tau(\kappa), \ldots, \tau^{n-1}(\kappa)\}$ is a normal basis, then the minimal polynomial of κ over $\mathbb{F}_q$ is of degree n. The proof that a normal basis of $\mathbb{F}_{q^n}$ over $\mathbb{F}_q$ always exists is based upon Dedekind's Independence Theorem which is presented next. Furthermore, it needs deeper methods from the theory of modules, therefore, we postpone the rest of the proof to Section 6.9.

Dedekind's Independence Theorem *Let G be a group and let $\varphi_0, \ldots, \varphi_{n-1}$ be pairwise distinct homomorphisms from G into the multiplicative group $\mathbb{F}^*$ of a field $\mathbb{F}$. Then $\varphi_0, \ldots, \varphi_{n-1}$ are linearly independent over $\mathbb{F}$.*

  3.3.6

Proof: By induction on n, we have to prove that

$$\sum_{i \in n} \kappa_i \varphi_i = 0, \qquad \kappa_i \in \mathbb{F},\ i \in n,$$

  3.3.7

implies that $\kappa_i = 0$ for $i \in n$.

If $n = 1$, then $\kappa_0 \varphi_0(1) = \kappa_0$, whence $\kappa_0 = 0$. Assume that $n > 1$ and that the induction hypothesis holds true for $n - 1$. Since the φ_i are pairwise distinct, there exists some $g \in G$ such that $\varphi_0(g) \neq \varphi_{n-1}(g)$. We have

$$0 = \sum_{i \in n} \kappa_i \varphi_i(g \cdot x) = \sum_{i \in n} \kappa_i \varphi_i(g) \varphi_i(x) = \left(\sum_{i \in n} \kappa_i \varphi_i(g) \varphi_i \right)(x), \qquad x \in G.$$

  3.3.8

From 3.3.7 we also obtain

$$\sum_{i \in n} \kappa_i \varphi_0(g) \varphi_i = 0.$$

Subtracting the last equation from 3.3.8 we obtain that

$$\sum_{i=1}^{n-1} \kappa_i (\varphi_i(g) - \varphi_0(g)) \varphi_i = 0.$$

By induction $\kappa_i(\varphi_i(g) - \varphi_0(g)) = 0$ for $1 \leq i \leq n - 1$, thus $\kappa_{n-1} = 0$, due to the particular choice of g. Inserting this into 3.3.7 and applying the induction hypothesis once again we derive that also $\kappa_i = 0$ for $i \in n - 1$.   □

Exercises

E.3.3.1   **Exercise** Let σ be the Frobenius automorphism $\kappa \mapsto \kappa^p$. Use Dedekind's Independence Theorem to show that $\langle \sigma \rangle$ is a linearly independent set in the vector space $\mathbb{F}_{p^n}^{\mathbb{F}_{p^n}}$. From this, deduce that $|\langle \sigma \rangle| \leq n$. Hint: σ is $\mathbb{F}_p$-linear.

E.3.3.2   **Exercise** Let α be a primitive element of $\mathbb{F}_{q^n}$. Show that the minimal polynomial of α over $\mathbb{F}_q$ is of degree n.

3.4   3.4 Enumeration under Group Actions, Lyndon Words

In this section, we will study the concept of group actions in more depth. We will also see more applications. Recall that we have already met two important group actions, namely the action of the linear isometry group on $\mathbb{F}_{q^n}$ and the action of the Galois group $\mathrm{Gal}\,[\,\mathbb{F}_{q^n} : \mathbb{F}_q\,]$ on the field $\mathbb{F}_{q^n}$. Further actions can be derived from these, like the action on the set of subspaces of $\mathbb{F}_{q^n}$ in the first case or the action on the set of roots of a polynomial with coefficients in $\mathbb{F}_q$ in the latter. In any case, the orbits of these actions are of interest. In the following, we will see several enumerative results in this area. In a first step, we will see how to count the overall number of orbits of a finite group acting on a finite set. After that, we will restrict attention to orbits of a particular type (this will be made precise later). One example of this will be the orbits of maximal length. It will turn out that one can determine the orbits of any given type provided one has a good understanding of the structure of the group which acts (also this will be made precise). As an application of this theory, we will count certain words of finite length over finite alphabets. This will later turn out to be related to the number of irreducible polynomials of a given degree over a given finite field.

To begin with, we have the following fundamental result on group actions, the proof of which is Exercise 3.4.1.

3.4.1   **Lemma** *For an action* $_G X$ *the following holds:*

— *The stabilizer*

$$G_x := \{g \in G \mid gx = x\}$$

of the point $x \in X$ *in G is a subgroup of G.*
— *For arbitrary* $x \in X$, *the mapping*

$$\theta_x : G/G_x \to G(x) \ : \ gG_x \mapsto gx, \quad g \in G,$$

is a bijection between the set

$$G/G_x = \{gG_x \mid g \in G\}$$

of left cosets (cf. Exercise 1.4.6) of the stabilizer of x and the orbit $G(x)$ of x.

— *For finite groups G we have the identity*

$$|G(x)| = |G : G_x| = |G/G_x| = |G|/|G_x|,$$

which means that the length of an orbit is the index of the stabilizer of any one of its elements. $\square$

An important notion is that of fixed points, as introduced in 1.6.13. Recall that $x \in X$ is called fixed point of the group element $g \in G$ if $gx = x$. The set of all fixed points of g is indicated by X_g.

If U is a subgroup of G, let X_U denote the set of all elements $x \in X$ which contain U in their stabilizer, i.e.

$$X_U := \{x \in X \mid U \leq G_x\} = \{x \in X \mid gx = x \text{ for all } g \in U\}.$$

This is the set of invariants of U on X (cf. 1.6.13).

It turns out that the number of orbits can be determined provided the number of fixed points is known for each element $g \in G$:

The Lemma of Cauchy–Frobenius *If $_GX$ is a finite group action, then the number of orbits of G on X is the average number of fixed points, i.e.*   **3.4.2**

$$|G\backslash\backslash X| = \frac{1}{|G|} \sum_{g \in G} |X_g|.$$

Proof:

$$\sum_{g \in G} |X_g| = \sum_{g \in G} \sum_{x:gx=x} 1 = \sum_{x \in X} \sum_{g:gx=x} 1 = \sum_{x \in X} |G_x|$$

$$\stackrel{3.4.1}{=} |G| \sum_{x \in X} \frac{1}{|G(x)|} \stackrel{1.4.6}{=} |G| \sum_{\omega \in G\backslash\backslash X} \sum_{x \in \omega} \frac{1}{|\omega|} = |G||G\backslash\backslash X|.$$

The last identity follows from the fact that each orbit ω contributes the value $\frac{1}{|\omega|}|\omega| = 1$ to the sum. $\square$

There are further methods which allow to count orbits with particular properties, for example, particular stabilizers of their elements. The basic result is

3.4.3   **Lemma** *For each action $_GX$ we have:*

— *If $g \in G$ and $x \in X$, then the stabilizers of x and of gx are conjugate subgroups:*

$$G_{gx} = gG_xg^{-1}.$$

— *Hence, the stabilizers of the elements in the orbits $G(x)$ form a full conjugacy class of subgroups (cf. Exercise 3.4.4):*

$$\left\{G_{x'} \mid x' \in G(x)\right\} = \left\{G_{gx} \mid g \in G\right\} = \left\{gG_xg^{-1} \mid g \in G\right\} =: \widetilde{G_x}. \qquad \square$$

For this reason we consider the set of all subgroups of G,

3.4.4
$$L(G) := \{U \mid U \leq G\}.$$

It is a partially ordered set, where the order $\leq$ is given by set theoretic inclusion. For U and V in $L(G)$ we write $U \leq V$ if $U \subseteq V$, which means that U is actually a subgroup of V. Moreover, any two subgroups in $L(G)$ have an infimum and a supremum with respect to this order. In fact, for subgroups U and V of G we have

$$U \wedge V := U \cap V \text{ and } U \vee V := \langle U \cup V \rangle,$$

the intersection of the two subgroups and the subgroup generated by the union of them. Together with these compositions $L(G)$ is a lattice, the *subgroup lattice*

$$(L(G), \wedge, \vee)$$

of G. For U a subgroup of G, we say that an orbit is of *type* $\widetilde{U}$ if the elements of that orbit have stabilizers which are conjugate to U. Recall from 3.4.3 that the stabilizers of elements of an orbit $G(x)$ form a full conjugacy class of subgroups, so that this concept makes sense. Let

$$G\backslash\backslash_{\widetilde{U}}X := \left\{G(x) \mid x \in X, G_x \in \widetilde{U}\right\}.$$

be the set of G-orbits which are of type $\widetilde{U}$. For example, $G\backslash\backslash_{\widetilde{1}}X$ and $G\backslash\backslash_{\widetilde{G}}X$ denote the sets of G-orbits of largest and of smallest size, respectively. Let us now evaluate the number of orbits of a given type. We use a similar method as in the determination of the number of irreducible polynomial over finite fields in the proof of 3.2.25. Recall that we applied a method called Möbius inversion to the lattice of integers ordered with respect to divisibility. Here, we are going to apply Möbius inversion to the lattice of subgroups of the finite group G. We start with the zeta function of $L(G)$, which is defined by

$$\zeta(U, V) := \begin{cases} 1 & \text{if } U \leq V, \\ 0 & \text{otherwise.} \end{cases}$$

Correspondingly, the zeta matrix of $L(G)$ is

$$Z(L(G)) := Z(L(G), \leq) := (\zeta(U, V))_{U, V \in L(G)},$$

whereas the Möbius matrix is defined to be

$$M(L(G)) := M(L(G), \leq) := (\mu(U, V))_{U, V \in L(G)} := Z(L(G))^{-1}.$$

We are now in a position to show that the number of orbits of a given type can be expressed in terms of numbers of fixed points:

Lemma *If U is a subgroup of the finite group G which acts on a finite set X, then the number of orbits of G on X of type $\widetilde{U}$ is*   **3.4.5**

$$|G\backslash\backslash_{\widetilde{U}} X| = \frac{|\widetilde{U}|}{|G/U|} \sum_{V \leq G} \mu(U, V)|X_V|,$$

where μ denotes the Möbius function of the lattice of subgroups of G.

Proof: Let T be a transversal of $G\backslash\backslash X$, i.e. a complete set of representatives of the G-orbits on X. Then the cardinality of the set of elements $x \in X$ which contain U in their stabilizer G_x is equal to

$$|X_U| = \sum_{x:U \leq G_x} 1 = \sum_{V:U \leq V \leq G} \sum_{x:V=G_x} 1 = \sum_{V:U \leq V \leq G} \frac{1}{|\widetilde{V}|} \sum_{x:G_x \in \widetilde{V}} 1$$

$$= \sum_{V:U \leq V \leq G} \frac{|G/V|}{|\widetilde{V}|} \sum_{t \in T:G_t \in \widetilde{V}} 1 = \sum_{V:U \leq V \leq G} \frac{|G/V|}{|\widetilde{V}|}|G\backslash\backslash_{\widetilde{V}} X|.$$

Using the zeta function of the poset $(L(G), \leq)$, the last equation can be rewritten as

$$|X_U| = \sum_{V \leq G} \zeta(U, V) \frac{|G/V|}{|\widetilde{V}|}|G\backslash\backslash_{\widetilde{V}} X|.$$

By Möbius inversion (cf. Exercise 3.2.16), this equation is equivalent to the assertion.   □

A special case is the number of orbits with trivial stabilizer $G_x = \{1\}$. These are the orbits of maximal length. According to the above lemma, their number is

$$|G\backslash\backslash_{\widetilde{1}} X| = \frac{1}{|G|} \sum_{V \leq G} \mu(1, V)|X_V|.$$ **3.4.6**

Our next goal is to show that the expression for $m_q(n)$ which was derived in 3.2.26 is a special case of this identity. For this purpose, recall that if G acts on X, then by 1.4.7 for any set Y there is an induced action of G on the set of mappings $Y^X = \{f \mid f : X \to Y\}$,

$$G \times Y^X \to Y^X : (g, f) \mapsto f \circ \overline{g}^{-1}.$$

It takes the pair (g, f) and maps it to the function from X to Y whose value at x is $f(g^{-1}x)$. This means that (g, f) is mapped to $f \circ \overline{g}^{-1}$, where $\overline{g}$ is the permutation induced by g on X (cf. 1.4.5). Iterating this formula we see that $f \in Y^X$ is a fixed point of g if and only if

$$f(x) = f(g^{-1}x) = f(g^{-2}x) = \ldots = f(gx) = f(x), \qquad x \in X.$$

3.4.7   **Corollary** *Let $_GX$ be a group action and Y a set. Consider the induced action of G on Y^X. The fixed points of $g \in G$ are the mappings which are constant on orbits of $\langle g \rangle$. In other words, f is fixed under g if and only if f is constant on the cyclic factors of the permutation $\overline{g}$ induced by g on X. Hence, for finite X, Y, and G we obtain from the Lemma of Cauchy–Frobenius that*

$$|G \backslash\backslash Y^X| = \frac{1}{|G|} \sum_{g \in G} |Y|^{|\langle g \rangle \backslash\backslash X|}. \qquad \Box$$

In order to verify that this formula yields the desired number of irreducible polynomials, we show that the action of the Galois group on $\mathbb{F}_{q^n}$ is essentially the same as a particular action of the cyclic group on a set of mappings Y^X, where "essentially the same" is understood in the following sense:

3.4.8   **Definition (similar actions)** Two actions $_GX$ and $_GY$ of a group G on sets X and Y, respectively, are called *similar* if there exists a bijection $\varphi \colon X \to Y$ which commutes with the actions, i.e.

$$\varphi(gx) = g\varphi(x), \qquad g \in G, \, x \in X.$$

We indicate similarity of actions by

$$_GX \approx {}_GY. \qquad \diamond$$

It is left to the reader to prove some basic facts about similar group actions:

3.4.9   **Lemma**

1. *If $_GX$ is an action, then for any $x \in X$, the action of G on the orbit $G(x)$ is similar to the action of G on the set of left cosets of the stabilizer G_x,*

$$_GG(x) \approx_G (G/G_x).$$

2. *Similar actions have the same numbers and sizes of orbits.* $\qquad \Box$

Here we are interested in the following situation:

Example The Galois group $\langle \tau : \kappa \mapsto \kappa^q \rangle$ of $\mathbb{F}_{q^n}$ gives rise to the action

$$\langle \tau \rangle \left(\mathbb{F}_{q^n} \right).$$

If β denotes a primitive element of $\mathbb{F}_{q^n}$, say $\mathbb{F}_{q^n}^* = \langle \beta \rangle$, then any nonzero element κ of $\mathbb{F}_{q^n}$ is of the form $\kappa = \beta^i$, for a suitable $i \in \{1, \ldots, q^n - 1\}$. Hence, we obtain a bijection $\varphi : \mathbb{F}_{q^n} \to Y^X := q^n$ by mapping $\kappa = \beta^i \in \mathbb{F}_{q^n}^*$ onto the q-adic decomposition of the exponent $i \geq 1$,

$$\kappa \mapsto (a_{n-1}, \ldots, a_0)_q, \text{ where } i = \sum_{j \in n} a_j q^j, \quad a_j \in q,$$

while the zero element is mapped onto the sequence of zeros,

$$0 \mapsto (0, \ldots, 0)_q.$$

Since $\tau(\kappa) = \tau(\beta^i) = (\beta^i)^q = \beta^{iq}$ and the q-adic decomposition of $iq \bmod q^n$ is given by $a_{n-1} + \sum_{j=0}^{n-2} a_j q^{j+1}$, the cyclic group $\langle \tau \rangle$ acts on the set of these sequences by cyclic shift. Thus we obtain an action

$$\langle \tau \rangle \left(q^n \right)$$

of the group $\langle \tau \rangle$ by putting

$$\tau(a_{n-1}, \ldots, a_0)_q := (a_{n-2}, \ldots, a_0, a_{n-1})_q.$$

Since

$$\varphi(\tau\kappa) = (a_{n-2}, \ldots, a_0, a_{n-1})_q = \tau(a_{n-1}, \ldots, a_0)_q = \tau\varphi(\kappa),$$

the actions are similar, i.e.

$$\langle \tau \rangle \left(\mathbb{F}_{q^n} \right) \approx_{\langle \tau \rangle} \left(q^n \right).$$

$\diamondsuit$

Using 3.4.9 we obtain the following result:

Corollary *The orbits of the Galois group $\langle \tau \rangle$ on the field $\mathbb{F}_{q^n}$ are mapped under φ*
bijectively onto the orbits of the cyclic group $\langle \tau \rangle$ on the set of mappings q^n:

$$\langle \tau \rangle \backslash\backslash \mathbb{F}_{q^n} \to \langle \tau \rangle \backslash\backslash q^n.$$

$\square$

Using a normal basis of $\mathbb{F}_{q^n}$ over $\mathbb{F}_q$, it is even possible to find a vector space isomorphism $\psi : \mathbb{F}_{q^n} \to \mathbb{F}_q^n$ and two similar actions of $\langle \tau \rangle$ on $\mathbb{F}_{q^n}$ and $\mathbb{F}_q^n$. We want to describe the action on $\mathbb{F}_q^n$ as an action of a group G on Y^X of the form 1.4.7. Consider the Galois group $G := \text{Gal}\,[\mathbb{F}_{q^n} : \mathbb{F}_q] = \langle \tau \rangle$, and let X be a normal basis $\{\kappa, \tau(\kappa), \ldots, \tau^{n-1}(\kappa)\}$ of $\mathbb{F}_{q^n}$ over $\mathbb{F}_q$, which is a particular orbit of some $\kappa \in \mathbb{F}_{q^n}$ under the Galois group. For Y we take $\mathbb{F}_q$, obtaining

$$Y^X = \mathbb{F}_q^{\{\kappa, \tau(\kappa), \ldots, \tau^{n-1}(\kappa)\}} \simeq \mathbb{F}_q^n \simeq \mathbb{F}_{q^n}.$$

Since X is a normal basis, each element $\alpha \in \mathbb{F}_{q^n}$ can be uniquely expressed as a linear combination of the elements of X, i.e.

3.4.12
$$\alpha = \sum_{i \in n} \alpha_i \tau^i(\kappa) \quad \text{with} \quad \alpha_i \in \mathbb{F}_q, \ i \in n,$$

and

$$\tau(\alpha) = \tau\left(\sum_{i \in n} \alpha_i \tau^i(\kappa)\right) = \sum_{i \in n} \alpha_i \tau^{i+1}(\kappa) = \alpha_{n-1}\kappa + \sum_{i \in n-1} \alpha_i \tau^{i+1}(\kappa).$$

Thus we obtain a group action of $\langle \tau \rangle$ on $\mathbb{F}_q^n$ by putting

$$\tau(\alpha_0, \ldots, \alpha_{n-1}) := (\alpha_{n-1}, \alpha_0, \ldots, \alpha_{n-2}).$$

The coefficient vectors of the conjugates $\tau^j(\alpha) = \alpha^{q^j}$ for $j \geq 1$ are given by

$$(\alpha_{n-j}, \alpha_{n-j+1}, \ldots, \alpha_{n-1}, \alpha_0, \ldots, \alpha_{n-j-1}).$$

Hence, they can be obtained by a cyclic shift of the coefficient vector

$$(\alpha_0, \ldots, \alpha_{n-1})$$

of α.

3.4.13   **Corollary** *Let τ be the Frobenius automorphism of $\mathbb{F}_{q^n}$ over $\mathbb{F}_q$. Let ψ be the vector space isomorphism which takes each element $\alpha \in \mathbb{F}_{q^n}$ to the coefficient vector of α with respect to the normal basis $\{\kappa, \tau(\kappa), \ldots, \tau^{n-1}(\kappa)\}$. Then the actions of the Galois group $\langle \tau \rangle$ on $\mathbb{F}_{q^n}$ and on $\mathbb{F}_q^n$ are similar, and*

$$\psi(\tau(\alpha)) = \tau(\psi(\alpha)), \qquad \alpha \in \mathbb{F}_{q^n}.$$

The elements of the orbit $\langle \tau \rangle(\alpha)$, i.e. the conjugates of α, are in bijection to the cyclic shifts of the coefficient vector $\psi(\alpha)$.   □

The next result shows how to determine the smallest field extension of $\mathbb{F}_q$ which contains α provided the coefficient vector of α with respect to a normal basis of $\mathbb{F}_{q^n}$ over $\mathbb{F}_q$ is given.

3.4.14   **Lemma** *The element $\alpha \in \mathbb{F}_{q^n}$, given by 3.4.12, belongs to the subfield $\mathbb{F}_{q^m}$ of $\mathbb{F}_{q^n}$ if and only if $\alpha_i = \alpha_{i+m \bmod n}$ for $i \in n$.*

Proof: Let m be a divisor of n. The element $\alpha \in \mathbb{F}_{q^n}$ belongs to the subfield $\mathbb{F}_{q^m}$ if and only if $\tau^m(\alpha) = \alpha$. This is equivalent to

$$\sum_{i \in n} \alpha_i \tau^{i+m}(\kappa) = \sum_{i \in n} \alpha_i \tau^i(\kappa),$$

whence

$$\sum_{j=0}^{m-1} \alpha_{j-m+n} \tau^j(\kappa) + \sum_{j=m}^{n-1} \alpha_{j-m} \tau^j(\kappa) = \sum_{i \in n} \alpha_i \tau^i(\kappa),$$

which yields $\alpha_i = \alpha_{i+m \bmod n}$ for $i \in n$, since the coefficients with respect to a basis are uniquely determined.   □

If α belongs to $\mathbb{F}_{q^m}$ with $m < n$ and necessarily m a divisor of n, we say that the coefficient vector $(\alpha_0, \ldots, \alpha_{n-1})$ has nontrivial cyclic symmetries, since $\alpha_i = \alpha_{i+m \bmod n}$. In this case the stabilizer of α contains $\langle \tau^m \rangle$. Consequently, the element α given by 3.4.12 belongs to $\mathbb{F}_q$ if and only if $\alpha_0 = \ldots = \alpha_{n-1}$. The element α belongs to no proper subfield of $\mathbb{F}_{q^n}$ if and only if the coefficient vector has no cyclic symmetries. In this case we call it *acyclic*.

Now we want to describe the natural correspondence between monic, irreducible polynomials of degree n over $\mathbb{F}_q$ and orbits of maximal length of the Galois group $G = \mathrm{Gal}[\mathbb{F}_{q^n} : \mathbb{F}_q]$. If α is the root of an irreducible polynomial of degree n over $\mathbb{F}_q$, then the orbit $G(\alpha)$ contains the n conjugates of α. Whence it is an orbit of length n, i.e. it is an orbit of maximal length. Conversely, if $G(\beta)$ is an orbit of length n for some $\beta \in \mathbb{F}_{q^n}$, then the powers $\beta, \beta^q = \tau(\beta), \ldots, \beta^{q^{n-1}} = \tau^{n-1}(\beta)$ are pairwise distinct. Thus β belongs to no proper subfield of $\mathbb{F}_{q^n}$ and its minimal polynomial is, therefore, a monic, irreducible polynomial of degree n over $\mathbb{F}_q$.

The Galois group $G = \langle \tau \rangle$ is cyclic of order n, and the induced permutation $\overline{\tau}$ of the normal basis X is a cycle of length n. Hence, $\langle \overline{\tau} \rangle$ is a cyclic group of order n isomorphic to the Galois group. We indicate a group generated by a cyclic permutation of length n by C_n. The action of $\langle \tau \rangle$ on $\mathbb{F}_{q^n}$ is similar to the action of C_n on $\mathbb{F}_q^n$ (cf. Exercise 3.4.8). The group C_n contains for each divisor d of n exactly one subgroup U of order d. This subgroup is generated by a permutation consisting of n/d cycles of length d. There exist exactly $\phi(d)$ generators of U, where ϕ denotes the *Euler function*,

$$\phi(d) := |\{0 \le i < d \mid \gcd(i, d) = 1\}|. \qquad \text{3.4.15}$$

Some properties of the Euler function are collected in Exercise 3.4.9. Hence, by the Lemma of Cauchy–Frobenius, the number of orbits of the Galois group on $\mathbb{F}_{q^n}$ is equal to

$$|\langle \tau \rangle \backslash\backslash \mathbb{F}_{q^n}| = \frac{1}{n} \sum_{d \mid n} \phi(d) q^{n/d}. \qquad \text{3.4.16}$$

The lattice of subgroups of the Galois group is isomorphic to the lattice of divisors of n. This means that the Möbius function of the lattice of subgroups of the Galois group coincides with the number theoretic Möbius function. Hence, by 3.4.6, the number of orbits of $\langle \tau \rangle$ of maximal length is equal to

$$|\langle \tau \rangle \backslash\backslash_{\hat{1}} \mathbb{F}_{q^n}| = \frac{1}{n} \sum_{d \mid n} \mu(d) q^{n/d} = m_q(n). \qquad \text{3.4.17}$$

The last equation is obtained from the proof of 3.2.25.

More generally, let C_n be a cyclic group of order n. We usually identify C_n with the permutation group of the set $n = \{0, 1, \ldots, n-1\}$, which is generated by the cyclic permutation $(0, 1, \ldots, n - 1)$. We are especially interested in the orbits of maximal length of C_n on m^n. The numbers

3.4.18
$$l_{mn} := |C_n \backslash\backslash_{\tilde{1}} m^n| = \frac{1}{n} \sum_{d|n} \mu(d) m^{n/d}$$

are called *Dedekind numbers*. The first few of these numbers are shown in the following table:

Table 3.2 Dedekind numbers

$m \backslash n$	1	2	3	4	5	6	7
1	1	0	0	0	0	0	0
2	2	1	2	3	6	9	18
3	3	3	8	18	48	116	312
4	4	6	20	60	204	670	2340
5	5	10	40	150	624	2580	11160
6	6	15	70	315	1554	7735	39990
7	7	21	112	588	3360	19544	117648

3.4.19   **Corollary** *The number $m_q(n)$ of irreducible polynomials in $\mathbb{F}_q[x]$ of degree n is equal to l_{qn} as in 3.4.18, the Dedekind number of asymmetric sequences of length n over an alphabet of size q.* $\square$

3.4.20   **The (co)lexicographical order** Assume that $(X, \leq)$ is a totally ordered set. For $n \in \mathbb{N}$ the functions $f \in X^n$ can be considered as *words*

$$f = f(0) \ldots f(n - 1)$$

of *length* n over the *alphabet* X. The set X^n, of all words of length n over X (or all vectors of length n over X, or all functions from n to X), is totally ordered by the *lexicographical order*. For two words $x_0 \ldots x_{n-1}$ and $y_0 \ldots y_{n-1}$ with $x_i, y_i \in X$, $i \in n$, this order puts $x_0 x_1 \ldots x_{n-1} < y_0 y_1 \ldots y_{n-1}$ if there exists $i \in n$ such that $x_j = y_j$ for $j \in i$ and $x_i < y_i$. In order to compare words which are not of the same length, we extend the lexicographical order in the following way. For words $\alpha = x_0 \ldots x_{m-1}$ and $\beta = y_0 \ldots y_{n-1}$ we define $\alpha < \beta$ if

— either there exists $i \in \{0, \ldots, \min\{m - 1, n - 1\}\}$ such that $x_j = y_j$ for $j \in i$ and $x_i < y_i$,

— or $n > m$ and $\beta = x_0 \ldots x_{m-1} y_m \ldots y_{n-1}$.

The *colexicographical order*, is defined by $x_0 x_1 \ldots x_{n-1} < y_0 y_1 \ldots y_{n-1}$ (with $x_i, y_i \in X$, $i \in n$) if there exists $i \in n$ such that $x_j = y_j$ for $i < j \leq n - 1$ and $x_i < y_i$. The set X^n is totally ordered by the colexicographical order. $\diamond$

Now we consider m^n, the set of all words of *length n* over the totally ordered *alphabet m*. It is totally ordered by means of the lexicographical order
(cf. Exercise 3.4.10). We write $f \leq C_n(f)$ if $f \leq g$ for all $g \in C_n(f)$. Thus, an
element f with $f \leq C_n(f)$ is the *lexicographically least* element in its orbit under
the cyclic group. It is called a *necklace*. A necklace whose orbit has length n
is called *Lyndon word* of length n over the alphabet m. These words form the
canonical transversal

$$L_n(m) := \left\{ f \in m^n \mid |C_n(f)| = n,\ f \leq C_n(f) \right\}.$$

The union of all these sets

$$L(m) := \bigcup_{n \in \mathbb{N}^*} L_n(m)$$

is called the *Lyndon set* over the alphabet m.

Moreover, we consider the set of all words over the alphabet m,

$$m^* := \bigcup_{n \in \mathbb{N}} m^n.$$

The *length* of the word $x_0 \ldots x_{n-1}$ in m^* is n. The *empty word* ϵ is the unique
word of length zero in m^*. The *concatenation* of the words $x_0 \ldots x_{m-1}$ and
$y_0 \ldots y_{n-1}$ yields the word $x_0 \ldots x_{m-1}y_0 \ldots y_{n-1}$ in m^*. The set m^* together
with the concatenation is a semigroup with neutral element ϵ.

A word $\beta \in m^*$ is called a *prefix* of $\alpha \in m^*$ if there is a word $\gamma \in m^*$ such
that α is the concatenation $\beta\gamma$. Correspondingly, $\gamma \in m^*$ is called a *suffix* of
$\alpha \in m^*$ if there is a word $\beta \in m^*$ such that $\alpha = \beta\gamma$. A suffix or prefix of
α is called *proper* if its length is at least 1 and less than the length of α. Any
nonempty prefix of a Lyndon word is called a *pre-Lyndon word*.

Next we want to present another characterization of Lyndon words.

Lemma *A word $\alpha \in m^n$ is a Lyndon word if and only if α is strictly less than each*   **3.4.21**
of its proper suffixes.

Proof: Assume that $\alpha = x_0 \ldots x_{n-1}$ is a Lyndon word. Then for any $j \in$
$\{1, \ldots, n-1\}$ we have $\alpha < x_j \ldots x_{n-1}x_0 \ldots x_{j-1}$, whence the first $n - j$ positions satisfy $x_0 \ldots x_{n-1-j} \leq x_j \ldots x_{n-1}$. We claim that this inequality always
holds with $<$. Assume, on the contrary, that $x_0 \ldots x_{n-1-j} = x_j \ldots x_{n-1}$ for
some j. Then there exists some $k \geq n - j$ such that $x_i = x_{i+j \bmod n}$ for $i \in k$ and
$x_k < x_{k+j \bmod n} = x_{k+j-n}$. But then

$$x_0 \ldots x_{k+j-n} > x_{n-j} \ldots x_k$$

which yields $\alpha > x_{n-j} \ldots x_{n-1}x_0 \ldots x_{n-j-1}$. This is a contradiction to α being a
Lyndon word. Consequently, α is strictly less than its proper suffix $x_j \ldots x_{n-1}$.

Conversely, assume that α is less than all its proper suffixes, then evidently, α is less than all its cyclic shifts, and α is a Lyndon word.   □

3.4.22   **Lemma** *Assume that ℓ_1 and ℓ_2 are Lyndon words. The concatenation $\ell_1\ell_2$ is a Lyndon word if and only if $\ell_1 < \ell_2$.*

Proof: Assume that $\ell_1 < \ell_2$. According to 3.4.21, we have to show that $\ell_1\ell_2$ is strictly less than each of its proper suffixes. Assume on the contrary that there exists a proper suffix β of $\ell_1\ell_2$ such that $\beta \leq \ell_1\ell_2$, therefore $\beta < \ell_1\ell_2$. Then either β is a suffix of ℓ_2 and $\ell_1 < \ell_2 \leq \beta$ or β is of the form $\alpha\ell_2$, where α is a proper suffix of ℓ_1, whence $\ell_1 < \alpha\ell_2 = \beta$. Consequently, in both situations we have $\ell_1 < \beta < \ell_1\ell_2$, and β can be written as $\beta = \ell_1\gamma$, where the suffix γ of β is a proper suffix of ℓ_2. (Otherwise $\beta = \ell_1\ell_2$ is not a proper suffix of $\ell_1\ell_2$.) Hence, $\ell_2 < \gamma$ and $\ell_1\ell_2 < \ell_1\gamma = \beta$ which is a contradiction to our assumption.

The remaining parts of the proof are left to the reader as Exercise 3.4.11.   □

Now we consider an arbitrary word $f \in m^n$. Since each letter $i \in m$ forms a Lyndon word of length 1, there exists a uniquely determined longest left factor $l_0(f) = f(0)\ldots$ of f which is a Lyndon word. The same is true for the remaining part w of $f = l_0(f)w$, whence there is a unique decomposition

3.4.23 $$f = l_0(f)l_1(f)\ldots l_{\lambda(f)-1}(f)$$

of f into Lyndon words such that in the right part $l_i(f)\ldots l_{\lambda(f)-1}(f)$ of f the Lyndon word $l_i(f)$ is the maximal left Lyndon factor. This decomposition is called the *Lyndon decomposition* of f, and the number $\lambda(f)$ of factors is called the *Lyndon length* of f. The Lyndon decomposition of f can also be described as follows:

3.4.24   **Lyndon's Theorem** *The factors $l_i(f)$ of the decomposition 3.4.23 of a word $f \in m^n$ are lexicographically decreasing, i.e.*

$$l_0(f) \geq l_1(f) \geq \ldots \geq l_{\lambda(f)-1}(f).$$

Conversely, every decomposition of f into (weakly) decreasing Lyndon words is the Lyndon decomposition of f.   □

In the next section we will discuss algorithms for listing all necklaces, Lyndon words and pre-Lyndon words of length n over the alphabet m.

Exercises

E.3.4.1   **Exercise** Prove 3.4.1.

Exercise Let G be a group. Prove that G acts on itself by *conjugation*, i.e.   **E.3.4.2**

$$G \times G \to G \,:\, (g, x) \mapsto gxg^{-1}$$

is a group action. The orbit $G(x)$ is called the *conjugacy class* of x. The stabilizer G_x is the *centralizer* of x. When is this group action transitive? When is each orbit of size 1?

Exercise Let G be a finite group acting transitively on the finite set X. Show   **E.3.4.3**
that for arbitrary $x \in X$ we have

$$|G_x \backslash\backslash X| = \frac{1}{|G|} \sum_{g \in G} |X_g|^2.$$

Exercise Let G be a group and $L(G) := \{U \mid U \le G\}$ its set of subgroups.   **E.3.4.4**
Prove that G acts on $L(G)$ by *conjugation*, i.e.

$$G \times L(G) \to L(G) \,:\, (g, U) \mapsto gUg^{-1} = \left\{ gxg^{-1} \,\middle|\, x \in U \right\}$$

is a group action. The orbit $G(U)$ is called the *conjugacy class*

$$\tilde{U} := \left\{ gUg^{-1} \,\middle|\, g \in G \right\}$$

of U. The stabilizer G_U is the *normalizer* $N_G(U)$ of U. A subgroup U is called *normal* if $|G(U)| = 1$. The fact that U is a normal subgroup of G is indicated by $U \trianglelefteq G$.

Exercise Prove 3.4.3.   **E.3.4.5**

Exercise Prove that $L(G)$ is indeed a lattice.   **E.3.4.6**

Exercise Prove 3.4.9.   **E.3.4.7**

Exercise Let $C_n := \langle (0, 1, \ldots, n-1) \rangle$ be the permutation group which is gen-   **E.3.4.8**
erated by a single cycle of length n. Describe the orbit $C_n(f)$ of a vector $f \in Y^n$
for the action of C_n on Y^n given by 1.4.7.

Exercise Let ϕ be the Euler function defined by 3.4.15. Show that for any prime   **E.3.4.9**
p and for $n \in \mathbb{N}^*$ the following assertions hold:

— $\phi(p) = p - 1$.

— $\phi(p^n) = p^{n-1}(p-1) = p^n - p^{n-1} = p^n(1 - \frac{1}{p})$.

— If a and b are positive integers which are relatively prime, then $\phi(a \cdot b) = \phi(a) \cdot \phi(b)$.

— $\phi(n) = n \cdot \prod_{p|n}(1 - \frac{1}{p})$.

— For $n \geq 2$ the order of the multiplicative group of $\mathbb{Z}_n^*$ is given by $\phi(n)$.

— $\sum_{d|n} \phi(d) = n$.

E.3.4.10 **Exercise** Assume that $(X, \leq)$ is a totally ordered set. Show that both the lexicographical and the colexicographical order are total orders on the set X^n.

E.3.4.11 **Exercise** Prove the missing details in 3.4.22.

E.3.4.12 **Exercise** Prove 3.4.24.

E.3.4.13 **Exercise** Compute the Lyndon decomposition of

$$2718281828459045235360287471352662497757$$

using the natural order on the alphabet $\{0, 1, \ldots, 9\}$.

3.5 3.5 Construction of Irreducible Polynomials

So far we have described finite fields as extensions over their prime fields. The structure of prime fields of prime characteristic p is clear, since they are isomorphic to the residue class rings $\mathbb{Z}_p = \mathbb{Z}/p\mathbb{Z}$. A finite field $\mathbb{F}_q$ with $q = p^n$ and $n \geq 2$ was described as

— a residue class ring $\mathbb{Z}_p[x]/I(f)$, where $f \in \mathbb{Z}_p[x]$ is a monic, irreducible polynomial of degree n (cf. 3.1.6),

— a $\mathbb{Z}_p$-vector space with basis $\{1, \alpha, \ldots, \alpha^{n-1}\}$, where α is a root of a monic, irreducible polynomial over $\mathbb{Z}_p$ of degree n (cf. 3.1.7),

— a $\mathbb{Z}_p$-vector space with normal basis $\{\kappa, \sigma(\kappa) = \kappa^p, \ldots, \sigma^{n-1}(\kappa) = \kappa^{p^{n-1}}\}$, where κ is a normal element and σ is the Frobenius automorphism. According to 3.3.5, κ is a root of a monic, irreducible polynomial over $\mathbb{Z}_p$ of degree n (cf. 3.4.12),

— the union $\{0\} \cup \mathbb{F}_q^*$, where $\mathbb{F}_q^*$ is a multiplicative group generated by a primitive element α, whence α is a root of a monic, irreducible polynomial over $\mathbb{Z}_p$ of degree n, and the multiplicative order of α is equal to $q - 1$ (cf. 3.2.8).

Consequently, for all computations in finite fields $\mathbb{F}_q$ it is necessary to know irreducible polynomials of given degree over $\mathbb{F}_p$ or, more generally, over a given finite field $\mathbb{F}_{p^m}$.

Now we want to discuss several methods for constructing irreducible polynomials. (A detailed description of these methods can be found in chapters 3 and 4 of [131].) We determine them as minimal polynomials, and we describe a method for computing all irreducible polynomials over $\mathbb{F}_q$ of degree n when a normal basis of $\mathbb{F}_{q^n}$ over $\mathbb{F}_q$ is known. We also discuss two factoring algorithms which determine all irreducible factors of a given polynomial in $\mathbb{F}_q[x]$. Furthermore, we present methods for randomly generating irreducible polynomials over $\mathbb{F}_q$ of given degree, by generating monic polynomials which later on must be tested for irreducibility over $\mathbb{F}_q$. Finally, we discuss how to find primitive elements in $\mathbb{F}_q$. The algorithms and ideas presented are taken mainly from [136] and [131].

Before going into details, a short remark about computing the n-fold sum or n-fold product (cf. Exercise 1.6.6) of an element can be very useful and helpful for computations in finite fields. When restricting to nonnegative multiples, then these computations can be defined in a semigroup with neutral element (cf. [136, pages 17ff]). Let $(S, +, 0)$ be a semigroup with neutral element 0, then the method of *repeated doubling and adding* allows the computation of $a + n \cdot b$ for $a, b \in S, n \in \mathbb{N}$ in the following way. Let

$$R(a, b, n) := a + n \cdot b,$$

then R satisfies

$$
\begin{aligned}
R(a, b, 0) &= a, \\
R(0, b, n) &= n \cdot b, \\
R(a, b, 2n) &= R(a, 2 \cdot b, n), \\
R(a, b, n+1) &= R(a + b, b, n).
\end{aligned}
$$

Using the last two rules, either the integer n can be reduced by 1 if n is odd, or it can be be divided by 2 if n is even, so that finally we arrive at $n = 0$. Then, according to the first rule, the result can be read from the first parameter.

For example $2 + 15 \cdot 9 = R(2, 9, 15) = R(11, 9, 14) = R(11, 18, 7) = R(29, 18, 6) = R(29, 36, 3) = R(65, 36, 2) = R(65, 72, 1) = R(137, 72, 0) = 137$. Of course, this method is even more helpful when S is a finite semigroup.

Correspondingly, the method of *repeated squaring and multiplying* can be applied for computing ab^n for elements a, b of a semigroup $(S, \cdot, 1)$ and $n \in \mathbb{N}$.

Let
$$R_M(a, b, n) := ab^n$$
then R_M satisfies
$$
\begin{array}{rcl}
R_M(a, b, 0) &=& a, \\
R_M(1, b, n) &=& b^n, \\
R_M(a, b, 2n) &=& R_M(a, b^2, n), \\
R_M(a, b, n+1) &=& R_M(ab, b, n).
\end{array}
$$
When we apply this method for computing powers in finite fields of characteristic p we also use
$$
\begin{array}{rcl}
R_M(a, b, pn) &=& R_M(a, b^p, n), \\
R_M(a, b, p+n) &=& R_M(ab^p, b, n),
\end{array}
$$
since computing the p-th power is usually easier done than computing an arbitrary power.

Since minimal polynomials are by definition irreducible, one obtains irreducible polynomials by computing minimal polynomials of algebraic elements (cf. [131, pages 102ff]).

3.5.1   **Example** Assume that $\{1, \alpha, \ldots, \alpha^{n-1}\}$, with $\alpha \in \mathbb{F}_{q^n}$, is a basis of $\mathbb{F}_{q^n}$ over $\mathbb{F}_q$. In order to find the minimal polynomial $f \in \mathbb{F}_q[x]$ of $\beta \in \mathbb{F}_{q^n}$ we express the powers $\beta^0, \beta^1, \ldots, \beta^n$ with respect to this basis, obtaining
$$\beta^i = \sum_{j \in n} \kappa_{ij} \alpha^j, \qquad \kappa_{ij} \in \mathbb{F}_q, \ j \in n, \ 0 \le i \le n.$$
We determine f as
$$f(x) = \sum_{i=0}^{n} \lambda_i x^i \in \mathbb{F}_q[x],$$
so that f is monic, and of least degree with $f(\beta) = 0$. The second property leads to the homogeneous system of linear equations
$$\sum_{i=0}^{n} \lambda_i \kappa_{ij} = 0, \qquad j \in n,$$
for the unknown $\lambda_0, \ldots, \lambda_n$. The coefficient matrix $K = (\kappa_{ij})$ is an $(n+1) \times n$-matrix of rank r with $1 \le r \le n$. Hence, the dimension of the space of solutions of this system is $s = n + 1 - r$, thus $1 \le s \le n$. Therefore, we can prescribe s values of $\lambda_0, \ldots, \lambda_n$. The remaining ones are then uniquely determined. If $s = 1$ we set $\lambda_n = 1$, if $s > 1$ we set $\lambda_n = \ldots = \lambda_{n-s+2} = 0$ and $\lambda_{n-s+1} = 1$. This way we obtain a monic polynomial f of smallest degree such that $f(\beta) = 0$.
$\diamond$

There is another approach using the Frobenius automorphism for the determination of the minimal polynomial.

Example [131, pages 103ff] Assume that β belongs to $\mathbb{F}_{q^n}$, and let $\tau \colon \kappa \mapsto \kappa^q$ **3.5.2**
be the Frobenius automorphism of $\mathbb{F}_{q^n}$ over $\mathbb{F}_q$. We compute the values $\tau(\beta)$,
$\tau^2(\beta), \ldots$ until we find the least positive integer m such that $\tau^m(\beta) = \beta$. This
integer m is the degree of the minimal polynomial f of β over $\mathbb{F}_q$, and f is
given by

$$f(x) = \prod_{i \in m} (x - \tau^i(\beta)).$$

Moreover, the elements $\beta, \tau(\beta), \ldots, \tau^{m-1}(\beta)$ are the m distinct conjugates of β.
$\diamond$

If we have a normal basis $\{\kappa, \tau(\kappa), \ldots, \tau^{n-1}(\kappa)\}$ of $\mathbb{F}_{q^n}$ over $\mathbb{F}_q$, where τ is
the Frobenius automorphism of $\mathbb{F}_{q^n}$ over $\mathbb{F}_q$, then it is very easy to compute all
irreducible polynomials of degree n over $\mathbb{F}_q$. Generalizing 3.4.14, these polynomials occur as the minimal polynomials of those elements $\alpha \in \mathbb{F}_{q^n}$ which
have an acyclic coefficient vector in the representation

$$\alpha = \sum_{i \in n} \alpha_i \tau^i(\kappa), \qquad \alpha_i \in \mathbb{F}_q.$$

Knowing all irreducible polynomials of degree at most k is essential for describing the conjugacy classes of the linear groups $\mathrm{GL}_k(q)$ (cf. Section 6.3).
As shown in the previous section, an irreducible polynomial of degree n corresponds to the n distinct conjugates $\tau^j(\alpha)$ for $j \in n$ which form an orbit of
maximal length n of C_n on $\mathbb{F}_q^n$.

Hence, in order to list all irreducible polynomials of degree n over $\mathbb{F}_q$, we
list all Lyndon words of length n over an alphabet of size $q = |\mathbb{F}_q|$, identify
these Lyndon words with coefficient vectors $(\alpha_0, \ldots, \alpha_{n-1})$ with respect to the
given normal basis and compute polynomials of the form

$$f(x) = \prod_{j \in n} \left(x - \sum_{i \in n} \alpha_{i-j \bmod n} \tau^i(\kappa) \right)$$

which are necessarily monic, irreducible, and of degree n.

Before describing an algorithm for listing all Lyndon words, we need some
more facts about pre-Lyndon words over the alphabet m. (For more details
see [111, Section 7.2.1.1].) The *n-extension* of a nonempty word $\alpha \in m^k, k \geq 1$, is
the concatenation of $\lfloor n/k \rfloor$ copies of α with the prefix α' of α of length $n \bmod k$.
For example the 10-extension of $\alpha = 123$ is 1231231231.

Lemma *Let α be a pre-Lyndon word of length $n \geq 1$.*   **3.5.3**

1. *If the final letter of α is increased by 1, then the resulting word is a Lyndon word.
   (Here we assume that the alphabet contains sufficiently many elements, so that the
   last letter of a given word α can be increased by 1.)*

2. *The word α is the n-extension of the first factor ℓ_0 in its Lyndon decomposition.*

3. *The word α cannot be the n-extension of two different Lyndon words.*

Proof: 1. Since each word of length 1 is a Lyndon word, the first assertion is true for words of length 1, and we can assume that α is of length greater than 1. Since α is a pre-Lyndon word there exists some $\omega \in m^*$ such that $\alpha\omega$ is a Lyndon word. Let α' be the word obtained from α by increasing its last letter by 1. Moreover, we write α as the concatenation of two nonempty words $\alpha = \beta\gamma$, and, correspondingly, $\alpha' = \beta\gamma'$. Thus γ' is a proper suffix of α', and we want to show that $\alpha' < \gamma'$.

Let θ be a prefix of α of the same length as γ, then $\theta \leq \alpha\omega < \gamma\omega$ since $\gamma\omega$ is a proper suffix of the Lyndon word $\alpha\omega$. Consequently, $\theta \leq \gamma < \gamma'$. Since θ is also a prefix of α', we deduce that $\alpha' < \gamma'$. The last inequality holds for all proper suffixes γ' of α', whence, by 3.4.21, it is shown that α' is a Lyndon word.

2. Let ℓ_0 be the first factor in the Lyndon decomposition of α, so that $\alpha = \ell_0\beta = x_0 \ldots x_{n-1}$. The length of ℓ_0 is indicated by r. If $r = n$ then α is the Lyndon word ℓ_0. So we assume that $r < n$. Since α is a pre-Lyndon word, there exists some $\omega \in m^*$ such that $\alpha\omega = \ell_0\beta\omega$ is a Lyndon word. Hence, $\ell_0\beta\omega < \beta\omega$, from which we want to deduce that $x_j = x_{j+r}$ for $j \in n - r$. For $j = 0$ we first obtain $x_0 \leq x_r$. Therefore $x_0 = x_r$, since otherwise $\ell_0 x_r$ would be a prefix of α which is a Lyndon word of length $r + 1$. This is a contradiction to the fact that in the decomposition 3.4.23 the word ℓ_0 is the longest prefix of α which is a Lyndon word.

Assume now that $x_i = x_{i+r}$ for $i \in j$ and $j < n - r - 1$. We have to prove that also $x_j = x_{j+r}$ is satisfied. From the fact that $\ell_0\beta\omega < \beta\omega$ and the assumptions on x_i we obtain that $x_j \leq x_{j+r}$. Assuming that $x_j < x_{j+r}$ and putting $j_0 = \lfloor j/r \rfloor r$, it follows from the first assertion of this lemma that $x_{j_0} \ldots x_{j+r}$ is a Lyndon word. By construction we have $\ell_0 < x_{j_0} \ldots x_{j+r}$, which yields together with repeated applications of 3.4.22 that $x_0 \ldots x_{j+r}$ is (as the concatenation of some copies of ℓ_0 and $x_{j_0} \ldots x_{j+r}$) a Lyndon word of length greater than r which is a prefix of α. Again this is a contradiction to the assumed Lyndon decomposition of α. Consequently, α is the n-extension of ℓ_0.

3. In order to prove the last assertion, we assume that α is the n-extension of two different Lyndon words ℓ_0 and ℓ_0'. Without loss of generality, the length of ℓ_0 is greater than the length of ℓ_0'. Then ℓ_0 is the concatenation of some copies of ℓ_0' and a proper prefix θ of ℓ_0'. Note that this proper prefix must necessarily appear in ℓ_0, for otherwise ℓ_0 would have a nontrivial cyclic symmetry. Thus $\ell_0 = \ell_0' \ldots \ell_0' \theta$. Since θ is a proper prefix of ℓ_0' we have $\theta < \ell_0'$. Since θ is a

proper suffix of ℓ_0 we have $\ell_0 < \theta$. These inequalities lead to the contradiction $\theta < \ell'_0 < \ell_0 < \theta$. $\qquad\qquad\qquad\qquad\qquad\qquad\qquad\qquad\qquad\qquad\square$

As an immediate consequence we obtain

Corollary *A word of length $n \geq 1$ is a pre-Lyndon word if and only if it is the* **3.5.4**
n-extension of a Lyndon word of length $k \leq n$. This Lyndon word is uniquely deter-
mined. $\qquad\qquad\qquad\qquad\qquad\qquad\qquad\qquad\qquad\qquad\qquad\qquad\qquad\quad\square$

This corollary describes a one-to-one correspondence between Lyndon words of length not greater than n and pre-Lyndon words of length n.

Now we are in a position to describe an algorithm for generating all pre-Lyndon words of length n (see also [57]). It generates all pre-Lyndon words $x_0 \ldots x_{n-1}$ over the alphabet m in the lexicographical order and indicates the index j in any pre-Lyndon word such that $x_0 \ldots x_j$ is the first factor in the Lyndon decomposition of $x_0 \ldots x_{n-1}$.

Algorithm (pre-Lyndon [111]) 3.5.5

    Input: Two positive integers n and m.
    Output: The list of all pre-Lyndon words of length n over the alphabet m.

(1) Set $x_0 := \ldots := x_{n-1} := 0$ and $j := 0$. (The word x_0 is a pre-Lyndon word.)

(2) Print $x_0 \ldots x_{n-1}$ and indicate $x_0 \ldots x_j$ as a Lyndon word.

(3) Set $j := n - 1$.
    If $x_j = m - 1$ decrease j until ($j < 0$ or $x_j < m - 1$).

(4) If $j < 0$ terminate the algorithm, otherwise increase x_j by 1. (The word $x_0 \ldots x_j$ is a Lyndon word of length $j + 1$ by the first assertion of 3.5.3.)

(5) Compute the n-extension of $x_0 \ldots x_j$, i.e. for k from $j + 1$ to $n - 1$ set $x_k := x_{k-(j+1)}$. (The word $x_0 \ldots x_{n-1}$ is a pre-Lyndon word by the second assertion of 3.5.3.)
    Goto (2).

In (3) and (4) we compute the smallest Lyndon word of length not greater than n which is (in the lexicographical order) greater than the previously computed pre-Lyndon word. $\qquad\qquad\qquad\qquad\qquad\qquad\qquad\qquad\qquad\qquad\qquad\square$

This algorithm can also be used for generating all necklaces of length n over m, since a pre-Lyndon word is a necklace if and only if it is the n-extension of a Lyndon word of length d where d is a divisor of n.

The output of this algorithm for $n = 4$ and $m = 3$ is

$$\begin{array}{llllllll}
\underline{0000} & \underline{0011} & \underline{0022} & \underline{0111} & \underline{0122} & \underline{0212} & 1111 & 1212 \\
\underline{0001} & \underline{0012} & 0101 & \underline{0112} & 0202 & 0220 & 1112 & 1221 \\
\underline{0002} & \underline{0020} & \underline{0102} & \underline{0120} & 0210 & \underline{0221} & 1121 & 1222 \\
\underline{0010} & \underline{0021} & \underline{0110} & \underline{0121} & \underline{0211} & 0222 & 1122 & 2222,
\end{array}$$

where the letters of the longest prefix, which is a Lyndon word, are underlined. In addition to the Lyndon words of length 4, the following pre-Lyndon words describe necklaces: 0000, 0101, 0202, 1111, 1212, and 2222.

Another method for constructing irreducible polynomials over $\mathbb{F}_q$ is to factor a given polynomial in $\mathbb{F}_q[x]$. (For more details see chapter 4 of [131] and chapter 12 of [136].) This means, for a given polynomial $f \in \mathbb{F}_q[x]$ we determine the factorization

$$f(x) = \prod_{i \in t} f_i^{m_i}$$

into pairwise distinct irreducible polynomials $f_i \in \mathbb{F}_q[x]$ with $m_i \in \mathbb{N}^*$ for $i \in t$. We describe two different algorithms. The first one is deterministic. It applies to polynomials which are defined over fields of small order. The second algorithm uses probabilistic methods. It is more suitable for larger fields. Here, a field $\mathbb{F}_q$ is considered large if q is substantially greater than the degree of the polynomial to be factored.

Both algorithms can only be applied to polynomials which are square free, i.e. which do not admit multiple factors. For this reason, we first describe the square free factorization. It is based on properties of the derivative of a polynomial and it yields a decomposition

$$\prod_{i \in t} f_i$$

without repeated factors.

Let $f(x) = \sum_{i=0}^{n} \kappa_i x^i$ be a polynomial in $\mathbb{F}[x]$. Then the formal *derivative* f' of f is defined by

$$f'(x) = \sum_{i=1}^{n} i\kappa_i x^{i-1}.$$

For $f, g \in \mathbb{F}[x]$ the following rules are satisfied:

3.5.6

$$(f + g)' = f' + g', \qquad (fg)' = f'g + fg', \qquad (f \circ g)' = (f' \circ g)g'.$$

The derivative of f permits us to determine whether f has multiple roots or not.

Lemma [131, 1.68 Theorem] *The element $\alpha \in \mathbb{F}$ is a multiple root of $f \in \mathbb{F}[x] \setminus \{0\}$*   **3.5.7**
if and only if α is a root of both f and f'. $\square$

Since, when computing the derivative of f, we have to form i-fold sums of the coefficients of f, the following lemma is very important.

Lemma *Let p be the characteristic of $\mathbb{F}_q$.*   **3.5.8**

1. *Assume that $f(x) = \sum_{i=0}^{n} \kappa_i x^i$ is a polynomial over $\mathbb{F}_q$. Then $f' = 0$ if and only if $\kappa_i = 0$ for $i \neq 0 \bmod p$. This in turn is equivalent to the existence of polynomials $g, h \in \mathbb{F}_q[x]$ such that $f(x) = g(x^p) = h(x)^p$.*

2. *If $f \in \mathbb{F}_q[x]$ is irreducible, then $f' \neq 0$.* $\square$

By using the derivative f' of f, we determine not only multiple roots of f but, in general, multiple irreducible factors of f.

Theorem [136, page 150] *Assume that $q = p^n$, $f, g \in \mathbb{F}_q[x]$, and g is irreducible.*   **3.5.9**
Consider $m \in \mathbb{N}^*$ such that g^m is a divisor of f and g^{m+1} does not divide f. Then g^{m-1} is a divisor of $\gcd(f, f')$. Moreover, g^m is a divisor of $\gcd(f, f')$ if and only if $m \equiv 0 \bmod p$.

Proof: Define $h \in \mathbb{F}_q[x]$ such that $f = g^m h$, then g is not a divisor of h. Furthermore, $f' = mg^{m-1}g'h + g^m h' = g^{m-1}(mg'h + gh')$. Hence, g^{m-1} is a divisor of $\gcd(f, f')$.

Moreover, g^m divides $\gcd(f, f')$ if and only if g is a divisor of $mg'h$. Since $g' \neq 0$ (cf. 3.5.8), $\deg(g') < \deg(g)$, g is irreducible, and since g is not a divisor of h, it follows from 3.2.4 that g is not a divisor of $g'h$. Therefore, $mg'h$ is divisible by g if and only if $m \equiv 0 \bmod p$. $\square$

Immediately we get

Corollary [136, page 150] *Let $q = p^n$ and $f \in \mathbb{F}_q[x]$. Then $f/\gcd(f, f')$ is the*   **3.5.10**
product of all irreducible factors of f whose multiplicities are not divisible by p. $\square$

In order to factor an arbitrary polynomial $f \in \mathbb{F}_q[x]$, we may assume that f is monic. We begin by computing $d = \gcd(f, f')$. If $d = 1$ then f has no repeated factors. If $d = f$ then necessarily $f' = 0$ and, consequently, there exists a polynomial $g \in \mathbb{F}_q[x]$ such that $f(x) = g(x)^p$ for a suitable polynomial $g \in \mathbb{F}_q[x]$. Hence, the factors of f are the factors of g. In order to factor g we have to check whether it has no repeated factors. If d is a nontrivial factor of f, then we factor f/d and d separately. The polynomial f/d is squarefree, whereas d may still have repeated factors, which can then be handled by the same method (apply induction on the degree).

The next theorem describes a first step in factorizing a polynomial:

3.5.11

Theorem [131, 4.1 Theorem] *If* $f \in \mathbb{F}_q[x]$ *is monic and* $h \in \mathbb{F}_q[x]$ *satisfies* $h^q \equiv h \bmod \mathrm{I}(f)$, *then*

$$f(x) = \prod_{\kappa \in \mathbb{F}_q} \gcd(f(x), h(x) - \kappa).$$

Proof: Since $\gcd(f(x), h(x) - \kappa)$ is a divisor of f, and $h(x) - \kappa$ are relatively prime for different $\kappa \in \mathbb{F}_q$, we derive that $\prod_{\kappa \in \mathbb{F}_q} \gcd(f(x), h(x) - \kappa)$ is a divisor of f.

Conversely, from $h^q \equiv h \bmod \mathrm{I}(f)$ we deduce that f is a divisor of

3.5.12

$$h(r)^q - h(r) = \prod_{\kappa \in \mathbb{F}_q} (h(r) - \kappa)$$

This equation follows from 3.2.3 by substituting $h(x)$ for x. Then f is the greatest common divisor $\gcd(f(x), \prod_{\kappa \in \mathbb{F}_q}(h(x) - \kappa))$ and, consequently, f is a divisor of $\prod_{\kappa \in \mathbb{F}_q} \gcd(f(x), h(x) - \kappa)$, which finishes the proof. $\square$

The last theorem does not always yield the complete factorization of f, since $\gcd(f(x), h(x) - \kappa)$ may be reducible. If $h(x) \equiv \kappa \bmod \mathrm{I}(f)$ for some $\kappa \in \mathbb{F}_q$, then this factorization is trivial. If the factorization is not trivial, then h is called an *f-reducing* polynomial. In the sequel, we present a method to construct *f*-reducing polynomials. Any $h \in \mathbb{F}_q[x]$ with $0 < \deg h < \deg f$ and $h^q \equiv h \bmod \mathrm{I}(f)$ is *f*-reducing.

Before we do that, we present the Chinese Remainder Theorem, which enables us to solve simultaneous equivalence relations. Recall that the composition $R \circ S$ of two relations $R, S \subseteq X \times X$ is defined as

$$R \circ S := \{(x, z) \in X \times X \mid \exists y \in X : (x, y) \in S, (y, z) \in R\}.$$

Now we are able to prove

3.5.13

The Chinese Remainder Theorem, set-theoretic version *Let* $R_0, \ldots, R_{t-1}$ *be equivalence relations on a set* X *and let* φ *be given by*

$$\varphi: X \to \underset{i \in t}{\times} X/R_i : x \mapsto \varphi(x) := ([x]_{R_0}, \ldots, [x]_{R_{t-1}}),$$

where $[x]_{R_i}$ *is the equivalence class of* x *with respect to the relation* R_i. *Then* φ *is surjective if and only if* $(R_0 \cap \ldots \cap R_{i-1}) \circ R_i = X \times X$ *for all* $i \in t$.

Proof: 1. We assume that φ is surjective. We have to prove that $(R_0 \cap \ldots \cap R_{i-1}) \circ R_i \supseteq X \times X$ is true for each $i \in t$. Choose any $(x', x'') \in X \times X$ and $i \in t$. Then there exists an element $x \in X$ such that

$$\varphi(x) = ([x'']_{R_0}, \ldots, [x'']_{R_{i-1}}, [x']_{R_i}, \ldots, [x']_{R_{t-1}}),$$

whence

$$(x, x'') \in \bigcap_{j \in i} R_j \text{ and } (x', x) \in R_i.$$

Consequently $(x', x'') \in (R_0 \cap \ldots \cap R_{i-1}) \circ R_i$. Since this holds true for any $(x', x'') \in X \times X$, the first part of the proof is finished.

2. Assume, conversely, that $(R_0 \cap \ldots \cap R_{i-1}) \circ R_i \supseteq X \times X$ is satisfied for each $i \in t$. By induction on t, we prove that φ is surjective. If $t = 1$ then φ is the natural projection from X onto X/R_0 which is surjective. If $t > 1$ then we assume that the mapping

$$\tilde{\varphi} \colon X \to \underset{i \in t-1}{\times} X/R_i \ : \ x \mapsto \tilde{\varphi}(x) := ([x]_{R_0}, \ldots, [x]_{R_{t-2}})$$

is surjective. Thus, for any $x_0, \ldots, x_{t-2}$ there exists an element $x \in X$ such that $\tilde{\varphi}(x) = ([x_0]_{R_0}, \ldots, [x_{t-2}]_{R_{t-2}})$. Consider furthermore $x_{t-1} \in X$. Since

$$(x_{t-1}, x) \in X \times X = (R_0 \cap \ldots \cap R_{t-2}) \circ R_{t-1},$$

there exists some $y \in X$ such that

$$(x_{t-1}, y) \in R_{t-1} \text{ and } (y, x) \in R_0 \cap \ldots \cap R_{t-2}.$$

Hence,

$$\varphi(y) = ([y]_{R_0}, \ldots, [y]_{R_{t-1}}) =$$
$$([x]_{R_0}, \ldots, [x]_{R_{t-2}}, [x_{t-1}]_{R_{t-1}}) = ([x_0]_{R_0}, \ldots, [x_{t-2}]_{R_{t-2}}, [x_{t-1}]_{R_{t-1}})$$

and φ is surjective. $\qquad\square$

This theorem is often applied in rings where the equivalence relation is induced by an ideal in the ring. Two elements r, s of the ring are called equivalent with respect to the ideal I if $r - s \in I$. The relation $\equiv_n$ on $\mathbb{Z}$ defined in Exercise 3.1.3 is an example for this. Here, the ideal is $I = I(n)$. Furthermore, we remind the reader that the sum of two ideals I and J is the set $I + J = \{i + j \mid i \in I, \ j \in J\}$, whereas the product IJ is the ideal generated by all elements of the form ij for $i \in I$ and $j \in J$. (See Exercises 3.5.7 and 3.5.8.) Two ideals I and J in R are called *relatively prime* if $I + J = R$.

Lemma *Let R be a ring with* 1.   3.5.14

1. *If I, J are ideals in R and S_I, S_J are the induced equivalence relations on R, then*

$$S_I \circ S_J = R \times R \Longleftrightarrow I + J = R,$$

   which means that I and J are relatively prime.

2. *If J and $I_0, \ldots, I_{t-1}$ are ideals and $I_j + J = R$ for all $j \in t$, then*

$$I_0 \cdots I_{t-1} + J = I_0 \cap \ldots \cap I_{t-1} + J = R.$$

Proof: Assume that $S_I \circ S_J = R \times R$. Then for any $r \in R$ the pair $(r, 0)$ belongs to $S_I \circ S_J$, whence there exists an element $s \in R$ such that $r - s \in J$ and $s - 0 \in I$. Consequently, $r = s - 0 + r - s \in I + J$. Since r was chosen arbitrarily, $I + J = R$.

On the other hand, if $I + J = R$ then there exist $i \in I$ and $j \in J$ such that $1 = i + j$. For $(r, s) \in R \times R$ let $v = ri + sj$. Then

$$r - v = r(1 - i) - sj = rj - sj \in J \quad \text{and} \quad v - s = ri + s(j - 1) = ri + si \in I.$$

For this reason $(r, v) \in S_J$ and $(v, s) \in S_I$, which means that $(r, s) \in S_I \circ S_J$. Since (r, s) was chosen arbitrarily, $S_I \circ S_J = R \times R$.

The second part is proved by induction on t. The case $t = 1$ is clear. Assume that $t > 1$. Since $I_0 \cdots I_{t-1} \subseteq I_0 \cap \ldots \cap I_{t-1}$, it suffices to show that $I_0 \cdots I_{t-1} + J = R$. For the induction step we assume that $I_0 \cdots I_{t-2} + J = R$. Then there exist $i \in I_0 \cdots I_{t-2}$ and $j \in J$ such that $i + j = 1$. Since $I_{t-1} + J = R$, there exist $r \in I_{t-1}$ and $s \in J$ such that $r + s = 1$. By multiplication we derive

$$1 = (i + j)(r + s) = (ir + is) + (jr + js) \in I_0 \cdots I_{t-1} + J.$$

Since $I_0 \cdots I_{t-1} + J$ is an ideal in R, the proof is finished.   □

Now we formulate

3.5.15   **The Chinese Remainder Theorem, ring-theoretic version** *Let R be a ring with 1 and let $I_0, \ldots, I_{t-1}$ denote ideals in R. Assume that the map φ is given by*

$$\varphi \colon R \to \underset{j \in t}{\times} R/I_j \; : \; r \mapsto \varphi(r) := (r + I_0, \ldots, r + I_{t-1}).$$

With addition and multiplication on $\times_{j \in t} R/I_j$ defined componentwise, the map φ is a ring homomorphism with

$$\ker(\varphi) = \bigcap_{j \in t} I_j.$$

Moreover, φ is surjective if and only if the ideals are pairwise relatively prime, i.e. $I_i + I_j = R$ for all $i, j \in t$ with $i \neq j$.   □

Now we come back to the construction of f-reducing polynomials.

3.5.16   **Berlekamp's algorithm** [131, pages 149ff] Assume that $f = f_0 \cdots f_{t-1}$ is the product of t distinct monic, irreducible polynomials f_i over $\mathbb{F}_q$. The mapping

$$\varphi \colon \mathbb{F}_q[x] \to \underset{j \in t}{\times} \mathbb{F}_q[x]/\mathrm{I}(f_j) \; : \; g \mapsto \varphi(g) := (g + \mathrm{I}(f_0), \ldots, g + \mathrm{I}(f_{t-1}))$$

is a surjective ring homomorphism with $\ker(\varphi) = \mathrm{I}(f)$. According to the Chinese Remainder Theorem, for each $(\kappa_0, \ldots, \kappa_{t-1}) \in \mathbb{F}_q^t$ there exists a unique polynomial $h \in \mathbb{F}_q[x]$ such that $h(x) \equiv \kappa_i \bmod \mathrm{I}(f_i)$ for $i \in t$ and $\deg h < \deg f$. Therefore, $h(x)^q \equiv \kappa_i^q = \kappa_i \equiv h(x) \bmod \mathrm{I}(f_i)$ for $i \in t$, thus

$$h^q \equiv h \bmod \mathrm{I}(f), \qquad \deg h < \deg f. \qquad\qquad \textbf{3.5.17}$$

Conversely, if $h \in \mathbb{F}_q[x]$ is a solution of 3.5.17, then, since 3.5.12 is satisfied, for each irreducible factor f_i of f there exists some $\kappa_i \in \mathbb{F}_q$ such that f_i divides $h(x) - \kappa_i$. Thus all solutions of 3.5.17 satisfy

$$h(x) \equiv \kappa_i \bmod \mathrm{I}(f_i), \qquad i \in t,$$

for some $(\kappa_0, \dots, \kappa_{t-1}) \in \mathbb{F}_q^t$. Consequently, there are q^t solutions of 3.5.17.

In order to solve 3.5.17, we reduce it to a system of linear equations. Let $n = \deg f$. We construct an $n \times n$-matrix $L = (\lambda_{ij})_{i,j \in n}$ with $\lambda_{ij} \in \mathbb{F}_q$ by expressing the powers x^{iq} modulo $f(x)$ as

$$x^{iq} \equiv \sum_{j \in n} \lambda_{ij} x^j \bmod \mathrm{I}(f), \qquad i \in n.$$

Then $h(x) = \sum_{i \in n} \mu_i x^i \in \mathbb{F}_q[x]$ is a solution of 3.5.17 if and only if the coefficient vector $(\mu_0, \dots, \mu_{n-1})$ is a solution of

$$(\mu_0, \dots, \mu_{n-1}) \cdot L = (\mu_0, \dots, \mu_{n-1}).$$

Indeed, h is a solution of 3.5.17 if and only if

$$\sum_{i \in n} \mu_i x^i \equiv \sum_{i \in n} \mu_i^q x^{iq} \bmod \mathrm{I}(f)$$
$$= \sum_{i \in n} \mu_i \sum_{j \in n} \lambda_{ij} x^j$$
$$= \sum_{j \in n} \Big(\sum_{i \in n} \mu_i \lambda_{ij} \Big) x^j.$$

Hence, h is a solution of 3.5.17 if and only if $\mu_j = \sum_{i \in n} \mu_i \lambda_{ij}$ for $j \in n$. This system of linear equations can be written as the homogeneous system

$$(\mu_0, \dots, \mu_{n-1}) \cdot (L - I_n) = 0, \qquad\qquad \textbf{3.5.18}$$

where I_n is the unit matrix. We have just shown that 3.5.18 has q^t solutions, whence the dimension of $\ker(L - I_n)$ is equal to t, and so the rank of $L - I_n$ is $n - t$.

In order to factor f, we determine the rank r of the matrix $L - I_n$. The number of irreducible factors of f is then $t = n - r$, which yields a stopping rule for the algorithm. After that, we determine a basis of the space of solutions of 3.5.18 and compute the corresponding polynomials $h_0, \dots, h_{t-1}$. Without loss of generality we label the polynomials in such a way that $h_0(x) = 1$ is the solution of 3.5.17 corresponding to the solution $(1, 0, \ldots, 0)$ of 3.5.18. Then $0 < \deg(h_i) < n$ for $1 \leq i < t$, and all these polynomials are f-reducing.

If $t = 1$, then f is irreducible. If $t \geq 2$, we first take the f-reducing polynomial $h_1(x)$ and calculate $\gcd(f(x), h_1(x) - \kappa)$ for all $\kappa \in \mathbb{F}_q$. This yields a nontrivial factorization of f by 3.5.11. If the use of h_1 did not split f into t factors we calculate $\gcd(g(x), h_2(x) - \kappa)$ for all $\kappa \in \mathbb{F}_q$ and all nontrivial factors g obtained so far. This procedure must be continued until we have found t factors of f.

This way we find all factors of f. Assume that f_1 and f_2 are two distinct irreducible factors, then there exist $(\kappa_0^{(1)}, \ldots, \kappa_{t-1}^{(1)})$ and $(\kappa_0^{(2)}, \ldots, \kappa_{t-1}^{(2)})$ in $\mathbb{F}_q^t$ such that

$$h_j(x) \equiv \kappa_j^{(i)} \bmod \mathrm{I}(f_i), \qquad j \in t, \; i = 1, 2.$$

We claim that $(\kappa_0^{(1)}, \ldots, \kappa_{t-1}^{(1)}) \neq (\kappa_0^{(2)}, \ldots, \kappa_{t-1}^{(2)})$. Suppose, on the contrary, that $\kappa_j^{(1)} = \kappa_j^{(2)}$ for $j \in t$, then for any solution h of 3.5.17 there exists some $\kappa \in \mathbb{F}_q$ such that

$$h(x) \equiv \kappa \bmod \mathrm{I}(f_i), \qquad i = 1, 2.$$

But at the beginning of 3.5.16 we have shown that there exists, for instance, a solution h of 3.5.17 with

$$h(x) \equiv 0 \bmod \mathrm{I}(f_1) \quad \text{and} \quad h(x) \equiv 1 \bmod \mathrm{I}(f_2),$$

which is a contradiction. □

The methods described above also provide the number of irreducible factors of f.

3.5.19   **Corollary** *Assume that $f \in \mathbb{F}_q[x]$ is a polynomial of degree n which does not have multiple factors. Let $L = (\lambda_{ij})_{i,j \in n}$ be the $n \times n$-matrix determined by*

$$x^{iq} \equiv \sum_{j \in n} \lambda_{ij} x^j \bmod \mathrm{I}(f), \qquad i \in n.$$

Then $\deg f - \mathrm{rank}(L - I_n)$ is the number of irreducible factors of f. □

Now we formulate a useful test for irreducibility of a polynomial f.

3.5.20   **Corollary** *Let $f \in \mathbb{F}_q[x]$ be a polynomial of degree n and let L be the $n \times n$-matrix $(\lambda_{ij})_{i,j \in n}$ determined by*

$$x^{iq} \equiv \sum_{j \in n} \lambda_{ij} x^j \bmod \mathrm{I}(f), \qquad i \in n.$$

Then f is irreducible if and only if f is square free and $\mathrm{rank}(L - I_n) = n - 1$. □

Based on this test we can randomly determine irreducible polynomials over $\mathbb{F}_q$ of degree n. We randomly produce a monic polynomial of degree n over $\mathbb{F}_q$ by choosing arbitrary coefficients $\kappa_0, \ldots, \kappa_{n-1}$. Then we test whether or not the polynomial $f(x) = \sum_{i=0}^{n-1} \kappa_i x^i + x^n$ is irreducible. If it is not, we move on to the next polynomial. There exist exactly q^n monic polynomials of degree n over $\mathbb{F}_q$. According to [136, page 145], approximately q^n/n of them are irreducible. So, on average we will have to test n polynomials in order to find an irreducible one.

Berlekamp's algorithm is useful only for small fields $\mathbb{F}_q$. The reason for this is as follows. For each f-reducing polynomial h we have to compute the greatest common divisor of $f(x)$ and $h(x) - \kappa$ for all $\kappa \in \mathbb{F}_q$. Since the factors occurring in 3.5.11 are relatively prime, and since f consists of t different factors, there will be at most t values $\kappa \in \mathbb{F}_q$ for which $\gcd(f(x), h(x) - \kappa)$ is different from 1. Now we give a characterization of those κ for which $\gcd(f(x), h(x) - \kappa) \neq 1$. (See also [131, pages 160ff].)

Let K be the set of those $\kappa \in \mathbb{F}_q$ for which $\gcd(f(x), h(x) - \kappa) \neq 1$, then from 3.5.11 we deduce that

$$f(x) = \prod_{\kappa \in K} \gcd(f(x), h(x) - \kappa), \qquad\qquad \textbf{3.5.21}$$

where $h \in \mathbb{F}_q[x]$ is an f-reducing polynomial. Hence, f is a divisor of

$$\prod_{\kappa \in K} (h(x) - \kappa).$$

Let G be the polynomial defined by

$$G(y) := \prod_{\kappa \in K} (y - \kappa) \in \mathbb{F}_q[y],$$

then $f(x)$ divides $G(h(x))$ and $G(y)$ can be characterized by

---

Theorem [131, 4.8 Theorem] *Consider $f \in \mathbb{F}_q[x]$ and $h \in \mathbb{F}_q[x]$ an f-reducing*   **3.5.22**
polynomial. Then $G(y)$ is the unique monic polynomial of least degree which satisfies $g(y) \in \mathbb{F}_q[y]$ *and* $f(x)$ *divides* $g(h(x))$.

Proof: We already know that $G \in \mathbb{F}_q[y]$ is monic, and that $f(x)$ is a divisor of $G(h(x))$. The set

$$\{g \in \mathbb{F}_q[y] \mid f(x) \text{ divides } g(h(x))\}$$

is an ideal in $\mathbb{F}_q[y]$, whence it is a principal ideal of the form $I(G_0)$. Consequently, G_0 is a divisor of G and

$$G_0(y) = \prod_{\kappa \in K_0} (y - \kappa)$$

for some subset K_0 of K. Furthermore, $f(x)$ is a divisor of

$$G_0(h(x)) = \prod_{\kappa \in K_0} (h(x) - \kappa),$$

thus

$$f(x) = \prod_{\kappa \in K_0} \gcd(f(x), h(x) - \kappa).$$

Comparing this with 3.5.21, we conclude that $K_0 = K$ and $G_0 = G$. $\square$

Assume that $f = f_0 \cdots f_{t-1}$ is the product of t distinct monic, irreducible polynomials $f_i \in \mathbb{F}_q[x]$. Let $h \in \mathbb{F}_q[x]$ be an f-reducing polynomial and let K and $G \in \mathbb{F}_q[y]$ be as above. Assume that the cardinality of K is equal to m. From 3.5.21 it follows that $m \leq t$. Next we introduce coefficients of G, so that

$$G(y) = \prod_{\kappa \in K} (y - \kappa) = \sum_{i=0}^{m} \mu_i y^i \in \mathbb{F}_q[y]$$

with $\mu_m = 1$. Since $f(x)$ is a divisor of $G(h(x))$, we have

$$\sum_{i=0}^{m} \mu_i h(x)^i \equiv 0 \bmod \mathrm{I}(f).$$

Because $\mu_m = 1$, this is a nontrivial linear dependence relation of the residues of $1, h(x), \ldots, h(x)^m$ modulo $f(x)$. From 3.5.22 it follows that this dependence relation is unique and that the residues of $1, h(x), \ldots, h(x)^{m-1}$ modulo $f(x)$ are linearly independent over $\mathbb{F}_q$.

The Zassenhaus algorithm (cf. [131, pages 157ff]) describes a method for finding G.

3.5.23   **Zassenhaus algorithm** Assume that $f = f_0 \cdots f_{t-1}$ is the product of t distinct monic irreducible polynomials f_i over $\mathbb{F}_q$ and let $h \in \mathbb{F}_q[x]$ be an f-reducing polynomial.

We compute the residues modulo $f(x)$ of $1, h(x), h^2(x), \ldots$ until we find the least power $h^m(x)$ which is linearly dependent on the previously computed residues of powers of h. In fact, $m \leq t$ where t can be determined by the Berlekamp algorithm. Let

$$\sum_{i=0}^{m} \mu_i h(x)^i \equiv 0 \bmod \mathrm{I}(f), \quad \mu_i \in \mathbb{F}_q, \; \mu_m \neq 0$$

be the first dependency relation occurring in this way. The monic polynomial

$$G(y) = \sum_{i=0}^{m} \frac{\mu_i}{\mu_m} y^i \in \mathbb{F}_q[y]$$

has as roots in $\mathbb{F}_q$ precisely the elements of K (which is the set of those $\kappa \in \mathbb{F}_q$ for which $\gcd(f(x), h(x) - \kappa) \neq 1$.) Determine them either by trial and error, or by a root finding method which we present next. $\square$

To find the roots of a polynomial, we introduce q-polynomials over $\mathbb{F}_{q^n}$ (cf. [131, page 108]). A polynomial of the form

$$f(x) = \sum_{i=0}^{m} \kappa_i x^{q^i} \in \mathbb{F}_{q^n}[x]$$

is called a *q-polynomial* over $\mathbb{F}_{q^n}$ or a *linearized polynomial*. The second name is motivated by

---

Lemma *Let $f \in \mathbb{F}_{q^n}[x]$ be a q-polynomial. Then for any λ_1, λ_2 in an arbitrary field extension of $\mathbb{F}_{q^n}$ and for any $\mu \in \mathbb{F}_q$ we have* **3.5.24**

$$f(\lambda_1 + \lambda_2) = f(\lambda_1) + f(\lambda_2)$$

and

$$f(\mu \lambda_1) = \mu f(\lambda_1). \qquad \square$$

Next we describe a method to determine the roots of a q-polynomial.

Example [131, pages 110ff] Let **3.5.25**

$$f(x) = \sum_{i=0}^{m} \kappa_i x^{q^i} \in \mathbb{F}_{q^n}[x]$$

be a q-polynomial over $\mathbb{F}_{q^n}$. We want to find all roots of f in a finite extension $\mathbb{F}$ over $\mathbb{F}_{q^n}$. Assume that $\mathbb{F}$ is an s-dimensional vector space over $\mathbb{F}_q$ and let $\{\beta_0, \ldots, \beta_{s-1}\}$ be an $\mathbb{F}_q$-basis of $\mathbb{F}$. According to 3.5.24, the mapping

$$f : \mathbb{F} \to \mathbb{F} : \beta \mapsto f(\beta)$$

is $\mathbb{F}_q$-linear. Consequently, it is represented by an $s \times s$-matrix $L = (\lambda_{ij})_{i,j \in s}$ over $\mathbb{F}_q$, where

$$f(\beta_i) = \sum_{j \in s} \lambda_{ij} \beta_j, \qquad \lambda_{ij} \in \mathbb{F}_q, \ i \in s.$$

Each element $\beta \in \mathbb{F}$ can be written uniquely as

$$\beta = \sum_{i \in s} \mu_i \beta_i, \qquad \mu_i \in \mathbb{F}_q, \ i \in s.$$

Then

$$f(\beta) = \sum_{i \in s} \mu_i f(\beta_i) = \sum_{i \in s} \mu_i \left(\sum_{j \in s} \lambda_{ij} \beta_j \right) = \sum_{j \in s} \left(\sum_{i \in s} \mu_i \lambda_{ij} \right) \beta_j.$$

Therefore, $f(\beta) = 0$ if and only if

$$\sum_{i \in s} \mu_i \lambda_{ij} = 0, \qquad j \in s.$$

This is a homogeneous system of linear equations

$$(\mu_0, \ldots, \mu_{s-1}) \cdot L = (0, \ldots, 0).$$

If r denotes the rank of L, then this system has q^{s-r} solutions $(\mu_0, \ldots, \mu_{s-1}) \in \mathbb{F}_q^s$. Each of these solutions gives a root $\beta = \sum_{i \in s} \mu_i \beta_i$ of f. In other words, we can find the roots of f in $\mathbb{F}$ by solving a homogeneous system of linear equations. $\diamond$

Assume that $f \in \mathbb{F}_{q^n}[x]$ is a q-polynomial and let $\kappa \in \mathbb{F}_{q^n}$. Then the polynomial $f(x) - \kappa$ is called an *affine q-polynomial*. (See [131, page 112].) Using the notation from above, $\beta \in \mathbb{F}$ is a root of $f(x) - \kappa$ if and only if $f(\beta) = \kappa$ which yields the system of linear equations

$$(\mu_0, \ldots, \mu_{s-1}) \cdot L = (\gamma_0, \ldots, \gamma_{s-1}),$$

where $\kappa = \sum_{i \in s} \gamma_i \beta_i$ with uniquely determined $\gamma_i \in \mathbb{F}_q$ for $i \in s$.

Since it is easier to determine the roots of an affine q-polynomial than of an arbitrary polynomial $f \in \mathbb{F}_{q^n}[x]$, we propose the following method to determine the roots of f in an extension field $\mathbb{F}$ of $\mathbb{F}_{q^n}$. First determine a nonzero affine q-polynomial which is divisible by f. This polynomial is called an *affine multiple* of f. Then compute the roots of the affine multiple of f, and check each root whether it is also a root of f.

We only have to describe how to obtain an affine multiple of f.

3.5.26   **Example** [131, page 112] Assume that $f \in \mathbb{F}_{q^n}[x]$ is of degree $m \geq 1$. In order to find a nonzero affine multiple of f, we compute for $i \in m$ the unique polynomial $r_i \in \mathbb{F}_{q^n}[x]$ with $\deg(r_i) < m$ such that $x^{q^i} \equiv r_i(x) \bmod I(f)$. Then we determine $\kappa_i \in \mathbb{F}_{q^n}$, not all equal to 0, in such a way that

$$\sum_{i \in m} \kappa_i r_i(x)$$

is constant. If we write r_i in the form $r_i(x) = \sum_{j \in m} \lambda_{ij} x^j$ with $\lambda_{ij} \in \mathbb{F}_{q^n}$ for $i, j \in m$, then we have to determine κ_i so that the polynomial

$$\sum_{i \in m} \kappa_i \left(\sum_{j \in m} \lambda_{ij} x^j \right) = \sum_{j \in m} \left(\sum_{i \in m} \kappa_i \lambda_{ij} \right) x^j$$

is constant, which yields

$$\sum_{i \in m} \kappa_i \lambda_{ij} = 0, \qquad 1 \leq j \leq m - 1.$$

This is the homogeneous system of linear equations

$$(\kappa_0, \ldots, \kappa_{m-1}) \cdot L = (0, \ldots, 0)$$

for the unknown κ_i, where L is the $m \times (m-1)$-matrix of the elements λ_{ij} for $i \in m$ and $1 \leq j \leq m - 1$. Since $\text{rank}(L) \leq m - 1$, this system has nontrivial solutions, and among those we determine a solution with $\kappa_{m-1} = 1$. Such a solution yields

$$\sum_{i \in m} \kappa_i r_i(x) = \kappa$$

for some $\kappa \in \mathbb{F}_{q^n}$. Therefore we have

$$\sum_{i \in m} \kappa_i x^{q^i} \equiv \sum_{i \in m} \kappa_i r_i(x) = \kappa \bmod \text{I}(f),$$

and, consequently, $\sum_{i \in m} \kappa_i x^{q^i} - \kappa$ is a nonzero affine multiple of f over $\mathbb{F}_{q^n}$. ◇

At the end of this section we want to discuss how to find a primitive element of $\mathbb{F}_q$. In order to do this, we pick an element of the multiplicative group $\mathbb{F}_q^*$ at random and compute its order. A general algorithm for computing the order of an element of a finite group is the following ([136, page 171]).

Algorithm (Order of a group element) Assume that G is a finite group of order   3.5.27

$$|G| = \prod_{i \in t} p_i^{m_i}$$

with pairwise distinct primes p_i and integers $m_i \geq 1$. Let G be written multiplicatively. The order of $g \in G$ is a divisor of $|G|$. We determine $\text{ord}(g)$ as the least positive integer n such that $g^n = 1$ and n is a divisor of $|G|$.

  Input: The order of the group G, its prime divisors p_i and their multiplicities m_i for $i \in t$, and one element $g \in G$.

  Output: The order of g.

(1) Set $n := |G|$.

(2) Set $i := 0$.

(3) Set $j := 0$.

(4) While ($g^{n/p_i} = 1$ and $j < m_i - 1$), set $n := n/p_i$ and increase j by 1.

(5) If $g^{n/p_i} = 1$, set $n := n/p_i$.

(6) If $i < t - 1$ increase i by 1 and goto (3).
     Otherwise $\text{ord}(g) =: n$. Terminate the algorithm.   □

Now we can describe the method for finding a primitive element of $\mathbb{F}_q^*$.

3.5.28   **Algorithm (Primitive element of $\mathbb{F}_q$ [136, page 172])**

(1) Pick at random an element $\alpha \in \mathbb{F}_q^*$ and compute its order.

(2) While $\mathrm{ord}(\alpha) < q - 1$, pick an element $\beta \in \mathbb{F}_q^*$ at random and compute its order.
    If $\mathrm{ord}(\beta)$ is not a divisor of $\mathrm{ord}(\alpha)$, then β is not contained in $\langle \alpha \rangle$. In this situation we compute an element $\gamma \in \mathbb{F}_q^*$ such that $\mathrm{ord}(\gamma) = \mathrm{lcm}(\mathrm{ord}(\alpha), \mathrm{ord}(\beta))$ and we set $\alpha := \gamma$.

(3) Output the element α. It is of order $q - 1$, whence it is a primitive element of $\mathbb{F}_q$.

If we assume that α is not yet a primitive element of $\mathbb{F}_q^*$, then there are $q - 1 - \mathrm{ord}(\alpha)$ elements β which do not belong to $\langle \alpha \rangle$. Therefore, the probability for choosing β not in $\langle \alpha \rangle$ is equal to

$$\frac{q - 1 - \mathrm{ord}(\alpha)}{q - 1} = 1 - \frac{\mathrm{ord}(\alpha)}{q - 1} \geq \frac{1}{2}.$$

If we have found some $\beta \notin \langle \alpha \rangle$, then we have to compute $\gamma \in \mathbb{F}_q^*$ such that $\mathrm{ord}(\gamma) = \mathrm{lcm}(\mathrm{ord}(\alpha), \mathrm{ord}(\beta))$. This can be done in the following way: Assume that $q - 1 = \prod_{i \in t} p_i^{m_i}$ with pairwise distinct primes p_i and positive integers m_i for $i \in t$. Moreover, we assume that

$$\mathrm{ord}(\alpha) = \prod_{i \in t} p_i^{r_i} \quad \text{and} \quad \mathrm{ord}(\beta) = \prod_{i \in t} p_i^{s_i}$$

with $0 \leq r_i, s_i \leq m_i$ for $i \in t$. If we put

$$a_i := \prod_{j \neq i} p_j^{r_j} \quad \text{and} \quad b_i := \prod_{j \neq i} p_j^{s_j}$$

then $\mathrm{ord}(\alpha^{a_i}) = p_i^{r_i}$ and $\mathrm{ord}(\beta^{b_i}) = p_i^{s_i}$. Let $I = \{i \in t \mid r_i \geq s_i\}$ and $J = \{i \in t \mid r_i < s_i\}$. If we determine γ as

$$\gamma := \prod_{i \in I} \alpha^{a_i} \prod_{j \in J} \beta^{b_j},$$

then we get, by Exercise 3.2.2,

$$\mathrm{ord}(\gamma) = \prod_{i \in I} p_i^{r_i} \prod_{j \in J} p_j^{s_j} = \mathrm{lcm}(\mathrm{ord}(\alpha), \mathrm{ord}(\beta)),$$

since $\mathbb{F}_q^*$ is an abelian group.

Since in this procedure $\mathrm{ord}(\gamma) > \mathrm{ord}(\alpha)$, the "while" loop will be performed at most $2 \log_2(q - 1)$ times on average (cf. [136, page 172]).

In order to apply 3.5.27 for computing the order of an element in $\mathbb{F}_q^*$, we must know the factorization of $q - 1 = p^n - 1$, which may be hard to obtain even for small values of p and n. However, there exist tables containing such factorizations, see [31]. In general, for doing these computations we need an arithmetic which enables us to use integers of arbitrary precision. $\qquad$ □

In Table 3.3 we have collected several primitive polynomials for extension fields of small degrees over $\mathbb{F}_p$ for $p = 2, 3, 5, 7$:

Table 3.3 Primitive polynomials for extension fields of small degrees over $\mathbb{F}_p$ for $p = 2, 3, 5, 7$

$\mathbb{F}_2:$	$\mathbb{F}_3:$	$\mathbb{F}_5:$
$x^2 + x + 1$	$x^2 + x + 2$	$x^2 + x + 2$
$x^3 + x^2 + 1$	$x^3 - x + 1$	$x^3 + x^2 + 2$
$x^4 + x^3 + 1$	$x^4 + x^3 + 2$	$x^4 + x^3 + 3x + 2$
$x^5 + x^2 + 1$	$x^5 - x + 1$	$x^5 - x + 2$
$x^6 + x^5 + 1$	$x^6 + x^5 + 2$	$x^6 + x^5 + 2$
$x^7 + x^6 + 1$	$x^7 + x^2 - x + 1$	$x^7 + x^6 + 2$
$x^8 + x^4 + x^3 + x^2 + 1$	$x^8 + x^3 - 1$	$x^8 + x^2 + 2x - 2$
$x^9 + x^4 + 1$	$x^9 - x^3 + x^2 + 1$	$x^9 + x^2 + 2x - 2$
$x^{10} + x^3 + 1$	$x^{10} + x^3 + x - 1$	$x^{10} + x^9 + 3x + 2$
$x^{11} + x^2 + 1$	$x^{11} + x^2 - x + 1$	$\mathbb{F}_7:$
$x^{12} + x^6 + x^4 + x + 1$	$x^{12} - x^4 + x^3 - x^2 - x - 1$	$x^2 + x + 3$
$x^{13} + x^4 + x^3 + x + 1$	$x^{13} - x + 1$	$x^3 + 3x + 2$
$x^{14} + x^5 + x^3 + x + 1$	$x^{14} + x^{13} + 2$	$x^4 + x^3 + x + 3$

Based on the results of the present section we are able to design an arithmetic for arbitrary finite fields $\mathbb{F}_q$ for given $q = p^n$. The arithmetic in the prime field $\mathbb{F}_p$ is just the arithmetic of the residue class ring $\mathbb{Z}/p\mathbb{Z}$. A primitive element in $\mathbb{F}_p$ can be determined by 3.5.28.

For computations in $\mathbb{F}_q$ we have to find a monic irreducible polynomial of degree n over $\mathbb{F}_p$. Hence, we randomly generate monic polynomials of degree n over $\mathbb{F}_p$ and check whether they are irreducible by an application of 3.5.20. If we have found an irreducible polynomial it can be used to determine an arithmetic in $\mathbb{F}_q$. Again with 3.5.28 we find a primitive element of $\mathbb{F}_q$. The minimal polynomial of the primitive element is a primitive polynomial. It can be computed as indicated in 3.5.1. Further down, in Section 6.9 we will describe how to determine a normal basis of $\mathbb{F}_q$ over $\mathbb{F}_p$. This normal basis together with all Lyndon words of length n over the alphabet $\mathbb{F}_p$, generated according to 3.5.5, can be used to determine all irreducible polynomials of degree n over $\mathbb{F}_p$.

In the next section we discuss two different possibilities to represent elements of finite fields and their advantages for computations.

Exercises

E.3.5.1   **Exercise** Let $\alpha \in \mathbb{F}_{64}$ be a root of the irreducible polynomial $x^6 + x + 1 \in \mathbb{F}_2[x]$. By using the method described in 3.5.1 show that the minimal polynomial of $\beta = \alpha^3 + \alpha^4$ over $\mathbb{F}_2$ is equal to $x^3 + x^2 + 1$. (See [131, 3.42 Example].)

E.3.5.2   **Exercise** Compute the minimal polynomials over $\mathbb{F}_2$ of all the elements of $\mathbb{F}_{16}$ with the method described in 3.5.2. In order to represent the elements of $\mathbb{F}_{16}$, assume that $\alpha \in \mathbb{F}_{16}$ is a root of the irreducible polynomial $x^4 + x + 1 \in \mathbb{F}_2[x]$. (See [131, 3.43 Example].)

E.3.5.3   **Exercise** Prove that $f \in m^n$ is a necklace if and only if f is the n-extension of a Lyndon word of length d dividing n.

E.3.5.4   **Exercise** Prove the derivation rules 3.5.6.

E.3.5.5   **Exercise** Prove 3.5.7.

E.3.5.6   **Exercise** Prove 3.5.8.

E.3.5.7   **Exercise** Assume that I and J are ideals in a ring R. Verify that the sum $I + J :=$ $\{i + j \mid i \in I, \, j \in J\}$ is an ideal in R.

E.3.5.8   **Exercise** Consider ideals I and J in a ring R. Let IJ be the set of all finite sums of products ij, where $i \in I$ and $j \in J$. Show that IJ is an ideal in R.

E.3.5.9   **Exercise** Prove 3.5.15.

E.3.5.10   **Exercise** Assume that f and g are two distinct irreducible polynomials over $\mathbb{F}_q$. Prove that the ideals $I(f)$ and $I(g)$ are relatively prime.

E.3.5.11   **Exercise** Prove *The Chinese Remainder Theorem, number-theoretic version*. Let $n > 1$ be an integer, where

$$n = \prod_{i \in t} p_i^{m_i}, \qquad m_i > 0, \, i \in t,$$

is the unique decomposition of n into pairwise distinct primes p_i. Moreover, assume that φ is given by

$$\varphi : \mathbb{Z} \to \underset{i \in t}{\times} \mathbb{Z}_{p_i^{m_i}} \; : \; z \mapsto \varphi(z) := (z \bmod p_0^{m_0}, \ldots, z \bmod p_{t-1}^{m_{t-1}})$$

with addition and multiplication on $\times_{i \in t} \mathbb{Z}_{p_i^{m_i}}$ defined componentwise. Prove that the mapping φ is a surjective ring homomorphism with

$$\ker(\varphi) = n\mathbb{Z}.$$

Exercise Apply Berlekamp's algorithm 3.5.16 in order to factor the polynomial $x^8 + x^6 + x^4 + x^3 + 1$ over $\mathbb{F}_2$. (See [131, 4.2 Example].)   **E.3.5.12**

Exercise Factorize the polynomial $f(x) = x^6 - 3x^5 + 5x^4 - 9x^3 - 5x^2 + 6x + 7$ over $\mathbb{F}_{23}$. Hint: Use Berlekamp's algorithm to show that this polynomial has three factors and that $h(x) = x^3 + 2x^2 + 4x$ is f-reducing. Then apply the algorithm of Zassenhaus 3.5.23 to obtain the elements κ of $\mathbb{F}_{23}$ for which $\gcd(f(x), h(x) - \kappa) \neq 1$. (See [131, 4.7 Example and 4.9 Example].)   **E.3.5.13**

Exercise Prove 3.5.24.   **E.3.5.14**

3.6 Representations of Field Elements   **3.6**

We continue with two further ways of describing elements of finite fields. In contrast to the above representation as residue classes, these methods yield *canonical labelings* of the field elements with values in $q = \{0, \ldots, q-1\}$. Then a single element of a field can be replaced just by its label.

In order to do this, we have two essentially different possibilities. In the *multiplicative representation* the nonzero elements of $\mathbb{F}_q$ are given as powers of a primitive element α. The multiplication of two nonzero elements of the field is done by just adding the corresponding exponents and taking into account that $\alpha^{q-1} = 1$,

$$\alpha^i \cdot \alpha^j = \begin{cases} \alpha^{i+j} & \text{if } i+j < q-1, \\ \alpha^{i+j-(q-1)} & \text{otherwise.} \end{cases}$$

The addition of two elements α^i and α^j of $\mathbb{F}_q$ reduces to the multiplication

$$\alpha^i + \alpha^j = \alpha^i \cdot (1 + \alpha^{j-i}),$$

which is easily carried out as soon as the exponent of the factor $1 + \alpha^{j-i}$ is known. For this purpose, we introduce a function $Z \colon q \to q$, called the *Zech logarithm* with respect to the basis α. It is defined in the following way: If $1 + \alpha^i \neq 0$, then $Z(i)$ is the unique element of $q-1$ so that $\alpha^{Z(i)} = 1 + \alpha^i$. If $1 + \alpha^i = 0$, then $Z(i) := q - 1$. The multiplicative representation of the elements in $\mathbb{F}_q$ is defined as follows:

3.6.1   **Definition (multiplicative representation)** The zero element of the field gets the number $q - 1$, while the elements of the multiplicative group $\mathbb{F}_q^*$ are labeled from 0 to $q - 2$, the label of α^i being its exponent i:

$$\varphi : \mathbb{F}_q \to \mathbb{N} \ : \ \alpha^i \mapsto i \text{ and } 0 \mapsto q - 1. \qquad \diamond$$

3.6.2   **Example** The field $\mathbb{F}_4$ is generated by a root α of the primitive polynomial $x^2 + x + 1$. In the following, we present composition tables in multiplicative notation, the bijection between $\mathbb{F}_4$ and $\{0, 1, 2, 3\}$ defined by the multiplicative notation φ, and the table of the Zech-logarithm. For example, the exponent $3 = 4 - 1 = q - 1$ represents the zero element.

Table 3.4 Multiplicative representation of $\mathbb{F}_4$

+	3	0	1	2
3	3	0	1	2
0	0	3	2	1
1	1	2	3	0
2	2	1	0	3

·	0	1	2
0	0	1	2
1	1	2	0
2	2	0	1

$\mathbb{F}_4$	0	1	α	$1 + \alpha$
φ	3	0	1	2
i	0	1	2	3
$Z(i)$	3	2	1	0

$\diamond$

For the generation of the elements of a finite field in their multiplicative representation we need a suitable primitive polynomial. See Table 3.3.

Now we describe a second way of labeling the elements of a finite field. The description of finite fields, used in 3.1.6, gives them as residue class rings

$$\mathbb{F}_q = \mathbb{F}_{p^m} = \mathbb{F}_p[x]/\mathrm{I}(f) = \left\{ \sum_{i \in m} k_i \alpha^i \ \middle| \ k_i \in \mathbb{Z}_p \right\}, \ \alpha = x + \mathrm{I}(f),$$

where $f \in \mathbb{F}_p[x]$ is an irreducible polynomial of degree m. An element κ in $\mathbb{F}_q$ is then displayed as a linear combination of the elements of the $\mathbb{F}_p$-basis $\{1, \alpha, \dots, \alpha^{m-1}\}$ of $\mathbb{F}_q$ in the form

$$\kappa = k_0 + k_1 \alpha + \dots + k_{m-1} \alpha^{m-1}, \qquad k_i \in \mathbb{F}_p.$$

The coordinate vector $(k_0, k_1, \dots, k_{m-1})$ can be interpreted as the p-adic decomposition $(k_{m-1}, \dots, k_0)_p$ of the decimal number $k_0 + k_1 p + \dots + k_{m-1} p^{m-1}$. Hence we associate κ with this number.

3.6.3   **Definition (additive representation)** If $\{1, \alpha, \dots, \alpha^{m-1}\}$ is an $\mathbb{F}_p$-basis of $\mathbb{F}_q$, then the *additive representation* of $\kappa = k_0 + k_1 \alpha + \dots + k_{m-1} \alpha^{m-1}$ with $k_i \in \mathbb{F}_p$ is given by the mapping

3.6.4 $$\psi : \mathbb{F}_q \to \{0, 1, \dots, q - 1\} \ : \ \kappa \mapsto k_0 + k_1 p + \dots + k_{m-1} p^{m-1}$$

where k_i is the standard representative of its residue class.   ⋄

For example, the identity element 1 gets the number 1, while the zero element gets the number 0.

Example The finite field obtained from the root α of the irreducible polynomial   **3.6.5**
$x^2 + x + 1$ is $\mathbb{F}_4 = \{0, 1, \alpha, \alpha^2 = \alpha + 1\}$. It has the basis $\{1, \alpha\}$. Its additive representation is $\psi(0) = 0$, $\psi(1) = 1$, $\psi(\alpha) = 2$, and $\psi(\alpha + 1) = 3$. If we replace the field elements by their numbers, we obtain the following composition tables:

Table 3.5 Additive representation of $\mathbb{F}_4$

+	0	1	2	3
0	0	1	2	3
1	1	0	3	2
2	2	3	0	1
3	3	2	1	0

·	1	2	3
1	1	2	3
2	2	3	1
3	3	1	2

⋄

3.7 Projective Geometry

3.7

Linear codes are closely related to finite projective spaces. The reason is the notion of isometry which says that (up to isometry) the columns of a generator matrix can be multiplied by nonzero scalar factors. Hence we can consider a generator matrix without zero column as a selection of one-dimensional subspaces in $\mathbb{F}_q^k$. The set of all these one-dimensional subspaces is known as a projective space which leads us to the following definition.

Definition (projective space) Let k be a positive integer and $v \in \mathbb{F}_q^k \setminus \{0\}$.   **3.7.1**
We denote by $\langle v \rangle^*$ the set $\langle v \rangle$ without the zero vector, the *punctured* one-dimensional space, as we sometimes call it. The *projective space* corresponding to $\mathbb{F}_q^k$, abbreviated as $\mathrm{PG}_{k-1}(q)$, consists of all one-dimensional subspaces of $\mathbb{F}_q^k$,

$$\mathrm{PG}_{k-1}(q) := \left\{ \langle v \rangle \mid v \in \mathbb{F}_q^k \setminus \{0\} \right\}.$$

A closely related set – which is motivated by the concept of group actions – is

$$\mathrm{PG}_{k-1}^*(q) := \left\{ \langle v \rangle^* \mid v \in \mathbb{F}_q^k \setminus \{0\} \right\} = \mathbb{F}_q^* \backslash\backslash \mathbb{F}_q^k \setminus \{0\} = \left\{ \mathbb{F}_q^*(v) \mid v \in \mathbb{F}_q^k \setminus \{0\} \right\},$$

the set of all punctured one-dimensional subspaces. In addition to this, we write $\mathrm{PG}(\mathbb{F}_q^k)$ for the set of all subspaces:

$$\mathrm{PG}(\mathbb{F}_q^k) := \left\{ U \mid U \leq \mathbb{F}_q^k \right\}.$$

Let U be a subspace of dimension r, then the *projective dimension* of U is defined to be $r - 1$. The subspaces of $\mathbb{F}_q^k$ of dimension one are called (projective) *points*. They are the elements of $\mathrm{PG}_{k-1}(q)$. Furthermore, subspaces of $\mathbb{F}_q^k$ of dimension $2, 3,$ or $k - 1$ are called *lines*, *planes*, or *hyperplanes*, respectively. Therefore, the projective dimensions of projective points, lines, planes, and hyperplanes in $\mathrm{PG}(\mathbb{F}_q^k)$ are $0, 1, 2,$ and $k - 2$ respectively. The null-space $\{0\}$ of $\mathbb{F}_q^k$ should be thought of as the "empty element" of $\mathrm{PG}(\mathbb{F}_q^k)$ and has projective dimension -1. The set $\mathrm{PG}(\mathbb{F}_q^k)$ carries an incidence relation. We say that a subspace U is *incident* with a subspace V in $\mathrm{PG}(\mathbb{F}_q^k)$ if U is a subspace of V in the vector space $\mathbb{F}_q^k$, for short $U \leq V$. Thus, $\mathrm{PG}(\mathbb{F}_q^k)$ together with the incidence relation is sometimes called the *projective geometry* over $\mathbb{F}_q^k$. The cardinality of $\mathrm{PG}_{k-1}(q)$ which is the number of projective points in $\mathrm{PG}(\mathbb{F}_q^k)$ is denoted as

3.7.2
$$\theta_{k-1}(q) := \frac{q^k - 1}{q - 1}.$$

◇

We have introduced the projective space over $\mathbb{F}_q^k$ in three different ways. All these notions will be used in later chapters. For instance in the context of enumeration we replace generator matrices of (n, k)-codes over $\mathbb{F}_q$ by mappings from n to $\mathrm{PG}_{k-1}^*(q)$. In the Chapters 8 and 9 some notions from geometry are applied, thus we rather use $\mathrm{PG}_{k-1}(q)$ and $\mathrm{PG}(\mathbb{F}_q^k)$.

Now we study a particular class of mappings from $\mathrm{PG}(\mathbb{F}_q^k)$ onto itself.

3.7.3   Definition (collineation) A *collineation* of $\mathrm{PG}(\mathbb{F}_q^k)$ is a bijective map λ from $\mathrm{PG}(\mathbb{F}_q^k)$ to $\mathrm{PG}(\mathbb{F}_q^k)$ which preserves incidence. For $U, V \in \mathrm{PG}(\mathbb{F}_q^k)$ this means that if $U \leq V$ then $\lambda(U) \leq \lambda(V)$.   ◇

According to Exercise 1.4.13, the group of linear maps $\mathrm{GL}_k(q)$ acts on $\mathbb{F}_q^k$. It takes subspaces to subspaces and preserves incidence, whence, it induces collineations of $\mathrm{PG}(\mathbb{F}_q^k)$.

From Exercise 1.4.13 we deduce that

3.7.4
$$\mathrm{GL}_k(q) \times \mathrm{PG}_{k-1}(q) \to \mathrm{PG}_{k-1}(q) : (A, \langle v \rangle) \mapsto \langle v \cdot A^\top \rangle$$

is the natural action of $\mathrm{GL}_k(q)$ on $\mathrm{PG}_{k-1}(q)$.

Let $\mathcal{Z}_k$ be the *center* of $\mathrm{GL}_k(q)$, i.e.

3.7.5
$$\mathcal{Z}_k := \left\{ A \in \mathrm{GL}_k(q) \mid A \cdot B = B \cdot A \text{ for all } B \in \mathrm{GL}_k(q) \right\},$$

then according to Exercise 3.7.2

$$\mathcal{Z}_k = \left\{ \kappa \cdot I_k \mid \kappa \in \mathbb{F}_q^* \right\}.$$

An element $A \in \mathrm{GL}_k(q)$ fixes all points of $\mathrm{PG}_{k-1}(q)$ if and only if A belongs to the center of $\mathrm{GL}_k(q)$. Hence, such multiplications by elements of the center do not matter, they can be factored out. For this reason, we introduce the *projective linear group* as the factor group

$$\mathrm{PGL}_k(q) := \mathrm{GL}_k(q)/\mathscr{Z}_k.$$

The details are presented in the next

Lemma   3.7.6

1. *The intersection of the stabilizers of all elements of* $\mathrm{PG}_{k-1}(q)$ *under the action of* $\mathrm{GL}_k(q)$, *defined in 3.7.4, is*

$$\bigcap_{y \in \mathrm{PG}_{k-1}(q)} \mathrm{GL}_k(q)_y = \left\{ \kappa \cdot I_k \mid \kappa \in \mathbb{F}_q^* \right\} = \mathscr{Z}_k.$$

2. *If* $_G X$ *is a group action, then*

$$N := \bigcap_{x \in X} G_x$$

is a normal subgroup of G. *It is also called the* **pointwise stabilizer** *of* X *or the* **kernel** *of the action of* G *on* X. *The action of* G *on* X *descends to an action of the factor group* G/N *on* X *via*

$$(G/N) \times X \to X \; : \; (gN, x) \mapsto gx.$$

For each $x \in X$ *the orbits* $G(x)$ *and* $(G/N)(x)$ *coincide, whence*

$$(G/N)\backslash\backslash X = G\backslash\backslash X.$$

3. *Combining the first two assertions, the projective general linear group* $\mathrm{PGL}_k(q)$ *acts in a natural way on the projective space* $\mathrm{PG}_{k-1}(q)$, *the orbits* $\mathrm{GL}_k(q)(y)$ *and* $\mathrm{PGL}_k(q)(y)$ *for* $y \in \mathrm{PG}_{k-1}(q)$ *are the same, whence*

$$\mathrm{PGL}_k(q)\backslash\backslash \mathrm{PG}_{k-1}(q) = \mathrm{GL}_k(q)\backslash\backslash \mathrm{PG}_{k-1}(q). \qquad \square$$

Hence, $\mathrm{PGL}_k(q)$ is the group which is induced by $\mathrm{GL}_k(q)$ on $\mathrm{PG}_{k-1}(q)$. We indicate the elements of $\mathrm{PGL}_k(q)$ either as cosets $A\mathscr{Z}_k$ or as orbits $\mathbb{F}_q^*(A)$ (cf. Exercise 3.7.4), or to simplify notation, we sometimes write them as matrices.

According to 3.4.8 two group actions $_G X$ and $_G Y$ are called similar if there exists a bijective mapping $\varphi \colon X \to Y$ so that

$$\varphi(gx) = g\varphi(x), \qquad g \in G, \; x \in X.$$

3.7.7   **Lemma** *The action of* $\mathrm{GL}_k(q)$ *on* $\mathrm{PG}_{k-1}(q)$ *and the action of* $\mathrm{GL}_k(q)$ *on* $\mathrm{PG}_{k-1}^*(q)$ *are similar. Consequently,* $\mathrm{PGL}_k(q)$ *is the group which is induced by* $\mathrm{GL}_k(q)$ *on* $\mathrm{PG}_{k-1}(q)$ *and on* $\mathrm{PG}_{k-1}^*(q)$. $\qquad\square$

The points of a projective space are indicated in different ways, depending on the situation where they occur. Sometimes we describe them as orbits $\mathbb{F}_q^*(v)$ of vectors $v \in \mathbb{F}_q^k \setminus \{0\}$, sometimes as points $P(v) := \langle v \rangle$ of a projective space, sometimes, in order to simplify notation, just as vectors.

Now we want to describe the projective semilinear group. Let V be a k-dimensional vector space over $\mathbb{F}_q$, without loss of generality, $V = \mathbb{F}_q^k$. Semilinear mappings were introduced in 1.5.7. Here we are interested in semilinear mappings σ which map V onto V. They are bijective. Any semilinear map is uniquely described by an automorphism α of $\mathbb{F}_q$ and by the values $\sigma(e^{(i)})$ on the standard basis $\{e^{(0)}, \ldots, e^{(k-1)}\}$ of V, since

$$\sigma\left(\sum_{i\in k} v_i e^{(i)}\right) = \sum_{i\in k} \alpha(v_i)\sigma(e^{(i)}), \qquad v_i \in \mathbb{F}_q, \ i \in k.$$

Assume that $q = p^r$. In Section 3.3 we have seen that the automorphism group of $\mathbb{F}_q$ is the Galois group $\mathrm{Gal} := \mathrm{Gal}\,[\,\mathbb{F}_q : \mathbb{F}_p\,]$, a cyclic group of order r. It is generated by the Frobenius automorphism $\tau(\kappa) := \kappa^p$, $\kappa \in \mathbb{F}_q$.

Let α be a nontrivial automorphism of $\mathbb{F}_q$. It acts componentwise on vectors in $\mathbb{F}_q^k$ or matrices in $\mathbb{F}_q^{k\times n}$. The induced mapping is a (not necessarily linear) bijection of $\mathbb{F}_q^k$ or $\mathbb{F}_q^{k\times n}$, respectively. If α acts componentwise on matrices in $\mathrm{GL}_k(q)$, then the induced mapping is a group automorphism. (See Exercise 3.7.5.)

Since σ is bijective, the set $\{\sigma(e^{(0)}), \ldots, \sigma(e^{(k-1)})\}$ is again a basis of $\mathbb{F}_q^k$. In other words, σ induces a change of the basis of V. According to Exercise 1.2.3, this change of bases can be expressed by a regular matrix $A \in \mathrm{GL}_k(q)$. We have

$$\sigma(e^{(i)}) = e^{(i)} \cdot A^\top, \qquad i \in k.$$

Thus σ is described as the pair (A, α) with $A \in \mathrm{GL}_k(q)$, $\alpha \in \mathrm{Gal}$ and where

3.7.8
$$\sigma(v) = \sum_{i\in k} \alpha(v_i)\sigma(e^{(i)}) = \alpha(v) \cdot A^\top.$$

If q is prime, i.e. if $r = 1$, then the Frobenius automorphism is the identity mapping. In this case, every semilinear bijection is described by a regular matrix, whence it is linear. However, if $r > 1$, not every semilinear bijection is linear, since there exist nontrivial automorphism α of $\mathbb{F}_q$.

3.7.9   **Theorem** *The semilinear bijections of* $\mathbb{F}_q^k$ *form a group, the* general semilinear group

$$\Gamma\mathrm{L}_k(q) := \{(A, \alpha) \mid A \in \mathrm{GL}_k(q), \ \alpha \in \mathrm{Gal}\}\,.$$

It is the semidirect product $GL_k(q) \rtimes Gal$ *(with the normal subgroup on the left). The identity element is the pair* (I_k, id), *where* $id = \tau^0$, *the identity element in* Gal. *The multiplication of two elements of* $\Gamma L_k(q)$ *is given by*

$$(A_2, \alpha_2)(A_1, \alpha_1) = (A_2 \cdot \alpha_2(A_1), \alpha_2\alpha_1), \qquad (A_1, \alpha_1), (A_2, \alpha_2) \in \Gamma L_k(q). \quad \square$$

Notice that in the last equation, the symbol $\alpha_2(A_1)$ denotes the application of the automorphism $\alpha_2 \in Gal$ to the entries of the matrix A_1.

The natural action of the general semilinear group on the projective space is the following

$$\Gamma L_k(q) \times PG_{k-1}(q) \to PG_{k-1}(q) \ : \ ((A, \alpha), \langle v \rangle) \mapsto \langle \alpha(v) \cdot A^\top \rangle. \qquad \textbf{3.7.10}$$

Since each element of $\Gamma L_k(q)$ maps a subspace of $\mathbb{F}_q^k$ onto a subspace, and preserves incidence, each semilinear bijection of $\mathbb{F}_q^k$ induces a collineation of the corresponding projective geometry $PG(\mathbb{F}_q^k)$. Moreover the pointwise stabilizer of $PG_{k-1}(q)$ in $\Gamma L_k(q)$ is $\{(A, id) \mid A \in \mathcal{Z}_k\}$, whence it is isomorphic to the center $\mathcal{Z}_k$ of $GL_k(q)$. Therefore, the group of semilinear bijections induces the action of the *projective semilinear group*

$$P\Gamma L_k(q) := \Gamma L_k(q) / \mathcal{Z}_k$$

on $PG_{k-1}(q)$. Its elements are of the form $(\mathbb{F}_q^*(A), \alpha) = (A\mathcal{Z}_k, \alpha)$ for $A \in GL_k(q)$ and $\alpha \in Gal$. Since the actions of $\Gamma L_k(q)$ on $PG_{k-1}(q)$ and on $PG_{k-1}^*(q)$ are similar, $P\Gamma L_k(q)$ is also the group induced by $\Gamma L_k(q)$ on $PG_{k-1}^*(q)$.

The projective semilinear group is the semidirect product

$$P\Gamma L_k(q) = PGL_k(q) \rtimes Gal, \qquad \textbf{3.7.11}$$

with the normal subgroup on the left. The identity element is $(I_k\mathcal{Z}_k, id)$, where $id = \tau^0$, the identity element in Gal. The multiplication of two elements $(A_1\mathcal{Z}_k, \alpha_1), (A_2\mathcal{Z}_k, \alpha_2) \in P\Gamma L_k(q)$ is given by

$$(A_2\mathcal{Z}_k, \alpha_2)(A_1\mathcal{Z}_k, \alpha_1) = ((A_2 \cdot \alpha_2(A_1))\mathcal{Z}_k, \alpha_2\alpha_1).$$

The inverse element of $(A\mathcal{Z}_k, \alpha)$ is $((\alpha^{-1}(A))^{-1}\mathcal{Z}_k, \alpha^{-1})$.

Since Gal is a cyclic group generated by the Frobenius-automorphism τ, for each $\alpha \in Gal$ there exists some $i \in r$ (recall that $q = p^r$) such that $\alpha = \tau^i$. Therefore in order to simplify notation, in Section 9.9 the elements of $P\Gamma L_k(q)$ are indicated as pairs (A, i), where $A \in GL_k(q)$ and i is an integer modulo r.

In 1.5.10 we have described a semilinear isometry σ of $\mathbb{F}_q^n$ as $\sigma = (\psi, (\alpha; \pi))$ where $\alpha \in Aut(\mathbb{F}_q) = Gal$ and $(\psi; \pi)$ is a linear isometry of $\mathbb{F}_q^n$. Thus $(\psi; \pi)$ can be identified with an element of the full monomial group $M_n(q)$ which is a subgroup of $GL_n(q)$.

3.7.12   **Theorem** *If $n \geq 3$, then the group of isometries of $\mathbb{F}_q^n$ which map subspaces onto subspaces is the semidirect product*

$$M_n(q) \rtimes \mathrm{Gal} = \{(A, \alpha) \mid A \in M_n(q),\ \alpha \in \mathrm{Gal}\},$$

with the normal subgroup on the left. It is a subgroup of $\Gamma L_n(q)$.   □

In Chapter 6 we will derive that the linear and semilinear isometry classes of linear (n,k)-codes can be described as orbits of $\mathrm{PGL}_k(q) \times S_n$ or $\mathrm{P\Gamma L}_k(q) \times S_n$ on the set of all generator matrices.

The *Fundamental Theorem of Projective Geometry* (cf. [4]) shows that any collineation of $\mathrm{PG}(\mathbb{F}_q^k)$, $k \geq 3$, is induced by a semilinear bijection.

Exercises

E.3.7.1   **Exercise** Verify 3.7.2.

E.3.7.2   **Exercise** Prove that the center $\mathcal{Z}_k$ of $\mathrm{GL}_k(\mathbb{F})$, which is the set of all matrices in $\mathrm{GL}_k(\mathbb{F})$ which commute with every matrix in $\mathrm{GL}_k(\mathbb{F})$, is given by

$$\{\kappa \cdot I_k \mid \kappa \in \mathbb{F}^*\},$$

where I_k denotes the $k \times k$ unit matrix. Hint: If A belongs to the center of $\mathrm{GL}_k(\mathbb{F})$, then A commutes with the elementary matrices $B_{i_0,j_0,1}^{(2)}$ from Exercise 1.7.3.

E.3.7.3   **Exercise** Prove 3.7.6.

E.3.7.4   **Exercise** Prove that the elements of $\mathrm{PGL}_k(q)$ are the $\mathbb{F}_q^*$-orbits of regular matrices under the action

$$\mathbb{F}_q^* \times \mathrm{GL}_k(q) \to \mathrm{GL}_k(q)\ :\ (\lambda, A) \mapsto \lambda A.$$

E.3.7.5   **Exercise** Let α be an automorphism of $\mathbb{F}_q$. Prove that the mapping

$$\mathbb{F}_q^{k \times n} \to \mathbb{F}_q^{k \times n}\ :\ (a_{ij})_{ij} \mapsto (\alpha(a_{ij}))_{ij}$$

is bijective. Moreover show that the mapping

$$\mathrm{GL}_k(q) \to \mathrm{GL}_k(q)\ :\ (a_{ij})_{ij} \mapsto (\alpha(a_{ij}))_{ij}$$

is a group isomorphism.

E.3.7.6   **Exercise** Prove 3.7.9.

E.3.7.7   **Exercise** Prove that 3.7.10 is well-defined and that it is a group action.

Chapter 4
Cyclic Codes

4

4   Cyclic Codes

4 Cyclic Codes

A very important class of codes is the class of cyclic codes, which are invariant under cyclic shifts

$$(c_0, c_1, \ldots, c_{n-1}) \mapsto (c_{n-1}, c_0, c_1, \ldots, c_{n-2})$$

of the coordinates. It is a subclass of the more general class of codes which are *invariant under a prescribed action of a finite group G*. We briefly introduce this more general class, and then we restrict attention to situations when G is the cyclic group of order n, acting on $n = \{0, \ldots, n-1\}$, and the action on the code C of length n is the cyclic shift. Particular classes of cyclic codes are the BCH-codes, the Reed–Solomon-codes, and the quadratic-residue-codes. They are of great practical relevance and will be introduced.

Further codes will be derived from these, the generalized Reed–Solomon-codes, the Goppa-codes and the Alternant-codes. In general, these codes are not necessarily cyclic, but they are isometric to cyclic codes. Moreover, we will revisit the Reed–Muller-codes over prime fields, since puncturing such codes yields cyclic ones.

Codes (over $\mathbb{F}_q$) that are invariant under an action of a prescribed finite group G are in particular the left ideals in the group algebra $\mathbb{F}_q^G$ of G, such codes will be called *group algebra codes*. In the present chapter we show that cyclic codes are group algebra codes. Moreover we will see that the group algebra of the cyclic group is the factor ring

$$\mathbb{F}_q[x]/I(x^n - 1),$$

and so cyclic codes are best studied in the algebraic setting of a polynomial factor ring. In fact, the cyclic codes of length n are in 1-1-correspondence with the ideals of this residue class ring.

Each cyclic code possesses a generator and a check polynomial. We derive the Structure Theorem for cyclic codes, and we show that particular classes of cyclic codes can also be described by idempotent generators. The variety of a cyclic code is the set of roots of its generator polynomial. We also present relations between the lattice of cyclic codes of length n and the lattice of their varieties.

Finally we present some encoding and decoding methods. We describe the use of shift registers for the encoding of cyclic codes. A decoding algorithm for cyclic codes using the syndrome of the received vector together with permutations of its entries is described in Section 4.12. In Section 4.13 we introduce the method of error-correcting pairs of subspaces. Finally in the last section we discuss the method of majority logic decoding which can be used not only for cyclic codes. We will illustrate it by an application to the binary Reed–Muller-codes.

4.1 Cyclic Codes as Group Algebra Codes

To begin with, we introduce an algebraic structure that contains many interesting codes which are invariant under a given finite group G. If we take for G the cyclic group of order n, then the corresponding structure will contain all the cyclic codes of length n over $\mathbb{F}_q$.

4.1.1 **Definition (the group algebra)** The *group algebra* of a finite group G over the field $\mathbb{F}$ is the set
$$\mathbb{F}^G := \{f \mid f \colon G \to \mathbb{F}\}$$
of all the mappings f from G to $\mathbb{F}$, together with the pointwise addition,
$$(f + \tilde{f})(g) := f(g) + \tilde{f}(g), \qquad f, \tilde{f} \in \mathbb{F}^G, \, g \in G,$$
the scalar multiplication
$$(\alpha f)(g) := \alpha \cdot f(g), \qquad \alpha \in \mathbb{F}, \, f \in \mathbb{F}^G, \, g \in G,$$
and the *convolution* product as multiplication:
$$(f\tilde{f})(g) := \sum_{x,y:xy=g} f(x)\tilde{f}(y), \qquad f, \tilde{f} \in \mathbb{F}^G, \, g \in G. \qquad \diamond$$

$\mathbb{F}^G$ is clearly a vector space of dimension $|G|$ over $\mathbb{F}$ with respect to the pointwise addition and scalar multiplication of functions as defined above. In order to describe a basis, we identify $g \in G$ with the element $1_{\mathbb{F}}g \in \mathbb{F}^G$ which is 1 at g and 0 everywhere else. The group algebra has the set of all these mappings $g = 1_{\mathbb{F}}g$, $g \in G$, as a linear basis. This is the *canonical basis* of $\mathbb{F}^G$. Therefore, the elements of the group algebra can be written as *formal sums*
$$f = \sum_{g \in G} \alpha_g g, \text{ where } \alpha_g := f(g).$$

The formal sum notation expresses the elements of the group algebra as linear combinations of the elements in the canonical basis. Moreover, in terms of formal sums, the addition is
$$\left(\sum_{g \in G} \alpha_g g\right) + \left(\sum_{g \in G} \beta_g g\right) = \sum_{g \in G} (\alpha_g + \beta_g)g, \qquad \alpha_g, \beta_g \in \mathbb{F},$$

while the convolution is just the long multiplication of these sums:

$$\left(\sum_{g\in G}\alpha_g g\right)\cdot\left(\sum_{g\in G}\beta_g g\right)=\sum_{k\in G}\gamma_k k,\quad \gamma_k:=\sum_{g,h:gh=k}\alpha_g\cdot\beta_h,\quad \alpha_g,\beta_g\in\mathbb{F}.$$

Summarizing, $\mathbb{F}^G$ is a vector space over $\mathbb{F}$, a ring with identity element $f=1_G$, and even an $\mathbb{F}$-algebra, since

$$\alpha(f\tilde{f})=(\alpha f)\tilde{f}=f(\alpha\tilde{f}),\qquad \alpha\in\mathbb{F},\ f,\tilde{f}\in\mathbb{F}^G.$$

Subspaces $C\leq\mathbb{F}^G$ which are invariant under left multiplication by elements $f\in\mathbb{F}^G$ are called *left ideals* of $\mathbb{F}^G$. Likewise, subspaces C of $\mathbb{F}^G$ which are invariant under right multiplication by elements $f\in\mathbb{F}^G$ are called *right ideals* of $\mathbb{F}^G$. Subspaces which are invariant under both left and right multiplication by elements $f\in\mathbb{F}^G$ are called *two-sided ideals* or simply *ideals*.

The left multiplication of a left ideal $C\subseteq\mathbb{F}^G$ by $g\in\mathbb{F}^G$, an element of the canonical basis of $\mathbb{F}^G$, is a linear mapping $D(g):C\to C$ since

$$g(\alpha f+\beta\tilde{f})=\alpha gf+\beta g\tilde{f}\qquad \alpha,\beta\in\mathbb{F},\ f,\tilde{f}\in C.$$

Hence the left ideals are invariant under the linear action of G via left multiplication by group elements

$$G\times C\to C\ :\ (g,c)\mapsto gc. \qquad\qquad \textbf{4.1.2}$$

This motivates the following

Definition (group algebra codes) The left ideals C of a group algebra $\mathbb{F}_q^G$ are called *group algebra codes*. $\diamond$ **4.1.3**

We are now in a position to introduce the main topic of the present chapter, the cyclic codes, and we show that they are group algebra codes:

Definition (cyclic codes) A linear code C of length n is called *cyclic* if it is invariant under the following action of the cyclic group $G:=\langle(0,\ldots,n-1)\rangle$, generated by the permutation $\pi:=(0,\ldots,n-1)$ of n. The action is defined by the mapping   **4.1.4**

$$G\times C\to C\ :\ (\pi^i,c)\mapsto \pi^i c:=(c_{\pi^{-i}(0)},\ldots,c_{\pi^{-i}(n-1)}).$$

Equivalently, C is cyclic, if the set of codewords of C is invariant under cyclic shifts of coordinates:

$$\forall\, c=(c_0,\ldots,c_{n-1})\in C\ :\ \pi c=(c_{n-1},c_0,c_1,\ldots,c_{n-2})\in C. \qquad \diamond$$

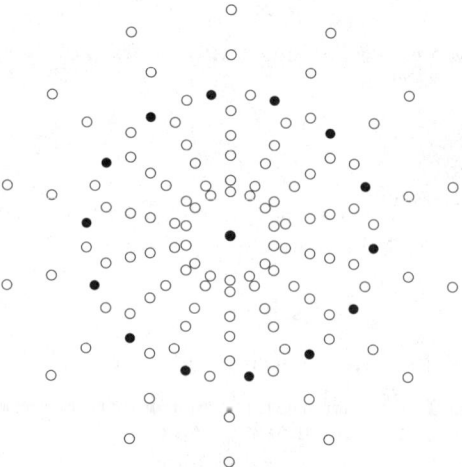

Fig. 4.1 The $(7,4)$-Hamming-code is cyclic

4.1.5   **Example** The binary $(7,4)$-Hamming-code is cyclic, as we will see now. It is easy to deduce from the generator matrix given in 1.3.6 that also the following matrix generates this code:

$$\Gamma := \begin{pmatrix} 1 & 1 & 0 & 1 & 0 & 0 & 0 \\ 0 & 1 & 1 & 0 & 1 & 0 & 0 \\ 0 & 0 & 1 & 1 & 0 & 1 & 0 \\ 0 & 0 & 0 & 1 & 1 & 0 & 1 \end{pmatrix}.$$

We note that the basis vectors, the rows of the generator matrix, are cyclic shifts of $(1,1,0,1,0,0,0)$, and that also the other three cyclic shifts, $(1,0,0,0,1,1,0)$, $(0,1,0,0,0,1,1)$ and $(1,0,1,0,0,0,1)$ are contained in this code. Hence the code consists of the linear combinations of the cyclic shifts of $(1,1,0,1,0,0,0)$, and so it must be cyclic (cf. also Exercises 4.1.2 and 4.1.1). This can be visualized as follows: We may project the Hamming space $H(7,2)$ into the complex plane as follows: Send the elements of the standard basis $e^{(0)}, \dots, e^{(6)}$ of $H(7,2)$ to the 7-th roots of unity, i.e. to the complex numbers of the form $e^{(2\pi i/7) \cdot j}$ with $j \in 7$. This defines a map, provided we identify the elements 0 and 1 of $\mathbb{F}_2$ with the complex numbers 0 and 1, respectively. In this projection, the Hamming-code corresponds to the black dots in Fig. 4.1. Notice that the center dot is actually two codewords of the Hamming-code. This is because both the zero vector $\mathbf{0}$ and the all-one vector $\mathbf{1}$ map to $0 \in \mathbb{C}$ under this map (since $\sum_{j \in 7} e^{(2\pi i/7) \cdot j} = 0$).

Later on we will see that all binary Hamming-codes can be described as cyclic codes, which is not always the case for ternary Hamming-codes (Exercise 4.1.3). ◇

In order to show that cyclic codes are group algebra codes, we note the following:

Remarks (on the algebra of the cyclic group)   4.1.6

— Consider the cyclic group G of order n, generated by the permutation $\pi := (0, \ldots, n-1)$,
$$G := \left\{ \pi^0, \pi^1, \ldots, \pi^{n-1} \right\}.$$
Its group algebra over $\mathbb{F}_q$ consists of the formal sums
$$\sum_{i \in n} \alpha_i \pi^i, \qquad \alpha_i \in \mathbb{F}_q, \; i \in n,$$
and so we obtain the following isomorphism between this group algebra and the vector space $\mathbb{F}_q^n$,
$$\psi \colon \mathbb{F}_q^n \to \mathbb{F}_q^G \; : \; (v_0, \ldots, v_{n-1}) \mapsto \sum_{i \in n} v_i \pi^i.$$

— Consider a code $C \leq \mathbb{F}_q^n$. An application of ψ gives
$$\psi(C) = \left\{ \psi(c) = \sum_{i \in n} c_i \pi^i \; \middle| \; c = (c_0, \ldots, c_{n-1}) \in C \right\}.$$
It is easy to check that
$$\psi(\pi c) = \pi \psi(c),$$
and so the code C is cyclic if and only if its image $\psi(C)$ is invariant under left multiplication by π.

— Since $\psi(C)$ is a subspace, this means that $\psi(C)$ is an *ideal* in the group algebra $\mathbb{F}_q^G$. It is a two-sided ideal, since the cyclic group G is abelian and, therefore, any ideal in the group algebra is two-sided.

— Conversely, the inverse image of each ideal in $\mathbb{F}_q^G$ is invariant under cyclic shift of coordinates, and so it is a cyclic code.   ◇

Corollary *Cyclic codes are group algebra codes.*   □   4.1.7

In the following, we will restrict our attention to cyclic codes. Nevertheless we remark that general group algebra codes have also been studied. For example, [132], [133], [205], [206], [109] discuss group algebra codes which come from symmetric groups.

Exercises

Exercise List all the codewords of the binary $(7,4)$-Hamming-code generated   E.4.1.1
by the matrix Γ given in 4.1.5 and check again that this code is indeed cyclic.

E.4.1.2   **Exercise** Prove that a linear (n,k)-code C is cyclic, if there is a codeword c in C such that $\pi c, \ldots, \pi^{n-1}c \in C$ and $c, \pi c, \ldots, \pi^{k-1}c$ are linearly independent, i.e. they are the rows of a generator matrix of C.

E.4.1.3   **Exercise** Show that the second order ternary Hamming-code is *not* cyclic.

E.4.1.4   **Exercise** Assume that G is the dihedral group of order 8. Implement a computer program or use MAPLE, in order to evaluate the parameters of all its group algebra codes over $\mathbb{F}_2$ which are of the form

$$\mathbb{F}_2^G \cdot f := \{\tilde{f} \cdot f \mid \tilde{f} \in \mathbb{F}_2^G\}, \qquad f \in \mathbb{F}_2^G.$$

E.4.1.5   **Exercise** Prove that $f \in \mathbb{F}^G$ and g an element of the canonical basis of $\mathbb{F}^G$ satisfy

$$gf(x) = f(g^{-1}x), \qquad x \in G.$$

Consequently, 4.1.2 describes an action of the group G on the left ideal C of $\mathbb{F}^G$.

E.4.1.6   **Exercise** Let G denote a finite group. For $f, \tilde{f} \in \mathbb{F}^G$ we denote by $[f, \tilde{f}]_g$ the coefficient of g in the convolution product $f\tilde{f}$. Prove the following:

— The mapping

$$\mathbb{F}^G \times \mathbb{F}^G \to \mathbb{F} : (f, \tilde{f}) \mapsto [f, \tilde{f}]_1$$

is an $\mathbb{F}$-bilinear form on $\mathbb{F}^G$.

— It is nondegenerate, i.e. for all $f \in \mathbb{F}^G$, $f \neq 0$, there exists $\tilde{f} \in \mathbb{F}^G$ so that $[f, \tilde{f}]_1 \neq 0$.

— It is symmetric, i.e. $[f, \tilde{f}]_1 = [\tilde{f}, f]_1$ for all $f, \tilde{f} \in \mathbb{F}^G$.

— It is associative, i.e. $[f\tilde{f}, \hat{f}]_1 = [f, \tilde{f}\hat{f}]_1$ for all $f, \tilde{f}, \hat{f} \in \mathbb{F}^G$.

— For $f, \tilde{f} \in \mathbb{F}^G$ and $g \in G$ we have $[\tilde{f}, f]_g = [g^{-1}\tilde{f}, f]_1$.

E.4.1.7   **Exercise** Characterize the annihilators of left, right, and two-sided ideals in the group algebra by proving:

— If L denotes a left ideal of $\mathbb{F}^G$, then its *right annihilator* is

$$\mathrm{Rann}(L) := \{f \in \mathbb{F}^G \mid L \cdot f = 0\} = \{f \in \mathbb{F}^G \mid \forall \tilde{f} \in L : [f, \tilde{f}]_1 = 0\}.$$

— If R denotes a right ideal of $\mathbb{F}^G$, then its *left annihilator* is

$$\mathrm{Lann}(R) := \{f \in \mathbb{F}^G \mid f \cdot R = 0\} = \{f \in \mathbb{F}^G \mid \forall\, \tilde{f} \in R \, : \, [f, \tilde{f}]_1 = 0\}.$$

— For each two-sided ideal I in $\mathbb{F}^G$ and its *annihilator*, we obtain

$$\mathrm{Ann}(I) := \{f \in \mathbb{F}^G \mid I \cdot f = f \cdot I = 0\} = \{f \in \mathbb{F}^G \mid \forall\, \tilde{f} \in I \, : \, [f, \tilde{f}]_1 = 0\}.$$

— If L is a left-ideal and R a right-ideal, then $\mathrm{Rann}(L)$ is a right-ideal and $\mathrm{Lann}(R)$ a left-ideal of $\mathbb{F}^G$. Moreover,

$$\mathbb{F}^G = L \oplus \mathrm{Rann}(L) = R \oplus \mathrm{Lann}(R) = I \oplus \mathrm{Ann}(I),$$

so that we have for the $\mathbb{F}$-dimensions

$$|G| = \dim(L) + \dim(\mathrm{Rann}(L)) = \ldots = \dim(I) + \dim(\mathrm{Ann}(I)).$$

Both the set of left-ideals and the set of right-ideals in $\mathbb{F}^G$ form a lattice with respect to $+$ and $\cap$. The mapping $L \mapsto \mathrm{Rann}(L)$ is a lattice anti-isomorphism between the lattices of left and right-ideals. This means that for any two left ideals L_1 and L_2 we have that

$$\mathrm{Rann}(L_1 + L_2) = \mathrm{Rann}(L_1) \cap \mathrm{Rann}(L_2)$$

and

$$\mathrm{Rann}(L_1 \cap L_2) = \mathrm{Rann}(L_1) + \mathrm{Rann}(L_2).$$

Exercise Consider the map from G to G, defined by   E.4.1.8

$$g \mapsto \tilde{g} := g^{-1},$$

which is an anti-isomorphism. Extend this map linearly to a map from $\mathbb{F}^G$ to $\mathbb{F}^G$, such that for $f = \sum_{g \in G} \alpha_g g \in \mathbb{F}^G$ we have

$$\tilde{f} = \sum_{g \in G} \alpha_g g^{-1} = \sum_{g \in G} \alpha_{g^{-1}} g.$$

For subsets $Y \subseteq \mathbb{F}^G$, define

$$\tilde{Y} := \{\tilde{y} \mid y \in Y\}.$$

Prove that for each group algebra code C in $\mathbb{F}^G$ we have

$$C^{\perp} = \widetilde{\mathrm{Rann}(C)}.$$

4.2 Polynomial Representation of Cyclic Codes

So far we have seen that cyclic codes of length n over $\mathbb{F}_q$ are group algebra codes. Thus they are ideals in the group algebra $\mathbb{F}_q^G$, where G is the cyclic group of order n. As we will see now, this group algebra is isomorphic to the residue class ring of polynomials in $\mathbb{F}_q[x]$ modulo the ideal which is generated by the polynomial $x^n - 1$. We denote this ring as

$$\operatorname{Res}_{q,n} := \mathbb{F}_q[x]/I(x^n - 1).$$

The map

$$\varphi : \mathbb{F}_q^G \to \operatorname{Res}_{q,n} \ : \ \sum_{i \in n} v_i \pi^i \mapsto \sum_{i \in n} v_i x^i + I(x^n - 1)$$

induces a correspondence between the elements of the group algebra and residue classes of polynomials. This correspondence is clearly a vector space isomorphism. Moreover, the identity $\pi^i \cdot \pi^j = \pi^k$ in $\mathbb{F}_q^G$ translates into the equation

$$\left(x^i + I(x^n - 1)\right) \cdot \left(x^j + I(x^n - 1)\right) = x^k + I(x^n - 1)$$

in the residue class ring. Here, k is the residue modulo n of $i + j$ in both equations. This shows that φ is in fact an isomorphism of algebras. Combining φ with the vector space isomorphism

$$\psi : \mathbb{F}_q^n \to \mathbb{F}_q^G \ : \ (v_0, \ldots, v_{n-1}) \mapsto \sum_{i \in n} v_i \pi^i,$$

described in 4.1.6, any cyclic code can be embedded into the residue class ring as follows:

Corollary *The mapping* $\iota := \varphi \circ \psi$, *defined by*

$$\iota : \mathbb{F}_q^n \to \operatorname{Res}_{q,n} \ : \ v \mapsto v(x) + I(x^n - 1), \quad v(x) := \sum_{i \in n} v_i x^i,$$

establishes the bijection

$$C \mapsto \iota(C) = \{\, c(x) + I(x^n - 1) \mid c \in C \,\}$$

between the set of cyclic codes in $\mathbb{F}_q^n$ and the set of ideals in $\operatorname{Res}_{q,n}$. □

For this reason, it is necessary to study the ideals of $\operatorname{Res}_{q,n}$ in some detail. To begin with, let us recall the following facts from ring theory.

— By the Isomorphism Theorem for Rings (see 4.7.3), the ideals in the residue class ring $\operatorname{Res}_{q,n}$ correspond to the ideals in $\mathbb{F}_q[x]$ which contain $I(x^n - 1)$. This correspondence is induced by the map which takes a polynomial in $\mathbb{F}_q[x]$ to its residue class modulo $I(x^n - 1)$.

— Every ideal I in $\mathbb{F}_q[x]$ is principal, i.e. it is of the form

$$I = I(g) = \{fg \mid f \in \mathbb{F}_q[x]\},$$

for a suitable polynomial g (see Exercise 3.1.11). Such a polynomial g is called a generator of the ideal I. It is unique up to scalar multiples. To achieve uniqueness, one often requires that the generator be monic, in this case it is also the unique monic nonzero polynomial of least degree in the ideal.

— In addition, $I(g)$ contains $I(x^n - 1)$ if and only if g is a divisor of $x^n - 1$ (Exercise 4.2.6). Thus the ideals in $\mathbb{F}_q[x]$ which contain $I(x^n - 1)$ are in one-to-one correspondence to the monic divisors of $x^n - 1$.

— Each element in $I(g)$, $g \neq 0$, can be written in a unique way as a product fg, for some $f \in \mathbb{F}_q[x]$. This follows from the fact that $\mathbb{F}_q[x]$ has no zero divisors, i.e. the product of two nonzero polynomials in $\mathbb{F}_q[x]$ is again nonzero.

— In the ideal $I(g)/I(x^n - 1)$, there is only one way of writing any given residue class as the product $fg + I(x^n - 1)$ with $\deg f < n - \deg g$.

Corollary

4.2.2

1. *The cyclic codes of length n over $\mathbb{F}_q$ are in one-to-one correspondence to the ideals of the residue class ring* Res $_{q,n}$.

2. *The ideals in* Res $_{q,n}$ *in turn correspond one-to-one to the monic divisors of $x^n - 1$.*

3. *Each such ideal can be written as*

$$I(g)/I(x^n - 1) = \{fg + I(x^n - 1) \mid f \in \mathbb{F}_q[x], \ \deg f < n - \deg g\},$$

where g is a monic divisor of $x^n - 1$. □

Definition (generator polynomial, check polynomial) The monic divisor g of $x^n - 1$ which generates the image $\iota(C)$ of the cyclic code C is called *generator polynomial* of C. The polynomial $h := (x^n - 1)/g$ is called *check polynomial* of C.

4.2.3

◇

Now we recall the following:

— The residue class ring is the set

$$\text{Res}_{q,n} = \mathbb{F}_q[x]/I(x^n - 1) = \{\overline{f} := f + I(x^n - 1) \mid f \in \mathbb{F}_q[x]\}$$

with multiplication defined by

$$\overline{f_0} \cdot \overline{f_1} = \overline{f_0 \cdot f_1}, \quad \text{for all } f_0, f_1 \in \mathbb{F}_q[x].$$

— The residue classes modulo $I(x^n - 1)$ of two polynomials $f_0, f_1 \in \mathbb{F}_q[x]$ are equal if and only if $f_0 - f_1$ is divisible by $x^n - 1$. Using the notation of Exercise 3.1.10, we may write

$$\overline{f_0} = \overline{f_1} \iff f_0 \equiv f_1 \bmod I(x^n - 1).$$

— By the Division Theorem for polynomials (Exercise 3.1.6), any $f \in \mathbb{F}_q[x]$ can be written uniquely as

$$f = s \cdot (x^n - 1) + r,$$

with $r, s \in \mathbb{F}_q[x]$ and either $r = 0$ or $0 \leq \deg r < n$. The polynomials s and r are called quotient and remainder upon dividing f by $x^n - 1$, respectively.

— Let $\mathrm{rem}_n(f)$ denote the remainder of f upon division by $x^n - 1$. It is clear that

$$\mathrm{rem}_n(f_0) = \mathrm{rem}_n(f_1) \iff \overline{f_0} = \overline{f_1}.$$

This shows that there is a one-to-one correspondence between the elements of the residue class ring $\mathrm{Res}_{q,n}$ and the set of possible remainders. We call $\mathrm{rem}_n(f)$ the *canonical representative* of the residue class of f.

Thus, the reader should carefully note the next

4.2.4   **Remarks** In the following sections of this chapter,

— a codeword c means, first of all, a vector

$$c = (c_0, \ldots, c_{n-1}) \in \mathbb{F}_q^n.$$

— On the other hand, we may also identify c with an element of the residue class ring,

$$c = c(x) + I(x^n - 1) \in \mathrm{Res}_{q,n},$$

where $c(x)$ is the uniquely defined polynomial $\sum_{i \in n} c_i x^i$ of degree less than n, the canonical representative of this particular residue class.

Therefore, a cyclic code C of length n can be regarded both as a subspace of $\mathbb{F}_q^n$ and as an ideal in the residue class ring $\mathrm{Res}_{q,n}$. It should be clear from the context which of the two interpretations is meant.   ◇

Theorem *Consider the cyclic code $C \leq \mathrm{Res}_{q,n}$ with generator polynomial $g = \sum_{i=0}^{t} g_i x^i$ of degree $t \leq n$ and check polynomial $h = (x^n - 1)/g = \sum_{i=0}^{n-t} h_i x^i$ of degree $n - t$. Then*   4.2.5

1. *The dimension of C is*

$$k = n - t = n - \deg g = \deg h.$$

   An $\mathbb{F}_q$-basis of C is the set

$$\left\{ \overline{x^i g} = x^i g + \mathrm{I}(x^n - 1) \mid i \in n - t \right\}.$$

2. *A generator matrix of C is the $(n - t) \times n$-matrix*

$$\Gamma := \begin{pmatrix} g_0 & g_1 & \cdots & \cdots & g_{t-1} & g_t & 0 & \cdots & & 0 \\ 0 & g_0 & g_1 & \cdots & \cdots & g_{t-1} & g_t & \cdots & & 0 \\ \vdots & \vdots & \ddots & \ddots & \cdots & \cdots & & \ddots & \ddots & \vdots \\ 0 & 0 & \cdots & g_0 & g_1 & \cdots & & \cdots & g_{t-1} & g_t \end{pmatrix}.$$

3. *The annihilator of C is*

$$\mathrm{Ann}(C) = \mathrm{I}(h)/\mathrm{I}(x^n - 1).$$

4. *The dual code $C^\perp$ is also cyclic. A generator matrix of $C^\perp$ and, therefore, also a check matrix of C is the $t \times n$-matrix*

$$\Delta := \begin{pmatrix} h_k & h_{k-1} & \cdots & \cdots & h_1 & h_0 & 0 & \cdots & 0 \\ 0 & h_k & h_{k-1} & \cdots & \cdots & h_1 & h_0 & \cdots & 0 \\ \vdots & \vdots & \ddots & \ddots & \cdots & \cdots & \ddots & \ddots & \vdots \\ 0 & 0 & \cdots & h_k & h_{k-1} & \cdots & \cdots & h_1 & h_0 \end{pmatrix}.$$

5. *$C^\perp$ is generated by $\widehat{h} := x^k h(x^{-1})$. The unique generator polynomial of $C^\perp$ is $\widehat{h}/h(0)$.*

Proof: 1. In order to prove the first assertion, we compare degrees and see that the codewords

$$\overline{g}, \overline{xg}, \ldots, \overline{x^{n-t-1}g}$$

are linearly independent elements of the vector space $\mathrm{Res}_{q,n}$. It remains to show that they generate the code C. To this end, consider a codeword $c = \overline{fg} \in C$, for some $f \in \mathbb{F}_q[x]$. From

$$\mathrm{rem}_n(fg) = \sum_{i \in n} c_i x^i$$

we deduce

$$c = \sum_{i \in n} c_i x^i + \mathrm{I}(x^n - 1).$$

We have already mentioned that $f \in \mathbb{F}_q[x]$ is uniquely determined if we impose the condition that $\deg f \leq n - t - 1$, and we know that this condition is no restriction of generality. In fact, this f of smallest degree is the unique remainder which we obtain when we divide *any* f with $c = \overline{fg}$ by the check polynomial (Exercise 4.2.1). Hence, c is a linear combination of the elements $x^i g, i \in n - t$.

2. The second assertion is a direct consequence of the proof of the first one.

3. In order to verify the assertion on the annihilator we note that for each $a = \tilde{f} h \in I(h)$ the following is true:

$$ag = \tilde{f} h g = \tilde{f}(x^n - 1) \equiv 0 \bmod I(x^n - 1).$$

Conversely, consider an $a \in \mathbb{F}_q[x]$ such that $ag \equiv 0 \bmod I(x^n - 1)$. There exists a polynomial $f \in \mathbb{F}_q[x]$ such that $ag = f(x^n - 1) = fhg$ and so

$$(a - fh)g = 0.$$

According to Exercise 3.1.1, the polynomial ring $\mathbb{F}_q[x]$ does not contain any zero divisors. Hence $a = fh \in I(h)$.

4. Any $c(x) = fg$ in C satisfies

$$c(x)h = fgh \equiv 0 \bmod I(x^n - 1).$$

According to Exercise 4.2.2, the coefficient of x^m, $m \in n$, in the canonical representative $\mathrm{rem}_n(c(x)h)$ is

$$\sum_{i \in n} c_i h_{(m-i) \bmod n} = 0,$$

and this implies $c \cdot \Delta^\top = 0$. Hence, the code generated by the $n - k$ linearly independent rows of the matrix Δ is contained in $C^\perp$. Since both codes have the same dimension, they are in fact equal. Thus, Δ is a generator matrix of $C^\perp$. On the other hand, by reversing the above argument we see that Δ is a generator matrix of the cyclic code generated by $x^k h(x^{-1})$. Thus, $C^\perp$ is cyclic with generator polynomial $x^k h(x^{-1}) = \sum_{i=0}^k h_{k-i} x^i$. This proves the last two assertions.   □

If $\deg h = k$, the polynomial

$$\widehat{h} := x^k h(x^{-1})$$

is called *reciprocal* of h. If $h(0)$ is nonzero, then $\widehat{h}/h(0)$ is monic (cf. Exercise 4.2.10).

Being ideals in a ring, cyclic codes can be added and intersected. We have the following result (Exercise 4.2.11):

Theorem   *Let C and C' be cyclic codes of length n over $\mathbb{F}_q$ with generator polynomials g and g', respectively.*   **4.2.6**

- *The intersection $C \cap C'$ is an ideal, i.e. a cyclic code. It is generated by the least common multiple*

$$g = \mathrm{lcm}(g, g').$$

- *The sum $C + C'$ is the ideal generated by the union $C \cup C'$ (see Exercise 3.5.7). Its generator polynomial is the greatest common divisor*

$$g = \gcd(g, g').$$

- *The set of cyclic codes of length n over $\mathbb{F}_q$ together with the operations $\cap$ and $+$ forms a lattice. The map from the set of monic divisors of $x^n - 1$ to ideals in $\mathrm{Res}_{q,n}$, given by $g \mapsto I(g)$, is a lattice anti-isomorphism.*   □

Examples Let us describe all binary cyclic codes of length 7. By 4.2.2, this   **4.2.7**
amounts to listing all ideals of $\mathrm{Res}_{2,7} = \mathbb{F}_2[x]/I(x^7 - 1)$. For this, we consider the set of all possible monic divisors of the polynomial $x^7 - 1$. To begin with, the polynomial $x^7 - 1 = x^7 + 1$ factors over $\mathbb{F}_2$ into monic irreducible polynomials as follows:

$$x^7 - 1 = \underbrace{(x+1)}_{f_0} \underbrace{(x^3 + x + 1)}_{f_1} \underbrace{(x^3 + x^2 + 1)}_{f_2}.$$

The 3 irreducible factors determine $2^3 = 8$ cyclic codes (if $\{0\}$ is included).

- The polynomial $g := f_0 f_1 f_2$ generates $\{0\}$.

- Let W_7 denote the cyclic code which is generated by

$$f_1 f_2 = (x^3 + x + 1)(x^3 + x^2 + 1) = x^6 + x^5 + x^4 + x^3 + x^2 + x + 1.$$

Its generator matrix is

$$\begin{pmatrix} 1 & 1 & 1 & 1 & 1 & 1 & 1 \end{pmatrix}$$

and, hence, W_7 is the $(7, 1)$ repetition code.

- The cyclic code S_3 with generator polynomial

$$f_0 f_1 = (x+1)(x^3 + x + 1) = x^4 + x^3 + x^2 + 1$$

is a $(7, 3)$-code with generator matrix

$$\begin{pmatrix} 1 & 0 & 1 & 1 & 1 & 0 & 0 \\ 0 & 1 & 0 & 1 & 1 & 1 & 0 \\ 0 & 0 & 1 & 0 & 1 & 1 & 1 \end{pmatrix},$$

a matrix which is the check matrix of the third order binary Hamming-code, and so S_3 is a binary simplex-code.

– The cyclic code S_3' with generator polynomial

$$f_0 f_2 = (x+1)(x^3 + x^2 + 1) = x^4 + x^2 + x + 1$$

is also a $(7,3)$-code which is isometric to S_3.

– The cyclic code P_7 with generator polynomial $f_0 = x + 1$ is a $(7,6)$-code. From its generator polynomial we obtain a generator matrix that can be transformed, using elementary row transformations, into the systematic generator matrix

$$\begin{pmatrix} 1 & 0 & 0 & 0 & 0 & 0 & 1 \\ 0 & 1 & 0 & 0 & 0 & 0 & 1 \\ 0 & 0 & 1 & 0 & 0 & 0 & 1 \\ 0 & 0 & 0 & 1 & 0 & 0 & 1 \\ 0 & 0 & 0 & 0 & 1 & 0 & 1 \\ 0 & 0 & 0 & 0 & 0 & 1 & 1 \end{pmatrix}.$$

Hence, P_7 is isometric to a parity check code. It consists of all even weight vectors in $\mathbb{F}_2^7$.

– The cyclic code H_3 generated by $f_1 = x^3 + x + 1$ is a $(7,4)$-code with generator matrix

$$\begin{pmatrix} 1 & 1 & 0 & 1 & 0 & 0 & 0 \\ 0 & 1 & 1 & 0 & 1 & 0 & 0 \\ 0 & 0 & 1 & 1 & 0 & 1 & 0 \\ 0 & 0 & 0 & 1 & 1 & 0 & 1 \end{pmatrix}.$$

According to 4.2.5, $H_3^\perp$ has the generator polynomial

$$x^4 + x^2 + x + 1,$$

which says that $H_3^\perp$ is the simplex-code S_3; hence H_3 is a Hamming-code.

– H_3' with generator polynomial $f_2 = x^3 + x^2 + 1$ is isometric to H_3, whence it is also a Hamming-code.

– The trivial factor $g = 1$ of $x^7 - 1$ is the generator polynomial of the full code $\text{Res}_{2,7}$ with generator matrix I_7.

Figure 4.2 shows the lattice of all binary cyclic codes of length 7 and the corresponding lattice of divisors of $x^7 - 1$ ("upside down", because of the third assertion in 4.2.6).

The codes S_3 and S_3' or H_3 and H_3' are isometric. This shows that there exist distinct cyclic codes which are isometric. Conversely, it is easy to find noncyclic codes which are permutationally isometric to cyclic ones (cf. Exercise 4.2.7). In other words, the isometry class of a cyclic code usually contains codes which are not cyclic.

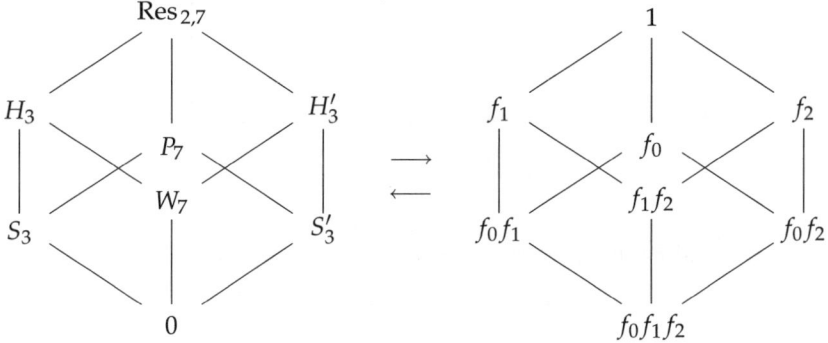

Fig. 4.2 The lattice of binary cyclic codes of length 7

◇

Systematic encoding of cyclic codes The generator matrix of a cyclic code as   **4.2.8**
given in 4.2.5 is not systematic. Here we will present two methods to encode a cyclic (n, k)-code in a systematic way (cf. [104, pages 80ff]). We will use the last k symbols of a codeword as the information symbols. That is, we divide the symbols of a codeword $c \neq 0$ into *check symbols* and *information symbols*,

$$c = (\underbrace{c_0, \ldots, c_{n-k-1}}_{\text{check symbols}}, \underbrace{c_{n-k}, \ldots, c_{n-1}}_{\text{information symbols}}).$$

For any choice of $(c_{n-k}, \ldots, c_{n-1}) \in \mathbb{F}_q^k$ we determine $(c_0, \ldots, c_{n-k-1}) \in \mathbb{F}_q^{n-k}$ so that $(c_0, \ldots, c_{n-1})$ belongs to C. There are two possible approaches.

1. In order to use the generator polynomial for encoding, we start with $\tilde{c}(x) = c_{n-k}x^{n-k} + \ldots + c_{n-1}x^{n-1}$. Let g be the generator polynomial of C. By the division algorithm, there exist uniquely determined polynomials r, s in $\mathbb{F}_q[x]$ such that $\tilde{c} = sg + r$ with $r = 0$ or $\deg r < \deg g = n - k$. If we set $c := \tilde{c} - r$, then g divides c and $\deg c < n$, i.e. c belongs to C. Moreover, this encoding is systematic in the last k coordinates, as the coefficient of x^j in $c(x)$ is c_j, for $n - k \leq j \leq n - 1$.

2. A second method for systematic encoding uses the check matrix Δ of C. We know that the vector $(c_0, \ldots, c_{n-1})$ belongs to C if and only if $c \cdot \Delta^\top = 0$.

Inserting the particular form of Δ as described in 4.2.5, we get

$$(c_0, \ldots, c_{n-1}) \cdot \begin{pmatrix} h_k & 0 & \cdots & 0 \\ h_{k-1} & h_k & \cdots & 0 \\ \vdots & h_{k-1} & \ddots & \vdots \\ h_1 & \vdots & \ddots & h_k \\ h_0 & h_1 & & h_{k-1} \\ 0 & h_0 & \ddots & \vdots \\ \vdots & \vdots & \ddots & h_1 \\ 0 & 0 & \cdots & h_0 \end{pmatrix} = 0 \,.$$

This yields the homogeneous system of linear equations

$$\begin{array}{rcl} h_k c_0 + h_{k-1}c_1 + \ldots + h_0 c_k & = & 0 \\ h_k c_1 + h_{k-1}c_2 + \ldots + h_0 c_{k+1} & = & 0 \\ \cdots & & \\ h_k c_{n-k-1} + h_{k-1}c_{n-k} + \ldots + h_0 c_{n-1} & = & 0 \end{array}$$

Since $h_k \neq 0$, we are able to determine the check symbol c_{n-k-1} from the last equation. The next to last equation then determines c_{n-k-2}. Proceeding in this way, we are able to determine all check symbols c_i, $i \in n - k$ (in reverse order).   □

It remains to discuss the factorization of $x^n - 1$ into the product of monic irreducible factors. To begin with, we assume that p denotes the characteristic of $\mathbb{F}_q$ and we define positive integers r and s by

$$n = p^s r, \text{ where } p \nmid r.$$

4.2.9   **Lemma** *The roots of the polynomial $x^r - 1$ are all nonzero and simple. Thus, the monic irreducible factors f_i in $x^r - 1 = \prod_{i \in I} f_i$ are pairwise distinct. Moreover, we have*

$$x^n - 1 = (x^r - 1)^{p^s} = \prod_{i \in I} (f_i)^{p^s} \,.$$

Proof: The roots of $x^r - 1$ are clearly nonzero. In order to show that they are simple, we consider the formal derivative of $x^r - 1$ over $\mathbb{F}_q$,

$$(x^r - 1)' = rx^{r-1}.$$

Since p is not a divisor of r, the only root of this derivative is zero, and this root is an $(r-1)$–fold root. Hence, $x^r - 1$ and its derivative do not have common roots, so that, according to 3.5.7, the polynomial $x^r - 1$ has only simple roots. The second statement follows directly from 3.2.12.   □

Definition (variety of a polynomial) Consider a nonzero polynomial $f \in \mathbb{F}_q[x]$.   **4.2.10**
The set of roots of f (in a suitable field extension of $\mathbb{F}_q$ or in the algebraic
closure $\overline{\mathbb{F}_q}$, cf. 3.2.23) is called the *variety* $V(f)$ of f.   ◇

For more details on $V(f)$ see Exercise 4.2.18. Let U_r denote the set of roots
of the polynomial $x^r - 1$. The elements of U_r are known as the r-th *roots of
unity*.

Let r be an integer which is relatively prime to q. Now we describe a constructive way of factoring $f = x^r - 1$ into irreducible polynomials over $\mathbb{F}_q$,
which is useful, at least for small r. During this factorization we may find that
not all roots of f are contained in the field $\mathbb{F}_q$. In this case, we have to work in
the splitting field of f, which is $\mathbb{F}_{q^m}$, for a suitable positive integer m. In this
situation it will be important to use the Galois group (cf. 3.3.1)

$$\mathrm{Gal} := \mathrm{Gal}\left[\mathbb{F}_{q^m} : \mathbb{F}_q\right].$$

This is the cyclic group of order m which is generated by the Frobenius automorphism

$$\sigma : \mathbb{F}_{q^m} \to \mathbb{F}_{q^m} \; : \; \alpha \mapsto \alpha^q.$$

According to 3.2.15, this group acts on the variety $V(f)$ of $f \in \mathbb{F}_q[x]$ in the
following way:

$$\mathrm{Gal} \times V(f) \to V(f) \; : \; (\sigma^i, \alpha) \mapsto \sigma^i(\alpha).$$

The orbit of the root α of f under Gal is

$$\mathrm{Gal}(\alpha) = \left\{ \alpha, \alpha^q, \alpha^{q^2}, \dots \right\}.$$

Lemma *Consider $r > 1$, coprime to q, and let $m := \mathrm{ord}_r(q)$ denote the smallest*   **4.2.11**
positive integer such that $q^m \equiv 1 \bmod r$. Then:

— *The polynomial $x^r - 1$ splits over $\mathbb{F}_{q^m}$ into linear factors:*

$$x^r - 1 = \prod_{i \in r} (x - \xi^i),$$

where $\xi \in U_r \subseteq \mathbb{F}_{q^m}$ denotes a primitive r-th root of unity.

— *The orbit of the root $\xi^i \in \mathbb{F}_{q^m}$ of $x^r - 1$ under Gal is*

$$\mathrm{Gal}(\xi^i) = \left\{ \xi^i, \xi^{iq}, \xi^{iq^2}, \dots, \xi^{iq^{t-1}} \right\},$$

*where t is the smallest positive integer such that $iq^t \equiv i \bmod r$. It consists of at
most m elements.*

— *The minimal polynomial (recall 3.1.5) of ξ^i over $\mathbb{F}_q$ is*

$$M_{\xi^i} := \prod_{\alpha \in \mathrm{Gal}(\xi^i)} (x - \alpha).$$

▬ *The factorization of $x^r - 1$ into irreducible factors over $\mathbb{F}_q[x]$ is*

$$x^r - 1 = \prod_{\zeta^i \in \mathcal{T}} M_{\zeta^i},$$

where $\mathcal{T}$ denotes a transversal of the orbits of the Galois group on U_r.

Proof: First we show that there exists an $m \in \mathbb{N}^*$ with $q^m \equiv 1 \bmod r$. Since $\gcd(r, q) = 1$, Bézout's Identity (Exercise 3.1.2) yields the existence of integers a and b such that $ar + bq = 1$. Hence, $bq \equiv 1 \bmod r$, which shows that the residue class $\bar{q}$ of q is a unit in the residue class ring $\mathbb{Z}_r$. Consequently, $\bar{q}$ generates a cyclic subgroup $\langle \bar{q} \rangle$ of the group of units in $\mathbb{Z}_r$. If m denotes the order of this group,

$$m := \mathrm{ord}_r(q) := |\langle \bar{q} \rangle|,$$

then $q^m \equiv 1 \bmod r$. This m is the smallest positive integer for which $\bar{q}^m = \bar{1} \in \mathbb{Z}_r$ or, equivalently, $q^m \equiv 1 \bmod r$.

The multiplicative group of the finite field $\mathbb{F}_{q^m}$ is cyclic. Hence, a generator β of this group is of order $q^m - 1$. By assumption r divides $q^m - 1$ and, therefore, $\zeta := \beta^{(q^m - 1)/r}$ is a primitive r-th root of unity. This element of $\mathbb{F}_{q^m}^*$ is, together with each of its powers, a root of $x^r - 1$, i.e. $\prod_{i \in r}(x - \zeta^i)$ is a divisor of $x^r - 1$. Since both these polynomials are monic and of the same degree, we deduce that

$$x^r - 1 = \prod_{i \in r}(x - \zeta^i).$$

Since the orbit $\mathrm{Gal}(\zeta^i)$ is finite, we may assume that t is the least integer $t \geq 1$ for which $\zeta^{iq^t} = \zeta^i$, whence $iq^t \equiv i \bmod r$. It is clear that $\mathrm{Gal}(\zeta^i)$ contains at most m elements, since $|\mathrm{Gal}| = m$.

Partitioning the set of roots of $x^r - 1$ into disjoint orbits of the Galois group, we obtain the factorization

$$x^r - 1 = \prod_{\zeta^i \in \mathcal{T}} M_{\zeta^i}.$$

Here $\mathcal{T}$ denotes a transversal of the orbits of the Galois group on the variety

$$U_r := \left\{ \kappa \in \mathbb{F}_{q^m} \mid \kappa^r = 1 \right\}.$$

By 3.3.4, the factor M_{ζ^i} is the minimal polynomial of ζ^i over $\mathbb{F}_q$.   □

Thus, the problem of factoring $x^r - 1$ in $\mathbb{F}_q[x]$ reduces to the problem of finding the zeros of the irreducible divisors of $x^r - 1$ over $\mathbb{F}_q$ and of evaluating their orbits under Gal. When we express the roots as powers ζ^i of a primitive root ζ of unity of order r, then we may *restrict attention to the exponents i of the powers ζ^i*, obtaining the orbits of the Galois group on r.

Definition (*q*-cyclotomic cosets modulo *r*) Let q be a prime power and let   4.2.12
$r \geq 2$ be an integer which is relatively prime to q. For an integer $i \in r$, the
q-cyclotomic coset modulo r containing i is the set

$$\{i,\ iq \bmod r,\ iq^2 \bmod r,\ iq^3 \bmod r, \ldots, iq^{t-1} \bmod r\} \subseteq r,$$

where t is the least positive integer such that $iq^t \equiv i \bmod r$. It is in fact an orbit
of the Galois group acting on r in the following way:

$$\mathrm{Gal} \times r \to r \ :\ (\sigma^j, i) \mapsto q^j \cdot i \bmod r.$$

Hence we may denote this set as

$$\mathrm{Gal}(i)$$

(or $\mathrm{Gal}(iq^j)$ for any $j \leq r - 1$). It is customary to assume that i is the least
element among all elements of $\mathrm{Gal}(i)$.   ◇

We hope that the orbit $\mathrm{Gal}(i)$ will not be mixed up with the orbit $\mathrm{Gal}(\zeta^i)$. Here
is an example:

Example Let us write $x^{15} - 1$ as a product of irreducible polynomials over   4.2.13
$\mathbb{F}_2$. Since $2^4 \equiv 1 \bmod 15$, $x^{15} - 1$ factorizes into linear factors over $\mathbb{F}_{2^4}$. The
2-cyclotomic cosets modulo 15 are

$$\mathrm{Gal}(0) = \{0\},\ \mathrm{Gal}(1) = \{1,2,4,8\},\ \mathrm{Gal}(3) = \{3,6,12,9\},$$

$$\mathrm{Gal}(5) = \{5,10\},\ \mathrm{Gal}(7) = \{7,14,13,11\}.$$

(Compare this with 3.2.14.) If ζ denotes a primitive 15-th root of unity, we
obtain the following irreducible divisors of $x^{15} - 1$:

$$\begin{aligned}
f_0 &= x - 1,\\
f_1 &= (x - \zeta)(x - \zeta^2)(x - \zeta^4)(x - \zeta^8),\\
f_2 &= (x - \zeta^3)(x - \zeta^6)(x - \zeta^{12})(x - \zeta^9),\\
f_3 &= (x - \zeta^5)(x - \zeta^{10}),\\
f_4 &= (x - \zeta^7)(x - \zeta^{14})(x - \zeta^{13})(x - \zeta^{11}).
\end{aligned}$$

In order to see what these polynomials really are, we need to construct the
field $\mathbb{F}_{16}$. Without restriction, let ζ denote a root of the primitive polynomial
$x^4 + x + 1 \in \mathbb{F}_2[x]$. As $\zeta^4 = \zeta + 1$, we can write the powers of ζ as linear
combinations of the elements of the $\mathbb{F}_2$-basis $\{1, \zeta, \zeta^2, \zeta^3\}$ in the following way

(see 3.2.9):

$$\begin{aligned}
\zeta^4 &= \zeta+1, & \zeta^5 &= \zeta^2+\zeta, \\
\zeta^6 &= \zeta^3+\zeta^2, & \zeta^7 &= \zeta^3+\zeta+1, \\
\zeta^8 &= \zeta^2+1, & \zeta^9 &= \zeta^3+\zeta, \\
\zeta^{10} &= \zeta^2+\zeta+1, & \zeta^{11} &= \zeta^3+\zeta^2+\zeta, \\
\zeta^{12} &= \zeta^3+\zeta^2+\zeta+1, & \zeta^{13} &= \zeta^3+\zeta^2+1, \\
\zeta^{14} &= \zeta^3+1.
\end{aligned}$$

For example, $\zeta^5 + \zeta^{10} = 1$, and $\zeta^5 \cdot \zeta^{10} = 1$. Hence,

$$f_3 = x^2 + x(\zeta^5 + \zeta^{10}) + \zeta^5 \cdot \zeta^{10} = x^2 + x + 1.$$

In a similar fashion, we obtain representations of the remaining f_i over $\mathbb{F}_2$. The following table shows the cyclotomic cosets together with the orbits of the Galois group on the set of roots of unity and the resulting minimal polynomials:

Cyclotomic set $\mathrm{Gal}(i)$	Orbit $\mathrm{Gal}(\zeta^i)$	Minimal polynomial M_{ζ^i}
$\{0\}$	$\{1\}$	$x+1$
$\{1,2,4,8\}$	$\{\zeta,\zeta^2,\zeta^4,\zeta^8\}$	x^4+x+1
$\{3,6,12,9\}$	$\{\zeta^3,\zeta^6,\zeta^{12},\zeta^9\}$	$x^4+x^3+x^2+x+1$
$\{5,10\}$	$\{\zeta^5,\zeta^{10}\}$	x^2+x+1
$\{7,14,13,11\}$	$\{\zeta^7,\zeta^{14},\zeta^{13},\zeta^{11}\}$	x^4+x^3+1 ◇

Recall that we can write $n = p^s \cdot r$ for some prime p and some integers s and r, where $p \nmid r$. Two special cases are of particular interest. These are the cyclic codes of *p-regular length* ($n = r$) and those of *p-power length* ($n = p^s$). The case of general block length will be discussed later. The foregoing results may be summarized as follows.

4.2.14   Corollary *Let p denote the characteristic of $\mathbb{F}_q$.*

- *If $p \nmid n$, then the set of all the cyclic codes of length n over $\mathbb{F}_q$ forms a Boolean lattice with 2^l elements, where l is the number of irreducible monic divisors of $x^n - 1$.*

- *If $n = p^s$, then $x^n - 1 = (x-1)^n$, and the cyclic codes of length n over $\mathbb{F}_q$ form the chain*

$$\mathbb{F}_q[x]/I(x^n - 1) = C_0 \supset C_1 \supset \ldots \supset C_{n-1} \supset C_n = 0,$$

where C_i is generated by the polynomial $(x-1)^i$.   □

Definition (the variety of a cyclic code) Consider a cyclic code C over $\mathbb{F}_q$ with   4.2.15
generator polynomial g. The variety of g will also be called the *variety of C over*
$\mathbb{F}_q$ and indicated as follows:

$$V(C) := V(g).$$

Every element of $V(C)$ is called a *root* of C over $\mathbb{F}_q$.   ◇

This is justified, since each $\alpha \in V(g)$ is a root of every $c(x)$, $c \in C$, and conversely, each root of all the codewords is a root of g, because $\bar{g} = g + I(x^n - 1)$
is contained in C.

Theorem *Let C denote a cyclic (n, k)-code over $\mathbb{F}_q$, where n is relatively prime to q,*   4.2.16
with generator polynomial g, and let m be $\mathrm{ord}_n(q)$. According to 4.2.11, $\mathbb{F}_{q^m}$ contains
the variety of g, and we have:

1. *The dimension k of C is $k = n - |V(C)|$.*

2. *In the case $V(C) = \{\alpha_0, \dots, \alpha_{n-k-1}\}$, the matrix*

$$\tilde{\Delta} = \begin{pmatrix} 1 & \alpha_0 & \alpha_0^2 & \cdots & \alpha_0^{n-1} \\ 1 & \alpha_1 & \alpha_1^2 & \cdots & \alpha_1^{n-1} \\ & \cdots & & \cdots & \\ 1 & \alpha_{n-k-1} & \alpha_{n-k-1}^2 & \cdots & \alpha_{n-k-1}^{n-1} \end{pmatrix}$$

is a check matrix of a code $\tilde{C}$ over $\mathbb{F}_{q^m}$ whose intersection with $\mathbb{F}_q^n$ is C,

$$\tilde{C} \downarrow \mathbb{F}_q = \tilde{C} \cap \mathbb{F}_q^n = C.$$

3. *The variety of $C^{\perp}$ consists of the inverses of the nonroots of C in U_n, i.e.*

$$V(C^{\perp}) = \left\{ \xi^{-i} \mid \xi^i \notin V(C), i \in n \right\},$$

where ξ is a primitive root of unity of order n.

Proof: 1. is clear from the fact that $k = n - \deg g$ and $\deg g = |V(g)| = |V(C)|$.
2. In order to prove the second statement we recall the definition of $V(C)$ and
deduce that

$$c \cdot \tilde{\Delta}^{\top} = (c(\alpha_0), \dots, c(\alpha_{n-k-1})) = (0, \dots, 0) = 0,$$

for each $c \in C$, and so $C \subseteq \tilde{C}$. Conversely, if $\tilde{c} \in \tilde{C} \cap \mathbb{F}_q^n$, then $\tilde{c}(\alpha_j) = 0$, for
each $\alpha_j \in V(C)$, which implies $\tilde{c} \in C$.
3. The final assertion is immediately clear from the fifth item of 4.2.5.   □

This theorem is very useful for applications. For example, the dimensions of cyclic codes of length n over $\mathbb{F}_q$ can be determined easily provided that n and q are relatively prime. According to 4.2.16, it suffices to calculate the varieties of all divisors of the polynomial $x^n - 1$ over $\mathbb{F}_q$. In fact, we need not even compute the divisors themselves. The varieties of the irreducible factors are given by the q-cyclotomic cosets modulo n, which can be computed easily. The variety of a divisor of $x^n - 1$ is a union of varieties of irreducible polynomials, i.e. a union of q-cyclotomic cosets modulo n.

4.2.17   **Example** We evaluate the dimensions of all binary cyclic codes of length 23. For this purpose we calculate first the dimensions of the maximal ones. They are generated by irreducible divisors of $x^{23} - 1$ over $\mathbb{F}_2$, whose zeros form varieties of U_{23} over $\mathbb{F}_2$, the orbits $\mathrm{Gal}(\zeta^s)$ of the Galois group. These orbits can be obtained from the 2-cyclotomic cosets modulo 23:

$$\{0\},$$
$$\{1, 2, 4, 8, 16, 9, 18, 13, 3, 6, 12\},$$
$$\{5, 10, 20, 17, 11, 22, 21, 19, 15, 7, 14\}.$$

Therefore, the varieties of the irreducible factors of $x^{23} - 1$ over $\mathbb{F}_2$ are

$$\mathrm{Gal}(1) = \{1\},$$
$$\mathrm{Gal}(\zeta) = \{\zeta^1, \zeta^2, \zeta^4, \zeta^8, \zeta^{16}, \zeta^9, \zeta^{18}, \zeta^{13}, \zeta^3, \zeta^6, \zeta^{12}\},$$
$$\mathrm{Gal}(\zeta^5) = \{\zeta^5, \zeta^{10}, \zeta^{20}, \zeta^{17}, \zeta^{11}, \zeta^{22}, \zeta^{21}, \zeta^{19}, \zeta^{15}, \zeta^7, \zeta^{14}\}.$$

Hence, there are exactly three maximal binary cyclic codes of length 23. According to 4.2.16, they are of dimension 22, 12 and 12. From 4.2.6 and Exercise 4.2.18 we obtain, by forming unions of these varieties, the dimensions of all other binary cyclic codes of length 23. The dimensions of the seven binary cyclic codes different from $\{0\}$ of length 23 are therefore 1, 11, 11, 12, 12, 22, and 23.   ◇

4.2.6 and 4.2.16 imply another important result on these lattices.

4.2.18   **Corollary** *The mapping $C \mapsto V(C)$ is an anti-isomorphism between the lattice of cyclic codes of length n over $\mathbb{F}_q$ and the lattice of the varieties of U_n over $\mathbb{F}_q$. (A subset V of U_n is a variety over $\mathbb{F}_q$ if $V = V(f)$ for some $f \in \mathbb{F}_q[x]$.)*   □

Exercises

E.4.2.1   **Exercise** Show that for $0 \neq c \in C \leq \mathrm{Res}_{q,n}$ (with generator polynomial g) there exists a unique polynomial f of degree $\deg f < \dim(C)$ such that $c = \overline{fg}$.

Exercise Assume that $f = \sum_{i=0}^{m} f_i x^i$ is a polynomial of degree m, and let $n \geq 1$. Prove by induction on the degree of f that

$$f \equiv \sum_{j \in n} \left(\sum_{i: i \equiv j \bmod n} f_i \right) x^j \bmod \mathrm{I}(x^n - 1).$$

E.4.2.2

Exercise Assume that R is a ring and I is an ideal in R. Show that the set $R/I = \{r + I \mid r \in R\}$ together with the two compositions

$$(r_1 + I) + (r_2 + I) = (r_1 + r_2) + I, \ (r_1 + I)(r_2 + I) = (r_1 r_2) + I, \quad r_1, r_2 \in R$$

is a ring, the *factor ring* of R modulo I.

E.4.2.3

Exercise Show that for any ideal I of the ring R, the canonical projection $\pi \colon R \to R/I$ is a surjective ring homomorphism.

E.4.2.4

Exercise Assume that $\varphi \colon R \to S$ is a ring homomorphism and that J is an ideal in S. Prove that $\varphi^{-1}(J)$ is an ideal in R and $\ker \varphi$ is an ideal contained in $\varphi^{-1}(J)$. If φ is surjective and I an ideal in R, show that $\varphi(I)$ is an ideal in S.

E.4.2.5

Exercise Let R be an integral domain (i.e. a commutative ring different from $\{0\}$ with 1 and without zero divisors). Show that the ideal $\mathrm{I}(r)$ is contained in the ideal $\mathrm{I}(s)$ for $r, s \in R$ if and only if s divides r. Hence, the ideals in $\mathbb{F}_q[x]/\mathrm{I}(f)$ are of the form $\mathrm{I}(g)/\mathrm{I}(f)$ where g divides f.

E.4.2.6

Exercise Construct a code which is not cyclic and permutationally isometric to the code S_3 from 4.2.7.

E.4.2.7

Exercise Show that if $g(x) \in \mathbb{F}_q[x]$ is the generator polynomial of a cyclic code then $g(0) \neq 0$.

E.4.2.8

Exercise Let C be a binary cyclic code of odd length n. Let g be the generator polynomial of C, and let $h = (x^n - 1)/g$. Prove that the following are equivalent:

E.4.2.9

$-$ $\mathbf{1}_n \in C$.

$-$ C contains a word of odd weight.

$-$ $g(1) \neq 0$.

$-$ $h(1) = 0$.

E.4.2.10   **Exercise** Let $f = \sum_{i=0}^{k} a_i x^i$ be a polynomial of degree k. Check the following properties of the reciprocal:

— $\widehat{f}$ has degree k if $a_0 = f(0) \neq 0$. In this case, the leading coefficient of $\widehat{f}$ is $a_0 = f(0)$ and, therefore, $\widehat{f}/f(0)$ is monic.

— Let $\alpha \in \mathbb{F}$ be a field element. Prove the following equivalence:

$$f(\alpha) = 0 \iff \widehat{f}(\alpha^{-1}) = 0.$$

E.4.2.11   **Exercise** Prove 4.2.6.

E.4.2.12   **Exercise** Evaluate the factorization of $x^8 - 1 \in \mathbb{F}_3[x]$ into monic irreducible polynomials, and derive generator matrices for the codes generated by products of degree four of these factors.

E.4.2.13   **Exercise** Describe the annihilator of the repetition code of length n.

E.4.2.14   **Exercise** Give for every binary cyclic code of length 9 its dimension, a generator matrix, a check matrix, and its dual code.

E.4.2.15   **Exercise** Factor $x^{15} - 1 \in \mathbb{F}_4[x]$ into irreducible polynomials.

E.4.2.16   **Exercise** Is the binary code generated by

$$\begin{pmatrix} 1 & 1 & 1 & 1 & 0 & 0 & 0 \\ 0 & 1 & 1 & 1 & 1 & 0 & 0 \\ 0 & 0 & 1 & 1 & 1 & 1 & 0 \\ 0 & 0 & 0 & 1 & 1 & 1 & 1 \end{pmatrix}$$

cyclic?

E.4.2.17   **Exercise** Show that the mapping $C \mapsto \mathrm{Ann}(C)$ is an anti-automorphism of the lattice of cyclic codes of length n over $\mathbb{F}_q$. In particular, for all cyclic codes C_0, C_1 of length n over $\mathbb{F}_q$ prove that

$$\mathrm{Ann}(C_0 + C_1) = \mathrm{Ann}(C_0) \cap \mathrm{Ann}(C_1)$$

and

$$\mathrm{Ann}(C_0 \cap C_1) = \mathrm{Ann}(C_0) + \mathrm{Ann}(C_1).$$

Exercise For each $f \in \mathbb{F}_q[x]$ we introduce the notation   E.4.2.18

$$V(f) := \left\{ \alpha \in \overline{\mathbb{F}_q} \mid f(\alpha) = 0 \right\}$$

for the corresponding *variety* over $\mathbb{F}_q$, where $\overline{\mathbb{F}_q}$ denotes the algebraic closure of $\mathbb{F}_q$ (cf. 3.2.23). A nonempty variety V over $\mathbb{F}_q$ (i.e. a subset V of $\overline{\mathbb{F}_q}$ which is a variety of a polynomial, $V = V(f)$, for some $f \in \mathbb{F}_q[x]$) is called *irreducible*, if it cannot be written as the disjoint union $V_1 \cup V_2$ of two proper subvarieties V_1, V_2 over $\mathbb{F}_q$. Prove that, for each $f_1, f_2 \in \mathbb{F}_q[x]$, the following holds:

- $f_1 \mid f_2 \iff V(f_1) \subseteq V(f_2)$.
- $V(f_1) \cap V(f_2) = V(h) \iff h = \kappa \cdot \gcd(f_1, f_2)$ for a suitable $\kappa \in \mathbb{F}_q^*$.
- $V(f_1) \cup V(f_2) = V(g) \iff g = \kappa \cdot \operatorname{lcm}(f_1, f_2)$ for a suitable $\kappa \in \mathbb{F}_q^*$.
- $V(f_i)$ is irreducible over $\mathbb{F}_q$ if and only if f_i is irreducible over $\mathbb{F}_q$.
- Each nonempty variety over $\mathbb{F}_q$ is the disjoint union of irreducible varieties over $\mathbb{F}_q$.
- Each nonempty variety over $\mathbb{F}_q$ is the disjoint union of orbits of Galois groups over $\mathbb{F}_q$, whence it is closed under the Frobenius automorphism $\alpha \mapsto \alpha^q$.

Exercise Assume that $\xi \in \mathbb{F}_{q^m}$ is a primitive n-th root of unity. Show that   E.4.2.19

$$|\operatorname{Gal}(\xi^i)| = |\operatorname{Gal}(i)|$$

is the number of different cyclic shifts of the vector

$$(i_{m-1}, \dots, i_0),$$

defined by the q-adic expansion of i, which means

$$i = i_{m-1}q^{m-1} + \dots + i_1 q + i_0, \qquad i_j \in q, \ j \in m.$$

4.3 BCH-Codes and Reed–Solomon-Codes   4.3

One of the most important classes of cyclic codes was introduced by R.C. Bose and D.K. Ray-Chauduri in [24] and independently also by A. Hocquenghem in [90]. This class of codes is known as the BCH-codes. A subclass of these codes are the Reed–Solomon-codes, which are due to I.S. Reed and G. Solomon [168]. Codes of these classes can be constructed easily from their varieties, and they have good error correcting qualities. For their decoding an efficient procedure is known, which will be described later on.

As before, we denote by n the length of the codewords, and we assume that it is *not divisible by the characteristic p of the field $\mathbb{F}_q$*. The order $\mathrm{ord}_n(q)$ of q in the group of units of $\mathbb{Z}_n$ is again indicated by m, so that, in particular, $q^m \equiv 1 \bmod n$. From a primitive element β of $\mathbb{F}_{q^m}$, we obtain the primitive n-th root of unity

$$\zeta := \beta^{(q^m-1)/n},$$

and so the set of all n-th roots of unity in $\mathbb{F}_{q^m}$ is

$$U_n = \langle \zeta \rangle = \left\{ \kappa \in \mathbb{F}_{q^m} \mid \kappa^n = 1 \right\}.$$

A subset $W \subseteq U_n$ will be called *consecutive* (with respect to ζ), if there exist integers $b \geq 0$ and $\delta \geq 2$, such that

$$W = \left\{ \zeta^b, \zeta^{b+1}, \ldots, \zeta^{b+\delta-2} \right\}.$$

The introduction of BCH-codes is motivated by the following result:

4.3.1   **The BCH-bound** *Let C be a cyclic code of length n over $\mathbb{F}_q$ where q is coprime to n. Assume further that the variety of C contains $\delta - 1$ consecutive powers of ζ, a primitive n-th root of unity, where $\delta \geq 2$. Then the minimum distance of C is at least δ. In formal terms,*

$$W := \left\{ \zeta^b, \ldots, \zeta^{b+\delta-2} \right\} \subseteq V(C) \Longrightarrow \mathrm{dist}(C) \geq \delta.$$

Proof: We want to prove this assertion by an application of 1.3.9. For this purpose we consider the $(\delta - 1) \times n$-matrix

$$\widetilde{\Delta} := \begin{pmatrix} 1 & \zeta^b & \zeta^{2b} & \cdots & \zeta^{(n-1)b} \\ 1 & \zeta^{b+1} & \zeta^{2(b+1)} & \cdots & \zeta^{(n-1)(b+1)} \\ \cdots & \cdots & \cdots & \cdots & \cdots \\ 1 & \zeta^{b+\delta-2} & \zeta^{2(b+\delta-2)} & \cdots & \zeta^{(n-1)(b+\delta-2)} \end{pmatrix}.$$

Note that this matrix is a matrix over the extension field $\mathbb{F}_{q^m}$ containing ζ, where $m := \mathrm{ord}_n(q)$, and that for each $c \in C$ we have

$$c \cdot \widetilde{\Delta}^\top = (c(\zeta^b), \ldots, c(\zeta^{b+\delta-2})) = (0, \ldots, 0) = 0.$$

We show that each subset of $\delta - 1$ columns of the matrix $\widetilde{\Delta}$ is linearly independent over $\mathbb{F}_q$. In order to verify this, we consider a submatrix consisting of $\delta - 1$ columns of $\widetilde{\Delta}$:

$$\begin{pmatrix} \zeta^{i_1 b} & \zeta^{i_2 b} & \cdots & \zeta^{i_{\delta-1} b} \\ \zeta^{i_1(b+1)} & \zeta^{i_2(b+1)} & \cdots & \zeta^{i_{\delta-1}(b+1)} \\ \cdots & & \cdots & \\ \zeta^{i_1(b+\delta-2)} & \zeta^{i_2(b+\delta-2)} & \cdots & \zeta^{i_{\delta-1}(b+\delta-2)} \end{pmatrix}$$

with $0 \leq i_1 < i_2 < \ldots < i_{\delta-1} \leq n - 1$. Its determinant is $\zeta^{(i_1+i_2+\ldots+i_{\delta-1})b}$ times the determinant of the Vandermonde matrix

$$
\begin{pmatrix}
1 & 1 & \ldots & 1 \\
\zeta^{i_1} & \zeta^{i_2} & \ldots & \zeta^{i_{\delta-1}} \\
\zeta^{2i_1} & \zeta^{2i_2} & \ldots & \zeta^{2i_{\delta-1}} \\
& \ldots & & \ldots \\
\zeta^{(\delta-2)i_1} & \zeta^{(\delta-2)i_2} & \ldots & \zeta^{(\delta-2)i_{\delta-1}}
\end{pmatrix}.
$$

Hence the determinant is different from 0 since the ζ^{ij} are pairwise distinct. This shows that any $\delta - 1$ columns of $\widetilde{\Delta}$ are linearly independent over $\mathbb{F}_{q^m}$. Thus, $\widetilde{\Delta}$ is a check matrix of a code $\widetilde{C}$ over $\mathbb{F}_{q^m}$ which has minimum distance

$$
\mathrm{dist}(\widetilde{C}) \geq \delta.
$$

Moreover, since $c \cdot \widetilde{\Delta}^\top = 0$, for all $c \in C$, we obtain the inclusion $C \subseteq \widetilde{C}$, and so we also have

$$
\mathrm{dist}(C) \geq \mathrm{dist}(\widetilde{C}) \geq \delta,
$$

as stated.   □

BCH-codes are defined to be the maximal cyclic codes containing a prescribed consecutive set W in their variety:

Definition (BCH-codes, designed distance, Reed–Solomon-codes) Let $W =$ 4.3.2
$\{\zeta^b, \ldots, \zeta^{b+\delta-2}\}$ be a consecutive subset of U_n for some $b \geq 0$ and some δ with $n > \delta \geq 2$. Define the polynomial

$$
g := \mathrm{lcm}\left\{ M_{\zeta^{b+i}} \mid i \in \delta - 1 \right\},
$$

where $M_{\zeta^{b+i}}$ is the minimal polynomial of ζ^{b+i} over $\mathbb{F}_q$. The code C with generator polynomial g is called *the BCH-code generated by W*. The value δ is the *designed distance* of C since $\mathrm{dist}(C) \geq \delta$ by 4.3.1. If $n = q^m - 1$, the code is called *primitive*, since in this case ζ is also a primitive element for $\mathbb{F}_{q^m}$. Moreover, if $b = 1$ we say that C is a BCH-code *in the narrow sense*. BCH-codes of length $n = q - 1$ over $\mathbb{F}_q$ are called *Reed–Solomon-codes*.   ◇

Reed–Solomon-codes are particularly easy to create since in case $q - 1 = n$ the minimal polynomials are all linear. Namely, in this case the field $\mathbb{F}_q$ contains all n-th roots of unity and, therefore, $M_{\zeta^i} = (x - \zeta^i)$ for all i.

Example Let us design a BCH-code C which can correct 2 errors. For this,   4.3.3
we need minimum distance at least 5, i.e. we put the designed distance to be $\delta = 5$. We decide to use a Reed–Solomon-code with $n = q - 1$. Since we want

$q - 1 = n > \delta = 5$, we choose $q = 7$ and $n = 6$. A primitive element modulo 7 is $\beta = 3$, so we may take

$$W := \{3, 3^2, 3^3, 3^4\} = \{3, 2, 6, 4\} \subset \mathbb{F}_7.$$

The Reed–Solomon-code C generated by the consecutive set W has generator polynomial

$$
\begin{aligned}
g &= (x - 3)(x - 3^2)(x - 3^3)(x - 3^4) \\
&= (x - 3)(x - 2)(x - 6)(x - 4) \\
&= x^4 + 6x^3 + 3x^2 + 2x + 4.
\end{aligned}
$$

It is a $(6, 6 - 4) = (6, 2)$-code with generator matrix

$$
\Gamma = \begin{pmatrix} 4 & 2 & 3 & 6 & 1 & 0 \\ 0 & 4 & 2 & 3 & 6 & 1 \end{pmatrix}.
$$

The check polynomial of C is

$$
h = \frac{x^6 - 1}{x - 1} = (x - 1)(x - 3^5) = x^2 + x + 5
$$

and, therefore, a check matrix of C is

$$
\Delta = \begin{pmatrix} 1 & 1 & 5 & 0 & 0 & 0 \\ 0 & 1 & 1 & 5 & 0 & 0 \\ 0 & 0 & 1 & 1 & 5 & 0 \\ 0 & 0 & 0 & 1 & 1 & 5 \end{pmatrix}.
$$

◇

The generator polynomial g of a BCH-code with designed distance δ is the least common multiple of the minimal polynomials over $\mathbb{F}_q$ of the elements in the consecutive set $W = \{\zeta^b, \zeta^{b+1}, \ldots, \zeta^{b+\delta-2}\}$. Hence, it is the polynomial of least degree over $\mathbb{F}_q$ with $\zeta^b, \zeta^{b+1}, \ldots, \zeta^{b+\delta-2}$ as roots. Consequently, c is an element of C if and only if

$$c(\zeta^b) = \ldots = c(\zeta^{b+\delta-2}) = 0.$$

In addition, since $m = \text{ord}_n(q)$ we have that n divides $q^m - 1$, i.e. we know that the primitive n-th root ζ belongs to $\mathbb{F}_{q^m}$. The minimal polynomial M_ζ of ζ over $\mathbb{F}_q$ is of degree m. Hence, similarly as in 3.1.9, the set $\{1, \zeta, \ldots, \zeta^{m-1}\}$ is a basis of $\mathbb{F}_{q^m}$ over $\mathbb{F}_q$, and each element $\alpha \in \mathbb{F}_{q^m}$ can be written in a unique way as

$$\alpha = \sum_{i \in m} \kappa_i \zeta^i, \qquad \kappa_i \in \mathbb{F}_q, \ i \in m.$$

Indeed, we may identify α with the coefficient vector $(\kappa_0, \ldots, \kappa_{m-1}) \in \mathbb{F}_q^m$ with respect to this basis. If we replace in the matrix

$$
\widetilde{\Delta} := \begin{pmatrix}
1 & \xi^b & \xi^{2b} & \cdots & \xi^{(n-1)b} \\
1 & \xi^{b+1} & \xi^{2(b+1)} & \cdots & \xi^{(n-1)(b+1)} \\
& \cdots & & \cdots & \\
1 & \xi^{b+\delta-2} & \xi^{2(b+\delta-2)} & \cdots & \xi^{(n-1)(b+\delta-2)}
\end{pmatrix},
$$

which occurs in the proof of 4.3.1, each component by the transposed of its coefficient vector with respect to the basis $\{1, \xi, \ldots, \xi^{m-1}\}$, then $\widetilde{\Delta}$ contains a check matrix of C. We actually get a check matrix of C if we choose a maximal set of independent rows of the extended matrix $\widetilde{\Delta}$.

The BCH-bound is a lower bound for the minimum distance of a BCH-code. Besides that, there is also a bound for the dimension:

Theorem *The dimension k of the BCH-code C generated by the consecutive set W of order $\delta - 1$ satisfies the inequality*   **4.3.4**

$$
k \geq n - m(\delta - 1) = n - m \cdot |W|.
$$

Proof: The least common multiple g of the minimal polynomials of the elements of W consists of at most $\delta - 1$ different factors, and each of them is of degree at most m, since m is the maximal orbit length of the Galois group (see 4.2.11). Thus $\deg g \leq m \cdot (\delta - 1)$. This, together with $k = n - |V(g)|$ (see 4.2.16) gives the desired estimate for the dimension k of C. $\square$

The BCH-bound $d \geq \delta$ is not always tight. In fact, $d > \delta$ happens frequently. The most prominent example of this situation is maybe that of the Golay-codes of length 11 and 23, which we will discuss in Section 4.4. It is an important (and sometimes difficult!) problem to determine the true minimum distance of BCH-codes. Several attempts have been made in developing better lower bounds. The easiest such improvement is to apply the BCH-bound to the longest consecutive set of roots in the variety of C. For example, if C is a ternary code and if $m = \mathrm{ord}_n(q) > 1$, then there is a 3-cyclotomic coset containing 1 and 3. Thus, whenever $W = \{\xi, \xi^2\}$ is a consecutive set of roots of C, then also $\{\xi, \xi^2, \xi^3\}$ is contained in $V(C)$. The optimal bound, i.e. the BCH-bound which comes from the longest consecutive set of roots in $V(C)$ is called the *Bose-distance*. We note that there are also results which show that under certain conditions the BCH-bound is sharp, i.e., the true minimum distance of the code agrees with the BCH-bound.

4.3.5   **Examples**

1. The following table gives the parameters of several binary BCH-codes of length 15, described in terms of a primitive 15-th root of unity ζ and the minimal polynomials of some of its powers:

generator polynomial	k	δ	d
1	15	1	1
M_ζ	11	3	3
$M_\zeta M_{\zeta^3}$	7	5	5
$M_\zeta M_{\zeta^3} M_{\zeta^5}$	5	7	7
$M_\zeta M_{\zeta^3} M_{\zeta^5} M_{\zeta^7}$	1	15	15

2. Now we consider the binary cyclic codes of length 23. The polynomial $x^{23} - 1$ decomposes over $\mathbb{F}_2$ in the following way into irreducible factors:

$$(x+1)(x^{11} + x^9 + x^7 + x^6 + x^5 + x + 1)(x^{11} + x^{10} + x^6 + x^5 + x^4 + x^2 + 1).$$

Because of $2^{11} \equiv 1 \bmod 23$, the roots of these polynomials are contained in $\mathbb{F}_{2^{11}}$. Let ζ denote a root of

$$g := x^{11} + x^9 + x^7 + x^6 + x^5 + x + 1.$$

According to 4.2.17, $\mathrm{Gal}(\zeta)$ contains the consecutive set $\{\zeta, \zeta^2, \zeta^3, \zeta^4\}$, and thus the binary cyclic code of length 23 generated by g has designed distance $\delta = 5$. The same holds true for the code which is generated by $x^{11} + x^{10} + x^6 + x^5 + x^4 + x^2 + 1$. In Section 4.4, we will show that both codes are permutationally isometric quadratic-residue-codes with minimum distance $d = 7$. $\Diamond$

Now we show that all binary Hamming-codes are BCH-codes.

4.3.6   **Theorem** *Let ζ denote a primitive $n = (2^m - 1)$-th root of unity. Then the consecutive set $W := \{\zeta, \zeta^2\}$ generates the m-th order binary Hamming-code. Thus, binary Hamming-codes are cyclic, they are in fact narrow sense BCH-codes.*

Proof: 1. According to Exercise 4.2.19 the degree of the minimal polynomial M_ζ is m, and so $\{1, \zeta, \ldots, \zeta^{m-1}\}$ is linearly independent over $\mathbb{F}_2$ and therefore an $\mathbb{F}_2$-basis of $\mathbb{F}_{2^m}$. Hence we can express the powers of ζ in terms of this basis, say

$$\zeta^j = \sum_{i \in m} h_{ij} \zeta^i, \qquad j \in n.$$

The coefficients in these equations form the $m \times n$-matrix

$$\Delta := (h_{ij})_{i \in m, j \in n}.$$

Consider the coefficient vectors of these powers of ξ when written as linear combinations of ξ^i, $i \in m$ with coefficients in $\mathbb{F}_2$. Since ξ is primitive, the powers ξ^i are pairwise distinct and so, since $n = 2^m - 1$, they are just all the binary representations of the positive integers $1, \ldots, n$. Therefore, the matrix Δ is a check matrix of the m-th order binary Hamming-code C.

2. Now we show that the code C is cyclic and that it has the generator polynomial

$$g := (x - \xi)(x - \xi^2)(x - \xi^{2^2}) \cdots (x - \xi^{2^{m-1}}) = \prod_{j \in m}(x - \xi^{2^j}).$$

This polynomial is the minimal polynomial of ξ over $\mathbb{F}_q$, since its roots form the orbit of ξ under the Galois group. Moreover, $c = (c_0, \ldots, c_{n-1})$ is contained in C if and only if $\Delta \cdot c^\top = 0$, which means that for each i the following holds:

$$\sum_{j \in n} h_{ij}c_j = 0, \qquad i \in m.$$

Multiplying both sides by ξ^i and summing over i yields

$$0 = \sum_{i \in m}\sum_{j \in n} h_{ij}c_j\xi^i = \sum_j c_j \sum_i h_{ij}\xi^i = \sum_j c_j\xi^j = c(\xi),$$

which shows that every codeword, when considered as a polynomial, has ξ as a root. This last implication is indeed reversible, since $1, \xi, \xi^2, \ldots, \xi^{m-1}$ is an $\mathbb{F}_2$-basis for $\mathbb{F}_{2^m}$. Thus

$$c \in C \iff c(\xi) = 0,$$

and hence $C = \mathrm{I}(M_\xi)/\mathrm{I}(x^n - 1)$. Since $\deg M_\xi = m$ and $\dim(C) = n - m$, we conclude that M_ξ is indeed the generator polynomial of C.

3. From the foregoing we deduce that the m-th order binary Hamming-code is cyclic with variety

$$V(C) = \left\{\xi, \xi^2, \xi^4, \ldots, \xi^{2^m - 1}\right\},$$

and it is generated by the consecutive set

$$W := \left\{\xi, \xi^2\right\},$$

as stated. $\qquad\qquad\qquad\qquad\qquad\qquad\qquad\qquad\qquad\qquad\qquad\qquad\qquad\square$

Examples The $(5,3)$ second order Hamming-code over $\mathbb{F}_4$ is cyclic. If ξ denotes a primitive 15-th root of unity, then **4.3.7**

$$g := (x - \xi^5)(x - \xi^{10}) = x^2 + x + 1$$

is a generator polynomial of this code (this polynomial has been computed in 4.2.13). On the other hand, the second order ternary Hamming-code is not cyclic, according to Exercise 4.1.3. $\qquad\qquad\qquad\qquad\qquad\qquad\qquad\diamond$

More generally, the following holds:

4.3.8   **Theorem**   *Let β be a primitive element for $\mathbb{F}_{q^m}$, put $\zeta := \beta^{q-1}$ and assume that $n = (q^m - 1)/(q - 1)$. The linear code C of length n over $\mathbb{F}_q$, the check matrix $\Delta = (h_{ij})$ of which is defined by the equations*

$$\zeta^j = \sum_{i \in m} h_{ij}\zeta^i, \qquad j \in n,$$

is isometric to the m-th order q-ary Hamming-code, provided m and $q - 1$ are relatively prime. Hence such Hamming-codes are BCH-codes generated by consecutive sets $W = \{\zeta\}$ and with varieties

$$V(C) = \left\{\zeta, \zeta^q, \ldots, \zeta^{q^{m-1}}\right\}.$$

Proof: 1. According to Exercise 4.2.19, the degree of the minimal polynomial M_ζ is m, and so $\{1, \zeta, \ldots, \zeta^{m-1}\}$ is linearly independent and therefore an $\mathbb{F}_q$-basis of $\mathbb{F}_{q^m}$. Hence we can in fact represent each ζ^j as an $\mathbb{F}_q$-linear combination of the $\zeta^i, i \in m$. Thus, the matrix

$$\Delta := (h_{ij})_{i \in m, j \in n}$$

is defined.

2. Because of $n = (q^m - 1)/(q - 1)$ we need only show (in order to prove that Δ is a check matrix of an m-th order q-ary Hamming-code) that the columns of Δ are pairwise linearly independent. If this were not the case, say

$$\zeta^i = \alpha \zeta^j,$$

for a suitable $\alpha \in \mathbb{F}_q^*$ and some $j < i \in n$, then $\zeta^{i-j} \in \mathbb{F}_q^*$ and so there were some $k \in q$ for which $\zeta^{i-j} = \beta^{nk}$ and, therefore, $(q-1)(i-j) \equiv nk \bmod q^m - 1$. Because of $0 < i - j < n$ we could even deduce that

$$(q-1)(i-j) = nk.$$

Now we derive also that n and $q - 1$ are relatively prime. As

$$
\begin{aligned}
n &= \frac{q^m - 1}{q - 1} \\
&= q^{m-1} + q^{m-2} + \ldots + q + 1 \\
&= m + (q^{m-1} - 1) + (q^{m-2} - 1) + \ldots + (q - 1) \\
&= m + (q - 1)\sum_{i \in m}\sum_{j \in i} q^j,
\end{aligned}
$$

each divisor of $q - 1$ and n divides m, and each divisor of $q - 1$ and m divides n (see Exercise 4.3.2). Hence, since m and $q - 1$ are supposed to be coprime, the same holds for n and $q - 1$. Thus, since $(q - 1)(i - j) = nk$, every divisor of n is a divisor of $i - j$, in particular n itself, which contradicts the choice of i and j. Hence Δ is in fact a check matrix of an m-th order q-ary Hamming-code.

3. As in the proof of 4.3.6, we can easily check that

$$c \in C \iff c(\xi) = 0.$$

Hence, C is generated by M_ξ and, therefore, the m-th order q-ary Hamming-code is a BCH-code with variety

$$V(C) = \left\{ \xi, \xi^q, \ldots, \xi^{q^{m-1}} \right\},$$

whence generated by $W = \{\xi\}$.   □

Now we describe a group of automorphisms of the parity extension of a primitive BCH-code. The *affine linear group*

$$\mathrm{AGL}_1(q) := \{ \sigma_{\kappa,\lambda} : \gamma \mapsto \kappa\gamma + \lambda \mid \kappa \in \mathbb{F}_q^*, \lambda \in \mathbb{F}_q \}$$

acts *transitively* on $\mathbb{F}_q$, i.e. for any two elements $\gamma, \beta \in \mathbb{F}_q$ there exist $\kappa, \lambda \in \mathbb{F}_q$ with $\sigma_{\kappa,\lambda}(\gamma) = \beta$. The elements of the group are called *affine transformations* on $\mathbb{F}_q$. For example, the inverse of $\sigma_{\kappa,\lambda}$ is

$$\sigma_{\kappa,\lambda}^{-1} = \sigma_{\kappa^{-1}, -\kappa^{-1}\lambda}.$$

Theorem   *The parity extension*   **4.3.9**

$$P(C) := \left\{ (c_0, \ldots, c_{n-1}, c_\infty) \;\middle|\; (c_0, \ldots, c_{n-1}) \in C, \; c_\infty := -\sum_{i=0}^{n-1} c_i \right\}$$

of a primitive BCH-code C of length $n = q^m - 1$ over $\mathbb{F}_q$ has a group of automorphisms which is isomorphic to $\mathrm{AGL}_1(q^m)$.

Proof: We prove the statement for a narrow sense BCH-code C with designed distance δ. The proof in the general case is similar. Hence, $V(C)$ contains the consecutive set $\{\xi, \xi^2, \ldots, \xi^{\delta-1}\}$, where ξ denotes a primitive element of $\mathbb{F}_{q^m}^*$. The parity check coordinate of $P(C)$ will be labeled by ∞. Thus, a vector $c = (c_0, \ldots, c_{n-1}, c_\infty) \in \mathbb{F}_q^{n+1}$ is contained in $P(C)$, if

1. $\sum_{i \in n} c_i \xi^{ij} = 0$ for $1 \le j \le \delta - 1$ and

2. $\sum_{i \in n} c_i + c_\infty = 0$.

The field $\mathbb{F}_{q^m}$ can be identified with the set of coordinates $\{0, \ldots, n-1\} \cup \{\infty\}$ via

$$\alpha \mapsto \log_\xi \alpha =: \log \alpha,$$

where we put $\log_\zeta 0 := \infty$, i.e. $\zeta^\infty := 0$. Then the conditions for $c \in P(C)$ read as follows:

1. $\sum_{\alpha \in \mathbb{F}_{q^m}} c_{\log \alpha} (\zeta^{\log \alpha})^j = 0$ for $1 \leq j \leq \delta - 1$ and

2. $\sum_{\alpha \in \mathbb{F}_{q^m}} c_{\log \alpha} = 0$.

It is easy to check that the seemingly additional summand for $\alpha = 0$ in the first condition vanishes. The second condition is certainly invariant under the action of $\mathrm{AGL}_1(q^m)$. We now prove the invariance of the first condition. For this purpose consider $\sigma \in \mathrm{AGL}_1(q^m)$ with

$$\sigma := \sigma_{\kappa, \lambda}^{-1} = \sigma_{\kappa^{-1}, -\kappa^{-1}\lambda}.$$

Then, for $1 \leq j \leq \delta - 1$, we obtain

$$\begin{aligned}
\sum_{\alpha \in \mathbb{F}_{q^m}} c_{\log \sigma(\alpha)} (\zeta^{\log \alpha})^j &= \sum_{\alpha \in \mathbb{F}_{q^m}} c_{\log \alpha} (\zeta^{\log \sigma^{-1}(\alpha)})^j \\
&= \sum_{\alpha \in \mathbb{F}_{q^m}} c_{\log \alpha} (\kappa \alpha + \lambda)^j \\
&= \sum_{\alpha \in \mathbb{F}_{q^m}} \sum_{l=0}^{j} \binom{j}{l} \kappa^l \lambda^{j-l} c_{\log \alpha} \alpha^l \\
&= \sum_{l=0}^{j} \binom{j}{l} \kappa^l \lambda^{j-l} \sum_{\alpha \in \mathbb{F}_{q^m}} c_{\log \alpha} (\zeta^{\log \alpha})^l = 0,
\end{aligned}$$

since the inner sum is zero, by assumption. $\square$

Based on this theorem we want to derive a result on the minimum distance of binary primitive BCH-codes. We still need the following

4.3.10   **Lemma** *Let C be a binary linear code of length n. Assume that $P(C)$, the parity extension of C, possesses a group of automorphisms which acts transitively on its components. Then the minimum weight of C is odd.*

Proof: Denote by A_i (resp. A_i') the number of codewords of weight i in C (resp. $P(C)$). The number of pairs $(l, c) \in (n \cup \{\infty\}) \times P(C)$ with $\mathrm{wt}(c) = 2i$ and $c_l = 1$ is $2i \cdot A_{2i}'$. The number of vectors $c \in P(C)$ such that $\mathrm{wt}(c) = 2i$ and $c_\infty = 1$ is A_{2i-1}. Since the automorphism group of $P(C)$ is transitive on the set of coordinates, for each $l \in n \cup \{\infty\}$ we have

$$|\{(l, c) \mid c \in P(C),\ \mathrm{wt}(c) = 2i,\ c_l = 1\}| =$$

$$|\{c \mid c \in P(C),\ \mathrm{wt}(c) = 2i,\ c_\infty = 1\}| = A_{2i-1},$$

so that $2i \cdot A'_{2i} = (n+1)A_{2i-1}$, whence

$$\frac{2i \cdot A'_{2i}}{n+1} = A_{2i-1}.$$

Furthermore, since C is a binary code, $P(C)$ is even. If d' is the minimum distance of $P(C)$, then the equation above gives $A_{d'-1} > 0$. Thus, according to the construction of $P(C)$, the minimum weight of C is equal to $d' - 1$ and odd. $\qquad \square$

Consequently, we obtain

Corollary *The minimum distance of a primitive binary BCH-code is odd.* $\qquad \square$ **4.3.11**

In certain cases, it is equal to the designed distance:

Theorem *The primitive, narrow sense, binary BCH-code of length $n = 2^m - 1$ with*   **4.3.12**
designed distance $\delta = 2t + 1$ has minimum distance $d = \delta$, provided that

$$\sum_{i=0}^{t+1} \binom{2^m - 1}{i} > 2^{mt}.$$

Proof: The generator polynomial g of such a code is the least common multiple of $\delta - 1$ minimal polynomials the degree of which is bounded above by m, the order of the Galois group. If $q = 2$ then i and $2i$ are in the same 2-cyclotomic coset modulo n, and hence $M_{\zeta^{2i}} = M_{\zeta^i}$ for all i. Thus

$$\mathrm{lcm}\left\{ M_{\zeta^1}, M_{\zeta^2}, \ldots, M_{\zeta^{2t}} \right\} = \mathrm{lcm}\left\{ M_{\zeta^1}, M_{\zeta^3}, \ldots, M_{\zeta^{2t-1}} \right\},$$

and therefore $k = n - \deg g \geq n - mt$. If $d := \mathrm{dist}(C)$ were not equal to $\delta = 2t + 1$, then by 4.3.11 $d \geq 2t + 3$. Such a code would correct $t + 1$ errors, and so the Hamming-bound would give

$$\sum_{i=0}^{t+1} \binom{2^m - 1}{i} \leq 2^{n-k} \leq 2^{mt},$$

in contradiction to the assumption. $\qquad \square$

Theorem *Let C be a narrow sense q-ary BCH-code of length n with designed distance*   **4.3.13**
δ. If δ divides n then $\mathrm{dist}(C) = \delta$.

Proof: By definition, $\zeta, \zeta^2, \ldots, \zeta^{\delta-1}$ are roots of C, where ζ is again a primitive n-th root of unity. Write $n = \delta s$ for some integer s. Then $\zeta^{is} \neq 1$ for all $0 < i < \delta$. In the expression

$$x^n - 1 = (x^s - 1)(x^{(\delta-1)s} + \ldots + x^{2s} + x^s + 1),$$

the roots $\zeta, \zeta^2, \ldots, \zeta^{\delta-1}$ must, therefore, all be roots of the second factor. Thus $x^{(\delta-1)s} + \ldots + x^{2s} + x^s + 1 + I(x^n - 1)$ is a codeword of C of weight δ, so $\delta \leq \text{dist}(C) \leq \delta$. $\square$

4.3.14   **Examples** In the following table we give several binary BCH-codes together with their designed distances.

BCH-code	$t = \lfloor (\delta - 1)/2 \rfloor$	δ
$(31, 26)$	1	3
$(31, 21)$	2	5
$(31, 16)$	3	7
$(31, 11)$	4	9

For $t \in \{1, 2, 3\}$ we have the following inequality

$$\sum_{i=0}^{t+1} \binom{31}{i} > 2^{5t}$$

and, therefore, the designed distances of the first three codes are equal to their minimum distances.

In addition, we present the binary BCH-codes of length 21. In this case, $m = 6$. Let β be a primitive element for $\mathbb{F}_{2^6} = \mathbb{F}_{64}$, where β is a root of $x^6 + x^5 + 1$ over $\mathbb{F}_2$. By means of 2-cyclotomic cosets modulo 21, we can compute the minimal polynomials of powers of ζ. We obtain

cyclotomic coset	M_{ζ^i}
$\{0\}$	$x + 1$
$\{1, 2, 4, 8, 11, 16\}$	$x^6 + x^5 + x^4 + x^2 + 1$
$\{3, 6, 12\}$	$x^3 + x + 1$
$\{5, 10, 13, 17, 19, 20\}$	$x^6 + x^4 + x^2 + x + 1$
$\{7, 14\}$	$x^2 + x + 1$
$\{9, 15, 18\}$	$x^3 + x^2 + 1$

The BCH-codes are

δ	g	$\deg g$	$\text{wt}(g)$	(n, k, d)	optimal?
1	1	0	1	$(21, 21, 1)$	yes
3	M_ζ	6	5	$(21, 15, 3)$	no
5	$M_\zeta M_{\zeta^3}$	9	7	$(21, 12, 5)$	yes
7	$M_\zeta M_{\zeta^3} M_{\zeta^5}$	15	11	$(21, 6, 7)$	no
9	$M_\zeta M_{\zeta^3} M_{\zeta^5} M_{\zeta^7}$	17	9	$(21, 4, 9)$	no
11	$M_\zeta M_{\zeta^3} M_{\zeta^5} M_{\zeta^7} M_{\zeta^9}$	20	21	$(21, 1, 21)$	yes

The minimum distances of the codes with $\delta = 3$ and $\delta = 7$ follow from 4.3.13. The B-construction implies that there is no $(21, 12, 6)$-code, and hence the code with $\delta = 5$ is an optimal $(21,12,5)$-code. The minimum distance of the code with $\delta = 9$ is 9 since the generator polynomial

$$g = M_\xi M_{\xi^3} M_{\xi^5} M_{\xi^7} = x^{17} + x^{15} + x^{14} + x^{10} + x^8 + x^7 + x^3 + x + 1$$

has weight 9.

For the construction of optimal $(21,15,4)$, $(21,6,8)$ and $(21,4,10)$-codes, see Exercise 4.3.8. ◇

Recall that Reed–Solomon-codes are BCH-codes of length $n = q - 1$. Even though these codes require larger field sizes, they are of enormous practical importance. One reason for this may be that they are defined so easily. The paper [168] by I.S. Reed and G. Solomon is considered to be a major breakthrough in coding theory. Today, the Reed–Solomon-codes are ubiquitous. Every compact disc player uses Reed–Solomon-codes for error-correction. We will have to say more on that in Chapter 5. At this point, we only mention that two codes which are defined over $\mathbb{F}_{2^8}$ play an important role. These codes, with parameters $(32,28,5)$ and $(28,24,5)$ are obtained from a $(255,251,5)$-Reed–Solomon-code over $\mathbb{F}_{2^8}$ by successive shortening. The encoding with respect to these two codes is completely explained in Section 5.4.

In the case when $n = q - 1$, i.e. in the case of Reed–Solomon-codes,

$$x^n - 1 = \prod_{i \in n}(x - \xi^i),$$

where ξ is a primitive element of $\mathbb{F}_q^*$, and each linear factor $x - \xi^i$ belongs to $\mathbb{F}_q[x]$. To begin with the discussion of these codes, we show that they are maximum distance separable:

Theorem *Any Reed–Solomon-code is MDS.*   **4.3.15**

Proof: An (n,k,d)-Reed–Solomon-code with designed distance δ has a generator polynomial of the form

$$g = (x - \xi^b) \cdots (x - \xi^{b+\delta-2}).$$

From 4.2.5 we obtain that

$$k = n - \deg g = n - \delta + 1 \geq n - d + 1.$$

The Singleton-bound implies the converse inequality. □

Corollary *For every positive integer $k \leq q - 1$, there exists a $(q-1,k)$-MDS-code*   **4.3.16**
over $\mathbb{F}_q$. □

4.3.17   **Example** The generator polynomial of a $(255, 251, 5)$-Reed–Solomon-code over $\mathbb{F}_{2^8}$ is given by

$$g = \prod_{i=1}^{4} (x - \zeta^i),$$

where ζ is a primitive element of $\mathbb{F}_{2^8}^*$, whence a primitive 255-th root of unity. Using the shortening procedure (cf. 2.2.17), we obtain a $(254, 250)$-code over $\mathbb{F}_{2^8}$ with minimum distance $d' \geq 5$. The Singleton-bound yields $d' = 5$. Further successive shortening gives MDS-codes with parameters $(32, 28, 5)$ and $(28, 24, 5)$. $\diamond$

Now we consider the parity extensions (cf. 2.2.2) of Reed–Solomon-codes.

4.3.18   **Theorem** *The parity extension $P(C)$ of an (n, k)-Reed–Solomon-code C over $\mathbb{F}_q$ with generator polynomial*

$$g = (x - \zeta)(x - \zeta^2) \cdots (x - \zeta^{n-k})$$

is MDS.

Proof: We know that C is an MDS-code, and so we can assume that $c = \sum_{i \in n} c_i x^i + I(x^n - 1) \in C$ is an element of minimum weight $d = n - k + 1$. There exists a polynomial $f \in \mathbb{F}_q[x]$ with $c = fg + I(x^n - 1)$. In the parity extension $P(C)$ of C, c is extended by the coordinate c_∞, defined by

$$-c_\infty = \sum_{i \in n} c_i = c(1).$$

We distinguish two cases:

1. If $k = n$, then $\mathrm{dist}(C) = 1$ and the minimum distance of $P(C)$ equals 2, therefore $P(C)$ is an MDS-code.

2. We assume now that $1 \leq k < n$, so that 1 is not among the roots of g, i.e. $g(1) \neq 0$. We claim that $c(1) \neq 0$. Otherwise, if $c(1) = 0$ then also $f(1)g(1) = c(1) = 0$. From $g(1) \neq 0$ it follows that $f(1) = 0$. Hence $c(x)$ is a multiple of $(x - 1)g = (x - \zeta^0)(x - \zeta^1) \cdots (x - \zeta^{n-k})$. By the BCH-bound, the weight of c is at least $n - k + 2 = d + 1$, a contradiction. $\square$

Exercises

E.4.3.1   **Exercise** Let $C = I(g)/I(x^n - 1)$ be the cyclic code of length n over $\mathbb{F}_q$ which is generated by $g \in \mathbb{F}_q[x]$. Assume that n is relatively prime to q. Factor g into irreducible polynomials as $g = f_0 \cdot f_1 \cdots f_{l-1}$ with $f_i \in \mathbb{F}_q[x]$. For $i \in l$, let β_i

be a root of f_i. Define the $l \times n$-matrix

$$\Delta' = \begin{pmatrix} 1 & \beta_0 & \beta_0^2 & \cdots & \beta_0^{n-1} \\ 1 & \beta_1 & \beta_1^2 & \cdots & \beta_1^{n-1} \\ \vdots & & & & \vdots \\ 1 & \beta_{l-1} & \beta_{l-1}^2 & \cdots & \beta_{l-1}^{n-1} \end{pmatrix} = (\beta_i^j)_{i \in l, j \in n}.$$

Then $c \in \mathbb{F}_q^n$ is in C if and only if $c \cdot \Delta'^\top = 0$. That is, Δ' is a check matrix of a code $\tilde{C}$ over some larger field containing $\beta_0, \ldots, \beta_{l-1}$ that restricts to C, i.e. $\tilde{C} \cap \mathbb{F}_q^n = C$.

Exercise Let a, b, q and r be integers with $a = qb + r$. Show that $\gcd(a, b) = \gcd(b, r)$.   E.4.3.2

Exercise Consider the ternary cyclic $(8, 4)$-code C with generator polynomial   E.4.3.3

$$g = x^4 + 2x^3 + 2x + 2 = (x^2 + 2x + 2)(x^2 + 1).$$

Denote by ζ a root of the primitive polynomial $x^2 + x + 2 \in \mathbb{F}_3[x]$. Check that this code has variety

$$V(C) = \{\zeta^2, \zeta^6, \zeta^5, \zeta^7\}$$

and conclude that C is an $(8, 4, 4)$-code.

Exercise Using 4.3.12, evaluate a generator matrix of a binary cyclic code with parameters $(63, 51, 5)$.   E.4.3.4

Exercise Show that the affine linear group $\mathrm{AGL}_1(q)$ acts *doubly transitive* on $\mathbb{F}_q$, i.e. for $\alpha, \beta, \gamma, \delta \in \mathbb{F}_q$ such that $\alpha \neq \beta$ and $\gamma \neq \delta$ there exist $\kappa, \lambda \in \mathbb{F}_q, \kappa \neq 0$, with $\sigma_{\kappa, \lambda}(\alpha) = \gamma$ and $\sigma_{\kappa, \lambda}(\beta) = \delta$.   E.4.3.5

Exercise Construct the elements of a $(6, 2, 5)$-Reed–Solomon-code over $\mathbb{F}_7$.   E.4.3.6

Exercise Show that the dual of a Reed–Solomon-code is again a Reed–Solomon-code.   E.4.3.7

Exercise Construct optimal binary codes with parameters   E.4.3.8

1. $(21, 15, 4)$,
2. $(21, 6, 8)$,
3. $(21, 4, 10)$.

Why are these codes optimal?

Hints: For 1., take the Reed–Muller-code $RM_{4,3}^2$, which is a $(32, 26, 4)$-code. Shorten this code at 11 positions. For 2., apply the $(u \mid u + v)$ construction to a $(10, 5, 4)$-code and a repetition code of length 10. The resulting code of length 20 may be extended by a zero position. A $(10, 5, 4)$-code can be constructed using the $(u \mid u + v)$ construction for a $(5, 4, 2)$-code with a repetition code. For 3., construct a $(20, 4, 10)$-code and extend it by a zero position. A $(20, 4, 10)$-code can be obtained from the (u, v) construction applied to a $(8, 4, 4)$-code and a $(12, 4, 6)$-code. A $(12, 4, 6)$-code results from the $(u \mid u + v)$ construction applied to a $(6, 3, 3)$-code and a repetition code. A $(6, 3, 3)$-code can be obtained as a shortened subcode of a $(7, 4, 3)$-code. For the upper bounds, apply the Griesmer-bound.

4.4 Quadratic-Residue-Codes, Golay-Codes

In this section, we will construct a class of cyclic codes of length n, assuming that n is an odd prime with $\gcd(n, q) = 1$. Recall from Exercise 3.1.3 that the *residue class ring* of integers modulo n is

$$\mathbb{Z}_n := \{\overline{0}, \overline{1}, \ldots, \overline{n-1}\} = \mathbb{Z}/I(n),$$

where

$$\overline{z} = z + I(n),$$

the equivalence class of the integer z modulo the ideal

$$I(n) = \{z \cdot n \mid z \in \mathbb{Z}\} = n\mathbb{Z} \subseteq \mathbb{Z},$$

consisting of the multiples of n. For any integer z, we denote by

$$\mathrm{rem}_n(z)$$

the canonical representative of its residue class, which means the unique integer r with $z = sn + r$ where $s \in \mathbb{Z}$ and $r \in n$. This r is called the *smallest remainder* of z modulo n. Also, since $n > 2$, we let

$$\mathrm{asr}_n(z)$$

be the unique integer r with $z = sn + r$ where $s \in \mathbb{Z}$ and $|r| \leq (n-1)/2$. This r is called the *absolutely smallest remainder* of z modulo n. We always have

$$z \equiv \mathrm{rem}_n(z) \equiv \mathrm{asr}_n(z) \bmod n, \text{ and } z + I(n) = \mathrm{rem}_n(z) + I(n) = \mathrm{asr}_n(z) + I(n).$$

Example If $n = 7$, the smallest remainders modulo 7 are $0, 1, 2, \ldots, 6$. The $\quad$ **4.4.1**
absolutely smallest remainders modulo 7 are $-3, -2, -1, 0, 1, 2, 3$. We have
$\mathrm{rem}_7(25) = 4$ and $\mathrm{asr}_7(25) = -3$. Also, $25 \equiv 4 \equiv -3 \bmod 7$. $\qquad\qquad$ $\diamond$

Definition (square, nonsquare modulo n) Let n be a prime and i an integer $\quad$ **4.4.2**
which is not divisible by n. Then i is called a *square (modulo n)*, if there exists an integer z such that $z^2 \equiv i \bmod n$. Otherwise, i is called a *nonsquare*
(modulo n). The multiples of n are neither squares nor nonsquares modulo n.
The residue classes $\bar{i}$ of squares (resp. nonsquares) are called *quadratic residues*
(resp. *quadratic non-residues*). Let Q (resp. N) be the set of quadratic residues
(resp. quadratic non-residues) modulo n,

$$Q := \left\{ \bar{i} \in \mathbb{Z}_n^* \mid \exists z \in \mathbb{Z} : \overline{z^2} = \bar{i} \right\},$$

while

$$N := \left\{ \bar{i} \in \mathbb{Z}_n^* \mid \nexists z \in \mathbb{Z} : \overline{z^2} = \bar{i} \right\} = \mathbb{Z}_n^* \setminus Q. \qquad\qquad \diamond$$

It is clear that the product of two squares is a square, and that, therefore,
the quadratic residues form a subgroup of the multiplicative group $(\mathbb{Z}_n^*, \cdot)$ of
(the field!) $\mathbb{Z}_n$. Moreover, the following holds:

Corollary *Let n be an odd prime, and assume that Q and N are the sets of squares* $\quad$ **4.4.3**
and nonsquares modulo n. Then

— *Q is a subgroup of index 2 in $\mathbb{Z}_n^*$. N is a coset of this subgroup, in fact*

$$\mathbb{Z}_n^* = Q \,\dot{\cup}\, N, \text{ and } |Q| = |N| = (n-1)/2.$$

— *If β is a primitive element for $\mathbb{Z}_n$ then $Q = \langle \beta^2 \rangle$. In particular, each $\alpha \in \mathbb{Z}_n^*$*
satisfies

$$\alpha \in Q \iff \alpha^{(n-1)/2} = 1.$$

— *The following identities hold for the complex products of Q and N,*

$$\begin{aligned} Q \cdot Q &= Q, \\ Q \cdot N &= N \cdot Q = N, \\ N \cdot N &= Q. \end{aligned} \qquad\qquad \square$$

The proofs are easy and left as Exercise 4.4.1. A more detailed description of
Q is contained in

Lemma *The quadratic residues modulo n, n an odd prime, form the set* $\quad$ **4.4.4**

$$Q = \left\{ \overline{\mathrm{rem}_n(i^2)} \;\middle|\; 1 \le i \le \frac{n-1}{2} \right\} \subset \mathbb{Z}_n^*.$$

Proof: The congruence

$$(n - a)^2 \equiv a^2 \bmod n$$

shows that all quadratic residues are contained in this set. Moreover, if $i^2 \equiv j^2 \bmod n$, then the prime n divides the difference $i^2 - j^2 = (i + j)(i - j)$ and, therefore, at least one of the two factors. Since $1 \leq i, j \leq (n-1)/2$, this implies that $i = j$.   □

4.4.5   **Example** For example, the squares modulo 7 are $1^2 = 1$, $2^2 = 4$, and $3^2 \equiv 2 \bmod 7$. The nonsquares modulo 7 are therefore 3, 5, 6.   ◇

Let $\zeta \in \mathbb{F}_{q^m}$ be a primitive n-th root of unity over $\mathbb{F}_q$, where $m := \mathrm{ord}_n(q)$, so that $q^m \equiv 1 \bmod n$. Then $x^n - 1$ splits into

4.4.6
$$x^n - 1 = \prod_{i \in n}(x - \zeta^i)$$

over $\mathbb{F}_{q^m}$. Now we partition the set of roots ζ^i into three subsets, according to the exponents i. The root $\zeta^0 = 1$ forms one of these sets, the second and third are defined as

$$\{\zeta^i \mid i \text{ is a square modulo } n\}, \text{ and } \{\zeta^i \mid i \text{ is a nonsquare modulo } n\}.$$

The quadratic-residue-codes will be defined as cyclic codes whose varieties are combinations from these three sets.

The following concept from Number Theory permits to decide the question of whether $z \in \mathbb{Z}$ is a square modulo n or not. To actually compute square roots modulo n, the probabilistic but efficient algorithm of Tonelli and Shanks can be used. For a description, see [39].

4.4.7   **Definition (Legendre-symbol)** Let n be any prime number (including 2), and denote by

$$\nu_n : \mathbb{Z} \to \mathbb{Z}_n \ : \ z \mapsto \bar{z}$$

the canonical homomorphism which maps an integer onto its residue class modulo n. Moreover, we consider the canonical epimorphism which has Q as its kernel, i.e.

$$\lambda : \mathbb{Z}_n^* \to \{1, -1\} \ : \ \bar{z} \mapsto \begin{cases} 1 & \text{if } \bar{z} \in Q, \\ -1 & \text{if } \bar{z} \in N. \end{cases}$$

We extend the function λ by defining its value to be zero if $\bar{z} = \bar{0}$. The composition of these two mappings is the mapping

$$\lambda \circ \nu_n : \mathbb{Z} \to \{0, 1, -1\} \ : \ a \mapsto \left(\frac{a}{n}\right),$$

where, for $a \in \mathbb{Z}$ we have

$$\left(\frac{a}{n}\right) := \begin{cases} 0 & \text{if } a \text{ is divisible by } n, \\ 1 & \text{if } a \text{ is a square modulo } n, \\ -1 & \text{otherwise.} \end{cases}$$

$\left(\frac{a}{n}\right)$ is called the *Legendre-symbol* associated to a (with respect to n).   ⬦

Euler's Lemma *For each integer a and every odd prime n, the following is true*   **4.4.8**

$$\left(\frac{a}{n}\right) \equiv a^{(n-1)/2} \bmod n.$$

Proof: Assume $a \equiv \alpha^r \bmod n$ where α is a primitive element of $\mathbb{Z}_n^*$ (the case $a \equiv 0 \bmod n$ is trivial). Then

$$a^{(n-1)/2} \equiv 1 \bmod n \iff \alpha^{r(n-1)/2} = 1$$
$$\iff (n-1) \text{ divides } r\frac{(n-1)}{2}$$
$$\iff r \text{ is even.}$$

The last condition is equivalent to $\left(\frac{a}{n}\right) = 1$.   □

The following lemma allows the evaluation of the Legendre-symbol:

Gauss' Criterion *Let n denote an odd prime, and assume that $n \nmid a \in \mathbb{Z}^*$. Let*   **4.4.9**

$$\mu := \left|\left\{ \text{asr}_n(ia) < 0 \,\Big|\, 1 \le i \le \frac{n-1}{2} \right\}\right|$$

be the number of absolutely smallest residues of $a, 2a, 3a, \ldots, (n-1)a/2$ modulo n which are negative. Then

$$\left(\frac{a}{n}\right) = (-1)^\mu.$$

Proof: Let $r_i := |\text{asr}_n(ia)|$ then r_i is positive and there exist $\epsilon_i \in \{-1,1\}$ so that $r_i = \epsilon_i \text{asr}_n(ia)$. As i ranges from 1 to $(n-1)/2$, the number of minus signs which occur in this way is equal to μ. We claim that $r_i \neq r_j$ if $i \neq j$ and $1 \le i,j \le (n-1)/2$. For, if $r_i = r_j$ then $\epsilon_i ia \equiv r_i = r_j \equiv \epsilon_j ja \bmod n$, and since n does not divide a it is clear that n divides $i\epsilon_i - j\epsilon_j$. But $-(n-1) \le i\epsilon_i - j\epsilon_j \le n-1$ and, therefore, $i\epsilon_i - j\epsilon_j = 0$, thus $i = j$. It follows that the two sets

$$\{1,2,\ldots,(n-1)/2\} \quad \text{and} \quad \{r_1,r_2,\ldots,r_{(n-1)/2}\}$$

coincide. Multiplying the congruences $ia \equiv \epsilon_i r_i \bmod n$ for $i = 1,\ldots,(n-1)/2$ together yields

$$((n-1)/2)! \, a^{(n-1)/2} \equiv (-1)^\mu ((n-1)/2)! \bmod n.$$

Canceling the term $((n-1)/2)!$ (which is prime to n) leads to $a^{(n-1)/2} \equiv (-1)^\mu \bmod n$. The assertion now follows from Euler's Lemma.   □

The most important properties of the Legendre-symbol are collected in the following

4.4.10   **Lemma** *Let n be an odd prime. For integers a and b, the following is true:*

1. $\left(\dfrac{a^2}{n}\right) = 1,$

2. $a \equiv b \bmod n \implies \left(\dfrac{a}{n}\right) = \left(\dfrac{b}{n}\right),$

3. $\left(\dfrac{ab}{n}\right) = \left(\dfrac{a}{n}\right)\left(\dfrac{b}{n}\right),$

4. $\left(\dfrac{-1}{n}\right) = (-1)^{(n-1)/2},$

5. $\left(\dfrac{2}{n}\right) = (-1)^{(n^2-1)/8}.$

6. *If m and n are distinct odd primes, then*

$$\left(\frac{m}{n}\right)\left(\frac{n}{m}\right) = (-1)^{(m-1)(n-1)/4}.$$

This equation is called the Law of Quadratic Reciprocity.

Proof: The first three assertions follow directly from Euler's Lemma. The fourth and fifth assertion can be obtained from Gauss' Criterion and Euler's Lemma. For a proof of the Law of Quadratic Reciprocity we refer to the literature on basic Number Theory or Algebra (e.g. [97], [100]).   □

We remark that 5. can be restated as

$$\left(\frac{2}{n}\right) = \begin{cases} 1 & \text{if } n \equiv \pm 1 \bmod 8, \\ -1 & \text{if } n \equiv \pm 3 \bmod 8. \end{cases}$$

The use of these rules is illustrated by the following

4.4.11   **Example** Is -42 a square modulo 53? The answer to this question can be obtained in the following way:

$$\left(\frac{-42}{53}\right) \overset{3.}{=} \left(\frac{-1}{53}\right)\left(\frac{2}{53}\right)\left(\frac{3}{53}\right)\left(\frac{7}{53}\right) \overset{4.,5.}{=} -\left(\frac{3}{53}\right)\left(\frac{7}{53}\right) \overset{6.}{=} -\left(\frac{53}{3}\right)\left(\frac{53}{7}\right)$$

$$\overset{2.}{=} -\left(\frac{2}{3}\right)\left(\frac{4}{7}\right) \overset{3.}{=} -\left(\frac{2}{3}\right)\underbrace{\left(\frac{2}{7}\right)\left(\frac{2}{7}\right)}_{\overset{3.,1.}{=}1} \overset{5.}{=} -(-1)^{\frac{9-1}{8}} = 1.$$

Hence -42 is a square modulo 53, and indeed we find that

$$-42 \equiv 8^2 \bmod 53. \qquad \diamond$$

In the following, we will need to impose the additional assumption that the prime power q (which we assumed to be relatively prime to n) is a square modulo n. This ensures that the set $\{\xi^a \mid \bar{a} \in Q\}$ is a variety over $\mathbb{F}_q$.

Lemma *Let n be an odd prime, let ξ be a primitive n-th root of unity over $\mathbb{F}_q$, and assume that q is a square modulo n.* **4.4.12**

— *The polynomials*

$$g_Q := \prod_{\bar{a} \in Q}(x - \xi^a) \quad \text{and} \quad g_N := \prod_{\bar{b} \in N}(x - \xi^b),$$

which depend on the chosen root of unity ξ, are contained in $\mathbb{F}_q[x]$.

— *The sets $\{\xi^a \mid \bar{a} \in Q\}$ and $\{\xi^b \mid \bar{b} \in N\}$ are varieties over $\mathbb{F}_q$.*

— *The polynomial $x^n - 1$ factorizes over $\mathbb{F}_q$ as*

$$x^n - 1 = (x - 1) \cdot g_Q \cdot g_N.$$

Proof: The coefficients of the polynomials g_Q and g_N are elementary symmetric functions of the roots ξ^a, $\bar{a} \in Q$, and ξ^b, $\bar{b} \in N$, respectively. Because of our assumptions on q, it follows from 4.4.3 that $\overline{qa} \in Q$ for all $\bar{a} \in Q$ and $\overline{qb} \in N$ for all $\bar{b} \in N$. Hence, the coefficients of both these polynomials remain fixed under the Frobenius automorphism $\alpha \mapsto \alpha^q$, thus they are elements of $\mathbb{F}_q$. This proves the first two statements. The last one follows from 4.4.6 and the fact that each element of $\mathbb{F}_n^*$ is either a square or a nonsquare. $\qquad\square$

Definition (quadratic-residue-codes) Let n be an odd prime such that q is a **4.4.13** square modulo n. The *quadratic-residue-codes* (or *QR-codes* for short) of length n over $\mathbb{F}_q$ are the cyclic codes which are defined as ideals in $\text{Res}_{q,n}$ as follows:

$$
\begin{aligned}
C_Q(n,q) &:= I(g_Q)/I(x^n - 1) \\
C_Q^1(n,q) &:= I((x-1)g_Q)/I(x^n - 1) \\
C_N(n,q) &:= I(g_N)/I(x^n - 1) \\
C_N^1(n,q) &:= I((x-1)g_N)/I(x^n - 1).
\end{aligned}
$$

We notice that these codes like their generating polynomials may depend on the choice of the primitive root of unity ξ. $\qquad \diamond$

We have $V(C_Q(n,q)) = \{\xi^i \mid \bar{i} \in Q\}$ and $V(C_N(n,q)) = \{\xi^i \mid \bar{i} \in N\}$. From 4.2.6 we obtain the following inclusions:

4.4.14

$$C_Q^1(n,q) \subseteq C_Q(n,q) \quad \text{and} \quad C_N^1(n,q) \subseteq C_N(n,q).$$

Using 4.2.5 and 4.4.3, we determine the dimensions of the cyclic codes as follows.

4.4.15 **Corollary** *Let n be an odd prime and assume that q is a square modulo n. Then*

  — *$C_Q(n,q)$ and $C_N(n,q)$ are $(n, (n+1)/2)$-codes,*

  — *$C_Q^1(n,q)$ and $C_N^1(n,q)$ are $(n, (n-1)/2)$-codes.*   □

4.4.16 **Examples** The binary Hamming- and simplex-codes of length 7 (cf. 4.2.7) are quadratic residue codes:

$$H_3 = C_Q(7,2), \ H_3' = C_N(7,2), \ S_3 = C_Q^1(7,2), \ \text{and } S_3' = C_N^1(7,2). \quad \diamond$$

By 4.4.10, binary QR-codes exist only for $n \equiv \pm 1 \bmod 8$, since only in these cases 2 is a square modulo n. In fact, the class of QR-codes is not as rich as it might seem at first sight.

4.4.17 **Theorem** *The QR-codes $C_Q(n,q)$ and $C_N(n,q)$ are permutationally isometric.*

Proof: We identify the coordinates $0, \ldots, n-1$ of vectors in $\mathbb{F}_q^n$ with the elements of the field $\mathbb{F}_n$ (remember that n is a prime). For $a \in \mathbb{F}_n^*$ the mapping σ_a defined by $i \mapsto a \cdot i$ is a bijection on $\mathbb{F}_n$. We show that for a nonsquare b the map σ_b in fact induces an isometry from $C_Q(n,q)$ to $C_N(n,q)$, which we denote again by σ_b:

$$c = \sum_{i \in \mathbb{F}_n} c_i x^i \mapsto \sigma_b(c) = \sum_{i \in \mathbb{F}_n} c_{\sigma_b(i)} x^i = \sum_{i \in \mathbb{F}_n} c_{bi} x^i.$$

To begin with, for each polynomial $c = \sum_{i \in \mathbb{F}_n} c_i x^i \in \mathbb{F}_q[x]$ and $a \in \mathbb{Z}$ we have

$$
\begin{aligned}
c(\xi^a) &= \sum_{i \in \mathbb{F}_n} c_i \xi^{ai} \\
&= \sum_{i \in \mathbb{F}_n} c_{\sigma_b(i)} \xi^{a\sigma_b(i)} \\
&= \sum_{i \in \mathbb{F}_n} c_{\sigma_b(i)} (\xi^{ab})^i.
\end{aligned}
$$

Because of $\{\overline{ba} \mid \bar{a} \in Q\} = N$, we obtain the following relation between the varieties of c and $\sigma_b(c) := \sum_{i \in \mathbb{F}_n} c_{\sigma_b(i)} x^i$:

$$\{\xi^a \mid \bar{a} \in Q\} \subseteq V(c) \iff \{\xi^a \mid \bar{a} \in N\} \subseteq V(\sigma_b(c)).$$

Thus, $c \in C_Q(n,q)$ if and only if $\sigma_b(c) \in C_N(n,q)$. Hence, the mapping

$$C_Q(n,q) \to C_N(n,q) \; : \; c \mapsto \sigma_b(c)$$

is an $\mathbb{F}_q$-isomorphism which is clearly a permutational isometry. $\square$

Corollary *The QR-codes $C_Q^1(n,q)$ and $C_N^1(n,q)$ are permutationally isometric.* $\square$ **4.4.18**

Example As already pointed out in 4.4.16, the QR-codes H_3 and H_3' respec-   **4.4.19**
tively S_3 and S_3' are permutationally isometric. $\diamond$

Theorem *Assume that $n \equiv \pm 1 \bmod 8$. The binary codes $C_Q^1(n,2)$ and $C_N^1(n,2)$*   **4.4.20**
are even.

Proof: By 4.4.18 the codes $C_Q^1(n,2)$ and $C_N^1(n,2)$ are isometric. The variety of
$C_Q^1(n,2)$ contains 1, therefore, $c(1) = 0$ for all $c \in C_Q^1(n,2)$. In other words
$\sum_{i \in n} c_i = 0$. Thus c, has even weight. $\square$

Since -1 is a square modulo n if and only if $n \equiv 1 \bmod 4$ by 4.4.10, the last
assertion of 4.2.16 implies the following:

Theorem   **4.4.21**

 — *If $n \equiv 1 \bmod 4$, then $C_Q(n,q)^{\perp} = C_N^1(n,q)$ and $C_N(n,q)^{\perp} = C_Q^1(n,q)$.*

 — *If $n \equiv -1 \bmod 4$, then $C_Q(n,q)^{\perp} = C_Q^1(n,q)$ and $C_N(n,q)^{\perp} = C_N^1(n,q)$.* $\square$

Now we are going to construct a group of automorphisms of the parity
extension of the binary QR-code $C_Q(n,2)$. This needs a few preparations. The
projective line (see 3.7.1) of $\mathbb{F}_n^2$,

$$PG_1^*(n) = \{\langle v \rangle^* \mid v \in \mathbb{F}_n^2 \setminus \{0\}\} = \{\langle (1,0) \rangle^*\} \cup \{\langle (\kappa,1) \rangle^* \mid \kappa \in \mathbb{F}_n\},$$

can be identified with the set $\mathbb{F}_n \cup \{\infty\}$ as follows:

$$PG_1^*(n) \to \mathbb{F}_n \cup \{\infty\} \; : \; \langle v \rangle^* \mapsto \begin{cases} \kappa & \text{if } \langle v \rangle^* = \langle (\kappa,1) \rangle^*, \\ \infty & \text{if } \langle v \rangle^* = \langle (1,0) \rangle^*. \end{cases}$$

Each element $(a,b) \in \langle (\kappa,1) \rangle^*$ satisfies $\kappa = \frac{a}{b} = ab^{-1}$. This motivates the
following two compositions

$$\begin{aligned} (a,b) + (c,d) &:= (ad + bc, bd), & (a,b),(c,d) \in \mathbb{F}_n^2 \setminus \{0\}, \\ (a,b) \cdot (c,d) &:= (ac, bd), & (a,b),(c,d) \in \mathbb{F}_n^2 \setminus \{0\}, \end{aligned}$$

which extend addition and multiplication of $\mathbb{F}_n$ to $\mathbb{F}_n \cup \{\infty\}$. In particular we
obtain

$$\kappa + \infty = \infty + \kappa = \infty, \qquad \kappa \in \mathbb{F}_n,$$

and

$$\kappa \cdot \infty = \infty \cdot \kappa = \infty \cdot \infty = \infty, \qquad \kappa \in \mathbb{F}_n^*.$$

For $\kappa \in \mathbb{F}_n^*$ we have $\kappa^{-1} = \frac{1}{\kappa}$ which is identified with $\langle(1,\kappa)\rangle^*$. Analogously we set $0^{-1} = \infty$ and $\infty^{-1} = 0$.

Using this identification we obtain from the action of the general linear group $\mathrm{GL}_2(n)$ on the projective line (recall 3.7.4),

$$\mathrm{GL}_2(n) \times \mathrm{PG}_1^*(n) \to \mathrm{PG}_1^*(n) \ : \ (A, \langle v \rangle^*) \mapsto \langle v \cdot A^\top \rangle^*,$$

actions of subgroups and of factor groups of $\mathrm{GL}_2(n)$. For example, a subgroup is the *special linear group*

$$\mathrm{SL}_2(n) := \{A \in \mathrm{GL}_2(n) \mid \det(A) = 1\}.$$

A factor group is the *projective linear group*

$$\mathrm{PGL}_2(n) = \mathrm{GL}_2(n) / \mathcal{Z}_2 = \{A\mathcal{Z}_2 \mid A \in \mathrm{GL}_2(n)\}$$

which was introduced in Section 3.7. ($\mathcal{Z}_2$ denotes the center of $\mathrm{GL}_2(n)$, $\mathcal{Z}_2 = \{\kappa \cdot I_2 \mid \kappa \in \mathbb{F}_n^*\}$.) Thus, $\mathrm{PGL}_2(n)$ arises from $\mathrm{GL}_2(n)$ by identifying matrices that are scalar multiples of each other. Since we are dealing with subspaces $\langle v \rangle^*$, we can replace the action of $\mathrm{GL}_2(n)$ by the action of $\mathrm{PGL}_2(n)$,

$$A\langle v \rangle^* = \langle v \cdot A^\top \rangle^* = \langle v \cdot (A\mathcal{Z}_2)^\top \rangle^* = A\mathcal{Z}_2\langle v \rangle^*.$$

A factor group of the special linear group is

$$\mathrm{PSL}_2(n) := \mathrm{SL}_2(n) / (\mathrm{SL}_2(n) \cap \mathcal{Z}_2),$$

the *projective special linear group*. Let $\tilde{\mathcal{Z}}_2 = \mathrm{SL}_2(n) \cap \mathcal{Z}_2$. The transformation induced by

$$A\tilde{\mathcal{Z}}_2 = \begin{pmatrix} \alpha_0 & \alpha_1 \\ \beta_0 & \beta_1 \end{pmatrix} \tilde{\mathcal{Z}}_2 \in \mathrm{PSL}_2(n)$$

on $\mathrm{PG}_1^*(n)$ is

$$A\tilde{\mathcal{Z}}_2 \colon \langle(1,0)\rangle^* \mapsto \langle(\alpha_0, \beta_0)\rangle^*, \ \langle(\kappa,1)\rangle^* \mapsto \langle(\kappa\alpha_0 + \alpha_1, \kappa\beta_0 + \beta_1)\rangle^*.$$

Using the identification of $\mathrm{PG}_1^*(n)$ with $\mathbb{F}_n \cup \{\infty\}$ we obtain the following map induced on $\mathbb{F}_n \cup \{\infty\}$,

$$A\tilde{\mathcal{Z}}_2 \colon \mathbb{F}_n \cup \{\infty\} \to \mathbb{F}_n \cup \{\infty\} \ : \ \kappa \mapsto \begin{cases} \frac{\alpha_0}{\beta_0} & \text{if } \kappa = \infty, \ \beta_0 \neq 0, \\ \infty & \text{if } \kappa = \infty, \ \beta_0 = 0, \\ \frac{\kappa\alpha_0 + \alpha_1}{\kappa\beta_0 + \beta_1} & \text{if } \kappa \neq \infty, \ \kappa\beta_0 + \beta_1 \neq 0, \\ \infty & \text{if } \kappa \neq \infty, \ \kappa\beta_0 + \beta_1 = 0. \end{cases}$$

Lemma *Let α denote a primitive element of $\mathbb{F}_n$. The group $\mathrm{PSL}_2(n)$ on $\mathrm{PG}_1^*(n)$ is*   **4.4.22**
generated by the permutations

$$\rho: z \mapsto z+1, \ \sigma: z \mapsto \alpha^2 z, \ and \ \tau: z \mapsto -z^{-1}.$$

Proof: The three mappings can be represented in the following way as elements of $\mathrm{PSL}_2(n)$.

$$\rho = \begin{pmatrix} 1 & 1 \\ 0 & 1 \end{pmatrix} \tilde{\mathbb{Z}}_2, \quad \sigma = \begin{pmatrix} \alpha & 0 \\ 0 & \alpha^{-1} \end{pmatrix} \tilde{\mathbb{Z}}_2, \quad \tau = \begin{pmatrix} 0 & -1 \\ 1 & 0 \end{pmatrix} \tilde{\mathbb{Z}}_2.$$

Thus $\langle \rho, \sigma, \tau \rangle$ is a subgroup of $\mathrm{PSL}_2(n)$.

Let $A = \begin{pmatrix} a & b \\ c & d \end{pmatrix}$ be an element of $\mathrm{GL}_2(n)$ with $\det(A) = ad - bc = 1$. We want to prove, that $A\tilde{\mathbb{Z}}_2$ belongs to $\langle \rho, \sigma, \tau \rangle$. If there are exactly two nonzero components of A, then either $b = c = 0$ and $ad = 1$, or $a = d = 0$ and $bc = -1$. Thus, either $d = a^{-1}$ or $c = -b^{-1}$ with $a, b \in \mathbb{F}_n^*$. Hence $A\tilde{\mathbb{Z}}_2$ can be represented as σ^k or $\tau\sigma^k$ for suitable k.

If there are exactly three nonzero components of A, then without loss of generality $c = 0$ and $d = a^{-1}$. It is easy to check that $\rho^\ell = \begin{pmatrix} 1 & \ell \\ 0 & 1 \end{pmatrix} \tilde{\mathbb{Z}}_2$. Since $\mathbb{F}_n$ is a prime field we obtain all matrices $\begin{pmatrix} 1 & \lambda \\ 0 & 1 \end{pmatrix}$, $\lambda \in \mathbb{F}_n^*$, in this way. The multiplication $\sigma^k \cdot \begin{pmatrix} 1 & \lambda \\ 0 & 1 \end{pmatrix}$ yields all $\begin{pmatrix} \kappa & \lambda \\ 0 & \kappa^{-1} \end{pmatrix} \tilde{\mathbb{Z}}_2$ with $\kappa, \lambda \in \mathbb{F}_n^*$. The other matrices with exactly one 0 can be derived as

$$\tau \cdot \begin{pmatrix} \kappa & \lambda \\ 0 & \kappa^{-1} \end{pmatrix} = \begin{pmatrix} 0 & -\kappa^{-1} \\ \kappa & \lambda \end{pmatrix} \tilde{\mathbb{Z}}_2, \quad \begin{pmatrix} \kappa & \lambda \\ 0 & \kappa^{-1} \end{pmatrix} \cdot \tau = \begin{pmatrix} \lambda & -\kappa \\ \kappa^{-1} & 0 \end{pmatrix} \tilde{\mathbb{Z}}_2,$$

$$\tau \cdot \begin{pmatrix} \kappa & \lambda \\ 0 & \kappa^{-1} \end{pmatrix} \cdot \tau = \begin{pmatrix} -\kappa^{-1} & 0 \\ \lambda & -\kappa \end{pmatrix} \tilde{\mathbb{Z}}_2.$$

If all four components of A are different from zero, then any three components determine the last one uniquely. For instance $c = (ad - 1)b^{-1}$. Simple computations show that $A\tilde{\mathbb{Z}}_2$ can be expressed as the product of two matrices each containing three nonzero entries:

$$\begin{pmatrix} a & b \\ c & d \end{pmatrix} \tilde{\mathbb{Z}}_2 = \begin{pmatrix} bd^{-1} & 1 \\ 0 & b^{-1}d \end{pmatrix} \cdot \begin{pmatrix} b^{-1} & 0 \\ d^{-1}(ad-1) & b \end{pmatrix} \tilde{\mathbb{Z}}_2. \qquad \square$$

The following theorem shows that $\mathrm{PSL}_2(n)$ is a group of automorphisms of the parity extension of the QR-code $C_Q(n,q)$. We identify, similarly as in the proof of 4.3.9, the parity coordinate with ∞. Then $\mathrm{PSL}_2(n)$ permutes the components of the codewords labeled by the elements of $\mathbb{F}_n \cup \{\infty\}$, i.e. by the elements of the projective line $\mathrm{PG}_1^*(n)$, in such a way that the code remains invariant. For a proof of the following result, we refer to [92], [93].

4.4.23   **Theorem of Gleason and Prange** *The group* $\mathrm{PSL}_2(n)$ *is a group of automorphisms of the parity extension of the QR-code* $C_Q(n,q)$.   □

In the binary case one shows that row vectors of the generator matrix Γ of the parity extension $P(C_Q(n,2))$ are transformed by the generators ρ, σ, and τ of $\mathrm{PSL}_2(n)$ into elements of $P(C_Q(n,2))$. Since $\mathrm{PSL}_2(n)$ acts transitively on the projective line $\mathrm{PG}_1^*(n)$ (see Exercise 4.4.6), we obtain from the Theorem of Gleason and Prange, together with 4.3.10, the important

4.4.24   **Corollary** *The minimum weight of* $C_Q(n,2)$ *is odd.*   □

4.4.25   **Theorem** *The minimum weights of* $C_Q(n,q)$ *and* $C_Q^1(n,q)$ *are related as*

$$\mathrm{dist}(C_Q^1(n,q)) = \mathrm{dist}(C_Q(n,q)) + 1.$$

Proof: A codeword $c \in C_Q(n,q)$ is contained in $C_Q^1(n,q)$ if and only if $c(1) = 0$, i.e.,

$$(c_0,\ldots,c_{n-1}) \in C_Q^1(n,q) \iff (c_0,\ldots,c_{n-1},0) \in P(C_Q(n,q)).$$

Therefore, we can consider each codeword $c \in C_Q^1(n,q)$ of minimum weight $d := \mathrm{dist}(C_Q^1(n,q))$ as a codeword of $P(C_Q(n,q))$ such that $c_\infty = 0$. From 4.4.14 we derive that

$$d \geq \mathrm{dist}(C_Q(n,q)).$$

Now assume

$$d = \mathrm{dist}(C_Q(n,q)).$$

Let c be a codeword of $C_Q^1(n,q)$ of weight d. Consider c as an element of $P(C_Q(n,q))$ with $c_\infty = 0$. Since the group $\mathrm{PSL}_2(n)$ acts transitively on the projective line $\mathrm{PG}_1^*(n)$, there exists a permutation in $\mathrm{PSL}_2(n)$, that maps c onto a codeword c', also of minimum weight, such that $c'_\infty \neq 0$. But in this case, the vector $(c'_0,\ldots,c'_{n-1})$ is of weight $d-1$, and is contained in $C_Q(n,q)$, which is a contradiction.   □

Now we are going to derive an important lower bound for the minimum weight of QR-codes.

4.4.26   **The Square-Root-bound** *The minimum distance d of the QR-code $C_Q(n,q)$ satisfies*

1. $d^2 \geq n$,

2. $d^2 - d + 1 \geq n$ *if* $n \equiv -1 \bmod 4$,

3. $d \equiv 3 \bmod 4$ *if* $n \equiv -1 \bmod 8$ *and* $q = 2$.

Proof: 1. Let $c = fg_Q \in C_Q(n,q)$ be a codeword of minimum weight and assume that $\overline{b} \in N$. The proof of 4.4.17 shows that the image of c under the isometry induced by $\sigma_{b^{-1}}$ gives the polynomial $\sigma_{b^{-1}}(c(x)) \equiv c(x^b) \equiv f_b g_N \mod I(x^n - 1)$, with a suitable $f_b \in \mathbb{F}_q[x]$. Therefore, $\sigma_{b^{-1}}(c)$ is an element of minimum weight in $C_N(n,q)$. Correspondingly $c(x)c(x^b) \in C_Q(n,q) \cap C_N(n,q)$. Because of 4.2.6, $C_Q(n,q) \cap C_N(n,q)$ is generated by the polynomial

$$g_Q g_N = \sum_{i \in n} x^i,$$

and hence it is an n-fold repetition code. There are two cases to consider:

— If $c(x)c(x^b) \not\equiv 0 \mod I(x^n - 1)$, then

$$c(x)c(x^b) \equiv \beta \cdot (1 + x + \ldots + x^{n-1}) \mod I(x^n - 1)$$

for some $\beta \in \mathbb{F}_q^*$. Since both c and $c(x^b)$ are of weight d, and $c(x)c(x^b)$ is of weight n, the inequality $n \leq d^2$ is satisfied (cf. Exercise 4.4.8).

— If $c(x)c(x^b) \equiv 0 \mod I(x^n - 1)$, then $c(x)$ or $c(x^b)$ is divisible by $x - 1$. This means that a codeword of minimum weight in $C_Q(n,q)$ or in $C_N(n,q)$ is contained even in $C_Q^1(n,q)$ or in $C_N^1(n,q)$. But this cannot happen, since then $\text{dist}(C_Q^1(n,q)) \leq \text{dist}(C_Q(n,q))$ contrary to 4.4.25.

2. If $n \equiv -1 \mod 4$, then, according to 4.4.10, -1 is a nonsquare modulo n. Hence we can put $b = -1$. This gives

$$c(x)c(x^{-1}) \equiv \beta \cdot (1 + x + \ldots + x^{n-1}) \mod I(x^n - 1)$$

for some suitable $\beta \in \mathbb{F}_q$. If $\beta \neq 0$, then the number of nonvanishing coefficients on the left hand side is at most $d(d-1) + 1$ (cf. Exercise 4.4.9), and so the stated inequality holds if $\beta \neq 0$. The case $\beta = 0$ does not arise, as pointed out in the preceding part of the proof.

3. Let $c \in C_Q(n,2) \setminus \{0\}$ be a codeword of minimal weight. Assume that $i_0, \ldots, i_{d-1}$ are the nonzero coordinates of $c \in C_Q(n,2)$, whence $c = \sum_{u \in d} x^{i_u}$. According to 4.4.24 the minimum distance d of $C_Q(n,2)$ is odd. Hence

$$1 + x + \ldots + x^{n-1} \equiv c(x)c(x^{-1})$$
$$= \sum_{(u,v) \in d \times d} x^{i_u - i_v} = 1 + \sum_{\substack{(u,v) \in d \times d \\ u \neq v}} x^{i_u - i_v} \mod I(x^n - 1).$$

On the right-hand side of this equation, the summands for (u,v) and (y,z) cancel if $i_u - i_v = i_y - i_z$. But then also $i_v - i_u = i_z - i_y$, $i_u - i_y = i_v - i_z$, and $i_y - i_u = i_z - i_v$. Therefore, the number of pairs of terms that cancel is a multiple of 4, i.e. $n = d^2 - d + 1 - 4s$ for some integer $s \geq 0$. From $d \equiv 1 \mod 4$ we would immediately obtain that $n \equiv 1 \mod 4$, in contradiction to $n \equiv -1 \mod 8$. Hence $d \equiv 3 \mod 4$. $\qquad\square$

The Square-Root-bound is sharp for the $(7,4)$-Hamming-code, since any $(7,4,d)$-QR-code satisfies $d \geq 3$. An improvement of the Square-Root-bound for self-dual QR-codes is reported in [84, third edition].

In 2.3.12 the binary Golay-code G_{24} of length 24 was introduced. Now we consider further Golay-codes G_{23} and G_{11}, which are particular QR-codes.

4.4.27   **Definition (Golay-codes)** The QR-code $G_{23} = C_Q(23,2)$ is called *the binary Golay-code* of length 23. The QR-code $G_{11} = C_Q(11,3)$ is called *the ternary Golay-code*.   ◇

4.4.28   **Theorem**

1. *The Golay-code G_{23} is a perfect $(23,12,7,2)$-code and its parity extension $P(G_{23})$ is a $(24,12,8,2)$-code.*

2. *The Golay-code G_{11} is a perfect $(11,6,5,3)$-code.*

Proof: 1. It is easily checked that the squares modulo 23 are

$$Q = \{1,2,4,8,16,9,18,13,3,6,12\}.$$

Thus, G_{23} is the binary QR-code $C_Q(23,2)$ with variety

$$V(G_{23}) = \{\zeta, \zeta^2, \zeta^3, \zeta^4, \zeta^6, \zeta^8, \zeta^9, \zeta^{12}, \zeta^{13}, \zeta^{16}, \zeta^{18}\}.$$

Hence, by 4.2.16, its dimension is $k = 12$ and by the BCH-bound 4.3.1, its minimum distance d is at least 5. Using part 3 of the Square-Root-bound 4.4.26, we obtain that $d \equiv 3 \bmod 4$, so that $d \geq 7$. But the minimum distance cannot be greater than 7, since the parameters $(23,12,7,2)$ attain the Hamming-bound, i.e. the code is perfect (cf. Exercise 2.1.2). Therefore, we have shown that G_{23} is a perfect $(23,12,7,2)$-code. It follows that the extended code $P(G_{23})$ has parameters $(24,12,8,2)$.

2. G_{11} is the ternary QR-code $C_Q(11,3)$ with generator polynomial

$$g = x^5 - x^3 + x^2 - x - 1.$$

If ζ is a root of g, then its variety is

$$V(G_{11}) = \{\zeta, \zeta^3, \zeta^4, \zeta^5, \zeta^9\}.$$

Hence, from 4.2.16 we deduce that its dimension is equal to 6, and the BCH-bound 4.3.1 implies that its minimum distance is $d \geq 4$. In order to determine the exact value of d, we consider a generator matrix of G_{11} which is formed from the following codewords.

$$
\begin{aligned}
(x^4 + x^3 + x^2 + x - 1) \cdot g &= x^9 + x^8 + x^6 + x^5 + x^4 + 1, \\
(x^4 + x^3 + x^2 - x) \cdot g &= x^9 + x^8 - x^6 - x^5 + x, \\
(-x^4) \cdot g &= -x^9 + x^7 - x^6 + x^5 + x^4, \\
(-x^4 + x^3 - x^2) \cdot g &= -x^9 + x^8 + x^6 - x^4 + x^2, \\
(x^5 + x^4) \cdot g &= x^{10} + x^9 - x^8 + x^5 - x^4, \\
(-x^3) \cdot g &= -x^8 + x^6 - x^5 + x^4 + x^3.
\end{aligned}
$$

Reordering the coefficients of these polynomials according to increasing powers of x, we get the matrix

$$
\Gamma = \begin{pmatrix}
1 & 0 & 0 & 0 & 1 & 1 & 1 & 0 & 1 & 1 & 0 \\
0 & 1 & 0 & 0 & 0 & -1 & -1 & 0 & 1 & 1 & 0 \\
0 & 0 & 0 & 0 & 1 & 1 & -1 & 1 & 0 & -1 & 0 \\
0 & 0 & 1 & 0 & -1 & 0 & 1 & 0 & 1 & -1 & 0 \\
0 & 0 & 0 & 0 & -1 & 1 & 0 & 0 & -1 & 1 & 1 \\
0 & 0 & 0 & 1 & 1 & -1 & 1 & 0 & -1 & 0 & 0
\end{pmatrix}.
$$

Column permutations lead to a systematic generator matrix for a code which is isometric, and hence has the same minimum distance:

$$
\Gamma' = \begin{pmatrix}
1 & 0 & 0 & 0 & 0 & 0 & 1 & 1 & 1 & 1 & 1 \\
0 & 1 & 0 & 0 & 0 & 0 & 0 & -1 & -1 & 1 & 1 \\
0 & 0 & 1 & 0 & 0 & 0 & 1 & 1 & -1 & 0 & -1 \\
0 & 0 & 0 & 1 & 0 & 0 & -1 & 0 & 1 & 1 & -1 \\
0 & 0 & 0 & 0 & 1 & 0 & -1 & 1 & 0 & -1 & 1 \\
0 & 0 & 0 & 0 & 0 & 1 & 1 & -1 & 1 & -1 & 0
\end{pmatrix}.
$$

This matrix satisfies

$$
\Gamma' \cdot \Gamma'^{\mathsf{T}} = \begin{pmatrix}
0 & 0 & 0 & 0 & 0 & 0 \\
0 & -1 & -1 & -1 & -1 & -1 \\
0 & -1 & -1 & -1 & -1 & -1 \\
0 & -1 & -1 & -1 & -1 & -1 \\
0 & -1 & -1 & -1 & -1 & -1 \\
0 & -1 & -1 & -1 & -1 & -1
\end{pmatrix}.
$$

Let $f \in \mathbb{F}_3^6$ be a message and $c = f \cdot \Gamma'$ be the corresponding codeword. Then, using the result of Exercise 1.3.18, the following is true over $\mathbb{F}_3 = \mathbb{Z}_3$:

$$
c \cdot c^{\mathsf{T}} = f \cdot \Gamma' \cdot \Gamma'^{\mathsf{T}} \cdot f^{\mathsf{T}} = \sum_{i=1}^{5} \sum_{j=1}^{5} -f_i f_j = -\left(\sum_{i=1}^{5} f_i\right)^2.
$$

The right hand side is clearly the negative of a square, and hence either 0 or -1 modulo 3. Writing the elements of $\mathbb{F}_3$ as $-1, 0, +1 \in \mathbb{Z}$, then the weight of c is obtained as $\mathrm{wt}(c) = c \cdot c^\top$ over $\mathbb{Z}$. This implies that $\mathrm{wt}(c) \not\equiv 1 \bmod 3$, i.e. $\mathrm{wt}(c) \neq 4$. Hence G_{11} has minimum weight $d \geq 5$. But the minimum distance cannot be greater than 5, since the parameters $(11, 6, 5, 3)$ attain the Hamming-bound, i.e. the code is perfect (cf. Exercise 2.1.2). □

We remark that the Golay-codes are the only nontrivial perfect codes which can correct more than one error ([191, 207]). The parity extension $P(G_{23})$ possesses an interesting combinatorial interpretation. It can be shown that the vectors of minimum weight in $P(G_{23})$ are exactly the blocks of the Steiner system $S(5, 8, 24)$. According to E. Witt [204], the automorphism group of this Steiner system is the Mathieu group $\mathcal{M}_{24}$. From this it follows that $\mathcal{M}_{24}$ is the (full) automorphism group of $P(G_{23})$. The Golay codes also demonstrate that the BCH-bound 4.3.1 may not be sharp. It is actually sometimes quite hard to determine the minimum distance of a BCH-code. If the n unit vectors of the Hamming space $H(n, q)$ are projected onto the n-th roots of unity in the complex number plane $\mathbb{C}$, then the two Golay codes G_{11} and G_{23} result in the point sets which are drawn in Fig. 4.3.

Exercises

E.4.4.1   **Exercise** Prove 4.4.3.

E.4.4.2   **Exercise** Is 99 a square modulo 101? Is 311 a square modulo 1001?

E.4.4.3   **Exercise** Let n be a prime. If a, b are integers with $ab \equiv 1 \bmod n$, show that
$$\left(\frac{a}{n} \right) = \left(\frac{b}{n} \right).$$

E.4.4.4   **Exercise** Verify 4.4.21.

E.4.4.5   **Exercise** Let n be a prime. Consider the following relation $\sim$ on $\mathbb{F}_n^2 \setminus \{(0,0)\}$,
$$(\alpha, \beta) \sim (\gamma, \delta) :\Longleftrightarrow \exists\, \lambda \in \mathbb{F}_n^* : \alpha\lambda = \gamma \text{ and } \beta\lambda = \delta.$$
Prove the following statements:

— $\sim$ is an equivalence relation.

— It has exactly $n + 1$ classes, and the set
$$\{(1,0), (\kappa, 1) \mid \kappa \in \mathbb{F}_n\}$$
is a system of representatives, the *homogeneous coordinates* of $\mathrm{PG}_1^*(n)$.

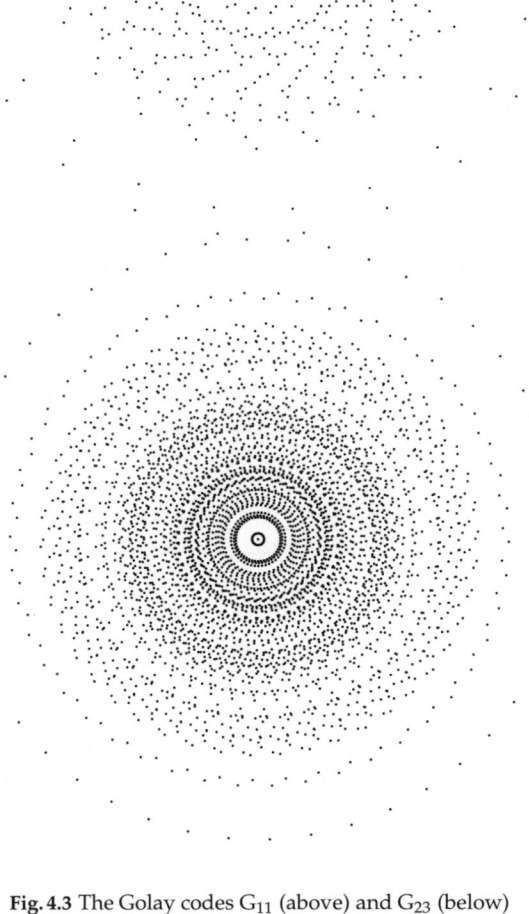

Fig. 4.3 The Golay codes G_{11} (above) and G_{23} (below)

E.4.4.6   **Exercise** Show that the group $PSL_2(n)$ acts transitively on $PG_1^*(n)$. Is it even *doubly-transitive?* (I.e., for $\alpha, \beta, \gamma, \delta \in PG_1^*(n)$ with $\alpha \neq \beta$ and $\gamma \neq \delta$, does there exist a transformation $\phi \in PSL_2(n)$ such that $\phi(\alpha) = \gamma$ and $\phi(\beta) = \delta$.)

E.4.4.7   **Exercise** Prove 4.4.23 for the binary quadratic-residue-codes $C_Q(n, 2)$.

E.4.4.8   **Exercise** Let c and c' be elements of $\mathrm{Res}_{q,n}$. Show that $\mathrm{wt}(c \cdot c') \leq \mathrm{wt}(c)\,\mathrm{wt}(c')$.

E.4.4.9   **Exercise** Let c and c' be elements of $\mathrm{Res}_{q,n}$. Assume that $c'(x) = c(x^{-1})$. Show that $\mathrm{wt}(c \cdot c') \leq \mathrm{wt}(c)(\mathrm{wt}(c) - 1) + 1$.

E.4.4.10   **Exercise** Use the generator polynomial of G_{11} given in the proof of 4.4.28 for the evaluation of a generator matrix of G_{11} and apply the attached software to check that its minimum distance is in fact 5.

4.5 Idempotents and the Discrete Fourier Transform

We have seen in 4.2.2 that cyclic codes correspond to ideals in the polynomial residue class ring $\mathrm{Res}_{q,n} = \mathbb{F}_q[x]/I(x^n - 1)$. In particular, we can *multiply* two codewords since we know that they correspond to polynomials in that ring. In this section, we are going to study the multiplicative structure of codes in more detail.

Besides its generator polynomial g, a cyclic code C may have other generators. Some of them have the additional property that they are *idempotent*, i.e. they satisfy $e^2 = e \neq 0$ (as an equation in $\mathrm{Res}_{q,n}$). An element which is both a generator and an idempotent is called an *idempotent generator*. It turns out that such elements exist and are unique provided that the very mild condition $\gcd(n, q) = 1$ is satisfied. We recall from the foregoing:

- By 4.2.9, the condition $\gcd(n, q) = 1$ implies that $x^n - 1$ has no multiple roots. Hence also $\gcd(g, h) = 1$, where $h = (x^n - 1)/g$ denotes the check polynomial. Thus, the ideals $I(g)$ and $I(h)$ are relatively prime. The Chinese Remainder Theorem (see Section 3.5) establishes the ring isomorphism

$$\mathbb{F}_q[x]/I(x^n - 1) \simeq (\mathbb{F}_q[x]/I(g)) \times (\mathbb{F}_q[x]/I(h)),$$

    defined by

$$f + I(x^n - 1) \mapsto (f + I(g), f + I(h)).$$

— Let
$$e := e(x) + I(x^n - 1)$$

be the inverse image of $(0 + I(g), 1 + I(h))$ under this ring isomorphism. We may take $e(x)$ as the canonical representative of $e(x) + I(x^n - 1)$ of degree less than n. From the definition of e, we obtain:

$$e(x) \equiv 0 \bmod I(g), \quad e(x) \equiv 1 \bmod I(h).$$

We conclude that there exist polynomials s and t such that $e(x) = sg = 1 + th$. Thus $e^2 = sg(1 + th) = e + stgh$. This gives

$$e^2 = e \neq 0,$$

i.e. e is an idempotent which is contained in C.

— e is even a *generating unit*: Assume that $c \in C$, say $c = \overline{fg}$. Then

$$ec = \overline{(1 + th)fg} = \overline{fg} = c,$$

which shows that e is a unit for C, as $ec = c$, for every $c \in C$. Moreover, as $e(x)$ is a multiple of g, we have

$$C = e \cdot C \subseteq e \cdot \mathrm{Res}_{q,n} = sg \cdot \mathrm{Res}_{q,n} = g \cdot \mathrm{Res}_{q,n} \subseteq C,$$

whence
$$C = e \cdot \mathrm{Res}_{q,n}.$$

This means that e is in fact a generating idempotent of the code, in the sense of ring theory (see the Exercises).

— In order to prove the uniqueness of such a generating unit, we consider two idempotents e and e'. Being units, they satisfy

$$e' = ee' = e'e = e.$$

— As each codeword is a multiple of e, this idempotent is even a generator, in the sense of coding theory. Hence, each cyclic code C contains exactly one such idempotent generator.

This, together with Exercise 4.5.3, yields the following

Corollary *Let C be a cyclic code of length n over $\mathbb{F}_q$ with generator polynomial g and assume that $\gcd(n, q) = 1$. Then there is a unique idempotent $e \in C$ which generates the ideal C,*   **4.5.1**

$$C = e \cdot C = e \cdot \mathrm{Res}_{q,n}.$$

The idempotent $e = e(x) + I(x^n - 1)$ *is uniquely determined as the solution of the congruences*

$$e(x) \equiv 0 \bmod I(g), \quad e(x) \equiv 1 \bmod I(h),$$

with $\deg e(x) < n$. *Moreover, it is a generator of the code,*

$$C = I(g(x))/I(x^n - 1) = I(e(x))/I(x^n - 1).$$

In addition, $1 - e$ *is the generating idempotent of the code generated by the check polynomial* $h = (x^n - 1)/g$. *The generating idempotent* e *determines the generator polynomial* g *of* C *as*

$$g = \gcd(e(x), x^n - 1). \qquad \square$$

4.5.2   **Example (Continuation of 4.2.7)** The binary cyclic codes of length 7 have the following generator polynomials and generating idempotents:

code C	$\dim(C)$	g	e
$\{0\}$	0	$x^7 - 1$	0
$\mathbb{F}_2^7$	7	1	1
W_7	1	$1 + x + x^2 + \ldots + x^6$	$1 + x + x^2 + \ldots + x^6$
P_7	6	$1 + x$	$x + x^2 + \ldots + x^6$
H_3	4	$1 + x + x^3$	$x + x^2 + x^4$
H_3'	4	$1 + x^2 + x^3$	$x^3 + x^5 + x^6$
S_3	3	$1 + x^2 + x^3 + x^4$	$1 + x^3 + x^5 + x^6$
S_3'	3	$1 + x + x^2 + x^4$	$1 + x + x^2 + x^4$

$\diamond$

The generating idempotent $e = \overline{e(x)}$, where $e(x) = \sum_{i \in n} e_i x^i$, gives the $k \times n$ generator matrix

4.5.3

$$\Gamma(e) = \begin{pmatrix} e_0 & e_1 & \cdots & e_{n-1} \\ e_{n-1} & e_0 & \cdots & e_{n-2} \\ \vdots & \vdots & \ddots & \vdots \\ e_{n-k+1} & e_{n-k+2} & \cdots & e_{n-k} \end{pmatrix}$$

for C, since $e(x)$ is a generating polynomial as well (the proof is similar to that which gives the generator matrix $\Gamma(g)$ obtained from the generator polynomial g). Of course, $\Gamma(g)$ and $\Gamma(e)$ may be different, examples are easily obtained from 4.5.2.

The next result extends 4.2.6.

4.5.4   **Lemma** *Let* C *and* C' *be cyclic codes of length* n *over* $\mathbb{F}_q$ *with generating idempotents* e, e', *respectively. Then*

1. $C \cap C'$ *has the generating idempotent* $e \cdot e'$.

2. $C + C'$ *has the generating idempotent* $e + e' - e \cdot e'$.

Proof: 1. It is easy to check that $e \cdot e'$ is an idempotent which is contained in $C \cap C'$. It remains to show that it generates $C \cap C'$. Let $c = u \cdot e$ be an element of $C \cap C'$ for some u. Since e' is unity in C', we deduce $c = c \cdot e' = u \cdot e \cdot e' \in I(e(x)e'(x))/I(x^n - 1)$.

2. It is clear that $e + e' - e \cdot e' \in C + C'$. Now let $\tilde{c} := c + c' \in C + C'$ be arbitrary, with $c \in C$ and $c' \in C'$. Applying $c \cdot e = c$ and $c' \cdot e' = c'$, we obtain

$$\tilde{c} \cdot (e + e' - e \cdot e') = c + c \cdot e' - c \cdot e' + c' \cdot e + c' - c' \cdot e$$
$$= c + c'$$
$$= \tilde{c},$$

so that $e + e' - e \cdot e'$ generates $C + C'$. An easy calculation shows that $e + e' - e \cdot e'$ is an idempotent. $\qquad\qquad\square$

We know from the previous section that binary QR-codes exist for $n \equiv \pm 1 \bmod 8$. The next result describes generating idempotents for these codes:

Lemma *Let n be a prime which is congruent to ± 1 modulo 8 and Q be the set of quadratic residues modulo n. Put*   **4.5.5**

$$e(x) = \sum_{i \in n\,:\,\bar{i} \in Q} x^i.$$

For a suitable choice of a primitive n-th root of unity ξ we have for g_Q and the corresponding codes C_Q and C_Q^1:

- *For $n \equiv 1 \bmod 8$, $e = \overline{e(x)}$ is the generating idempotent of the binary QR-code $C_Q^1(n, 2)$.*

- *For $n \equiv -1 \bmod 8$, $e = \overline{e(x)}$ is the generating idempotent of the binary QR-code $C_Q(n, 2)$.*

Proof: 1. To begin with we show that e is idempotent. This follows easily from $q = 2$ and the fact that 2 is a square modulo n :

$$e^2(x) = \sum_{\bar{i} \in Q} x^{2i} = \sum_{\bar{i} \in Q} x^i = e(x) \neq 0,$$

and so

$$e^2 = \overline{e(x)}^2 = \overline{e(x)^2} = \overline{e(x)} = e \neq 0.$$

2. Let ξ be a primitive n-th root of unity, $\bar{b} \in N$ a quadratic non-residue modulo n. For ξ^b we have, as $QN = N$,

$$e(\xi) + e(\xi^b) = \sum_{\bar{i} \in Q} \xi^i + \sum_{\bar{i} \in Q} \xi^{ib} = 1 + \sum_{i=0}^{n-1} \xi^i = \underbrace{(\xi^n - 1)(\xi - 1)^{-1}}_{=0} + 1 = 1,$$

and so

$$e(\xi) = 0 \iff e(\xi^b) = 1.$$

3. Assume that $e(\xi) = 0$ (for otherwise we could replace ξ by another primitive n-th root of unity $\xi^{b'}$ with $\overline{b}' \in N$). For each $\overline{a} \in Q$ we have $e(\xi^a) = \sum_{\overline{i} \in Q} \xi^{i \cdot a} = e(\xi) = 0$. This implies that

$$e(x) \equiv 0 \bmod \mathrm{I}(g_Q), \text{ thus } e \in C_Q(n,2).$$

Moreover, $e(\xi^b) + 1 = e(\xi) = 0$ for each $\overline{b} \in N$, and hence

$$e(x) \equiv 1 \bmod \mathrm{I}(g_N), \text{ thus } e + 1 \in C_N(n,2).$$

— If $n \equiv 1 \bmod 8$, then $\mathrm{wt}(e) = (n-1)/2 \equiv 0 \bmod 2$, and so $e(1) = 0$, i.e. e is even and, therefore, contained in $C_Q^1(n,2)$, the even weight subcode of $C_Q(n,2)$. This gives

$$e(x) \equiv 0 \bmod \mathrm{I}((x-1)g_Q) \text{ and } e(x) \equiv 1 \bmod \mathrm{I}(g_N).$$

By 4.5.1 (noting that $\gcd(n,q) = 1$), this implies that e is the generating idempotent of $C_Q^1(n,2) = \mathrm{I}((x-1)g_Q)/\mathrm{I}(x^n - 1)$.

— In the case $n \equiv -1 \bmod 8$ we have $\mathrm{wt}(e) = (n-1)/2 \equiv 1 \bmod 2$ and so $e(1) = 1$. From this we deduce that $e(x) \equiv 1 \bmod \mathrm{I}((x-1)g_N)$. This together with $e(x) \equiv 0 \bmod \mathrm{I}(g_Q)$ implies that e is the generating idempotent of $C_Q(n,2) = \mathrm{I}(g_Q)/\mathrm{I}(x^n - 1)$. $\square$

The question of determining the zeros of a polynomial in a cyclic code of length n can be answered by evaluating the polynomial at all possible n-th roots of unity. This is exactly what the following construction does. In addition, the results of all these evaluations are taken as the coefficients of another polynomial (possibly in a larger field). The construction is based on the Discrete Fourier Transform, which is in fact the following further application of the Chinese Remainder Theorem 3.5.15:

4.5.6   **Corollary** *Assume again that $\gcd(n,q) = 1$, that $m = \mathrm{ord}_n(q)$, so that $\mathbb{F}_{q^m}$ contains a primitive n-th root of unity ξ, and $x^n - 1 = \prod_{\overline{i} \in n}(x - \xi^i)$ in $\mathbb{F}_{q^m}[x]$. Since the ideals $\mathrm{I}(x - \xi^i)$ are pairwise relatively prime, we have the following isomorphism:*

$$\mathbb{F}_{q^m}[x]/\mathrm{I}(x^n - 1) \simeq \underset{i \in n}{\times} \mathbb{F}_{q^m}[x]/\mathrm{I}(x - \xi^i),$$

defined by

$$f + \mathrm{I}(x^n - 1) \mapsto (f + \mathrm{I}(x - \xi^0), f + \mathrm{I}(x - \xi^1), \dots, f + \mathrm{I}(x - \xi^{n-1})).$$

In addition, the right hand side is equal to

$$(f(\xi^0) + \mathrm{I}(x - \xi^0), \dots, f(\xi^{n-1}) + \mathrm{I}(x - \xi^{n-1})),$$

and so we can rephrase the isomorphism as follows:

$$\{f \in \mathbb{F}_{q^m}[x] \mid \deg f < n\} \simeq \mathbb{F}_{q^m}^n \; : \; f \mapsto (f(\xi^0), \ldots, f(\xi^{n-1})).$$

On the left hand side addition and multiplication are modulo $\mathrm{I}(x^n - 1)$*, while on the right hand side they are pointwise.*   □

Definition (Discrete Fourier Transform, MS-polynomial) Assuming again   **4.5.7**
that $\gcd(n, q) = 1$, we obtain a primitive root of unity ξ, and so the mapping

$$f + \mathrm{I}(x^n - 1) \mapsto (f(\xi^0), f(\xi^1), \ldots, f(\xi^{n-1}))$$

is an $\mathbb{F}_q$-isomorphism of algebras. It is called the *Discrete Fourier Transform* and abbreviated as DFT. It can be considered as the right multiplication

$$f + \mathrm{I}(x^n - 1) \mapsto (f_0, \ldots, f_{n-1}) \cdot \Phi_n,$$

where $f = \sum_{i=0}^{n-1} f_i x^i$ and where the representing matrix Φ_n, called the n-th order *Fourier matrix*, is given by

$$\Phi_n := \begin{pmatrix} 1 & 1 & 1 & \cdots & 1 \\ 1 & \xi & \xi^2 & \cdots & \xi^{n-1} \\ 1 & \xi^2 & \xi^4 & \cdots & \xi^{2(n-1)} \\ \vdots & \vdots & \vdots & \vdots & \vdots \\ 1 & \xi^{n-1} & \xi^{2(n-1)} & \cdots & \xi^{(n-1)(n-1)} \end{pmatrix} = (\xi^{ij})_{i,j \in n}.$$

The image vector

$$F := (F_0, \ldots, F_{n-1}) := (f(\xi^0), \ldots, f(\xi^{n-1})) \in \mathbb{F}_{q^m}^n$$

is the *Fourier vector* of f and the corresponding polynomial

$$F(y) := \sum_{i \in n} F_i y^{n-i} \in \mathbb{F}_{q^m}[y]$$

is called the *Discrete Fourier Transform* or the *Mattson–Solomon polynomial* (MS-polynomial, for short) of f, cf. [145].   ◇

Summarizing, we have obtained the

Corollary *The mapping* $f \mapsto F$ *is an* $\mathbb{F}_{q^m}$*-isomorphism between the* $\mathbb{F}_{q^m}$*-algebras*   **4.5.8**
$\{f \in \mathbb{F}_{q^m}[x] \mid \deg f < n\}$*, with addition and multiplication modulo* $\mathrm{I}(x^n - 1)$*, and*
$\mathbb{F}_{q^m}^n$*, with pointwise addition and multiplication (the* Hadamard-product*),*

$$\mathrm{Res}_{q^m, n} \simeq \mathbb{F}_{q^m}^n \; : \; f \mapsto F.$$ □

For an application of the DFT we need the inverse matrix (cf. Exercise 4.5.6)

$$\Phi_n^{-1} = \frac{1}{n}(\xi^{-ij})_{i,j \in n}.$$

Application of the MS-polynomial is based on the

4.5.9   **Lemma** *The Mattson–Solomon-polynomial has the following properties:*

1. $F(y)$ is the MS-polynomial of a polynomial $f(x) \in \mathbb{F}_q[x]$ if and only if

$$\forall\, j \in n\; :\; F_{qj \bmod n} = F_j^q.$$

2. $F(y)$ is the MS-polynomial of a codeword $f = f(x) + I(x^n - 1)$ of the cyclic code $C \leq \mathrm{Res}_{q,n}$ if and only if $F_i = 0$ for all $\xi^i \in V(C)$ and $F_{qj \bmod n} = F_j^q$ for all j.

3. For $f \in \mathrm{Res}_{q,n}$ we have $\mathrm{wt}(f) = n - s$, where s is the number of n-th roots of unity which are zeros of the MS-polynomial F of f. For short:

$$\mathrm{wt}(f) = n - |V(F) \cap U_n|.$$

4. If $F(y)$ is the MS-polynomial of $f(x)$ then $\mathrm{wt}(f) \geq n - \deg F$.

Proof: 1. Let $F(y)$ be the MS-polynomial of $f(x) \in \mathbb{F}_q[x]$. Since the coefficients of $f(x)$ are in $\mathbb{F}_q$, we have

$$F_j^q = f(\xi^j)^q = f(\xi^{qj}) = F_{qj \bmod n}.$$

On the other hand, let us now assume that $F_{qj \bmod n} = F_j^q$ for all j. First we note that $1/n$ is in the prime field and hence $(1/n)^q = 1/n$ (recall that $1/n$ is a shorthand for the unique solution c of the congruence $nc \equiv 1 \bmod p$, where p is the characteristic of $\mathbb{F}_q$). Secondly, we recall that over $\mathbb{F}_q$ we have $(\gamma + \delta)^q = \gamma^q + \delta^q$. Therefore we compute (indices modulo n)

$$
f_j^q = \left(\frac{1}{n} F(\xi^j)\right)^q = \frac{1}{n} \sum_{i \in n} F_i^q \xi^{(n-i)jq} = \frac{1}{n} \sum_{i \in n} F_{qi} \xi^{-qij}
$$
$$
= \frac{1}{n} \sum_{i \in n} F_i \xi^{-ij} = \frac{1}{n} \sum_{i \in n} F_i \xi^{(n-i)j} = \frac{1}{n} F(\xi^j) = f_j
$$

for all j. This shows that f_j is in $\mathbb{F}_q$ and, therefore, we have $f(x) \in \mathbb{F}_q[x]$.

2. $f(x)$ is an element of the cyclic code C if and only if $f(\xi^j) = 0$ for all $\xi^j \in V(C)$. This means that $F_j = 0$ for $\xi^j \in V(C)$.

3. This follows from the definition of the MS-polynomial, since simple computations show that $F(\xi^j) = nf_j$, $j \in n$. Therefore $F(\xi^j) = 0$ if and only if $f_j = 0$.

4. This is a consequence of the third assertion since the polynomial F can have no more than $\deg(F)$ zeros. $\qquad\square$

This result implies the following facts about idempotents:

Theorem *Let n and q be relatively prime, as before, and let ζ denote a primitive n-th*   **4.5.10**
root of unity over $\mathbb{F}_q$.

1. *The element $e \in \mathrm{Res}_{q,n}$ is idempotent if and only if $e(\zeta^i) \in \{0,1\}$ for all $i \in n$.*

2. *A polynomial $e(x) \in \mathrm{Res}_{q,n}$ is the generating idempotent of $C = I(g)/I(x^n - 1)$*
   if and only if

$$e(\zeta^i) = \begin{cases} 0 & \text{if } \zeta^i \in V(C), \\ 1 & \text{if } \zeta^i \notin V(C), \end{cases} \qquad i \in n.$$

Proof: 1. The Discrete Fourier Transform turns e into $E = (E_0, \ldots, E_{n-1})$ and
the equation $e^2 = e$ into

$$E * E = (E_0, \ldots, E_{n-1})^2 = (E_0^2, \ldots, E_{n-1}^2) = E,$$

where $*$ is the Hadamard-product. Thus $E_i^2 = E_i$, which is equivalent to $E_i = e(\zeta^i) \in \{0,1\}$.

2. Assume that $e(\zeta^i) = 0$ for $\zeta^i \in V(C)$ and $e(\zeta^i) = 1$ for $\zeta^i \notin V(C)$. Since
the idempotent polynomial $e(x)$ and the generator polynomial $g(x)$ have the
same n-th roots of unity as zeros and $g(x)$ has no other roots, we have that $e(x)$
is a multiple of $g(x)$. Moreover by 4.5.1, $g(x)$ is the greatest common divisor
of $e(x)$ and $x^n - 1$. Thus, by Bézout's Identity (cf. Exercise 3.1.6). there exist
polynomials s and t such that $g = se + t(x^n - 1)$. When read modulo $I(x^n - 1)$,
this equation shows that $g(x)$ is a multiple of $e(x)$. Hence e and g generate the
same code. By part 1 we know that e is an idempotent. Hence e is a generating
idempotent. The converse follows using Exercise 4.5.3.   □

The next result shows how to determine the dimension of a code from *any*
of its generating polynomials.

Theorem *Let $C = I(f)/I(x^n - 1)$, $f \in \mathbb{F}_q[x]$ a polynomial with $\deg f < n$. Then*   **4.5.11**
the dimension of C is equal to the number of nonzero Fourier coefficients F_i of f.

Proof: The elements $\overline{f}, \overline{xf}, \ldots, \overline{x^{n-1}f}$ certainly generate the ideal C as a vector
space. If we write the coefficients of these elements into the rows of a matrix,
we obtain

$$\Theta_n(f) = \begin{pmatrix} f_0 & f_1 & \cdots & f_{n-2} & f_{n-1} \\ f_{n-1} & f_0 & \cdots & f_{n-3} & f_{n-2} \\ \vdots & \vdots & \ddots & \vdots & \vdots \\ f_1 & f_2 & \cdots & f_{n-1} & f_0 \end{pmatrix}.$$

The rows of $\Theta_n(f)$ generate the code. Nevertheless, the matrix $\Theta_n(f)$ is in general not a generator matrix, as it may contain linearly dependent rows. In any case, the rank of $\Theta_n(f)$ is equal to the dimension of the code. Now, consider Φ_n, the matrix of the Fourier Transform. Let $\Phi_{*,j}$ be the j-th column of Φ_n for $j \in n$. The i-th entry in the vector $\Theta_n(f) \cdot \Phi_{*,j}$ is

$$\sum_{k \in n} f_{k-i}\xi^{kj} = \xi^{ij} \sum_{k \in n} f_{k-i}\xi^{(k-i)j} = \xi^{ij} \sum_{k \in n} f_k \xi^{kj} = f(\xi^j)\xi^{ij}.$$

This shows that $\Theta_n(f) \cdot \Phi_{*,j} = F_j \Phi_{*,j}$, i.e. that Φ_n is a matrix of eigenvectors for $\Theta_n(f)$ with corresponding eigenvalues $F_0, F_1, \ldots, F_{n-1}$. Since Φ_n is non-singular, the columns of Φ_n form a basis of eigenvectors, and hence $\Theta_n(f)$ is diagonalizable. The rank of $\Theta_n(f)$ is, therefore, the number of nonzero eigenvalues, i.e. the number of nonzero Fourier coefficients F_i.   □

Using the Discrete Fourier Transform, the Reed–Solomon-codes can be described as follows. Recall that the Reed–Solomon-code C with designed distance δ, base b and length $n = q - 1$ is the code generated by

$$g := (x - \xi^b) \cdots (x - \xi^{b+\delta-2}) \in \mathbb{F}_q[x],$$

and that in this case the primitive n-th root of unity ξ is a primitive element of the field $\mathbb{F}_q$.

4.5.12   **Theorem** *Let C be a Reed–Solomon-code with designed distance δ and base b over $\mathbb{F}_q$. The dual code $C^\perp$ is generated by*

$$g^\perp := \prod_{i=-n-b+1}^{-b-\delta+1} (x - \xi^i) = \prod_{i=n-b+1}^{2n-b-\delta+1} (x - \xi^i)$$

and, therefore, $C^\perp$ has base $n - b + 1$ and designed distance $\delta^\perp = n - \delta + 2$.

Proof: The parity check polynomial of C is

$$h(x) = \frac{x^n - 1}{g} = \frac{\prod_{i=0}^{n-1}(x - \xi^i)}{\prod_{i=b}^{b+\delta-2}(x - \xi^i)} = \prod_{i=b+\delta-1}^{b+n-1}(x - \xi^i),$$

which is of degree $n - \delta + 1 = k$. The reciprocal polynomial is

$$\widehat{h} = x^k h(x^{-1}) = x^k \prod_{i=b+\delta-1}^{b+n-1}(x^{-1} - \xi^i) = \prod_{i=b+\delta-1}^{b+n-1}(1 - \xi^i x),$$

and the constant term of h is

$$h(0) = \prod_{i=b+\delta-1}^{b+n-1}(-\xi^i).$$

So, the dual code has the generator polynomial

$$\frac{\widehat{h}(x)}{h(0)} = \prod_{i=b+\delta-1}^{b+n-1} (x - \zeta^{-i}) = \prod_{i=-n-b+1}^{-b-\delta+1} (x - \zeta^{i}) = \prod_{i=n-b+1}^{2n-b-\delta+1} (x - \zeta^{i}).$$

Therefore, it is a Reed–Solomon-code for base $n - b + 1$ and designed distance $n - \delta + 2$. $\qquad\qquad\qquad\qquad\qquad\qquad\qquad\qquad\qquad\qquad\qquad\qquad\qquad\qquad\square$

Thus,

$$V(C^{\perp}) = \left\{ \zeta^{i} \mid n - b + 1 \leq i \leq 2n - b - \delta + 1 \right\}$$

and hence, according to Exercise 4.3.1, and since $k = \dim(C) = n - \delta + 1$,

$$\Delta = (\zeta^{ij})_{n-b+1 \leq i \leq n-b+k,\, j \in n}$$

is a check matrix for the dual code.

Corollary *The Reed–Solomon-code of length n and dimension k over $\mathbb{F}_q$, generated*   **4.5.13**
by the consecutive set
$$W := \{\zeta^{b}, \ldots, \zeta^{b+\delta-2}\},$$
has the following submatrix of the Fourier matrix as generator matrix:
$$\Gamma := (\zeta^{ij})_{n-b+1 \leq i \leq n-b+k,\, j \in n}.$$

In particular, the narrow-sense Reed–Solomon-code of length n and dimension k over
$\mathbb{F}_q$ *is generated by*
$$\Gamma := (\zeta^{ij})_{i \in k,\, j \in n},$$

and so it is equal to

$$\left\{ (f(1), f(\zeta), \ldots, f(\zeta^{n-1})) = (F_0, \ldots, F_{n-1}) \,\middle|\, f \in \mathbb{F}_q[x],\ f = 0 \text{ or } \deg f < k \right\}.$$
$$\square$$

In other words, the Discrete Fourier Transform can be used as a natural encoding method for narrow-sense Reed–Solomon-codes.

Example The narrow-sense Reed–Solomon-code of length 7 with designed dis-   **4.5.14**
tance 5 has dimension 3. It is generated by

$$g(x) = (x - \zeta)(x - \zeta^{2})(x - \zeta^{3})(x - \zeta^{4}) = \zeta^{3} + \zeta x + x^{2} + \zeta^{3}x^{3} + x^{4},$$

where ζ is the usual primitive n-th root of unity, a primitive element of $\mathbb{F}_8$ with $\zeta^{3} = \zeta + 1$. From 4.5.13 we obtain that this linear code over $\mathbb{F}_8$ is generated by

$$\Gamma := \begin{pmatrix} 1 & 1 & 1 & 1 & 1 & 1 & 1 \\ 1 & \zeta & \zeta^{2} & \zeta^{3} & \zeta^{4} & \zeta^{5} & \zeta^{6} \\ 1 & \zeta^{2} & \zeta^{4} & \zeta^{6} & \zeta & \zeta^{3} & \zeta^{5} \end{pmatrix}.$$

$\diamond$

The fact that narrow-sense Reed–Solomon-codes consist of Fourier Transforms

$$(f(1), f(\xi), \ldots, f(\xi^{n-1}))$$

of polynomials f of degree $< k$ motivates the following notation and generalization of Reed–Solomon-codes: For $1 \le k \le n$ we denote the subspaces of all polynomials of degree strictly less than k by

$$\mathbb{F}_q[x]_{<k} := \bigoplus_{i=0}^{k-1} \mathbb{F}_q \cdot x^i.$$

4.5.15 **Definition (Generalized Reed–Solomon-code)** Let $n \le q$ be a positive integer, $\kappa = (\kappa_0, \kappa_1, \ldots, \kappa_{n-1})$ an n-tuple of pairwise distinct elements of $\mathbb{F}_q$ and $\beta = (\beta_0, \beta_1, \ldots, \beta_{n-1})$ an n-tuple of nonzero elements of $\mathbb{F}_q$. For k with $1 \le k \le n$ we define the *Generalized Reed–Solomon-code* $\mathrm{GRS}_k(\kappa, \beta)$, a code which need not be cyclic, as

$$\left\{ (f(\kappa_0)\beta_0, \ f(\kappa_1)\beta_1, \ \ldots, \ f(\kappa_{n-1})\beta_{n-1}) \ \middle| \ f(x) \in \mathbb{F}_q[x]_{<k} \right\}.$$

In formal terms, using the matrix (of rank k)

$$\Phi_{(\kappa)} := (\kappa_j^i)_{i \in k, j \in n},$$

and the notation

$$f(x) \cdot \Phi_{(\kappa)} := (f_0, \ldots, f_{k-1}) \cdot \Phi_{(\kappa)} = (f(\kappa_0), f(\kappa_1), \ldots, f(\kappa_{n-1})),$$

$\mathrm{GRS}_k(\kappa, \beta)$ is generated by the matrix of rank k

$$\Gamma := \Phi_{(\kappa)} \cdot \mathrm{diag}(\beta_0, \ldots, \beta_{n-1}) = \begin{pmatrix} \beta_0 & \beta_1 & \cdots & \beta_{n-1} \\ \kappa_0\beta_0 & \kappa_1\beta_1 & \cdots & \kappa_{n-1}\beta_{n-1} \\ \kappa_0^2\beta_0 & \kappa_1^2\beta_1 & \cdots & \kappa_{n-1}^2\beta_{n-1} \\ \vdots & & & \vdots \\ \kappa_0^{k-1}\beta_0 & \kappa_1^{k-1}\beta_1 & \cdots & \kappa_{n-1}^{k-1}\beta_{n-1} \end{pmatrix},$$

in the following way:

$$\mathrm{GRS}_k(\kappa, \beta) = \left\{ f(x) \cdot \Gamma \ \middle| \ f(x) \in \mathbb{F}_q[x]_{<k} \right\}. \qquad \diamond$$

4.5.16 **Example** Let $n = q - 1$ and let ξ be a primitive element for $\mathbb{F}_q$. Put $\kappa_j = \xi^j$ and $\beta_j = \kappa_j^{n-b+1}$ for $j \in n$. Then

$$\Phi_{(\kappa)} = (\xi^{ij})_{i \in k, \, j \in n}$$

and the multiplication by

$$\mathrm{diag}(\beta_0,\ldots,\beta_{n-1}) = \mathrm{diag}\big(\zeta^{0\cdot(n-b+1)},\ldots,\zeta^{(n-1)\cdot(n-b+1)}\big)$$

gives the generator matrix

$$\Gamma = (\zeta^{ij})_{n-b+1\le i\le n-b+k,\,j\in n}.$$

Hence, by 4.5.13, this particular Generalized Reed–Solomon-code $\mathrm{GRS}_k(\kappa,\beta)$ is in fact a Reed–Solomon-code. $\diamond$

Theorem *Generalized Reed–Solomon-codes are MDS-codes.*   **4.5.17**

Proof: Since $\deg(f) < k$ and since the β_i are all nonzero, a codeword

$$c = \big(f(\kappa_0)\beta_0, f(\kappa_1)\beta_1,\ldots,f(\kappa_{n-1})\beta_{n-1}\big)$$

in $\mathrm{GRS}_k(\kappa,\beta)$ has at most $k-1$ zeros, i.e. $\mathrm{wt}(c) \ge n-k+1$. This shows that the minimum distance is at least $n-k+1$ and, therefore, the code is MDS, since $k = \dim(\mathrm{GRS}_k(\kappa,\beta))$. $\square$

Since Generalized Reed–Solomon-codes are MDS, they can be encoded systematically at any k positions.

Theorem *A systematic generator matrix for the code* $\mathrm{GRS}_k(\kappa,\beta)$ *is* $\Gamma = (I_k \mid M)$,   **4.5.18**
where $M = (m_{ij})_{i\in k,\,j\in n-k}$, *with*

$$m_{ij} = \frac{\beta_{k+j}\prod_{u\in k\setminus\{i\}}(\kappa_{k+j} - \kappa_u)}{\beta_i \prod_{u\in k\setminus\{i\}}(\kappa_i - \kappa_u)}.$$

Proof: We consider first the case where $\beta = 1_n$ is the all-one vector. For $i \in k$, let $v^{(i)}$ be the i-th row of $(I_k \mid M)$. By 4.5.15 we need to find a polynomial $f^{(i)} \in \mathbb{F}_q[x]$ of degree at most $k-1$ such that $v^{(i)} = f^{(i)} \cdot \Phi_{(\kappa)}$, i.e.

$$f^{(i)}(\kappa_j) = \begin{cases} 1 & \text{if } i = j \\ 0 & \text{if } i \neq j,\ j < k. \end{cases}$$

By the Lagrange Interpolation Theorem, $f^{(i)}$ exists and is unique. In fact, we have

$$f^{(i)}(x) = \frac{\prod_{u\in k\setminus\{i\}}(x - \kappa_u)}{\prod_{u\in k\setminus\{i\}}(\kappa_i - \kappa_u)},\qquad i \in k.$$

Hence $m_{ij} = f^{(i)}(\kappa_{k+j})$ is of the stated form provided that $\beta_j = 1$ for all j. Otherwise, we multiply the j-th column by β_j and divide the elements of the i-th row by β_i to get the systematic generator matrix Γ as stated. $\square$

The dual of a Generalized Reed–Solomon-code is again a Generalized Reed–Solomon-code:

4.5.19   **Theorem** *The dual code of* $\mathrm{GRS}_k(\kappa,\beta)$ *is* $\mathrm{GRS}_{n-k}(\kappa,\gamma)$ *where* $\gamma = (\gamma_0,\dots,\gamma_{n-1})$ *with*

$$\gamma_j = \frac{1}{\beta_j \prod_{i\in n, i\neq j}(\kappa_j - \kappa_i)}, \qquad j \in n.$$

Proof: Let $\Gamma = (I_k \mid M)$ be a systematic generator matrix of a $\mathrm{GRS}_k(\kappa,\beta)$ with $M = (m_{ij})$ as in 4.5.18. In the same fashion as before one can show (cf. Exercise 4.5.9) that $\Delta = (N \mid I_{n-k})$ is a systematic generator matrix of $\mathrm{GRS}_{n-k}(\kappa,\gamma)$, where $N = (n_{ij})$ with

$$n_{ij} = \frac{\gamma_j \prod_{u\in n-k\setminus\{i\}}(\kappa_j - \kappa_{k+u})}{\gamma_{k+i} \prod_{u\in n-k\setminus\{i\}}(\kappa_{k+i} - \kappa_{k+u})}, \quad i \in n-k,\ j \in k.$$

Substituting the values of γ_j, we compute

$$
\begin{aligned}
n_{ij} &= \frac{\beta_{k+i} \prod_{v\in n, v\neq k+i}(\kappa_{k+i} - \kappa_v) \prod_{u\in n-k\setminus\{i\}}(\kappa_j - \kappa_{k+u})}{\beta_j \prod_{v\in n, v\neq j}(\kappa_j - \kappa_v) \prod_{u\in n-k\setminus\{i\}}(\kappa_{k+i} - \kappa_{k+u})} \\
&= \frac{\beta_{k+i} \prod_{v\in k}(\kappa_{k+i} - \kappa_v)}{\beta_j (\kappa_j - \kappa_{k+i}) \prod_{v\in k\setminus\{j\}}(\kappa_j - \kappa_v)} \\
&= -\frac{\beta_{k+i} \prod_{v\in k\setminus\{j\}}(\kappa_{k+i} - \kappa_v)}{\beta_j \prod_{v\in k\setminus\{j\}}(\kappa_j - \kappa_v)} \\
&= -m_{ji}.
\end{aligned}
$$

This shows that $N = -M^{\top}$, and hence by Exercise 1.3.9, Δ is a check matrix for the code generated by Γ. Thus, $\mathrm{GRS}_{n-k}(\kappa,\gamma)$ is the dual of $\mathrm{GRS}_k(\kappa,\beta)$.   □

This Generalized Reed–Solomon-codes $\mathrm{GRS}_k(\kappa,\beta)$ can be rephrased by replacing the sequence β by a polynomial: We again fix a positive integer $n \leq q$ and a sequence $\kappa = (\kappa_0,\dots,\kappa_{n-1})$ of pairwise distinct elements of $\mathbb{F}_q$, and we form the corresponding polynomial

$$h := \prod_{i\in n}(x - \kappa_i).$$

The Chinese Remainder Theorem establishes the $\mathbb{F}_q$-algebra isomorphism

$$\phi\colon \mathbb{F}_q[x]/I(h) \to \mathbb{F}_q^n \ :\ f + I(h) \mapsto (f(\kappa_0),\dots,f(\kappa_{n-1})).$$

This map restricts to an isomorphism between the two groups of units, which are the polynomials $g \in F_q[x]/I(h)$ with $g(\kappa_i) \neq 0$, for $i \in n$ on one side and $(F_q^*)^n$, the sequences of length n over F_q whose entries are all nonzero on the other side (with the Hadamard product as multiplication). The map ϕ takes a polynomial $g \in \mathbb{F}_q[x]/I(h)$ which is a unit in this ring, i.e. with $g(\kappa_i) \neq 0$, to the sequence $(g(\kappa_0), \ldots, g(\kappa_{n-1}))$. In the other direction, given a sequence $c = (c_0, \ldots, c_{n-1})$ with $c_i \neq 0$, for $i \in n$, we can obtain the inverse image under the map ϕ from Lagrange's interpolation formula. Namely, we have

$$\phi^{-1} : \mathbb{F}_q^n \to \mathbb{F}_q[x]/I(h) \; : \; c \mapsto \sum_{i \in n} \frac{c_i}{\ell(\kappa_i)} \prod_{j \in n \setminus \{i\}} (x - \kappa_j) + I(h),$$

where ℓ is the unique polynomial of degree less than n with

$$\ell(\kappa_i) := \prod_{j \in n \setminus \{i\}} (\kappa_i - \kappa_j) \; \text{ for each } i \in n.$$

The residue class $\ell + I(h)$ is a unit in $\mathbb{F}_q[x]/I(h)$. It serves in the following as a normalizing factor. For each unit $g + I(h)$ we form, as already announced, the following linear code which is isometric to $\mathrm{GRS}_k(\kappa, \mathbf{1}_n)$,

$$\mathrm{GRS}_k(\kappa, g) = \mathrm{GRS}_k(\kappa, \beta)$$

where $\beta = (\beta_0, \ldots, \beta_{n-1})$ is the sequence whose entries are $\beta_i = g(\kappa_i)/\ell(\kappa_i)$, i.e.

$$\mathrm{GRS}_k(\kappa, g) = \left\{ \left(\frac{c_0 g(\kappa_0)}{\ell(\kappa_0)}, \ldots, \frac{c_{n-1} g(\kappa_{n-1})}{\ell(\kappa_{n-1})} \right) \; \middle| \; c \in \mathrm{GRS}_k(\kappa, \mathbf{1}_n) \right\}.$$

Theorem Let $\kappa = (\kappa_0, \ldots, \kappa_{n-1})$ be a sequence of pairwise distinct elements of $\mathbb{F}_q$,   **4.5.20**
where $n \leq q$ and $h := \prod_{i \in n}(x - \kappa_i)$. Consider $g \in \mathbb{F}_q[x]_{<n}$ with $g(\kappa_i) \neq 0$ for all $i \in n$.

— We have:

$$\mathrm{GRS}_k(\kappa, g) = \left\{ c \in \mathbb{F}_q^n \; \middle| \; \exists f \in \mathbb{F}_q[x]_{<k} : \sum_{i \in n} c_i \prod_{j \neq i} (x - \kappa_j) \equiv fg \bmod I(h) \right\}.$$

— If g is of degree $n - k$, then

$$\mathrm{GRS}_k(\kappa, g) = \left\{ c \in \mathbb{F}_q^n \; \middle| \; \sum_{i \in n} c_i \prod_{j \neq i} (x - \kappa_j) \equiv 0 \bmod I(g) \right\}.$$

This set can also be characterized as the set of all $c \in \mathbb{F}_q^n$ which satisfy

$$\sum_{i \in n} c_i (x - \kappa_i)^{-1} = 0$$

in $\mathbb{F}_q[x]/I(g)$.

Proof: 1. A vector $c \in \mathbb{F}_q^n$ is contained in $\mathrm{GRS}_k(\kappa, g)$ if and only if there exists a polynomial $f \in \mathbb{F}_q[x]_{<k}$ such that

$$c = \left(\frac{f(\kappa_0) g(\kappa_0)}{\ell(\kappa_0)}, \ldots, \frac{f(\kappa_{n-1}) g(\kappa_{n-1})}{\ell(\kappa_{n-1})} \right).$$

This is equivalent to

$$\begin{aligned}(c_0 \ell(\kappa_0), \ldots, c_{n-1} \ell(\kappa_{n-1})) &= (f(\kappa_0) g(\kappa_0), \ldots, f(\kappa_{n-1}) g(\kappa_{n-1})) \\ &= \phi(fg + I(h)),\end{aligned}$$

i.e.

$$fg + I(h) = \phi^{-1}(c_0 \ell(\kappa_0), \ldots, c_{n-1} \ell(\kappa_{n-1})),$$

and so

$$fg \equiv \sum_{i \in n} c_i \prod_{j \neq i} (x - \kappa_j) \bmod I(h).$$

2. To each codeword $c \in \mathrm{GRS}_k(\kappa, g)$, there exist polynomials $f \in \mathbb{F}_q[x]_{<k}$ and $s \in \mathbb{F}_q[x]$ with

$$\sum_{i \in n} c_i \prod_{j \neq i} (x - \kappa_j) = fg + sh.$$

If the degree of g is $n - k$, then fg is of degree less than n, which implies that

$$\sum_{i \in n} c_i \prod_{j \neq i} (x - \kappa_j) = fg,$$

and hence

$$\sum_{i \in n} c_i \prod_{j \neq i} (x - \kappa_j) \equiv 0 \bmod I(g).$$

Since $h + I(g)$ is a unit in $\mathbb{F}_q[x] / I(g)$, we obtain via multiplication by its inverse that

$$\sum_{i \in n} c_i (x - \kappa_i)^{-1} = 0$$

in $\mathbb{F}_q[x] / I(g)$. Since these arguments can be reversed, we obtain the assertion. $\qquad\square$

The MDS-codes of the second part of this theorem are called *Goppa-MDS-codes*. Goppa-codes are obtained from Goppa-MDS-codes by restriction. These codes were discovered in the early 1970's by V.D. Goppa (cf. [71, 72, 10]). We will discuss them in the following section.

Exercises

E.4.5.1   **Exercise** Consider a ring R with identity element. An element $e \in R \setminus \{0\}$ is called *idempotent* if it satisfies $e^2 = e$. *Central* idempotents are idempotents e that commute with every ring element. Furthermore, there is the notion of *orthogonal* idempotents, i.e. idempotents e, e' with $ee' = e'e = 0$. Lastly, we introduce *primitive* idempotents e, for which there do *not* exist orthogonal idempotents e', e'' such that $e = e' + e''$. Idempotents that are both central and primitive are called *centrally-primitive*.

Assume that an idempotent $e \in R$ generates the left ideal $L := Re$. Prove the following assertions:

— Besides being a generator of L, e is also a generating *unit* in L, i.e. $xe = x$ for all $x \in L$.

— If $e \neq 1$, then both Re and $R(1-e)$ are nontrivial left ideals. Moreover, both e and $1 - e$ are orthogonal idempotents, and we obtain the following direct decomposition of R into left ideals

$$R = Re \oplus R(1 - e).$$

Exercise Show that direct decompositions of rings with identity into left ideals   E.4.5.2
correspond to decompositions of the identity element into orthogonal idempotents in the following way:

1. If

$$R = \bigoplus_{i \in I} L_i$$

is a decomposition of R into a direct sum of left ideals $L_i \neq 0$, and

$$1 = \sum_{i \in I} e_i,$$

with $e_i \in L_i$, is the corresponding decomposition of the identity element, then these decompositions satisfy the following conditions:
   — The index set I is finite.

   — For $i, j \in I, i \neq j$, the elements e_i and e_j are orthogonal idempotents, and $L_i = Re_i$ for $i \in I$.

   — If L_i is indecomposable, then e_i is primitive.

   — If all the left ideals L_i are two-sided ideals, then the idempotents e_i are central.

   — If all the L_i are two-sided and L_{i_0} is indecomposable (as a two-sided ideal), then e_{i_0} is centrally-primitive.

2. If, conversely,

$$1 = \sum_{i \in I} e_i$$

is a decomposition of the identity element into pairwise orthogonal idempotents e_i, then we have, for the left ideals $L_i := Re_i$:

- $R = \bigoplus_{i \in I} L_i$.

- If e_i is primitive, then L_i is indecomposable.

- If e_i is central, then L_i is a two-sided ideal.

- If e_i is centrally-primitive, then L_i is indecomposable (as a two-sided ideal).

Moreover, such decompositions of the identity into centrally-primitive orthogonal idempotents are uniquely determined.

E.4.5.3   **Exercise** Let C be a cyclic code of length n over $\mathbb{F}_q$ with generating idempotent e. Show that the generator polynomial of C is $g = \gcd(x^n - 1, e)$.

E.4.5.4   **Exercise** Give another proof of 4.5.1 using the Euclidean Algorithm.

E.4.5.5   **Exercise** Let C denote a cyclic code of length n over $\mathbb{F}_q$ with generating idempotent e. Prove that $1 - x^n e(x^{-1})$ is a generating idempotent of $C^{\perp}$.

E.4.5.6   **Exercise** Prove that the inverse of the Fourier matrix is

$$\Phi_n^{-1} = \frac{1}{n}(\xi^{-ij})_{i,j \in n}.$$

Describe a fast multiplication of polynomials, which enables us to obtain the product of two polynomials in $\mathbb{F}_q[x]$ from the Hadamard-product of the corresponding Discrete Fourier Transforms. (An analysis of the complexity of this method shows that it is more efficient than direct multiplication. For a detailed description we recommend [1], [38], or [159].)

E.4.5.7   **Exercise** *Phase property* of DFT: Let ξ be a primitive n-th root of unity and $c = (c_0, \dots, c_{n-1})$ a vector in $\mathbb{F}_q^n$. Let

$$d = (c_s, c_{s+1}, \dots, c_{s+n-1 \bmod n})$$

be the vector which is obtained from c by s cyclic shifts of the entries. Then we have, for the corresponding Fourier-transforms $\hat{c} = c \cdot \Phi_n$ and $\hat{d} = d \cdot \Phi_n$,

$$\hat{d}_i = \hat{c}_i \xi^{-is}, \qquad i \in n.$$

Exercise *Scaling* via DFT: For the discrete Fourier Transform $\hat{d} := d \cdot \Phi_n$ of the   **E.4.5.8**
vector

$$d = (c_0, c_{s \bmod n}, \ldots, c_{s(n-1) \bmod n})$$

obtained from $c = (c_0, \ldots, c_{n-1})$ by a scaling factor $s \in \mathbb{N}^*$, where $\gcd(n, s) = 1$ (which permutes the entries), the following holds true:

$$\hat{d}_i = \hat{c}_{is^{-1} \bmod n}, \qquad i \in n.$$

Exercise Verify that the matrix $(N \mid I_{n-k})$ with $N = (n_{ij})$ as defined in 4.5.19   **E.4.5.9**
generates $\mathrm{GRS}_{n-k}(\kappa, \beta)$.

Exercise Let g and h be as in 4.5.20. Show that $g + I(h)$ is a unit of $\mathbb{F}_q[x]/I(h)$ **E.4.5.10**
and that $h + I(g)$ is a unit of $\mathbb{F}_q[x]/I(g)$. Compute the inverse of $(x - \kappa_i) + I(g)$.

Exercise Give a necessary and sufficient condition for two Generalized Reed–   **E.4.5.11**
Solomon-codes $\mathrm{GRS}_k(\kappa, g)$ and $\mathrm{GRS}_k(\tilde{\kappa}, \tilde{g})$ to be linearly isometric.

Exercise Consider $\kappa = (0, 1, 2, \alpha, 2\alpha)$, where α is a root of the irreducible poly-   **E.4.5.12**
nomial $x^2 + x + 2 \in \mathbb{F}_3[x]$. Construct the Generalized Reed–Solomon-code
$\mathrm{GRS}_3(\kappa, g)$, where $g \in \mathbb{F}_9[x]$ is a suitable polynomial of degree 3.

4.6 Alternant-Codes, Goppa-Codes   4.6

The class of Alternant-codes which we will consider now is obtained by restricting Generalized Reed–Solomon-codes to subfields. So again fix a positive
integer n, a sequence κ of n pairwise distinct elements of $\mathbb{F}_{q^m}$, and consider the
corresponding polynomial

$$h = \prod_{i \in n}(x - \kappa_i) \in \mathbb{F}_{q^m}[x].$$

According to the preceding section, for every polynomial $g \in \mathbb{F}_{q^m}[x]$ with
$g(\kappa_i) \neq 0$ for $0 \leq i \leq n - 1$, i.e. for every unit $g + I(h)$ in $\mathbb{F}_{q^m}[x]/I(h)$, there is
the (n, k)-MDS-code $\mathrm{GRS}_k(\kappa, g)$.

Definition (Alternant-codes) The restriction of a Generalized Reed–Solomon-   **4.6.1**
code $\mathrm{GRS}_k(\kappa, g)$ over $\mathbb{F}_{q^m}$ to the subfield $\mathbb{F}_q$ is called *Alternant-code* over $\mathbb{F}_q$,
denoted as

$$\mathrm{Alt}_{k,q}(\kappa, g) := \mathrm{GRS}_k(\kappa, g) \cap \mathbb{F}_q^n. \qquad \diamond$$

We obtain from 4.5.20 that

$$\text{Alt}_{k,q}(\kappa, g) = \left\{ c \in \mathbb{F}_q^n \;\middle|\; \exists f \in \mathbb{F}_{q^m}[x]_{<k} \;:\; \sum_{i \in n} c_i \prod_{j \neq i}(x - \kappa_j) \equiv fg \bmod \mathrm{I}(h) \right\}.$$

In order to obtain bounds for the parameters of such codes, we prove

4.6.2   **Lemma** *If C is an (n,k,d)-code over $\mathbb{F}_{q^m}$, and if C' is the (n,k',d')-code C' which is obtained by restriction of C to $\mathbb{F}_q$, we have the inequalities*

$$d' \geq d \quad \text{and} \quad m(n-k) \geq n - k' \geq n - k.$$

Proof: As C' is an $\mathbb{F}_q$-subspace of C, it follows that $d \leq d'$ and $k' \leq k$. Assume that Δ is a check matrix of C. Choose a basis of $\mathbb{F}_{q^m}$ over $\mathbb{F}_q$ and express the entries of Δ in terms of this basis, thereby obtaining an $m(n-k) \times n$-matrix over $\mathbb{F}_q$, the rank of which lies between $n - k$ and $m(n-k)$. The rows of this matrix span the dual code of C', and so the second inequality is also true.   □

Thus we obtain from 4.5.17

4.6.3   **Corollary** *The Alternant-code $\text{Alt}_{k,q}(\kappa, g)$ is an (n,k',d')-code over $\mathbb{F}_q$, with*

$$k \geq k' \geq n - m(n-k) \quad \text{and} \quad d' \geq n - k + 1.$$ □

The following theorem proves the existence of Alternant-codes with a prescribed lower bound for the minimum distance.

4.6.4   **Theorem** *Let $n = q^m - 1$ and assume that δ and r are positive integers such that*

$$\sum_{i=1}^{\delta-1} \binom{n}{i}(q-1)^i < (q^m - 1)^r.$$

Then there exists an Alternant-code over $\mathbb{F}_q$ of length n, dimension $k \geq n - mr$, and minimum distance $d \geq \delta$.

Proof: The proof is based on an enumerative argument. We count the number of Alternant-codes of a particular type which contain a vector of weight less than δ.

For this purpose we consider a Generalized $(n, n-r)$-Reed–Solomon-code $\text{GRS}_{n-r}(\kappa, \mathbf{1}_n)$ over $\mathbb{F}_{q^m}$. To each polynomial g with $g(\kappa_i) \neq 0$, $i \in n$, we associate the vector $g = (g(\kappa_0)/\ell(\kappa_0), \ldots, g(\kappa_{n-1})/\ell(\kappa_{n-1})) \in \mathbb{F}_{q^m}^n$. Each vector $g \in \mathbb{F}_{q^m}^n \setminus \{0\}$ defines an Alternant-code $\text{Alt}_{n-r,q}(\kappa, g)$ obtained by restriction to $\mathbb{F}_q$. There are $(q^m - 1)^n$ Generalized Reed–Solomon-codes $\text{GRS}_{n-r}(\kappa, g)$ coming from $\text{GRS}_{n-r}(\kappa, \mathbf{1}_n)$ and, according to 4.6.3, they determine Alternant-codes of dimension at least $n - mr$.

A nonzero vector c of weight less than δ can occur in at most $(q^m - 1)^{n-r}$ such Alternant-codes. The reason is that $\mathrm{GRS}_{n-r}(\kappa, \mathbf{1}_n)$ is an MDS-code: If $c \neq 0$ is vector of weight $w < \delta$ which occurs in such an Alternant-code, then c is obtained from a vector $\tilde{c} \in \mathrm{GRS}_{n-r}(\kappa, \mathbf{1}_n)$ of weight w:

$$c = (\tilde{c}_0 g(\kappa_0)\ell(\kappa_0)^{-1}, \dots, \tilde{c}_{n-1} g(\kappa_{n-1})\ell(\kappa_{n-1})^{-1}).$$

Therefore $w \geq n - (n-r) + 1$ and consequently $n - w \leq (n-r) - 1 < n - r$. The same vector c occurs in at most $(q^m - 1)^{n-w} < (q^m - 1)^{n-r}$ such Alternant-codes. They are obtained from those vectors g which can have arbitrary nonzero values in all those positions where $\tilde{c}_i = 0$. Moreover, the number of nonzero vectors in $\mathbb{F}_q^n$ which are of weight less than δ is

$$\sum_{i=1}^{\delta-1} \binom{n}{i}(q-1)^i.$$

Hence, the number of the above Alternant-codes containing a vector $c \neq 0$ of weight less than δ is not greater than

$$(q^m - 1)^{n-r} \sum_{i=1}^{\delta-1} \binom{n}{i}(q-1)^i.$$

By assumption,

$$(q^m - 1)^{n-r} \sum_{i=1}^{\delta-1} \binom{n}{i}(q-1)^i < (q^m - 1)^n.$$

This implies that at least one (n, k, d)-Alternant-code exists which satisfies $k \geq n - mr$ and $d \geq \delta$. $\square$

Now we briefly discuss the classical Goppa-codes. An introduction to the theory of algebraic-geometric Goppa-codes can be found in [193, 192, 196]. For further reading, we refer to [94, 187].

Definition (Goppa-codes) The restriction of a Generalized Reed–Solomon-code $\mathrm{GRS}_k(\kappa, g)$ over $\mathbb{F}_{q^m}$ with $\deg g = n - k$ to $\mathbb{F}_q$ is called a *Goppa-code*. It is indicated by $\mathrm{GO}_q(\kappa, g)$. The polynomial $g \in \mathbb{F}_{q^m}[x]$ is called *Goppa-polynomial*.   **4.6.5**

$\diamond$

According to 4.5.20, the Goppa-code $\mathrm{GO}_q(\kappa, g)$ has the form

$$\mathrm{GO}_q(\kappa, g) = \left\{ c \in \mathbb{F}_q^n \ \Big|\ \sum_{i \in n} c_i \prod_{j \neq i}(x - \kappa_j) \equiv 0 \bmod \mathrm{I}(g) \right\},$$

or (cf. 4.6.10),

$$\mathrm{GO}_q(\kappa, g) = \left\{ c \in \mathbb{F}_q^n \ \Big|\ \sum_{i \in n} \frac{c_i}{x - \kappa_i} \equiv 0 \bmod \mathrm{I}(g) \right\}.$$

From 4.6.2 we can deduce the following result:

4.6.6   **Corollary** *A Goppa-code* $GO_q(\kappa, g)$ *is an* (n, k, d)*-code with*

$$d \geq 1 + \deg g \quad \text{and} \quad k \geq n - m \cdot \deg g. \qquad \square$$

4.6.7   **Theorem** *A binary Goppa-code* $GO_2(\kappa, g)$ *has minimum distance* $d \geq 1 + 2 \deg g$, *provided its Goppa-polynomial has no multiple roots.*

Proof: Let $c = (c_0, \dots, c_{n-1})$ be a codeword in $GO_2(\kappa, g)$ of weight $w > 0$. Define the polynomial

$$f_c(x) = \prod_{i \in n, c_i = 1} (x - \kappa_i) = \sum_{i \in w+1} \gamma_i x^i$$

of degree w. The formal derivative f_c' of f_c can be obtained from the product rule as

$$f_c'(x) = \sum_{i \in n, c_i = 1} \prod_{j \in n \setminus \{i\}, c_j = 1} (x - \kappa_j) = \sum_{i \in n} c_i \prod_{j \in n \setminus \{i\}, c_j = 1} (x - \kappa_j).$$

It is a divisor of $\tilde{f}_c(x) = \sum_{i \in n} c_i \prod_{j \neq i} (x - \kappa_j)$, indeed

$$\tilde{f}_c(x) = f_c'(x) \prod_{j \in n, c_j = 0} (x - \kappa_j).$$

Since c is in the Goppa-code, the polynomial g divides $\tilde{f}_c(x)$. By assumption $g(\kappa_i) \neq 0$, $i \in n$, and therefore g and $\prod_j (x - \kappa_j)$ are relatively prime. Hence g divides f_c'. On the other hand, since we are in characteristic 2, we deduce that

$$
\begin{aligned}
f_c'(x) &= \sum_{i \in w+1} i \gamma_i x^{i-1} \\
&= \sum_{i \equiv 1 \bmod 2} \gamma_i x^{i-1} \\
&= \sum_{i \in u+1} \gamma_{2i+1} x^{2i} \\
&= \left(\lambda_0 + \lambda_1 x + \dots + \lambda_u x^u \right)^2
\end{aligned}
$$

where $u = \lfloor (w-1)/2 \rfloor$ and where $\lambda_i = \sigma^{-1}(\gamma_{2i+1})$ for $0 \leq i \leq u$. Here σ denotes the Frobenius automorphism (that is, $\lambda_i^2 = \gamma_{2i+1}$). This shows that $f_c'(x)$ is a square. Thus, since g divides $f_c'(x)$ and since g has no multiple roots, we deduce that in fact g^2 divides $f_c'(x)$. This shows that

$$2 \deg g \leq \deg f_c' = w - 1 \leq d - 1. \qquad \square$$

Example The polynomial $g = x^2 + x + 1 \in \mathbb{F}_2[x]$ is separable, i.e. has no   **4.6.8**
multiple roots. In order to construct a Goppa-code, we consider the field $\mathbb{F}_8$.
We use the irreducible polynomial $x^3 + x^2 + 1$ to generate this field. Let α
denote a root of this polynomial, so that $\alpha^3 = \alpha^2 + 1$. We label the elements of
$\mathbb{F}_8$ as

$$
\begin{aligned}
\kappa_0 &= 0, \\
\kappa_1 &= \alpha^0 = 1, \\
\kappa_2 &= \alpha^1 = \alpha, \\
\kappa_3 &= \alpha^2 = \alpha^2, \\
\kappa_4 &= \alpha^3 = \alpha^2 + 1, \\
\kappa_5 &= \alpha^4 = \alpha^3 + \alpha = \alpha^2 + \alpha + 1, \\
\kappa_6 &= \alpha^5 = \alpha^3 + \alpha^2 + \alpha = \alpha + 1, \\
\kappa_7 &= \alpha^6 = \alpha^2 + \alpha.
\end{aligned}
$$

The Generalized Reed–Solomon-code $\mathrm{GRS}_6(\kappa, g)$, with $\kappa = (\kappa_0, \ldots, \kappa_7)$ is the
set of $c \in \mathbb{F}_8^8$ such that

$$
\sum_{i \in 8} \frac{c_i}{x - \kappa_i} \equiv 0 \mod g.
$$

In order to find a check matrix for this code, we note that modulo $g(x)$ we have

$$
\frac{1}{x} \equiv x + 1, \quad \frac{1}{x+1} \equiv x, \quad \frac{1}{x+\alpha} \equiv \alpha^3 x + \alpha, \quad \frac{1}{x+\alpha^2} \equiv \alpha^6 x + \alpha^2,
$$

$$
\frac{1}{x+\alpha^3} \equiv \alpha^6 x + \alpha, \quad \frac{1}{x+\alpha^4} \equiv \alpha^5 x + \alpha^4, \quad \frac{1}{x+\alpha^5} \equiv \alpha^3 x + \alpha^4,
$$

$$
\frac{1}{x+\alpha^6} \equiv \alpha^5 x + \alpha^2 \mod g(x),
$$

which can be verified easily (for a way to actually compute these polynomials,
we refer to 5.2.8). Thus,

$$
\begin{aligned}
\tilde{\Delta} &= \begin{pmatrix} 1 & 0 & \alpha & \alpha^2 & \alpha & \alpha^4 & \alpha^4 & \alpha^2 \\ 1 & 1 & \alpha^3 & \alpha^6 & \alpha^6 & \alpha^5 & \alpha^3 & \alpha^5 \end{pmatrix} \\[4pt]
&= \begin{pmatrix} 1 & 0 & \alpha & \alpha^2 & \alpha & \alpha^2+\alpha+1 & \alpha^2+\alpha+1 & \alpha^2 \\ 1 & 1 & \alpha^2+1 & \alpha^2+\alpha & \alpha^2+\alpha & \alpha+1 & \alpha^2+1 & \alpha+1 \end{pmatrix}
\end{aligned}
$$

is a check matrix. Using the basis of $\mathbb{F}_8$ over $\mathbb{F}_2$ consisting of the elements
$1, \alpha, \alpha^2$, we can rewrite this matrix over $\mathbb{F}_2$ as

$$
\begin{pmatrix}
1 & 0 & 0 & 0 & 0 & 1 & 1 & 0 \\
0 & 0 & 1 & 0 & 1 & 1 & 1 & 0 \\
0 & 0 & 0 & 1 & 0 & 1 & 1 & 1 \\
1 & 1 & 1 & 0 & 0 & 1 & 1 & 1 \\
0 & 0 & 0 & 1 & 1 & 1 & 0 & 1 \\
0 & 0 & 1 & 1 & 1 & 0 & 1 & 0
\end{pmatrix}.
$$

This matrix has rank 6. Using Gaussian elimination, it can be brought into the form

$$\Delta = \begin{pmatrix} 1 & 0 & 0 & 0 & 0 & 1 & 0 & 1 \\ 0 & 1 & 0 & 0 & 0 & 1 & 0 & 1 \\ 0 & 0 & 1 & 0 & 0 & 1 & 0 & 0 \\ 0 & 0 & 0 & 1 & 0 & 1 & 0 & 0 \\ 0 & 0 & 0 & 0 & 1 & 0 & 0 & 1 \\ 0 & 0 & 0 & 0 & 0 & 0 & 1 & 1 \end{pmatrix},$$

so that a generator matrix of the Goppa-code is

$$\Gamma = \begin{pmatrix} 1 & 1 & 1 & 1 & 0 & 1 & 0 & 0 \\ 1 & 1 & 0 & 0 & 1 & 0 & 1 & 1 \end{pmatrix}.$$

It is easy to check that this code has minimum distance 5, as predicted by the previous theorem. Thus, we have found an $(8,2,5)$-code. As we can see from 2.2.13, this code is optimal. ◇

4.6.9   **Theorem** *BCH-codes in the narrow sense are Goppa-codes.*

Proof: Let C be a BCH-code in the narrow sense over $\mathbb{F}_q$ with length $n = q^m - 1$ and designed distance δ. Consider a primitive n-th root of unity ζ. Then $c \in \mathbb{F}_q^n$ is contained in C if and only if

$$\sum_{i \in n} c_i(\zeta^j)^i = 0, \qquad 1 \leq j \leq \delta - 1,$$

holds true.

For $\kappa = (1, \zeta^{-1}, \zeta^{-2}, \ldots, \zeta^{-(n-1)})$ and $g = x^{\delta-1}$ consider the Goppa-code $GO_q(\kappa, g)$. Then $c \in \mathbb{F}_q^n$ is contained in $GO_q(\kappa, g)$ if and only if

$$\sum_{i \in n} c_i \frac{x^n - 1}{x - \zeta^{-i}} \equiv 0 \bmod I(g).$$

Using the identity

$$\frac{x^n - 1}{x - \zeta^{-i}} = \zeta^i \frac{1 - (\zeta^i x)^n}{1 - \zeta^i x} = \zeta^i \sum_{l \in n} (\zeta^i x)^l = \sum_{l \in n} \zeta^{i(l+1)} x^l,$$

this condition can be rephrased as

$$\sum_{i \in n} c_i \sum_{l \in n} \zeta^{i(l+1)} x^l \equiv 0 \bmod I(g),$$

which is equivalent to

$$\sum_{l \in n} \left(\sum_{i \in n} c_i (\zeta^{l+1})^i \right) x^l \equiv 0 \bmod I(g).$$

Thus c is an element of $GO_q(\kappa, g)$ if and only if

$$\sum_{i \in n} c_i(\zeta^{l+1})^i = 0, \qquad 0 \le l \le \delta - 2,$$

which means that c is an element of C.   □

One can show that not every BCH-code is a Goppa-code (Exercise 4.6.2).

The check matrix of a code that restricts to a Goppa-code $GO_q(\kappa, g)$ can be deduced directly from its definition. For this purpose we form the inverse of $x - \kappa_i$ in $\mathbb{F}_q[x]/I(g)$,

$$(x - \kappa_i)^{-1} = -\frac{g(x) - g(\kappa_i)}{x - \kappa_i} g(\kappa_i)^{-1}. \tag{4.6.10}$$

This shows that $c \in \mathbb{F}_q^n$ is contained in $GO_q(\kappa, g)$ if and only if

$$\sum_{i \in n} c_i \frac{g(x) - g(\kappa_i)}{x - \kappa_i} g(\kappa_i)^{-1} = 0 \tag{4.6.11}$$

is true in $\mathbb{F}_q[x]/I(g)$. For degree reasons, this equation can be considered as an equation over $\mathbb{F}_q[x]$. The Goppa-polynomial $g = \sum_{i=0}^r g_i x^i$ is of degree $r := n - k$ and it satisfies

$$\frac{g(x) - g(\kappa_i)}{x - \kappa_i} = \sum_{j=1}^r g_j \sum_{l=1}^j x^{l-1} \kappa_i^{j-l} = \sum_{l=1}^r \left(\sum_{j=l}^r g_j \kappa_i^{j-l} \right) x^{l-1}, \qquad i \in n.$$

An application of 4.6.11 together with comparison of coefficients shows that $c \in \mathbb{F}_q^n$ is contained in $GO_q(\kappa, g)$ if and only if $c \cdot \widetilde{\Delta}^\top = 0$, where

$$\widetilde{\Delta} = \begin{pmatrix} g_r g(\kappa_0)^{-1} & \cdots & g_r g(\kappa_{n-1})^{-1} \\ (g_{r-1} + \kappa_0 g_r) g(\kappa_0)^{-1} & \cdots & (g_{r-1} + \kappa_{n-1} g_r) g(\kappa_{n-1})^{-1} \\ \vdots & & \vdots \\ \left(\sum_{i=1}^r \kappa_0^{i-1} g_i \right) g(\kappa_0)^{-1} & \cdots & \left(\sum_{i=1}^r \kappa_{n-1}^{i-1} g_i \right) g(\kappa_{n-1})^{-1} \end{pmatrix}.$$

This matrix can be written as a product

$$\widetilde{\Delta} = \begin{pmatrix} g_r & 0 & \cdots & 0 \\ g_{r-1} & g_r & \cdots & 0 \\ & & \ddots & 0 \\ g_1 & g_2 & \cdots & g_r \end{pmatrix} \begin{pmatrix} g(\kappa_0)^{-1} & \cdots & g(\kappa_{n-1})^{-1} \\ \kappa_0 g(\kappa_0)^{-1} & \cdots & \kappa_{n-1} g(\kappa_{n-1})^{-1} \\ \vdots & & \vdots \\ \kappa_0^{r-1} g(\kappa_0)^{-1} & \cdots & \kappa_{n-1}^{r-1} g(\kappa_{n-1})^{-1} \end{pmatrix}.$$

Since the first of the two matrices is invertible, the latter is a check matrix of a code $\widetilde{C}$ that restricts to $GO_q(\kappa, g)$. Hence, we have proved the following

4.6.12   **Theorem** *The code $\tilde{C}$ with check matrix*

$$\begin{pmatrix} g(\kappa_0)^{-1} & \cdots & g(\kappa_{n-1})^{-1} \\ \kappa_0 g(\kappa_0)^{-1} & \cdots & \kappa_{n-1} g(\kappa_{n-1})^{-1} \\ \vdots & & \vdots \\ \kappa_0^{r-1} g(\kappa_0)^{-1} & \cdots & \kappa_{n-1}^{r-1} g(\kappa_{n-1})^{-1} \end{pmatrix} = \Phi_{(\kappa)} \cdot \mathrm{diag}(\beta),$$

with $\kappa = (\kappa_0, \ldots, \kappa_{n-1})$ and $\beta = (g(\kappa_0)^{-1}, \ldots, g(\kappa_{n-1})^{-1})$ restricts to a Goppa-code, i.e. $\tilde{C} \cap \mathbb{F}_q^n = \mathrm{GO}_q(\kappa, g)$ with $r = \deg g$.   □

Exercises

E.4.6.1   **Exercise** Describe the Alternant-code over $\mathbb{F}_3$ that corresponds to the Generalized Reed–Solomon-code $\mathrm{GRS}(\kappa, g)$ of Exercise 4.5.12.

E.4.6.2   **Exercise** Let C be the binary cyclic code C of length 15 which is generated by $x^2 + x + 1$. Show that C is a BCH-code but not a Goppa-code.

E.4.6.3   **Exercise** Construct a generator matrix of the binary $(31, 16, 7)$-Goppa-code with Goppa-polynomial $g = x^3 + x + 1$.

4.7

4.7 The Structure Theorem

In order to prepare a proof of the Structure Theorem for cyclic codes in general, we have to discuss the *module structure* of these codes. They are left ideals of the group algebra of the cyclic group, and left ideals in a group algebra $\mathbb{F}G$ are subspaces that have the group algebra $\mathbb{F}G$ as ring of operators. This means that cyclic codes are $\mathbb{F}G$-left modules. For this reason we briefly recall the basics from module theory:

  — Assume that R is a ring with identity element 1_R. An *R-left module* is an additively written abelian group $(M, +)$, together with an outer multiplication from the left by the elements of the ring R:

$$R \times M \to M : (r, m) \mapsto rm,$$

which satisfies the following equations:

$$r(m + m') = rm + rm', \; (r + r')m = rm + r'm, \; r(r'm) = (rr')m, \; 1_R m = m,$$

for all $r, r' \in R$ and $m, m' \in M$. We indicate this by writing

$$_R M.$$

Correspondingly, we can define *R-right modules* and develop the theory of *R*-right modules.

— An important example, for our purposes here, is $R := \mathbb{F}^G$ and $M := L$, a left ideal in the group algebra. The outer multiplication is simply the multiplication in the group algebra:

$$\mathbb{F}^G \times L \to L : (a, b) \mapsto ab.$$

It follows immediately from the definitions of addition and multiplication in the group algebra that L is an $\mathbb{F}^G$-left module. Hence, cyclic codes of length n over $\mathbb{F}$ are $\mathbb{F}^G$-left modules in the group algebra $\mathbb{F}^G$ of the cyclic group $G = \langle (0, \ldots, n-1) \rangle$.

— A nonempty subset U of the R-left module M is called a *submodule* of M if U, together with the induced addition and multiplication, is itself an R-left module. In order to prove that a nonempty subset U of M is a submodule of the R-left module M, it is enough to prove that $u_0 + u_1 \in U$ for all $u_0, u_1 \in U$ and that $ru \in U$ for all $r \in R$ and $u \in U$. If U is a submodule of the R-left module M, then the set consisting of all cosets modulo U

$$m + U := \{ m + u \mid u \in U \}$$

forms, together with the two compositions

$$
\begin{aligned}
(m_0 + U) + (m_1 + U) &:= (m_0 + m_1) + U, & m_0, m_1 \in M, \\
r(m + U) &:= rm + U, & r \in R,\ m \in M,
\end{aligned}
$$

an R-left module. It is called the *factor module* and indicated by

$$M/U := \{ m + U \mid m \in M \}.$$

— The *R-linear closure* of a subset T of the R-left module M is the following set of all finite linear combinations

$$\{ r_0 t_0 + \ldots + r_{n-1} t_{n-1} \mid n \in \mathbb{N},\ r_i \in R,\ t_i \in T,\ i \in n \}.$$

The empty linear combination is the zero element 0_M of M. We denote the R-linear closure of T by $\langle T \rangle$. An R-left module M is called *finitely generated* if there exists a finite subset T of M such that $M = \langle T \rangle$. If $|T| = 1$ then $\langle T \rangle$ is a *cyclic* module.

— Correspondingly, a vector space V over $\mathbb{F}$ is called *cyclic* if there exists some $v \in V$ so that $V = \mathbb{F}v = \{ \kappa v \mid \kappa \in \mathbb{F} \}$.

— If $(U_i)_{i \in I}$ is a family of submodules of M, then the sum $\sum_{i \in I} U_i$ of these submodules is the R-linear closure of $T := \bigcup_{i \in I} U_i$,

$$\sum_{i \in I} U_i := \left\langle \bigcup_{i \in I} U_i \right\rangle.$$

This is the smallest submodule of M which contains U_i for all $i \in I$. The finite sum $U_0 + \ldots + U_{s-1}$ of submodules of M consists of all sums $u_0 + \ldots + u_{s-1}$ with $u_i \in U_i$, for $i \in s$. The sum $U_0 + \ldots + U_{s-1}$ of submodules of M is called *direct* if for all $u_i \in U_i$ the implication

$$u_0 + \ldots + u_{s-1} = 0 \Longrightarrow u_0 = \ldots = u_{s-1} = 0$$

holds true. The direct sum will be indicated as

$$U_0 \oplus \ldots \oplus U_{s-1}.$$

— The R-left module M is called *simple* if it has exactly two submodules, namely $\{0_M\}$ and M. A submodule $U \neq M$ of the R-left module M is called *maximal* if for any submodule V of M with $U \subseteq V \subseteq M$ either $U = V$ or $V = M$ holds. Maximal submodules can be characterized in the following way: U is a maximal submodule of M if and only if M/U is a simple module.

— Modules are a generalization of the notion of vector spaces. Whereas a vector space is defined over a field, a module is defined over a ring. The right kind of mappings between modules are the module homomorphisms. Let M and N be two R-left modules. A mapping $f : M \to N$ is an R-module *homomorphism* if f satisfies the conditions

$$f(m_0 + m_1) = f(m_0) + f(m_1), \quad f(rm) = rf(m), \quad r \in R, \, m, m_0, m_1 \in M.$$

If $f : M \to N$ is an R-module homomorphism, then its *kernel*

$$\ker f := \{m \in M \mid f(m) = 0_N\}$$

is a submodule of M. Moreover, if U is a submodule of M, then the mapping

$$\pi : M \to M/U \, : \, m \mapsto m + U$$

is a surjective R-module homomorphism (we also call it an R-module *epimorphism*), with kernel U. It is called the *natural projection* of M to M/U. Two R-left modules M and N are called *isomorphic* if there exists a bijective R-module homomorphism (we also call it an R-module *isomorphism*) $f : M \to N$. When M and N are isomorphic R-modules we also write

$$M \simeq_R N \quad \text{or} \quad M \simeq N.$$

The following is the basic structure theorem for modules. Its specialization to vector spaces is known to the reader from Linear Algebra:

Homomorphism Theorem for Modules *Consider two R-left modules M and N,*   4.7.1
an R-module homomorphism $f: M \to N$, *and* $\pi: M \to M/\ker f$, *the natural pro-*
jection. Then there is a unique R-module homomorphism $\overline{f}: M/\ker f \to N$, *such*
that $f = \overline{f} \circ \pi$, *namely*

$$\overline{f}(m + \ker f) = f(m).$$

Moreover, $\overline{f}$ *is injective (we also call it an R-module* monomorphism*), and* $M/\ker f$
is isomorphic to $f(M)$.

Proof: First we prove that the function $\overline{f}$ defined by $\overline{f}(m + \ker f) := f(m)$ has
the desired properties. It is well-defined since for $m_0, m_1 \in M$ the following
chain of implications is satisfied:

$$m_0 + \ker f = m_1 + \ker f \Rightarrow m_0 - m_1 \in \ker f \Rightarrow f(m_0 - m_1) = 0.$$

Thus $f(m_0) - f(m_1) = 0$ or, equivalently, $f(m_0) = f(m_1)$.

   It is easy to prove that $\overline{f}$ is an R-module homomorphism. In order to show
that it is injective, we verify that its kernel is just $0 + \ker f$. This is evidently
true, since $m + \ker f \in \ker \overline{f}$ is equivalent to $f(m) = 0$, which means that m
belongs to $\ker f$. Finally, from the definition of $\overline{f}$ it is clear that $f = \overline{f} \circ \pi$ is
satisfied.

   If $\overline{\overline{f}}: M/\ker f \to N$ is an R-module homomorphism with $f = \overline{\overline{f}} \circ \pi$, then
necessarily $\overline{f}$ is uniquely determined by $\overline{f}(m + \ker f) = (\overline{f} \circ \pi)(m) = f(m)$.

   If f is surjective, then $\overline{f}$ is also surjective, whence it is an R-module isomorphism. $\qquad\square$

Since, according to 4.2.6, the cyclic codes form a lattice with respect to $\cap$ and
$+$, the following theorem applies to cyclic codes:

First Isomorphism Theorem for Modules   *If* U, V *are submodules of the R-left*   4.7.2
module M, then

$$(U + V)/V \simeq_R U/(U \cap V).$$

Proof: It is easy to prove that the mapping $\mu: U \to (U + V)/V$ defined by
$\mu(u) := u + V$ is a surjective R-module homomorphism. Moreover, its kernel
is equal to $U \cap V$. Thus, by the Homomorphism Theorem for Modules the
mapping

$$\overline{\mu}: U/(U \cap V) \to (U + V)/V \; : \; \overline{\mu}(u + (U + V)) = \mu(u) = u + V$$

is an isomorphism. $\qquad\square$

Since every ideal in a ring R is an R-left module, we directly obtain the

4.7.3 **First Isomorphism Theorem for Rings** *Let I, J be ideals in a ring R. Then*

$$(I + J)/J \simeq_R I/(I \cap J).$$ □

Continuing the examination of modules we state the

4.7.4 **Second Isomorphism Theorem for Modules** *If $U \subseteq V$ are submodules of the R-left module M then*

$$M/V \simeq_R (M/U)/(V/U).$$

Proof: First we show that the mapping

$$f: M/U \to M/V \; : \; m + U \mapsto m + V$$

is an R-module epimorphism. It is well-defined, since for all $m_0, m_1 \in M$ the following chain of implications is true:

$$m_0 + U = m_1 + U \Rightarrow m_0 - m_1 \in U \subseteq V \Rightarrow m_0 + V = m_1 + V.$$

It is evidently surjective, whence $f(M/U) = M/V$. Let

$$\pi: M/U \to (M/U)/\ker f$$

be the natural projection. The Homomorphism Theorem 4.7.1 shows that there exists exactly one R-left module monomorphism

$$\overline{f}: (M/U)/\ker f \to M/V$$

such that $f = \overline{f} \circ \pi$, where f is the epimorphism $M/U \to M/V$ introduced above. Since f is surjective, $\overline{f}$ is even an isomorphism. Finally, to confirm the assertion, we have to show that $\ker f = V/U$. For $m \in M$ the element $m + U$ belongs to $\ker f$ if and only if $m + V = 0_{M/V}$ which is equivalent to $m \in V$. Thus $\overline{f}$ is an R-left module isomorphism between $(M/U)/(V/U)$ and M/V.   □

We directly obtain the

4.7.5 **Second Isomorphism Theorem for Rings** *Let I, J be ideals in a ring R with $I \subseteq J$. Then J/I is an ideal in R/I for which the following is true:*

$$(R/I)/(J/I) \simeq_R R/J.$$ □

Modular Law *If U, V, W are submodules of the R-left module M, and if $V \subseteq W$*   **4.7.6**
then

$$(U + V) \cap W = (U \cap W) + (V \cap W).$$

Proof: Assume that $u + v = w$, for $u \in U$, $v \in V$, $w \in W$. Since $V \subseteq W$ we
have $u = w - v \in W$, whence $u + v \in (U \cap W) + V = (U \cap W) + (V \cap W)$.

Conversely, since $(U \cap W) \subseteq U$ and $(V \cap W) \subseteq V$, the sum $(U \cap W) + (V \cap W)$ is contained in $U + V$ and in $W + W \subseteq W$.   □

This implies for group algebra codes, and in particular for cyclic ones:

Corollary *For cyclic codes $C_0, C_1, C_2 \subseteq \mathbb{F}^G$, $G = \langle (0, \dots, n-1) \rangle$, the following is*   **4.7.7**
true:

$$(C_0 + C_1)/C_1 \simeq_{\mathbb{F}G} C_0/(C_0 \cap C_1).$$

Moreover, if $C_0 \subseteq C_1 \subseteq C_2$, then

$$C_2/C_1 \simeq_{\mathbb{F}G} (C_2/C_0)/(C_1/C_0)$$

and, for any C_0, C_1, C_2 such that $C_1 \subseteq C_2$, we have

$$(C_0 + C_1) \cap C_2 = (C_0 \cap C_2) + (C_1 \cap C_2).$$ □

The next problem we want to answer is the question how modules, and in particular cyclic codes, can be composed of submodules and factor modules. For this purpose we introduce the following notion. A sequence $\mathcal{M} = (M_i)_{0 \leq i \leq n}$ of submodules of an R-left module M is called a *normal series* of M, if

$$M = M_0 \supseteq M_1 \supseteq \dots \supseteq M_n = \{0_M\}.$$

The factor modules M_i/M_{i+1} for $i \in n$ are the *factors* of $\mathcal{M}$. Normal series of binary cyclic codes are, for example (cf. 4.2.7),

$$\mathbb{F}_2^7 \supset P_7 \supset S_3 \supset \{0\} \quad \text{and} \quad \mathbb{F}_2^7 \supset H_3' \supset W_7 \supset \{0\}.$$

A normal series $\mathcal{M}' = (M_j')_{0 \leq j \leq m}$ is a *refinement* of the normal series $\mathcal{M} = (M_i)_{0 \leq i \leq n}$ if $\mathcal{M}$ is a subseries (or subsequence) of $\mathcal{M}'$. Two normal series $\mathcal{M} = (M_i)_{0 \leq i \leq n}$ and $\mathcal{M}' = (M_j')_{0 \leq j \leq m}$ of M are called *isomorphic* if $n = m$ and if there exists a permutation $\pi \in S_n$ such that

$$M_i/M_{i+1} \simeq_R M_{\pi(i)}'/M_{\pi(i)+1}', \quad i \in n.$$

(Recall that S_n consists of all permutations of the set $n := \{0, 1 \dots, n-1\}$.)

Theorem (Schreier) *Any two normal series of an R-left module M have isomorphic*   **4.7.8**
refinements.

Proof: Let $\mathcal{M} = (M_i)_{0 \leq i \leq n}$ and $\mathcal{M}' = (M'_j)_{0 \leq j \leq m}$ be two normal series of M. If $n = 1$ or $m = 1$ then the assertion is trivial. Hence, we assume that both $n > 1$ and $m > 1$. We introduce further submodules of M by

$$M_{i,j} := M_i + (M_{i-1} \cap M'_j), \qquad 1 \leq i \leq n, \, 0 \leq j \leq m,$$
$$M'_{j,i} := M'_j + (M'_{j-1} \cap M_i), \qquad 1 \leq j \leq m, \, 0 \leq i \leq n.$$

Then

$$M_{i-1} = M_{i,0} \supseteq M_{i,1} \supseteq \ldots \supseteq M_{i,m} = M_i, \qquad 1 \leq i \leq n,$$
$$M'_{j-1} = M'_{j,0} \supseteq M'_{j,1} \supseteq \ldots \supseteq M'_{j,n} = M'_j, \qquad 1 \leq j \leq m.$$

From these submodules we form the two normal series

$$M = M_{1,0} \supseteq \ldots \supseteq M_{1,m} = M_{2,0} \supseteq M_{2,1} \supseteq \ldots \supseteq M_{n,m} = \{0_M\}$$
$$M = M'_{1,0} \supseteq \ldots \supseteq M'_{1,n} = M'_{2,0} \supseteq M'_{2,1} \supseteq \ldots \supseteq M'_{m,n} = \{0_M\}$$

which have the same length. In order to show that the two series are isomorphic, we verify that $M_{i,j-1}/M_{i,j} \simeq_R M'_{j,i-1}/M'_{j,i}$ for $1 \leq i \leq n$ and $1 \leq j \leq m$. The following is derived from the definition of $M_{i,j}$, the fact that $M_{i-1} \cap M'_j \subseteq M_{i-1} \cap M'_{j-1}$, by 4.7.2, and by 4.7.6: (In order to make the notation more clear, we sometimes use fractions to indicate factor modules.)

$$
\begin{aligned}
M_{i,j-1}/M_{i,j} &= \frac{M_i + (M_{i-1} \cap M'_{j-1})}{M_i + (M_{i-1} \cap M'_j)} \\[1em]
&= \frac{M_i + (M_{i-1} \cap M'_j) + (M_{i-1} \cap M'_{j-1})}{M_i + (M_{i-1} \cap M'_j)} \\[1em]
&\simeq_R \frac{M_{i-1} \cap M'_{j-1}}{[M_i + (M_{i-1} \cap M'_j)] \cap (M_{i-1} \cap M'_{j-1})} \\[1em]
&= \frac{M_{i-1} \cap M'_{j-1}}{(M_i \cap M'_{j-1}) + (M_{i-1} \cap M'_j)}
\end{aligned}
$$

In the same way we show that

$$M'_{j,i-1}/M'_{j,i} \simeq_R \frac{M'_{j-1} \cap M_{i-1}}{(M'_j \cap M_{i-1}) + (M'_{j-1} \cap M_i)},$$

which completes the proof. □

A normal series $(M_i)_{0 \leq i \leq n}$ of M is said to be a *composition series* of M if M_{i+1} is maximal in M_i for $i \in n$. The two normal series mentioned above,

$$\mathbb{F}_2^7 \supset P_7 \supset S_3 \supset \{0\} \quad \text{and} \quad \mathbb{F}_2^7 \supset H'_3 \supset W_7 \supset \{0\},$$

are composition series. This follows from their definition in 4.2.7. For example,

$$P_7 = I(x+1)/I(x^7 - 1) \supset I((x^3 + x + 1)(x+1))/I(x^7 - 1) = S_3$$

and so there is no ideal J such that $P_7 \supset J \supset S_3$, since $x^3 + x + 1$ is irreducible. Moreover, by Schreier's Theorem, these two composition series are isomorphic. In fact it follows from the Second Isomorphism Theorem 4.7.4 that

$$
\begin{aligned}
\mathbb{F}_2^7/P_7 &= \left(\mathrm{I}(1)/\mathrm{I}(x^7-1)\right)/\left(\mathrm{I}(x+1)/\mathrm{I}(x^7-1)\right) \simeq \mathrm{I}(1)/\mathrm{I}(x+1),\\
P_7/S_3 &= \left(\mathrm{I}(x+1)/\mathrm{I}(x^7-1)\right)/\left(\mathrm{I}((x^3+x+1)(x+1))/\mathrm{I}(x^7-1)\right)\\
&\simeq \mathrm{I}(x+1)/\mathrm{I}((x^3+x+1)(x+1)),\\
S_3/\{0\} &= \left(\mathrm{I}((x^3+x+1)(x+1))/\mathrm{I}(x^7-1)\right)/\left(\mathrm{I}(x^7-1)/\mathrm{I}(x^7-1)\right)\\
&\simeq \mathrm{I}((x^3+x+1)(x+1))/\mathrm{I}(x^7-1),\\
\mathbb{F}_2^7/H_3' &= \left(\mathrm{I}(1)/\mathrm{I}(x^7-1)\right)/\left(\mathrm{I}(x^3+x^2+1)/\mathrm{I}(x^7-1)\right)\\
&\simeq \mathrm{I}(1)/\mathrm{I}(x^3+x^2+1),\\
H_3'/W_7 &= \left(\mathrm{I}(x^3+x^2+1)/\mathrm{I}(x^7-1)\right)/\left(\mathrm{I}((x^3+x^2+1)(x^3+x+1))/\mathrm{I}(x^7-1)\right)\\
&\simeq \mathrm{I}(x^3+x^2+1)/\mathrm{I}((x^3+x^2+1)(x^3+x+1)),\\
W_7/\{0\} &= \left(\mathrm{I}((x^3+x^2+1)(x^3+x+1))/\mathrm{I}(x^7-1)\right)/\left(\mathrm{I}(x^7-1)/\mathrm{I}(x^7-1)\right)\\
&\simeq \mathrm{I}((x^3+x^2+1)(x^3+x+1))/\mathrm{I}(x^7-1).
\end{aligned}
$$

Recall from Exercises 3.5.10 and 4.7.11 that for relatively prime polynomials $f, g \in \mathbb{F}[x]$ we have

$$\mathrm{I}(f) + \mathrm{I}(g) = \mathrm{I}(1) \quad \text{and} \quad \mathrm{I}(fg) = \mathrm{I}(f) \cap \mathrm{I}(g).$$

In this case, the First Isomorphism Theorem 4.7.3 implies that

$$(\mathrm{I}(f)/\mathrm{I}(fg)) \simeq_{\mathbb{F}[x]} (\mathrm{I}(1)/\mathrm{I}(g)).$$

In particular, the two composition series

$$\mathbb{F}_2^7/P_7 \simeq W_7/\{0\}, \quad P_7/S_3 \simeq H_3'/W_7, \quad S_3/\{0\} \simeq \mathbb{F}_2^7/H_3'.$$

are isomorphic, in accordance with Schreier's Theorem.

If $(M_i)_{0 \le i \le n}$ is a composition series of the R-left module M, then we say that M is a module of *finite length* $n = \ell_R(M)$. In the sequel, all modules will have finite length.

Lemma *If M is an R-left module of finite length and U a submodule of M, then both U and M/U are of finite length and*

4.7.9

$$\ell_R(M) = \ell_R(U) + \ell_R(M/U).$$

Proof: Let $\mathcal{M}$ be a composition series of M. From 4.7.8 we get that $\mathcal{M}$ and the normal series $M \supseteq U \supseteq \{0_M\}$ have isomorphic refinements. By deleting repeated submodules, we obtain two composition series of M, at least one of which contains the submodule U. Assume that this series is given by

$$M = M_0 \supset M_1 \supset \ldots \supset M_k = U \supset M_{k+1} \supset \ldots \supset M_n = \{0_M\}.$$

Consequently, U has the composition series

$$U = M_k \supset M_{k+1} \supset \ldots \supset M_n = \{0_M\}.$$

From 4.7.4 it follows that

$$(M_i/U)/(M_{i+1}/U) \simeq_R M_i/M_{i+1}, \qquad i \in k,$$

whence M_{i+1}/U is maximal in M_i/U. Thus,

$$M/U = M_0/U \supset M_1/U \supset \ldots \supset M_k/U = U/U = \{0_{M/U}\}$$

is a composition series of M/U. Summarizing, we have proved that U is of length $n - k$ and M/U is of length k. $\qquad\square$

4.7.10   **Lemma**   *If $U_0, \ldots, U_{k-1}$ are submodules of an R-left module M, which is of finite length, then also $\sum_{i \in k} U_i$ is of finite length and*

$$\ell_R\left(\sum_{i \in k} U_i\right) \leq \sum_{i \in k} \ell_R(U_i).$$

Proof: Since U_j and $\sum_{i \in k} U_i$ are submodules of M, they are of finite length. For $j \in k + 1$ let $V_j = \sum_{i \in j} U_i$, then

$$\sum_{i \in k} U_i = V_k \supseteq V_{k-1} \supseteq \ldots \supseteq V_0 = \{0_M\}$$

is a normal series of $\sum_{i \in k} U_i$. Moreover, for $i \in k$ let

$$U_i = U_{i,0} \supset U_{i,1} \supset \ldots \supset U_{i,n_i} = \{0_M\}$$

be a composition series of U_i of length $\ell_R(U_i) = n_i$. In the normal series of $\sum_{i \in k} U_i$ we replace the consecutive submodules

$$V_i \supseteq V_{i-1}$$

for $1 \leq i \leq k$ by

$$V_{i-1} + U_{i,0} \supseteq V_{i-1} + U_{i,1} \supseteq \ldots \supseteq V_{i-1} + U_{i,n_i}.$$

Furthermore, we identify $V_{i-1} + U_{i,0}$ with $V_i + U_{i+1,n_{i+1}}$, for $1 \leq i \leq k - 1$. This way, we obtain a refinement of a composition series of $\sum_{i \in k} U_i$ which consists of $n_0 + \ldots + n_{k-1}$ modules. The factors of this refined series are all either simple or trivial since for $1 \le i \le k$ and $j \in n_i$ we have

$$\frac{V_{i-1} + U_{i,j}}{V_{i-1} + U_{i,j+1}} = \frac{V_{i-1} + U_{i,j+1} + U_{i,j}}{V_{i-1} + U_{i,j+1}}$$

$$\cong_R \frac{U_{i,j}}{(V_{i-1} + U_{i,j+1}) \cap U_{i,j}} = \begin{cases} U_{i,j}/U_{i,j+1} \\ U_{i,j}/U_{i,j} \end{cases}$$

depending on whether $(V_{i-1} + U_{i,j+1}) \cap U_{i,j}$ is equal to $U_{i,j+1}$ or $U_{i,j}$. Deleting repeated modules in the series, we obtain a composition series of $\sum_{i \in k} U_i$ whose length is at most $n_0 + \ldots + n_{k-1}$. $\square$

The rings R which we consider here are in many cases polynomial rings over fields, so we can make additional assumptions: We suppose that R is a commutative ring with 1 which contains no zero divisors, and has at least two elements, so that $0 \ne 1$. This means that the ring R is an *integral domain*. Moreover, we assume that each ideal can be generated by a single element, i.e. the ring is a *principal ideal domain*; an example is the ring of integers $(\mathbb{Z}, +, \cdot)$, another one is $\mathbb{F}[x]$, as we know.

Recall that $a \in R$ is a *divisor* of $b \in R$ if there exists $c \in R$ such that $ac = b$, this is indicated by $a \mid b$. An element $u \in R \setminus \{0\}$ which is a divisor of 1 is called a *unit*. Two elements $a, b \in R$ are called *relatively prime* if the common divisors of a and b are all units. The *associates* of $a \in R$ are those $b \in R$ which are of the form $b = au$ where u is a unit in R. An element $p \in R \setminus \{0\}$ which is not a unit is called a *prime element* of R if, for any $a, b \in R$, $p \mid ab$ implies $p \mid a$ or $p \mid b$. An element $p \in R \setminus \{0\}$ which is not a unit is called *irreducible* if every divisor of p is either a unit or an associate of p. Any prime element is irreducible. In a principal ideal domain, the converse is true, i.e. p is prime if and only if p is irreducible.

Theorem *If ζ is a prime element in a principal ideal domain R, and n denotes a positive integer, then the factor ring $\overline{R} := R/\mathrm{I}(\zeta^n)$ contains exactly $n + 1$ ideals, and these ideals form the chain*   **4.7.11**

$$\overline{R} \supset \mathrm{I}(\overline{\zeta}) \supset \mathrm{I}(\overline{\zeta^2}) \supset \ldots \supset \mathrm{I}(\overline{\zeta^n}) = \{\overline{0}\},$$

where $\overline{r} := r + \mathrm{I}(\zeta^n)$ for all $r \in R$.

Here $\mathrm{I}(r)$ denotes the ideal generated by r in R, $\mathrm{I}(r) = Rr$, whereas $\mathrm{I}(\overline{r})$ is the ideal $\overline{R}\overline{r}$ in $\overline{R}$. Moreover,

$$\mathrm{I}(\overline{\zeta^j}) = \overline{R}\overline{\zeta^j} = R\zeta^j/\mathrm{I}(\zeta^n) = \mathrm{I}(\zeta^j)/\mathrm{I}(\zeta^n).$$

Proof: Let $\pi \colon R \to \overline{R} \colon r \mapsto \overline{r} = r + \mathrm{I}(\zeta^n)$ be the canonical projection, and let J be an ideal in $\overline{R}$. Then $\pi^{-1}(J)$ is an ideal in R (see Exercise 4.2.5), whence it is of the form $I(q)$ for some $q \in R$. Since π is surjective, we have $J = \pi(\pi^{-1}(J)) = \pi(I(q)) = I(\overline{q})$. Choose $m \in \mathbb{N}^*$ maximal with the property $\zeta^m \mid q$. Hence, there exists $r \in R$ such that $q = \zeta^m r$ and ζ is not a divisor of r. Applying Exercise 4.7.10, we deduce that ζ and r are relatively prime. Thus ζ^n and r are also relatively prime, and from Exercise 4.7.11 we deduce that there exist $x, y \in R$ such that $rx + \zeta^n y = 1$. Consequently, $1 - rx \in I(\zeta^n)$ and $\overline{1} = \overline{rx}$. This shows that $\overline{r}$ is a unit in $\overline{R}$ and, finally, we get that $I(\overline{q}) = I(\overline{\zeta^m r}) = I(\overline{\zeta^m})$. Since J was an ideal in $\overline{R}$ chosen arbitrarily, each ideal in $\overline{R}$ is of the form $I(\overline{\zeta^\ell})$ for some $\ell \in n + 1$. Obviously these ideals form a chain as indicated above.   □

From Exercise 4.7.2 we know that any ideal of a ring R with 1 is both an R-left and R-right module. In particular, the additive group $(R, +)$ itself can be considered as both an R-left and an R-right module.

4.7.12   **Theorem** *If the R-left module R and the R-right module R both have exactly one composition series (of finite length) then every finitely generated R-left module M is a direct sum of cyclic submodules.*

Proof: For $m \in M$ the cyclic submodule $\langle \{m\} \rangle = Rm$ of M is isomorphic to a factor module of the R-left module R. In order to prove this, let

$$\psi : R \to Rm \; : \; r \mapsto rm$$

be the natural R-module epimorphism, then, according to 4.7.1, Rm is isomorphic to $R/\ker \psi$. Since R is of finite length, so is Rm.

Now assume that $M = \langle \{m_0, \ldots, m_{k-1}\} \rangle$, then $M = Rm_0 + \ldots + Rm_{k-1}$. According to 4.7.10, it is also of finite length and

$$\ell_R(M) \le \sum_{i \in k} \ell_R(Rm_i).$$

In addition, if the elements $m_0, \ldots, m_{k-1}$ of M are chosen in such a way that $\sum_{i \in k} \ell_R(Rm_i)$ is minimal, then M is the direct sum $Rm_0 \oplus \ldots \oplus Rm_{k-1}$. We only have to prove that this sum is direct. Assume on the contrary that

$$0 = \sum_{i \in k} r_i m_i$$

and that the set $I := \{i \in k \mid r_i m_i \ne 0\}$ is not empty. Since the R-right module R has exactly one composition series, among the R-right modules $r_i R, i \in I$, there is a largest one. Without loss of generality we take this to be $r_0 R$. Consequently, there exist $q_i \in R$ for $i \in I$ such that $r_i = r_0 q_i$. Let $q_i = 0$, for $i \notin I$. Then

$$0 = \sum_{i \in k} r_i m_i = \sum_{i \in k} r_0 q_i m_i = r_0 \left(m_0 + \sum_{i=1}^{k-1} q_i m_i \right) =: r_0 m_0'.$$

Since

$$m_0 = m_0' - \sum_{i=1}^{k-1} q_i m_i \in \langle \{m_0', m_1, \ldots, m_{k-1}\} \rangle,$$

the set $\{m_0', m_1, \ldots, m_{k-1}\}$ also generates M. Now we prove that $\ell_R(Rm_0') < \ell_R(Rm_0)$. This yields a contradiction to the fact that the generators $m_0, \ldots, m_{k-1}$ were chosen such that $\sum_{i \in k} \ell_R(Rm_i)$ is minimal. Hence, our assumption $I \neq \emptyset$ does not hold and, consequently, M is the direct sum of the cyclic modules Rm_i for $i \in k$.

Let $\psi : R \to Rm_0 : r \mapsto rm_0$ and $\psi' : R \to Rm_0' : r \mapsto rm_0'$ be the natural R-module epimorphisms. Since $\ker \psi$ and $\ker \psi'$ are R-submodules of R and the R-left module R has exactly one composition series, we have $\ker \psi \subseteq \ker \psi'$ or $\ker \psi' \subseteq \ker \psi$. From the fact that $r_0 m_0 \neq 0$ and $r_0 m_0' = 0$ we deduce that $r_0 \notin \ker \psi$ and $r_0 \in \ker \psi'$, whence $\ker \psi \subset \ker \psi'$. Consequently, $\ell_R(\ker \psi) < \ell_R(\ker \psi')$. According to 4.7.1, Rm_0 and Rm_0' are isomorphic to $R/\ker \psi$ and $R/\ker \psi'$, respectively, and for that reason, together with 4.7.9, we obtain

$$\ell_R(Rm_0') = \ell_R(R) - \ell_R(\ker \psi') < \ell_R(R) - \ell_R(\ker \psi) = \ell_R(Rm_0). \quad \square$$

For instance the ring $\overline{R}$ from Theorem 4.7.11 satisfies the assumptions of the last theorem.

Now we are going to describe the structure of the set of cyclic codes in $\mathbb{F}_q^n$.

We again assume that p is the characteristic of the field and that $n = p^s r$, where $p \nmid r$. Furthermore, we suppose that $x^r - 1 = \prod_{i \in I} f_i$ is the decomposition of $x^r - 1$ into irreducible polynomials over $\mathbb{F}_q$. According to 4.2.9, we have

$$x^n - 1 = \prod_{i \in I} (f_i)^{p^s} = \prod_{i \in I} f_i(x^{p^s}).$$

Since the irreducible factors f_i of $x^r - 1$ are pairwise different, they and therefore also the ideals which are generated by their powers $(f_i)^{p^s} = f_i(x^{p^s})$ are relatively prime. Thus, the Chinese Remainder Theorem implies

---

Corollary *The residue class ring $\mathbb{F}_q[x]/\mathrm{I}(x^n - 1)$ has the decomposition*   **4.7.13**

$$\mathbb{F}_q[x]/\mathrm{I}(x^n - 1) \simeq \underset{i \in I}{\times} \mathbb{F}_q[x]/\mathrm{I}(f_i^{p^s}). \qquad \square$$

The structure of the summands in this decomposition is shown in the next

---

Theorem *If f denotes an irreducible factor of $x^r - 1$, then:*   **4.7.14**

— *The residue class ring $\mathbb{F}_q[x]/\mathrm{I}(f)$ is a field and also an $\mathbb{F}_q$-vector space of dimension $\deg f$.*

— *The residue class ring* $\mathbb{F}_q[x]/\mathrm{I}(f^{p^s})$ *possesses a* unique *composition series*

$$\mathbb{F}_q[x]/\mathrm{I}(f^{p^s}) \supset \mathrm{I}(f)/\mathrm{I}(f^{p^s}) \supset \ldots \supset \mathrm{I}(f^{p^s-1})/\mathrm{I}(f^{p^s}) \supset 0.$$

All factors of this series are simple submodules and also minimal ideals of the ring $\mathbb{F}_q[x]/\mathrm{I}(f^{p^s})$.

Proof: The first assertion was already proved in 3.1.6 and 3.1.7. The second assertion was shown in 4.7.11. Here we want to give a second proof: First we note that because f is irreducible, the divisors of f^{p^s} are of the form f^j, $0 \le j \le p^s$. The introductory remark in Section 4.2 (or Exercise 4.2.6) implies that every ideal of $\mathbb{F}_q[x]/\mathrm{I}(f^{p^s})$ has the form $\mathrm{I}(f^j)/\mathrm{I}(f^{p^s})$, $0 \le j \le p^s$. For each of the factors of the given series we deduce from the Second Isomorphism Theorem for Rings 4.7.5 that

$$\left(\mathrm{I}(f^j)/\mathrm{I}(f^{p^s})\right)/\left(\mathrm{I}(f^{j+1})/\mathrm{I}(f^{p^s})\right) \simeq \mathrm{I}(f^j)/\mathrm{I}(f^{j+1}), \qquad j \in p^s.$$

The mapping

$$\mathbb{F}_q[x] \to \mathrm{I}(f^j)/\mathrm{I}(f^{j+1}) \; : \; g \mapsto gf^j + \mathrm{I}(f^{j+1}), \qquad j \in p^s,$$

is a ring epimorphism with kernel $\mathrm{I}(f)$, and hence, by the Homomorphism Theorem,

$$\mathbb{F}_q[x]/\mathrm{I}(f) \simeq \mathrm{I}(f^j)/\mathrm{I}(f^{j+1}), \qquad j \in p^s.$$

Thus each factor of the series is a field, and hence a minimal ideal of a residue class ring $\mathbb{F}_q[x]/\mathrm{I}(f^{p^s})$.   □

In every ring the *minimal ideals* (minimal with respect to set theoretic inclusion) and the *maximal* ideals are of particular importance as well as the *indecomposable* ones, i.e. the ideals that cannot be expressed as direct sums of different nonzero ideals. A cyclic code (of length n) is called *indecomposable* if it is indecomposable as an ideal in $\mathbb{F}_q[x]/\mathrm{I}(x^n - 1)$.

According to 4.2.6, the lattice of divisors of $x^n - 1$ over $\mathbb{F}_q$ is anti-isomorphic to the lattice of cyclic codes of length n. This fact permits the characterization of minimal and maximal cyclic codes:

4.7.15   **Corollary** *A cyclic code of length* $n = p^s \cdot r$, *with* $p \nmid r$, *over a finite field of characteristic* p *with generator polynomial* g *is minimal (resp. maximal) if* g *is a maximal (resp. minimal) divisor of*

$$x^n - 1 = \prod_{i \in I}(f_i)^{p^s},$$

which means that $g = (x^n - 1)/f_i$ *(resp.* $g = f_i$*), for some* i. *Hence, if* l *is the number of distinct irreducible factors of* $x^r - 1$ *then there are exactly* l *maximal (resp. minimal) cyclic codes of length* n.   □

Example (Continuation of 4.2.7) The minimal binary cyclic codes of length 7   **4.7.16**
are the repetition code W_7 and the two simplex-codes S_3 and S_3'. The parity
check code P_7 and the two Hamming-codes H_3 and H_3' are the maximal binary
cyclic codes of length 7. This agrees with the fact that $x^7 - 1 \in \mathbb{F}_2[x]$ has 3
different irreducible factors. $\diamond$

Summarizing, we obtain the announced structural description of cyclic codes:

The Structure Theorem for Cyclic Codes *For $0 \leq j \leq p^s$ and $i \in l$ let*   **4.7.17**

$$C_{i,j} := \mathrm{I}(g_{i,j})/\mathrm{I}(x^n - 1)$$

denote the cyclic codes of length n over $\mathbb{F}_q$ with generator polynomials

$$g_{i,j} := f_0^{p^s} \cdots f_{i-1}^{p^s} \cdot f_i^j \cdot f_{i+1}^{p^s} \cdots f_{l-1}^{p^s}.$$

These codes have the following structure:

1. *The decomposition*
$$\mathbb{F}_q[x]/\mathrm{I}(x^n - 1) = \bigoplus_{i \in l} C_{i,0}$$
   is a decomposition of $\mathbb{F}_q[x]/\mathrm{I}(x^n - 1)$ into indecomposable ideals. Each $C_{i,0}$ is called a block *of $\mathbb{F}_q[x]/\mathrm{I}(x^n - 1)$.*

2. *Each of these blocks possesses the unique composition series*
$$C_{i,0} \supset C_{i,1} \supset \ldots \supset C_{i,p^s-1} \supset C_{i,p^s} = 0, \qquad i \in l.$$

3. *Every cyclic code C of length n over $\mathbb{F}_q$ can be decomposed as*
$$C = \bigoplus_{i \in l} (C_{i,0} \cap C).$$

4. *Every indecomposable cyclic code of length n over $\mathbb{F}_q$ is of the form $C_{i,j}$, where $j \in p^s$ and $i \in l$. In particular, each minimal cyclic code of length n over $\mathbb{F}_q$ is of the form C_{i,p^s-1}.*

5. *Every cyclic (n,k)-code C over $\mathbb{F}_q$ can be decomposed as*
$$C = \bigoplus_{i \in \mathcal{I}} C_{i,j_i},$$
   into indecomposable cyclic codes C_{i,j_i}, where $\mathcal{I} \subseteq l$ and $j_i \in p^s$ for each $i \in \mathcal{I}$, in which case it has the $\mathbb{F}_q$-dimension
$$k = \sum_{i \in \mathcal{I}} (p^s - j_i) \cdot \deg f_i.$$

Proof: The Chinese Remainder Theorem provides the ring isomorphism

$$\mathbb{F}_q[x]/\mathrm{I}(x^n - 1) \to \underset{i \in l}{\times}\, \mathbb{F}_q[x]/\mathrm{I}(f_i^{p^s}),$$

which is defined by

$$g + \mathrm{I}(x^n - 1) \mapsto (g + \mathrm{I}(f_0^{p^s}), \ldots, g + \mathrm{I}(f_{l-1}^{p^s})).$$

The inverse image of $(0, \ldots, 0, \mathrm{I}(f_i^j)/\mathrm{I}(f_i^{p^s}), 0, \ldots, 0)$ under this mapping is the cyclic code $C_{i,j}$. In particular, the cyclic code $C_{i,0}$ is the inverse image of

$$(0, \ldots, 0, \mathbb{F}_q[x]/\mathrm{I}(f_i^{p^s}), 0, \ldots, 0), \qquad i \in l.$$

By 4.7.14, the $\mathbb{F}_q$-algebra $\mathbb{F}_q[x]/\mathrm{I}(f_i^{p^s})$ has the unique composition series

$$\mathbb{F}_q[x]/\mathrm{I}(f_i^{p^s}) \supset \mathrm{I}(f_i)/\mathrm{I}(f_i^{p^s}) \supset \ldots \supset \mathrm{I}(f_i^{p^s-1})/\mathrm{I}(f_i^{p^s}) \supset 0, \qquad i \in l.$$

Since a decomposable ideal has more than one composition series, it follows that $\mathbb{F}_q[x]/\mathrm{I}(f_i^{p^s})$ and hence $C_{i,0}$ is indecomposable. The inverse image of this composition series furnishes the composition series of $C_{i,0}$. The first assertion follows from 4.7.18, which states that the ring $\mathbb{F}_q[x]/\mathrm{I}(x^n - 1)$ is isomorphic to the sum of the blocks $C_{i,0}$.

Now we consider an arbitrary cyclic code C of length n over $\mathbb{F}_q$ with generator polynomial g. According to 4.2.9, g can be written as a product of irreducible polynomials in the form

$$g = \prod_{i \in l} f_i^{j_i}$$

with uniquely determined exponents j_i with $0 \le j_i \le p^s$ for $i \in l$. By 4.2.6, the intersection $C_{i,0} \cap C$ of cyclic codes is again cyclic with generator polynomial

$$\mathrm{lcm}(g_{i,0}, g) = f_i^{j_i} g_{i,0} = g_{i,j_i},$$

and so $C \cap C_{i,0} = C_{i,j_i}$. Also by 4.2.6, the sum

$$\hat{C} = \sum_{i \in l} C_{i,j_i}$$

is cyclic, with generator polynomial

$$\hat{g} = \gcd\{f_i^{j_i} g_{i,0} \mid i \in l\} = \prod_{i \in l} f_i^{j_i} = g.$$

This shows that $\hat{C} = C$. Moreover, the sum $\sum_{i \in l} C_{i,j_i}$ is direct, since the sum of the $C_{i,0}$ is direct and $C_{i,j}$ is contained in $C_{i,0}$. This proves the third assertion.

For $j_i = p^s$ we have $C_{i,j_i} = 0$, whence C can be written as a direct sum of indecomposable cyclic codes C_{i,j_i} whose dimension is $(p^s - j_i) \cdot \deg f_i$ by 4.2.5. This proves the assertion on the dimensions.

Finally, let C be an indecomposable cyclic code of length n over $\mathbb{F}_q$. Because of the third assertion, C must be contained in a block $C_{i,0}$, and so $C = C_{i,j}$ for some j, according to the second assertion. Conversely, we already know that each code of the form $C_{i,j}$ is indecomposable. Particular indecomposable codes are the minimal cyclic codes. They are of the form C_{i,p^s-1}. □

Because of their importance, let us summarize the assertions of this theorem once again:

- $\mathbb{F}_q[x]/I(x^n - 1)$ possesses a decomposition into a direct sum

$$\mathbb{F}_q[x]/I(x^n - 1) = \bigoplus_{i \in I} C_{i,0}$$

of indecomposable cyclic codes $C_{i,0}$.

- Each block $C_{i,0}$ has a unique decomposition series.

- Every indecomposable cyclic code of length n over $\mathbb{F}_q$ is, up to isomorphism, contained in a block $C_{i,0}$. (The isomorphism is due to 4.7.13.)

- Up to isomorphism, every cyclic code C of length n over $\mathbb{F}_q$ has a direct sum decomposition

$$C = \bigoplus_{i \in \mathcal{I}} C_i,$$

where $\mathcal{I} \subseteq I$ is an index set and C_i is a cyclic code contained in the block $C_{i,0}$.

We rephrase the Structure Theorem for codes of p-regular length:

Corollary *Assume that $p \nmid n$.*   4.7.19

- *Every indecomposable cyclic code of length n over $\mathbb{F}_q$ is minimal and isomorphic to a block $C_{i,0}$, $i \in I$.*

- *Every block $C_{i,0}$ is isomorphic to a field.*

- *A cyclic code of length n over $\mathbb{F}_q$ is a direct sum of minimal cyclic codes.*

- *Up to isomorphism, every cyclic code of length n over $\mathbb{F}_q$ is a direct summand of $\mathbb{F}_q[x]/I(x^n - 1)$.*

- *Let $q^m \equiv 1$ mod n. Then every minimal cyclic code of length n over $\mathbb{F}_{q^m}$ is one-dimensional.* □

The proof of this is recommended as Exercise 4.7.18. An R-left module M is said to be *semi-simple* if it can be decomposed as a direct sum of simple R-modules. We say that the ring R is *semi-simple* if it is semi-simple as a left module over itself, that is

$$R \simeq_R \bigoplus_{i \in I} L_i$$

where L_i is a simple submodule of R, i.e. a minimal left ideal of R.

The last corollary means that the residue class ring $R = \mathbb{F}_q[x]/I(x^n - 1)$ is semi-simple for $p \nmid n$. That is, every cyclic code of length n over $\mathbb{F}_q$ is a direct summand of R. Therefore, these cyclic codes are sometimes also called *semi-simple*. More generally, for $p \nmid n$ each R-left module is a direct summand of R (as an R-left module). By contrast, the residue class ring $\mathbb{F}_q[x]/I(x^n - 1)$, with $p \mid n$, is not semi-simple, by 4.7.17.

4.7.20   **Example** In the case $q = p$ and $n = p - 1$ we have the decomposition

$$x^n - 1 = \prod_{i=1}^{p-1} (x - i)$$

over $\mathbb{F}_p$. Hence, by 4.7.17, the residue class ring $\mathbb{F}_p[x]/I(x^n - 1)$ is isomorphic to the direct sum

$$\mathbb{F}_p[x]/I(x^{p-1} - 1) = \bigoplus_{i \in p-1} I(g_i)/I(x^{p-1} - 1)$$

of cyclic codes with generator polynomials

$$g_i(x) = \frac{x^{p-1} - 1}{x - i - 1}.$$

All these direct summands are Reed–Solomon-codes of dimension 1 and minimum distance $n = p - 1$. ◇

4.7.21   **Examples (Continuation of 4.2.7 and 4.7.16)**

1. Using the notation from 4.2.7 for the binary cyclic codes of length 7, we obtain:

$$\begin{aligned}
\mathbb{F}_2[x]/I(x^7 - 1) &= W_7 \oplus S_3 \oplus S_3' \\
H_3 &= W_7 \oplus S_3 \\
H_3' &= W_7 \oplus S_3' \\
P_7 &= S_3 \oplus S_3'.
\end{aligned}$$

2. Now we consider the binary cyclic codes of length $n = 28 = 7 \cdot 4$. In this case we get

$$\mathbb{F}_2[x]/I(x^{28} - 1) = C_{0,0} \oplus C_{1,0} \oplus C_{2,0}$$

where

$$\begin{aligned}
C_{0,0} &= I(f_1(x^4)f_2(x^4))/I(x^{28} - 1) \\
&= I((x^{12} + x^4 + 1)(x^{12} + x^8 + 1))/I(x^{28} - 1),
\end{aligned}$$

$$
\begin{aligned}
C_{1,0} &= I(f_0(x^4)f_2(x^4))/I(x^{28}-1) \\
&= I((x^4+1)(x^{12}+x^8+1))/I(x^{28}-1), \\
C_{2,0} &= I(f_0(x^4)f_1(x^4))/I(x^{28}-1) \\
&= I((x^4+1)(x^{12}+x^4+1))/I(x^{28}-1).
\end{aligned}
$$

Each block $C_{i,0}$ has a unique composition series of length 4.   ◇

From Exercise 4.5.2 together with 4.7.17 we obtain a characterization of all cyclic codes that possess an idempotent generator:

Corollary *Assume that C is a cyclic code of length n over $\mathbb{F}_q$. Then C possesses an idempotent generator if and only if C is a direct summand of $\mathbb{F}_q[x]/I(x^n-1)$. Using the notation of 4.7.17, the code C has an idempotent generator if and only if $C = \bigoplus_{i\in\mathcal{I}} C_{i,0}$, where $\mathcal{I}$ is a nonempty subset of l. In particular, in the case when $p \nmid n$, every cyclic code of length n over $\mathbb{F}_q$ possesses an idempotent generator.*   **4.7.22**

Proof: Since C is cyclic of length n over $\mathbb{F}_q$, it is of the form $I(g)/I(x^n-1)$, where $g \in \mathbb{F}_q[x]$ is a divisor of x^n-1.

If C possesses an idempotent generator e, then $C = (\mathbb{F}_q[x]/I(x^n-1)) \cdot e$. If $e = 1$, then $C = \mathbb{F}_q[x]/I(x^n-1)$. Otherwise, $1 = e + (1-e)$ is a nontrivial decomposition of the identity element, and C is a direct summand of $\mathbb{F}_q[x]/I(x^n-1)$ according to the second part of Exercise 4.5.2. Thus by the first part of 4.7.17, it is a direct sum of blocks $C_{i,0}$.

Conversely, if C is a direct summand of $\mathbb{F}_q[x]/I(x^n-1)$, then there exists C' such that $\mathbb{F}_q[x]/I(x^n-1) = C \oplus C'$. Consequently, $1 = e + e'$ with $e \in C$ and $e' \in C'$, and according to the first part of Exercise 4.5.2 the idempotent e is even an idempotent generator of $C = (\mathbb{F}_q[x]/I(x^n-1)) \cdot e$.

If $p \nmid n$, then according to 4.7.19 each cyclic code of length n over $\mathbb{F}_q$ is equivalent to a direct summand of $\mathbb{F}_q[x]/I(x^n-1)$.   □

Example (Continuation of 4.7.21 and 4.5.2) Let us now consider binary cyclic codes of length $n = 28$. In the notation of 4.7.21 we have   **4.7.23**

$$
\mathbb{F}_2[x]/I(x^{28}-1) = C_{0,0} \oplus C_{1,0} \oplus C_{2,0}.
$$

Putting $y = x^4$, the idempotent generators of indecomposable binary cyclic codes of length 28 are:

code	idempotent generator
$C_{0,0}$	$y^6 + y^5 + y^4 + y^3 + y^2 + y + 1$
$C_{1,0}$	$y^4 + y^2 + y + 1$
$C_{2,0}$	$y^6 + y^5 + y^3 + 1$

◇

Exercises

E.4.7.1   **Exercise** Prove that any nonempty subset U of an R-left module M is a submodule of M if and only if

$$u_1 + u_2 \in U \quad \text{and} \quad ru \in U, \qquad \forall\, u_1, u_2, u \in U, r \in R.$$

E.4.7.2   **Exercise** Prove that any ideal of a ring R with 1 is both an R-left and an R-right module.

E.4.7.3   **Exercise** Let M be an R-left module and $T \subset M$. Prove that the R-linear closure of T is a submodule of M.

E.4.7.4   **Exercise** Assume that U is a submodule of the R-left module M. Prove that the factor module M/U is indeed an R-left module. Moreover, verify that the natural projection $\pi \colon M \to M/U$ is a surjective R-module homomorphism.

E.4.7.5   **Exercise** Prove that a submodule U of the R-left module M is maximal if and only if the factor module M/U is simple.

E.4.7.6   **Exercise** Let $f \colon M \to N$ be an R-module homomorphism. Prove that the image $f(M)$ is a submodule of N, and that the kernel of f is a submodule of M.

E.4.7.7   **Exercise** Prove that the factors of a refinement of a composition series are either trivial or simple.

E.4.7.8   **Exercise** Prove that any prime element p of a integral domain is irreducible.

E.4.7.9   **Exercise** Prove that any irreducible element p of a principal ideal domain is prime.

E.4.7.10   **Exercise** Let p be an irreducible element of an integral domain R. If p is not a divisor of $a \in R$, then p and a are relatively prime.

E.4.7.11   **Exercise** Prove that for arbitrary elements a, b of the principal ideal domain R the following facts are equivalent:

— a and b are relatively prime,

— $R = Ra + Rb$,

— there exist $r, s \in R$ such that $1 = ra + sb$.

Exercise Assume that the decomposition of $x^n - 1$ into irreducible factors over   **E.4.7.12**
$\mathbb{F}_q$ is given by
$$x^n - 1 = \prod_{i \in l} (f_i)^{p^s}.$$
Show that the ideals $\mathrm{I}(f_i^{p^s})$ for $i \in l$ are pairwise relatively prime.

Exercise For $f \in \mathbb{F}[x]$ show that $\mathbb{F}[x]/\mathrm{I}(f)$ is an $\mathbb{F}$-algebra.   **E.4.7.13**

Exercise Show that a decomposable ideal possesses more than a single com-   **E.4.7.14**
position series.

Exercise Describe the m-th order binary Hamming-code as a direct sum of   **E.4.7.15**
minimal cyclic codes.

Exercise Show that the polynomials $g_{i,j}$ and $(x^r - 1)^j g_{i,0}$ generate the same   **E.4.7.16**
cyclic code of length n $(0 \le j \le p^s)$.

Exercise Give all indecomposable binary cyclic codes of length $n = 9$.   **E.4.7.17**

Exercise Verify 4.7.19.   **E.4.7.18**

Exercise Assume that $C_{i_1, p^s - 1}, \dots, C_{i_u, p^s - 1}$ is a complete system of representa-   **E.4.7.19**
tives of the isomorphism classes (not to be mixed up with the isometry classes!)
of minimal cyclic codes as ideals in $\mathbb{F}_q[x]/\mathrm{I}(x^n - 1)$. Prove that there are exactly $u(p^s - 1)$ isomorphism classes of indecomposable cyclic codes of length
n over $\mathbb{F}_q$.

4.8 Codes of *p*-Power Block Length   4.8

In this section we consider cyclic codes of block length $n = p^s$ over $\mathbb{F}_q$, where p
is the characteristic of the field $\mathbb{F}_q$. Thus $x^n - 1 = x^{p^s} - 1 = (x - 1)^{p^s}$, i.e. the

divisors of $x^n - 1$ are of the form $(x-1)^t, t \in p^s + 1$, and the cyclic codes correspond to the ideals

$$I(x-1)^t := I((x-1)^t) = (I(x-1))^t.$$

This case was briefly mentioned in Corollary 4.2.14.

4.8.1 **Theorem**

1. *The set of all cyclic codes of length $n = p^s$ over $\mathbb{F}_q$ forms the composition series*

$$\mathbb{F}_q[x]/I(x^{p^s} - 1) = C_0 \supset C_1 \supset \ldots \supset C_{p^s-1} \supset C_{p^s} = 0,$$

   where the code C_t has generator polynomial $(x-1)^t$.

2. *The code C_t is of dimension $k = p^s - t$ and has the two $\mathbb{F}_q$-bases*

$$\left\{ x^i(x-1)^t + I(x^{p^s} - 1) \mid i \in p^s - t \right\}$$

   and

$$\left\{ (x-1)^j + I(x^{p^s} - 1) \mid t \leq j \in p^s \right\}.$$

   Putting $x_i := x^{p^i}$ the last basis turns out to be the set

$$\left\{ (x_0 - 1)^{b_0} \cdots (x_{s-1} - 1)^{b_{s-1}} + I(x^{p^s} - 1) \mid \sum_{i \in s} b_i p^i \geq t, \ b_i \in p, \ i \in s \right\}.$$

   This basis is called the Jennings basis *of C_t.*

3. *The dual code of C_t is C_{p^s-t}.*

4. *The code C_{p^s-1} is a repetition code.*

Proof: 1. According to 4.2.14, all cyclic codes of length p^s over $\mathbb{F}_q$ are of the form $C_t = I(x-1)^t/I(x^{p^s} - 1)$. From the Second Isomorphism Theorem for Rings (4.7.5) we deduce that

$$\begin{aligned} C_t/C_{t+1} &= (I(x-1)^t/I(x^{p^s} - 1))/(I(x-1)^{t+1}/I(x^{p^s} - 1)) \\ &\simeq I(x-1)^t/I(x-1)^{t+1}, \end{aligned}$$

i.e. the factors of successive terms in the chain have dimension 1, and so the chain cannot be refined, it is a composition series.

2. From 4.2.5 it follows that C_t has $\mathbb{F}_q$-dimension $p^s - t$. Since

$$\deg(x^i(x-1)^t) = t+i \quad \text{and} \quad \deg((x-1)^j) = j,$$

the sets in question are linearly independent. Both sets have the right number of elements, namely $p^s - t$, and so they are bases. In characteristic p, we have the identity

$$(x - 1)^j = (x_0 - 1)^{b_0} \cdots (x_{s-1} - 1)^{b_{s-1}},$$

for any integer $j \in p^s$ with base-p representation

$$j = b_0 + b_1 p + \ldots + b_{s-1} p^{s-1}, \quad b_i \in p, \ i \in s.$$

This means that the second basis $\{(x - 1)^j + I(x^{p^s} - 1) \mid t \leq j \leq p^s - 1\}$ equals the Jennings basis of C_t.

3. The dual code $C_t^{\perp}$ has dimension t and is also cyclic because of 4.2.5. By a dimension argument, we deduce that $C_t^{\perp}$ is in fact C_{p^s-t}.

4. From the previous item it follows that the code C_{p^s-1} is the dual code of $C_1 = I(x - 1)$. From 4.2.5 we deduce that C_{p^s-1} is generated by $x^{p^s} h(x^{-1})/h(0)$ where h is the check polynomial of C_1. Therefore

$$h(x) = \frac{x^{p^s} - 1}{x - 1} = x^{p^s-1} + \ldots + x + 1,$$

and C_{p^s-1} is generated by $x^{p^s-1} + \ldots + x + 1$. Hence it is the p^s-fold repetition code over $\mathbb{F}_q$.   □

We remark that the Jennings basis is originally due to Lombardo–Radice who studied the radical of the group ring of a finite group whose order is divisible by p over a field of characteristic p. It was later picked up by Jennings [102]. As a general reference, we refer to the article by Assmus and Key in the Handbook of Coding Theory [163] and the references listed there.

Corollary   *The residue class ring* $\mathbb{F}_q[x]/I(x^{p^s} - 1)$ *possesses the Jennings basis*   **4.8.2**

$$\left\{ (x_0 - 1)^{b_0} \cdots (x_{s-1} - 1)^{b_{s-1}} + I(x^{p^s} - 1) \ \middle| \ b_i \in p, \ i \in s \right\}. \qquad \square$$

The minimum distance of such codes of p-power block length was first evaluated by S. Berman (cf. [11]). We will present a different proof of his result, using *visible sets*, as introduced by H.N. Ward in [197]:

Definition (visible sets) A nonempty set S of vectors in $\mathbb{F}_q^n$ is called *visible* over   **4.8.3**
$\mathbb{F}_q$, if every linear code C which is generated by a nonempty subset T of S over $\mathbb{F}_q$ satisfies

$$\text{dist}(C) = \min\{\text{wt}(v) \mid v \in T\}. \tag*{**4.8.4**}$$

That is, the minimum distance of C is the weight of at least one element in T. A code C with this property is said to be *visible* (with respect to S).   ◇

Thus, for visible codes the complexity of evaluating the minimum distance depends only on the size of the given visible set. Hence, we will derive such a set for *certain linear codes of prime block length p over* $\mathbb{F}_q$, *p the characteristic of* $\mathbb{F}_q$. The proof is due to E.F. Assmus and H.F. Mattson (cf. [8]).

4.8.5   **Lemma** *The subset*

$$S := \{(x-1)^t + \mathrm{I}(x^p - 1) \mid t \in p\} \subseteq \mathbb{F}_q[x]/\mathrm{I}(x^p - 1)$$

is visible over $\mathbb{F}_q$. *Moreover, the linear code generated by a nonempty* $T \subseteq S$ *has the minimum distance*

$$\mathrm{dist}(\langle T \rangle) = \min \{t \mid (x-1)^t \in T\} + 1.$$

Proof: Consider a nonempty subset T of S and denote by C the code generated by T. We use induction on t, the smallest exponent such that $(x-1)^t$ is contained in the generating set T.

1. If $t = 0$, then $1 \in T$, which means that $(1, 0, \ldots, 0) \in C$. (Recall that the basis of the residue class ring consists of the powers $x^0 = 1, x^1, \ldots, x^{p-1}$ of x.) Thus $\mathrm{dist}(C) = 1 = t + 1$, as stated.

2. If $t > 0$, we derive a lower and an upper bound for $\mathrm{dist}(C)$. The upper bound is easy since C contains the vector $(x-1)^t + \mathrm{I}(x^p - 1)$, which has weight $t + 1$ as $(x-1)^t$ is the sum of $t + 1$ monomials $\alpha_i x^i, i \in p, \alpha_i \neq 0$. This shows that

$$\mathrm{dist}(C) \leq t + 1.$$

In order to derive a lower bound, we consider $0 \neq a = a(x) + \mathrm{I}(x^p - 1) \in C$. It is a linear combination of vectors $(x-1)^i + \mathrm{I}(x^p - 1)$, $i \geq t$, and therefore $a = b \cdot (x-1)^t + \mathrm{I}(x^p - 1)$, for a suitable polynomial b. Choose an exponent $i \in p$, for which the shifted vector

$$c := x^i \cdot a(x) + \mathrm{I}(x^p - 1) = c(x) + \mathrm{I}(x^p - 1)$$

has nonzero constant term c_0. We note that $\mathrm{wt}(a) = \mathrm{wt}(c)$. That c is not necessarily an element of C but of an isometric code does not matter. If $c(x) = \sum_{i \in p} c_i x^i$, then the formal derivative $c'(x)$ is of the form $c'(x) = \sum_{i \in p} i c_i x^{i-1} = \sum_{i=1}^{p-1} i c_i x^{i-1}$. Therefore, as $c_0 \neq 0$, $\mathrm{wt}(c) = \mathrm{wt}(c') + 1$. Moreover, $c'(x) \neq 0$, since otherwise $c_i = 0$ for $1 \leq i \leq p-1$ and consequently $a(x) = \alpha x^{p-i}$, for some $\alpha \in \mathbb{F}_q^*$, but this contradicts the assumption $t > 0$. Hence, $0 \neq c'(x) = d \cdot (x-1)^{t-1}$, for a suitable polynomial d. It means that $0 \neq c' \in C_{t-1}$, since according to 4.8.1 this code consists of all polynomial multiples of $(x-1)^{t-1}$. Thus, the induction hypothesis applies to C_{t-1}, and we obtain the inequality

$$\mathrm{wt}(a) = \mathrm{wt}(c) \geq \mathrm{dist}(C_{t-1}) + 1 = t + 1.$$

It gives the desired lower bound $\mathrm{dist}(C) \geq t + 1$, and the proof is complete. $\square$

The weight of a polynomial $(x - 1)^t$, for arbitrary t, is given in

---

Lemma Let $m = \sum_{i \in s} a_i p^i$ be the base-p representation of m (with $a_i \in p$). Then   **4.8.6**

$$\mathrm{wt}\left((x - 1)^m \right) = \prod_{i \in s} (a_i + 1).$$

Proof: Write $m = a_0 + a_1 p + \ldots + a_{s-1} p^{s-1}$ where $a_i \in p$ for $i \in s$. It follows from Exercise 4.8.2 that

$$(x - 1)^m = \sum_{n=0}^{m} (-1)^{m-n} \binom{m}{n} x^n \equiv \sum_{n=0}^{m} (-1)^{m-n} \prod_{i \in s} \binom{a_i}{b_i} x^n \quad \mathrm{mod}\ p,$$

where $n = b_0 + b_1 p + \ldots + b_{s-1} p^{s-1}$, with $b_i \in p$ for $i \in s$. The coefficient of x^n is nonzero modulo p precisely if $b_i \leq a_i$ for all $i \in s$. Since there are $a_i + 1$ choices for b_i, we get $\prod_{i \in s} (a_i + 1)$ nonzero coefficients in the expansion of $(x - 1)^m$.   □

We want to know when a generating system S of C is visible (without explicitly knowing the minimum distance of C). It will turn out to be important that the generating system is closed under the tensor product. For this reason we mention the following result of H.N. Ward ([197]).

Theorem Consider $m \in \mathbb{N}^*$. For $i \in m$, let S_i denote a visible subset of $\mathbb{F}_q^{n_i}$. Then   **4.8.7**

$$\left\{ v^{(0)} \otimes \ldots \otimes v^{(m-1)} \;\middle|\; v^{(i)} \in S_i, \, i \in m \right\} \subseteq \mathbb{F}_q^{n_0} \otimes \ldots \otimes \mathbb{F}_q^{n_{m-1}}$$

is visible over $\mathbb{F}_q$.

Proof: It is clear that it suffices to prove the theorem for $m = 2$. The general case then follows by a trivial induction which we will skip. So we consider two visible sets $S_0 \subseteq \mathbb{F}_q^{n_0}$ and $S_1 \subseteq \mathbb{F}_q^{n_1}$. Let T be a nonempty subset of

$$S_0 \otimes S_1 := \{ v \otimes w \mid v \in S_0, w \in S_1 \}$$

and C the linear code generated by T. We have to prove that

$$\mathrm{dist}(C) = \min \left\{ \mathrm{wt}(v \otimes w) \mid v \otimes w \in T \right\}.$$

1. To begin with, we note that we can express a nonzero element $a \in C$ as $a = \sum_{v \in S_0} v \otimes c(v)$, where $c(v)$ denotes a linear combination of the vectors in

$$S_1(v) := \{ w \in S_1 \mid v \otimes w \in T \} \subseteq S_1.$$

Moreover, we can discard the vectors $v \in S_0$ for which $S_1(v)$ is empty. In other words, we can restrict attention to the subset $S_0^* \subseteq S_0$ of all vectors $v \in S_0$ that appear as left factor of some element $v \otimes w \in T$, obtaining

$$a = \sum_{v \in S_0^*} v \otimes c(v).$$

We note that S_0^* is a nonempty visible set.

2. In order to evaluate the weight of a we introduce the vectors

$$a^{(k)} := \sum_{v \in S_0^*} v_k \cdot c(v) \in \mathbb{F}_q^{n_1}, \quad k \in n_0.$$

By definition of tensor multiplication, the weight of a is the sum of the weights of these vectors. More precisely,

$$\mathrm{wt}(a) = \sum_{k \in N} \mathrm{wt}(a^{(k)}), \quad \text{where } N := \left\{ k \in n_0 \mid a^{(k)} \neq 0 \right\}.$$

3. The main point of the proof is to show that, for each $k \in N$,

$$a^{(k)} \in \left\langle \bigcup_{v \in S_0^* \setminus W} S_1(v) \right\rangle, \quad \text{for } W := \{ v \in S_0^* \mid \mathrm{wt}(v) > |N| \}.$$

This is clear if $N = n_0$, since in this case $S_0^* \setminus W = S_0$. If $N \subset n_0$, we consider the $(n_0 \times |W|)$-matrix V which contains the transposed of the elements of W in its columns, together with its $(|M| \times |W|)$-submatrix V_M consisting of the rows with numbers i in $M := n_0 \setminus N$. By assumption, V_M is not empty. These matrices define linear mappings

$$f_0 : \mathbb{F}_q^{|W|} \to \mathbb{F}_q^{n_0} \ : \ b \mapsto b \cdot V^\top = \sum_{v \in W} b_v \cdot v$$

and

$$f_1 : \mathbb{F}_q^{|W|} \to \mathbb{F}_q^{|M|} \ : \ b \mapsto b \cdot V_M^\top = \sum_{v \in W} b_v \cdot \pi(v),$$

where π means the projection of $\mathbb{F}_q^{n_0}$ onto $\mathbb{F}_q^{|M|}$, and the vector b is indicated as the sequence $(b_v)_{v \in W}, b_v \in \mathbb{F}_q$. We should like to show that $\ker(f_0) = \ker(f_1)$.

Obviously $\ker(f_0) \subseteq \ker(f_1)$. If there were vectors $b \in \mathbb{F}_q^{|W|}$, contained in $\ker(f_1)$ but not in $\ker(f_0)$, there were elements $b_v \in \mathbb{F}_q$, for $v \in W$, such that

$$\sum_{v \in W} b_v \cdot v_j = 0 \text{ for all } j \in M = n_0 \setminus N,$$

while

$$\sum_{v \in W} b_v \cdot v_k \neq 0 \ \text{ for some } \ k \in N.$$

This would imply that $w := \sum_{v \in W} b_v \cdot v \neq 0$ were contained in the subspace generated by W and $\mathrm{wt}(w) \leq |N|$. But this contradicts the visibility of S_0^*.

Hence the kernels are in fact equal, and so the rows of V are linear combinations of the rows of the submatrix V_M. It follows, that there are elements $\gamma_{k,j} \in \mathbb{F}_q, k \in N, j \in M$, such that

$$v_k = \sum_{j \in M} \gamma_{k,j} \cdot v_j \quad \text{for all } v \in W.$$

Let $k \in N$. Since $a^{(j)} = 0$, for $j \in M$, we can write

$$a^{(k)} = a^{(k)} - \sum_{j \in M} \gamma_{k,j} \cdot a^{(j)}.$$

Therefore,

$$\begin{aligned} a^{(k)} &= \sum_{v \in S_0^*} v_k \cdot c(v) - \sum_{j \in M} \gamma_{k,j} \cdot \Big(\sum_{v \in S_0^*} v_j \cdot c(v) \Big) \\ &= \sum_{v \in S_0^*} \Big(v_k - \sum_{j \in M} \gamma_{k,j} v_j \Big) \cdot c(v) \\ &= \sum_{v \in S_0^* \setminus W} \Big(v_k - \sum_{j \in M} \gamma_{k,j} v_j \Big) \cdot c(v), \end{aligned}$$

which completes the proof of

$$a^{(k)} \in \Big\langle \bigcup_{v \in S_0^* \setminus W} S_1(v) \Big\rangle.$$

4. In order to finish the proof of the theorem, let w be a vector of minimum weight in this union $\bigcup_{v \in S_0^* \setminus W} S_1(v)$. Since S_1 is visible, $\mathrm{wt}(a^{(k)}) \geq \mathrm{wt}(w)$. If we take a vector v from $S_0^* \setminus W$ such that $v \otimes w \in T$, then $\mathrm{wt}(v) \leq |N|$, obtaining

$$\mathrm{wt}(a) = \sum_{k \in N} \mathrm{wt}(a^{(k)}) \geq |N| \cdot \mathrm{wt}(w) \geq \mathrm{wt}(v) \cdot \mathrm{wt}(w) = \mathrm{wt}(v \otimes w).$$

This shows the stated visibility.   □

We apply this result to cyclic codes of p-power length.

Theorem *The Jennings basis of the residue class ring $\mathbb{F}_q[x]/\mathrm{I}(x^{p^s} - 1)$ is visible over $\mathbb{F}_q$.*   **4.8.8**

Proof: The s-fold tensor product

$$\otimes^s (\mathbb{F}_q[x]/\mathrm{I}(x^p - 1))$$

of the residue class ring $\mathbb{F}_q[x]/\mathrm{I}(x^p - 1)$ with itself has as $\mathbb{F}_q$-basis the set

$$\{x^{i_0} \otimes \ldots \otimes x^{i_{s-1}} \mid i_j \in p, \ j \in s\}.$$

According to 4.8.5 and 4.8.7, the subset

$$\Big\{ (x-1)^{b_0} \otimes \ldots \otimes (x-1)^{b_{s-1}} \ \Big| \ b_j \in p, \ j \in s \Big\}$$

of the tensor space $\otimes^s (\mathbb{F}_q[x]/\mathrm{I}(x^p - 1))$ is visible over $\mathbb{F}_q$. Choosing the powers of x, $x_0, \ldots, x_{s-1}$ as in 4.8.1, the mapping

$$\otimes^s (\mathbb{F}_q[x]/\mathrm{I}(x^p - 1)) \to \mathbb{F}_q[x]/\mathrm{I}(x^{p^s} - 1),$$

defined by associating canonical representatives

$$x^{b_0} \otimes \ldots \otimes x^{b_{s-1}} \mapsto x_0^{b_0} \cdots x_{s-1}^{b_{s-1}},$$

is a linear isometry that maps the visible set onto the Jennings basis of $\mathbb{F}_q[x]/I(x^{p^s} - 1)$. Hence, also the Jennings basis of $\mathbb{F}_q[x]/I(x^{p^s} - 1)$ is visible over $\mathbb{F}_q$. $\square$

Now we are able to evaluate the minimum distance of cyclic codes of p-power length.

4.8.9 **Theorem of Berman** *Let $t = a_0 + a_1 p + \ldots + a_{s-1} p^{s-1}$ with $a_j \in p$, $j \in s$, be the base-p representation of the integer $t \in p^s - 1$. Let i be the unique index such that $a_i < p - 1$ and $a_{i+1} = \ldots = a_{s-1} = p - 1$. Then*

$$\text{dist}(C_t) = \begin{cases} (a_i + 1)p^{s-i-1} & \text{if } (a_0, \ldots, a_{i-1}) = (0, \ldots, 0), \\ (a_i + 2)p^{s-i-1} & \text{if } (a_0, \ldots, a_{i-1}) \neq (0, \ldots, 0). \end{cases}$$

Proof: By 4.8.8, the Jennings basis of C_t is a subset of the visible Jennings basis of the residue class ring $\mathbb{F}_q[x]/I(x^{p^s} - 1)$. Hence, the minimum distance of C_t is the weight of an element in its Jennings basis, i.e. an element of the form $(x - 1)^j$, where $t \le j \le p^s - 1$. Let $j = b_0 + b_1 p + \ldots + b_{s-1} p^{s-1}$, $b_l \in p$, $l \in s$. It follows from 4.8.6 that in order to minimize the weight of a basis vector, we must minimize the value of the function $\text{wt}\left((x - 1)^j\right) = \prod_{l \in s}(b_l + 1)$ taken over all j with $t \le j \le p^s - 1$. Therefore, $b_{i+1} = \ldots = b_{s-1} = p - 1$. If $a_0 = \ldots = a_{i-1} = 0$, we may take $b_0 = \ldots = b_{i-1} = 0$ and $b_i = a_i$, so that

$$j = \sum_{l \in s} b_l p^l = a_i p^i + (p - 1)p^{i+1} + \ldots + (p - 1)p^{s-1} \ge t,$$

and hence

$$\text{wt}\left((x - 1)^j\right) = \prod_{l \in s}(b_l + 1) = (a_i + 1)(a_{i+1} + 1) \cdots (a_{s-1} + 1) = (a_i + 1)p^{s-i-1}.$$

Otherwise, if at least one coefficient $a_k > 0$ for $k < i$, we may take $b_0 = \ldots = b_{i-1} = 0$ and $b_i = a_i + 1$, so that

$$j = \sum_{l \in s} b_l p^l = (a_i + 1)p^i + (p - 1)p^{i+1} + \ldots + (p - 1)p^{s-1} \ge t$$

and hence

$$\text{wt}\left((x - 1)^j\right) = \prod_{l \in s}(b_l + 1) = (a_i + 2)(a_{i+1} + 1) \cdots (a_{s-1} + 1) = (a_i + 2)p^{s-i-1}.$$

One can verify that no $j \ge t$ with $b_i = a_i$ leads to a smaller weight. $\square$

Corollary *Every cyclic code of length p over $\mathbb{F}_q$ is an MDS-code.*   4.8.10

Proof: By 4.8.1, every cyclic code of length p over $\mathbb{F}_q$ is of the form C_t, $t \in p$. Its dimension is $p - t$, and its minimum weight is $t + 1$ because of 4.8.9.   □

Corollary *For every cyclic code C_t of length p^s over $\mathbb{F}_q$ there is a $\hat{t} \in \{t, \ldots, p^s - 1\}$*   4.8.11
such that

$$\mathrm{dist}(C_t) = \mathrm{wt}\left((x-1)^{\hat{t}}\right).$$ □

A decoding method for binary cyclic codes whose length is a power of 2 is presented in Section 4.14.

Exercises

Exercise Determine the dimension and the minimum distance of all binary   E.4.8.1
cyclic codes of length 32.

Exercise Let p be prime, and let $m = a_0 + a_1 p + \ldots + a_{s-1}p^{s-1}$, $n = b_0 + b_1 p +$ E.4.8.2
$\ldots + b_{s-1}p^{s-1}$ where $a_i, b_i \in p$ for $i \in s$. Show Lucas' theorem that

$$\binom{m}{n} \equiv \prod_{i \in s} \binom{a_i}{b_i} \quad \mathrm{mod}\ p.$$

Hint: Consider first the case that $m = cp + a$ and $n = dp + b$, where $a, b \in p$. Let x be an indeterminate. Then $(1 + x)^p \equiv 1 + x^p \mod p$. Thus (modulo p) we have

$$(1+x)^m = (1+x)^{cp}(1+x)^a \equiv (1+x^p)^c(1+x)^a = \sum_{i=0}^{c}\binom{c}{i}x^{pi} \cdot \sum_{j=0}^{a}\binom{a}{j}x^j.$$

Compare coefficients of $x^n = x^{dp+b}$ on both sides to get

$$\binom{m}{n} \equiv \binom{c}{d}\binom{a}{b} \quad \mathrm{mod}\ p.$$

Finish the proof by induction.

4.9 Bounds for the Minimum Distance   4.9

This section contains various basic results on the minimum distance of cyclic codes. As before, we assume that $n = p^s r$, where $p \nmid r$. Moreover, we suppose that

$$x^r - 1 = \prod_{i \in I} f_i$$

is the decomposition of $x^r - 1$ into pairwise distinct irreducible polynomials over $\mathbb{F}_q$. In 4.7.17, we introduced the cyclic code $C_{i,j}$ which is generated by

$$g_{i,j} := f_0^{p^s} \cdots f_{i-1}^{p^s} \cdot f_i^j \cdot f_{i+1}^{p^s} \cdots f_{l-1}^{p^s}.$$

According to Exercise 4.7.16, the code $C_{i,j}$ is also generated by $(z-1)^j g_{i,0}$, where $z = x^r$. In particular, the code generated by $g_{i,0}$ is a direct summand of the residue class ring $\mathbb{F}_q[x]/\mathrm{I}(x^n - 1)$, and by Exercise 4.5.2 it is generated by a primitive idempotent e_i. The primitive idempotent e_i and, therefore, every idempotent in $\mathbb{F}_q[x]/\mathrm{I}(x^n - 1)$ is contained in the subalgebra $\mathbb{F}_q[y]/\mathrm{I}(y^r - 1)$ of $\mathbb{F}_q[x]/\mathrm{I}(x^n - 1)$, where $y = x^{p^s}$. By Exercise 4.9.1, the cyclic code with generator polynomial $g_{i,j}$ is also generated by $(z-1)^j e_i$.

To begin with, we consider a special class of cyclic codes, which contains the indecomposable cyclic codes. In the sequel we always use $n = p^s r$, where $p \nmid r$, $y = x^{p^s}$ and $z = x^r$.

4.9.1   **Theorem** *Suppose that e is an idempotent of the residue class ring $\mathbb{F}_q[x]/\mathrm{I}(x^n - 1)$, and that $j \in p^s$. The cyclic code C of length n over $\mathbb{F}_q$ with generator polynomial*

$$(z-1)^j e$$

is linearly isometric to the product code $Y \otimes Z$, where Y is the cyclic code of length r over $\mathbb{F}_q$ with generating idempotent e and Z is the cyclic code of length p^s over $\mathbb{F}_q$ with generator polynomial $(z-1)^j$. The parameters of C are

$$\dim(C) = \dim(Y) \cdot \dim(Z) \quad and \quad \mathrm{dist}(C) = \mathrm{dist}(Y) \cdot \mathrm{dist}(Z).$$

The proof of this theorem relies on two further results, the verification of which is recommended to the reader as an exercise.

4.9.2   **Lemma** *To each polynomial $h \in \mathbb{F}_q[x]$ of degree at most $n - 1$ there exist polynomials $h^{(0)}, \ldots, h^{(r-1)} \in \mathbb{F}_q[z]$, each of degree at most $p^s - 1$, such that h can be written uniquely in the form*

$$h(x) \equiv \sum_{i \in r} y^i h^{(i)}(z) \bmod \mathrm{I}(x^n - 1). \qquad \square$$

4.9.3   **Lemma** *The canonical mapping*

$$\psi \colon \mathbb{F}_q[y]/\mathrm{I}(y^r - 1) \otimes \mathbb{F}_q[z]/\mathrm{I}(z^{p^s} - 1) \to \mathbb{F}_q[x]/\mathrm{I}(x^n - 1) \ : \ a \otimes b \mapsto ab$$

is an $\mathbb{F}_q$-isomorphism. Choosing the $\mathbb{F}_q$-basis

$$\left\{ y^i \otimes z^j \ \middle|\ i \in r, \ j \in p^s \right\}$$

of the tensor product, the mapping ψ is a linear isometry with respect to the Hamming metric. $\qquad \square$

Proof of 4.9.1: It follows from 4.9.2 that there are uniquely determined polynomials $h^{(0)}, \ldots, h^{(r-1)} \in \mathbb{F}_q[z]$ with $\deg h^{(i)} < p^s$ such that each element c of C can be written in the form

$$c(x) = \sum_{i \in r} y^i h^{(i)}(z)(z-1)^j e = \psi \left(\sum_{i \in r} y^i \otimes h^{(i)}(z)(z-1)^j \right).$$

Hence

$$\psi(Y \otimes Z) = C.$$

Consequently, by 4.9.3 C is linearly isometric to the product code $Y \otimes Z$. The statements on the dimension and the minimum distance of C now follow directly from Exercise 2.3.5.   □

Corollary *Every codeword $c = ab$ of a cyclic code of length n with $a \in \mathbb{F}_q[y]$ of degree less than r and $b \in \mathbb{F}_q[z]$ of degree less than p^s has the weight*   **4.9.4**

$$\mathrm{wt}(c) = \mathrm{wt}(a) \cdot \mathrm{wt}(b).$$ □

Example The cyclic code of length $p - 1$ over $\mathbb{F}_p$ with generator polynomial   **4.9.5**
$g = (y - 1) \cdots (y - (p - k - 1)), k \in p - 1$, is a $(p - 1, k)$-Reed–Solomon-code.
The generating idempotent e is a sum of k primitive idempotents (cf. 4.7.20).
Moreover, the cyclic code generated by $(z - 1)^j$ over $\mathbb{F}_p$ is a $(p, p - j)$-MDS-code. From 4.9.1 we deduce that the cyclic code generated by

$$(x^p - 1) \cdots (x^p - (p - k - 1))(x^{p-1} - 1)^j$$

over $\mathbb{F}_p$ has length $n = p(p - 1)$, dimension $k(p - j)$ and minimum distance $(p - k)(j + 1)$.   ◇

Example (Continuation of Example 4.7.23) An indecomposable binary cyclic   **4.9.6**
code $C_{i,j}$ of length $28 = 2^2 \cdot 7$ has the generator

$$(z - 1)^j e_i,$$

where $0 \le j \le 3$. According to 4.7.23, e_i is one of the following three primitive idempotents:

$$\begin{aligned}
e_0 &= 1 + y + y^2 + y^3 + y^4 + y^5 + y^6, \\
e_1 &= 1 + y^3 + y^5 + y^6, \\
e_2 &= 1 + y + y^2 + y^4.
\end{aligned}$$

If Y_i denotes the binary cyclic code of length 7 with generating idempotent e_i and if Z_j is the binary cyclic code of length 4 with generator polynomial

$(z - 1)^j$, then $C_{i,j}$ is linearly isometric to the product code $Y_i \otimes Z_j$. The evaluation of the parameters of $C_{i,j}$ is left to the reader (Exercise 4.9.4). $\diamond$

According to the Structure Theorem for cyclic codes, every cyclic code C of
length n over $\mathbb{F}_q$ can be written as a direct sum

4.9.7
$$C = \bigoplus_{i \in \mathcal{I}} C_{i,j_i}, \quad \mathcal{I} \subseteq l, \quad j_i \in p^s,$$

of indecomposable cyclic codes C_{i,j_i}.

In the following, we denote by Y_i the cyclic code of length r with generating
idempotent e_i. Also, Z_j is the cyclic code of length p^s with generator polynomial $(z - 1)^j$. From 4.9.1 we deduce that $C_{i,j}$ is equivalent to the product code
$Y_i \otimes Z_j$.

Now we want to obtain a lower bound for the minimum distance of C.

4.9.8   **Theorem** *Let C denote the cyclic code of length n over $\mathbb{F}_q$ defined in 4.9.7. Then*
$$\dim(C) = \sum_{i \in \mathcal{I}} \dim(Y_i) \cdot \dim(Z_{j_i})$$

and
$$\mathrm{dist}(C) = \min \left\{ \mathrm{wt}((z-1)^j) \cdot \mathrm{dist}(C_j) \mid \min \{j_i \mid i \in \mathcal{I}\} \le j \le p^s - 1 \right\},$$

where C_j indicates the cyclic code
$$C_j := \bigoplus_{i \in \mathcal{I}, \, j \ge j_i} Y_i$$

of length r over $\mathbb{F}_q$.

In the proof we will use the *augmentation mapping*, the algebra epimorphism
$$\eta : \mathbb{F}_q[z]/\mathrm{I}(z^{p^s} - 1) \to \mathbb{F}_q,$$

defined by
$$\sum_{i \in p^s} \kappa_i z^i + \mathrm{I}(z^{p^s} - 1) \mapsto \sum_{i \in p^s} \kappa_i.$$

This epimorphism can be extended as follows:

4.9.9   **Lemma** *The mapping*
$$\phi : \mathbb{F}_q[x]/\mathrm{I}(x^n - 1) \to \mathbb{F}_q[y]/\mathrm{I}(y^r - 1),$$

defined by
$$\sum_{i \in r} y^i h^{(i)}(z) + \mathrm{I}(x^n - 1) \mapsto \sum_{i \in r} y^i \eta(h^{(i)}(z)) + \mathrm{I}(y^r - 1)$$

is an algebra epimorphism. Its kernel is the ideal generated by $z - 1$.

Proof: It is clear that ϕ is an epimorphism of algebras. Also, the principal ideal I which is generated by $z - 1$ is contained in the kernel of ϕ.

Conversely, by 4.9.2 the canonical representative of a residue class in the residue class ring $\mathbb{F}_q[x]/I(x^n - 1)$ is of the form $\sum_{i \in r} y^i h^{(i)}$, with $h^{(i)} \in \mathbb{F}_q[z]$ and $\deg(h^{(i)}) < p^s$.

Since $\{(z-1)^j \mid j \in p^s\}$ is a basis of $\mathbb{F}_q[z]/I(z^{p^s} - 1)$, by 4.8.1, the set

$$B := \left\{ y^i(z-1)^j \mid i \in r, \, j \in p^s \right\}$$

is linearly independent, contained in I, and of size

$$|B| = r(p^s - 1) = n - r = \dim(\ker(\phi)).$$

This proves the statement. $\square$

Proof of 4.9.8:   The assertion on the dimension follows from 4.9.1. The remainder of the proof is split into four steps:

1. For each j with $\min\{j_i \mid i \in \mathcal{I}\} \leq j \leq p^s - 1$ the inequality

$$\mathrm{dist}(C) \leq \mathrm{wt}((z-1)^j) \cdot \mathrm{dist}(C_j)$$

holds true. If $c^{(j)}$ is a nonzero element of C_j then $c := (z-1)^j c^{(j)}$ belongs to the subcode

$$\bigoplus_{i \in \mathcal{I}, \, j \geq j_i} C_{i,j}$$

of C and

$$\mathrm{wt}(c) = \mathrm{wt}((z-1)^j) \cdot \mathrm{wt}(c^{(j)})$$

by 4.9.4.

2. For each nonzero $c \in C$ there exists an index j with

$$\min\{j_i \mid i \in \mathcal{I}\} \leq j \leq p^s - 1$$

and a polynomial $c^{(z)} \in \mathbb{F}_q[x]$ of degree less than n, which is not divisible by $z - 1$, such that

$$c = (z-1)^j c^{(z)}.$$

From the definition of C it follows that c is of the form

$$c = \sum_{i \in \mathcal{I}} (z-1)^{k_i} f^{(i)} e_i$$

where $k_i \geq j_i$ and $f^{(i)} \in \mathbb{F}_q[x]$ are not divisible by $z - 1$. Putting

$$j := \min \left\{ k_i \mid (z-1)^{k_i} f^{(i)} e_i \neq 0, \, i \in \mathcal{I} \right\},$$

we derive that $c = (z - 1)^j c^{(z)}$ with $c^{(z)} = \sum_{i \in \mathcal{I}} (z - 1)^{k_i - j} f^{(i)} e_i$. Due to the choice of j, exactly one summand of $c^{(z)}$ is not divisible by $z - 1$, and so the same holds also for $c^{(z)}$.

3. Assume that $c = (z - 1)^j c^{(z)} \in C$, as in the previous step. Define $l, j \leq l \leq p^s - 1$, such that

$$\mathrm{wt}((z - 1)^l) = \min\left\{\mathrm{wt}((z - 1)^k) \mid j \leq k \leq p^s - 1\right\}$$

and put $c^{(l)} := (z - 1)^l \phi(c^{(z)})$. Then

$$c^{(l)} \in C \text{ and } 0 < \mathrm{wt}(c^{(l)}) \leq \mathrm{wt}(c).$$

The second part of 4.9.9 implies that $c^{(l)} \neq 0$, since $c^{(z)}$ is not contained in the kernel of the epimorphism ϕ. The image $\phi(c^{(z)})$ is contained in C_j and so $c^{(l)} \in C$. By 4.9.2, $c^{(z)}$ can be expressed in the form

$$c^{(z)} = \sum_{i \in r} y^i h^{(i)}.$$

Then

4.9.10
$$\mathrm{wt}(c) = \sum_{i \in r} \mathrm{wt}((z - 1)^j h^{(i)}),$$

since the supports of the summands $y^i (z - 1)^j h^{(i)}$, $i \in r$, of c are disjoint. For each nonzero polynomial $(z - 1)^j h^{(i)}$ the following is true, by the choice of l and by 4.8.9:

$$\mathrm{wt}((z - 1)^j h^{(i)}) \geq \mathrm{wt}((z - 1)^l).$$

Let N_c be the number of indices i for which $h^{(i)} \neq 0$. We deduce from 4.9.10 that

4.9.11
$$\mathrm{wt}(c) \geq N_c \cdot \mathrm{wt}((z - 1)^l).$$

On the other hand, we obtain from 4.9.4 that

$$\mathrm{wt}(c^{(l)}) = \mathrm{wt}((z - 1)^l) \cdot \mathrm{wt}(\phi(c^{(z)})),$$

where $\mathrm{wt}(\phi(c^{(z)}))$ is the number of indices i with $\eta(h^{(i)}) \neq 0$. Since this number is at most N_c, we obtain from 4.9.11 that $\mathrm{wt}(c) \geq \mathrm{wt}(c^{(l)})$.

4. Lastly, we prove that

$$\mathrm{dist}(C) \geq \min\{\mathrm{wt}((z - 1)^j) \cdot \mathrm{dist}(C_j) \mid \min\{j_i \mid i \in \mathcal{I}\} \leq j \leq p^s - 1\}.$$

Consider a codeword $c \in C$ of minimal weight. From the third step we know that there exists a word $c^{(l)} \in C$ which has the same weight as c. This proves the stated inequality. $\square$

The evaluation of the minimum distance of C using the given formula, due to G. Castagnoli et al. (cf. [36]), is difficult, since often for semi-simple codes just lower bounds of the minimum distance are known.

Example (Continuation of Example 4.9.6) We evaluate the parameters of the   **4.9.12**
binary cyclic codes $C = C_{1,2} \oplus C_{2,3}$ described in 4.9.6. According to 4.9.8

$$\dim(Y_1) \cdot \dim(Z_2) + \dim(Y_2) \cdot \dim(Z_3) = 2 \cdot 3 + 1 \cdot 3 = 9$$

and $\mathrm{dist}(C)$ is the minimum of

$$\mathrm{wt}((z-1)^2) \cdot \mathrm{dist}(Y_1) = 2 \cdot 4 = 8$$

and

$$\mathrm{wt}((z-1)^3) \cdot \mathrm{dist}(Y_1 \oplus Y_2) = \mathrm{wt}((z-1)^3) \cdot \mathrm{dist}(P_7) = 4 \cdot 2 = 8,$$

where $Y_1 = H_3$ and $Y_2 = H_3'$ denote the $(7,3)$-simplex-codes, the sum of which is the parity check code P_7 (for the notation, see 4.2.7). This shows that C is a $(28,9,8)$-code.   ◇

The following result shows that the minimum distance of a cyclic code is given by the minimum distance of a certain subcode. This subcode is itself a product code, consisting of a semi-simple cyclic code and a cyclic code whose length is a power of the prime p.

Theorem *Let C denote a cyclic code of length n over $\mathbb{F}_q$, given by 4.9.7. Then*   **4.9.13**
there exists an idempotent $e \in \mathbb{F}_q[x]/\mathrm{I}(x^n - 1)$ and an integer $j \in p^s$, such that the
minimum distance of C is given by the minimum distance of a cyclic subcode of C
with generator $(z-1)^j e$.

Proof: By 4.9.8, C contains a vector c of minimal weight of the form $c = (z - 1)^j w$. Here, w is a vector of minimal weight in C_j, defined in 4.9.8. Hence, by 4.9.4,

$$\mathrm{dist}(C) = \mathrm{wt}((z-1)^j) \cdot \mathrm{wt}(w).$$

On the one hand, due to the choice of j and by 4.8.11, the minimum distance of the cyclic code Z_j is the weight of $(z-1)^j$. On the other hand, since C_j is the direct sum of the $C_{i,j}$, a generating idempotent is

$$e = \sum_{i \in \mathcal{I}, j \geq j_i} e_i.$$

That is, $\mathrm{dist}(C) = \mathrm{dist}(C_j \otimes Z_j)$. The image of $C_j \otimes Z_j$ under ψ (defined in 4.9.3) is contained in C, by the choice of j and e in C, and it is generated by $(z-1)^j e$.   □

Recall that the *relative minimum distance* of an (n, k, d)-code is the ratio d/n. From the following statement, also due to G. Castagnoli et al. [36], it will become clear that cyclic codes of composed length are not better than semi-simple cyclic codes.

4.9.14   **Theorem** *Let C denote a cyclic code of length n over $\mathbb{F}_q$ as defined in 4.9.7. Then there exists a cyclic code of length r over $\mathbb{F}_q$, whose rate and relative minimum distance are at least the rate and the relative minimum distance of C.*

Proof: Putting $j := \max\{j_i \mid i \in \mathcal{I}\}$, then 4.9.8 implies that

$$\frac{\dim(C)}{n} = \sum_{i \in \mathcal{I}} \frac{\dim(C_{i,j_i})}{n} = \sum_{i \in \mathcal{I}} \frac{\dim(Y_i)}{r} \cdot \frac{\dim(Z_{j_i})}{p^s}$$

$$\leq \sum_{i \in \mathcal{I}} \frac{\dim(Y_i)}{r} = \frac{\dim(C_j)}{r}$$

and

$$\frac{\text{dist}(C)}{n} \leq \frac{\text{wt}((z-1)^j)}{p^s} \cdot \frac{\text{dist}(C_j)}{r} \leq \frac{\text{dist}(C_j)}{r}.$$

Therefore, the cyclic code C_j has the desired properties.   □

Exercises

E.4.9.1   **Exercise** If e_i is the primitive idempotent of $C_{i,0}$, prove that $(x-1)^j e_i$ is another generating polynomial of the cyclic code $C_{i,j}$.

E.4.9.2   **Exercise** Verify 4.9.2.

E.4.9.3   **Exercise** Check 4.9.3.

E.4.9.4   **Exercise** Evaluate the parameters of all indecomposable binary cyclic codes of length 28.

E.4.9.5   **Exercise** Give lower bounds for the minimum distance of the binary cyclic codes of length 92.

4.10 Reed–Muller-Codes

In Section 2.4, Reed–Muller-codes have been introduced as subspaces of the algebra $\mathcal{B}_m^q$ consisting of the polynomial functions from $\mathbb{F}_q^m$ to $\mathbb{F}_q$. We begin by describing a connection between Reed–Muller-codes and cyclic codes. After that, we prove that the Reed–Muller-codes $\mathrm{RM}_{m,t}^p$ are group algebra codes and we evaluate their dimensions and minimum distances. A decoding algorithm for binary Reed–Muller-codes will be presented at the end of Section 4.14.

First we show that the code which is obtained by puncturing a Reed–Muller-code at a position is cyclic. Consider the Reed–Muller-code $\mathrm{RM}_{m,t}^q$, choose a primitive element $\xi \in \mathbb{F}_{q^m}$ and represent each element of $\mathbb{F}_{q^m}^*$ as a linear combination of the members of the $\mathbb{F}_q$-basis $\{1, \xi, \ldots, \xi^{m-1}\}$ of $\mathbb{F}_{q^m}$:

$$\xi^i = \sum_{j \in m} \kappa_{j,i} \xi^j, \qquad \kappa_{j,i} \in \mathbb{F}_q, \quad i \in q^m.$$

According to 4.2.11, the minimal polynomial of ξ over $\mathbb{F}_q$ has the form

$$M_\xi(x) = x^m + \sum_{i \in m} \lambda_i x^i.$$

By definition, its *companion matrix*

$$Y_\xi := \begin{pmatrix} 0 & 0 & \ldots & 0 & 0 & -\lambda_0 \\ 1 & 0 & \ldots & 0 & 0 & -\lambda_1 \\ 0 & 1 & \ldots & 0 & 0 & -\lambda_2 \\ \vdots & \vdots & \ddots & \vdots & \vdots & \vdots \\ 0 & 0 & \ldots & 1 & 0 & -\lambda_{m-2} \\ 0 & 0 & \ldots & 0 & 1 & -\lambda_{m-1} \end{pmatrix}$$

satisfies the identity

$$(\kappa_{0,i}, \ldots, \kappa_{m-1,i}) \cdot Y_\xi^\top = (\kappa_{0,i+1}, \ldots, \kappa_{m-1,i+1}),$$

4.10.1

where the index $i + 1$ is understood modulo $q^m - 1$. The codeword $f \in \mathrm{RM}_{m,t}^q$ is the q^m-tuple

$$(f(0), f(1), f(\xi), f(\xi^2), \ldots, f(\xi^{q^m-2})),$$

where

$$f(\xi^i) := f(\kappa_{0i}, \ldots, \kappa_{m-1,i}).$$

The polynomial function $g \in \mathcal{B}_m^q$, defined by

$$g(x_0, \ldots, x_{m-1}) := f((x_0, \ldots, x_{m-1}) \cdot Y_\xi^\top),$$

is also contained in $\mathrm{RM}_{m,t}^q$. By 4.10.1, the associated q^m-tuple is

$$(f(0), f(\xi), f(\xi^2), \ldots, f(\xi^{q^m-2}), f(1)).$$

This shows that the code which is obtained from $\mathrm{RM}_{m,t}^q$ by puncturing the first component is cyclic. In 2.4.7, the parameters of binary Reed–Muller-codes have been calculated. In the following, we will generalize this result. We prove that Reed–Muller-codes over a prime field $\mathbb{F}_p$ are group algebra codes.

Let G denote an elementary-abelian p-group of order p^m. This means that G admits generators $g_0, \ldots, g_{m-1}$, each of order p, and each element g of G can be written uniquely in the form

$$g = g_0^{a_0} \cdots g_{m-1}^{a_{m-1}}, \qquad a_i \in p, \ i \in m.$$

We consider a particular basis of the group algebra $\mathbb{F}_p^G$.

4.10.2   **Theorem** *The group algebra $\mathbb{F}_p^G$ possesses the basis*

$$B := \left\{ \prod_{i \in m} (g_i - 1)^{a_i} \,\middle|\, a_i \in p, \ i \in m \right\}.$$

It is the Jennings basis *of $\mathbb{F}_p^G$.*

Proof: We use induction on $|G|$. The case $G = \{1\}$ is trivial, so we may assume that $|G| > 1$. It suffices to show that the given set is a system of $\mathbb{F}_p$-generators of $\mathbb{F}_p^G$. Denote by H the subgroup which is generated by $g_0, \ldots, g_{m-2}$. Then H is an elementary abelian p-group of order p^{m-1}, and the set

$$\left\{ 1, g_{m-1}, \ldots, g_{m-1}^{p-1} \right\}$$

is a transversal of the cosets of H in G. Hence, each element $a \in \mathbb{F}_p^G$ is of the form

$$a = \sum_{i \in p} A_i g_{m-1}^i, \qquad A_i \in \mathbb{F}_p^H.$$

The statement then follows from the identity

$$g_{m-1}^i = \left((g_{m-1} - 1) + 1\right)^i = \sum_{j=0}^{i} \binom{i}{j} (g_{m-1} - 1)^j$$

and the induction hypothesis about $\mathbb{F}_p^H$. $\square$

To each element $\prod_{i \in m} (g_i - 1)^{a_i}$ of this basis of $\mathbb{F}_p^G$ we associate its *Jennings weight*

$$\sum_{i \in m} a_i.$$

In the following we identify the integers $0, \ldots, p-1$ with the residue classes in $\mathbb{Z}_p$, thus we consider them as elements of the field $\mathbb{F}_p$. Now we introduce various ideals of the group algebra $\mathbb{F}_p^G$.

Lemma *The set*   4.10.3

$$I := \left\{ \sum_{g \in G} \alpha_g g \in \mathbb{F}_p^G \;\middle|\; \sum_{g \in G} \alpha_g = 0 \right\}$$

is an ideal, the augmentation ideal *of $\mathbb{F}_p^G$ with $\mathbb{F}_p$-basis*

$$B' := \{ g - 1 \mid g \in G, \; g \neq 1 \},$$

and

$$I = \left\{ a \in \mathbb{F}_p^G \mid a^p = 0 \right\}.$$

Proof: I is the kernel of the $\mathbb{F}_p$-algebra epimorphism

$$\mathbb{F}_p^G \to \mathbb{F}_p \;:\; \sum_{g \in G} \alpha_g g \mapsto \sum_{g \in G} \alpha_g$$

and, therefore, it is an ideal of $\mathbb{F}_p^G$. The set B' is clearly linearly independent and contained in I. Since I is a proper ideal of $\mathbb{F}_p^G$, a dimension argument shows that B' must be a basis, i.e. I is of codimension 1. Moreover, because of 3.2.12 each element $a = \sum_{g \in G} \alpha_g g \in I$ satisfies

$$a^p = \sum_{g \in G} \alpha_g^p g^p = \left(\sum_{g \in G} \alpha_g^p \right) 1 = \left(\sum_{g \in G} \alpha_g \right)^p 1 = 0.$$

Thus, I is contained in the proper subspace $\{ a \in \mathbb{F}_p^G \mid a^p = 0 \}$ of $\mathbb{F}_p^G$. Hence, the statement follows again from a dimension argument.   □

Theorem *Suppose that $0 \le t \le m(p-1)$.*   4.10.4

— *The set*

$$B_t := \left\{ \prod_{i \in m} (g_i - 1)^{a_i} \;\middle|\; a_i \in p, \; i \in m, \; \sum_{i \in m} a_i \ge t \right\}$$

is an $\mathbb{F}_p$-basis of I^t.

— *The* degree of nilpotency *of I is $m(p-1)+1$, i.e. $t = m(p-1)+1$ is the smallest positive integer for which $I^t = 0$.*

— *Let x be an indeterminate over $\mathbb{R}$. If $z_t = \dim(I^t/I^{t+1})$ for $0 \le t \le m(p-1)$, then*

$$\left(\sum_{j \in p} x^j \right)^m = \sum_{t=0}^{m(p-1)} z_t x^t$$ 4.10.5

and

$$\dim(I^t) + \dim(I^{m(p-1)+1-t}) = \dim(\mathbb{F}_p^G).$$ 4.10.6

— *The annihilator of I^t is $I^{m(p-1)+1-t}$.*

Proof: 1. By 4.10.2 the set B_0 is the Jennings basis B of $\mathbb{F}_p^G = I^0$.

2. In 4.10.3 we have determined a basis B' of I. Its elements are of the form $g - 1 = \prod_{i \in m} g_i^{a_i} - 1$ for $g \neq 1$. Consequently there exists at least one $i_0 \in m$ so that $a_{i_0} \neq 0$. Since (cf. Exercise 4.10.1)

$$g_i^{a_i} - 1 = (g_i - 1)^{a_i} - \sum_{j=1}^{a_i - 1} \binom{a_i}{j} (-1)^{a_i - j} (g_i^j - 1), \qquad 1 \leq a_i \leq p - 1,$$

we obtain by induction that $g_i^{a_i} - 1$ is a linear combination of elements of B_1. Assume that $a_i \neq 0$ for $i \in \{i_0, \ldots, i_{r-1}\}$. Then

$$\prod_{i \in m} g_i^{a_i} - 1 = \prod_{j \in r} g_{i_j}^{a_{i_j}} - 1$$

$$= \left(\prod_{j \in r-1} g_{i_j}^{a_{i_j}} - 1 \right) \left(g_{i_{r-1}}^{a_{i_{r-1}}} - 1 \right) + \left(\prod_{j \in r-1} g_{i_j}^{a_{i_j}} - 1 \right) + \left(g_{i_{r-1}}^{a_{i_{r-1}}} - 1 \right).$$

By induction we see that $g - 1$ is a linear combination of elements of B_1. Hence, B_1 is a generating set of I. Since B_1 is a subset of the Jennings basis B of $\mathbb{F}_p^G$ it is even a basis.

3. According to the definition of I^t (see Exercise 3.5.8, where the product of two ideals is defined) $B_t \subseteq I^t$ and, by 4.10.2, B_t is linearly independent. Since B_1 is a basis of I, the complex product $B_t = (B_1)^t$ is a basis of I^t. This proves the first statement.

4. The basis $B_{m(p-1)}$ of $I^{m(p-1)}$ contains only the element $\prod_{i \in m} (g_i - 1)^{p-1}$. It is the element of the Jennings basis of $\mathbb{F}_p^G$ with maximal Jennings weight. The elements of $I^{m(p-1)+1}$ are linear combinations of

$$\prod_{i \in m} (g_i - 1)^{a_i} \prod_{i \in m} (g_i - 1)^{p-1}$$

where at least one a_i is different from 0. Therefore, this term contains at least one factor of the form $(g_i - 1)^{b_i}$ where $b_i \geq p$. According to the last assertion of 4.10.3 we have $(g_i - 1)^{b_i} = 0$, whence this term vanishes and $I^{m(p-1)+1} = 0$. Hence, $m(p-1) + 1$ is the degree of nilpotency of I.

5. From the particular form of the basis B_t we deduce that z_t is the number of elements

$$\prod_{i \in m} (g_i - 1)^{a_i}$$

in the Jennings basis of $\mathbb{F}_p^G$ with Jennings weight t. On the other hand, the coefficient of x^t on the left hand side of 4.10.5 is the number of possibilities to choose m-tuples $(a_0, \ldots, a_{m-1})$ of integers with $a_i \in p$ and $\sum_{i \in m} a_i = t$. This proves 4.10.5. Moreover, we have

$$\sum_{t=0}^{m(p-1)} z_t x^t = \left(\sum_{j \in p} x^j \right)^m = \left(\sum_{j \in p} x^{(p-1)-j} \right)^m = x^{m(p-1)} \left(\sum_{j \in p} x^{-j} \right)^m$$

$$= x^{m(p-1)} \sum_{t=0}^{m(p-1)} z_t x^{-t} = \sum_{t=0}^{m(p-1)} z_{m(p-1)-t} x^t.$$

Comparing coefficients leads to $z_t = z_{m(p-1)-t}$, for $0 \leq t \leq m(p-1)$. Thus

$$\dim(I^{m(p-1)}) = z_{m(p-1)} = z_0 = \dim(\mathbb{F}_p^G/I) = 1.$$

Therefore, for $1 \leq t \leq m(p-1)$ we have

$$\dim(\mathbb{F}_p^G/I^t) = \dim(\mathbb{F}_p^G/I) + \sum_{s=1}^{t-1} \dim(I^s/I^{s+1})$$

$$= \dim(I^{m(p-1)}) + \sum_{s=1}^{t-1} \dim(I^{m(p-1)-s}/I^{m(p-1)-s+1})$$

$$= \dim(I^{m(p-1)-t+1}),$$

i.e. 4.10.6 is verified.

6. Lastly, from $I^t \cdot I^{m(p-1)+1-t} = I^{m(p-1)+1} = 0$ we deduce that $I^{m(p-1)+1-t}$ is contained in the annihilator of I^t. By Exercise 4.1.7, the dimension of I^t is the codimension of its annihilator. Thus, the last statement follows from 4.10.6. □

The Reed–Muller-codes over $\mathbb{F}_p$ form ideals in the group algebra $\mathbb{F}_p^G$, thus they are group algebra codes.

Theorem *Suppose that $0 \leq t \leq m(p-1)$. The linear extension φ of the mapping*   **4.10.7**

$$B_t \to RM^p_{m,m(p-1)-t} : \prod_{i \in m}(g_i - 1)^{a_i} \mapsto \prod_{i \in m} \prod_{l \in p-1-a_i} \frac{x_i - l}{p - 1 - a_i - l}$$

to I^t is an $\mathbb{F}_p$-isomorphism and even a linear isometry between I^t and $RM^p_{m,m(p-1)-t}$.
The fraction $\frac{x_i-l}{p-1-a_i-l}$ is understood as $(x_i - l)(p - 1 - a_i - l)^{-1} \in \mathbb{F}_p[x_i]$. If
$a_i = p - 1$ then the product $\prod_{l \in p-1-a_i} \frac{x_i-l}{p-1-a_i-l} = 1$, the empty product.

Proof: The product $\prod_{i \in m}(g_i - 1)^{a_i}$ is an element of the basis of I^t if and only if $\sum_{i \in m} a_i \geq t$. By definition

$$\deg(\varphi(\prod_{i \in m}(g_i - 1)^{a_i})) = \sum_{i \in m}(p - 1 - a_i) \leq m(p-1) - t.$$

Thus $\varphi(\prod_{i \in m}(g_i - 1)^{a_i})$ is an element of $RM^p_{m,m(p-1)-t}$. It is a linear combination of monomials of degree at most $m(p-1) - t$. Since

$$\varphi(\prod_{i \in m}(g_i - 1)^{a_i}) = \prod_{i \in m} \frac{x_i^{p-1-a_i}}{\prod_{l \in p-1-a_i}(p - 1 - a_i - l)} + f(x_0, \ldots, x_{m-1})$$

with $\deg f < \sum_{i \in m}(p - 1 - a_i)$, each element of the canonical basis of the code $RM^p_{m,m(p-1)-t}$ occurs in the image of a suitable element of the Jennings basis of I^t. This shows that φ is an isomorphism between I^t and $RM^p_{m,m(p-1)-t}$.

Now we have to prove that φ is even an isometry. The elements of $\mathbb{F}_p^G$ are of the form $\sum_{g \in G} \alpha_g g$ with $\alpha_g \in \mathbb{F}_p$. The elements $f \in B_m^p$ are identified with vectors

$$(f(0, \ldots, 0), \ldots, f(p-1, \ldots, p-1)) \in \mathbb{F}_p^{p^m}.$$

Let ψ be the bijective map

$$\psi: G \to \mathbb{F}_p^{p^m} \ : \ \prod_{i \in m} g_i^{a_i} \mapsto (p-1-a_0, \ldots, p-1-a_{m-1}).$$

We show that from $\prod_{i \in m}(g_i - 1)^{a_i} = \sum_{g \in G} \alpha_g g$ we obtain

$$\varphi\left(\prod_{i \in m}(g_i - 1)^{a_i}\right)(\psi(g)) = \alpha_g, \qquad g \in G.$$

This means that evaluating $\varphi(\prod_{i \in m}(g_i - 1)^{a_i})$ at $\psi(g)$ yields the coefficient of g in $\prod_{i \in m}(g_i - 1)^{a_i}$. Consider

$$\prod_{i \in m}(g_i - 1)^{a_i} = \prod_{i \in m}\left(\sum_{l=0}^{a_i} \binom{a_i}{l} g_i^l (-1)^{a_i - l}\right)$$

$$= \sum_{(j_0, \ldots, j_{m-1}) \in p^m} \prod_{i \in m} \binom{a_i}{j_i} g_i^{j_i} (-1)^{a_i - j_i}.$$

Then

4.10.8

$$\varphi\left(\prod_{i \in m}(g_i - 1)^{a_i}\right)(\psi(g_0^{j_0} \cdots g_{m-1}^{j_{m-1}})) = \prod_{i \in m} \prod_{l \in p-1-a_i} \frac{p-1-j_i-l}{p-1-a_i-l}.$$

If $j_{i_0} > a_{i_0}$ for some $i_0 \in m$, then $p - 1 - j_{i_0} < p - 1 - a_{i_0} \le p - 2 - a_{i_0}$ and therefore 4.10.8 is equal to 0. Assume that $0 \le j_i \le a_i$ for $i \in m$, then $j_i = a_i - r_i$ with $0 \le r_i \le a_i$ for $i \in m$. Since all these computations are done in $\mathbb{F}_p$ the last product in 4.10.8 can be written as

$$\prod_{l \in p-1-a_i} \frac{p-1-j_i-l}{p-1-a_i-l} = \prod_{l \in p-1-a_i} \frac{p-1-(a_i - r_i)-l}{p-1-a_i-l}$$

$$= \frac{\displaystyle\prod_{l=0}^{r_i-1}(p-1-a_i-l+r_i) \prod_{l=r_i}^{p-2-a_i}(p-1-a_i-(l-r_i))}{\displaystyle\prod_{l=0}^{p-2-a_i-r_i}(p-1-a_i-l) \prod_{l=p-1-a_i-r_i}^{p-2-a_i}(p-1-a_i-l)}$$

$$= \frac{\displaystyle\prod_{l=0}^{r_i-1}(-1)(a_i+(l+1)-r_i) \prod_{l=0}^{p-2-a_i-r_i}(p-1-a_i-l)}{\displaystyle\prod_{l=0}^{p-2-a_i-r_i}(p-1-a_i-l) \prod_{l=1}^{r_i} l}$$

$$= \frac{(-1)^{a_i-j_i} a_i (a_i - 1) \cdots (a_i - r_i + 1)}{(a_i - j_i)!}$$

$$= \binom{a_i}{a_i - j_i}(-1)^{a_i-j_i} = \binom{a_i}{j_i}(-1)^{a_i-j_i}.$$

Therefore, 4.10.8 is equal to

$$\prod_{i \in m} \binom{a_i}{j_i}(-1)^{a_i-j_i}.$$

In conclusion, we represent the elements of the Jennings basis of I^t as linear combinations of the standard basis of $\mathbb{F}_p^G$ and the elements of $\mathrm{RM}_{m,m(p-1)-t}^p$ as p^m-tuples over $\mathbb{F}_p$. Then for each element $\prod_{i \in m}(g_i - 1)^{a_i}$ of the Jennings basis the coefficient of the component g coincides with $\varphi(\prod_{i \in m}(g_i - 1)^{a_i})$ evaluated at $\psi(g)$. Since φ is an isomorphism for each $f \in \mathbb{F}_p^G$, $f = \sum_{g \in G} f(g)g$, we obtain $\varphi(f)(\psi(g)) = f(g)$ for $g \in G$, so that φ is a permutational isometry. $\square$

The minimum distance of Reed–Muller-codes can be evaluated using a visible generating set.

Theorem *The Jennings basis*   **4.10.9**

$$\left\{ \prod_{i \in m}(g_i - 1)^{a_i} \;\middle|\; a_i \in p \right\}$$

of $\mathbb{F}_p^G$ is visible.

Proof: Let $H = \langle g \rangle$ denote a cyclic group of order p. We consider the m-fold tensor power $\otimes^m \mathbb{F}_p^H$ of $\mathbb{F}_p^H$ with its canonical $\mathbb{F}_p$-basis

$$\{ g^{a_0} \otimes \ldots \otimes g^{a_{m-1}} \mid a_i \in p \}.$$

The mapping

$$\otimes^m \mathbb{F}_p^H \to \mathbb{F}_p^G \;:\; g^{a_0} \otimes \ldots \otimes g^{a_{m-1}} \mapsto g_0^{a_0} \cdots g_{m-1}^{a_{m-1}}$$

is an $\mathbb{F}_p$-isomorphism and even an isometry. Hence, we obtain the statement from 4.8.5 and 4.8.7. $\square$

Now we are able to describe the parameters of Reed–Muller-codes (cf. [11], [37], [121], [205]).

Theorem *The parameters of the Reed–Muller-code $\mathrm{RM}_{m,t}^p$ for $t = r(p-1) - s$ with*   **4.10.10**
$1 \leq r \leq m$ and $s \in p - 1$ are

$$(n, k, d, q) = \left(p^m, \sum_{i=(m-r)(p-1)+s}^{m(p-1)} z_i, (s+1)p^{m-r}, p \right),$$

where the integers z_i are given by 4.10.5.

Proof: It is clear that $\mathrm{RM}_{m,t}^p$ has length p^m. By 4.10.7, $\mathrm{RM}_{m,t}^p$ is isometric to the code $\mathrm{I}^{m(p-1)-t} = \mathrm{I}^{(m-r)(p-1)+s}$. This proves the assertion on the dimension of $\mathrm{RM}_{m,t}^p$. According to 4.10.4 and 4.10.9, $B_{(m-r)(p-1)+s}$ is a visible generating set for $\mathrm{I}^{(m-r)(p-1)+s}$. Hence it suffices to find a codeword of minimum weight in this basis. By Exercise 4.10.2, the Hamming weight of $\prod_{i \in m}(g_i - 1)^{a_i}$ is equal to $\prod_{i \in m}(a_i + 1)$. Hence, $\prod_{i \in m}(g_i - 1)^{a_i}$ with $a_0 = \ldots = a_{m-r-1} = p-1$ and $a_{m-r} = s$ is an element of minimum weight in $B_{(m-r)(p-1)+s}$.   □

Exercises

E.4.10.1   **Exercise** Prove that

$$g_i^{a_i} - 1 = (g_i - 1)^{a_i} - \sum_{j=1}^{a_i-1} \binom{a_i}{j}(-1)^{a-j}(g_i^j - 1)$$

holds true for a_i with $1 \le a_i \le p-1$.

E.4.10.2   **Exercise** Prove that

$$\mathrm{wt}\left(\prod_{i \in m}(g_i - 1)^{a_i}\right) = \prod_{i \in m}(a_i + 1).$$

E.4.10.3   **Exercise** Let C and C' be linear codes with given bases B and B'. Find an example of a linear isomorphism between C and C' which preserves the weights of all elements in the bases that is not an isometry.

E.4.10.4   **Exercise** Generalizing 2.4.8, prove that $\mathrm{RM}_{m,m(p-1)-t-1}^p$ is the dual code of $\mathrm{RM}_{m,t}^p$ for $0 \le t < m(q-1)$.

4.11

4.11 Encoding

The main advantage of cyclic codes is the fact that efficient encoding and decoding methods are available. In this section we will discuss an easily realizable circuit for the systematic encoding.

For the realization as circuits, three types of switches are used:

1. An *adder*, indicated by the symbol

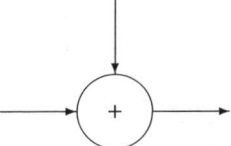

takes as input two field elements from different input streams and sends their sum to the output.

2. A *multiplier,* shown as

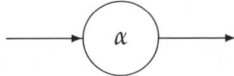

takes an element $\kappa \in \mathbb{F}_q$ from the input stream, multiplies it by $\alpha \in \mathbb{F}_q$, and outputs the product $\kappa\alpha$.

3. A *cell* is able to store field elements. We indicate it by the symbol

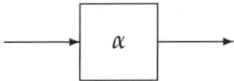

The cell reads an element from the input, stores this element for one clock cycle, and sends the element to the output stream afterwards (while at the same time reading another element from the input).

The multiplication of a message $f \in \mathbb{F}_q^k$ by the generator matrix corresponding to the generator polynomial g of the cyclic (n,k)-code over $\mathbb{F}_q$ in question is easily realized by an encoding circuit (see Exercise 4.11.1). Unfortunately, this encoder is not systematic, consequently it has the disadvantage that we need another transformation of the received vector in order to obtain the original message.

On the contrary, the use of a systematic encoding and decoding method allows us to take the first k components of the received vector as the original message. In the following, we will deduce a systematic encoder for cyclic codes. The underlying idea is based on the following observation.

Lemma *Let $C \leq \mathrm{Res}_{q,n}$ denote a cyclic (n,k)-code over $\mathbb{F}_q$ with generator polynomial g. Let $\mathrm{rem}_g(h)$ denote the unique remainder of h after division by g (cf. Exercise 3.1.6), then the map*   **4.11.1**

$$\varepsilon : \mathbb{F}_q[x]_{<k} \to C : f \mapsto x^{n-k}f - \mathrm{rem}_g(x^{n-k}f) + I(x^n - 1)$$

is a bijection.

Proof: 1. Consider an $f \in \mathbb{F}_q[x]_{<k}$. According to the Division Theorem there exist polynomials s and r with $x^{n-k}f = sg + r$ and either $\deg r < \deg g$ or $r = 0$. Consequently $r = \mathrm{rem}_g(x^{n-k}f)$ and

$$\varepsilon(f) = x^{n-k}f - \mathrm{rem}_g(x^{n-k}f) = sg \in C.$$

2. Let $f = \sum_{i \in k} f_i x^i$. Since g has degree $n - k$, $r = \sum_{i \in n-k} r_i x^i$. Hence $x^{n-k}f - \mathrm{rem}_g(x^{n-k})$ has the following n-tuple of coefficients:

$$(-r_0, \ldots, -r_{n-k-1}, f_0, \ldots, f_{k-1}).$$

This shows clearly that ε is injective, and so, since $|\mathbb{F}_q[x]_{<k}| = |C| = q^k$, it is even bijective.   ☐

Now we claim that the division of polynomials can be realized by a division shift register. Figure 4.4 shows an m-step *division shift register* with the monic *recoupling polynomial* $g = \sum_{i=0}^{m} g_i x^i$, which we abbreviate by $\mathrm{DR}(g)$; it has two input gates, indicated by A and B.

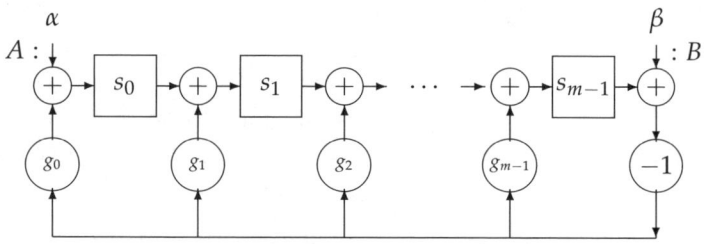

Fig. 4.4 Division shift register $\mathrm{DR}(g)$

4.11.2   **Lemma** *Assume that the shift register $\mathrm{DR}(g)$ is initialized with the coefficients of the polynomial $s = \sum_{i \in m} s_i x^i$ and that it receives the input $\alpha \in \mathbb{F}_q$ at A and $\beta \in \mathbb{F}_q$ at B. Then, after one clock cycle, $\mathrm{DR}(g)$ contains the coefficients of*

$$\mathrm{rem}_g(xs + \alpha + \beta x^m).$$

Proof: We assume that after one clock cycle the register contains the coefficients of $\tilde{s} = \sum_{i \in m} \tilde{s}_i x^i$. Then

$$\tilde{s}_0 = \alpha - g_0(\beta + s_{m-1})$$

and

$$\tilde{s}_i = s_{i-1} - g_i(\beta + s_{m-1}) \text{ for } 1 \leq i \leq m - 1.$$

This gives:

$$
\begin{aligned}
\tilde{s} &= \sum_{i \in m} \tilde{s}_i x^i \\
&= \sum_{i=1}^{m-1} (s_{i-1} - g_i(\beta + s_{m-1}))x^i + \alpha - g_0(\beta + s_{m-1}) \\
&= \sum_{i=1}^{m-1} s_{i-1}x^i + \alpha - (\beta + s_{m-1}) \sum_{i \in m} g_i x^i \\
&= x\left(\sum_{i=1}^{m-1} s_{i-1}x^{i-1}\right) + \alpha + (\beta + s_{m-1})x^m - (\beta + s_{m-1})g \\
&= xs + \alpha + \beta x^m - (\beta + s_{m-1})g \\
&= \text{rem}_g(xs + \alpha + \beta x^m). \qquad \square
\end{aligned}
$$

We are now in a position to describe the behavior of the division shift register $DR(g)$ with respect to successive input.

Corollary *Let g be a monic polynomial of degree m and assume that $DR(g)$ is initialized with zeros.* **4.11.3**

- *If we input at A the components of the vector $f \in \mathbb{F}_q^n$ as a sequence $f_{n-1}, \dots, f_0$ and at B always 0, then, after n successive clock cycles, the content of the division shift register $DR(g)$ is*

$$\text{rem}_g(f).$$

- *If we input at B the components of the vector $f \in \mathbb{F}_q^n$ as a sequence $f_{n-1}, \dots, f_0$ and at A always 0, then, after n clock cycles, the content of the division shift register $DR(g)$ is*

$$\text{rem}_g(x^m f). \qquad \square$$

This shows that we can construct a systematic encoder of C using a circuit that performs division of polynomials.

Corollary *Let $C \leq \text{Res}_{q,n}$ denote a cyclic (n, k)-code over $\mathbb{F}_q$ with generator polynomial g. The mapping* **4.11.4**

$$f \mapsto x^{n-k}f - \text{rem}_g(x^{n-k}f)$$

is a systematic encoding:

$$\mathbb{F}_q^k \to \mathbb{F}_q^n : (f_0, \dots, f_{k-1}) \mapsto (-r_0, \dots, -r_{n-k-1}, f_0, \dots, f_{k-1}),$$

corresponding to a generator matrix of the form $\Gamma = (A \mid I_k)$ of C. Being a cyclic code, $(I_k \mid A)$ generates the same code, and so,

$$\mathbb{F}_q^k \to \mathbb{F}_q^n : f \mapsto f - x^k \text{rem}_g(x^{n-k}f),$$

$$(f_0, \ldots, f_{k-1}) \mapsto (f_0, \ldots, f_{k-1}, -r_0, \ldots, -r_{n-k-1})$$

is the systematic encoding corresponding to $(I_k \mid A)$.

In order to determine the coefficients of $r = \mathrm{rem}_g(x^{n-k}f)$, *we start with the division shift register* $\mathrm{DR}(g)$ *initialized with zeros, and input at A always zeros, and at B successively the elements* $f_{k-1}, \ldots, f_0$.   □

4.11.5   **Example** The check digits of the binary $(7,4)$-Hamming-code with the generator polynomial $g = x^3 + x + 1$ can be obtained by the following division shift register $\mathrm{DR}(g)$:

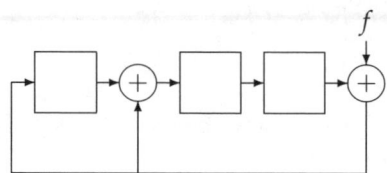

Fig. 4.5 The division shift register $\mathrm{DR}(g)$ for the encoding of the $(7,4)$-Hamming-code

◇

In 4.5.13 we have shown that the Discrete Fourier Transformation can be used as a natural encoding method for narrow-sense Reed–Solomon-codes.

Exercises

E.4.11.1   **Exercise** Consider a cyclic (n,k)-code C over $\mathbb{F}_q$ with generator polynomial g. Derive an encoder that realizes, for every message $f \in \mathbb{F}_q^k$, the multiplication $c = f \cdot \Gamma$. Use, for this purpose, the generator matrix Γ given in 4.2.5.

E.4.11.2   **Exercise** Encode the message (f_0, f_1) by using a narrow-sense $(6, 2, 5)$-Reed–Solomon-code over $\mathbb{F}_7$.

4.12

4.12 Permutation Decoding

We describe a decoding method due to MacWilliams [138] that uses automorphisms of a code to permute the entries of the received vector in such a way that there is no error on a given set of information bits. This method is particularly useful for cyclic codes since cyclic codes have all cyclic shifts of the coordinates as automorphisms.

Let C be a cyclic (n,k,d,q)-code with generator polynomial g, where the length n of C is not divisible by the characteristic of the field $\mathbb{F}_q$. As usual we denote by $y \in \mathbb{F}_q^n$ a received vector, and we assume that the number of errors that have occurred during the transmission is $t \leq (d-1)/2$.

The syndrome depends on the check matrix we use. Here we discuss two different ways to evaluate the syndrome. Our first method uses the *variety*

$$V(C) = \{\alpha_0, \ldots, \alpha_{n-k-1}\}$$

of the code C in question. As pointed out in 4.2.16, the matrix

$$\widetilde{\Delta} = \begin{pmatrix} 1 & \alpha_0 & \alpha_0^2 & \cdots & \alpha_0^{n-1} \\ 1 & \alpha_1 & \alpha_1^2 & \cdots & \alpha_1^{n-1} \\ & \cdots & & \cdots & \\ 1 & \alpha_{n-k-1} & \alpha_{n-k-1}^2 & \cdots & \alpha_{n-k-1}^{n-1} \end{pmatrix}$$

is a check matrix of a code $\widetilde{C}$ over an field extension of $\mathbb{F}_q$ that restricts to C, i.e. $C = \widetilde{C} \downarrow \mathbb{F}_q = \widetilde{C} \cap \mathbb{F}_q^n$.

Theorem *The syndrome $y \cdot \widetilde{\Delta}^\top$ of the received vector $y \in \mathbb{F}_q^n$ can be evaluated using $n-k$ simultaneous division shift registers $DR(x - \alpha_i)$, $i \in n - k$, each initialized with zeros. The coefficients of y are fed in reverse order, i.e. $y_{n-1}, \ldots, y_0$, into the input gate A (see Fig. 4.6) of each of these division shift registers.*　　4.12.1

Proof: The syndrome $y \cdot \widetilde{\Delta}^\top$ of y satisfies

$$y \cdot \widetilde{\Delta}^\top = (y(\alpha_0), \ldots, y(\alpha_{n-k-1})).$$

For each $\alpha \in \mathbb{F}_q$ we have that

$$y(\alpha) = \mathrm{rem}_{x-\alpha}(y),$$

where rem is as in 4.11.1. According to 4.11.3, $y(\alpha_i)$ can be evaluated using the circuit shown in Fig. 4.6.　　□

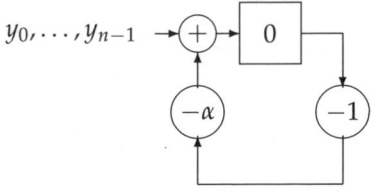

Fig. 4.6 Evaluation of $y(\alpha)$ with a division shift register

For the remainder of this section, let us fix an information set as described in Section 1.7, i.e. a set of k coordinates whose values determine the codeword uniquely. We assume that the first k positions in the code form such an information set. In this case, we may assume that the code has a *systematic generator matrix* $\Gamma = (I_k \mid A)$, so that $\Delta = (-A^\top \mid I_{n-k})$ is a check matrix of C by Exercise 1.3.9. The following result enables us to compute the syndrome $y \cdot \Delta^\top$ of a received vector y using a division shift register $\mathrm{DR}(g)$.

4.12.2 **Theorem** *If $\Gamma = (I_k \mid A)$ is a generator matrix and $\Delta = (-A^\top \mid I_{n-k})$ is the corresponding check matrix of the cyclic (n,k)-code C generated by g, the syndrome of a received vector y is the sequence of coefficients of the remainder $\mathrm{rem}_g(x^k y) = \sum_{i \in n-k} r_i x^i$,*

$$(y_0, \ldots, y_{n-1}) \cdot \Delta^\top = (r_0, \ldots, r_{n-k-1}).$$

Proof: 1. Multiplying $y(x)$ by x^k, the residue $\mathrm{rem}_{x^n-1}(x^k y)$ corresponds to the vector $(y_{n-k}, \ldots, y_{n-1}, y_0, \ldots, y_{n-k-1})$. Since the generator polynomial divides $x^n - 1$ we have $\mathrm{rem}_g(x^k y) = \mathrm{rem}_g(\mathrm{rem}_{x^n-1}(x^k y))$.

2. We divide $y(x)$ by x^k to obtain unique polynomials $f, s \in \mathbb{F}_q[x]$ such that

$$y = x^k s + f,$$

where $f = 0$ or $\deg f < k$. Moreover, $\deg s < n - k$. In terms of the sequences of coefficients this means that

$$(y_0, \ldots, y_{n-1}) = (f \mid s) := (f_0, \ldots, f_{k-1}, s_0, \ldots, s_{n-k-1}).$$

3. Replacing the sequence s of coefficients by the vector

$$\tilde{f} := (f_0, \ldots, f_{k-1}) \cdot A \in \mathbb{F}_q^{n-k}$$

we obtain $(f \mid \tilde{f}) \cdot \Delta^\top = -f \cdot A + \tilde{f} = -f \cdot A + f \cdot A = 0$. Hence,

$$f(x) + x^k \tilde{f}(x) + \mathrm{I}(x^n - 1) \in C.$$

Since C is cyclic, also

$$x^{n-k} f(x) + \tilde{f}(x) + \mathrm{I}(x^n - 1) \in C,$$

from which we deduce that $x^{n-k} f(x) + \tilde{f}(x)$ is a multiple of the generator polynomial g, and so

$$\tilde{f} = \mathrm{rem}_g(\tilde{f}) = -\mathrm{rem}_g(x^{n-k} f).$$

This gives that the sequence corresponding to

$$y \cdot \Delta^\top = (f \mid s) \cdot \Delta^\top = s - f \cdot A = s - \tilde{f}$$

is the sequence of coefficients of

$$s(x) + \text{rem}_g(x^{n-k}f) = \text{rem}_g(s + x^{n-k}f) = \text{rem}_g(x^k y),$$

since $s = \text{rem}_g(s)$ and $s + x^{n-k}f$ corresponds to $(s_0, \ldots, s_{n-k-1}, f_0, \ldots, f_{k-1}) = (y_k, \ldots, y_{n-1}, y_0, \ldots, y_{k-1})$. □

Permutation decoding is another way of using the syndrome $y \cdot \Delta^\top$ to correct errors. The idea is to apply a permutation σ from the automorphism group of the code to the received vector y in such a way that $\sigma(y)$ contains no errors in a fixed set of information places. We apply the parity check equations to $\sigma(y)$ and obtain a correct but possibly permuted codeword. By reversing the permutation we obtain the codeword transmitted originally. The basic result for this method is

Theorem *Consider a linear code C with a systematic generator matrix $\Gamma = (I_k \mid A)$ and the corresponding check matrix $\Delta = (-A^\top \mid I_{n-k})$. Assume that the codeword c was sent, that $y = c + e$ was received, and that $\text{wt}(e) = t$ with $2t + 1 \le \text{dist}(C)$. Then the weight of the syndrome $y \cdot \Delta^\top$ is not greater than t if and only if y is correct in all its information places. In formal terms* 4.12.3

$$\text{wt}(y \cdot \Delta^\top) \le t \iff e_0 = \ldots = e_{k-1} = 0.$$

Proof: The assumption $e_0 = \ldots = e_{k-1} = 0$ yields

$$y \cdot \Delta^\top = e \cdot \Delta^\top = (e_k, \ldots, e_{n-1})$$

and so $\text{wt}(y \cdot \Delta^\top) = \text{wt}(e) = t$.
   Conversely, if $(e_0, \ldots, e_{k-1}) \ne 0$ we may write

$$v := (e_0, \ldots, e_{k-1}), \quad w := (e_k, \ldots, e_{n-1}).$$

Since $y \cdot \Delta^\top = e \cdot \Delta^\top = -v \cdot A + w$, we can use the inequality from Exercise 4.12.1 in the form

$$\text{wt}(y \cdot \Delta^\top) \ge \text{wt}(v \cdot A) - \text{wt}(w).$$

As $v \cdot (I_k \mid A) = v + v \cdot A$ is an element of C, we have

$$\begin{aligned}
2t + 1 &\le \text{wt}(v \cdot A) + \text{wt}(v) \\
&\le \text{wt}(y \cdot \Delta^\top) + \text{wt}(v) + \text{wt}(w) \\
&= \text{wt}(y \cdot \Delta^\top) + \text{wt}(e) \\
&= \text{wt}(y \cdot \Delta^\top) + t,
\end{aligned}$$

from which $\text{wt}(y \cdot \Delta^\top) \ge t + 1$ follows. □

We are now in a position to formulate the announced algorithm due to MacWilliams:

4.12.4   **Algorithm (permutation decoding)** Consider a linear code C with minimum distance $d \geq 2t + 1$, systematic generator matrix $\Gamma = (I_k \mid A)$ and corresponding check matrix $\Delta = (-A^\top \mid I_{n-k})$. In order to correct up to t transmission errors with permutation decoding, for each potential error vector $e \in \mathbb{F}_q^n$ with $\mathrm{wt}(e) \leq t$ we must know an automorphism $\sigma_e \in S_n$ of C so that the first k components of $\sigma_e(e)$ are 0.

We apply all these automorphisms to the received vector $y \in \mathbb{F}_q^n$, obtain the permuted vectors $\sigma_e(y)$ and determine the corresponding syndromes $\sigma_e(y) \cdot \Delta^\top$.

— If $\mathrm{wt}(\sigma_e(y) \cdot \Delta^\top) \leq t$ for an automorphism σ_e, then the information places, i.e. the first k components, of $\sigma_e(y)$ are correct, and the error of $\sigma_e(y)$ is

$$e' = (0, \ldots, 0, e'_k, \ldots, e'_{n-1}) = (\mathbf{0}_k \mid \sigma_e(y) \cdot \Delta^\top).$$

By reverting this automorphism, we can reconstruct c as

$$c = \sigma_e^{-1}(\sigma_e(y) - e').$$

— If we cannot find such a permutation σ_e, we conclude that more than t errors have occurred and we cannot correct the error.   □

To apply this algorithm, it remains to find a suitable set of permutations in the automorphism group of the code that does the job. That means, we want a set of permutations σ which guarantees that for each error vector e the permuted vector $\sigma(e)$ has zeros in all its first k positions. Such a set is called a *PD-set* (for permutation decoding). Of course, we want a *small* PD-set. For cyclic codes, this can be accomplished with trap decoding.

4.12.5   **Trap decoding** Let C be a cyclic (n, k, d)-code. If

$$t \leq \left\lfloor \frac{d-1}{2} \right\rfloor \quad \text{and} \quad t \leq \frac{n-1}{k},$$

then every error vector e of weight t contains a consecutive set of at least k zeros. If $\pi = (0, \ldots, n-1)$ is the cyclic shift then the assumptions imply that there is an $i \in n$ such that $\pi^i(e)$ has k leading zeros. That is, it suffices to consider the cyclic group

$$G = \langle \pi \rangle$$

of order n. This form of permutation decoding using the cyclic group of order n is called *error trapping* or *trap decoding*. It permits the correction of up to t errors.   ◇

For the $(23,12)$-Golay-code, trap decoding enables us to correct only a single error. It is, therefore, not very efficient, but it allows the correction of certain burst errors (cf. Section 5.3).

In many cases a subset of the set of elements of the cyclic group suffices, as the following example shows.

Example The binary $(7,4)$-Hamming-code has minimum weight $d = 3$ and hence corrects 1 error. Since $t = 1 \leq (7-1)/4$, it admits a trap decoder. For each vector $e \in \mathbb{F}_2^7$ of weight 1 we determine a suitable permutation π^j in the cyclic group $G := \langle \pi \rangle$, $\pi := (0,\ldots,6)$, of order 7, so that the cycled vector $\pi^j(e)$ has 4 leading zeros. For each possible single bit error e, the following table shows a suitable exponent j and the corresponding cyclic shift $\pi^j(e)$ with 4 leading zeros.

4.12.6

e	j	$\pi^j(e)$
0000001	0	0000001
0000010	0	0000010
0000100	0	0000100
0001000	3	0000001
0010000	3	0000010
0100000	3	0000100
1000000	6	0000001

In this case, it suffices to consider the PD-set $\pi^0 = \mathrm{id}$, π^3, and π^6 rather than the full cyclic group of order 7. $\diamond$

It is, of course, an interesting problem to determine the minimal size of a set of permutations for permutation decoding. The following lower bound is due to D. M. Gordon [73], see also [94]:

Theorem *A lower bound for the size of a PD-set for a t-error-correcting (n,k)-code is*

4.12.7

$$\left\lceil \frac{n}{n-k} \left\lceil \frac{n-1}{n-k-1} \left[\cdots \left\lceil \frac{n-t+1}{n-k-t+1} \right\rceil \cdots \right] \right\rceil \right\rceil .$$

This bound is sharp in the sense that there are codes for which no smaller set of permutations can be used.

Proof: A minimal set of permutations necessary in the worst case needs to contain enough permutations π to move the set R of redundancy places (in systematic codes: $R = \{k,\ldots,n-1\}$) so that the resulting sets $\pi(R)$ cover all possible t-subsets of n. Let S be the set of sets $\pi(R)$ where π runs through such a minimal set of permutations. Then each element of S is of cardinality $n-k$.

We indicate the cardinality of S, which is also the number of permutations in a minimal set, by

$$N(t, n-k, n).$$

Consequently, the set of pairs

$$\{(i, T) \mid i \in T, \ T \in S\}$$

is of cardinality $(n-k)N(t, n-k, n)$. This set can be described as the disjoint union

$$\bigcup_{j \in n} \{(j, T) \mid j \in T, \ T \in S\}.$$

For $j \in n$ let S_j be the set $\{T \in S \mid j \in T\}$. Since the elements of S cover all possible t-subsets of n, and $j \in T$ for all $T \in S_j$, the elements of S_j cover all possible t-subsets of n which contain j. Thus, the elements of $\{T \setminus \{j\} \mid T \in S_j\}$ cover all possible $t-1$-subsets of $n \setminus \{j\}$. Hence

$$|S_j| \geq N(t-1, n-k-1, n-1), \qquad j \in n,$$

and, therefore,

$$(n-k) \cdot N(t, n-k, n) \geq n \cdot N(t-1, n-k-1, n-1)$$

holds true. Finally, the statement follows from

$$N(1, n-k-t+1, n-t+1) = \left\lceil \frac{n-t+1}{n-k-t+1} \right\rceil.$$

A PD-set for the binary $(24, 12, 8)$-Golay-code of size 14 can be found in [94]. This shows that the lower bound is sharp. $\square$

Let C be a cyclic (n, k)-code. A modification of trap decoding is the *Kasami-decoder*, which often allows the correction of a higher number of errors. It uses a set of polynomials $u^{(0)} = 0, u^{(1)}, \ldots, u^{(s)} \in \mathbb{F}_q[x]$ of degree less than k, called *covering polynomials* of C. We suppose that for each error vector e with $\mathrm{wt}(e) \leq t$ there exists a covering polynomial $u^{(i)}$, such that the j-fold cyclic shift $\pi^j(e)$ of e, for a suitable j, agrees with (the sequence of coefficients of) $u^{(i)}$ in the information positions, i.e. in the first k places. Kasami-decoding is based on the following result:

4.12.8   **Theorem**   *Let C be a linear code of minimum distance $d \geq 2t+1$ with a systematic generator matrix $\Gamma = (I_k \mid A)$ and the corresponding check matrix $\Delta = (-A^\top \mid I_{n-k})$. Assume that $c \in C$ was sent, that $y = c + e$ was received, and that $\mathrm{wt}(e) \leq t$. Let $u^{(i)}$ be a polynomial of degree less than k. The information positions of $y + u^{(i)}$ are correct if and only if*

$$\mathrm{wt}((y + u^{(i)}) \cdot \Delta^\top) \leq t - \mathrm{wt}(u^{(i)}).$$

In this case, the error vector is given by

$$e = \left(-u^{(i)} \mid (y + u^{(i)}) \cdot \Delta^\top\right).$$

Proof: Assume that the first k positions of $y + u^{(i)}$ are correct. Then $u^{(i)}$ corrects all errors among the first k positions of y. These are $\mathrm{wt}(u^{(i)})$ corrections. Since $\deg u^{(i)} < k$, at most $t - \mathrm{wt}(u^{(i)})$ errors occur in the final $n - k$ positions. Therefore, $y + u^{(i)} - c = (0 \mid e')$ where $e' \in \mathbb{F}_q^{n-k}$ is of weight at most $t - \mathrm{wt}(u^{(i)})$, and

$$(y + u^{(i)}) \cdot \Delta^\top = (y + u^{(i)} - c) \cdot \Delta^\top = (0 \mid e') \cdot \Delta^\top = e'.$$

Consequently, $\mathrm{wt}((y + u^{(i)}) \cdot \Delta^\top) \leq t - \mathrm{wt}(u^{(i)})$.

Conversely, we assume that $\mathrm{wt}((y + u^{(i)}) \cdot \Delta^\top) \leq t - \mathrm{wt}(u^{(i)})$. Writing $y + u^{(i)} - c$ as $(e'' \mid e')$ with $e'' \in \mathbb{F}_q^k$ and $e' \in \mathbb{F}_q^{n-k}$, we show that $e'' = 0$. Suppose that $\mathrm{wt}(e'') \neq 0$. We can restrict ourselves to covering polynomials of weight at most t. Due to the construction we express the weights of e'' and e' as $\mathrm{wt}(e'') = v + w$ and $\mathrm{wt}(e') = z$, where

$$v = \left|\left\{j \in k \,\middle|\, y_j \neq c_j \text{ and } u_j^{(i)} \neq c_j - y_j\right\}\right|,$$

$$w = \left|\left\{j \in k \,\middle|\, y_j = c_j \text{ and } u_j^{(i)} \neq 0\right\}\right|,$$

$$z = \left|\left\{j \in \{k, \ldots, n - 1\} \,\middle|\, y_j \neq c_j\right\}\right|.$$

These cardinalities satisfy $v + z \leq t$ and $w \leq \mathrm{wt}(u^{(i)})$. From

$$(y + u^{(i)}) \cdot \Delta^\top = (y + u^{(i)} - c) \cdot \Delta^\top = (e'' \mid e') \cdot \Delta^\top = e' - e'' \cdot A$$

we obtain that

$$\mathrm{wt}((y + u^{(i)}) \cdot \Delta^\top) \geq \mathrm{wt}(e'' \cdot A) - \mathrm{wt}(e').$$

Since A comes from a systematic generator matrix $\Gamma = (I_k \mid A)$ of a code with minimum distance $d \geq 2t + 1$, we have

$$\mathrm{wt}(e'' \cdot A) \geq d - \mathrm{wt}(e'') \geq 2t + 1 - (v + w)$$

and, consequently,

$$\mathrm{wt}((y + u^{(i)}) \cdot \Delta^\top) \geq 2t + 1 - v - w - z \geq 2t + 1 - t - \mathrm{wt}(u^{(i)}) > t - \mathrm{wt}(u^{(i)}).$$

We have just shown that when $e'' \neq 0$, then $\mathrm{wt}((y + u^{(i)}) \cdot \Delta^\top) \not\leq t - \mathrm{wt}(u^{(i)})$. Finally, if $\mathrm{wt}((y + u^{(i)}) \cdot \Delta^\top) \leq t - \mathrm{wt}(u^{(i)})$, then the error vector is equal to

$$y - c = (0 \mid e') - u^{(i)} = \left(-u^{(i)} \mid (y + u^{(i)}) \cdot \Delta^\top\right). \qquad \square$$

For cyclic codes, the Kasami-decoder computes the cyclic shifts $\pi^j(y)$ of the received vector y until the inequality

$$\text{wt}((\pi^j(y) + u^{(i)}) \cdot \Delta^\top) \leq t - \text{wt}(u^{(i)})$$

is satisfied for a covering polynomial $u^{(i)}$. Then it decodes y into the codeword

$$c = \pi^{n-j}\left(\pi^j(y) - (-u^{(i)} \mid (y + u^{(i)}) \cdot \Delta^\top)\right).$$

4.12.9 **Example** The set $\{0, x^5, x^6\}$ is a set of covering polynomials of the binary $(23, 12)$-Golay-code with generator polynomial $g = x^{11} + x^{10} + x^6 + x^5 + x^4 + x^2 + 1$. It permits us to correct up to $t = 3$ errors. $\diamond$

Exercises

E.4.12.1 **Exercise** Prove that if x and y are in $\mathbb{F}_q^n$, then $\text{wt}(x + y) \geq \text{wt}(x) - \text{wt}(y)$.

E.4.12.2 **Exercise** Realize a Kasami-decoder using a division shift registers.

E.4.12.3 **Exercise** Prove that $u^{(0)} = 0$, $u^{(1)} = x^5$, and $u^{(2)} = x^6$ is a set of covering polynomials for the code in 4.12.9.

4.13

4.13 Error-Correcting Pairs

For decoding BCH-, Reed–Solomon- and Goppa-codes, the method of *error-correcting pairs* was introduced by R. Pellikaan [161] and independently by R. Kötter [115].

Let C be an (n, k, d, q)-code and assume that the vector $y = c + e$ has been received. Moreover, we suppose that $I := \text{supp}(e)$ is contained in a subset $J \subseteq n$, where $|J| \leq d - 1$. Consider an $\mathbb{F}_q$-basis of $C^\perp$

$$\left\{w^{(0)}, \ldots, w^{(n-k-1)}\right\}.$$

4.13.1 **Lemma** *The error vector e is the unique solution $u \in \mathbb{F}_q^n$ of the system of linear equations*

$$\langle u, w^{(i)} \rangle = \langle y, w^{(i)} \rangle \quad \text{for all } i \in n - k,$$
$$u_j = 0 \quad \text{for all } j \in n \setminus J.$$

Proof: It is clear that e is a solution. Moreover, if u is a solution, then $u - e$ is contained in C, and $\text{wt}(u - e) \leq |J|$, so that we can deduce from our assumption on J that $u - e = 0$, and so $u = e$.   □

Hence, it remains to construct such a set J. For this purpose we assume that $\mathrm{wt}(e) \le t \le (d-1)/2$, i.e. no more than t errors have occurred during the transmission of c.

Consider two linear codes A and B of length n over an extension field $\mathbb{F}_{q^m}$ of $\mathbb{F}_q$ – later on we will see that such codes can easily be constructed if we are given a BCH-code in the strict sense, a Reed–Solomon-code or a classical Goppa-code. In the following we use the notation $a \cdot b$ for the Hadamard product of $a \in A$ and $b \in B$,

$$a \cdot b := (a_0 b_0, \dots, a_{n-1} b_{n-1}),$$

while $A \cdot B$ means the (Hadamard) complex product $\{a \cdot b \mid a \in A, b \in B\}$. The codes A and B are supposed to have the following four properties:

- $A \cdot B \subseteq C^{\perp}$, where the elements of C are considered as elements of $\mathbb{F}_{q^m}^n$. This condition guarantees that the vector space (corresponding to the received vector y)

$$K_y := \{a \in A \mid \langle y, a \cdot b \rangle = 0 \text{ for all } b \in B\}$$

coincides with the vector space (corresponding to the error vector e)

$$K_e := \{a \in A \mid \langle e, a \cdot b \rangle = 0 \text{ for all } b \in B\}.$$

K_y contains the set

$$A(I) := \{a \in A \mid a_i = 0 \text{ for all } i \in I\}$$

of all codewords in A, the I-coordinates of which are zero. This is obvious if we recall that $I = \mathrm{supp}(e)$, since for each $a \in A(I)$ and $b \in B$ we have

$$\langle y, a \cdot b \rangle = \langle e, a \cdot b \rangle = \sum_{i \in I} a_i b_i e_i = 0,$$

and so $a \in K_y$.

- $\dim(A) \ge t+1$: This condition guarantees that $K_y \ne \{0\}$. The set K_y contains $A(I)$, and $A(I)$ is the intersection of A with

$$D := \{v \in \mathbb{F}_{q^m}^n \mid v_i = 0 \text{ for all } i \in I\}.$$

Obviously, D is a vector space of dimension $\ge n - t$. If $A(I) = \{0\}$, then $A + D = A \oplus D$ is a direct sum and a subspace of $\mathbb{F}_{q^m}^n$ of dimension at least $n+1$ which is impossible. Therefore, $\dim(A(I)) \ge 1$.

- $\mathrm{dist}(B^{\perp}) \ge t+1$: This fact implies $K_y = A(I)$, since each $a \in K_y$ satisfies

$$0 = \langle y, a \cdot b \rangle = \langle e, a \cdot b \rangle = \langle e \cdot a, b \rangle, \quad \text{for all } b \in B.$$

Thus $e \cdot a \in B^\perp$. But $\mathrm{wt}(e \cdot a) \leq \mathrm{wt}(e) \leq t$, and, since $\mathrm{dist}(B^\perp) \geq t+1$, we have $e \cdot a = 0$. Hence, a is zero whenever e is nonzero, so that $a \in A(I)$.

— $\mathrm{dist}(A) + \mathrm{dist}(C) \geq n+1$: This condition assures that each nonzero $a \in A$ contains at most $d-1$ zeros, i.e. the set

$$J_a := \{i \in n \mid a_i = 0\}$$

consists of at most $(d-1)$ elements. Indeed, for each $0 \neq a \in A(I)$ we have

$$n - |J_a| = \mathrm{wt}(a) \geq \mathrm{dist}(A) \geq n - \mathrm{dist}(C) + 1,$$

therefore $|J_a| \leq \mathrm{dist}(C) - 1 = d - 1$.

4.13.2 **Definition (error-correcting pair)** Let C be a linear code over $\mathbb{F}_q$ of length n and minimum distance $d \geq 2t+1$. Any pair (A, B) of linear codes of length n over $\mathbb{F}_{q^m}$ is called *t-error-correcting pair for C* if the following four conditions are satisfied:

— $A \cdot B \subseteq C^\perp$,

— $\dim(A) \geq t+1$,

— $\mathrm{dist}(B^\perp) \geq t+1$ and

— $\mathrm{dist}(A) + \mathrm{dist}(C) \geq n+1$.   ◇

4.13.3 **Theorem** *Let C be a linear (n, k, d, q)-code, and let (A, B) be a t-error-correcting pair for C for some $1 \leq t \leq (d-1)/2$. Then it is possible to decode $y = c + e$ with $\mathrm{wt}(e) \leq t$ using the linear system from 4.13.1.*

Proof: We determine K_y, choose a nonzero element $a \in K_y$ and consider the set $J := J_a$ of coordinates where a is zero. Since we know that J contains the support I of the error vector and that it consists of at most $d-1$ elements, we obtain e by solving the system of linear equations.   □

Now we show how to decode Generalized Reed–Solomon-codes, BCH-codes with $b = 0$, as well as classical Goppa-codes using a pair of error-correcting codes.

4.13.4 **Example** Let $\kappa = (\kappa_0, \ldots, \kappa_{n-1})$ denote a sequence of pairwise different elements of $\mathbb{F}_q$. We consider, without restriction, the Generalized Reed–Solomon-code $C = \mathrm{GRS}_{2t}(\kappa, \mathbf{1}_n)^\perp$. According to 4.5.17 and 2.5.1, this code is an $(n, n - 2t)$-MDS-code. We claim that the Generalized Reed–Solomon-codes

$$A := \mathrm{GRS}_{t+1}(\kappa, \mathbf{1}_n) \quad \text{and} \quad B := \mathrm{GRS}_t(\kappa, \mathbf{1}_n)$$

form a t-error-correcting pair for C over $\mathbb{F}_q$. Because of 4.5.17, the codes A and B are $(n, t+1)$ and (n, t)-MDS-codes, respectively. It follows from 2.5.1 that $B^{\perp}$ is an $(n, n-t, t+1)$-MDS-code and hence $\dim(A) = t+1$, $\text{dist}(A) + \text{dist}(C) = (n-t) + (2t+1) \geq n+1$ and $\text{dist}(B^{\perp}) = t+1$. Now we choose $f \in \mathbb{F}_q[x]_{<t+1}$ and $g \in \mathbb{F}_q[x]_{<t}$. Then, by definition, $(f(\kappa_0), \ldots, f(\kappa_{n-1})) \in A$ and $(g(\kappa_0), \ldots, g(\kappa_{n-1})) \in B$. Since $\deg(fg) < 2t$, the Hadamard product $(fg(\kappa_0), \ldots, fg(\kappa_{n-1}))$ of these two codewords is contained in $\text{GRS}_{2t}(\kappa, \mathbf{1}_n) = C^{\perp}$. Consequently $A \cdot B \subseteq C^{\perp}$. Thus, all four conditions for a t-error-correcting pair are satisfied. $\diamond$

Example Let C be a BCH-code of length $n = q^m - 1$ over $\mathbb{F}_q$ with designed distance δ. If α is a primitive element of $\mathbb{F}_{q^m}$, this means that $W = \{1, \alpha, \ldots, \alpha^{\delta-2}\}$ is contained in the variety of C.   **4.13.5**

Consider $\kappa = (1, \alpha, \ldots, \alpha^{n-1})$ and $t = \lfloor (\delta-1)/2 \rfloor$. The Generalized Reed–Solomon-codes

$$A := \text{GRS}_{t+1}(\kappa, \mathbf{1}_n) \quad \text{and} \quad B := \text{GRS}_t(\kappa, \mathbf{1}_n)$$

form a t-error-correcting pair for C. We prove only the first condition, the remaining ones follow by arguments along the lines of the previous example.

As shown in 4.13.4, the complex product $A \cdot B$ is contained in the Generalized Reed–Solomon-code $\text{GRS}_{2t}(\kappa, \mathbf{1}_n)$. This code contains the basis elements $\kappa^{(i)} := (1, \alpha^i, \ldots, \alpha^{i(n-1)})$ for $0 \leq i \leq 2t - 1$. These vectors are in the dual code $C^{\perp}$, which follows from the construction of BCH-codes in 4.3.1 and from the assumption that $2t - 1 \leq \delta - 2$. This completes the proof. $\diamond$

Example Now we consider the Goppa-code $\text{GO}_q(\kappa, g)$ with Goppa-polynomial g of degree r. For $t = \lfloor r/2 \rfloor$ there is a t-error-correcting pair, given by   **4.13.6**

$$A := \text{GRS}_{t+1}(\kappa, \mathbf{1}_n)$$

and the code

$$B := \left\{ \left(f(\kappa_0)g(\kappa_0)^{-1}, \ldots, f(\kappa_{n-1})g(\kappa_{n-1})^{-1} \right) \mid f \in \mathbb{F}_q[x]_{<t} \right\}$$

which is isometric to the Generalized Reed–Solomon-code $\text{GRS}_t(\kappa, \mathbf{1}_n)$. (See Exercise 4.13.1). $\diamond$

Another classical decoding method for BCH- and Reed–Solomon-codes is the Berlekamp–Massey algorithm, for which we refer to [19] and [139], for example. See also Section 5.2 where we present a decoding algorithm for BCH-codes which enables us to correct both erasures and transmission errors.

Exercises

E.4.13.1   **Exercise** Check the statement in 4.13.6.

E.4.13.2   **Exercise** Consider the $(7,3)$-Reed–Solomon-code over $\mathbb{F}_8$ with generator polynomial $g = (x - \zeta)(x - \zeta^2)(x - \zeta^3)(x - \zeta^4)$, where ζ denotes a root of the primitive polynomial $x^3 + x + 1 \in \mathbb{F}_2[x]$. Choose a suitable error-correcting pair in order to correct the error in the received vector $y = (\zeta^3, 0, 0, \zeta, 1, \zeta^3, \zeta)$.

4.14

4.14 Majority Logic Decoding

Now we discuss a very practical and efficient method to decode certain linear codes. It can easily be realized using circuits, and it works particularly well with binary Reed–Muller-codes. The method goes back to Reed [167], and was then thoroughly discussed by Massey [144]. Apart from binary Reed–Muller-codes, we will show how binary cyclic codes of length 2^s can be decoded by this method.

Recall that if C is a linear (n,k)-code over $\mathbb{F}_q$, the elements in the dual code $C^\perp$ are parity checks, i.e.

$$h_0 c_0 + \ldots + h_{n-1} c_{n-1} = 0$$

for all $c = (c_0, \ldots, c_{n-1}) \in C$ and all $h = (h_0, \ldots, h_{n-1}) \in C^\perp$. Since $C^\perp$ has dimension $n - k$, this gives us q^{n-k} parity check equations. The problem is to decide which of these parity checks should be used in order to decode errors.

4.14.1   **Definition** Let C be a linear (n,k)-code over $\mathbb{F}_q$. A set of parity check equations $h^{(0)}, \ldots, h^{(r-1)} \in C^\perp$ is called *focused* on the i-th coordinate of C if

1. $h_i^{(k)} \neq 0$ for all $k \in r$ and

2. for all $j \neq i$ there is at most one $k \in r$ with $h_j^{(k)} \neq 0$.   ◇

4.14.2   **Example** The parity check matrix of the $(7,3,4)$-simplex-code is

$$\begin{pmatrix} 1 & 1 & 0 & 1 & 0 & 0 & 0 \\ 0 & 1 & 1 & 0 & 1 & 0 & 0 \\ 0 & 0 & 1 & 1 & 0 & 1 & 0 \\ 0 & 0 & 0 & 1 & 1 & 0 & 1 \end{pmatrix},$$

the generator matrix of the $(7,4,3)$-Hamming-code which we know from 4.2.7. Under cyclic shifts, we get 7 parity checks:

i	$h^{(i)}$
0	$(1,1,0,1,0,0,0)$
1	$(0,1,1,0,1,0,0)$
2	$(0,0,1,1,0,1,0)$
3	$(0,0,0,1,1,0,1)$
4	$(1,0,0,0,1,1,0)$
5	$(0,1,0,0,0,1,1)$
6	$(1,0,1,0,0,0,1)$

Out of these seven parity checks, $h^{(0)}$, $h^{(4)}$ and $h^{(6)}$ are focused on the 0-th coordinate. This is because c_0 appears in all of

$$0 = \langle h^{(0)}, c \rangle = c_0 + c_1 + c_3,$$
$$0 = \langle h^{(4)}, c \rangle = c_0 + c_4 + c_5,$$
$$0 = \langle h^{(6)}, c \rangle = c_0 + c_2 + c_6,$$

whereas the other coordinates appear at most once. We assume that $c \in C$ is sent, and that $y = c + e$ is received, where $e = (e_0, \ldots, e_{n-1})$ is the error vector (cf. 1.2.7). Its entries e_i are known as the *error bits*. Since

$$\sigma_i = \langle h^{(i)}, y \rangle = \langle h^{(i)}, c \rangle + \langle h^{(i)}, e \rangle = \langle h^{(i)}, e \rangle$$

we get

$$\sigma_0 = \langle h^{(0)}, y \rangle = e_0 + e_1 + e_3,$$
$$\sigma_4 = \langle h^{(4)}, y \rangle = e_0 + e_4 + e_5,$$
$$\sigma_6 = \langle h^{(6)}, y \rangle = e_0 + e_2 + e_6. \qquad \diamond$$

Theorem *Let C be a binary linear code. If $\{h^{(0)}, \ldots, h^{(r-1)}\}$ is a set of parity check*   **4.14.3**
equations focused on the i-th coordinate and if no more than $\lfloor r/2 \rfloor$ errors have occurred, i.e. if $y = c + e$ with $\mathrm{wt}(e) \leq \lfloor r/2 \rfloor$, then

$$e_i = \begin{cases} 0 & \text{if } \mathrm{wt}\left((\sigma_0, \ldots, \sigma_{r-1})\right) \leq r/2, \\ 1 & \text{otherwise,} \end{cases}$$

where $\sigma_i = \langle y, h^{(i)} \rangle$ for $i \in r$. That is, the true value of e_i is the value taken by the majority of the σ_i's. In case that r is even and there is a tie, we decode e_i as 0.

Proof: 1. If $e_i = 0$ then there are at most $r/2$ error bits $e_j, j \neq i$, which can affect at most $r/2$ of the σ_k's. That is, wt $\left((\sigma_0, \ldots, \sigma_{r-1})\right) \leq r/2$ and the decoding decision is correct.

2. If $e_i = 1$ then less than $r/2$ equations are affected by the $e_j, j \neq i$, i.e. $\sigma_k \neq 0$ for more than $r/2$ parity checks. In this case, wt $\left((\sigma_0, \ldots, \sigma_{r-1})\right) > r/2$ and e_i is decoded correctly. $\square$

4.14.4 **Corollary** *If r focused parity checks exist for every coordinate i, the binary code C can correct $\lfloor r/2 \rfloor$ errors.* $\square$

Majority logic decoding is particularly useful for cyclic codes. Once a focused set of parity checks has been found for say the 0-th coordinate, cyclic shift of these parity checks give focused sets for all other coordinates (using the fact that the dual code is cyclic as well). The described method is known under the name *one-step majority logic decoding*. It works well for the simplex-code. However, for the $(23, 12, 7)$-Golay-code (which is cyclic) only a single error can be corrected. This is because of the following result (the dual code of the Golay-code has minimum distance 8):

4.14.5 **Theorem** *For a binary linear code of length n, majority logic can correct at most*

$$\frac{n-1}{2(d^\perp - 1)}$$

errors, where $d^\perp$ is the minimum distance of the dual code.

Proof: Consider the parity checks $h^{(0)}, \ldots, h^{(r-1)}$ which are focused on the 0-th coordinate. Counting disjoint coordinates, this means that

$$1 + (|\mathrm{supp}(h^{(0)})| - 1) + \ldots + (|\mathrm{supp}(h^{(r-1)})| - 1) \leq n.$$

Since each parity check corresponds to a word in the dual code, the left hand side is bounded from below by $1 + r(d^\perp - 1)$, i.e. $r \leq (n-1)/(d^\perp - 1)$. The result follows from 4.14.4. $\square$

In order to correct more errors, multi-step majority decoding can be used.

4.14.6 **Definition** Let C be a linear (n, k)-code over $\mathbb{F}_q$ and let $C^\perp$ be the dual code. Fix a set $S \subseteq n$ of $l = |S|$ coordinates. A set of parity checks $h^{(0)}, \ldots, h^{(r-1)} \in C^\perp$ is called *focused* on S if

1. $h_i^{(k)} \neq 0$ for all $i \in S$, $k \in r$ and
2. for all $j \notin S$ there is at most one $k \in r$ with $h_j^{(k)} \neq 0$. $\diamond$

The following example describes the dual setting of 4.14.2.

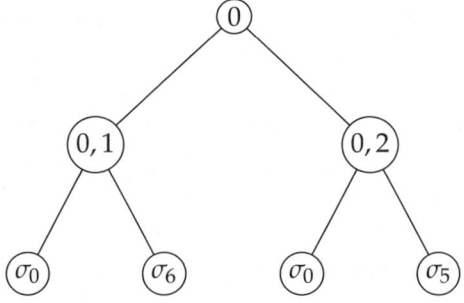

Fig. 4.7 A 2-step majority logic decoder for the $(7,4)$-Hamming-code

Example A parity check matrix of the $(7,4,3)$-Hamming-code is the generator   **4.14.7**
matrix of the $(7,3,3)$-simplex-code. Under cyclic shifts, we get 7 parity checks:

i	$h^{(i)}$
0	$(1,1,1,0,1,0,0)$
1	$(0,1,1,1,0,1,0)$
2	$(0,0,1,1,1,0,1)$
3	$(1,0,0,1,1,1,0)$
4	$(0,1,0,0,1,1,1)$
5	$(1,0,1,0,0,1,1)$
6	$(1,1,0,1,0,0,1)$

The parity checks $h^{(0)}$ and $h^{(6)}$ are focused on $\{0,1\}$, since

$$\sigma_0 = \langle h^{(0)}, y\rangle = e_0 + e_1 + e_2 + e_4,$$
$$\sigma_6 = \langle h^{(6)}, y\rangle = e_0 + e_1 + e_3 + e_6.$$

The parity checks $h^{(0)}$ and $h^{(5)}$ are focused on $\{0,2\}$, since

$$\sigma_0 = \langle h^{(0)}, y\rangle = e_0 + e_1 + e_2 + e_4,$$
$$\sigma_5 = \langle h^{(5)}, y\rangle = e_0 + e_2 + e_5 + e_6,$$

whereas the other coordinates appear at most once. Suppose there is one error.
By a majority rule, the first two parity checks determine the value of $e_0 + e_1$.
Similarly, the second pair of parity checks determines the value of $e_0 + e_2$. The
estimates $e_0 + e_1$ and $e_0 + e_2$ can be used to estimate by majority vote the value
of e_0. Thus we have a 2-step process to determine e_0 (cf. Fig. 4.7). Since C
is cyclic, the same procedure works for each of the other error coordinates,
provided the parity checks are cycled accordingly.   ◇

A decoder which has L levels of majority logic is called L-step majority logic decoder. The L levels form what is known as the *decoding tree*. We will see some examples below. As before, we find that $\lfloor r/2 \rfloor$ errors can be corrected as long as there are r checks at each stage in the decoding process. Even with L-step majority logic, we may not be able to correct the theoretically possible number of $\lfloor d/2 \rfloor$ errors. For instance, L-step majority logic applied to the $(23, 12, 7)$-Golay-code corrects only 2 errors. This is due to the following result:

4.14.8   **Theorem** *For a binary (n,k)-code C, L-step majority logic corrects at most*

$$\frac{n}{d^\perp} - \frac{1}{2}$$

errors, where $d^\perp$ is the minimum distance of the dual code.

Proof: Let $\{h^{(0)}, \ldots, h^{(r-1)}\}$ be a set of r parity checks focused on l coordinates S. For each $i \in S$, let m_i be the number of coordinates involved in $h^{(i)}$ outside S, i.e. $m_i = \text{wt}(h^{(i)}) - l$. Thus

4.14.9 $$m_i \geq d^\perp - l.$$

Also, since any two parity checks in a focused set have only l coordinates in common,

4.14.10 $$m_i + m_j \geq d^\perp \quad \text{for } i \neq j.$$

Set $M = \sum_{i \in r} m_i$, then

4.14.11 $$M \leq n - l.$$

Summing 4.14.9 over all i yields

4.14.12 $$rl + M \geq rd^\perp,$$

whereas summing 4.14.10 over all sets $\{i, j\}$, with $i \neq j$ yields

4.14.13 $$(r-1)M \geq \binom{r}{2} d^\perp.$$

Putting 4.14.11 and 4.14.12 together gives

4.14.14 $$n - M \geq l \geq \frac{rd^\perp - M}{r}$$

and hence

$$rd^\perp \leq rn - (r-1)M \leq rn - (r-1)rd^\perp/2$$

using 4.14.13. That is,

$$n \geq d^\perp(1 + \frac{r-1}{2}) = d^\perp \frac{r+1}{2}$$

so

$$r \leq \frac{2n}{d^\perp} - 1.$$

It follows from 4.14.4 that we can correct at most $\frac{n}{d^\perp} - \frac{1}{2}$ errors.   □

We will soon see that multi-step majority logic works well for decoding binary Reed–Muller-codes. Recall from 2.4.4 that the coordinates of the Reed–Muller-code $RM_{m,t}^2$ are labeled by m-tuples over $\mathbb{F}_2$, i.e. the elements of the vector space $\mathbb{F}_2^m$. The code itself is a subspace of the polynomial algebra $\mathcal{B}_m^2$ of all mappings from $\mathbb{F}_2^n$ to $\mathbb{F}_2$, considered as vector space over $\mathbb{F}_2$. The Reed–Muller-code $RM_{m,t}^2$ consists of those function in $\mathcal{B}_m^2$ which when expressed as reduced polynomials in $\mathbb{F}_2[x_0, \ldots, x_{m-1}]$ (cf. 2.4.3) have degree at most t.

Example In $RM_{4,2}^2$, we find the polynomial function $x_1 x_2$. This function has the following values (we write "$x_1 x_2$" for the function in order to distinguish the function from its arguments):   **4.14.15**

$x_3 x_2 x_1 x_0$	"$x_1 x_2$"$(x_3 x_2 x_1 x_0)$
0000	0
0001	0
0010	0
0011	0
0100	0
0101	0
0110	1
0111	1
1000	0
1001	0
1010	0
1011	0
1100	0
1101	0
1110	1
1111	1

When read as a row vector, the second column in this table is the corresponding element in the Reed–Muller-code. $\diamond$

According to 2.4.8, the dual code of $RM_{m,t}^2$ is $RM_{m,m-1-t}^2$, and from 2.4.7 we know that the minimum distance of $RM_{m,t}^2$ is 2^{m-t}. For a subset $S \subseteq \mathbb{F}_2^m$, let f_S be the characteristic function of S, i.e. the function which takes the value $f_S(x) = 1$ if x is in S and 0 otherwise. We will investigate which subsets S of $\mathbb{F}_2^m$ have the property that their characteristic function is in the Reed–Muller-code $RM_{m,t}^2$. We need some language from geometry in order to state the result.

Definition (Affine Geometry) The affine geometry, $AG(V)$, where V is a vector   **4.14.16**
space over a field F, consists of all cosets, $U + x$, of all subspaces U of V with incidence define through the natural inclusion relation. The dimension of $U +

Table 4.1 The 2-flats in $\mathbb{F}_2^3$

2-flat V	elements of V	as integer	characteristic vector
$x_0 = 0$	$000, 010, 100, 110$	$0, 2, 4, 6$	10101010
$x_1 = 0$	$000, 001, 100, 101$	$0, 1, 4, 5$	11001100
$x_1 + x_0 = 0$	$000, 011, 100, 111$	$0, 3, 4, 7$	10011001
$x_2 = 0$	$000, 001, 010, 011$	$0, 1, 2, 3$	11110000
$x_2 + x_0 = 0$	$000, 010, 101, 111$	$0, 2, 5, 7$	10100101
$x_2 + x_1 = 0$	$000, 001, 110, 111$	$0, 1, 6, 7$	11000011
$x_2 + x_1 + x_0 = 0$	$000, 011, 101, 110$	$0, 3, 5, 6$	10010110
$x_0 = 1$	$001, 011, 101, 111$	$1, 3, 5, 7$	01010101
$x_1 = 1$	$010, 011, 110, 111$	$2, 3, 6, 7$	00110011
$x_1 + x_0 = 1$	$001, 010, 101, 110$	$1, 2, 5, 6$	01100110
$x_2 = 1$	$100, 101, 110, 111$	$4, 5, 6, 7$	00001111
$x_2 + x_0 = 1$	$001, 011, 100, 110$	$1, 3, 4, 6$	01011010
$x_2 + x_1 = 1$	$010, 011, 100, 110$	$2, 3, 4, 6$	00111010
$x_2 + x_1 + x_0 = 1$	$001, 010, 100, 111$	$1, 2, 4, 7$	01101001

x is that of the defining subspace U, and if the latter has dimension k, we will also refer to a coset of U as a k-flat. Thus the *points* are all the vectors, the *lines* are 1-dimensional cosets, or 1-flats, the *planes* are the 2-dimensional cosets, or 2-flats, and so on, with the *hyperplanes* the cosets of dimension $n - 1$, where V is of dimension n over F. We also write $AG_n(F)$ for $AG(V)$, in analogy with the projective case. The affine geometry of these cosets is defined by the inclusion relation which specifies that, if $M = U + x$ and $N = W + y$ are cosets in $AG(V)$, then M *contains* N if $M \supseteq N$, from which it follows that W is a subspace of U. We write $AG_n(q)$ for $AG(\mathbb{F}_q^n)$ ◇

4.14.17   **Example** Let x_0, x_1, x_2 be the coordinates of $\mathbb{F}_2^3$. We use these coordinates and the corresponding integer representation interchangeably, i.e.

$$x_2 x_1 x_0 \longleftrightarrow 4x_2 + 2x_1 + x_0$$

where on the left we think of the elements of $\mathbb{F}_2$ as the integers 0 and 1, so that on the right we have an integer from 0 to 7. Consider 2-flats in $\mathbb{F}_2^3$. There are exactly 14 of them, they correspond to the solutions of all non-trivial linear equations in the three binary variables x_0, x_1, x_2, listed in the first column of Table 4.1. The second and third column list the elements of the flat V, and the last column the characteristic vector, which is just the vector holding the values of the characteristic function f_V, with entries in the following order

$$\left(f_V(0), f_V(1), \ldots, f_V(15)\right).$$

Coincidentally, the characteristic vectors listed in the rightmost column of the table are the minimum weight vectors in the $(8,4,4)$-code which is the extension of the $(7,4,3)$-Hamming-code (see Exercise 4.14.4). $\diamond$

Theorem *For $i \geq m - t$, the characteristic vector of any i-dimensional flat in* $\mathrm{AG}_m(2)$ *is a codeword of weight 2^i in* $\mathrm{RM}_{m,t}^2$. *In particular, the characteristic vector of an $(m - t)$-flat in* $\mathrm{AG}_m(2)$ *is a minimum weight vector of* $\mathrm{RM}_{m,t}^2$.   **4.14.18**

Proof: Every $(m - t)$-flat in $\mathrm{AG}_m(2)$ can be represented by t linearly independent inhomogeneous linear equations in the form

$$\sum_{j\in m} a_{ij}c_j = b_i, \quad i \in t.$$

The elements of the flat are the vectors $c = (c_0, \ldots, c_{m-1})$ satisfying these equations. To show that c is in $\mathrm{RM}_{m,t}^2$, we rewrite the condition as

$$\sum_{j\in m} a_{ij}c_j + b_i + 1 = 1, \quad i \in t.$$

In characteristic 2, these conditions can be replaced by the single equation

$$\prod_{i\in t}\left(\sum_{j\in m} a_{ij}c_j + b_i + 1\right) = 1.$$

This is because if one of the original equations does not hold then one of the right hand sides must be zero, which makes the product in the second equation become zero. Conversely, if c is a solution to the latter equation then again since we are in characteristic 2, each factor in the product must be equal to one, i.e. the former conditions hold for all $i \in r$. The fact that the latter condition is a polynomial equation of degree at most t implies that c is in $\mathrm{RM}_{m,t}^2$, i.e. each $(m - t)$-flat of $\mathrm{AG}_m(2)$ is contained in $\mathrm{RM}_{m,t}^2$. If $m - t$ is replaced by $i > m - t$, then fewer linear conditions are needed to define the affine subspace. By the same reasoning, we can represent the conditions by a single equation. Since that equation has degree less than t, the characteristic function of an i-dimensional affine subspace is contained in $\mathrm{RM}_{m,t}^2$ as well.

Since the characteristic vector of an $(m - t)$-flat has weight 2^{m-t}, the statement about the minimum weight vectors follows. $\square$

We note that the converse holds true as well. That is, the minimum weight vectors in $\mathrm{RM}_{m,t}^2$ are precisely the characteristic vectors of $(m - t)$-flats in the affine geometry $\mathrm{AG}_m(2)$.

Table 4.2 Basis elements of the Reed–Muller-code $\text{RM}^2_{4,2}$

1	1	1	1	1	1	1	1	1	1	1	1	1	1	1	1	1
x_0	0	1	0	1	0	1	0	1	0	1	0	1	0	1	0	1
x_1	0	0	1	1	0	0	1	1	0	0	1	1	0	0	1	1
x_2	0	0	0	0	1	1	1	1	0	0	0	0	1	1	1	1
x_3	0	0	0	0	0	0	0	0	1	1	1	1	1	1	1	1
$x_0 x_1$	0	0	0	1	0	0	0	1	0	0	0	1	0	0	0	1
$x_0 x_2$	0	0	0	0	0	1	0	1	0	0	0	0	0	1	0	1
$x_0 x_3$	0	0	0	0	0	0	0	0	0	1	0	1	0	1	0	1
$x_1 x_2$	0	0	0	0	0	0	1	1	0	0	0	0	0	0	1	1
$x_1 x_3$	0	0	0	0	0	0	0	0	0	0	1	1	0	0	1	1
$x_2 x_3$	0	0	0	0	0	0	0	0	0	0	0	0	1	1	1	1

4.14.19   **Example** Consider the binary Reed–Muller-code $\text{RM}^2_{4,t}$ for $t \in \{0,\dots,4\}$. Recall that the coordinates of these codes are labeled by the elements of $\mathbb{F}^4_2$, which we identify with the integers

$$0 = 0000, \quad 1 = 0001, \quad 2 = 0010, \quad \dots, \quad 15 = 1111,$$

where $x_3 x_2 x_1 x_0$ is the binary representation of the integer. Basis elements of the code are listed in Table 4.2. The horizontal lines separate basis elements according to the chain of subspaces

$$\text{RM}^2_{4,0} \leq \text{RM}^2_{4,1} \leq \text{RM}^2_{4,2}.$$

Assume we want to find parity checks focused on the characteristic vector of the one-dimensional subspace $\langle 0001 \rangle = \{0,1\}$. Since $(\text{RM}^2_{4,1})^\perp = \text{RM}^2_{4,2}$, we may take the 2-dimensional subspaces which contain $\langle 0001 \rangle$ (their characteristic vectors are in $\text{RM}^2_{4,2}$). These 2-dimensional subspaces are listed in Table 4.3. The leftmost column describes the 7 subspaces by means of generator matrices. The middle column lists the defining condition, which is translated from Boolean logic into $\mathbb{F}_2$-arithmetic using the following dictionary:

$$a \wedge b \equiv ab \bmod 2,$$
$$\overline{a \wedge b} \equiv 1 + ab \bmod 2,$$
$$\overline{a \vee b} \equiv (1+a)(1+b) \bmod 2,$$
$$a \vee b \equiv 1 + (1+a)(1+b) \bmod 2,$$
$$a = b \equiv 1 + a + b \bmod 2,$$
$$a \neq b \equiv a + b \bmod 2.$$

Table 4.3 Focused parity checks expressed in the basis of $RM_{4,2}^2$

subspace	condition	polynomial in $\mathcal{B}_4^2$	elements
$\begin{pmatrix} 0010 \\ 0001 \end{pmatrix}$	$x_2 = x_3 = 0$	$(1+x_2)(1+x_3)$ $= 1 + x_2 + x_3 + x_2x_3$	$0,1,2,3$
$\begin{pmatrix} 0100 \\ 0001 \end{pmatrix}$	$x_1 = x_3 = 0$	$(1+x_1)(1+x_3)$ $= 1 + x_1 + x_3 + x_1x_3$	$0,1,4,5$
$\begin{pmatrix} 1000 \\ 0001 \end{pmatrix}$	$x_1 = x_2 = 0$	$(1+x_1)(1+x_2)$ $= 1 + x_1 + x_2 + x_1x_2$	$0,1,8,9$
$\begin{pmatrix} 0110 \\ 0001 \end{pmatrix}$	$x_3 = 0, \ x_1 = x_2$	$(1+x_3)(1+x_1+x_2)$ $= 1 + x_1 + x_2 + x_3$ $+ x_1x_3 + x_2x_3$	$0,1,6,7$
$\begin{pmatrix} 1010 \\ 0001 \end{pmatrix}$	$x_2 = 0, \ x_1 = x_3$	$(1+x_2)(1+x_1+x_3)$ $= 1 + x_1 + x_2 + x_3$ $+ x_1x_2 + x_2x_3$	$0,1,10,11$
$\begin{pmatrix} 1100 \\ 0001 \end{pmatrix}$	$x_1 = 0, \ x_2 = x_3$	$(1+x_1)(1+x_2+x_3)$ $= 1 + x_1 + x_2 + x_3$ $+ x_1x_2 + x_1x_3$	$0,1,12,13$
$\begin{pmatrix} 1110 \\ 0001 \end{pmatrix}$	$x_1 = x_2 = x_3$	$(1+x_1+x_2)(1+x_2+x_3)$ $= 1 + x_1 + x_3 + x_2 + x_1x_3$ $+ x_1x_2 + x_2x_3$	$0,1,14,15$

Also, the fact that $x^2 \equiv x$ over $\mathbb{F}_2$ is used. For instance, the first subspace $\langle 0010, 0001 \rangle$ contains all vectors $x_3x_2x_1x_0$ with $x_3 = x_2 = 0$. In Boolean logic,

$$x_3 = 0 \ \text{ and } \ x_2 = 0,$$

which translates into

$$(1+x_3) \cdot (1+x_2)$$

using $\mathbb{F}_2$-arithmetic. This can be verified by looking at the elements. The subspace contains $0000, 0001, 0010$, and 0011 (i.e. $0, 1, 2$ and 3 in binary). This corresponds to the fact that the polynomial $(1+x_3)(1+x_2)$ evaluates to 1 precisely for those values. The corresponding vector in the code is

$$(1,1,1,1,0,0,0,0,0,0,0,0,0,0,0,0),$$

i.e. the vector which is one at exactly these 4 positions. In the table, the rightmost column lists the vectors contained in each of the subspaces (such as $0,1,2,3$ in the first row). The characteristic vectors of these subsets are the 7 parity checks which are focused on 0 and 1, as shown in Fig. 4.8. The fact that the characteristic vectors of the 7 subspaces are in $RM_{4,2}^2$ follows

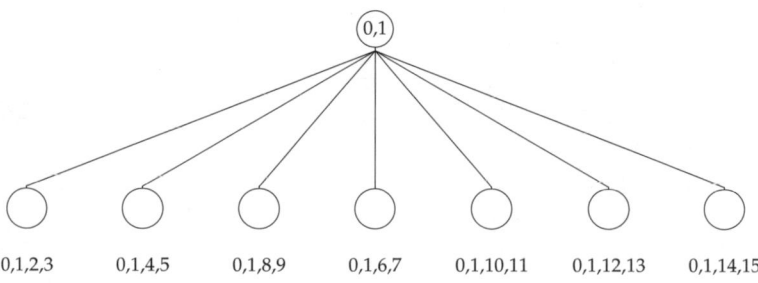

Fig. 4.8 Focused parity checks on $\langle 0001 \rangle = \{0,1\}$ in $\mathrm{RM}_{4,2}^2$

from 4.14.18 (and of course also from the table which shows that all polynomials are quadratic). ◇

4.14.20   **Lemma**   *In* $\mathrm{AG}_m(q)$, *each s-flat is contained in exactly* $(q^{m-s} - 1)/(q - 1)$ *flats of dimension* $s + 1$. *These* $(s + 1)$-*flats intersect pairwise exactly in the elements of the given s-flat.*

Proof: Consider first an s-dimensional linear subspace V of $\mathbb{F}_q^m$. There are q^m vectors overall, and each $w \notin V$ defines a linear subspace $W = \langle V, w \rangle$ of dimension $s + 1$ containing V. The same vector space W arises in $q^{s+1} - q^s$ different ways by choosing any one of the vectors $w \in W$ which are not in V. This amounts to $(q^m - q^s)/(q^{s+1} - q^s) = (q^{m-s} - 1)/(q - 1)$ linear subspaces W of dimension $s + 1$ containing V. It is clear that any two of these subspaces of dimension $s + 1$ intersect precisely in the given s-subspace V.

Next we consider the shifted subspaces $V + x$ and $W + x$, where W runs through the set of linear subspaces containing V. It is clear that each such $W + x$ contains $V + x$. Also, the collection of all these flats $W + x$ intersects pairwise exactly in $V + x$. □

4.14.21   **Theorem**   *The t-th order binary Reed–Muller-code* $\mathrm{RM}_{m,t}^2$ *admits a* $(t + 1)$-*step majority decoder that can correct* $\lfloor (2^{m-t} - 1)/2 \rfloor$ *errors.*

Proof: Recall from 2.4.8 that the dual code of $C = \mathrm{RM}_{m,t}^2$ is $C^\perp = \mathrm{RM}_{m,m-1-t}^2$. The characteristic vectors of $(t + 1)$-flats in $\mathrm{AG}_m(2)$ are minimum weight vectors in $C^\perp$. From 4.14.18 it follows that the characteristic function f_{V+x} is an element of $\mathrm{RM}_{m,t}^2$ if $\dim(V) \geq m - t$. Dually, f_{V+x} is an element of $C^\perp = \mathrm{RM}_{m,m-t-1}^2$ if $\dim(V) \geq t + 1$.

To set up a decoding tree for an arbitrary bit e_P, $P \in AG_m(2)$, we proceed as follows. We regard P as a 0-flat in $AG_m(2)$. By induction, we will create parity checks corresponding to s-flats in $AG_m(2)$ for $s = 0, 1, \ldots, t, t+1$. It is important to note that the majority logic decoding will proceed in the reverse order, i.e. starting from the nodes at level $t+1$ and working up the tree until an estimate for e_P is obtained (at level 0 in the tree). The parity check corresponding to an s-flat $V_s + P$ is the corresponding characteristic function f_{V_s+P}. This is simply the word whose support is $V_s + P$.

Now we describe how to find an estimate for the error bit e_P where P is a point in $AG_m(2)$. In the first step of decoding, for each node $V_t + P$ at level t of the decoding tree we use the parity checks corresponding to those $V_{t+1} + P$ which are focused on $V_t + P$. From 4.14.20 we know that these are $2^{m-t} - 1$ flats and they all belong to $RM_{m,m-(t+1)}^2 = C^\perp$, i.e. they are "true" parity checks, as required for a $t+1$-step majority logic decoder. They can be used to obtain an estimate for the error bits $\sum_{Q \in V_t+P} e_Q$. After having determined these estimates for all t-flats $V_t + P$, in the second step for each $V_{t-1} + P$ at the level $t-1$ of the decoding tree we use the error estimates of the parity checks corresponding to those $V_t + P$ which are focused on $V_{t-1} + P$. From these $2^{m-t+1} - 1$ estimates, which were computed in the first step, we obtain an estimate for the error bits $\sum_{Q \in V_{t-1}+P} e_Q$. Repeating this procedure we eventually obtain in the $(t+1)$-th step an error estimate for e_P.

To see that this decoder can correct $\lfloor (2^{m-t} - 1)/2 \rfloor$ errors, we have to argue as follows. From the construction of the decoding tree, it is clear that we have $2^{m-t} - 1$ parity checks corresponding to $(t+1)$-flats $V_{t+1} + P$ for each t-flat $V_t + P$ in the bottom level of the tree. Also, in the intermediate levels we have $2^{m-s} - 1 \geq 2^{m-t} - 1$ parity checks corresponding to $(s+1)$-flats for each s-flat. As remarked previously (before 4.14.8), this decoder can correct $\lfloor (2^{m-t} - 1)/2 \rfloor$ errors. □

The previous result is best possible since the minimum distance of the code is $d = 2^{m-t}$.

Example Figure 4.9 shows the decoding tree for the Reed–Muller-code $RM_{4,2}^2$. The tree is drawn from the left to the right. To simplify matters, not all lines are drawn. Note that the same parity checks are used over and over again. In order to save nodes, such a parity check has several ancestors. This way, the tree is not really a tree but rather a partially ordered set. The root node (at the left) corresponds to the 0-th coordinate. The parity checks focused on 0 are the incidence vectors of the 15 one-dimensional subspaces (generated by 1 through 15). Each one-dimensional subspace is contained in 7 two-dimensional subspaces, and there are 35 of them, corresponding to the second level. Each two-dimensional subspace is contained in 3 of the 15 three-dimensional subspaces.   **4.14.22**

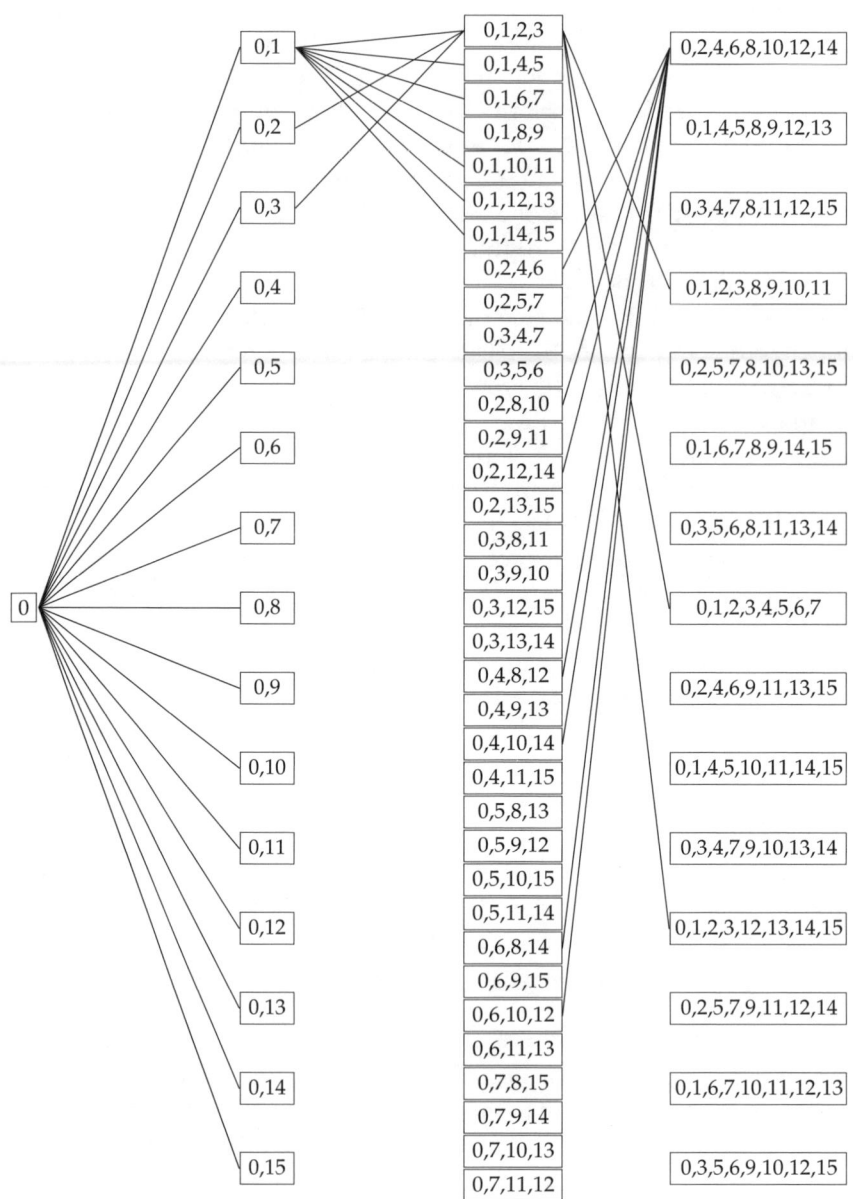

Fig. 4.9 A 3-step majority logic decoder for the Reed–Muller-code $RM_{4,2}^2$

We note that the decoding "tree" is in fact the projective geometry $PG_3(2)$ with the top element removed. In the picture, not all lines between nodes at levels $1, 2$ and 3 are shown. Only the descendants and ancestors of the first (i.e., top) node are shown. We note that the first node at level 1 has 7 descendants and that there are 3 descendants of the first node at the second level. Also, the first node at level 2 has 3 ancestors, whereas the first node at level 3 has 7 of them. The point is that the number of ancestors (and descendants) is constant for nodes at a given level. This decoder corrects one error since there are at least $r = 3$ descendants at each level. A decoding tree for $RM_{4,1}^2$ is obtained by removing the bottom level of the tree. The resulting decoder can correct 3 errors since there are at least $r = 7$ descendants at each level. $\diamond$

We remark that in the examples which we have seen, the subspaces used to build the decoding tree were all linear. That is, we never really needed any affine subspace which was not already a linear subspace. The reason for this is that the decoding trees presented were all decoding the error bit e_0. In fact, if we had chosen to decode a different bit e_P, say, where $P \in \mathbb{F}_2^n \setminus \{0\}$, then we really would have used "true" affine subspaces. Of course, we know that the binary Reed–Muller-code is cyclic, hence building a decoding tree for e_0 suffices do decode this code.

At the end of this section we show that cyclic codes of length 2^s over $\mathbb{F}_2$ can be decoded by using the majority logic decoding method.

Theorem *Using a one-step majority logic decoder for a binary cyclic code C of length 2^s, allows us to correct up to $\lfloor (\text{dist}(C) - 1)/2 \rfloor$ errors.*   **4.14.23**

Proof: From 4.8.1 we know that the code is of the form C_t with generator polynomial $(x - 1)^t$ for some $t \in \{1, \ldots, 2^s - 1\}$. Let

$$t = a_0 + a_1 2 + \ldots + a_{s-1} 2^{s-1}, \qquad a_i \in \{0, 1\}, \ i \in s,$$

be the binary representation of $t \in \{1, \ldots, 2^s - 1\}$. We distinguish four cases and in each of these we compute the minimum distance using 4.8.9.

1. If $1 \leq t < 2^{s-1}$, then $\text{dist}(C_t) = (a_{s-1} + 2) \cdot 2^0 = 2$. Such a code cannot correct a single error and hence there is nothing to show in this case.

2. If $t = 2^s - 1$, then C_{2^s-1} is a repetition code, which can be decoded. In this case, majority logic is the usual nearest neighbor decoding for repetition codes.

3. If $t = 2^{s-1} + \ldots + 2^{s-r}$, for $1 \leq r \leq s$, then $\text{dist}(C_t) = 2^r$. From 4.8.1 we know that the dual code of $C = C_t$ is

$$C_{2^s - t} = C_{2^s - 1 - t + 1} = C_{1 + 2 + \ldots + 2^{s-r-1} + 1} = C_{2^s - r}$$

with generator polynomial $(x-1)^{2^{s-r}}$. Since $x^i - 1 = (x-1)(x^{i-1} + \ldots + x + 1)$, the polynomial

$$x^{i2^{s-r}} - 1 = (x^i - 1)^{2^{s-r}} = (x-1)^{2^{s-r}}(x^{i-1} + \ldots + x + 1)^{2^{s-r}}$$

is in $C_t^{\perp}$. Thus,

$$\left\{ x^{i2^{s-r}} - 1 \,\middle|\, 1 \le i \le 2^r - 1 \right\}$$

is a set of $(2^r - 1)$ elements of $C_t^{\perp}$. From the form of these polynomials it is clear that they are focused on 1, i.e. on the 0-th coordinate of the code. It follows from 4.14.4 that we can correct $\lfloor (2^r - 1)/2 \rfloor = \lfloor (\mathrm{dist}(C) - 1)/2 \rfloor$ errors.

4. Since cases 1 and 2 have been dealt with, it remains to consider the case when $t = 2^{s-1} + \ldots + 2^{s-r+1} + 0 \cdot 2^{s-r} + \ldots$, with $1 \le r \le s$. We define $t' = 2^{s-1} + \ldots + 2^{s-r+1}$. Since case 3 is settled, we can assume that $t - t' > 0$. It follows that $\mathrm{dist}(C_t) = 2 \cdot 2^{s-(s-r)-1} = 2^r$. From

$$2^s - t < 2^s - t' = 2^s - 1 - t' + 1$$
$$= 1 + 2 + \ldots + 2^{s-1} - 2^{s-r+1} - \ldots - 2^{s-1} + 1$$
$$= 1 + 2 + \ldots + 2^{s-r} + 1 = 2^{s-r+1}$$

and 4.8.1 we deduce that the code C_{2^s-r+1} is contained in C_{2^s-t} which is the dual of C_t. Similarly as in the third case,

$$\left\{ x^{i2^{s-r+1}} - 1 \,\middle|\, 1 \le i \le 2^{r-1} - 1 \right\}$$

is a set of $2^{r-1} - 1$ elements of $C_t^{\perp}$ that are focused on 1. For each $i \in 2^{r-1}$, also

$$c^{(i)} := x^{i2^{s-r+1}}(x-1)^{2^s-t} + (x^{i2^{s-r+1}} - 1)$$

belongs to $C_t^{\perp}$. Since $1 \le 2^s - t < 2^{s-r+1}$, the support of $c^{(i)}$ consists of the terms 1 and elements of the form x^j with $i2^{s-r+1} < j < (i+1)2^{s-r+1}$. Therefore,

$$\left\{ x^{i2^{s-r+1}} - 1 \,\middle|\, 1 \le i \le 2^{r-1} - 1 \right\} \cup \left\{ c^{(i)} \,\middle|\, i \in 2^{r-1} \right\}$$

is a set of $(2^r - 1)$ elements of $C_t^{\perp}$ which are focused to 1. Since $\mathrm{dist}(C) = 2^r$, this allows one to decode $\lfloor (\mathrm{dist}(C) - 1)/2 \rfloor$ errors as claimed. $\square$

4.14.24 **Example** In Table 4.4, we show the parity checks for binary cyclic codes C of length 16 with $\mathrm{dist}(C) > 2$ which arise from the previous theorem. They are focused on 1. As pointed out previously, focused sets on the remaining coordinates of the code can be obtained by multiplying the checks from the previous proof by suitable powers of x modulo $x^{p^s} - 1$. $\diamond$

Table 4.4 Focused parity checks for binary cyclic codes of length 16

t	dist	parity checks
9	4	$(1+x)^7, 1+x^8, (1+x)^7 x^8 + (1+x^8)$
10	4	$(1+x)^6, 1+x^8, (1+x)^6 x^8 + (1+x^8)$
11	4	$(1+x)^5, 1+x^8, (1+x)^5 x^8 + (1+x^8)$
12	4	$1+x^4, 1+x^8, 1+x^{12}$
13	8	$(1+x)^3, 1+x^4, (1+x)^3 x^4 + (1+x^4),$
		$1+x^8, (1+x)^3 x^8 + (1+x^8),$
		$1+x^{12}, (1+x)^3 x^{12} + (1+x^{12})$
14	8	$1+x^2, 1+x^4, 1+x^6, 1+x^8,$
		$1+x^{10}, 1+x^{12}, 1+x^{14}$
15	16	$1+x, \ldots, 1+x^{15}$

Theorem 4.14.23 cannot be applied to cyclic codes of length p^s, where $p > 2$. This is because cyclic codes of prime length p are MDS according to 4.8.10 (cf. Exercise 4.14.3).

Exercises

Exercise Show that the m-th order binary Hamming-code admits an $(m-1)$-step majority logic decoder.   **E.4.14.1**

Exercise Starting with the following list of row-reduced echelon matrices over $\mathbb{F}_2$, determine all 35 two-dimensional subspaces of $\mathbb{F}_2^4$. Using the translation from Boolean algebra to $\mathbb{F}_2$-arithmetic described in 4.14.19, find the representing polynomials in $\mathcal{B}_4^2$ for the characteristic functions of these spaces. In the following list of matrices, a star ("$*$") represents an arbitrary element of $\mathbb{F}_2$.   **E.4.14.2**

$$\begin{pmatrix} 0 & 0 & 1 & 0 \\ 0 & 0 & 0 & 1 \end{pmatrix}, \begin{pmatrix} 0 & 1 & * & 0 \\ 0 & 0 & 0 & 1 \end{pmatrix}, \begin{pmatrix} 1 & * & * & 0 \\ 0 & 0 & 0 & 1 \end{pmatrix}, \begin{pmatrix} 0 & 1 & 0 & * \\ 0 & 0 & 1 & * \end{pmatrix},$$

$$\begin{pmatrix} 1 & * & 0 & * \\ 0 & 0 & 1 & * \end{pmatrix}, \begin{pmatrix} 1 & 0 & * & * \\ 0 & 1 & * & * \end{pmatrix}.$$

Hint: in the end, the subspaces you get should match those in level 2 of Fig. 4.9.

Exercise Discuss the question whether an MDS-code can be decoded using majority logic.   **E.4.14.3**

E.4.14.4   **Exercise**

1. Show that $\text{RM}^2_{m,1}$ is the dual of the extended m-th order binary Hamming-code.
2. Use 2.4.8 to show that $\text{RM}^2_{m,m-2}$ is the extended m-th order binary Hamming-code.

E.4.14.5   **Exercise** Let V be a finite set and $\mathcal{B}$ a collection of subsets of V. Assume that there are $v = |V|$ "points" in V and that $\mathcal{B} = \{B_0, \dots, B_{b-1}\}$ consists of b subsets called "blocks," each of size k. The pair $(V, \mathcal{B})$ is said to be a t-(v, k, λ) *design* if the following property holds: For each subset T of V of size t, there are exactly λ blocks $B_{i_0}, \dots, B_{i_{\lambda-1}}$ which contain T, i.e. $T \subseteq B_{i_j}$ for $j \in \lambda$. The quadruple t-(v, k, λ) is known as the *parameters of the design*.

1. Show that the entries in the third column of Table 4.1 form a 3-$(8, 4, 1)$ design with 14 blocks (the blocks correspond to the 4 entries per row).
2. Show that a t-(v, k, λ) design is at the same time also an s-(v, k, λ_s) design for $0 \le s \le \lambda$ and for the integer $\lambda_s = \lambda\binom{v-s}{t-s}/\binom{k-s}{t-s}$. Here, $\lambda_t = \lambda$.
3. Show that $b = \lambda\binom{v}{t}/\binom{k}{t}$.
4. Show that $vr = kb$, where $r := \lambda_1 = \lambda\binom{v-1}{t-1}/\binom{k-1}{t-1}$.
5. Show that the incidence relation between points and lines in a $\text{PG}_2(q)$ gives rise to a 2-$(q^2 + q + 1, q + 1, 1)$ design (but not all such designs arise from this construction). Can you "draw" the design which results from $\text{PG}_2(2)$? Hint: Chapter 8.
6. Show that the incidence relation between points and i-flats in $\text{AG}_n(q)$ gives rise to a
$$2\text{-}\left(q^n, q^i, \frac{(q^{n-1} - 1)(q^{n-2} - 1) \cdots (q^{n-i} - 1)}{(q^{i-1} - 1)(q^{i-2} - 1) \cdots (q - 1)}\right)$$
design. If $i > 1$ and $q = 2$, it is even a
$$3\text{-}\left(2^n, 2^i, \frac{(2^{n-2} - 1) \cdots (2^{n-i} - 1)}{(2^{i-2} - 1) \cdots (2 - 1)}\right)$$
design.
7. Show that the incidence relation between points and i-subspaces of $\text{PG}_n(q)$ gives rise to a
$$2\text{-}\left(\theta_n(q), \theta_i(q), \frac{(q^{n-1} - 1)(q^{n-2} - 1) \cdots (q^{n-i} - 1)}{(q^{i-1} - 1)(q^{i-2} - 1) \cdots (q - 1)}\right)$$
design, where $\theta_n(q)$ is as in 3.7.2.

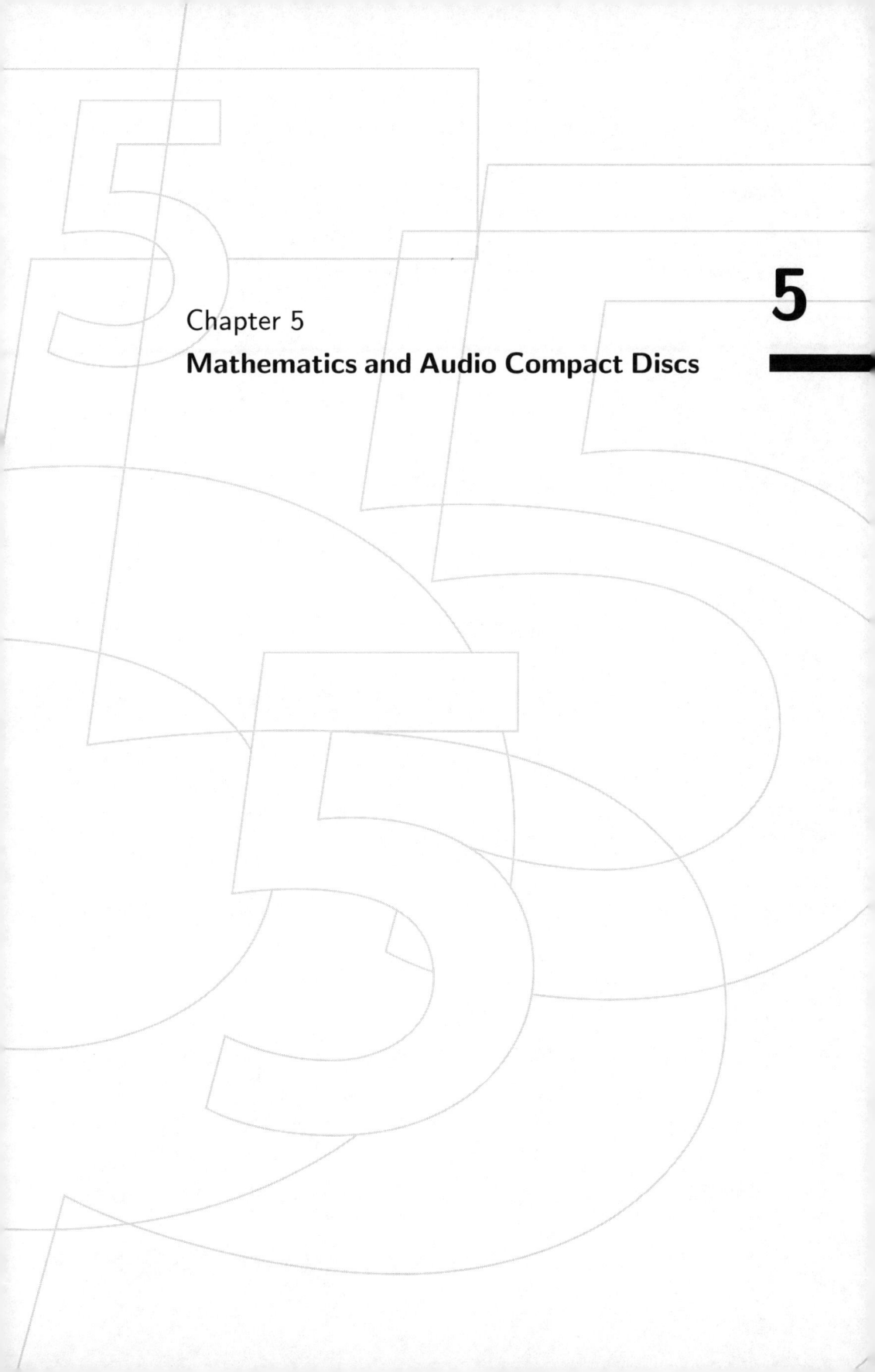

Chapter 5
Mathematics and Audio Compact Discs

5

5 **Mathematics and Audio Compact Discs**

5

5 Mathematics and Audio Compact Discs

In this chapter we give a short description of the mathematical background behind the technology used for compact discs. Since we are dealing with problems arising from a real-word application, we must adapt our assumptions to this particular situation.

In the first section we present a short introduction to digital audio transmission. Some facts about Fourier Series and Fourier Transforms are collected. We describe sampling and filtering of signals from a mathematical point of view, explain analog digital conversion, dither, pulse code modulation and, finally, we prove Shannon's Sampling Theorem, which gives an interpolation formula which expresses the value $f(x)$ of a signal at any time x in terms of its values $f(ns)$ at the discrete points ns for $n \in \mathbb{Z}, s > 0$.

First of all, as far as the error-correction in connection with compact discs is concerned, we should notice that errors are not uniformly distributed random errors. In fact, errors tend to occur in bursts, for instance due to manufacturing errors when the compact disc was produced, or due to surface errors arising from scratches or fingerprints while handling the disc. Such a burst error is actually a string of errors happening within a short period of time. In the third section we will investigate how to detect and correct burst errors from a general point of view.

We have already mentioned that in fact two linear codes are applied for the encoding process in the production of a compact disc. This method is known as interleaving of two codes, which is thoroughly described in the third section. This interleaving process allows us to correct burst errors. Since we are using two codes, one of them can be used for error detection. We already know that the error detection rate of a code is larger than its error correction rate. In case the second code detects an error in the received vector, it marks all the corresponding components sent to the first code as erasures (these are errors where the position of the error but not its value is known) and the first code can be applied to correct both errors and erasures. For this reason, we investigate the correction of erasures with linear codes in the second section. In particular, we present an algorithm for correcting erasures with BCH-codes.

In the third section we meet product codes (cf. 2.3.15) again. We describe how they can be used for correction of transmission errors and erasures. Especially products of cyclic codes are analyzed.

Finally, in the last two sections we present all the important facts about the methods used for error detection and error correction both in audio compact discs and in CD-ROM. In particular, the CIRC encoding and decoding is described in detail. Furthermore, we explain how to use interpolation in order to deal with errors which can not be eliminated by CIRC. Also the pit/land structure of a track on a CD and the EFM are mentioned.

The second and third section describe independently from the application in compact discs some interesting coding and decoding methods. The first section is mainly devoted to digital audio, the fourth to a detailed description of error detection and correction in audio compact discs. The fifth section gives a short overview how the CD-ROM standard extends the standard of digital audio discs.

5.1 Fourier Transform, Shannon's Sampling Theorem

In this section we explain how an acoustic signal is transformed into digital data. Usually this process is called *sampling*. In our setting the acoustic signal is a mapping $f: \mathbb{R} \to \mathbb{R}$, where $f(x)$ is the sound pressure at the time x. When dealing with audio data, it is important to analyze which frequencies occur in the signal. For this reason, we give a short introduction to the theory of Fourier Transforms, which describes how to express a signal f as a sum of functions $x \mapsto e^{2\pi i \xi x}$ for $\xi \in \mathbb{R}$. We limit our presentation to the essential facts, since a more complete discussion would require detailed knowledge of the theory of Lebesgue integration, which is beyond the scope of this text. Our presentation of the Fourier analysis is based on [64] and [30]. Technical details about the compact disc system are taken from Pohlmann's book [164]. Further facts about acoustics and audio engineering are taken from [164], [202] and [175].

Let $I \subseteq \mathbb{R}$ be an interval and let $p \geq 1$ be a real number. (The reader should take care that in this section p does not indicate a prime number!) We denote the set of all measurable functions $f: I \to \mathbb{C}$ for which $|f(x)|^p$ is integrable on I by $L^p(I)$. Then $L^p(I)$ together with the norm

$$\|f\|_p := \left(\int_I |f(x)|^p \, dx \right)^{1/p}, \qquad f \in L^p(I),$$

is a complete normed vector space (cf. Exercise 5.1.2 and [64, 15.2.3 Proposition]).

In this setting, $\|f\|_p = 0$ if and only if $f = 0$ almost everywhere on I, in other words, $f(x) = 0$ for all $x \in I \setminus M$, where M is a subset of I of measure 0. The set $\mathcal{N} := \left\{ f \in L^p(I) \mid \|f\|_p = 0 \right\}$ is a subspace of $L^p(I)$. In order to be more precise, instead of $L^p(I)$ we should actually consider the factor space $L^p(I)/\mathcal{N}$.

The set $L^2(I)$ is also equipped with an inner product defined by

$$\langle f, g \rangle = \int_I \overline{f(x)} g(x)\, dx, \qquad f, g \in L^2(I),$$

where $\overline{f(x)}$ means the complex conjugate of $f(x)$. It is easy to show that $L^2(I)$ is a Hermitian inner product space and that

$$\|f\|_2 = \sqrt{\langle f, f \rangle}, \qquad f \in L^2(I).$$

Consequently, $L^2(I)$ is a Hilbert space.

Different notions of convergence are considered. Let $(f_n)_{n \in \mathbb{N}}$ be a sequence with $f_n \in L^p(I)$ and let $f \in L^p(I)$. We say that $(f_n)_{n \in \mathbb{N}}$ *converges uniformly* to f if

$$\lim_{n \to \infty} \sup_{x \in I} |f_n(x) - f(x)| = 0,$$

which can also be expressed as

$$\forall \epsilon > 0,\ \exists N_\epsilon,\ \forall n > N_\epsilon,\ \forall x \in I\ :\ |f_n(x) - f(x)| < \epsilon.$$

The sequence $(f_n)_{n \in \mathbb{N}}$ *converges pointwise* to f if

$$\forall x \in I,\ \forall \epsilon > 0,\ \exists N_{\epsilon,x},\ \forall n > N_{\epsilon,x}\ :\ |f_n(x) - f(x)| < \epsilon.$$

The sequence $(f_n)_{n \in \mathbb{N}}$ converges to f in $L^p(I)$ if

$$\lim_{n \to \infty} \|f_n - f\|_p = 0.$$

Convergence in $L^1(I)$ is also called *mean convergence*, whereas convergence in $L^2(I)$ is known as *mean quadratic convergence* or *convergence "in energy"*. If $(f_n)_{n \in \mathbb{N}}$ converges uniformly to f in $L^1(I)$ and I is a finite interval, then integration and limit can be interchanged so that

$$\lim_{n \to \infty} \int_I f_n(x)\, dx = \int_I f(x)\, dx.$$

The *support* of a function $f : \mathbb{R} \to \mathbb{C}$ is the closure of the set of all elements $x \in \mathbb{R}$ for which $f(x) \neq 0$:

$$\mathrm{supp}(f) = \mathrm{cl}\,\{x \in \mathbb{R} \mid f(x) \neq 0\}.$$

A function $f : \mathbb{R} \to \mathbb{C}$ has period $a > 0$ if $f(x) = f(x + a)$ for all $x \in \mathbb{R}$. Especially for $p = 2$ and $I = (0, a)$, we consider the set

$$L^2_{\mathrm{per}}(0, a) := \left\{ f : \mathbb{R} \to \mathbb{C} \ \middle|\ f \text{ has period } a \text{ and } \int_0^a |f(x)|^2\, dx < \infty \right\}$$

which, together with the norm

$$\|f\|_2 := \left(\int_0^a |f(x)|^2\, dx \right)^{1/2},$$

is a normed vector space.

At first we approximate functions $f \in L^2_{\text{per}}(0, a)$ by

---

5.1.1   **Trigonometric polynomials** A *trigonometric polynomial* of degree N in $L^2_{\text{per}}(0, a)$, $a > 0$, is an expression of the form

$$p(x) := \sum_{n=-N}^{N} c_n e^{2\pi i n x / a}$$

with $c_n \in \mathbb{C}$. We denote by T_N the set of all trigonometric polynomials of degree N. Let e_n be the trigonometric polynomial $x \mapsto e^{2\pi i n x / a}$. Then

$$\langle e_n, e_m \rangle = \begin{cases} a & \text{if } n = m, \\ 0 & \text{if } n \neq m. \end{cases} \quad \text{and} \quad \|e_n\|_2 = \sqrt{a}.$$

Thus, the set $\{e_n \mid -N \le n \le N\}$ forms an orthogonal basis of T_N, and T_N is a $(2N + 1)$-dimensional space. Moreover, for $p = \sum_{n=-N}^{N} c_n e_n \in T_N$ we have $\langle e_n, p \rangle = c_n \|e_n\|_2^2 = c_n a$, whence

$$c_n = \frac{1}{a} \langle e_n, p \rangle = \frac{1}{a} \int_0^a p(x) e^{-2\pi i n x / a} \, dx, \qquad -N \le n \le N.$$

Using the fact that

$$e^{ix} = \cos(x) + i \sin(x), \qquad x \in \mathbb{R},$$

any $p \in T_N$ can be expressed as

$$
\begin{aligned}
p(x) &= c_0 + \sum_{n=1}^{N} \left(c_n e^{2\pi i n x / a} + c_{-n} e^{-2\pi i n x / a} \right) \\
&= c_0 + \sum_{n=1}^{N} \left((c_n + c_{-n}) \cos(2\pi n x / a) + i(c_n - c_{-n}) \sin(2\pi n x / a) \right).
\end{aligned}
$$

If we put $a_n := c_n + c_{-n}$ and $b_n := c_n - c_{-n}$ for $n \ge 0$, we obtain

$$p(x) = \frac{a_0}{2} + \sum_{n=1}^{N} \left(a_n \cos(2\pi n x / a) + i b_n \sin(2\pi n x / a) \right), \qquad x \in \mathbb{R},$$

with

$$a_n = \frac{2}{a} \int_0^a p(x) \cos(2\pi n x / a) \, dx$$

and

$$b_n = \frac{2}{a} \int_0^a p(x) \sin(2\pi n x / a) \, dx. \qquad \square$$

Theorem [64, page 30] *Let N be a positive integer, and assume that a > 0. For*   **5.1.2**
$f \in L^2_{\text{per}}(0,a)$ *there exists exactly one trigonometric polynomial*

$$f_N(x) := \sum_{n=-N}^{N} c_n e^{2\pi i n x / a}$$

such that $\|f - f_N\|_2 = \min \{\|f - p\|_2 \mid p \in T_N\}$. *The coefficients of* f_N *are given
by*

$$c_n = c_n(f) = \frac{1}{a} \int_0^a f(x) e^{-2\pi i n x / a} \, dx, \qquad -N \le n \le N.$$ **5.1.3**

For all $N \in \mathbb{N}$ *Bessel's inequality*

$$\sum_{n=-N}^{N} |c_n|^2 \le \frac{1}{a} \int_0^a |f(x)|^2 \, dx$$

holds true. Consequently,

$$\sum_{n=-\infty}^{\infty} |c_n|^2 < +\infty$$

and $c_n = c_n(f) \to 0$ *as* $|n| \to \infty$.   □

Fourier Series Deeper methods from the theory of the Lebesgue integral show   **5.1.4**
[64, 4.3.1 Theorem] that for $f \in L^2_{\text{per}}(0,a)$, $a > 0$,

$$\|f_N - f\|_2 \to 0 \text{ as } N \to \infty.$$

In other words,

$$
\begin{aligned}
f(x) &= \sum_{n=-\infty}^{\infty} c_n e^{2\pi i n x / a} \\
&= \frac{a_0}{2} + \sum_{n=1}^{\infty} (a_n \cos(2\pi n x / a) + i b_n \sin(2\pi n x / a))
\end{aligned}
$$ **5.1.5**

almost everywhere in $\mathbb{R}$, since this is an equality in $L^2_{\text{per}}(0,a)$. From this representation of f we obtain *Parseval's equality*

$$\sum_{n=-\infty}^{\infty} |c_n|^2 = \frac{1}{a} \int_0^a |f(x)|^2 \, dx.$$

The coefficients c_n are called *Fourier coefficients* of f. The right hand side of
5.1.5 is the *Fourier series* of f.   □

In addition, we take for granted the following result on Fourier series:

Theorem [64, 5.3.1 Theorem] *Assume that f has period a > 0, is continuous on* $\mathbb{R}$,   **5.1.6**
differentiable on $[0,a]$ *with exception of possibly a finite number of points, and that* f'
is piecewise continuous. Then the Fourier series of f converges uniformly to f on $\mathbb{R}$.

□

5.1.7   **Example** We have just explained how to approximate a periodic function with trigonometric polynomials. For example consider the periodic function f with period $a = 2\pi$ defined on $I = [-\pi, \pi)$ by

$$f(t) := \begin{cases} -1 & \text{if } -\pi \leq t < 0, \\ +1 & \text{if } 0 \leq t < \pi. \end{cases}$$

The three approximations for $N = 1, 3, 5$ are given by

$$f_1(t) = \frac{4}{\pi} \sin(t)$$

$$f_3(t) = \frac{4}{\pi} (\sin(t) + \frac{1}{3} \sin(3t))$$

$$f_5(t) = \frac{4}{\pi} (\sin(t) + \frac{1}{3} \sin(3t) + \frac{1}{5} \sin(5t)).$$

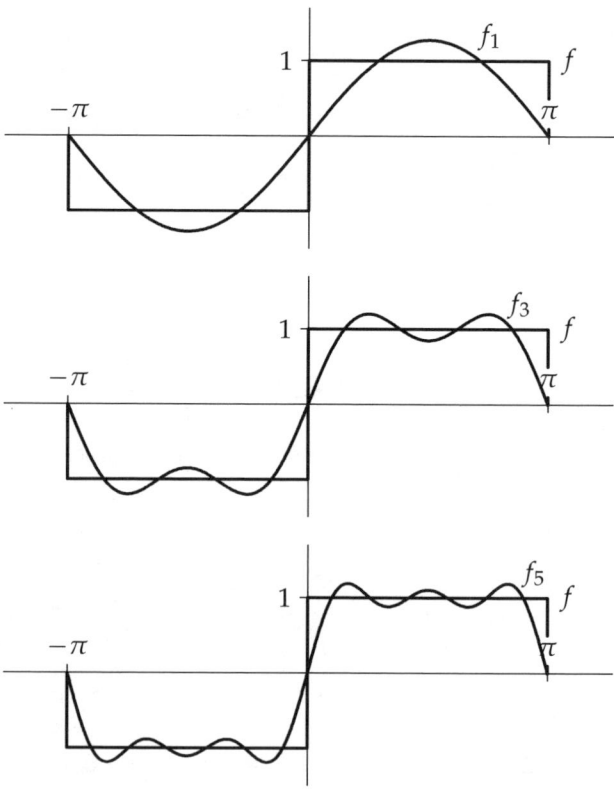

Fig. 5.1 Approximation by trigonometric polynomials

◇

The Discrete Fourier Transform Let f be a periodic function with period $a > 0$ **5.1.8**
and let N be a positive integer. Let $x_k := ka/N, k \in N$, be an evenly spaced
subdivision of the interval $[0, a]$. Assume that the values $y_k := f(x_k), k \in N$,
are known. Furthermore we assume that the Fourier series of f converges
pointwise to f and that

$$f(x) = \frac{1}{2}(f(x^+) + f(x^-))$$

holds true at points x of discontinuity. Here we have $f(x^+) := \lim_{t \to x, \, t > x} f(t)$
and $f(x^-) := \lim_{t \to x, \, t < x} f(t)$. In order to simplify our notation, we assume
that N is odd. For $-(N-1)/2 \le n \le (N-1)/2$, the coefficients of the Fourier
expansion of f can be approximated by using the trapezoid formula as

$$c_n' = \frac{1}{a} \sum_{k \in N} y_k \frac{a}{N} e^{-2\pi i n k a/(Na)} = \frac{1}{N} \sum_{k \in N} y_k e^{-2\pi i n k/N}.$$

It can be shown that these c_n' are the Fourier coefficients of the trigonometric
polynomial $p \in T_{(N-1)/2}$ which interpolates f at the points ka/N for $k \in N$. If
we put

$$Y_n := \begin{cases} c_n' & \text{if } 0 \le n \le (N-1)/2, \\ c_{n-N}' & \text{if } (N+1)/2 \le n \le N-1, \end{cases}$$

then we obtain the two equivalent formulae of the *Discrete Fourier Transform*

$$y_k = \sum_{n \in N} Y_n e^{2\pi i n k/N}, \qquad k \in N,$$

$$Y_n = \sum_{k \in N} y_k e^{-2\pi i k n/N}, \qquad n \in N. \qquad \square$$

The Fourier Transform [64, 17.1.3 Theorem] *The Fourier Transform $\hat{f}$ of $f \in$*   **5.1.9**
$L^1(\mathbb{R})$ is defined by

$$\hat{f}(\xi) := \int_{\mathbb{R}} f(x) e^{-2\pi i \xi x} \, dx.$$

The Fourier Transform $\hat{f}$ of $f \in L^1(\mathbb{R})$ is continuous and bounded with

$$\sup_{x \in \mathbb{R}} |\hat{f}(x)| \le \|f\|_1 \quad \text{and} \quad \lim_{x \to \pm \infty} \hat{f}(x) = 0. \qquad \square$$

The term 2π in the exponent is often omitted in the definition of the Fourier
Transform. In this case, the integral is multiplied by the normalization factor
$1/\sqrt{2\pi}$.

Comparing $\hat{f}$ with the Fourier coefficients 5.1.3, we deduce that the Fourier
Transform allows one to pass from the time domain of the signal f to the fre-
quency domain. The value $|\hat{f}(\xi)|$ is considered to represent the amplitude of
the frequency ξ in the signal f.

Some properties of $\hat{f}$ are collected in Exercise 5.1.4 and in the following

5.1.10   **Lemma** [64, 17.2.1 Theorem]

1. *If $x^k f \in L^1(\mathbb{R})$ for $0 \le k \le n$, then $\hat{f}$ is n times differentiable and the k-th derivative of $\hat{f}$ is*

$$\hat{f}^{(k)}(\xi) = \hat{f}_k(\xi), \qquad 1 \le k \le n,$$

for $f_k(x) := (-2\pi i x)^k f(x)$.

2. *If $f \in L^1(\mathbb{R})$ is n times continuously differentiable and all the derivatives $f^{(k)}$, $1 \le k \le n$, are in $L^1(\mathbb{R})$, then*

$$\widehat{f^{(k)}}(\xi) = (2\pi i \xi)^k \hat{f}(\xi), \qquad 1 \le k \le n.$$

3. *If $f \in L^1(\mathbb{R})$ has bounded support, then $\hat{f}$ is infinitely many times differentiable. We also write $\hat{f} \in C^\infty(\mathbb{R})$.*   □

It is not true in general that the Fourier Transform of $f \in L^1(\mathbb{R})$ is again in $L^1(\mathbb{R})$ as the following example shows.

5.1.11   **Example** Consider, for instance, the function

$$f(x) := \begin{cases} e^{-x} & \text{if } x \ge 0, \\ 0 & \text{if } x < 0, \end{cases}$$

then

$$\hat{f}(\xi) = \int_0^\infty e^{-2\pi i \xi x - x}\, dx = \lim_{R \to \infty} \int_0^R e^{-(2\pi i \xi + 1)x}\, dx$$

$$= \lim_{R \to \infty} -\frac{e^{-(2\pi i \xi + 1)R}}{2\pi i \xi + 1} + \frac{1}{2\pi i \xi + 1} = \frac{1}{2\pi i \xi + 1},$$

which is not integrable.   ◇

Therefore, it is interesting to determine under which conditions $\hat{f} \in L^1(\mathbb{R})$. The next lemma gives sufficient conditions on f.

5.1.12   **Lemma** [64, 18.1.2 Theorem] *If f is twice continuously differentiable and if f, f' and f'' are in $L^1(\mathbb{R})$, then $\hat{f} \in L^1(\mathbb{R})$.*   □

5.1.13   **The Inverse Fourier Transform** [64, 18.1.1 Theorem] *If both f and $\hat{f}$ belong to $L^1(\mathbb{R})$, then*

$$f(x) = \int_{\mathbb{R}} \hat{f}(\xi) e^{2\pi i \xi x}\, d\xi$$

for all points x, where f is continuous. This integral is called the Inverse Fourier Transform *of f.*   □

A function $f \in L^1(\mathbb{R})$ is called *band limited* if the support of its Fourier Transform $\hat{f}$ is bounded. In other words, if there exists a limiting value $\lambda_c > 0$ such that $\text{supp}(\hat{f}) \subseteq [-\lambda_c, \lambda_c]$, which means that in the signal f no frequencies greater than λ_c occur.

Consequence of the Paley–Wiener Theorem [64, 31.5.2 Theorem and p. 360]   5.1.14
If $f \neq 0$ is band limited, then $f \in C^\infty(\mathbb{R})$. Hence, f vanishes on no interval of positive length. The assumption that $f \neq 0$ is band limited implies that f is analytic with $\mathrm{supp}(f) = \mathbb{R}$.

If f has bounded support (which means bounded in time), then f cannot be band limited since $\hat{f}$ is a C^∞-function.   □

Example Consider the band limited function f with   5.1.15

$$\hat{f}(\xi) = \begin{cases} 1 & \text{if } |\xi| \leq \lambda_c, \\ 0 & \text{if } |\xi| > \lambda_c. \end{cases}$$

Since f is band limited, it is analytic, thus continuous. By the Inverse Fourier Transform we get

$$f(x) = \int_{\mathbb{R}} \hat{f}(\xi) e^{2\pi i \xi x} \, d\xi = \int_{-\lambda_c}^{\lambda_c} e^{2\pi i \xi x} \, d\xi = \left. \frac{e^{2\pi i \xi x}}{2\pi i x} \right|_{-\lambda_c}^{\lambda_c}$$

$$= \frac{1}{\pi x} \cdot \frac{e^{2\pi i \lambda_c x} - e^{-2\pi i \lambda_c x}}{2i} = 2\lambda_c \frac{\sin(2\pi \lambda_c x)}{2\pi \lambda_c x}.$$

This is closely related to the *cardinal sine function* sinc : $\mathbb{R} \to \mathbb{R}$ which is defined by

$$\mathrm{sinc}(x) := \begin{cases} \sin(x)/x & \text{if } x \neq 0, \\ 1 & \text{if } x = 0. \end{cases}$$

It has zeros at $k\pi$ for $k \in \mathbb{Z} \setminus \{0\}$. The set of local extrema of $\mathrm{sinc}(x)$ corresponds to its intersections with the cosine function $\cos(x)$. The main peak of $\mathrm{sinc}(x)$ has width 2π whereas all other peaks have width π.

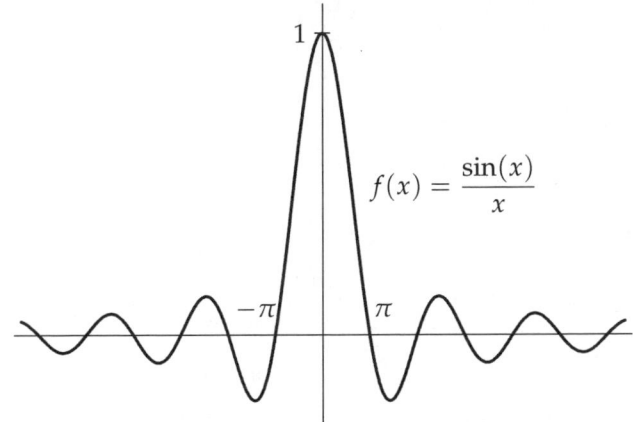

Fig. 5.2 The cardinal sine function

If we consider $\mathrm{sinc}(2\pi \lambda_c x)$ as a function of λ_c we notice that as λ_c increases, the width of the peaks decreases.   ◇

5.1.16   **Sampling** Assume that $f : \mathbb{R} \to \mathbb{C}$ is a signal and $s > 0$ is a real number. *Sampling* f every s time units means to replace f by the sequence $(f(ns))_{n \in \mathbb{Z}}$. The *sampling rate* or *sampling frequency* is given by $1/s$.

For instance, when watching a film, we are presented a sequence of 15 to 20 pictures per second. From this sampling we get the impression of smoothly moving pictures. But when seeing a turning wheel equipped with spokes, we realize that sometimes the wheel seems to rotate in the right direction, sometimes it is standing still, and sometimes it is turning in the converse direction. This is caused by the sampling of the pictures. Consider, for instance, a turning wheel with 4 spokes, which has a periodic movement with period $\pi/2$. If it is turning slowly enough, i.e. if we take more than two pictures within one period we get the right impression. For example, consider a wheel which is turning anti-clockwise. If it is turning from one sample to the next by $\pi/6$, we get:

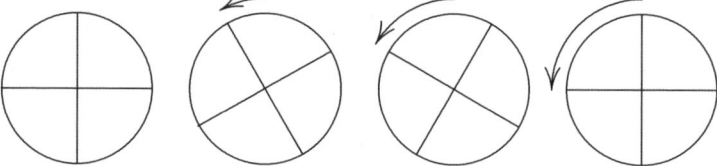

Turning between two samplings by $\pi/4$ produces only two different pictures so that we have the impression of a standing wheel which has 8 spokes.

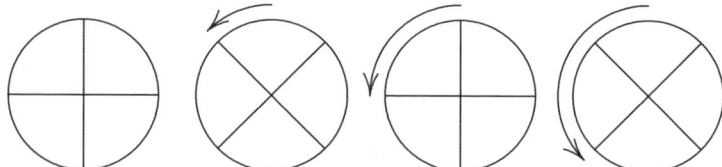

If the wheel is turning even faster, for instance by $\pi/3$ from one sampling to the next, then we have the impression that the wheel is rotating in the opposite direction.

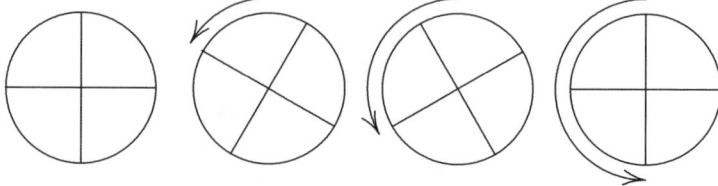

If the wheel is turning even faster we will realize similar phenomena as just described. From this example we deduce that on the one hand some information of the original signal is lost by sampling. But we will see that the loss of information can be neglected for practical purposes, provided the sampling rate is sufficiently high. On the other hand, digital processing offers many advantages. For instance, real time processing of the data can be done between the arrival of two consecutive samples. In addition, sampling and digital processing also allow us to compress the signal in such a way that it can later be reconstructed without loss of essential data. This technique is known as source coding. For further details we refer to [81].

Shannon's formula is an interpolation formula which expresses the value $f(x)$ of a signal at any time x in terms of its values $f(ns)$ at the discrete points ns, $n \in \mathbb{Z}$, $s > 0$. This theorem is attributed to C.E. Shannon, but was already discovered earlier by E.T. Whittaker [200], J.M. Whittaker [201], Kotel'nikov [114], or Nyquist [160]. For a historical background see also [103] or [141].

Shannon's Sampling Theorem [179] *Assume that $f \in L^1(\mathbb{R})$ is a band limited signal with* $\mathrm{supp}(\hat{f}) \subseteq [-\lambda_c, \lambda_c]$. *Consider a sampling rate* $1/s \geq 2\lambda_c$. *If the Fourier Transform $\hat{f}$ is piecewise continuously differentiable on the closed interval* $[-\lambda_c, \lambda_c]$, *then*   **5.1.17**

$$f(x) = \sum_{n=-\infty}^{\infty} f(ns) \frac{\sin \pi(x/s - n)}{\pi(x/s - n)}, \qquad x \in \mathbb{R}.$$ **5.1.18**

Proof: We extend the restriction of $\hat{f}$ to the interval $[-1/(2s), 1/(2s))$, which contains $[-\lambda_c, \lambda_c)$, to a continuous, periodic function $\tilde{f}$ with period $a = 1/s$ defined by

$$\tilde{f}(x + n/s) := \hat{f}(x) \text{ for } x \in [-1/(2s), 1/(2s)), \ n \in \mathbb{Z}.$$

Since $\tilde{f}$ is continuous and piecewise continuously differentiable, according to 5.1.6 the Fourier series of $\hat{f}$ converges uniformly. Consequently, for each $\xi \in [-1/(2s), 1/(2s)) = [-a/2, a/2)$ we have

$$\hat{f}(\xi) = \sum_{n=-\infty}^{\infty} c_n e^{2\pi i n \xi/a},$$

where

$$c_n = \frac{1}{a} \int_{-a/2}^{a/2} \hat{f}(\xi) e^{-2\pi i n \xi/a} \, d\xi = s \int_{\mathbb{R}} \hat{f}(\xi) e^{2\pi i \xi \cdot (-ns)} \, d\xi = s f(-ns)$$

by 5.1.13. Another application of the Fourier Inversion formula yields

$$f(x) = \int_{\mathbb{R}} \hat{f}(\xi) e^{2\pi i \xi x} \, d\xi = \int_{-1/(2s)}^{1/(2s)} \sum_{n=-\infty}^{\infty} c_n e^{2\pi i n \xi s} e^{2\pi i \xi x} \, d\xi.$$

Since the Fourier series converges uniformly, we are allowed to interchange the sequence of integration and summation, obtaining

$$f(x) = s \sum_{n=-\infty}^{\infty} f(-ns) \int_{-1/(2s)}^{1/(2s)} e^{2\pi i \xi(ns+x)} \, d\xi$$

$$= \sum_{n=-\infty}^{\infty} sf(ns) \int_{-1/(2s)}^{1/(2s)} e^{2\pi i \xi(x-ns)} \, d\xi = \sum_{n=-\infty}^{\infty} f(ns) \frac{\sin \pi(x/s - n)}{\pi(x/s - n)}$$

by an application of 5.1.15.   □

The assumptions of the last theorem are satisfied, for instance, in the situation when $f \in L^1(\mathbb{R})$ is band limited and the function $x \mapsto xf(x)$ is also integrable, since then by 5.1.10 $\hat{f}$ is continuously differentiable.

In order to describe a band limited signal f with $\mathrm{supp}(\hat{f}) \subseteq [-\lambda_c, \lambda_c]$ by a sampling with rate $1/s$ we must choose s so that $1/s \geq 2\lambda_c$. This critical value $2\lambda_c$ is called the *Nyquist rate*. Obviously, the largest frequency which can be described properly using a sampling rate $1/s$ is $1/(2s)$. This frequency is also known as the *Nyquist frequency*.

The following picture (see [30, page 62]) shows that if the frequency of a signal is too high for a given sampling rate, then this frequency is not properly described by the sampling. For instance, taking 10 samples per time unit yields $s = 1/10$ and a sampling rate of 10. In this situation we cannot distinguish between the two functions $f_1(x) = \sin(8\pi x)$ and $f_2(x) = \sin(28\pi x)$. The first one produces exactly 4 sine waves per time unit the second one 14 sine waves. If the time unit is a second, then we have a sampling rate of 10 Hz and two frequencies of 4 Hz and 14 Hz, respectively. Since $10 > 2 \cdot 4$, the sampling rate is high enough for describing f_1. According to 5.1.17, the frequency of f_2 is too high for the given sampling rate, so f_2 cannot be properly reconstructed from the sampling.

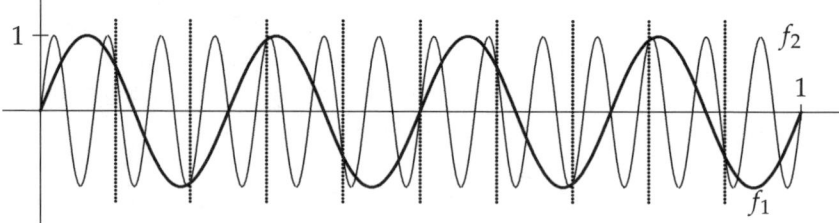

Fig. 5.3 Aliasing

The two functions f_1 and f_2 coincide for all x of the form $k/10$, $k \in \mathbb{Z}$. Hence, sampling f_2 at a sampling rate of 10 produces f_1. This phenomenon is called *aliasing*. In general, if $1/s$ is the sampling frequency and $f > 1/(2s)$ is sampled, then new frequencies f' appear, where $f' = n/s \pm f$ for $n \in \mathbb{Z}$.

In signal processing, $L^2(\mathbb{R})$ models the space of signals which are functions of a continuous variable (usually time) and which have finite energy. So far the Fourier Transform has only been defined for functions in $L^1(\mathbb{R})$, and $L^2(\mathbb{R})$ is not included in $L^1(\mathbb{R})$. In [64, Lesson 22] it is explained how to extend the

Fourier Transform in a natural way to $L^2(\mathbb{R})$. The Fourier Transform in $L^2(\mathbb{R})$ has the major advantage that $\hat{f} \in L^2(\mathbb{R})$ whenever $f \in L^2(\mathbb{R})$.

Filters In order to apply Shannon's Sampling Theorem, we must have a band limited signal in order to determine the correct sampling rate $1/s$. Thus, the natural signal, which is usually not band limited, must be filtered.

5.1.19

Assume that f is a time limited, piecewise continuously differentiable signal, then by the Inverse Fourier Transform

$$f(x) = \int_{\mathbb{R}} \hat{f}(\xi)e^{2\pi i \xi x}\,d\xi, \qquad x \in \mathbb{R}.$$

Usually a *filter* is described by its *transfer function* $\hat{h}$ for some $h \in L^2(\mathbb{R})$. Applying the filter with transfer function $\hat{h}$ to the signal f should produce (cf. [30, page 202ff]) the signal

$$\tilde{f}(x) = \int_{\mathbb{R}} \hat{f}(\xi)\hat{h}(\xi)e^{2\pi i \xi x}\,d\xi, \qquad x \in \mathbb{R}.$$

The amplitudes $|\hat{f}(\xi)|$ of the original signal are multiplied by the amplitudes $|\hat{h}(\xi)|$ and the phases $\arg(\hat{f}(\xi))$ are changed by adding the phase $\arg(\hat{h}(\xi))$. A complex number $c \neq 0$ may be presented as $c = x + iy$ with $x, y \in \mathbb{R}$, or $c = |c|e^{i\arg(c)}$, where $|c| = \sqrt{x^2 + y^2}$ is a positive real number called the complex modulus of c and $\arg(c)$ is a real number in the interval $(-\pi, \pi]$. It is called the argument of c and can be computed as

$$\arg(x + iy) = \begin{cases} \tan^{-1}(y/x) & \text{if } x \neq 0, \\ -\pi/2 & \text{if } x = 0 \text{ and } y < 0, \\ \pi/2 & \text{if } x = 0 \text{ and } y > 0. \end{cases}$$

An *ideal low-pass filter* suppresses all frequencies greater than a limiting frequency λ_c. It produces a signal $\tilde{f}$ which may be delayed by $x_0 \geq 0$, where the frequencies in $[-\lambda_c, \lambda_c]$ are not changed and all the frequencies outside this interval are canceled. Thus, the transfer function of an ideal low-pass filter is given by

$$\hat{h}(\xi) = \begin{cases} A_0 e^{-2\pi i \xi x_0} & \text{if } |\xi| \leq \lambda_c, \\ 0 & \text{if } |\xi| > \lambda_c, \end{cases}$$

with $A_0 > 0$, and the filtered signal is

$$\tilde{f}(x + x_0) = A_0 \int_{-\lambda_c}^{\lambda_c} \hat{f}(\xi)e^{2\pi i \xi x}\,d\xi, \qquad x \in \mathbb{R}.$$

Moreover, we derive that $\tilde{f}$ is band limited, infinitely many times differentiable, but it is not time limited, i.e. its support is the whole real line. From 5.1.15 we deduce that

$$h(x) = 2A_0\lambda_c \frac{\sin(2\pi\lambda_c(x - x_0))}{2\pi\lambda_c(x - x_0)}.$$

A filter is called *realizable* if for each signal f and each $x_0 \in \mathbb{R}$ the implication

$$\Big(f(x) = 0 \text{ for all } x < x_0\Big) \implies \Big(\tilde{f}(x) = 0 \text{ for all } x < x_0\Big)$$

holds true. From the explicit form of h given above we derive that an ideal low-pass filter is actually not realizable. The best we can expect is to find realizable filters whose transfer functions approximate the transfer function of an ideal filter. For instance, *Butterworth filters* are realizable approximations of ideal low-pass filters.

5.1.20   **Impulse** Similarly as in 5.1.15 we compute the Fourier Transform of the time limited constant signal

$$\delta_\epsilon(x) := \begin{cases} 1 & \text{if } |x| \leq \epsilon/2, \\ 0 & \text{if } |x| > \epsilon/2, \end{cases} \qquad \epsilon > 0.$$

For $\epsilon = 1/m$, $m \in \mathbb{N}^*$, we obtain

$$m\widehat{\delta_{1/m}}(\zeta) = m \int_{\mathbb{R}} \delta_{1/m}(x) e^{-2\pi i \zeta x}\, dx = \operatorname{sinc}\left(\frac{\pi \zeta}{m}\right).$$

In the following picture this function is plotted for $m = 1$ (solid), $m = 10$ (dashed), $m = 100$ (dotted), and $m = 1000$ (solid).

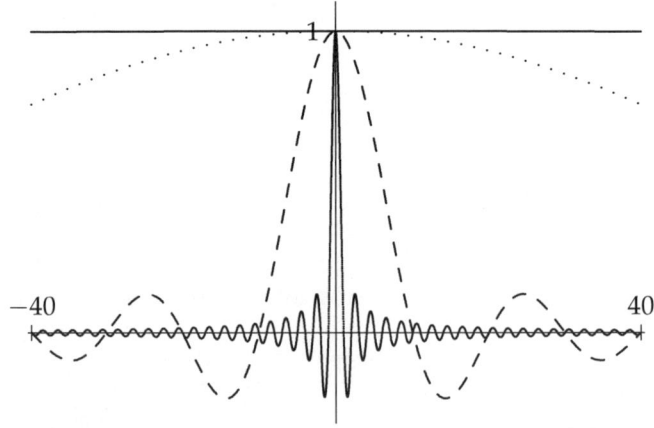

Given a limiting frequency λ_c we can find an integer m large enough so that $m\widehat{\delta_{1/m}}$ is approximately equal to 1 on the whole interval $[-\lambda_c, \lambda_c]$. Equivalently, if the signal $\delta_{1/m}$ is short enough, we call it an *impulse*, then $m\widehat{\delta_{1/m}}$ is approximately equal to 1 on this interval. This fact can be used for the

---

5.1.21   **Reconstruction of the original signal from a sampled signal** In order to reconstruct the original signal from a sampled signal we cannot use the formula

5.1.18 of the Sampling Theorem, since computing $f(x)$ for $x \in \mathbb{R}$ would require the knowledge of the sampled data $f(ns)$ for all $n \in \mathbb{Z}$, where f is a
function whose support is the whole real line.

If we assume that we know only finitely many sampled values, for instance
$f(ns)$ for $-M \le n \le N$, then we can use the sequence of impulses

$$f_{\epsilon,s}(x) := \sum_{n=-M}^{N} f(ns)\delta_\epsilon(x - ns)$$

as an input signal to an ideal low-pass filter described in 5.1.19 with limit-
ing frequency $\lambda_c = 1/(2s)$. If we choose ϵ small enough so that $(1/\epsilon)\hat{\delta}_\epsilon$ is
approximately 1 on $[-\lambda_c, \lambda_c]$, then the output of the ideal low-pass filter is
approximately

$$
\begin{aligned}
\widetilde{f_{\epsilon,s}}(x) &= \int_{-\lambda_c}^{\lambda_c} \widehat{f_{\epsilon,s}}(\xi)\hat{h}(\xi)e^{2\pi i \xi x}\, d\xi \\
&= A_0 \sum_{n=-M}^{N} f(ns) \int_{-\lambda_c}^{\lambda_c} \hat{\delta}_\epsilon(\xi)e^{2\pi i \xi(x-(x_0+ns))}\, d\xi \\
&\approx 2\lambda_c A_0 \epsilon \sum_{n=-M}^{N} f(ns)\frac{\sin \pi((x-x_0)/s - n)}{\pi((x-x_0)/s - n)}, \qquad x \in \mathbb{R}.
\end{aligned}
$$

Up to the factor $2\lambda_c A_0 \epsilon$ and the delay by x_0 this is an approximation of f,
converging to $2\lambda_c A_0 \epsilon f(x - x_0)$ if $N, M \to \infty$.

Digital audio transmission As an input we have the time limited audio sig-   **5.1.22**
nal f. Using an approximation of an ideal low-pass filter with maximum
frequency λ_c we obtain a band limited signal $\tilde{f}$. Using a sampling rate
$1/s \ge 2\lambda_c$ we obtain the sequence of samples $\tilde{f}(ns)$ for $-M \le n \le N$. An
analog-digital converter replaces each sample $\tilde{f}(ns)$ by a digital codeword.
These codewords are sent through the channel. The receiver produces a sequence of samples $F(ns)$ for $0 \le n \le N + M$ by using a digital-analog converter. Applying an ideal low-pass filter we smooth the sequence of impulses
$\sum_{n=0}^{N+M} F(ns)\delta_\epsilon(x - ns)$ and obtain an approximation of the input signal.

Sound pressure and decibels *Sound power* or *acoustic power* P is a measure of   **5.1.23**
sonic energy per time unit. It is measured in watts, abbreviated by W. The ratio
of two sound powers P_1 and P_0 is usually described in decibels by

$$dB = 10 \log_{10}(P_1/P_0).$$

If no second power P_0 is indicated, then the reference sound power in air is
usually taken to be $10^{-12}\,\text{W} = 0\,\text{dB}$. The decibel is a dimensionless "unit".

Since most audio engineers work with voltages and since power is proportional to the square of voltage, we also have

$$dB = 10 \log_{10}(V_1^2 / V_0^2) = 20 \log_{10}(V_1 / V_0).$$

The smallest change of sound power detectable for the human ear corresponds to one decibel. A change of three decibel is noticeable to most people. 10 dB seems to be approximately twice as loud.

The *sound intensity* is defined as the sound power P per unit area. The usual context is the measurement of sound intensity in the air at a listener's location. It is expressed in W/m². *Sound pressure p* or *acoustic pressure* is the measurement in Pascal (Pa $=$ N/m²) of the average sound wave pressure variations as the sound wave passes by a fixed point. We have

$$p = F/A,$$

where p is the sound pressure in Pascal Pa, F is the force measured in Newton N, and A is the area measured in m².

Sound is usually measured by microphones and they respond with voltage approximately proportional to the sound pressure p. Thus, we also have

$$dB = 20 \log_{10}(V_1 / V_0) = 20 \log_{10}(p_1 / p_0).$$

Unless specified otherwise, the reference level for air is chosen as 20 micro Pascal. This is about the limit of sensitivity of the human ear in the most sensitive range of frequency.

If two sound intensities satisfy $p_1 = 2p_0$, then the signal of p_1 is twice as loud as the signal of p_0. This yields a ratio of $20 \log_{10}(2) \approx 6.02059913$ dB.

5.1.24   **Analog-digital converter** So far we have been dealing with sound sampling and we have seen that when the sampling rate is high enough it is possible to reconstruct totally a band limited signal from the discrete samples. At each sampling the amplitude of a sound signal is measured. The amplitude is an analog signal which takes infinitely many values. When these values are stored in a digital system, only a finite number of discrete values or steps can be represented by digital numbers of finite length. So, for each measured amplitude value a digital value must be found which approximates the original value as good as possible. *Quantization* is the technique of approximating an analog amplitude by discrete numbers. (For more details see [164, pages 27ff].) By convention, the signal is attached with a sign. This means, that half of the digital values are used for positive amplitudes and half of them for 0 and negative amplitudes. When producing a compact disc each sampling is quantized by a 16 bit word. Hence, there are totally $2^{16} = 65536$ values which can be used to approximate all occurring amplitudes. In general, at each sampling small approximation errors occur. These errors are not bigger than half of a step between two consecutive digital values.

The number of bits used to represent a single sampling value, i.e. the word length, determines the resolution of the quantization.

The *signal-to-noise ratio*, often abbreviated by S/N, is an engineering term for the ratio between the magnitude of a signal and the magnitude of background noise. It is often expressed in terms of the decibel scale. If the incoming signal strength in microvolts is V_s, and the noise level, also in microvolts, is V_n, then the signal-to-noise ratio in decibels is given by the formula

$$S/N(dB) = 20 \log_{10}(V_s/V_n).$$

Signal-to-noise ratios are closely related to the concept of dynamic range. Whereas dynamic range measures the ratio between noise and the greatest undistorted signal on a channel, S/N measures the ratio between noise and an arbitrary signal on the channel, not necessarily the most powerful signal possible. Because of this, measuring signal-to-noise ratios requires the selection of a representative or reference signal. In audio engineering, this reference signal is usually a sine wave, i.e. a plain tone, at a recognized and standardized magnitude (cf. [202]).

In connection with digital audio, the noise is the error signal caused by the quantization of the signal. The *signal-to-error ratio*, in short S/E ratio, is the number of available digital values divided by the maximal quantization error. Above we have just seen that the quantization error is not greater than 1/2 bit. For instance, the S/E ratio of 16 bit audio is $2^{17} = 131\,072$. Every added bit doubles this ratio and also the number of possible digital values.

Usually the amplitude of a signal is attached with a sign. Therefore, using a quantization with n bits, the range of the digital signal lies between $-(2^{n-1} - 1)$ and 2^{n-1}. When a voltage amplitude of V_{max} is used, then a quantization step is given by

$$\Delta = \frac{V_{max}}{2^{n-1}}.$$

In terms of decibels, the S/E ratio depends on the word length n as

$$S/E(dB) = 20 \log_{10} \frac{V_{max}}{\Delta/2} = 20 \log_{10}(2^n) = 20n \log_{10}(2) \approx 6.02n.$$

This formula yields for 16-bit audio a S/E ratio of about 96dB. For more details see [156].

Pulse code modulation The amplitude of the signal is stored in *pulse code modulation*, PCM (cf. [164, pages 35ff]). This means that the analog signal is   **5.1.25**

represented by the sequence of binary values produced by the analog-digital converter. In PCM format the quantized value of every sampling is stored. In 16-bit audio format we can represent an audio signal by $2^{16} = 65\,536$ discrete levels. As mentioned above, the signal is attached with a sign. Hence, there are $32\,768$ binary values for representing positive amplitudes and $32\,767$ binary values for negative amplitudes.

PCM was invented by A.H. Reeves in 1939 (American Patents 2272070, 1942-2, see [156]) and was analyzed and developed as a modulation system from the point of view of communication theory by C.E. Shannon [180]. For 16-bit audio with sampling rate $44\,100$ Hz the demand on the storage device and speed of the transmission channel is $88\,200$ Bytes/sec. (Usually 8 bits are combined into one byte, thus each sampling produces 2 bytes of audio information. For an audio compact disc the left and right channel are sampled separately, which yields the above-mentioned number of audio bytes per second for each channel.) This is a "brute force" approach, which is not the most effective way of using the storage device and transmission channel.

"Sampling and quantization are the two fundamental design elements for audio digitization." ([164])

5.1.26   **Dither** As a matter of fact, even though the quantization error of 16-bit audio is quite small, it is obvious that when the signal amplitude decreases, the relative error increases. If the signal level is approximately as big as the signal difference corresponding to two consecutive digital values, then these errors could be audible. In other words, quantization not only loses information, but also causes unexpected problems. The following picture shows a low level sine signal and its quantization which is a square wave. As a square wave, according to 5.1.7, its trigonometric approximation is rich of odd harmonics extending far beyond the sampling frequency. (Harmonics are sine waves the frequencies of which are positive integer multiples of the wave with the smallest frequency.) For this reason, as we already know, low-pass filters must be used in order to obtain a band-limited output signal.

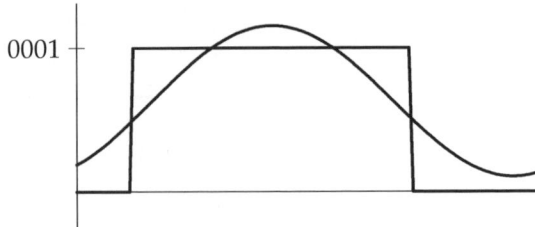

0001

Fig. 5.4 Quantization of a low level sine signal

After quantization of a low amplitude signal the resulting signal differs extremely from the original signal. This effect is also known as *granulation noise*. In high amplitude audio signals this effect is usually not audible.

Granulation noise can be removed by, surprisingly, adding small amounts of analog noise. This noise is called *dither* (cf. [164, pages 32ff]).

Adding low level noise to the previous signal we obtain:

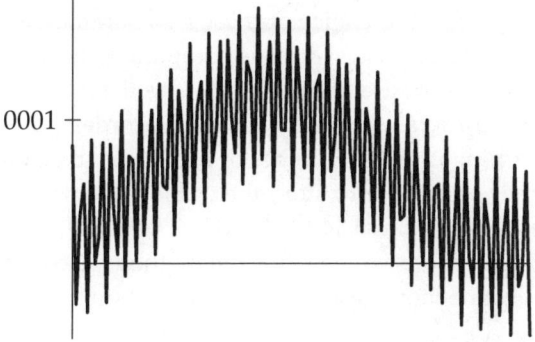

Fig. 5.5 Adding dither to a low level sine signal

After quantization it looks like:

Fig. 5.6 Quantization of the low level sine signal together with dither

Finally, taking the average amplitude of this quantized signal over a certain period of time we obtain

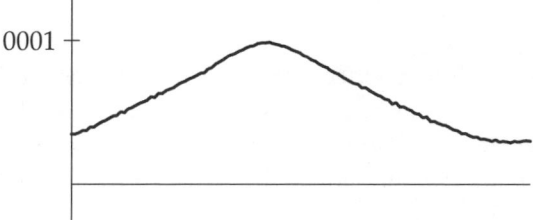

Fig. 5.7 The average amplitude of the quantized low level sine signal with added dither

which is quite similar to the original signal.

Since human beings are unable to hear frequencies which are beyond about 20 000 Hz, in the process of storing audio data on a compact disc all frequencies which are higher than 20 000 Hz are filtered out first. According to the Sampling Theorem, the sampling rate must be at least 40 000 samples per second in order to describe a band limited signal of maximal 20 000 Hz. For audio compact discs, 44 100 samples per second are taken both for the left and the right stereo channel. Each sample value is represented by 16 bits, this is a binary vector of length 16. The choice of these two parameters guarantees that the fidelity of the compact disc system is comparable to the best analog systems. For historical reasons the sampling rate was fixed to 44 100 Hz. Namely, early digital tape recorders used video cassette recorders for storage. Caused by the video refreshing rate and number of pixels per screen used in different video systems, they were able to store the audio information produced by a sampling frequency of maximal 44 100 Hz ([164, page 22]).

In the following sections we describe the methods used for error detection and correction in connection with audio compact discs.

Exercises

E.5.1.1   **Exercise** Show that $L^1(I)$ is a normed vector space.

E.5.1.2   **Exercise** Assume that $p > 1$ is a real number. Verify the details that $L^p(I)$ is a normed vector space.

Hints: For $f, g \in L^p(I)$, show that the inequality

$$|f(x) + g(x)|^p \le (|f(x)| + |g(x)|)^p \le 2^p \left(|f(x)|^p + |g(x)|^p\right)$$

holds almost everywhere on I, and deduce that $f + g \in L^p(I)$.

The proof of the triangle inequality for $\|.\|_p$ is based on *Hölder's Inequality*. Assume that $p, q > 1$ satisfy $1/p + 1/q = 1$. If φ belongs to $L^p(I)$ and ψ belongs to $L^q(I)$, then $\varphi\psi$ is an element of $L^1(I)$ and

$$\|\varphi\psi\|_1 \le \|\varphi\|_p \|\psi\|_q.$$

In this setting, the triangle inequality is also known as *Minkowski's Inequality*. Determine q such that $1/p + 1/q = 1$. Let $f, g \in L^p(I)$, then $h := |f + g|^{p-1}$ belongs to $L^q(I)$, and $h^q = |f + g|^{q(p-1)} = |f + g|^p = |f + g|h \le |fh| + |gh|$. Applying Hölder's inequality twice, once for $(\varphi, \psi) = (f, h)$ and once for $(\varphi, \psi) = (g, h)$, show that

$$\|f + g\|_p^p \le \left(\|f\|_p + \|g\|_p\right) \|f + g\|_p^{p/q}.$$

Exercise Prove that $L^2(I)$ is an Hermitian inner product space, i.e. the inner product is antilinear in the first argument, linear in the second argument and $\langle g, f \rangle = \overline{\langle f, g \rangle}$.

Exercise For $f_1, f_2 \in L^1(\mathbb{R})$ and $c_1, c_2 \in \mathbb{R}$ show that

$$\widehat{c_1 f_1 + c_2 f_2} = c_1 \hat{f}_1 + c_2 \hat{f}_2.$$

Consider $f \in L^1(\mathbb{R})$ and $a \in \mathbb{R}$. Prove that $g(x) := e^{-2\pi i a x} f(x)$ and $h(x) := f(x - a)$ are in $L^1(\mathbb{R})$ and verify

$$\hat{g}(\xi) = \hat{f}(\xi + a) \quad \text{and} \quad \hat{h}(\xi) = e^{-2\pi i a \xi} \hat{f}(\xi).$$

5.2 Correction of Erasures

After this short introduction of Fourier Analysis and digital audio we come back to the theory of error correcting codes.

Definition (erasures) An *erasure* is a transmission error where the exact position of the error in the received vector is known (but of course not the exact value of the error).   ◇

Erasures occur, for instance, when the receiver obtained a vector of which it could not read certain components, or when it is known to the decoder that certain positions in a received vector are not valid.

In the sequel when speaking about a transmission error, we use the term *random error* in order to distinguish it from an erasure. The reader can imagine that, in general, it is easier to correct erasures than random errors.

Theorem [104, 3.3.1 Satz] *Let C be an (n, k, d)-code over $\mathbb{F}_q$ and let t, u be non-negative integers. If $d \geq 2t + u + 1$, then it is possible to correct up to t errors and, additionally, u erasures with C.*

Proof: Assume that the receiver has obtained a vector $y \in \mathbb{F}_q^n$ containing exactly u erasures, which are located at $0 \leq i_0 < i_1 < \ldots < i_{u-1} \leq n - 1$, and in the remaining $n - u$ coordinates exactly t errors have occurred. Consider the set

$$V := \{ v \in \mathbb{F}_q^n \mid v_i = y_i \text{ for } i \notin \{i_0, \ldots, i_{u-1}\} \},$$

then there is exactly one $c \in C$ such that $d(c, v) \leq t$ for a suitable $v \in V$. From the assumptions it is clear that such a codeword c exists. We have to show that it is uniquely determined. Assuming that there exist $c^{(1)}, c^{(2)} \in C$, with $c^{(1)} \neq c^{(2)}$ and $v^{(1)}, v^{(2)} \in V$ such that $d(c^{(1)}, v^{(1)}) \leq t$ and $d(c^{(2)}, v^{(2)}) \leq t$, we obtain

$$d(c^{(1)}, c^{(2)}) \leq d(c^{(1)}, v^{(1)}) + d(v^{(1)}, v^{(2)}) + d(v^{(2)}, c^{(2)}) \leq t + u + t < d,$$

which is a contradiction to $\text{dist}(C) = d$. $\square$

5.2.3   **Example** If C is a binary code equipped with a decoding algorithm for random errors, then this algorithm can also be used for the correction of erasures.

Assume that $c \in C$ was sent and the received vector $y \in \mathbb{F}_2^n$ contains u erasures in the positions $i_0 < \ldots < i_{u-1}$. Moreover, assume that among the remaining components no more than t transmission errors have occurred and that $d \geq 2t + u + 1$. We consider two particular elements $v^{(0)}, v^{(1)}$ of V. The vector $v^{(0)}$ is obtained from y by replacing all erasures by 0, and $v^{(1)}$ is obtained by replacing them by 1. Since C is a binary code, one of two cases must occur: Either at least half the values at the erasure positions are zero or at least half the values are one. Consequently, in the first case $d(c, v^{(0)}) \leq t + \lfloor u/2 \rfloor$ and in the second case $d(c, v^{(1)}) \leq t + \lfloor u/2 \rfloor$.

Without loss of generality, assume that $d(c, v^{(0)}) \leq t + \lfloor u/2 \rfloor$. Then it is possible to decode $v^{(0)}$, and $v^{(0)}$ is decoded into the codeword originally sent. If $v^{(1)}$ cannot be decoded, then we are done. Assume that $v^{(1)}$ is decoded into $c' \in C$. If $c = c'$ we are also done. If $c' \neq c$ then

$$\left| \left\{ j \in n \mid c'_j \neq y_j, \ j \notin \{i_0, \ldots, i_{u-1}\} \right\} \right| > t,$$

thus c' cannot be obtained by filling the erasures and correcting up to t nonerased components of y. $\Diamond$

For codes over $\mathbb{F}_q$ with $q > 2$ this method is not very useful. In this case it would be necessary to compute all the q^u different vectors of V and find, according to 5.2.2, the existing and uniquely determined vector which can be decoded into a codeword so that at most t nonerased components are changed.

Hence, for BCH-codes we describe another method (cf. [104] and [68]). Let C be a BCH-code of length n over $\mathbb{F}_q$ with designed distance δ. By definition, there exists an integer b and a primitive n-th root of unity ξ such that the variety $V(C)$ contains the consecutive set $\{\xi^b, \ldots, \xi^{b+\delta-2}\}$. For $m = \text{ord}_n(q)$, the primitive n-th root ξ belongs to $\mathbb{F}_{q^m}$. Let t and u be nonnegative integers such that $\delta \geq 2t + u + 1$. Moreover, we assume that the codeword c was sent and that the vector $y \in \mathbb{F}_q^n$ was received. It contains exactly u erasures and we assume that during the transmission $w \leq t$ errors have occurred in the nonerased positions.

From y we derive the vector $\widetilde{y}$ in which all the erased positions are replaced by 0. Thus, we can express $\widetilde{y}$ as the sum

$$\widetilde{y} = c + e + a,$$

where e is the error vector describing the errors which occurred in the non-erased positions, and a is the vector which produces the value 0 at all the erased positions. Then the error vector of $\widetilde{y}$ equals $f := e + a$.

Using the check matrix $\widetilde{\Delta}$ over the extension field $\mathbb{F}_{q^m}$, which was introduced in the proof of 4.3.1, we compute the syndrome s of $\widetilde{y}$ as

$$s = \widetilde{y} \cdot \widetilde{\Delta}^\top.$$

We are interested in certain components of s, the *partial syndromes*, which are given by

$$s_j = \sum_{i \in n} \widetilde{y}_i \xi^{ij} = \widetilde{y}(\xi^j) = c(\xi^j) + f(\xi^j) = f(\xi^j), \qquad b \le j \le b + 2t + u - 1. \qquad \textbf{5.2.4}$$

As in 4.2.2 we consider the vectors $\widetilde{y}, c, f$ as polynomials. Since c is a codeword, $c(\xi^j) = 0$. This allows the computation of the *partial syndrome polynomial*

$$s(x) := \sum_{j \in 2t + u} s_{b+j} x^j.$$

From the received vector y we deduce that the erasures have occurred in positions $i_0, \ldots, i_{u-1}$. Moreover, we assume that the errors in the nonerased positions are located in $i_u, \ldots, i_{u+w-1}$. Actually these positions are not known in the beginning. In connection with these positions we have the *erasure location polynomial* which is given by

$$\lambda(x) := \prod_{j \in u} (1 - \xi^{i_j} x)$$

and the *error location polynomial* given by

$$\sigma(x) := \prod_{j=u}^{u+w-1} (1 - \xi^{i_j} x).$$

So far we know λ, but σ remains to be computed.

Now we turn our attention to the actual values of the errors. Once again, we recall that f_{i_j} for $j \in u + w$ denotes the error occurring at the position i_j when sending c and receiving $\widetilde{y}$. From 5.2.4, it follows that

$$s_\ell = \sum_{j \in u+w} f_{i_j} \xi^{i_j \ell}, \qquad b \le \ell \le b + 2t + u - 1. \qquad \textbf{5.2.5}$$

Let ω be the unique polynomial of degree less than $2t + u$ satisfying

$$\omega(x) \equiv s(x)\lambda(x)\sigma(x) \bmod \mathrm{I}(x^{2t+u}).$$

It is called the *error evaluation polynomial*.

5.2.6   **Lemma** *Using the notation from above we have*

$$\omega(x) = \sum_{j \in u+w} f_{i_j} \zeta^{i_j b} \prod_{r \neq j} (1 - \zeta^{i_r} x).$$

Proof: From 5.2.4 and 5.2.5, we get

$$s(x)\lambda(x)\sigma(x) = \left(\sum_{\ell \in 2t+u} \sum_{j \in u+w} f_{i_j} \zeta^{i_j(b+\ell)} x^\ell \right) \prod_{r \in u+w} (1 - \zeta^{i_r} x)$$

$$= \sum_{j \in u+w} f_{i_j} \zeta^{i_j b} \left(\sum_{\ell \in 2t+u} (\zeta^{i_j} x)^\ell \right) \prod_{r \in u+w} (1 - \zeta^{i_r} x)$$

$$= \sum_{j \in u+w} f_{i_j} \zeta^{i_j b} \frac{1 - (\zeta^{i_j} x)^{2t+u}}{1 - \zeta^{i_j} x} (1 - \zeta^{i_j} x) \prod_{r \neq j} (1 - \zeta^{i_r} x)$$

$$\equiv \sum_{j \in u+w} f_{i_j} \zeta^{i_j b} \prod_{r \neq j} (1 - \zeta^{i_r} x) \bmod I(x^{2t+u}).$$

This is a polynomial of degree at most $w + u - 1 < 2t + u$, whence it represents the polynomial ω.   □

Since we know the polynomial λ and since its degree is equal to u, we write $s\lambda$ in the form

$$s(x)\lambda(x) = T_1(x) + x^u T_2(x)$$

with $T_1, T_2 \in \mathbb{F}_{q^m}[x]$ and $\deg T_1 < u$. Then

$$s(x)\lambda(x)\sigma(x) - \sigma(x)T_1(x) = x^u \sigma(x)T_2(x),$$

whence

$$\omega(x) - \sigma(x)T_1(x) \equiv x^u \sigma(x)T_2(x) \bmod I(x^{2t+u}).$$

Consequently, x^u is a divisor of $\omega(x) - \sigma(x)T_1(x)$ and finally we obtain

$$\frac{\omega(x) - \sigma(x)T_1(x)}{x^u} \equiv \sigma(x)T_2(x) \bmod I(x^{2t}).$$

We denote the left-hand side of this congruence by $\Omega(x)$. It is a polynomial of degree at most $w - 1 \leq t - 1$.

5.2.7   **Lemma** *Let C be a BCH-code of length n over $\mathbb{F}_q$ with designed distance δ. The variety of C contains powers of ζ where $\zeta \in \mathbb{F}_{q^m}$ is a primitive n-th root of unity. Consider nonnegative integers t, u such that $u + 2t + 1 \leq \delta$. Assume that the product of the partial syndrome polynomial and the erasure location polynomial is written in the form $s(x)\lambda(x) = T_1(x) + x^u T_2(x)$ with $\deg T_1 < u$. Then there exist relatively* *prime polynomials* σ, $\Omega \in \mathbb{F}_{q^m}[x]$ *such that* $\deg \Omega \leq t - 1$, $\deg \sigma \leq t$, *and* $\Omega(x) \equiv \sigma(x) T_2(x) \bmod \mathrm{I}(x^{2t})$.

In addition, the polynomials σ and Ω are unique modulo scalars in the following sense. If σ_1 and Ω_1 are two further polynomials over $\mathbb{F}_{q^m}$ which are relatively prime and satisfy $\deg \Omega_1 \leq t - 1$, $\deg \sigma_1 \leq t$, and $\Omega_1(x) \equiv \sigma_1(x) T_2(x) \bmod \mathrm{I}(x^{2t})$, then there exists a constant $\kappa \in \mathbb{F}_{q^m}^*$ such that $\sigma_1 = \kappa \sigma$ and $\Omega_1 = \kappa \Omega$.

Proof: From the previous computations we already know that polynomials σ, Ω exist with $\deg \Omega \leq t - 1$, $\deg \sigma \leq t$, and $\Omega(x) \equiv \sigma(x) T_2(x) \bmod \mathrm{I}(x^{2t})$. Now we prove that they are relatively prime. If we assume that $\varphi \in \mathbb{F}_{q^m}[x]$ is a common divisor of σ and Ω, then it is also a divisor of $x^u \Omega(x)$, therefore, a divisor of $\omega - \sigma T_1$, thus a divisor of ω. From 5.2.6 it is clear that ω and σ are relatively prime, thus φ is a unit in $\mathbb{F}_{q^m}[x]$, in other words, a nonzero constant, thus σ and Ω are relatively prime.

If σ_1 and Ω_1 also have the desired properties, then

$$\Omega(x)\sigma_1(x) \equiv \sigma(x) T_2(x)\sigma_1(x) \equiv \sigma(x)\Omega_1(x) \bmod \mathrm{I}(x^{2t}).$$

Comparing degrees, we obtain that

$$\Omega\sigma_1 = \sigma\Omega_1.$$

Thus, σ_1 is a divisor of $\sigma\Omega_1$. Since σ_1 and Ω_1 are relatively prime, σ_1 is a divisor of σ. Correspondingly, we derive that σ is a divisor of σ_1, whence there exists a constant $\kappa \in \mathbb{F}_{q^m}^*$ such that $\sigma_1 = \kappa \sigma$. Finally, from $\Omega\kappa\sigma = \Omega\sigma_1 = \sigma\Omega_1$ we deduce that $\kappa\Omega = \Omega_1$.   □

So far we have proved that relatively prime polynomials σ and Ω of certain degrees exist, such that $\Omega(x) \equiv \sigma(x) T_2(x) \bmod \mathrm{I}(x^{2t})$. But we still don't know how to compute them. Next we describe a method for doing this which is based upon the

---

Extended Euclidean Algorithm [104, 3.2.12 Satz] *Consider nonzero polynomials*   **5.2.8**
$\varphi, \psi \in \mathbb{F}_q[x]$ *with* $\deg \varphi \geq \deg \psi$. *Put*

$$\begin{aligned} f_0 &:= 1, & g_0 &:= 0, & r_0 &:= \varphi, \\ f_1 &:= 0, & g_1 &:= 1, & r_1 &:= \psi, \end{aligned}$$

and recursively for $i \geq 1$ determine q_i, r_{i+1}, f_{i+1}, and g_{i+1} by

$$\begin{aligned} r_{i-1} &= q_i r_i + r_{i+1} \quad \text{with } \deg r_{i+1} < \deg r_i, \\ f_{i+1} &= f_{i-1} - q_i f_i, \\ g_{i+1} &= g_{i-1} - q_i g_i. \end{aligned}$$

Then

$$f_i \varphi + g_i \psi = r_i, \qquad i \geq 0,$$ **5.2.9**

and

5.2.10
$$\deg g_{i+1} \leq \deg \varphi - \deg r_i, \qquad i \geq 0.$$

Since $\deg r_{i+1} < \deg r_i$, *after finitely many steps this algorithm terminates with* $r_N \neq 0$ *and* $r_{N+1} = 0$. *Then* r_N *is equal to* $\gcd(\varphi, \psi)$. □

An application of the extended Euclidean Algorithm allows us to find polynomials σ and Ω with the properties described in 5.2.7.

5.2.11   **Sugiyama–Kasahara–Hirasawa–Namekawa algorithm** [188], [189] Let $\varphi(x) := x^{2t}$ and let $\psi(x)$ be the remainder of $T_2(x)$ after division by x^{2t}. Then there exist polynomials $f_N, g_N \in \mathbb{F}_{q^m}[x]$ such that

$$d(x) := \gcd(x^{2t}, T_2(x)) = \gcd(x^{2t}, \psi(x)) = f_N(x)x^{2t} + g_N(x)\psi(x).$$

If the assumptions of 5.2.7 are satisfied, then there exist relatively prime solutions Ω and σ of the congruence

$$\Omega(x) \equiv \sigma(x)\psi(x) \bmod \mathrm{I}(x^{2t})$$

with $\deg \Omega \leq t - 1$ and $\deg \sigma \leq t$. Thus, there exists a polynomial ϕ such that

$$\Omega(x) = \sigma(x)\psi(x) + \phi(x)x^{2t}$$

and d is a divisor of Ω. Hence $\deg d \leq t - 1$. The degrees of the polynomials r_i in 5.2.8 are strictly decreasing, and $\deg r_0 = \deg x^{2t} = 2t$. So, there exists an index $j_0 \in \{1, \dots, N\}$ such that $\deg r_{j_0-1} \geq t$ and $\deg r_{j_0} < t$. From 5.2.9 and 5.2.10 we obtain

$$f_{j_0}(x)x^{2t} + g_{j_0}(x)\psi(x) = r_{j_0}(x)$$

and $\deg g_{j_0} \leq 2t - \deg r_{j_0-1} \leq t$. Consequently, g_{j_0} and r_{j_0} satisfy the congruence

$$r_{j_0}(x) \equiv g_{j_0}(x)T_2(x) \bmod \mathrm{I}(x^{2t})$$

with $\deg g_{j_0} \leq t$ and $\deg r_{j_0} < t$. Since, furthermore, Ω and σ are required to be relatively prime we set

$$\Omega(x) := \frac{r_{j_0}(x)}{\gcd(r_{j_0}(x), g_{j_0}(x))} \quad \text{and} \quad \sigma(x) := \frac{g_{j_0}(x)}{\gcd(r_{j_0}(x), g_{j_0}(x))}.$$

If the assumptions of 5.2.7 are not satisfied, then it may happen that $T_2 \neq 0$ and $\deg d \geq t$. In this case we cannot find j_0 such that $\deg r_{j_0} < t$ and an uncorrectable error has occurred. □

So far we have computed the partial syndrome polynomial s, the error location polynomial σ, the erasure location polynomial λ, and the error evaluation polynomial

$$w(x) = \Omega(x)x^u + \sigma(x)T_1(x).$$

In the next step the error locations $i_u, \ldots, i_{u+w-1}$ are determined by finding the roots of σ. It is obvious that $\kappa \in \mathbb{F}_{q^m}$ is a root of σ if and only if there exists some $j \in \{u, \ldots, u+w-1\}$ such that $1 - \xi^{i_j}\kappa = 0$, which is equivalent to $\kappa = \xi^{-i_j}$. From the representation of the multiplicative inverse of the roots of σ as powers of ξ, we are able to deduce the error locations. In Section 3.5 we were briefly discussing a method how to determine the roots of a polynomial. For the present problem of finding the zeros of the error locator polynomial, we suggest another method. If we are able to determine the polynomial σ, then, according to 5.2.7, we can assume that $\sigma(0) = 1$. Hence, the roots of σ can be determined by

---

Chien search   5.2.12

Input: n the length of the code, $\xi \in \mathbb{F}_{q^m}$ a primitive n-th root of unity, and the error location polynomial $\sigma(x) = \sum_{i=0}^{w} \sigma_i x^i$ with $\sigma_0 = 1$.

Output: All roots of σ of the form ξ^j.

(1)   We introduce w variables $X_1, \ldots, X_w$. For $1 \le j \le w$ set $X_j := \sigma_j$ and set $i := 0$.

(2)   If

$$\sum_{j=1}^{w} X_j = -1,$$

   then output ξ^i.

(3)   If $i = n$, then STOP.
   Otherwise, set $i := i+1$ and for $1 \le j \le w$ set $X_j := X_j\xi^j$. Goto (2).   □

In the i-th step the sum in (2) stands for

$$\sum_{j=1}^{w} \sigma_j\xi^{ij} = \sigma(\xi^i) - 1,$$

whence ξ^i is a root of σ if and only if this sum equals -1.

Finally, we have to compute the error values f_{i_j} for $j \in u + w$ which describe the errors of $\tilde{y}$ in the i_j-th component. A useful method is given by the

---

Forney algorithm [55] *For $j \in u + w$ the error f_{i_j} can be computed as*   5.2.13

$$f_{i_j} = \frac{\xi^{-i_j b}\omega(\xi^{-i_j})}{\prod_{r \ne j}(1 - \xi^{i_r - i_j})} = -\frac{\xi^{-i_j(b-1)}\omega(\xi^{-i_j})}{\rho'(\xi^{-i_j})},$$

where $\rho := \sigma\lambda$.

Proof: From 5.2.6 we obtain

$$w(\xi^{-i_j}) = f_{i_j}\xi^{i_j b}\prod_{r\neq j}(1 - \xi^{i_r-i_j}),$$

which proves the first equality. Computing the formal derivative of ρ leads to

$$\rho'(x) = \frac{d}{dx}\prod_{j\in u+w}(1 - \xi^{i_j}x) = -\sum_{j\in u+w}\xi^{i_j}\prod_{r\neq j}(1 - \xi^{i_r}x).$$

Thus,

$$\rho'(\xi^{-i_j}) = -\xi^{i_j}\prod_{r\neq j}(1 - \xi^{i_r-i_j}),$$

which proves the second equality. □

Hence, we obtain the error values f_{i_j} for $j \in u + w$, from which together with the error- and erasure locations i_j, the error vectors e and a can easily be determined. By subtraction we obtain the originally sent codeword $c = \tilde{y} - e - a$.

5.2.14   **Algorithm (Decoding of errors and erasures with a BCH-code)** Let C be a BCH-code with designed distance δ and consecutive subset $\{\xi^b, \ldots, \xi^{b+\delta-2}\}$ of its variety, where $\xi \in \mathbb{F}_{q^m}$ is a primitive n-th root of unity.

Input: A vector y containing u erasures.

Output: A codeword c or an error message. If t denotes the number of errors in the nonerased positions and if $2t \leq \delta - 1 - u$, then the output is the codeword originally sent.

(1) Determine the vector $\tilde{y}$ by setting the erased components of y all equal to 0, and define $t := \lfloor(\delta - u - 1)/2\rfloor$.

(2) Compute the partial syndromes $s_j = \tilde{y}(\xi^j)$ for $b \leq j \leq b + 2t + u - 1$.

(3) Determine the partial syndrome polynomial

$$s(x) := \sum_{j\in 2t+u}s_{b+j}x^j.$$

If $s = 0$, then set $c := \tilde{y}$ and goto (12).

(4) Determine the locations of the erasures $i_0, \ldots, i_{u-1}$ and form the erasure location polynomial

$$\lambda(x) := \prod_{j\in u}(1 - \xi^{i_j}x).$$

(5) Write $s(x)\lambda(x)$ in the form $T_1(x) + x^u T_2(x)$ with $\deg T_1 < u$.

(6) Compute relatively prime polynomials $\Omega, \sigma \in \mathbb{F}_{q^m}[x]$ such that $\deg \Omega \leq t - 1$, $\deg \sigma \leq t$, and $\Omega(x) \equiv \sigma(x)T_2(x) \bmod I(x^{2t})$ as described in

5.2.11. If such polynomials do not exist, output an error message and STOP.

(7) Compute by Chien search the roots ζ^{-i_j} of σ for $j = u, \ldots, u + w - 1$, where $w \leq \deg \sigma \leq t$, and their multiplicative inverses. Then the error locations are $i_u, \ldots, i_{u+w-1}$.

(8) Set $\rho := \sigma \lambda$ and compute the formal derivative $\rho'(x)$.

(9) Compute the error evaluation polynomial

$$w(x) = \Omega(x)x^u + \sigma(x)T_1(x).$$

(10) For $j \in u + w$ determine the errors

$$f_{i_j} = -\frac{\zeta^{-i_j(b-1)}w(\zeta^{-i_j})}{\rho'(\zeta^{-i_j})}$$

as indicated by Forney's algorithm.

(11) Set

$$a(x) := \sum_{j \in u} f_{i_j} x^{i_j}, \qquad e(x) := \sum_{j=u}^{u+w-1} f_{i_j} x^{i_j},$$

and $c(x) := \tilde{y}(x) - a(x) - e(x)$.

(12) Output c.   □

This algorithm can also be used for decoding a BCH-code when no erasures occurred. In this case $u = 0$, $\tilde{y} = y$, $\lambda = 1$, $T_1 = 0$, and $w = \Omega$.

Algorithm (Decoding of errors with a BCH-code) Let C be a BCH-code   **5.2.15**
with designed distance δ and consecutive subset $\{\zeta^b, \ldots, \zeta^{b+\delta-2}\}$ of its variety, where $\zeta \in \mathbb{F}_{q^m}$ is a primitive n-th root of unity.
Input: A vector y.
Output: A codeword c or an error message. If t denotes the number of errors and if $2t \leq \delta - 1$, then the output is the codeword originally sent.

(1) Set $t := \lfloor (\delta - 1)/2 \rfloor$ and compute the partial syndromes $s_j = y(\zeta^j)$ for $b \leq j \leq b + 2t - 1$.

(2) Determine the partial syndrome polynomial

$$s(x) := \sum_{j \in 2t} s_{j+b} x^j.$$

If $s = 0$, then set $c := y$ and goto (8).

(3) Compute relatively prime polynomials $w, \sigma \in \mathbb{F}_{q^m}[x]$ such that $\deg w \le t - 1$, $\deg \sigma \le t$ and $w(x) \equiv \sigma(x) s(x) \mod I(x^{2t})$ as described in 5.2.11. If such polynomials do not exist, output an error message and STOP.

(4) Compute by Chien search the roots ξ^{-i_j} of σ for $j \in w$, where $w \le \deg \sigma \le t$, and their multiplicative inverses. Then the error locations are $i_0, \ldots, i_{w-1}$.

(5) Compute the formal derivative $\sigma'(x)$.

(6) For $j \in w$ determine the errors

$$f_{i_j} = - \frac{\xi^{-i_j(b-1)} w(\xi^{-i_j})}{\sigma'(\xi^{-i_j})}$$

as indicated by Forney's algorithm.

(7) Set

$$e(x) := \sum_{j \in w} f_{i_j} x^{i_j} \text{ and } c(x) := y(x) - e(x).$$

(8) Output c.   □

At the end of this section we want to discuss two numerical examples. First we illustrate the method of 5.2.3 for binary codes. It was taken from [104, 3.3.2 Beispiel].

5.2.16   **Example** Consider the binary $(15, 7, 5)$-code with generator polynomial $g(x) = x^8 + x^7 + x^6 + x^4 + 1$. According to 4.2.13, it is a BCH-code with $\delta = 5$, since g is the product of $x^4 + x + 1$ and $x^4 + x^3 + x^2 + x + 1$. Assume that

$$c = (1, 0, 0, 1, 1, 0, 1, 0, 1, 1, 1, 1, 0, 0, 0)$$

was sent and that the vector

$$y = (1, 0, \square, 1, 1, 0, 0, 0, 1, 1, 1, 1, 0, 0, \square)$$

was received. Thus, two erasures and one error have occurred. Now we set

$$v^{(0)} := (1, 0, 0, 1, 1, 0, 0, 0, 1, 1, 1, 1, 0, 0, 0)$$

and

$$v^{(1)} := (1, 0, 1, 1, 1, 0, 0, 0, 1, 1, 1, 1, 0, 0, 1).$$

Using a decoding algorithm for cyclic codes (for instance 5.2.15), we realize that $v^{(0)}$ is decoded to c, whereas $v^{(1)}$ cannot be decoded.   ◇

Now we consider a BCH-code over $\mathbb{F}_{11}$. A similar example can be found in [104, 3.3.4 Beispiel].

Example Let C be a BCH-code in the strict sense of length 10 with designed distance $\delta = 6$ over $\mathbb{F}_{11}$. The parameters were chosen so that $\zeta = 2$ is a primitive 10-th root of unity. Thus, it is not necessary to do computations in an extension field of $\mathbb{F}_{11}$.

5.2.17

The receiver has obtained the vector

$$y = (8, \boxed{}, 8, \boxed{}, 2, \boxed{}, 4, 2, 2, 1).$$

Since $u = 3$ erasures have occurred, we can detect at most $t = 1$ error in the nonerased positions. Initially, we set

$$\tilde{y} := (8, 0, 8, 0, 2, 0, 4, 2, 2, 1).$$

Then we compute the first five components of its syndrome

$$s_{j-1} = \sum_{i=0}^{9} \tilde{y}_i 2^{ij}, \qquad 1 \leq j \leq 5,$$

and obtain the partial syndrome polynomial

$$s(x) = 10x^4 + 5x^3 + 2x + 2.$$

The erasures occurred at the positions $i_0 = 1$, $i_1 = 3$ and $i_2 = 5$, whence

$$\lambda(x) = (1 - 2x)(1 - 2^3 x)(1 - 2^5 x) = 5x^3 + 6x^2 + 2x + 1.$$

Now we compute

$$s(x)\lambda(x) = 6x^7 + 8x^6 + 6x^5 + 8x^4 + 5x^3 + 5x^2 + 6x + 2$$

and determine

$$T_1(x) = 5x^2 + 6x + 2 \text{ and } T_2(x) = 6x^4 + 8x^3 + 6x^2 + 8x + 5 \equiv 8x + 5 \bmod I(x^2).$$

A solution of the congruence

$$\Omega(x) \equiv \sigma(x) T_2(x) \bmod I(x^2)$$

with $\deg \sigma \leq 1$ and $\deg \Omega = 0$ is easily computed as

$$\sigma(x) = x + 9 \quad \text{and} \quad \Omega(x) = 1.$$

The root of σ is $\kappa = 2 = \zeta$, whence $\kappa^{-1} = \zeta^{-1} = \zeta^9$ and the error location $i_3 = 9$. Now we determine

$$\rho(x) = \sigma(x)\lambda(x) = 5x^4 + 7x^3 + x^2 + 8x + 9,$$

$$\rho'(x) = 9x^3 + 10x^2 + 2x + 8$$

and

$$w(x) = \Omega(x)x^3 + \sigma(x)T_1(x) = 6x^3 + 7x^2 + x + 7.$$

This allows us to compute the following data

j	0	1	2	3
i_j	1	3	5	9
2^{-i_j}	6	7	10	2
$w(2^{-i_j})$	10	6	7	8
$\rho'(2^{-i_j})$	3	2	7	3
f_{i_j}	4	8	10	1

so that $a(x) = 4x + 8x^3 + 10x^5$, $e(x) = x^9$ and

$$c(x) = \tilde{y}(x) - a(x) - e(x) = 8 + 7x + 8x^2 + 3x^3 + 2x^4 + x^5 + 4x^6 + 2x^7 + 2x^8,$$

which results in the codeword

$$c = (8,7,8,3,2,1,4,2,2,0). \qquad \diamond$$

Exercises

E.5.2.1   **Exercise** Prove 5.2.8. First prove by induction that 5.2.9 is satisfied for $i \geq 0$. Show that $\deg q_i = \deg r_{i-1} - \deg r_i$ for $i \geq 1$. From the monotonicity of $\deg r_i$ deduce by induction that 5.2.10 is satisfied for $i \geq 0$. Finally, show by induction that $\gcd(r_{i+1}, r_i) = \gcd(r_i, r_{i-1}) = \gcd(\varphi, \psi)$ for $i \geq 1$.

E.5.2.2   **Exercise** Consider the code from 5.2.16 and the decoding method described in 5.2.3. Using the algorithm 5.2.15, show that both $v^{(0)}$ and $v^{(1)}$ constructed from

$$y = (0,0,0,0,1,0,0,\square,\square,0,0,0,0,0,0)$$

can be decoded into two distinct codewords $c^{(0)}$ and $c^{(1)}$. Assume that at most one error has occurred in the positions unaffected by erasures. Which of $c^{(0)}$ and $c^{(1)}$ was sent originally? (See [104, Übung 60].)

E.5.2.3   **Exercise** Let C be the BCH-code over $\mathbb{F}_7$ of length 6 with consecutive subset $\{\xi, \ldots, \xi^4\}$ of its variety and designed distance $\delta = 5$, where $\xi = 3 \in \mathbb{F}_7$. The vector $y = (3,3,\square,5,\square,2)$ was received. Which codeword was sent? (See [104, Übung 61].)

5.3 Burst Errors and Interleaving of Codes

If during the transmission of a codeword at least two errors have occurred, then the error vector can be seen as a burst error.

Definition (burst error) Let C be a code of length n. A *burst error* of length $\ell \leq n$ is an error vector e in which (after some cyclic rearrangement) all the nonzero entries (maybe mixed with some zero entries) occur in ℓ adjacent positions, say among the coordinates $e_i, \ldots, e_{i+\ell-1 \bmod n}$, with nonzero e_i and $e_{i+\ell-1 \bmod n}$.

5.3.1

A burst error of length ℓ is completely described by its *location*, which is the index i of the first nonzero component of e, and by its *pattern*, which can also be read as a polynomial $b(x) := e_i \cdot 1 + e_{i+1 \bmod n} x + \ldots + e_{i+\ell-1 \bmod n} x^{\ell-1}$ of degree $\ell - 1$. ◇

For example the two vectors

$$(0,0,0,0,1,0,1,1,0,0,0,0) \quad \text{and} \quad (1,0,0,0,0,0,0,0,0,1,1,0)$$

are both burst errors of length 4. The first burst has location 4 and pattern $1 + x^2 + x^3$, the second one has location 9 and pattern $1 + x + x^3$. In order to indicate that the second burst starts somewhere at the end of the vector and is continued at the beginning, we also call it a *wrap around burst*. However, neither the burst error nor its pattern and length are uniquely determined. The first vector is also a wrap around burst with location 6 and pattern $1 + x + x^{10}$, so a burst of length 11, or a wrap around burst with location 7 and pattern $1 + x^9 + x^{11}$. Of course, it makes sense to consider bursts of rather short length.

Again we use the term *random error* instead of *error* in the ordinary sense in order to distinguish it from a burst error. Random errors are sometimes seen as burst errors of length 1.

If the burst error has location i and pattern b of degree $\ell - 1$, then

$$e(x) \equiv x^i b(x) \bmod I(x^n - 1).$$

Theorem *A cyclic code with generator polynomial of degree t is able to detect all burst errors of length at most t.*

5.3.2

Proof: Let e be a burst error of length at most t, then $e(x) \equiv x^i b(x) \bmod I(x^n - 1)$ for some i, with $b \neq 0$ and $\deg b < t$. In order to show that e is not a codeword, we prove that e is not a multiple of the generator polynomial g.

On the contrary, if we suppose that $x^i b(x)$ belongs to the code, then there exists a polynomial a such that

$$x^i b(x) \equiv a(x) g(x) \bmod I(x^n - 1).$$

Hence, $x^n - 1$ is a divisor of $x^i b(x) - a(x)g(x)$, thus g is a divisor of $x^i b(x)$. Since $g(0) \neq 0$, we deduce that g is a divisor of b, which is impossible since $0 \leq \deg b < \deg g$. ☐

This fact motivates the method of *cyclic redundancy check* decoding (CRC-decoding) which is described below. It is applied for instance in computer networks, where it is important to detect errors. If a block of data was not transmitted correctly, then a transmission error is detected and the corresponding data must be transmitted again. CRC-decoding is also applied for error detection in digital audio applications.

5.3.3   **Algorithm (CRC-decoding [104, pages 86ff])**
Assume that C is a cyclic code of length n with generator polynomial g of degree $t = n - k$. Moreover, assume that the code was encoded systematically.

Input: A polynomial $y(x) = y_0 + \ldots + y_{n-1}x^{n-1} \in \mathbb{F}_q[x]$.

Output: A vector belonging to $\mathbb{F}_q^k$ or an error message. If a transmission error has occurred which is a burst of length at most t, then this error is definitely detected by this algorithm.

(1) By polynomial division determine $q, r \in \mathbb{F}_q[x]$ such that $y = qg + r$ with $r = 0$ or $\deg r < \deg g$.

(2) If $r \neq 0$ then output an error message.

(3) If $r = 0$ output the information symbols of y.   ☐

Especially the error detection rate of binary CRC-codes is quite high. Many burst errors of even bigger length can be detected by this method.

5.3.4   **Lemma** *Let C be a binary cyclic code with a generator polynomial of degree t and let $\ell > t$. Assume that all burst errors of length ℓ occur with the same probability. Then the probability that C detects a burst of length ℓ is equal to*

$$
\begin{aligned}
1 - 2^{-t+1} \qquad & \text{if } \ell = t + 1, \\
1 - 2^{-t} \qquad & \text{if } \ell > t + 1.
\end{aligned}
$$

Proof: Assume that e is a burst error with location i and pattern $b(x) = 1 + \ldots + x^{\ell-1}$. As we saw in the proof of 5.3.2, the burst e is not detected by the CRC-algorithm if and only if b is a multiple of g, thus $b = gf$ for some polynomial f of degree $\ell - 1 - t$. Moreover, from $b(0) = g(0) = 1$ we deduce that $f(0) = 1$ must be satisfied.

If $\ell = t + 1$, then there exists exactly one polynomial f with these properties, namely $f = 1$. If $\ell \geq t + 2$, then there are exactly $2^{\ell-t-2}$ polynomials f with these properties. By assumption, each burst pattern b of length ℓ occurs with the same probability. Since there are exactly $2^{\ell-2}$ burst patterns of length ℓ, we obtain the following probabilities for detecting a burst of length ℓ.

If $\ell = t + 1$, a burst of length ℓ is detected with the probability

$$\frac{2^{\ell-2} - 1}{2^{\ell-2}} = 1 - 2^{-\ell+2} = 1 - 2^{-t+1}.$$

If $\ell > t + 1$, a burst of length ℓ is detected with the probability

$$\frac{2^{\ell-2} - 2^{\ell-t-2}}{2^{\ell-2}} = 1 - 2^{-t}. \qquad\qquad \square$$

For the correction of burst and random errors we refer the reader to [134] and [68]. Now we analyze the burst error correction ability of linear codes.

Lemma *A linear code C is able to correct all bursts of length at most ℓ if and only if no codeword different from 0 is the sum of two bursts of length at most ℓ.*   **5.3.5**

Proof: Assume that C is able to correct all bursts of length up to ℓ and suppose that there exist two burst errors e and e' of length at most ℓ and $c \in C \setminus \{0\}$ such that $c = e + e'$. Then $0 + e' = c - e$. Since C is able to correct all bursts of length at most ℓ, the vector $0 + e$ is decoded into 0, and $c - e$ is decoded into c, whence $c = 0$, which is a contradiction to our assumption.

Conversely, we prove that if it is not possible to correct all bursts of length up to ℓ, then there exists a codeword different from 0 which is the sum of two bursts of length at most ℓ. By assumption, there exist $c, c' \in C$, $c \neq c'$ and two bursts e, e' of length at most ℓ such that $c + e = c' + e'$. Consequently, $e' - e = c - c' \in C \setminus \{0\}$ shows that the codeword $c - c'$ is the sum of e' and $-e$. $\square$

The Reiger-bound [169] *If C is a linear (n,k)-code over $\mathbb{F}_q$ which is able to correct all bursts of length at most ℓ, then $2\ell \leq n - k$.*   **5.3.6**

Proof: Since C corrects all bursts of length at most ℓ, according to 5.3.5, no codeword different from 0 is the sum of two bursts of length at most ℓ. Consequently $2\ell < n$, since each vector of length at most 2ℓ is either a burst of length at most ℓ or the sum of two bursts of length at most ℓ. Hence, it is possible to consider the vector space

$$T := \left\{ (v_0, v_1, \ldots, v_{2\ell-1}, 0, \ldots, 0) \in \mathbb{F}_q^n \;\middle|\; v_i \in \mathbb{F}_q, \, i \in 2\ell \right\},$$

which is a subspace of $\mathbb{F}_q^n$ of dimension 2ℓ. Each element of $T \setminus \{0\}$ is either a burst of length at most ℓ or can be written as the sum of two bursts of length at most ℓ. Therefore, $c = 0$ is the unique codeword belonging to T. For $v, v' \in T$,

$v \neq v'$, the cosets $v + C$ and $v' + C$ are distinct elements of $\mathbb{F}_q^n / C$. (Assuming on the contrary that $v + C = v' + C$, then $v - v' \in C \cap T$, thus $v - v' = 0$, whence $v = v'$ a contradiction to our assumption.) Hence, the syndromes of distinct elements of T are distinct. In other words, the mapping $T \to \mathbb{F}_q^{n-k}$, $v \mapsto v \cdot \Delta^\top$, where Δ is a check matrix of C, is injective. Thus $2\ell = \dim(T) \leq \dim(\mathbb{F}_q^{n-k}) = n - k$.   □

If an (n,k)-code is able to correct all burst errors of length at most ℓ, then its *burst error correcting efficiency* is defined as

$$\frac{2\ell}{n-k}.$$

From the Reiger-bound it is clear that the burst error correcting efficiency cannot be greater than 1. The minimum distance d of each MDS-code satisfies $d - 1 = n - k$. Hence, we obtain the following

5.3.7   **Corollary** *MDS-codes are able to correct all bursts of length at most $\lfloor (n-k)/2 \rfloor$. If $n - k$ is even, then their burst error correcting efficiency is equal to 1.*   □

Assume that C is a linear code encoded systematically so that the *last k* positions of a codeword are the information symbols. If g is the generator polynomial of C, then the syndrome s of $y \in \mathbb{F}_q^n$ satisfies $s(x) \equiv y(x) \bmod g(x)$. Moreover, it is easily verified that the syndrome of the cyclic shift $x^i y(x)$ is given by $x^i s(x) \bmod g(x)$. In order to correct correctable bursts, the method of error trapping (cf. 4.12.5) can be applied. By the Reiger-bound, a correctable burst is always of length $\ell \leq (n-k)/2$, whence it can always be rotated by a cyclic shift into the first $n - k$ positions.

Now we describe a method which allows us to deal with long burst errors. Let C be a linear (n,k)-code over $\mathbb{F}_q$ and let λ be a positive integer. Consider the set of all λ-tuples of codewords of C, i.e.

$$C^\lambda = \{ f \mid f : \lambda \to C \},$$

which is an $\mathbb{F}_q$ vector space. Its elements can be represented as matrices. Assume that $f(i) = c^{(i)} = (c_0^{(i)}, \dots, c_{n-1}^{(i)})$ for $i \in \lambda$, then f can be written as a matrix

$$f = \begin{pmatrix} c^{(0)} \\ c^{(1)} \\ \vdots \\ c^{(\lambda-1)} \end{pmatrix} = \begin{pmatrix} c_0^{(0)} & \cdots & c_{n-1}^{(0)} \\ c_0^{(1)} & \cdots & c_{n-1}^{(1)} \\ \vdots & \vdots & \vdots \\ c_0^{(\lambda-1)} & \cdots & c_{n-1}^{(\lambda-1)} \end{pmatrix},$$

where the codewords of C are the rows of this matrix. Now the main idea of interleaving is, to read this matrix columnwise from top to bottom and from left to right. For this reason, the symbols of λ consecutive codewords are mixed. Finally, we identify the matrix f with the vector

$$(c_0^{(0)}, c_0^{(1)}, \ldots, c_0^{(\lambda-1)}, \ldots, c_{n-1}^{(0)}, c_{n-1}^{(1)}, \ldots, c_{n-1}^{(\lambda-1)})$$

of length λn over $\mathbb{F}_q$. The identification of $\lambda \times n$ matrices over $\mathbb{F}_q$ with vectors of length λn over $\mathbb{F}_q$ is a vector space isomorphism

$$\phi : \mathbb{F}_q^{\lambda \times n} \to \mathbb{F}_q^{\lambda n}.$$

The image of C^λ under this isomorphism is indicated by $C^{(\lambda)}$.

Definition (λ-way interleave) The code $C^{(\lambda)}$ is called the λ-*way interleave* of the code C. $\diamond$ **5.3.8**

The proof of the next theorem is left to the reader.

Theorem *Let C be a linear (n, k, d, q)-code, and let λ be a positive integer. Then the λ-way interleave $C^{(\lambda)}$ is a linear $(\lambda n, \lambda k, d, q)$-code.* $\square$ **5.3.9**

Theorem *If the linear code C is able to correct all bursts of length at most ℓ, then the λ-way interleave $C^{(\lambda)}$ corrects all burst errors of length at most $\lambda\ell$.*   **5.3.10**

Proof: Assume that $e \in \mathbb{F}_q^{\lambda n}$ is a burst error of length at most $\lambda\ell$. Applying the inverse isomorphism ϕ^{-1}, this burst error is split over the λ rows of $\phi^{-1}(e)$. In each row at most ℓ consecutive symbols are effected, whence in each row the bursts can be corrected by C. Thus, e is corrected by $C^{(\lambda)}$.

Conversely, choose a burst of length $\ell + 1$ in $\mathbb{F}_q^n$ which cannot be corrected by C. If we replace ℓ consecutive symbols in an arbitrary row of $\phi^{-1}(c)$ for some $c \in C^{(\lambda)}$ by this burst, then we obtain a burst of length $\lambda\ell + 1$ in $C^{(\lambda)}$ which cannot be corrected. $\square$

Corollary *Let C be a code over $\mathbb{F}_{2^m}$. If C allows one to correct bursts of length at most ℓ, then C is able to correct bursts of maximal $(\ell - 1)m + 1$ bits, and $C^{(\lambda)}$ corrects burst errors of length up to $(\lambda\ell - 1)m + 1$ bits.* $\square$ **5.3.11**

Since C and $C^{(\lambda)}$ have the same minimum distance, interleaving does not increase the random error correction abilities of a code. As a matter of fact, depending on how the errors are distributed among the λ interleaves, more than $(d-1)/2$ errors might be corrected by $C^{(\lambda)}$.

Theorem *Let C be a cyclic code of length n with generator polynomial g over $\mathbb{F}_q$. Then $g(x^\lambda)$ is the generator polynomial of $C^{(\lambda)}$. Hence, the λ-way interleave is also cyclic.*   **5.3.12**

Proof: Let $c = (c_0, \ldots, c_{n\lambda-1})$ be an element of $C^{(\lambda)}$, then

$$\phi^{-1}(c) = \begin{pmatrix} c_0 & c_\lambda & \cdots & c_{(n-1)\lambda} \\ c_1 & c_{\lambda+1} & \cdots & c_{(n-1)\lambda+1} \\ \vdots & \vdots & \ddots & \vdots \\ c_{\lambda-1} & c_{2\lambda-1} & \cdots & c_{n\lambda-1} \end{pmatrix}.$$

Each row

$$c^{(i)} := (c_i, c_{\lambda+i}, \ldots, c_{(n-1)\lambda+i}), \quad i \in \lambda,$$

is a codeword of C, thus $g(x)$ is a divisor of

$$c^{(i)}(x) := c_i + c_{\lambda+i}x + \ldots + c_{(n-1)\lambda+i}x^{n-1}, \quad i \in \lambda.$$

The polynomial

$$c(x) = c_0 + c_1 x + \ldots + c_{\lambda n-1}x^{\lambda n-1}$$

is of the form

$$c(x) = c^{(0)}(x^\lambda) + xc^{(1)}(x^\lambda) + \ldots + x^{\lambda-1}c^{(\lambda-1)}(x^\lambda).$$

Since g is a divisor of $c^{(i)}(x)$ in $\mathbb{F}_q[x]/I(x^n - 1)$, it is obvious that $g(x^\lambda)$ is a divisor of $c(x)$ in $\mathbb{F}_q[x]/I(x^{\lambda n} - 1)$. Hence, each codeword of $C^{(\lambda)}$ is a multiple of $g(x^\lambda)$.

Conversely, we know that g is a divisor of $x^n - 1$ and that $k = \dim(C) = n - \deg g$. Thus, $g(x^\lambda)$ is a divisor of $x^{n\lambda} - 1$, and the cyclic code generated by $g(x^\lambda)$ is of dimension $n\lambda - \deg g(x^\lambda) = n\lambda - (n-k)\lambda = k\lambda$. This is the dimension of $C^{(\lambda)}$. Since $C^{(\lambda)}$ is contained in the cyclic code generated by $g(x^\lambda)$, and both codes have the same dimension, they are equal. Thus, $C^{(\lambda)}$ is the cyclic code generated by $g(x^\lambda)$. $\square$

The minimum distance of an interleaved code can be increased significantly by adding check symbols across interleaves. Usually product codes (cf. 2.3.15) are used for this construction.

Assume that C_i is an (n_i, k_i, d_i, q)-code, for $i = 1, 2$. Without loss of generality, we assume that the codes are systematically encoded, so that the first k_i symbols in each codeword form an information set. Then the product $C_1 \otimes C_2$ is an $(n_1 n_2, k_1 k_2, d_1 d_2, q)$-code. Its elements are represented as $n_1 \times n_2$-matrices over $\mathbb{F}_q$. First write down k_1 rows containing codewords of C_2. Then consider each of the n_2 columns as an information set of a codeword of C_1. For each column compute the remaining $n_1 - k_1$ check symbols and attach them at the bottom of the column. Finally, we obtain a matrix of the form

$k_1 \times k_2$ information symbols	$k_1 \times (n_2 - k_2)$ checks on rows
$(n_1 - k_1) \times k_2$ checks on columns	$(n_1 - k_1) \times (n_2 - k_2)$ checks on rows and columns

5.3.13

containing $k_1 k_2$ information symbols in the upper left corner. The code C_1 is also called the *column code* or *outer code*, whereas C_2 is the *row code* or *inner code*. As a matter of fact, all the rows of this matrix, i.e. also the last $n_1 - k_1$ rows, are codewords of C_2 (cf. Exercise 5.3.3). For the encoding process it is not important whether first rows and then columns, or first columns and then rows of the matrix 5.3.13 are determined. Usually the components of this matrix are finally read in columns. Thus from the matrix

$$
\begin{pmatrix}
c_{00} & c_{01} & \cdots & c_{0,n_2-1} \\
c_{10} & c_{11} & \cdots & c_{1,n_2-1} \\
\vdots & \vdots & \ddots & \vdots \\
c_{n_1-1,0} & c_{n_1-1,1} & \cdots & c_{n_1-1,n_2-1}
\end{pmatrix}
$$

we obtain the vector

$$(c_{00}, c_{10}, \ldots, c_{n_1-1,0}, c_{01}, c_{11}, \ldots, c_{n_1-1,1}, \ldots, c_{0,n_2-1}, c_{1,n_2-1}, \ldots, c_{n_1-1,n_2-1}).$$

There are various methods for decoding product codes. The conventional decoding is done in two steps, which are called *inner* and *outer decoding*. The inner decoding, also known as *row decoding*, is used both for error correction of short errors and for error detection. If errors were detected and not corrected in a row, then all the symbols of this row are marked as erasures. The outer decoder, also known as *column decoder*, is provided with information on erasures by the inner decoder, whence its main task is the correction of these erasures. Moreover, it is possible to use it for further error correction as well. If the outer decoder cannot correct the erasures, then in applications like compact discs an *error concealment* must be applied.

Example Consider the product code $C_1 \otimes C_2$ constructed from an extended binary Hamming-code C_1 with systematic generator matrix

5.3.14

$$
\Gamma_1 = \begin{pmatrix}
1 & 0 & 0 & 0 & 0 & 1 & 1 & 1 \\
0 & 1 & 0 & 0 & 1 & 0 & 1 & 1 \\
0 & 0 & 1 & 0 & 1 & 1 & 0 & 1 \\
0 & 0 & 0 & 1 & 1 & 1 & 1 & 0
\end{pmatrix},
$$

a binary cyclic code C_2 with generator polynomial

$$g_2(x) = (1+x)(1+x+x^3) = 1 + x^2 + x^3 + x^4$$

and systematic generator matrix

$$\Gamma_2 = \begin{pmatrix} 1 & 0 & 0 & 1 & 0 & 1 & 1 \\ 0 & 1 & 0 & 1 & 1 & 1 & 0 \\ 0 & 0 & 1 & 0 & 1 & 1 & 1 \end{pmatrix}.$$

We already know that $d_1 = d_2 = 4$, whence $C_1 \otimes C_2$ is a binary $(56, 12, 16)$-code. Since the generator polynomial g_2 is of degree 4, the code C_2 detects all bursts of length ≤ 4 and, moreover, it detects bursts of length greater than 4 with probability at least $7/8$. For encoding the information

$$111\ 101\ 010\ 011$$

we insert it in form of rows into the array

$$M_0 = \begin{pmatrix} 1 & 1 & 1 \\ 1 & 0 & 1 \\ 0 & 1 & 0 \\ 0 & 1 & 1 \end{pmatrix}.$$

Then, by using the systematic generator matrix Γ_2, we compute the four codewords of C_2 whose first three components are given by the rows of M_0, obtaining the rows of

$$M_1 = \begin{pmatrix} 1 & 1 & 1 & 0 & 0 & 1 & 0 \\ 1 & 0 & 1 & 1 & 1 & 0 & 0 \\ 0 & 1 & 0 & 1 & 1 & 1 & 0 \\ 0 & 1 & 1 & 1 & 0 & 0 & 1 \end{pmatrix}.$$

Finally, by using Γ_1, we compute the seven codewords of C_1 (written as columns) whose first four components are given by the columns of M_1. This way we obtain

$$M_2 = \begin{pmatrix} 1 & 1 & 1 & 0 & 0 & 1 & 0 \\ 1 & 0 & 1 & 1 & 1 & 0 & 0 \\ 0 & 1 & 0 & 1 & 1 & 1 & 0 \\ 0 & 1 & 1 & 1 & 0 & 0 & 1 \\ 1 & 0 & 0 & 1 & 0 & 1 & 1 \\ 1 & 1 & 0 & 0 & 1 & 0 & 1 \\ 0 & 0 & 1 & 0 & 1 & 1 & 1 \\ 0 & 0 & 0 & 0 & 0 & 0 & 0 \end{pmatrix}.$$

Reading M_2 column by column yields the vector

$$11001100\ 10110100\ 11010010\ 01111000\ 01100110\ 10101010\ 00011110.$$

After transmission we obtain the following vector

$$01011000\ 00100000\ 01010010\ 01111000\ 01100110\ 10101010\ 10011110.$$

For better readability the transmission errors are underlined. In this vector we can find a burst of length 17 in position 0 with pattern 10010100100101001 and a random error in position 48. Rewriting this vector as a matrix we obtain

$$M_2' = \begin{pmatrix} 0 & 0 & 0 & 0 & 0 & 1 & 1 \\ 1 & 0 & 1 & 1 & 1 & 0 & 0 \\ 0 & 1 & 0 & 1 & 1 & 1 & 0 \\ 1 & 0 & 1 & 1 & 0 & 0 & 1 \\ 1 & 0 & 0 & 1 & 0 & 1 & 1 \\ 0 & 0 & 0 & 0 & 1 & 0 & 1 \\ 0 & 0 & 1 & 0 & 1 & 1 & 1 \\ 0 & 0 & 0 & 0 & 0 & 0 & 0 \end{pmatrix}.$$

In the present example we use C_2 just for error detection. The first row of M_2' contains a burst of length 4 which will be detected by C_2, whence all entries of this row are marked as erasures. Also the errors in the two other rows are detected by C_2, and the elements in these rows are also marked as erasures. This way we obtain the matrix

$$M_2'' = \begin{pmatrix} \Box & \Box & \Box & \Box & \Box & \Box & \Box \\ 1 & 0 & 1 & 1 & 1 & 0 & 0 \\ 0 & 1 & 0 & 1 & 1 & 1 & 0 \\ \Box & \Box & \Box & \Box & \Box & \Box & \Box \\ 1 & 0 & 0 & 1 & 0 & 1 & 1 \\ \Box & \Box & \Box & \Box & \Box & \Box & \Box \\ 0 & 0 & 1 & 0 & 1 & 1 & 1 \\ 0 & 0 & 0 & 0 & 0 & 0 & 0 \end{pmatrix}.$$

Since $d_1 = 4$, it is possible to correct three erasures in each codeword of C_1, whence each column of M_2'' can be corrected. Consequently we are able to reconstruct the originally sent vector.

Just using interleaving of one code it would be impossible to correct these errors. However, if the errors were distributed so that a fourth row of M_2' would be infected, then our decoding strategy would fail. For this reason, usually the code C_2 is also used for correction of short errors. Moreover, further interleaving of codewords of the product code $C_1 \otimes C_2$ protects better against burst errors.   ◇

A similar example can be found in [104, 3.5.3 Beispiel].

5.3.15   **Example** For error correction in a DVD a product code of two Reed–Solomon-codes over $\mathbb{F}_{2^8}$ is used. The column code C_1 is a $(208, 192, 17)$-code with generator polynomial

$$g_1(x) = \prod_{i=0}^{15}(x + \xi^i),$$

where ξ is a root of the primitive polynomial $x^8 + x^4 + x^3 + x^2 + 1 \in \mathbb{F}_2[x]$. The row code C_2 is a $(182, 172, 11)$-code with generator polynomial

$$g_2(x) = \prod_{i=0}^{9}(x + \xi^i).$$

For further details see [49].   ◇

The error- and burst-correcting properties of product codes are described in

5.3.16   **Theorem** [134, pages 275ff] *Assume that C_i is a linear code of length n_i with minimum distance d_i which can correct all bursts of length at most ℓ_i, for $i = 1, 2$. Then:*

1. *The product code $C_1 \otimes C_2$ is capable of correcting*

$$t := \left\lfloor \frac{d_1 d_2 - 1}{2} \right\rfloor$$

   random errors.
2. *There exist decoding methods such that $C_1 \otimes C_2$ corrects all bursts of length up to $\max\{n_1\ell_2, n_2\ell_1\}$.*
3. *There exists a decoding algorithm for the product code $C_1 \otimes C_2$ which allows the correction of all random errors of weight at most t and of all burst errors of length at most $\max\{n_1 t_2, n_2 t_1\}$ with $t_i := \lfloor (d_i - 1)/2 \rfloor$.*

Proof: 1. The first assertion is trivial since the minimum distance of $C_1 \otimes C_2$ is $d = d_1 d_2$.

2. Assume that the elements of a codearray are transmitted in columns. If during the transmission a burst error of length at most $n_1\ell_2$ has occurred, after rearranging the received data in an $n_1 \times n_2$-array, the elements of the burst error lie in at most $\ell_2 + 1$ columns. Each row of the array is affected with a burst of length not greater than ℓ_2. Thus, these bursts can be corrected by C_2, whence $C_1 \otimes C_2$ corrects all bursts of length up to $n_1\ell_2$. If $n_2\ell_1 > n_1\ell_2$, then assume that the elements of a codearray are transmitted in rows. Similar arguments prove that under these assumptions $C_1 \otimes C_2$ is capable of correcting all bursts of length at most $n_2\ell_1$.

3. Without loss of generality, we assume that $n_1 t_2 \geq n_2 t_1$ and set $\ell := n_1 t_2$. If the received vector y contains a random error of weight at most t we apply

the decoding method of 1. Otherwise, we suppose that y contains a burst error e of length not greater than ℓ which is not a random error of weight at most t. Rearranging the elements of y as an $n_1 \times n_2$-array, the burst error affects at most $t_2 + 1$ columns, where each row of this array contains at most t_2 errors. Consequently, these errors can be corrected by C_2. Finally, we want to show that the syndromes of errors can be used in order to determine which decoding strategy should be applied. For doing this, we prove that the syndrome of e, a burst of length ℓ which is not a random error of weight at most t, is different from the syndromes of all correctable random errors. Supposing, on the contrary, that the syndrome of e coincides with the syndrome of a correctable random error e', then the difference $e - e'$ is a codeword of $C_1 \otimes C_2$. By assumption, $e - e' \neq 0$. Hence, each nonzero row of $e - e'$ is of weight at least d_2. Thus, it consists of at least $d_2 - t_2$ nonzero entries of e' and at most t_2 nonzero components of e. Since there are at most t random errors in e', there exist at most $\lfloor t/(d_2 - t_2) \rfloor$ rows of e' containing at least $d_2 - t_2$ nonzero entries. Thus,

$$\text{wt}(e - e') \leq \left\lfloor \frac{t}{d_2 - t_2} \right\rfloor t_2 + t \leq \frac{tt_2}{d_2 - t_2} + t = t \left(\frac{t_2}{d_2 - t_2} + 1 \right) = t \frac{d_2}{d_2 - t_2} < 2t,$$

since

$$\frac{d_2}{d_2 - t_2} = \frac{d_2}{d_2 - \lfloor (d_2 - 1)/2 \rfloor} \leq \frac{2d_2}{2d_2 - d_2 + 1} < 2.$$

Consequently, $e - e'$ is a nonzero codeword of $C_1 \otimes C_2$ of weight less than $2t < d$, which is a contradiction.   □

As already mentioned, there exist many other decoding methods for product codes. For instance, first determine all the rows and columns where errors have occurred, and flag all those entries lying both in infected rows and columns. Then use the conventional decoding strategy. But instead of erasing complete rows, just the flagged symbols in these rows will be erased.

The decoding process can be iterated. After having decoded rows and columns, start the decoding process once again. This method is especially useful for decoders with soft input and output. The word "soft" indicates that each data symbol is attached with a measure, usually an element of the real interval $[0, 1]$, indicating its reliability. See the vast literature on concatenated codes, e.g. [56], and on Turbo codes, e.g. [9], [12], and many others.

The effectiveness of the decoding method quite often depends on the situation where it is applied:

Example Consider the product code of two Hamming-codes. Its minimum **5.3.17**
distance is $3 \cdot 3 = 9$, whence it is possible to correct up to 4 errors. We will compare three different decoding methods in two different situations:
Method 1: First correct all correctable rows with the row code, and then correct all correctable columns with the column code.

Method 2: Use the row code for error detection and mark all infected rows as erasures. With the column code try to decode errors and erasures.

Situation 1: Assume that four errors are located in the positions

$$(i_1, j_1), \quad (i_1, j_2), \quad (i_2, j_1), \quad (i_2, j_2)$$

of the codearray with $i_1 \neq i_2$, $j_1 \neq j_2$. Method 1 introduces in the rows indexed with i_1 and i_2 and also in the columns indexed with j_1 and j_2 a third error, whence the error cannot be corrected. Method 2 allows one to realize that errors have occurred in the rows i_1 and i_2, whence these rows are erased. Since there are exactly two erasures in each column, the erased symbols can be computed, and the errors are corrected.

Situation 2: Now assume that the four errors have occurred in four different rows and columns, thus there are exactly four rows and four columns containing one error. Method 1 corrects the errors in the four erroneous rows, whence there are no errors left for the column decoding. Thus, all the errors were corrected. Method 2 allows one to realize that errors have occurred in four rows. The symbols in each of these rows are erased. Since there are four erased symbols in each column, the column decoder cannot correct these errors.

The errors in both situations can be corrected by using

Method 3: Correct each row using a decoder for the row code. For each row i, remember ν_i, the number of symbols corrected. Larger values of ν_i correspond to rows which are more likely to have been miscorrected. Uncorrectable rows are tagged with $\nu_i = \infty$ and all symbols in these rows are immediately erased. Then correct the columns using an errors-and-erasures correction method. For $j \in n_2$ attempt to decode column j. If decoding fails because the column is not correctable, or if decoding succeeds but changes a symbol in an unerased row, some of the row decodings were incorrect. In this case, erase the two least reliable unerased rows (rows with the largest values of ν_i), and repeat the decoding for this column. $\diamond$

There are many other ways of interleaving codewords. The construction of the direct product $C_1 \otimes C_2$ can also be described as follows: First interleave k_1 codewords of C_2. Then divide the interleaved vector into n_2 rows each of length k_1, extend each of these rows to a codeword of C_1 and append the additional symbols at the end of each column. In general, any combination of interleaving methods and encoding with respect to two (or more) codes is called *cross interleaving*.

Now we describe the interleaving applied for the error protection in compact discs. The method is called *cross interleaved Reed–Solomon-codes*, for short CIRC. It is a combination of three interleaving processes and encoding with respect to two Reed–Solomon-codes. CIRC involves another form of interleaving, namely, interleaving with delay $d \geq 1$, which is described below:

Ordinary n-fold interleaving of a code of length n yields blocks consisting of exactly n interleaved codewords, thus each block contains n^2 symbols. Moreover, each codeword is part of exactly one block. Interleaving with delay $d \geq 1$ is another method for interleaving a finite sequence of codewords $(c^{(r)})_{0 \leq r \leq N}$ with $c^{(r)} = (c_{rn}, c_{rn+1}, \ldots, c_{rn+(n-1)})$. As we will immediately see using interleaving with delay $d \geq 1$, a single codeword does not belong to a single block, as it is the case with product codes. Each block contains exactly n symbols which belong to n codewords. Each codeword in this sequence starts a new block and completes another block. With this method each codeword $c^{(r)}$ is spread over n different blocks.

Interleaving with delay d means that the codewords $c^{(r)}$ are inserted as certain diagonals of an array of n rows. For $i \in n$ put c_i, the i-th component of $c^{(0)}$, into the i-th row and the di-th column of this array. If the codeword $c^{(r)}$ is already inserted, then the components of $c^{(r+1)}$ are placed exactly one column to the right from the corresponding components of $c^{(r)}$. For instance, for $d = 1$ we obtain an array of the form:

$$
\begin{array}{ccccccccc}
c_0 & c_n & & \cdots & c_{(n-2)n} & c_{(n-1)n} & c_{nn} & c_{(n+1)n} & \cdots \\
& c_1 & c_{n+1} & \cdots & c_{(n-3)n+1} & c_{(n-2)n+1} & c_{(n-1)n+1} & c_{nn+1} & \cdots \\
& & c_2 & \ddots & & & & & \\
& & & \ddots & & & & & \\
& & & c_{n-2} & c_{n+n-2} & c_{2n+n-2} & c_{3n+n-2} & \cdots \\
& & & & c_{n-1} & c_{n+n-1} & c_{2n+n-1} & \cdots
\end{array}
$$

Of course blank fields at the beginning and at the end of this array must be filled with zeros. Finally, the interleaves are read as columns of the form

$$
\begin{pmatrix}
c_{rn} \\
c_{(r-1)n+1} \\
\vdots \\
c_{(r-(n-2))n+n-2} \\
c_{(r-(n-1))n+n-1}
\end{pmatrix}.
$$

In the general case for $d \geq 1$, *interleaving with delay d* yields blocks of the form

$$
\begin{pmatrix}
c_{rn} \\
c_{(r-d)n+1} \\
\vdots \\
c_{(r-(n-2)d)n+n-2} \\
c_{(r-(n-1)d)n+n-1}
\end{pmatrix}.
$$

5.3.18

For $i \in n$ the i-th component of the word $c^{(r)}$ stands in the $r + id$-th column. Thus, the components of a single codeword occur in n blocks distributed over

$(n-1)d+1$ blocks. To be more precise, the components occur in the blocks

$$
\begin{pmatrix}
c_{rn} \\
\quad c_{(r-d)n+1} \\
\quad\quad \vdots \\
c_{(r-(n-2)d)n+n-2} \\
c_{(r-(n-1)d)n+n-1}
\end{pmatrix}
,
\begin{pmatrix}
c_{(r+d)n} \\
c_{rn+1} \\
\vdots \\
c_{(r-(n-3)d)n+n-2} \\
c_{(r-(n-2)d)n+n-1}
\end{pmatrix}
,\dots,
\begin{pmatrix}
c_{(r+(n-1)d)n} \\
c_{(r+(n-2)d)n+1} \\
\vdots \\
c_{(r+d)n+n-2} \\
c_{rn+n-1}
\end{pmatrix}.
$$

By increasing d, symbols of a single codeword are spread over longer sequences of interleaved symbols, whence they are better protected against burst errors. On the other hand, deinterleaving becomes more difficult and time consuming, since more symbols must be read and kept in the memory before they can be collected to the original codewords of C.

Finally, at the end of this section we analyze certain relations between product codes and cyclic codes. These considerations are not necessary for understanding the encoding of compact discs, but they are interesting for their own sake. The reason is that the product of two cyclic codes is a cyclic code again, if the lengths of the codes are relatively prime. (See also [139, Ch. 18 §2].)

5.3.19   **Lemma** *Let A be an arbitrary alphabet and let n_1, n_2 be positive integers. By $A^{n_1 \times n_2}$ we denote the set of all $n_1 \times n_2$ matrices over A. If G_i is a subgroup of the symmetric group S_{n_i} for $i = 1, 2$, then the mapping*

$$
(G_1 \times G_2) \times A^{n_1 \times n_2} \to A^{n_1 \times n_2} : \left((\sigma, \pi), (a_{ij}) \right) \mapsto \left(a_{\sigma^{-1}(i), \pi^{-1}(j)} \right)
$$

defines an action of the direct product $G_1 \times G_2$ on $A^{n_1 \times n_2}$.   □

The proof of this lemma is left as an exercise for the reader.

Consider the alphabet $A = \mathbb{F}_q$, C_i a cyclic code of length n_i over $\mathbb{F}_q$ and G_i the cyclic group generated by the cycle $(0, 1, \dots, n_i - 1) \in S_{n_i}$ for $i = 1, 2$. Then for each $(\sigma, \pi) \in G_1 \times G_2$ we have

$$
(c_{ij}) \in C_1 \otimes C_2 \iff (c_{\sigma^{-1}(i), \pi^{-1}(j)}) \in C_1 \otimes C_2.
$$

In other words, $G_1 \times G_2$ is contained in the automorphism group of $C_1 \otimes C_2$. Moreover, if n_1 and n_2 are relatively prime, then the direct product $G_1 \times G_2$ is a cyclic group of order $n_1 n_2$.

Now we represent a codeword $c = (c_{ij}) \in C_1 \otimes C_2$ as

$$
c(x, y) = \sum_{i \in n_1} \sum_{j \in n_2} c_{ij} x^i y^j + \mathrm{I}(x^{n_1} - 1, y^{n_2} - 1) \in \mathbb{F}_q[x, y] / \mathrm{I}(x^{n_1} - 1, y^{n_2} - 1).
$$

Then $x \cdot c(x, y)$ and $y \cdot c(x, y)$ describe cyclic shifts of the rows and columns of $c = (c_{ij})$.

Assuming again that n_1 and n_2 are relatively prime, by the Chinese Remainder Theorem (cf. 3.5.15 and Exercise 3.5.11), we obtain that for each $(i,j) \in n_1 \times n_2$ there exists exactly one $\phi(i,j) \in n_1n_2$ such that

$$\phi(i,j) \equiv i \bmod n_1 \text{ and } \phi(i,j) \equiv j \bmod n_2. \qquad \textbf{5.3.20}$$

Lemma *Assume that n_1 and n_2 are relatively prime positive integers, and consider*   **5.3.21**
$a,b \in \mathbb{Z}$ such that $an_1 + bn_2 = 1$. Then:

1. *$\phi(i,j) \equiv jan_1 + ibn_2 \bmod n_1n_2$.*

2. *There exist integers $\tilde{a} > 0$ and $\tilde{b} \leq 0$ such that $\tilde{a}n_1 + \tilde{b}n_2 = 1$. If $n_1 > 1$ and $\tilde{a} > 0$, then $\tilde{b} < 0$. Moreover, $\gcd(a,b) = \gcd(\tilde{a},\tilde{b}) = 1$.*

3. *$I(x^n - 1, y^m - 1) = I(x^n - 1) + I(y^m - 1)$ for arbitrary $n, m > 0$.* $\square$

The proof is left to the reader.

We claim that it is possible to rewrite $c(x,y)$ in terms of a single variable z by replacing $x^i y^j$ by $z^{\phi(i,j)}$.

Lemma *Assume that n_1 and n_2 are relatively prime positive integers, and consider*   **5.3.22**
$a,b \in \mathbb{Z}$ such that $an_1 + bn_2 = 1$. Let $\varphi \colon \mathbb{F}_q[x,y] \to \mathbb{F}_q[z]/I(z^{n_1n_2} - 1)$ be the homomorphism defined by

$$x \mapsto \varphi(x) := z^{bn_2} + I(z^{n_1n_2} - 1), \quad y \mapsto \varphi(y) := z^{an_1} + I(z^{n_1n_2} - 1).$$

Then:

1. *$\varphi(x^i y^j) = z^{\phi(i,j)} + I(z^{n_1n_2} - 1)$ for all $(i,j) \in n_1 \times n_2$.*

2. *φ is surjective, $\ker \varphi = I(x^{n_1} - 1, y^{n_2} - 1)$ and*

$$\Phi \colon \mathbb{F}_q[x,y]/I(x^{n_1} - 1, y^{n_2} - 1) \to \mathbb{F}_q[z]/I(z^{n_1n_2} - 1)$$

$$\Phi(f(x,y) + I(x^{n_1} - 1, y^{n_2} - 1)) := \varphi(f(x,y))$$

is a ring-isomorphism.

Proof: 1. For $(i,j) \in n_1 \times n_2$ we have

$$\varphi(x^i y^j) = \left(z^{ibn_2} + I(z^{n_1n_2} - 1)\right)\left(z^{jan_1} + I(z^{n_1n_2} - 1)\right)$$

$$= z^{jan_1 + ibn_2} + I(z^{n_1n_2} - 1) = z^{\phi(i,j)} + I(z^{n_1n_2} - 1).$$

2. From the definition of φ it is obvious that $I(x^{n_1} - 1, y^{n_2} - 1) \subseteq \ker \varphi$. Assume, conversely, that $f(x,y) \in \mathbb{F}_q[x,y]$ belongs to $\ker \varphi$. It can be expressed as

$$f(x,y) = \sum_{i \in n_1} \sum_{j \in n_2} f_{ij} x^i y^j + \tilde{f}(x,y),$$

where $\tilde{f}(x,y) \in I(x^{n_1} - 1, y^{n_2} - 1)$. Since φ is a homomorphism, we deduce

$$0 + I(z^{n_1 n_2} - 1) = \varphi(f(x,y)) = \sum_{i \in n_1} \sum_{j \in n_2} f_{ij} z^{\phi(i,j)} + I(z^{n_1 n_2} - 1).$$

Since ϕ is a bijection between $n_1 \times n_2$ and $n_1 n_2$, all the coefficients f_{ij} vanish for $(i,j) \in n_1 \times n_2$ and, consequently, $f(x,y) = \tilde{f}(x,y) \in I(z^{n_1 n_2} - 1)$. Obviously, φ is surjective, whence Φ is an isomorphism. □

This way we obtain

$$\Phi(c(x,y) + I(x^{n_1} - 1, y^{n_2} - 1)) = \sum_{i \in n_1} \sum_{j \in n_2} c_{ij} z^{\phi(i,j)} + I(z^{n_1 n_2} - 1),$$

which allows us to determine $c(z)$ as

$$c(z) = \sum_{i \in n_1} \sum_{j \in n_2} c_{ij} z^{\phi(i,j)},$$

as was claimed.

If $c \in C_1 \otimes C_2$, then $zc(z) + I(z^{n_1 n_2} - 1) = \Phi(xy)\Phi(c(x,y)) = \Phi(xyc(x,y))$ and $xyc(x,y) \in C_1 \otimes C_2$, whence $C_1 \otimes C_2$ is an ideal in $\mathbb{F}_q[z]/I(z^{n_1 n_2} - 1)$. In other words, using an appropriate order of the canonical basis vectors of $C_1 \otimes C_2$, the product code is cyclic: We associate $c' \otimes c'' \in C_1 \otimes C_2$ with the vector

$$c := (c_{\phi^{-1}(0)}, \ldots, c_{\phi^{-1}(n_1 n_2 - 1)}),$$

where $c_{ij} = c_i' c_j''$ for $i \in n_1, j \in n_2$.

This proves the first assertion of

5.3.23   **Theorem** [34], [135] *Assume that C_i is a cyclic linear (n_i, k_i, d_i, q)-code with generator polynomial g_i and check polynomial h_i for $i = 1, 2$. Suppose that $n_1 > 1$ and n_2 are relatively prime and that $a > 0$ and $b \leq 0$ are integers such that $an_1 + bn_2 = 1$. Then:*

1. *The product code $C_1 \otimes C_2$ is a cyclic code.*

2. *The generator polynomial of $C_1 \otimes C_2$ is*

$$g(z) = \gcd\left(z^{n_1 n_2} - 1, \left(z^{\ell n_1 n_2} g_1(z^{bn_2})\right) g_2(z^{an_1})\right)$$

   with $\ell = 2(-b)$.

3. *The check polynomial of $C_1 \otimes C_2$ is*

$$h(z) = \gcd\left(z^{mn_1 n_2} h_1(z^{bn_2}), h_2(z^{an_1})\right)$$

   with $m = -b$.

4. *If e_i is the idempotent generator of C_i for $i = 1, 2$, then $\Phi(e_1 e_2)$ is the idempotent generator of $C_1 \otimes C_2$.*

Proof: 2. The integer ℓ is chosen so that $z^{\ell n_1 n_2} f(z^{bn_2})$ is a polynomial in z for all polynomials f of degree not greater than $2n_1$. Assume that $c = (c_{ij})$ belongs to $C_1 \otimes C_2$. Let $c(z)$ be the uniquely determined polynomial of degree less than $n_1 n_2$ associated with c. The i-th row of c belongs to C_2, whence adding suitable multiples of $y^{n_2} - 1$ we obtain a polynomial

$$\zeta_i(y) \equiv \sum_{j \in n_2} c_{ij} y^j \bmod I(y^{n_2} - 1), \qquad i \in n_1,$$

such that the generator polynomial $g_2(y)$ is a divisor of $\zeta_i(y)$ in $\mathbb{F}_q[y]$. Similarly, the j-th column of c belongs to C_1, whence adding suitable multiples of $x^{n_1} - 1$ we obtain a polynomial

$$\sigma_j(x) \equiv \sum_{i \in n_1} c_{ij} x^i \bmod I(x^{n_1} - 1), \qquad j \in n_2,$$

such that the generator polynomial $g_1(x)$ is a divisor of $\sigma_j(x)$ in $\mathbb{F}_q[x]$. It is always possible to find polynomials σ_j of degree less than $2n_1$. Consequently, after multiplying with $z^{\ell n_1 n_2}$ we obtain

$$c(z) \equiv \sum_{i \in n_1} \zeta_i(z^{an_1}) z^{ibn_2 + \ell n_1 n_2} \equiv \sum_{j \in n_2} \sigma_j(z^{bn_2}) z^{jan_1 + \ell n_1 n_2} \bmod I(z^{n_1 n_2} - 1).$$

Thus, $c(z)$ can be expressed in two ways

$$
\begin{aligned}
c(z) &= q(z)(z^{n_1 n_2} - 1) + \sum_{i \in n_1} \zeta_i(z^{an_1}) z^{ibn_2 + \ell n_1 n_2}, \\
c(z) &= \tilde{q}(z)(z^{n_1 n_2} - 1) + \sum_{j \in n_2} \sigma_j(z^{bn_2}) z^{jan_1 + \ell n_1 n_2}.
\end{aligned}
$$

For this reason $\gcd(z^{n_1 n_2} - 1, g_2(z^{an_1}))$ and $\gcd(z^{n_1 n_2} - 1, g_1(z^{bn_1}) z^{\ell n_1 n_2})$ are divisors of $c(z)$ for all $c \in C_1 \otimes C_2$. Hence,

$$L(z) := \mathrm{lcm}\big(\gcd(z^{n_1 n_2} - 1, g_2(z^{an_1})), \gcd(z^{n_1 n_2} - 1, g_1(z^{bn_1}) z^{\ell n_1 n_2})\big)$$

is a divisors of the generator polynomial $g(z)$.

Now assume that $c_1 = (c_0^{(1)}, \ldots, c_{n_1-1}^{(1)})$ and $c_2 = (c_0^{(2)}, \ldots, c_{n_2-1}^{(2)})$ are the codewords of C_1 and C_2 corresponding to the generator polynomials $g_1(x) = \sum_{i \in n_1} c_i^{(1)} x^i$ and $g_2(y) = \sum_{i \in n_2} c_i^{(2)} y^i$. Then

$$
\begin{aligned}
(c_1 \otimes c_2)(z) &\equiv \sum_{i \in n_1} \sum_{j \in n_2} c_i^{(1)} c_j^{(2)} z^{jan_1 + ibn_2 + \ell n_1 n_2} \bmod I(z^{n_1 n_2} - 1) \\
&\equiv z^{\ell n_1 n_2} \sum_{i \in n_1} c_i^{(1)} z^{ibn_2} \sum_{j \in n_2} c_j^{(2)} z^{jan_1} \bmod I(z^{n_1 n_2} - 1) \\
&\equiv z^{\ell n_1 n_2} g_1(z^{bn_2}) g_2(z^{an_1}) \bmod I(z^{n_1 n_2} - 1).
\end{aligned}
$$

After adding suitable multiples of $z^{n_1 n_2} - 1$ to $(c_1 \otimes c_2)(z)$, we deduce that $g(z)$ is a divisor of $(c_1 \otimes c_2)(z)$. Since $g(z)$ is also a divisor of $z^{n_1 n_2} - 1$, it follows that $g(z)$ divides $z^{\ell n_1 n_2} g_1(z^{bn_2}) g_2(z^{an_1})$, whence it is a divisor of

$$G(z) := \gcd\big(z^{n_1 n_2} - 1, z^{\ell n_1 n_2} g_1(z^{bn_2}) g_2(z^{an_1})\big).$$

Summarizing, so far we have deduced that $L(z) \mid g(z) \mid G(z)$. Finally, we want to prove that $L(z) = G(z)$. The polynomials $L(z)$ and $G(z)$ have the same irreducible factors. If $L(z)$ were a proper divisor of $G(z)$, then there exists an irreducible factor of $z^{n_1 n_2} - 1$ which is both a factor of $z^{\ell n_1 n_2} g_1(z^{b n_2})$ and $g_2(z^{a n_1})$ which occurs in $G(z)$ with a greater multiplicity than in $L(z)$. Then necessarily $n_1 n_2 = p^s n$, where p is the characteristics of $\mathbb{F}_q$, $s > 0$, and $\gcd(n, p) = 1$. Thus $z^{n_1 n_2} - 1 = (z^n - 1)^{p^s}$. Since n_1 and n_2 are relatively prime, either p is a divisor of n_1 or of n_2. If $p \mid n_1$, then $n_1 = p^s n_1'$ and $g_2(z^{a n_1}) = \left(g_2(z^{a n_1'})\right)^{p^s}$. Hence, each common factor of $z^{n_1 n_2} - 1$ and $g_2(z^{a n_1})$ occurs with the multiplicity p^s both in L and G. If $p \mid n_2$, then $n_2 = p^s n_2'$ and $z^{\ell n_1 n_2} g_1(z^{b n_2}) = \left(z^{\ell n_1 n_2'} g_1(z^{b n_2'})\right)^{p^s}$. Hence, each common factor of $z^{n_1 n_2} - 1$ and $z^{\ell n_1 n_2} g_1(z^{b n_2})$ occurs with the multiplicity p^s both in L and G. This proves the second assertion.

3. From the representation of the generator polynomial g in 2. and Bézout's Identity (cf. Exercise 3.1.6), we derive that

$$g(z) = f_1(z)(z^{n_1 n_2} - 1) + f_2(z) z^{\ell n_1 n_2} g_1(z^{b n_2}) g_2(z^{a n_1})$$

for some $f_1, f_2 \in \mathbb{F}_q[z]$. We want to prove that $h(z)$ is a divisor of $z^{m n_1 n_2} h_1(z^{b n_2})$ and of $h_2(z^{a n_1})$. Since $g(z) h(z) = z^{n_1 n_2} - 1$, it is enough to show that $z^{n_1 n_2} - 1$ is a divisor of $g(z) z^{m n_1 n_2} h_1(z^{b n_2})$ and of $g(z) h_2(z^{a n_1})$. The first assertion is proved by

$$g(z) z^{m n_1 n_2} h_1(z^{b n_2}) =$$
$$f_1(z)(z^{n_1 n_2} - 1) z^{m n_1 n_2} h_1(z^{b n_2}) + f_2(z) z^{\ell n_1 n_2} g_1(z^{b n_2}) g_2(z^{a n_1}) z^{m n_1 n_2} h_1(z^{b n_2}) =$$
$$f_1(z)(z^{n_1 n_2} - 1) z^{m n_1 n_2} h_1(z^{b n_2}) + f_2(z) g_2(z^{a n_1}) z^{(\ell + m + b) n_1 n_2} (1 - z^{-b n_1 n_2})$$

what follows from 4.2.3. Indeed, $z^{n_1 n_2} - 1$ is a factor of the first and of the second summand, since $b < 0$. Similarly, we prove that $z^{n_1 n_2} - 1$ is a divisor of $g(z) h_2(z^{a n_1})$.

Thus, $h(z)$ is a divisor of $H(z) := \gcd\left(z^{m n_1 n_2} h_1(z^{b n_2}), h_2(z^{a n_1})\right)$. Now we prove that $H(z)$ is a divisor of $z^{n_1 n_2} - 1$. If ζ is a root of H in a suitable extension field, then $\zeta \neq 0$, since $h_2(0) \neq 0$. Consequently $h_1(\zeta^{b n_2}) = 0 = h_2(\zeta^{a n_1})$. This implies that $\zeta^{b n_2}$ is a root of $z^{n_1} - 1$, and $\zeta^{a n_1}$ is a root if $z^{n_2} - 1$. Consequently $\zeta^{b n_1 n_2} = 1 = \zeta^{a n_1 n_2}$. In other words, $(\zeta^{n_1 n_2})^a = 1 = (\zeta^{n_1 n_2})^b$, from which we finally obtain that $1 = (\zeta^{n_1 n_2})^{(n_1 a + n_2 b)} = \zeta^{n_1 n_2}$. Hence, ζ is a root of $z^{n_1 n_2} - 1$. It is easy to prove that $h_2(z^{a n_1})$ divides $z^{a n_1 n_2} - 1$ and $z^{m n_1 n_2} h_1(z^{b n_2})$ divides $z^{-b n_1 n_2} - 1$. If ζ is a root of $H(z)$, then the minimal polynomial M_ζ of ζ over $\mathbb{F}_q$ is an irreducible factor of $z^{n_1 n_2} - 1$, of $z^{a n_1 n_2} - 1$ and of $z^{-b n_1 n_2} - 1$. Since a and b are relatively prime, the multiplicity of M_ζ in $H(z)$ is not greater than the multiplicity of M_ζ in $z^{n_1 n_2} - 1$. Therefore, $H(z)$ is a divisor of $z^{n_1 n_2} - 1$. The decomposition of $z^{n_1 n_2} - 1$ into linear factors is completely described in

Exercise 5.3.6. Each root of H can be expressed as the product $\alpha\beta$ of roots of 1 of order n_1 and n_2.

The common roots of $z^{mn_1n_2}h_1(z^{bn_2})$ and $z^{n_1n_2} - 1$ are of the form $\alpha\beta$ where α is a root of h_1 and $\beta^{n_2} = 1$. Similarly, the common roots of $h_2(z^{an_2})$ and $z^{n_1n_2} - 1$ are of the form $\alpha\beta$ where β is a root of h_2 and $\alpha^{n_1} = 1$.

Assume that $p \nmid n_1n_2$. Then there exist exactly k_1 distinct roots α of h_1 and k_2 distinct roots β of h_2. Moreover, $\alpha\beta$ is a root of H if and only if α is a root of h_1 and β is a root of h_2. Hence, each pair (α, β) of these roots determines uniquely a root $\alpha\beta$ of H. Consequently $\deg H = k_1k_2$ and, therefore, $H = h$, since h is the check polynomial of a code of dimension k_1k_2.

Assume that $p \mid n_2$. Then $\gcd(an_1, p) = 1$. From Exercise 5.3.7 we deduce: If β is a root of h_2 of multiplicity r and $\alpha^{n_1} = 1$, then $\alpha\beta$ is a root of $h_2(z^{an_1})$ of the same multiplicity r. Consequently, $\alpha\beta$ is a root of H if and only if α is a root of h_1 and β is a root of h_2. Moreover, the multiplicity of $\alpha\beta$ as a root of H is at most the multiplicity of β as a root of h_2. Hence, $\deg H \leq k_1k_2$ and, therefore, $H = h$.

Finally, assume that $p \mid n_1$. Then $\gcd(bn_2, p) = 1$. For $\zeta \neq 0$ it is easy to prove that ζ is a root of $z^{mn_1n_2}h_1(z^{bn_2})$ of multiplicity r if and only if ζ^{-1} is a root of $h_1(z^{-bn_2})$ of multiplicity r. Similar arguments as above show that also in this case $H = h$.

4. In order to prove the last assertion we derive

$$\Phi(e_1(x)e_2(y))^2 = \Phi(e_1(x)^2e_2(y)^2) = \Phi(e_1(x)e_2(y)),$$

whence $\Phi(e_1(x)e_2(y))$ is an idempotent element of $\mathbb{F}_q[z]/I(z^{n_1n_2} - 1)$. Assume that $f(z) \in \mathbb{F}_q[x]/I(z^{n_1n_2} - 1)$. Then there exists a unique $\tilde{f}(x,y) \in \mathbb{F}_q[x,y]/I(x^{n_1} - 1, y^{n_2} - 1)$ such that $f(z) = \Phi(\tilde{f}(x,y))$. Since n_1 and n_2 are relatively prime there exist $f_1(x) \in \mathbb{F}_q[x]/I(x^{n_1} - 1)$ and $f_2(y) \in \mathbb{F}_q[y]/I(y^{n_2} - 1)$ such that $\tilde{f}(x,y) = f_1(x)f_2(y)$. Since e_i is a generator of C_i, for $i = 1, 2$, there are $r(x) \in \mathbb{F}_q[x]/I(x^{n_1} - 1)$ and $s(y) \in \mathbb{F}_q[y]/I(y^{n_2} - 1)$ such that $f_1(x) = e_1(x)r(x)$ and $f_2(y) = e_2(y)s(y)$. Consequently,

$$f(z) = \Phi(\tilde{f}(x,y)) = \Phi(f_1(x)f_2(y))$$
$$= \Phi(e_1(x)r(x)e_2(y)s(y)) = \Phi(e_1(x)e_2(y))\Phi(r(x)s(y)),$$

which finishes the proof. □

Example [139, Ch. 18 §2] Let C_1 be the cyclic binary $(3,2,2)$-code with generator polynomial $g_1(x) = x + 1$, check polynomial $h_1(x) = x^2 + x + 1$, and idempotent generator $e_1(x) = x^2 + x$. And let C_2 be the cyclic binary $(5,4,2)$-code with generator polynomial $g_2(y) = y + 1$, check polynomial $h_2(y) = y^4 + y^3 + y^2 + y + 1$, and idempotent generator $e_2(y) = y^4 + y^3 + y^2 + y$.

5.3.24

Since 3 and 5 are relatively prime, $C_1 \otimes C_2$ is a cyclic $(15, 8, 4)$-code. With $a = 2$ and $b = -1$ we get $\ell = 2$, $m = 1$, and the generator polynomial

$$g(z) = \gcd\left(z^{15} - 1, z^{30} g_1(z^{-5}) g_2(z^6)\right) = z^7 + z^6 + z^5 + z^2 + z + 1.$$

Moreover, the check polynomial of $C_1 \otimes C_2$ is

$$h(z) = \gcd\left(z^{15} h_1(z^{-5}), h_2(z^6)\right) = z^8 + z^7 + z^5 + z^4 + z^3 + z.$$

Of course $h(z) = (z^{15} + 1)/g(z)$. Finally, the idempotent generator can be determined by

$$e(z) = \Phi(e_1(x)e_2(y)) = \Phi(x^2 y^4 + xy^4 + x^2 y^3 + xy^3 + x^2 y^2 + xy^2 + x^2 y + xy)$$
$$= z + z^2 + z^4 + z^7 + z^8 + z^{11} + z^{13} + z^{14}. \qquad \diamond$$

However, not all cyclic codes are products of cyclic codes (cf. [139, Ch. 18 §3]). Let C be a minimal binary cyclic (n, k)-code with $n = n_1 n_2$, $\gcd(n_1 n_2, 2) = 1$, $\gcd(n_1, n_2) = 1$, and $n_1 > 1$, $n_2 > 1$. Since C is minimal, its check polynomial h is irreducible, and its roots are of the form $\zeta, \zeta^2, \ldots, \zeta^{2^{k-1}}$, where ζ is an n-th root of 1 in $\mathbb{F}_{2^k}$. Moreover, we assume that h is a primitive polynomial, whence ζ is a primitive element of $\mathbb{F}_{2^k}^*$.

Since $\gcd(n_1, n_2) = 1$, there exist integers a, b such that $a n_1 + b n_2 = 1$. Define $\alpha := \zeta^{b n_2}$, $\beta := \zeta^{a n_1}$, and let k_1, k_2 be the least integers for which $\alpha \in \mathbb{F}_{2^{k_1}}$ and $\beta \in \mathbb{F}_{2^{k_2}}$. Clearly k_1 and k_2 are divisors of k. In fact, $k = \operatorname{lcm}(k_1, k_2)$. Based on these assumptions we can prove the next

5.3.25 **Theorem** [139, Ch. 18 §3] *Let C be a minimal binary cyclic (n, k)-code with $n = n_1 n_2$, $\gcd(n_1 n_2, 2) = 1$, $\gcd(n_1, n_2) = 1$, and $n_1 > 1$, $n_2 > 1$. There exist binary cyclic (n_i, k_i)-codes C_i such that $C = C_1 \otimes C_2$ if and only if $\gcd(k_1, k_2) = 1$.*

Proof: Since Φ is a ring isomorphism, we obtain from $c \in C$ an $n_1 \times n_2$-array (f_{ij}) by

$$\Phi^{-1}(c(z)) = \sum_{i \in n_1} \sum_{j \in n_2} f_{ij} x^i y^j + I(x^{n_1} - 1, y^{n_2} - 1) = f(x, y) + I(x^{n_1} - 1, y^{n_2} - 1).$$

Moreover, $\Phi^{-1}(z^{a n_1} + I(z^{n_1 n_2} - 1)) = y + I(x^{n_1} - 1, y^{n_2} - 1)$ and analogously $\Phi^{-1}(z^{b n_2} + I(z^{n_1 n_2} - 1)) = x + I(x^{n_1} - 1, y^{n_2} - 1)$. Since C is cyclic, $z^{a n_1} c(z) \in C$ for all $c \in C$, whence $y f(x, y) + I(x^{n_1} - 1, y^{n_2} - 1) = \Phi^{-1}(z^{a n_1} c(z)) \in \Phi^{-1}(C)$, and similarly $x f(x, y) + I(x^{n_1} - 1, y^{n_2} - 1) \in \Phi^{-1}(C)$. Therefore, the two sets

$$C_1 := \left\{ (f_{0j}, \ldots, f_{n_1 - 1, j}) \,\middle|\, j \in n_2, \ \sum_{i \in n_1} \sum_{j \in n_2} f_{ij} x^i y^j + I \in \Phi^{-1}(C) \right\}$$

and

$$C_2 := \left\{ (f_{i0}, \dots, f_{i,n_2-1}) \mid i \in n_1, \; \sum_{i \in n_1} \sum_{j \in n_2} f_{ij} x^i y^j + I \in \Phi^{-1}(C) \right\}$$

are cyclic codes of length n_1 and n_2, where $I = I(x^{n_1} - 1, y^{n_2} - 1)$.

For α and β as above, we have $\alpha^{n_1} = \beta^{n_2} = 1$ and $\zeta = \alpha\beta$. Hence, the roots of h are

$$\alpha\beta, (\alpha\beta)^2, (\alpha\beta)^4, \dots, (\alpha\beta)^{2^{k-1}}.$$

We still have to determine the dimensions of these codes. Our claim is that $\dim(C_i) = k_i$ for $i = 1, 2$, where k_1, k_2 are the least integers for which $\alpha \in \mathbb{F}_{2^{k_1}}$ and $\beta \in \mathbb{F}_{2^{k_2}}$. We prove that $\alpha, \alpha^2, \dots, \alpha^{2^{k_1}-1}$ are k_1 zeros of the check polynomial h_1 of C_1 and $\beta, \beta^2, \dots, \beta^{2^{k_2}-1}$ are k_2 zeros of the check polynomial h_2 of C_2. Let $\beta_0 \in \mathbb{F}_{2^{k_2}-1}$ be an n_2-th root of 1 not belonging to $\{\beta, \beta^2, \dots, \beta^{2^{k_2}-1}\}$, and let α_0 be any n_1-th root of 1. Then $\alpha_0\beta_0$ is a root of $g(z)$, since $g(z) = (z^n + 1)/h(z)$. Let $f(x, y) = r_0(y) + r_1(y)x + \dots + r_{n_1-1}(y)x^{n_1-1}$ correspond to a nonzero codeword of C, where r_i are codewords of C_2. Then

$$f(\alpha_0, \beta_0) = \sum_{i \in n_1} r_i(\beta_0)\alpha_0^i = 0.$$

This holds true for any root α_0 satisfying $\alpha_0^{n_1} = 1$. Since there exist n_1 different α_0, the values $r_i(\beta_0)$, $i \in n_1$, satisfy a system of n_1 linear homogeneous equations. The coefficient matrix is a Vandermonde matrix, thus it is regular and $r_i(\beta_0) = 0$ for $i \in n_1$. Therefore, β_0 is a root of each codeword of C_2, whence it is a root of the generator polynomial g_2, and consequently not a root of h_2.

Since $g(\alpha^{2^j}\beta^{2^j}) \neq 0$ for $j \in k_2$, there exist codewords $r_i \in C_2$ such that $r_i(\beta^{2^j}) \neq 0$. Whence, the roots of h_2 are exactly given by β^{2^j} for $j \in k_2$. Similarly, the roots of h_1 can be determined.

In conclusion we obtain: If $\gcd(k_1, k_2) = 1$, then $k = \text{lcm}(k_1, k_2) = k_1 k_2$ and $C = C_1 \otimes C_2$. Conversely, if $C = C_1 \otimes C_2$, then $k = \text{lcm}(k_1, k_2) = k_1 k_2$, whence $\gcd(k_1, k_2) = 1$. $\qquad\square$

Examples [139, Ch. 18 §3]   **5.3.26**

1. Let C be the cyclic binary $(15, 4, 8)$-code with generator polynomial $g(z) = z^{11} + z^8 + z^7 + z^5 + z^3 + z^2 + z + 1$ and idempotent generator $e(z) = zg(z)$. Let $n_1 = 3$ and $n_2 = 5$. The check polynomial of C is irreducible of degree 4 and its roots are of the form $\zeta, \zeta^2, \zeta^4, \zeta^8$ with $\zeta^{15} = 1$. Then $a = 2$, $b = -1$, $\alpha = \zeta^{-5} = \zeta^{10}$, $\beta = \zeta^6$, $\alpha^4 = \alpha$, $\beta^{16} = \beta$, $k_1 = 2$, and $k_2 = 4$. Since $\gcd(2, 4) \neq 1$, the product $C_1 \otimes C_2$ is different from C. Indeed C_1 is a $(3, 2, 2)$-code and C_2 is a $(5, 4, 2)$-code.

2. Let C be the $(21, 6, 8)$-product code of Exercise 5.3.8 where $\zeta, \zeta^2, \zeta^4, \zeta^8, \zeta^{16}$, $\zeta^{32} = \zeta^{11}$ are the roots of its check polynomial, with $\zeta^{21} = 1$. Then $n_1 = 3$, $n_2 = 7$, $a = 5$, $b = -2$, $\alpha = \zeta^{-14} = \zeta^7$, $\beta = \zeta^{15}$, $k_1 = 2$, $k_2 = 3$, and C is the product code of C_1 and C_2 as given in Exercise 5.3.8.   ◇

Exercises

E.5.3.1   **Exercise** Prove 5.3.9.

E.5.3.2   **Exercise** Prove 5.3.11.

E.5.3.3   **Exercise** Prove that all the rows of a matrix representing an element of the product code $C_1 \otimes C_2$ are elements of C_2.

E.5.3.4   **Exercise** Prove 5.3.19.

E.5.3.5   **Exercise** Assume that n_1 and n_2 are relatively prime positive integers, σ and π are given as in 5.3.19 and ϕ satisfies 5.3.20. Show that

$$\phi(\sigma(i), \pi(j)) \equiv \phi(i, j) + 1 \bmod n_1 n_2.$$

E.5.3.6   **Exercise** Let p be a prime and assume that n_1 and n_2 are relatively prime positive integers. Prove the following assertions.

1. If $\gcd(p, n_1 n_2) = 1$, then

$$z^{n_1 n_2} - 1 = \prod_{i \in n_1} \prod_{j \in n_2} (z - \alpha^i \beta^j),$$

where α and β are primitive roots of 1 of order n_1 and n_2, respectively.
2. If $p \mid n_1 n_2$ assume without loss of generality that p is a divisor of n_1, and $n_1 = n_1' p^s$ with $s > 0$ and $\gcd(n_1', p) = 1$. Then

$$z^{n_1 n_2} - 1 = \left(\prod_{i \in n_1'} \prod_{j \in n_2} (z - \alpha^i \beta^j) \right)^{p^s},$$

where α and β are primitive roots of 1 of order n_1' and n_2, respectively.

E.5.3.7   **Exercise** Let $h \in \mathbb{F}_q[x]$ be a divisor of $x^m - 1$ and $n \in \mathbb{N}$ with $\gcd(n, m) = 1$ and $\gcd(n, p) = 1$, where p is the characteristic of $\mathbb{F}_q$. Prove that all the roots of $h(x^n)$ are of the form $\alpha\beta$, where α is a root of h and $\beta^n = 1$. If, moreover, α occurs with the multiplicity r in h, then each $\alpha\beta$ occurs as a root of $h(x^n)$ with the same multiplicity r.

Exercise [139, Ch. 18 §2] Let C_1 be the cyclic $(3,2,2,2)$-code of 5.3.24 and let   **E.5.3.8**
C_2 be the cyclic $(7,3,4,2)$-code S_3 of 4.2.7. Compute the parameters, generator
and check polynomial and the idempotent generator of $C_1 \otimes C_2$.

5.4   More Details on Compact Discs   **5.4**

We already know that an audio compact disc contains 16-bit audio which is
sampled at a rate of 44 100 Hz. Each sampling is divided into two 8-bit vectors, and each of these vectors is considered as one byte and also as an element of $\mathbb{F}_{2^8}$. Thus, each sampling process yields 2 bytes of audio information. Since both the left and the right channel are sampled separately, each
sampling produces two bytes of audio information for the left and two bytes
of audio information for the right channel. Hence, every second we collect
$2 \cdot 2 \cdot 44\,100 = 176\,400$ bytes of audio information. This gives $10\,584\,000$ bytes
per minute and $635\,040\,000$ bytes or $5\,080\,320\,000$ bits per hour. As we will
see the total capacity required for storing this information on a compact disc
is approximately three times as big. If not otherwise specified we excerpt or
"quote" from the third chapter of [164]. "Storing audio information places
great demands on a digital medium. [...] Error correction, synchronization
and modulation are required for successful storage." "The compact disc was
developed in order to meet" different user demands as "random access, small
size, convenience to use, robustness, low cost, and ease of replication." The
specifications for the compact disc system "were jointly developed by Philips
and Sony and are defined in [...] the *Red Book*." (The *Red Book* is the 1980 document which provides the specifications for the standard compact disc (CD)
developed by Sony and Philips. According to legend, the document was in a
binder with red covers, originating the tradition for subsequent adaptations of
CD specifications to be referred to as variously colored books. The *Red Book* describes the compact discs physical specifications, such as the tracks, sector and
block layout, coding, and sampling. Sony and Philips referred to the discs as
CD-DA (digital audio), defined as a content medium for audio data digitized
at 44 100 samples per second and in a range of 65 536 possible values cf. [166]).
"It is also contained in the IEC standard *Compact Disc Digital Audio* [95]." A
compact disc allows us to store at least 74 minutes of stereo high fidelity audio. The disc must be made of transparent material with a refraction index of
1.55. "The optical system that reads the data from the disc uses a laser beam
with a 780 nanometer wavelength." All the information of a compact disc is
stored in the area between radius 23 mm and 58.5 mm. A lead-in and lead-out
area cover the innermost and outermost part of this area. They do not con-

tain any audio information. The audio data is stored between radius 25 mm and 58 mm. In general, the information of a compact disc is stored in a track in form of "a continuous spiral running from the inner circumference to the outer." "Viewed from the readout surface the disc rotates counter-clockwise." The distance between successive tracks is 1.6 micrometers. "There are 22 188 revolutions across the discs surface." The rotational speed of a compact disc varies on the position of the pickup. "The disc rotates at a speed of 500 rotations per minute when the pickup is reading the inner circumference, and as the pickup moves out, the rotational speed decreases to 200 rotations per minute. Thus, a *constant linear velocity*, CLV, is maintained." Depending on the disc, this velocity can vary between 1.2 and 1.4 meters per second. The CD-player automatically regulates the disc rotational speed to maintain a constant bit rate of 4.3218 Mb/sec (cf. 5.4.9). That kind of track is also called a CLV servo system; i.e. the player constantly reads synchronization words from the data and adjusts the speed accordingly. Audio data is stored in a frame format on the disc. Among other information, each frame contains exactly 24 audio bytes. Consequently, there are exactly 7350 frames per second. Further details about frames will be presented later (cf. 5.4.9). First we analyze the error detection and correction process used for compact discs. The analog to digital converter produces sequences of audio data in PCM format. This way we obtain finite sequences $(L_{i,A})_{0 \le i \le N}$, $(L_{i,B})_{0 \le i \le N}$, $(R_{i,A})_{0 \le i \le N}$, and $(R_{i,B})_{0 \le i \le N}$ of bytes representing the first (A) or the second (B) byte of the left (L) or right (R) channel at the i-th sampling for $0 \le i \le N$, where N is the number of the last sampling. With L_i or R_i we denote the i-th sampling of the left or right channel, i.e. the pair $(L_{i,A}, L_{i,B})$ or $(R_{i,A}, R_{i,B})$, respectively. The analog to digital converter produces two sequences $(L_i)_{0 \le i \le N}$ and $(R_i)_{0 \le i \le N}$. Now we want to describe the CIRC-process in more details. Even though this process is usually illustrated with diagrams, we try to use common mathematical notation.

5.4.1   **CIRC encoding** In *step A* this data is scrambled into a series of vectors containing 24 bytes. The vectors obtained from the first samplings are the rows of the following array:

$$
\begin{array}{cccccccccccc}
0 & 0 & 0 & 0 & 0 & 0 & L_1 & L_3 & L_5 & R_1 & R_3 & R_5 \\
0 & 0 & 0 & 0 & 0 & 0 & L_7 & L_9 & L_{11} & R_7 & R_9 & R_{11} \\
L_0 & L_2 & L_4 & R_0 & R_2 & R_4 & L_{13} & L_{15} & L_{17} & R_{13} & R_{15} & R_{17} \\
L_6 & L_8 & L_{10} & R_6 & R_8 & R_{10} & L_{19} & L_{21} & L_{23} & R_{19} & R_{21} & R_{23} \\
\end{array}
$$

$\cdots$ $\cdots$

In general, for $n \ge 0$ the n-th vector is built as

$$
\begin{array}{cccccc}
L_{6(n-2)} & L_{6(n-2)+2} & L_{6(n-2)+4} & R_{6(n-2)} & R_{6(n-2)+2} & R_{6(n-2)+4} \\
L_{6n+1} & L_{6n+3} & L_{6n+5} & R_{6n+1} & R_{6n+3} & R_{6n+5}.
\end{array}
$$

The first half of this vector contains the sampling values of even samples, the second half of odd samples. Of course, at the very beginning and at the very end empty fields must be filled with zero bytes. Using a $(28, 24)$ Reed–Solomon-code C_2 over $\mathbb{F}_{2^8}$, in *step B* this vector is encoded into the C_2-codeword

$$
\begin{array}{cccc}
L_{6(n-2),A} & L_{6(n-2),B} & L_{6(n-2)+2,A} & L_{6(n-2)+2,B} \\
L_{6(n-2)+4,A} & L_{6(n-2)+4,B} & R_{6(n-2),A} & R_{6(n-2),B} \\
R_{6(n-2)+2,A} & R_{6(n-2)+2,B} & R_{6(n-2)+4,A} & R_{6(n-2)+4,B} \\
Q_{n,0} & Q_{n,1} & Q_{n,2} & Q_{n,3} \\
L_{6n+1,A} & L_{6n+1,B} & L_{6n+3,A} & L_{6n+3,B} \\
L_{6n+5,A} & L_{6n+5,B} & R_{6n+1,A} & R_{6n+1,B} \\
R_{6n+3,A} & R_{6n+3,B} & R_{6n+5,A} & R_{6n+5,B},
\end{array}
$$

5.4.2

where the four new symbols (they are usually denoted by the letter Q) are inserted in the middle of the vector, i.e. between the audio information of the even and odd samples. (In general, these codewords are just sequences of 28 bytes. To increase the readability they were arranged in form of an array.) The code C_2 is obtained by shortening a $(255, 251, 5, 2^8)$-Reed–Solomon-code, which is a shortened BCH-code and also an MDS-code (cf. Exercise 5.4.1).

In *step C* these C_2-codewords are interleaved with delay $d = 4$. This yields, according to 5.3.18, vectors of the form

$$
\begin{array}{cccc}
L_{6(n-2),A} & L_{6(n-2-d),B} & L_{6(n-2-2d)+2,A} & L_{6(n-2-3d)+2,B} \\
L_{6(n-2-4d)+4,A} & L_{6(n-2-5d)+4,B} & R_{6(n-2-6d),A} & R_{6(n-2-7d),B} \\
R_{6(n-2-8d)+2,A} & R_{6(n-2-9d)+2,B} & R_{6(n-2-10d)+4,A} & R_{6(n-2-11d)+4,B} \\
Q_{n-12d,0} & Q_{n-13d,1} & Q_{n-14d,2} & Q_{n-15d,3} \\
L_{6(n-16d)+1,A} & L_{6(n-17d)+1,B} & L_{6(n-18d)+3,A} & L_{6(n-19d)+3,B} \\
L_{6(n-20d)+5,A} & L_{6(n-21d)+5,B} & R_{6(n-22d)+1,A} & R_{6(n-23d)+1,B} \\
R_{6(n-24d)+3,A} & R_{6(n-25d)+3,B} & R_{6(n-26d)+5,A} & R_{6(n-27d)+5,B}.
\end{array}
$$

Another Reed–Solomon-code C_1, a $(32, 28)$-code over $\mathbb{F}_{2^8}$, is used *in step D* to encode these vectors as C_1-codewords. Again we have to attach 4 bytes, this time they are appended at the end of the vector. In the literature these bytes are usually indicated with the letter P. We obtain

$$
\begin{array}{cccc}
L_{6(n-2),A} & L_{6(n-2-d),B} & L_{6(n-2-2d)+2,A} & L_{6(n-2-3d)+2,B} \\
L_{6(n-2-4d)+4,A} & L_{6(n-2-5d)+4,B} & R_{6(n-2-6d),A} & R_{6(n-2-7d),B} \\
R_{6(n-2-8d)+2,A} & R_{6(n-2-9d)+2,B} & R_{6(n-2-10d)+4,A} & R_{6(n-2-11d)+4,B} \\
Q_{n-12d,0} & Q_{n-13d,1} & Q_{n-14d,2} & Q_{n-15d,3} \\
L_{6(n-16d)+1,A} & L_{6(n-17d)+1,B} & L_{6(n-18d)+3,A} & L_{6(n-19d)+3,B} \\
L_{6(n-20d)+5,A} & L_{6(n-21d)+5,B} & R_{6(n-22d)+1,A} & R_{6(n-23d)+1,B} \\
R_{6(n-24d)+3,A} & R_{6(n-25d)+3,B} & R_{6(n-26d)+5,A} & R_{6(n-27d)+5,B} \\
P_{n,0} & P_{n,1} & P_{n,2} & P_{n,3}.
\end{array}
$$

5.4.3

Also this code is obtained by shortening a $(255, 251, 5, 2^8)$-Reed–Solomon code.

Finally, in *step E* all bytes in odd positions, i.e. in position $1, 3 \ldots, 31$, of this codeword are combined with the bytes in even positions, i.e. in position $0, 2, \ldots, 30$, of the preceding C_1-codeword, and the 8 bytes representing the Q- and P-check symbols are inverted, i.e. in the representation of these bytes as binary vectors the binary values 1 and 0 are exchanged. (This is indicated by overlining the corresponding P and Q symbols.) There are only technical reasons for this inversion. It assists data readout during areas with muted audio program.

In conclusion, we obtain a sequence of vectors

$$
\begin{array}{llll}
L_{6(n-3),A} & L_{6(n-2-d),B} & L_{6(n-3-2d)+2,A} & L_{6(n-2-3d)+2,B} \\
L_{6(n-3-4d)+4,A} & L_{6(n-2-5d)+4,B} & R_{6(n-3-6d),A} & R_{6(n-2-7d),B} \\
R_{6(n-3-8d)+2,A} & R_{6(n-2-9d)+2,B} & R_{6(n-3-10d)+4,A} & R_{6(n-2-11d)+4,B} \\
\overline{Q}_{n-1-12d,0} & \overline{Q}_{n-13d,1} & \overline{Q}_{n-1-14d,2} & \overline{Q}_{n-15d,3} \\
L_{6(n-1-16d)+1,A} & L_{6(n-17d)+1,B} & L_{6(n-1-18d)+3,A} & L_{6(n-19d)+3,B} \\
L_{6(n-1-20d)+5,A} & L_{6(n-21d)+5,B} & R_{6(n-1-22d)+1,A} & R_{6(n-23d)+1,B} \\
R_{6(n-1-24d)+3,A} & R_{6(n-25d)+3,B} & R_{6(n-1-26d)+5,A} & R_{6(n-27d)+5,B} \\
\overline{P}_{n-1,0} & \overline{P}_{n,1} & \overline{P}_{n-1,2} & \overline{P}_{n,3}.
\end{array}
$$

This completes the description of the encoding process for error-detection and error-correction. For technical reasons, the resulting sequence of bytes is encoded once again before the data is written onto the disc. This final encoding, called EFM, ensures that the stored information satisfies certain standards used for binary data written on optical discs. For further details see 5.4.7 and 5.4.9.

Let v be the row vector of 5.4.2. In step B the four parity bytes $Q_{n,0}, \ldots, Q_{n,3}$ are determined by the equation $v \cdot \Delta_2^\top = 0$, where the check matrix Δ_2 of C_2 is given by

$$
\Delta_2 := \begin{pmatrix}
1 & 1 & \ldots & 1 & 1 & 1 \\
\alpha^{27} & \alpha^{26} & \ldots & \alpha^2 & \alpha & 1 \\
\alpha^{54} & \alpha^{52} & \ldots & \alpha^4 & \alpha^2 & 1 \\
\alpha^{81} & \alpha^{78} & \ldots & \alpha^6 & \alpha^3 & 1
\end{pmatrix}
$$

for α a root of the primitive polynomial $x^8 + x^4 + x^3 + x^2 + 1 \in \mathbb{F}_{2^8}[x]$. Let w be the row vector of 5.4.3. In step D the four parity bytes $P_{n,0}, \ldots, P_{n,3}$ are determined by the equation $w \cdot \Delta_1^\top = 0$, where the check matrix Δ_1 of C_1 is given by

$$
\Delta_1 := \begin{pmatrix}
1 & 1 & \ldots & 1 & 1 & 1 \\
\alpha^{31} & \alpha^{30} & \ldots & \alpha^2 & \alpha & 1 \\
\alpha^{62} & \alpha^{60} & \ldots & \alpha^4 & \alpha^2 & 1 \\
\alpha^{93} & \alpha^{90} & \ldots & \alpha^6 & \alpha^3 & 1
\end{pmatrix}
$$

for the same α as above.

CIRC decoding The standard for compact discs does not explicitly describe a   **5.4.4**
CIRC decoding strategy. Different CD-players use different strategies, so the
quality of error correcting performance varies from player to player. Here we
describe one possible decoding strategy.

The decoder obtains vectors containing 32 bytes, 24 of them contain audio
information, the other 8 are check symbols added by C_1 and C_2. Odd numbered symbols are delayed by one vector and the parity symbols are inverted
in order to reverse *step E* of the encoding process. The code C_1 has minimum
distance $d_1 = 5$. It is used in order to correct a single error and to detect two or
three errors in a codeword. If it detects exactly one error, then the wrong byte
will be replaced by the corrected one. If it detects more than one error, then
all 28 information symbols of the corresponding C_1-codeword are marked as
erasures. Thus, C_1 is designed to correct short random errors and to detect
longer burst errors.

How large is the probability that C_1 does not detect an error? (Cf. [104].)
Assume that c was sent and $c + e$ was received. The error vector e is not detected if and only if $c + e$ belongs to a ball of radius 1 around a C_1-codeword c'
different from c. This probability is

$$\frac{|C_1 \setminus \{c\}| \cdot |\{v \in \mathbb{F}_q^n \mid \mathrm{dist}(v,c) \leq 1\}|}{|\mathbb{F}_q^n|}$$

for $q = 2^8$ and $n = 32$. This is

$$\frac{(q^{28} - 1)(1 + 32(q - 1))}{q^{32}} \leq \frac{32q - 31}{q^4} \approx 2^{-19}.$$

The interleaving of two consecutive C_1-codewords in *step E* allows one to break
short burst errors.

Then, in order to reverse *step D* and *step C*, the last four symbols of each
vector output by C_1, which are the check symbols, are deleted. Deinterleaving
collects the 28 symbols representing C_2-codewords. The code C_2 is first of all
used for correcting erasures. Since its minimum distance $d_2 = 5$, according
to 5.2.2, it is possible to correct up to 4 erasures per codeword. If it is not
possible to correct all erasures with C_2, then the erased symbols are passed to
an interpolation process.

In addition, C_2 can be used to correct a single error and to detect symbols
miscorrected by C_1. If miscorrected symbols are found, then all 24 audio symbols of the corresponding C_2-codeword are marked as erased and passed to
the interpolation process.

Thus, C_2 is designed for the correction of burst errors and short random
errors which were not corrected or miscorrected by C_1.

In worst case situations when the error is so massive that even interpolation fails, the audio signal is usually muted. In general, the brief silence is preferable to the burst of digital noise usually heard as a click.

What is the maximal size of a burst error which can still be corrected by CIRC? The interleaving with delay $d = 4$ in *step C* causes that each C_2-codeword is spread over 28 different C_1-codewords distributed among $27 \cdot 4 + 1 = 109$ consecutive C_1-codewords. This fact allows one to break long burst errors. Even if C_1 marked the symbols of 16 consecutive C_1-codewords as erasures, the original information can be reconstructed if no further errors have occurred in this data area. After deinterleaving, these $16 \cdot 28 = 448$ erasure marks are distributed over 124 different C_2-codewords, where at most 4 erasures occur in each of these codewords. Consequently, these erasures can be corrected. As mentioned above, there are exactly 24 audio symbols contained in a frame. These are the 24 audio bytes contained in a C_2-codeword. Since there are exactly 7350 frames per second which cover a track of 1200 mm, the CIRC decoding allows one to correct physical track errors of about 2.6 mm length. Thus, that kind of maximum length correctable error contains 384 audio bytes. The erased symbols are contained in 124 C_2-codewords, which are responsible for approximately 16.7 milliseconds of music.

The raw error bit rate of a CD is around 10^{-5} to 10^{-6}. This means that there is one wrong bit every 10^5 to 10^6 bits. Considering that an audio compact disc has an output of more than 4 million bits per second (cf. 5.4.9), the need for error correction is obvious. With error correction, perhaps 200 errors per second will be completely corrected. According to [164], the error rate after CIRC is between 10^{-10} and 10^{-11}. Nevertheless, the quality of error-correction varies from player to player, depending on the chosen CIRC decoding strategy.

5.4.5   **Interpolation** If it is impossible for the decoder to reconstruct a C_2-codeword, then the CD-player tries to interpolate the missing audio bytes from neighboring ones in case they are reliable. Because of the high correlation between music samples, an uncorrected error can be made virtually inaudible by synthesizing new data from surrounding data. Various interpolation schemes are used with different performance levels. In its simplest form the previous value is simply repeated. In first order interpolation the erased audio bytes are replaced by the mean value of the previous and the subsequent byte.

The interpolation process can be applied to determine the missing values even if the audio bytes of two consecutive C_2-codewords are marked as erasures, based on the scrambling in *step A*. Assume that the n-th and $(n+1)$-th C_2-codeword are erased, then the neighboring bytes of even samples occur in the $(n-3)$-th, $(n-2)$-th, and $(n-1)$-th codeword and the neighboring bytes of odd samples occur in the $(n+2)$-th, $(n+3)$-th, and $(n+4)$-th codeword.

What is the maximal size of a burst error which can still be reconstructed with interpolation? Even if C_1 marked the symbols of 48 consecutive C_1-codewords as erasures, the audio information can still be reconstructed by interpolation. After deinterleaving the received information we obtain a sequence of vectors $(y^{(n)})_n$ with $y^{(n)} = (y_{28n}, y_{28n+1}, \ldots, y_{28n+27})$. These $48 \cdot 28 = 1344$ erasure marks are distributed over 124 different vectors $y^{(n)}$. Assume that $y^{(m)}$ is the first vector in this sequence whose last component belongs to these erased rows. Hence, in the interleaving array the column the top entry of which has the index $m+108$ is the first erased column. Now we have to check that it is still possible to reconstruct all the audio information with error correction and interpolation. The vectors $y^{(m)}, \ldots y^{(m+3)}$ contain exactly one erasure which occurs in the last position. The vectors $y^{(m+4)}, \ldots y^{(m+7)}$ contain exactly two erasures which occur in the last two positions. The vectors $y^{(m+8)}, \ldots y^{(m+11)}$ contain exactly three erasures which occur in the last three positions. The vectors $y^{(m+12)}, \ldots y^{(m+15)}$ contain exactly four erasures which occur in the last four positions. Consequently, in all these vectors so far it is possible to fill the erased positions by correcting erasures with C_2.

For the following vectors we analyze how many erased bytes they contain, in which position they occur, and in which vectors and positions the neighboring audio bytes occur. Finally we will see that it is always possible to approximate the missing values by first order interpolation. The first column gives the index n of the vector $y^{(n)}$, the second shows the number of erased bytes in this vector, the third contains the position of the erased bytes, the next column contains the labels n' of the vectors $y^{(n')}$ which contain the audio bytes necessary for interpolation, and finally the last column contains the positions where these neighboring audio bytes occur. Careful investigation of this table proves that it is possible to reconstruct the missing information by interpolation.

n	erasures	pos.	n'	pos.
$m+16, \ldots, m+19$	5	$23 - 27$	$m+18, \ldots, m+22$	$6 - 11$
$m+20, \ldots, m+23$	6	$22 - 27$	$m+22, \ldots, m+26$	$6 - 11$
$m+24, \ldots, m+27$	7	$21 - 27$	$m+26, \ldots, m+30$	$0 - 11$
$m+28, \ldots, m+31$	8	$20 - 27$	$m+30, \ldots, m+34$	$0 - 11$
$m+32, \ldots, m+35$	9	$19 - 27$	$m+34, \ldots, m+38$	$0 - 11$
$m+36, \ldots, m+39$	10	$18 - 27$	$m+38, \ldots, m+42$	$0 - 11$
$m+40, \ldots, m+43$	11	$17 - 27$	$m+42, \ldots, m+46$	$0 - 11$
$m+44, \ldots, m+47$	12	$16 - 27$	$m+46, \ldots, m+50$	$0 - 11$
$m+48, \ldots, m+51$	12	$15 - 26$	$m+50, \ldots, m+54$	$0 - 11$
$m+52, \ldots, m+55$	12	$14 - 25$	$m+54, \ldots, m+58$	$0 - 11$

$m+56,\ldots,m+59$	12	$13-24$	$m+58,\ldots,m+62$	$0-11$
$m+60,\ldots,m+63$	12	$12-23$	$m+62,\ldots,m+66$	$0-9$
$m+64,\ldots,m+67$	12	$11-22$	$m+66,\ldots,m+70$	$0-9$
			$m+62,\ldots,m+65$	$24-27$
$m+68,\ldots,m+71$	12	$10-21$	$m+70,\ldots,m+74$	$0-5$
			$m+66,\ldots,m+69$	$24-27$
$m+72,\ldots,m+75$	12	$9-20$	$m+74,\ldots,m+78$	$0-5$
			$m+70,\ldots,m+73$	$22-27$
$m+76,\ldots,m+79$	12	$8-19$	$m+78,\ldots,m+81$	$0-5$
			$m+74,\ldots,m+77$	$22-27$
$m+80,\ldots,m+83$	12	$7-18$	$m+82,\ldots,m+85$	$0-5$
			$m+77,\ldots,m+81$	$22-27$

The remaining cases can be studied in a similar way. For instance, the study of the four columns $m+80,\ldots,m+83$ corresponds to the situation of $m+72,\ldots,m+75$.

Consequently, all these erasures can be filled by interpolation. Even if 48 C_1-codewords are erased, the last byte of the first codeword preceding this erased block and the first byte of the codeword following this block are erroneous, it is possible to reconstruct approximations of the erased bytes by interpolation. These 48 erased C_1-codewords contain the audio information of 48 frames. Each second the CD-player reads 7350 frames which cover 1200 mm of the track, whence interpolation is able to deal with burst errors of length up to 7.8 mm.

Finally we discuss how data is stored on a CD (cf. [164, pages 51ff]).

5.4.6   Pits and lands "A transparent plastic substrate forms most of a discs 1.2 mm thickness. Data is physically contained in *pits* which are impressed along its top surface and are covered with a very thin metal layer. Another thin plastic layer protects the metallized pit surface on top of which the identifying label is printed. A laser beam is used to read the data. It is applied from below and passes through the transparent substrate, is reflected at the metallized pit surface, and passes back." The beam of size 800 micrometers at the discs surface is focused to 1.7 micrometers on the metallized pit surface. Pits are very small, approximately 0.5 micrometers wide and 0.11 to 0.13 micrometers high. As we will see, they are of varying length. Thus, "the laser beam is focused to a point about three times larger than the pit width."

"When viewed from the lasers perspective, the pits appear as bumps." The areas between pits are called *lands*. Data are read from the compact disc by measuring the reflected light. Almost 90% of the laser beam are reflected by a land. Caused by the height of the pits, the refraction index of the transparent material, the wavelength of the laser beam, and the fact that the laser beam hits also land located around pits, almost no light is reflected from a pit. The remaining reflected light is used as a tracking signal. "The transition from pit to land or from land to pit, i.e. the change of the intensity of the reflected light, is considered as a binary one. When the laser beam hits areas within a land or within a pit it is interpreted as a sequence of zeros." For technical reasons pits and lands may not be too short or too long. To be more precise, a land or pit must have the length of at least 2 and at most 10 zeros. In other words, when reading binary data from a compact disc or writing it onto a compact disc, between two consecutive binary ones there must be a sequence of at least 2 and at most 10 zeros. Obviously, "binary data obtained from sampling and encoding does not satisfy these requirements. Thus, data provided by the CIRC encoder still must be changed before it can be written onto a compact disc." This is done by the channel encoder.

EFM The channel encoder uses *eight-to-fourteen modulation*, EFM, in order to change a byte, which is a binary vector of length 8, into a binary vector of length 14 which satisfies the requirements on the number of zeros between two consecutive binary ones (cf. [164, pages 77ff]). This process is usually done by table-lookup. Here is a small part of this table: **5.4.7**

00000000	01001000100000
00000001	10000100000000
00000010	10010000100000
00000011	10001000100000
00000100	01000100000000
00000101	00000100010000
00000110	00010000100000
00000111	00100100000000
⋮	⋮
11111000	01001000010010
11111001	10000000010010
11111010	10010000010010
11111011	10001000010010
11111100	01000000010010
11111101	00001000010010
11111110	00010000010010
11111111	00100000010010

It is easy to see (cf. Exercise 5.4.3) that 14 is the least length of vectors needed, in order to represent all 2^8 bytes by different vectors satisfying the requirements on the number of zeros between two consecutive binary ones. Actually there are 267 vectors with these properties. Two of them which are not used by EFM are used for subcode synchronization words (cf. 5.4.10).

5.4.8   **Merging bits** Now it can still happen that the concatenation of two vectors produced by EFM still does not satisfy the condition on the number of zeros between two consecutive binary ones. For this reason three *merging bits* are inserted between two vectors of length 14 (cf. [164, page 80]). Actually, there are only four possible choices for these bits, namely

$$000 \quad 100 \quad 010 \quad 001.$$

Two merging bits 00 are necessary to prevent consecutive binary ones. The third merging bit is added so that the average *digital sum* value is close to zero.

A bit pattern can also be interpreted as a rectangular wave. It admits values ± 1. A binary one causes a change of the sign. For instance, the vector 00000100010000 can be considered as

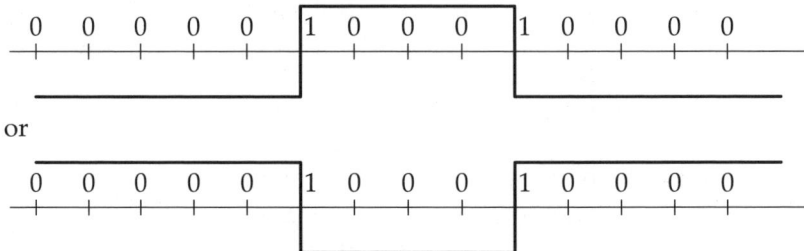

Fig. 5.8 Rectangular wave form of a bit pattern

The digital sum is determined by assigning $+1$ to the positive and -1 to the negative amplitude and summing these values for each bit. This is the integral of the rectangular wave form of the bit pattern when the distance between two bits is considered to be equal to 1.

Concatenating the above binary vector with itself, we may only use the merging bits containing a binary one, since otherwise there would be too many consecutive zeros. The three possibilities 100, 010, and 001 yield the following rectangular wave forms:

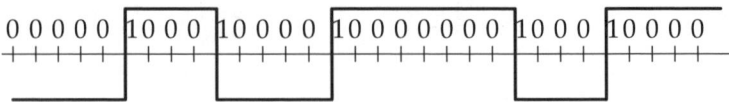

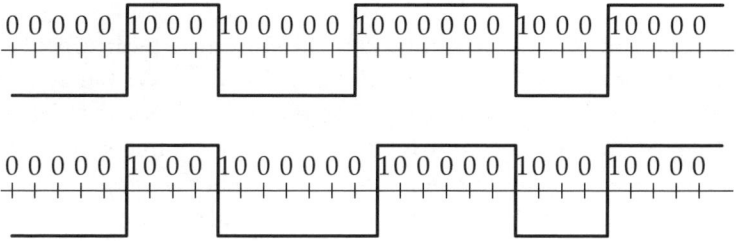

Fig. 5.9 Concatenation with merging bits

They have the digital sum values 3, 1, and −1. Depending on the digital sum value we started from, we insert those merging bits so that the final digital sum value is close to zero.

Frames As was already mentioned, the information on a compact disc is orga-
nized in form of frames (cf. [164, pages 82ff]). A frame consist of exactly 588
channel bits. Among other information it contains the audio information of
one C_1-codeword.

5.4.9

Each frame starts with a 24-bit synchronization pattern which is uniquely
distinguishable from all other possible data patterns. It is given by

$$100000000001000000000010,$$

and it is used to maintain a constant data readout rate. The rate of synchronization patterns influences the rotational speed of the disc. After three merging
bits a 14-bit subcode and another three merging bits are added. Then each byte
of a C_1-codeword encoded by EFM to a 14 bit sequence and another 3 merging
bits are added. This means that each byte of this codeword is enlarged to 14
bits by EFM and between two sequences produced by EFM three merging bits
are inserted.

Thus a frame is of the following form, where "Sync." indicates the synchronization patterns, "M." suitable merging bits, "Subc." the subcode, and
w_i the EFM-encoding of the i-th byte of the C_1-codeword. In the second row
the number of used channel bits is indicated.

	Sync.	M.	Subc.	M.	w_0	M.	...	w_{31}	M.
bits	24	3	14	3	14	3	$30 \cdot (14+3)$	14	3

In summary, we have 24 synchronization bits, one 14-bit subcode, $32 \cdot 14$ bits
produced by EFM from one C_1-codeword, and $34 \cdot 3$ merging bits in each
frame.

Since there are 7350 frames per second, 4 321 800 channel bits are read or
written per second. This gives 259 308 000 channel bits (32 413 500 channel
bytes) per minute and 15 558 480 000 bits (1 944 810 000 channel bytes) per hour.

5.4.10   **Subcode** The 14-bit *subcode* is produced by EFM from an 8-bit subcode. These 8 bits are usually referred to as P, Q, R, S, T, U, V, and W. On an audio CD only the P and Q bits are used. They should not be mixed with the P and Q symbols of the CIRC. The subcode holds information about the different tracks on a compact disc, display information, digital copy permission, and control information for different functions of the CD-player. (For more details see [164, pages 90ff].)

Each frame contains 8 subcode bits. In order to make better use of these bits and store useful information with them, the subcode bits of 98 frames are collected to form a *subcode block*. Each frame contributes one P-bit, Q-bit and so forth to the subcode block. Thus the 8 subcode bits are used as 8 different channels.

Since there are 7350 frames per second and the subcode of 98 frames is collected to one block, there are 75 subcode blocks per second.

In general, each subcode channel contains synchronization words, instructions, commands, data, and even some parity check symbols. Each subcode block starts with two synchronization words. These are two patterns not used by the EFM. In other words, the 14-bit representation of the subcode of the first two frames in a subcode block can uniquely be detected. The two synchronization words are given by

$$00100000000001 \quad \text{and} \quad 00000000010010.$$

Hence the first two bits of each channel are reserved for synchronization. After the first two bits of the Q-channel reserved for synchronization, there follow 4 control bits, 4 address bits describing 3 different modes, then 72 bits of data, and finally 16 bits for cyclic redundancy checking. This CRC is done in the following way. Consider the vector of the $4 + 4 + 72 = 80$ (control-, address-, data-) bits as a polynomial f of degree at most 79 over $\mathbb{F}_2$. By the division algorithm determine $q, r \in \mathbb{F}_2[x]$ so that

$$f(x) = \left(x^{16} + x^{12} + x^5 + 1\right)q(x) + r(x),$$

with $\deg r < 16$, and consider the coefficients of r (or as it is actually done, the inverted binary values of them) as the 16 check symbols. When the receiver obtains a Q-channel, the first two bits are stripped, the next 80 bits are read, the division algorithm is applied and the remainder is compared with the final 16 parity check bits. If they coincide the receiver assumes that the data was correct.

The P-channel designates the starting and stopping of tracks.

In the next section we describe some differences between CD-DA and CD-ROM.

Exercises

Exercise Let C be a Reed–Solomon-code of length $n = p^r - 1$, dimension $k = p^r - d$ and minimum distance d. For $1 \leq s < k$ prove the following facts about the shortened Reed–Solomon-code $C(s)$ of length $n - s$, which is obtained from C by taking all codewords which have zeros in the last s positions and deleting the last s positions:

1. If g is the generator polynomial of C, then

$$C(s) = \{fg \mid f \in \mathbb{F}_{p^r}[x], \ \deg f < k - s\}.$$

2. $C(s)$ is an $(n - s, k - s, d)$-code, thus it is an MDS-code.

Exercise Show that when the symbols of 17 consecutive C_1-codewords are erased, CIRC fails to correct this error. Show that when the symbols of 49 consecutive C_1-codewords are erased, the interpolation process fails to correct this error.

Exercise Let $a(n)$ be the number of binary vectors of length n such that between two consecutive binary ones there are at least 2 and at most 10 zeros. Show that $n = 14$ is the least integer such that $a(n) \geq 2^8$. Prove that $a(14) = 267$.

5.5 More Details on CD-ROM

CD-ROM and diverse other disc formats are thoroughly described in chapter 6 of [164]. Here we provide the reader with a short summary. "CD-ROM is the logical extension of the compact disc format towards the much broader application of information storage in general." The compact disc is used "as a read-only memory system" which can contain "any kind of program material." It is a "cost-effective way of distributing large amounts of information, especially information not requiring frequent updating," for instance, databases and mass storage for computer related applications.

Although the CD-ROM looks identically to an audio compact disc it uses a modified data format. "A CD-ROM identifies itself as differing from an audio compact disc through the Q-subcode channel."

The CD-ROM standard as specified in the *Yellow Book*, in the ISO/IEC standard *Information technology – Data interchange on read-only 120 mm optical data disks (CD-ROM)* [96], or also in [48], "does not link CD-ROM to a specific application." Unlike the audio CD standard, "it does not define the type of information that is stored" on a CD-ROM. "Furthermore, the layout of the information on the disc is not defined, and it does not indicate where and how to store the directory, how to identify the beginning or end of a file, or how to open a file." On an audio compact disc sampled digital audio is stored. Since usually neighboring samples are quite similar, it is possible to apply interpolation in order to reconstruct audio information which could not be properly decoded by CIRC. For obvious reasons, when storing arbitrary information on a CD-ROM the method of interpolation cannot be applied. Hence, the error correction and error detection of a CD-ROM must provide a higher data integrity than on an audio compact disc.

The smallest data area of an audio compact disc is a frame containing 24 bytes of audio information. A frame is too short for numerical applications and there is no way of addressing frames. Similarly as with the subcode channels, 98 frames are combined to form a *sector* of a CD-ROM. (See also [164, pages 215ff].) A sector is the basic data unit of a CD-ROM. In general a frame contains $98 \cdot 24 = 2352$ bytes of information. Since in an audio CD there are exactly 7350 frames per second, we have $7350/98 = 75$ sectors per second on a CD-ROM.

5.5.1   **Sector and sector modes** The first 12 bytes of each sector are used as a synchronization word. The next 4 bytes form a header field containing three address bytes and one mode byte. The *address bytes* indicate the minute (usually from 1 to 74), the second within this minute (from 0 to 59) and the sector within this second (from 1 to 75). For example, the three values $45 - 20 - 12$ indicate the 12-th sector in the 20-th second of the 45-th minute. This information is also found in the Q-subcode channel, but it speeds up and provides greater accuracy for searching. The *mode byte* indicates one of three different modes available for CD-ROM sectors. (See also [164, Fig. 6.2].)

Mode 0 just contains null data. Thus, after the synchronization word and the header it just contains 2336 zero-bytes.

Sync.	Header				null data
	Address			Mode 0	
	Min.	Sec.	Block		
12	1	1	1	1	2336

Fig. 5.10 CD-ROM sector mode 0

Mode 1 specifies, as described in Fig. 5.11, that 2048 of the remaining bytes are devoted to user data and the final 288 bytes are reserved for error detection EDC and error correction ECC. The error detection code is a CRC-code with

Sync.	Header				user data	EDC	Space	Auxiliary data	
	Address			Mode 1				ECC	
	Min.	Sec.	Block					P-parity	Q-parity
12	1	1	1	1	2048	4	8	172	104

Fig. 5.11 CD-ROM sector mode 1

respect to the polynomial

$$g(x) := \left(x^{16} + x^{15} + x^2 + 1\right)\left(x^{16} + x^2 + x + 1\right) \in \mathbb{F}_2[x].$$

The sequence of synchronization-, header-, and user data is considered as a binary polynomial. Dividing this polynomial by g, the division algorithm yields a remainder of degree less than 32, the coefficients of which are stored in the 4 EDC-bytes. Immediately after these bytes a sequence of 8 zero-bytes is appended as a space between the EDC and ECC bytes. The error correction encoding of a sector is carried out by a Reed–Solomon Product-like Code, RSPC. For more details see 5.5.2.

In **Mode 2** all remaining 2336 bytes are available for user data.

Sync.	Header				user data
	Address			Mode 2	
	Min.	Sec.	Block		
12	1	1	1	1	2336

Fig. 5.12 CD-ROM sector mode 1

The additional EDC and ECC of Mode 1 ensure a level of data integrity essential for storing arbitrary information. In Mode 1 each sector contains exactly 2 kB of information. Because of extended error correction Mode 1 has the most number of applications. The error rate is improved over that of an audio CD. Theoretically there will be one uncorrectable bit in every 10^{16} or 10^{17} bits. In Mode 2 it is possible to store more data on a CD-ROM, however with lower data integrity than in Mode 1. Therefore, it is mainly used for "gracefully degrading data such as video and audio."

RSPC encoding We assume that the bytes of a sector in Mode 1 are labeled   **5.5.2**
as $B_0, \ldots, B_{2351}$. The header-bytes, user data, EDC-bytes, and space bytes are

input to the RSPC encoder. These input bytes and the P-parity and Q-parity bytes make a total of 2340 bytes per sector. They are ordered in 1170 words $W_n, n \in 1170$. Each word consists of exactly 2 bytes, in more details

$$W_n = (B_{2n+12}, B_{2n+13}), \qquad n \in 1170.$$

Then we form two arrays of bytes, namely

$$(B_{2n+12})_{n \in 1170} \quad \text{and} \quad (B_{2n+13})_{n \in 1170}.$$

These two arrays are encoded separately. Let $V_n, n \in 1170$, denote the elements of such an array, i.e. either $V_n = B_{2n+12}$ or $V_n = B_{2n+13}$. The bytes V_n must be inserted into the diagram of Fig. 5.13 so that the first row contains from left to right the elements $V_0, V_1, \ldots, V_{42}$, the next one $V_{43}, V_{44}, \ldots, V_{85}$ and so on.

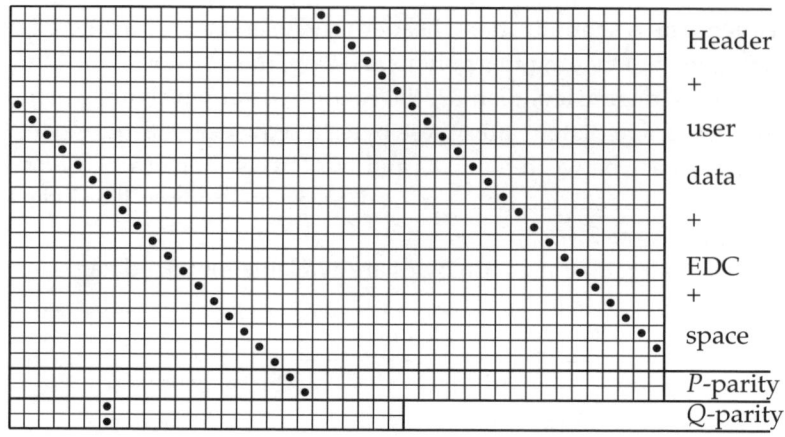

Header

+

user

data

+

EDC

+

space

P-parity
Q-parity

Fig. 5.13 RSPC encoding

The P-parity symbols are determined by a $(26, 24)$-Reed–Solomon-code over $\mathbb{F}_{2^8}$. It takes the first 24 entries of each column

$$(V_r, V_{r+43}, V_{r+2\cdot43}, \ldots, V_{r+23\cdot43}), \qquad r \in 44,$$

as an input and computes the symbols $V_{r+24\cdot43}$ and $V_{r+25\cdot43}$. A check matrix of this code is given by

$$\begin{pmatrix} 1 & 1 & \ldots & 1 & 1 & 1 \\ \alpha^{25} & \alpha^{24} & \ldots & \alpha^2 & \alpha & 1 \end{pmatrix}$$

for α a root of the primitive polynomial $x^8 + x^4 + x^3 + x^2 + 1 \in \mathbb{F}_{2^8}[x]$.

In order to determine the Q-parity symbols, a $(45, 43)$-Reed–Solomon-code over $\mathbb{F}_{2^8}$ is applied. Its codewords are diagonals as indicated by the black circles in the above diagram (cf. [68]). The i-th codeword, $i \in 26$, starts in the leftmost column of the i-th row of this array. The 45 symbols occurring in the first 26 rows are the information symbols. The parity check symbols of this code are input into the last two rows of this array. The Q-parity check symbols of the i-th codeword occur in the i-th column of this part of the array. A check matrix of this code is given by

$$\begin{pmatrix} 1 & 1 & \cdots & 1 & 1 & 1 \\ \alpha^{44} & \alpha^{43} & \cdots & \alpha^2 & \alpha & 1 \end{pmatrix}$$

for the same α as above. Actually, these two codes are obtained by shortening a $(255, 253, 3, 2^8)$-Reed–Solomon code (cf. Exercise 5.4.1).

Encoding and decoding After having determined the bytes filled into a sector (null data in Mode 0, user data, EDC and ECC in Mode 1, and just user data in Mode 2), the bytes $B_{12}, \dots, B_{2351}$ of a sector are *scrambled*. For more details see [48] and [96]. "A regular bit pattern fed into the EFM encoder can cause large values of the digital sum value in case the merging bits cannot reduce this value. The scrambler reduces this risk by converting the input bit stream with a shift register in a prescribed way." The bytes of a *scrambled sector* are mapped onto a series of consecutive frames. Each frame consists of exactly 24 bytes. However, the starting point of a sector is not necessarily the starting point of the frame. The byte B_0 of a sector can be insert as the $4n$-th byte of a frame, $0 \le n \le 5$. Consecutive bytes of the sector are placed in consecutive bytes of a frame. After the byte B_{2351} the byte B_0 of the next sector is inserted. A frame is, therefore, of the form

5.5.3

$$
\begin{array}{cccccc}
B_{4r} & B_{4r+1} & B_{4r+2} & B_{4r+3} & B_{4r+4} & B_{4r+5} \\
B_{4r+6} & B_{4r+7} & B_{4r+8} & B_{4r+9} & B_{4r+10} & B_{4r+11} \\
B_{4r+12} & B_{4r+13} & B_{4r+14} & B_{4r+15} & B_{4r+16} & B_{4r+17} \\
B_{4r+18} & B_{4r+19} & B_{4r+20} & B_{4r+21} & B_{4r+22} & B_{4r+23}.
\end{array}
$$

Next the byte order of each even-odd numbered pair of bytes in the frame is reversed. We obtain

$$
\begin{array}{cccccc}
B_{4r+1} & B_{4r} & B_{4r+3} & B_{4r+2} & B_{4r+5} & B_{4r+4} \\
B_{4r+7} & B_{4r+6} & B_{4r+9} & B_{4r+8} & B_{4r+11} & B_{4r+10} \\
B_{4r+13} & B_{4r+12} & B_{4r+15} & B_{4r+14} & B_{4r+17} & B_{4r+16} \\
B_{4r+19} & B_{4r+18} & B_{4r+21} & B_{4r+20} & B_{4r+23} & B_{4r+22}
\end{array}
$$

a so called *F1-frame*. Each F1-frame is then encoded by a conventional CIRC encoder (cf. 5.4.1). This yields an *F2-frame*, containing exactly 32 bytes. Adding one additional subcode byte to each F2-frame yields an *F3-frame*. Similarly as in an audio compact disc, there are eight different subcode bits referred to as P, Q, R, S, T, U, V, and W (cf. 5.4.10). The information stored in the subcode of 98 consecutive frames is collected to 8 subcode channels. A group of 98 F3-frames is also known as a *section*. Because of the delays during the CIRC encoding, sections have nothing to do with sectors. Finally, after sending the 33 bytes of an F3-frame to EFM (cf. 5.4.7) the data are written onto the CD.

Usually the error rate of a compact disc after CIRC decoding is approximately 10^{-10} to 10^{-11} errors per bit. The CIRC decoder also delivers information about bytes that could not be correctly decoded. Since we know the exact position of these bytes, they are erasures. The two Reed–Solomon-codes used in RSPC have minimum distance $d = 3$. Hence, each of them can be used to correct 1 error or 2 erasures per codeword. Combining both CIRC and RSPC decreases the bit error rate of a CD-ROM so that it will be between 10^{-16} and 10^{-17} (cf. [164]).

There exist various extension of the CD-ROM standard, for instance Compact Disc-Interactive, CD-I, described in the *Green Book*, Video compact discs defined in the *White Book*, or recordable discs, the standard of which can be found in the *Orange Book*. For more details see [164].

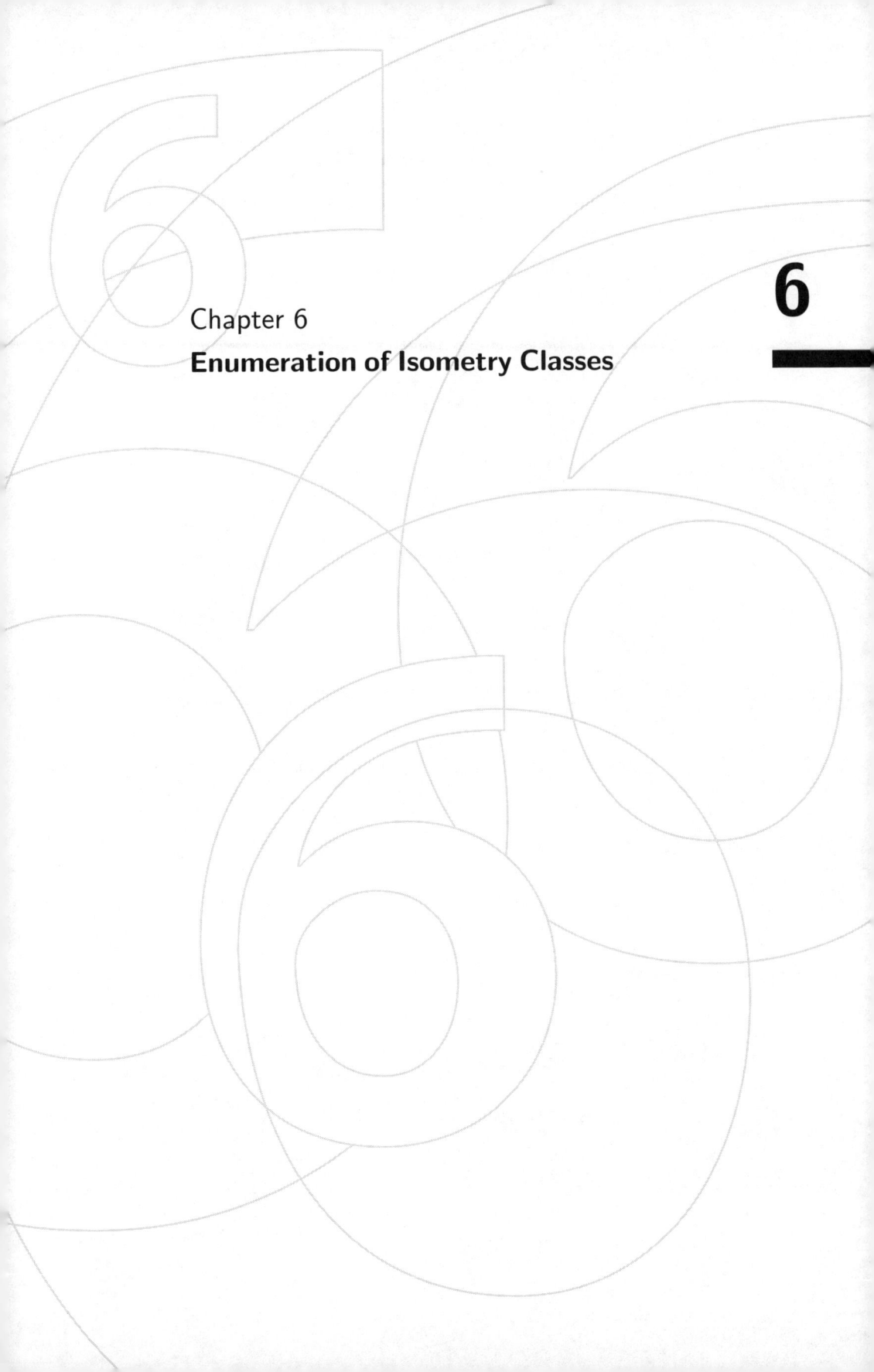

Chapter 6

Enumeration of Isometry Classes

6

6

Enumeration of Isometry Classes

6

6 Enumeration of Isometry Classes

We have gathered linear codes in classes of codes which are of the same quality with respect to error correction. Since the metric structure of a code determines its error correction properties we have introduced the notion of *isometric* codes and the just mentioned classes of codes are called *isometry classes*. Each of these classes is an orbit of an isometry group of $\mathbb{F}_q^n$. The linear isometry classes are orbits under the linear isometry group $M_n(q)$, the semilinear isometry classes orbits under the semilinear isometry group. This chapter is concerned with the enumeration of isometry classes of codes using methods from Combinatorics, in particular *Pólya's Theory of Enumeration*. This theory deals with the combinatorial properties of finite group actions. In particular, properties of the acting group like numbers of fixed points are used to get results about the number of orbits. The fundamental tool is the *Lemma of Cauchy-Frobenius*, which was introduced in 3.4.2 and refinements thereof. To count the isometry classes of codes we need detailed information about the isometry groups. Depending on whether we count linear isometry classes (in the first sections) or semilinear isometry classes (in Section 6.7) we have to study the projective linear or the projective semilinear groups over the appropriate finite fields.

An interesting and helpful notion introduced in Section 6.2 is the concept of *indecomposable* linear codes. Each code can be written in an essentially unique way as a sum of such codes. We derive the number of indecomposable linear codes, obtaining this way an idea of the complexity of the construction of all the isometry classes of indecomposable linear codes. Furthermore, a special class of indecomposable codes, the *critical indecomposable codes*, are described in detail in Section 6.5.

For the actual computation of the number of linearly nonisometric (n, k)-codes over $\mathbb{F}_q$, we need detailed information about the natural group action of the projective linear group $\mathrm{PGL}_k(q)$ on $\mathrm{PG}_{k-1}^*(q)$. Especially, we describe the conjugacy classes of the linear group $\mathrm{GL}_k(q)$ by using the *Jacobi normal form* of the automorphisms of $\mathbb{F}_q^k$. This approach is based on module theoretic considerations already introduced in Chapter 4. In Section 6.3 we derive a complete description of the cycle index for the natural action of $\mathrm{PGL}_k(q)$ on $\mathrm{PG}_{k-1}^*(q)$.

Numerical results concerning the enumeration of linear isometry classes of codes are displayed in Section 6.4. Extended tables, computed by SYMMET-RICA (cf. [190]), can be found online [58] or on the attached CD.

Closely related to the enumeration of nonisometric codes is the *random generation* of linear codes. The algorithm presented in Section 6.6 generates representatives of linear isometry classes which are distributed *uniformly at random* over all classes of $(n, \leq k)$-codes over $\mathbb{F}_p$ for given n, k and p. We use a quite general method which is due to Dixon and Wilf [46]. This method applies whenever the structure under consideration is defined as an orbit of a finite group acting on a finite set.

At the very end of this chapter in Section 6.8 we prove that every local isometry between two (n, k)-codes over $\mathbb{F}_q$ can be extended to a global isometry of $\mathbb{F}_q^n$. This demonstrates that the seemingly weaker condition of a local isometry is equivalent to our approach from Section 1.4 and Section 1.5. (See also [84, second edition, Section 9.1].)

Normal bases of a finite extension $\mathbb{F}_q$ over $\mathbb{F}_p$ have been introduced in Section 3.3. Finally, in Section 6.9 we prove that it is always possible to construct a normal basis of a finite field extension over a finite field. The proof uses methods from module theory introduced in Chapter 4 and Section 6.3.

6.1 Enumeration of Linear Isometry Classes

To begin with, we recall that two linear codes C and C' in $\mathbb{F}_q^n$ are said to be *linearly isometric* if there exists a linear isometry

$$\iota : \mathbb{F}_q^n \to \mathbb{F}_q^n$$

which maps C onto C'. The group of *all* linear isometries on $\mathbb{F}_q^n$, the *linear isometry group*, was indicated in Section 1.4 by

$$M_n(q).$$

It is the set of all $n \times n$-matrices over $\mathbb{F}_q$ which contain in each of their rows and columns exactly one nonzero element of $\mathbb{F}_q$. The application of a linear isometry to a generator matrix (via right multiplication) amounts to a permutation of its columns and/or a multiplication of columns by nonzero elements of $\mathbb{F}_q$. We have seen in 1.4.12 that $M_n(q)$ is isomorphic to a wreath product,

$$M_n(q) \simeq \mathbb{F}_q^* \wr_n S_n.$$

The linear isometry group acts on $\mathbb{F}_q^n$, whence also on its power set, and it has already been mentioned that the corresponding set of orbits,

$$\mathbb{F}_q^* \wr_n S_n \backslash\backslash 2^{\mathbb{F}_q^n},$$

is the set of isometry classes of block codes. Some of them are sets of subspaces, the linear isometry classes of linear codes. Using the notation

$$\mathcal{U}(n, q) := \left\{ U \mid U \leq \mathbb{F}_q^n \right\}$$

for the set of all subspaces of $\mathbb{F}_q^n$, we express the set of linear isometry classes of linear codes in $\mathbb{F}_q^n$ as

$$\mathbb{F}_q^* \wr_n S_n \backslash\!\backslash \mathcal{U}(n,q).$$

This set can still be refined since each linear isometry preserves both the dimension and the minimum distance of a code. For this reason, we introduce the following subsets of $\mathcal{U}(n,q)$

$$\mathcal{U}(n,k,q) := \left\{ U \leq \mathbb{F}_q^n \mid \dim(U) = k \right\}, \qquad 1 \leq k \leq n,$$

and

$$\mathcal{U}(n,k,d,q) := \left\{ U \leq \mathbb{F}_q^n \mid \dim(U) = k, \; \mathrm{dist}(U) = d \right\}.$$

Thus we obtain

The metric classification of linear codes   *The set of nontrivial linear isometry*   6.1.1
classes of linear codes of length n over $\mathbb{F}_q$ *is the set of orbits*

$$\mathbb{F}_q^* \wr_n S_n \backslash\!\backslash (\mathcal{U}(n,q) \setminus \{0\}) = \bigcup_{k=1}^{n} \bigcup_{d=1}^{d_{\max}(n,k,q)} \mathbb{F}_q^* \wr_n S_n \backslash\!\backslash \mathcal{U}(n,k,d,q).$$

Each transversal of the orbit set

$$\mathbb{F}_q^* \wr_n S_n \backslash\!\backslash \mathcal{U}(n,k,d,q)$$

is a complete system of pairwise linearly nonisometric linear (n,k,d,q)*-codes.*   □

Example   Considering the set of linear isometry classes of linear (n,k)-codes   6.1.2
instead of the set of all (n,k)-codes reduces dramatically the number of objects
to be classified. For instance, the numbers $\begin{bmatrix} n \\ k \end{bmatrix}(2)$ of k-dimensional subspaces
of $\mathbb{F}_2^n$ (cf. Exercise 6.1.3) are displayed in Table 6.1.

With the methods described in this section we will be able to determine
the numbers U_{nk2} given in Table 6.7. They are the numbers of linear isometry classes of binary (n,k)-codes. From these tables we deduce, for instance,
that there are more than 53 million 4-dimensional subspaces of $\mathbb{F}_2^{10}$ but only
516 linear isometry classes of binary $(10,4)$-codes. Later on (cf. Table 6.7) we
will see that there are only 276 isometry classes of $(10,4)$-codes without zero
columns. Using methods from Chapter 9, we will obtain that there are only 19
isometry classes of $(10,4)$-codes with optimal minimum distance $d = 4$.   ◇

If we want to apply the metric classification of linear codes for enumerative
or constructive purposes, we run into problems since the sets $\mathcal{U}(n,k,q)$ are
abstract sets of vector spaces. But we know from Linear Algebra that each
code possesses bases, k-tuples of linearly independent elements. They are the
generator matrices of a code. Still there is a problem concerning complexity.

Table 6.1 Values of $\begin{bmatrix} n \\ k \end{bmatrix}(2)$

$n\backslash k$	1	2	3	4	5
1	1	0	0	0	0
2	3	1	0	0	0
3	7	7	1	0	0
4	15	35	15	1	0
5	31	155	155	31	1
6	63	651	1 395	651	63
7	127	2 667	11 811	11 811	2 667
8	255	10 795	97 155	200 787	97 155
9	511	43 435	788 035	3 309 747	3 309 747
10	1 023	174 251	6 347 715	53 743 987	109 221 651
11	2 047	698 027	50 955 971	866 251 507	3 548 836 819
12	4 095	2 794 155	408 345 795	13 910 980 083	114 429 029 715

Table 6.2 Values of U_{nk2}

$n\backslash k$	1	2	3	4	5	6	7
1	1	0	0	0	0	0	0
2	2	1	0	0	0	0	0
3	3	3	1	0	0	0	0
4	4	6	4	1	0	0	0
5	5	10	10	5	1	0	0
6	6	16	22	16	6	1	0
7	7	23	43	43	23	7	1
8	8	32	77	106	77	32	8
9	9	43	131	240	240	131	43
10	10	56	213	516	705	516	213
11	11	71	333	1 060	1 988	1 988	1 060
12	12	89	507	2 108	5 468	7 664	5 468
13	13	109	751	4 064	14 724	29 765	29 765
14	14	132	1 088	7 641	39 006	117 169	173 035

Each (n,k)-code has *many bases,* except for very trivial cases. Hence, if we want to describe a subspace by a generator matrix, we are faced with a great variety of possibilities. So, instead of the abstract set of vector spaces we have to manage a big set of matrices. In 1.4.14 and Exercise 1.4.14 we have already introduced for $n \geq k \geq 1$ the set of all generator matrices of (n,k)-codes over $\mathbb{F}_q$ as the set

$$\mathbb{F}_q^{k \times n, k} := \left\{ \Gamma \mid \Gamma \in \mathbb{F}_q^{k \times n}, \ \mathrm{rank}(\Gamma) = k \right\}$$

of all $k \times n$-matrices over $\mathbb{F}_q$ of rank k. In the Exercises 1.4.14 and 1.4.16 we have described actions of the general linear group $\mathrm{GL}_k(q)$ and of the full monomial group $M_n(q)$ on $\mathbb{F}_q^{k \times n, k}$, and we have shown that these two actions commute. According to Exercise 1.4.10, two commuting group actions $_G X$ and $_H X$ induce an action of the direct product $G \times H$ on X.

Since, according to Exercise 1.4.14, exactly the left multiplications by elements $A \in \mathrm{GL}_k(q)$ transform a generator matrix of the space $C \in \mathcal{U}(n,k,q)$ into another generator matrix of C, the orbits of $\mathrm{GL}_k(q)$ on $\mathbb{F}_q^{k \times n, k}$ correspond to the subspaces of dimension k:

$$\mathcal{U}(n,k,q) = \mathrm{GL}_k(q) \backslash\!\backslash \mathbb{F}_q^{k \times n, k}.$$

The operations of elements of the linear isometry group commute with the operations of the elements of $\mathrm{GL}_k(q)$, and so the set of linear isometry classes of (n,k)-codes is equal to the set of orbits

$$(\mathrm{GL}_k(q) \times \mathbb{F}_q^* \wr_n S_n) \backslash\!\backslash \mathbb{F}_q^{k \times n, k} \qquad\qquad\qquad \textbf{6.1.3}$$

with respect to the action

$$(\mathrm{GL}_k(q) \times \mathbb{F}_q^* \wr_n S_n) \times \mathbb{F}_q^{k \times n, k} \longrightarrow \mathbb{F}_q^{k \times n, k},$$

defined by

$$((A, B), \Gamma) \mapsto A \cdot \Gamma \cdot B^\top.$$

Using Exercise 1.4.9 (which says that the set of orbits of a direct product is the set of orbits of one factor on the set of orbits of the other factor) we rephrase 6.1.3 as

$$\mathrm{GL}_k(q) \backslash\!\backslash \left(\mathbb{F}_q^* \wr_n S_n \backslash\!\backslash \mathbb{F}_q^{k \times n, k} \right). \qquad\qquad\qquad \textbf{6.1.4}$$

Because of the condition on the rank, the set $\mathbb{F}_q^{k \times n, k}$ is not easy to handle. We may thus prefer to work with the even larger set $\mathbb{F}_q^{k \times n}$ of *all* $k \times n$-matrices without any condition on the rank. In 6.1.15 and Exercise 6.1.6 it will be clear that it is possible to determine the number of isometry classes of linear (n,k)-codes from $|\mathrm{GL}_k(q) \backslash\!\backslash (\mathbb{F}_q^* \wr_n S_n \backslash\!\backslash \mathbb{F}_q^{k \times n})|$ and $|\mathrm{GL}_{k-1}(q) \backslash\!\backslash (\mathbb{F}_q^* \wr_n S_n \backslash\!\backslash \mathbb{F}_q^{(k-1) \times n})|$. The set $\mathbb{F}_q^{k \times n}$ can be reduced a bit, since matrices which contain zero columns are not of interest for coding theoretic purposes. (Such columns are redundant in coding theory, since the corresponding components are zero in each code-word, and so they give no information. Moreover, two generator matrices of the same code have the same number of columns of zeros, and these columns occur at the same column indices. Generator matrices of isometric codes also have the same number of columns of zeros, but these columns need not oc-cur at the same column indices.) For this reason we introduce the following notion:

6.1.5   **Definition (nonredundant code)** A linear code C is called *nonredundant* if its generator matrix Γ contains no zero column.   ◇

In fact, this condition is independent of the choice of the generator matrix Γ.

It is, therefore, reasonable to restrict attention to the set of all $k \times n$-matrices without zero columns. The advantage is that the set of *all $k \times n$-matrices over $\mathbb{F}_q$ which contain no zero column can be written as a set of mappings*

$$(\mathbb{F}_q^k \backslash \{0\})^n = \left\{ f \mid f \colon n \to \mathbb{F}_q^k \backslash \{0\} \right\}.$$

The generator matrix Γ of a nonredundant (n,k)-code is identified with the mapping $\Gamma \colon n \to \mathbb{F}_q^k \backslash \{0\}$ where $\Gamma(i)^\top$ is the i-th column of Γ.

Rewriting our problem in these terms shows that instead of the situation in 6.1.4 we are now faced with the set of orbits

6.1.6 $$\mathrm{GL}_k(q) \backslash\backslash \left(\mathbb{F}_q^* \wr_n S_n \backslash\backslash (\mathbb{F}_q^k \backslash \{0\})^n \right).$$

According to Exercise 1.4.9, the general linear group acts in the following way on $\mathbb{F}_q^* \wr_n S_n \backslash\backslash (\mathbb{F}_q^k \backslash \{0\})^n$:

$$\mathrm{GL}_k(q) \times \left(\mathbb{F}_q^* \wr_n S_n \backslash\backslash (\mathbb{F}_q^k \backslash \{0\})^n \right) \to \mathbb{F}_q^* \wr_n S_n \backslash\backslash (\mathbb{F}_q^k \backslash \{0\})^n,$$

6.1.7 $$(A, \mathbb{F}_q^* \wr_n S_n(f)) \mapsto \mathbb{F}_q^* \wr_n S_n(Af).$$

When writing Af, we identify the function $f \in (\mathbb{F}_q^k \backslash \{0\})^n$ with the corresponding $k \times n$-matrix $(f(0)^\top \mid \ldots \mid f(n-1)^\top)$. Then $Af = (A \cdot f(0)^\top \mid \ldots \mid A \cdot f(n-1)^\top)$ and, therefore, $Af(i) = (A \cdot f(i)^\top)^\top = f(i) \cdot A^\top$.

For this reason, we first investigate the action of a wreath product in more detail and explain how to split it into two group actions which are easier to handle (cf. [123], [124]).

6.1.8   **Lehmann's Lemma** *Let $_G X$ and $_H Y$ be two group actions. For the natural action of the wreath product $H \wr_X G$ on Y^X, defined in 1.4.9, we have:*

1. *If the mapping φ is given by*

$$\varphi \colon Y^X \to (H \backslash\backslash Y)^X \colon f \mapsto \varphi(f) \text{ where } \varphi(f)(x) = H(f(x)),$$

then the mapping

$$\Phi\colon H\wr_X G\backslash\backslash Y^X \to G\backslash\backslash((H\backslash\backslash Y)^X)\ :\ H\wr_X G(f) \mapsto G(\varphi(f))$$

is a bijection, where G acts canonically (cf. 1.4.7) on this set of functions.

2. *The orbit of $f \in Y^X$ under the action of $H \wr_X G$ is given by*

$$H\wr_X G(f) = \varphi^{-1}(\Phi(H\wr_X G(f))) = \varphi^{-1}(G(\varphi(f))).$$

Proof: 1. For $f_1, f_2 \in Y^X$ the following facts are equivalent:

$$\Phi(H\wr_X G(f_1)) = \Phi(H\wr_X G(f_2))$$
$$G(\varphi(f_1)) = G(\varphi(f_2))$$
$$\varphi(f_2) \in G(\varphi(f_1))$$
$$\varphi(f_2) = \varphi(f_1) \circ \overline{g} \text{ for some } g \in G$$
$$\varphi(f_2)(x) = \varphi(f_1)(gx) \text{ for some } g \in G \text{ and all } x \in X$$
$$H(f_2(x)) = H(f_1(gx)) \text{ for some } g \in G \text{ and all } x \in X$$
$$f_2(x) \in H(f_1(gx)) \text{ for some } g \in G \text{ and all } x \in X$$
$$f_2 = (\psi;g)f_1 \text{ for some } (\psi;g) \in H\wr_X G$$
$$f_2 \in H\wr_X G(f_1)$$
$$H\wr_X G(f_2) = H\wr_X G(f_1).$$

Reading these implications from the bottom to the top, we deduce that Φ is well-defined. Reading them the other way round it follows that Φ is injective. In order to prove that Φ is surjective, we first realize that φ is surjective, i.e. each $F \in (H\backslash\backslash Y)^X$ is of the form $\varphi(f) = F$ for some $f \in Y^X$. (The function f should be determined in such a way that for each $x \in X$ the value $f(x)$ belongs to $F(x)$, i.e. $F(x) = H(f(x))$.) If $\varphi(f) = F$, then

$$\Phi(H\wr_X G(f)) = G(\varphi(f)) = G(F),$$

whence Φ is also surjective.

2. In order to prove the second assertion, consider a function $F \in (H\backslash\backslash Y)^X$ and assume that $F = \varphi(f)$ for some $f \in Y^X$. Then

$$\varphi^{-1}(\{F\}) = \varphi^{-1}(\{\varphi(f)\}) = H\wr_X \{1\}(f)$$
$$= \left\{ \tilde{f} \in Y^X \,\middle|\, \tilde{f}(x) = \psi(x)f(x) \text{ for } \psi \in H^X \text{ and } x \in X \right\}.$$

Next we prove that

$$\varphi(f \circ \overline{g}) = \varphi(f) \circ \overline{g}, \qquad g \in G.$$

(The reader should realize that on the left hand side we are faced with the natural action of G on Y^X and on the right hand side with the natural action of G on $(H\backslash\backslash Y)^X$.) The action of G commutes with the application of φ, since $\varphi(f \circ \bar{g})(x) = H(f(g(x)))$ and $(\varphi(f) \circ \bar{g})(x) = \varphi(f)(gx) = H(f(gx))$ for all $x \in X$. Finally we obtain

$$
\begin{aligned}
H \wr_X G(f) &= \{(\psi; g)f \mid (\psi; g) \in H \wr_X G\} \\
&= \left\{x \mapsto \psi(x)f(g^{-1}x) \;\middle|\; \psi \in H^X,\, g \in G\right\} \\
&= \bigcup_{g \in G} \left\{x \mapsto \psi(x)f(g^{-1}x) \;\middle|\; \psi \in H^X\right\} \\
&= \bigcup_{g \in G} H \wr_X \{1\}\, (f \circ \bar{g}^{-1}) \\
&= \bigcup_{g \in G} \varphi^{-1}\left(\left\{\varphi(f \circ \bar{g}^{-1})\right\}\right) \\
&= \bigcup_{g \in G} \varphi^{-1}\left(\left\{\varphi(f) \circ \bar{g}^{-1}\right\}\right) \\
&= \varphi^{-1}\left(\bigcup_{g \in G} \left\{\varphi(f) \circ \bar{g}^{-1}\right\}\right) \\
&= \varphi^{-1}\left(\left\{\varphi(f) \circ \bar{g}^{-1} \;\middle|\; g \in G\right\}\right) \\
&= \varphi^{-1}\left(G(\varphi(f))\right) \\
&= \varphi^{-1}\left(\Phi(H \wr_X G(f))\right). \qquad\qquad \square
\end{aligned}
$$

An application of Lehmann's Lemma allows us to rewrite 6.1.6 in the form

6.1.9
$$
\mathrm{GL}_k(q)\backslash\backslash \left(S_n \backslash\backslash \left(\mathbb{F}_q^* \backslash\backslash (\mathbb{F}_q^k \backslash \{0\})\right)^n\right).
$$

This result shows the close connection between finite geometry and the theory of linear codes: The set of orbits of $\mathbb{F}_q^*$ on $\mathbb{F}_q^k \backslash \{0\}$ is the set of elements (also called points) of the $(k-1)$-dimensional projective space $\mathrm{PG}_{k-1}^*(q)$ (cf. Section 3.7). Hence, we actually investigate

6.1.10
$$
\mathrm{GL}_k(q)\backslash\backslash (S_n \backslash\backslash \mathrm{PG}_{k-1}^*(q)^n).
$$

Here the symmetric group S_n acts in a natural way on the domain of the mappings in $\mathrm{PG}_{k-1}^*(q)^n$. How does $\mathrm{GL}_k(q)$ act on the orbits $S_n \backslash\backslash \mathrm{PG}_{k-1}^*(q)^n$? From 6.1.7 we deduce that the application of $A \in \mathrm{GL}_k(q)$ to the $\mathbb{F}_q^* \wr_n S_n$-orbit of $f \in (\mathbb{F}_q^k \backslash \{0\})^n$ yields the orbit $\mathbb{F}_q^* \wr_n S_n(Af)$. If φ is the mapping defined as in Lehmann's Lemma, then the elements of $A(S_n(F))$ for $F \in \mathrm{PG}_{k-1}^*(q)^n$ are the elements in $\varphi(\mathbb{F}_q^* \wr_n S_n(Af))$ for some $f \in \varphi^{-1}(\{F\})$. We want to describe this set again as an orbit under a suitable group action. For this reason, in Section 3.7 we have deduced from Exercise 1.4.13 the natural action of $\mathrm{GL}_k(q)$ on $\mathrm{PG}_{k-1}^*(q)$ as described in 3.7.4. Here it is repeated once again.

$$\mathrm{GL}_k(q) \times \mathrm{PG}_{k-1}^*(q) \to \mathrm{PG}_{k-1}^*(q) \; : \; (A, \mathbb{F}_q^*(v)) \mapsto \mathbb{F}_q^*(v \cdot A^\top).$$

Lemma *Consider $A \in \mathrm{GL}_k(q)$ and let φ be given by*   **6.1.11**

$$\varphi \colon (\mathbb{F}_q^k \setminus \{0\})^n \to \mathrm{PG}_{k-1}^*(q)^n \; : \; f \mapsto \varphi(f) \text{ where } \varphi(f)(i) := \mathbb{F}_q^*(f(i)).$$

Then

$$\varphi\left(\mathbb{F}_q^* \wr_n S_n(Af)\right) = A(S_n(\varphi(f))), \qquad f \in (\mathbb{F}_q^k \setminus \{0\})^n,$$

where on the right hand side the action of $\mathrm{GL}_k(q)$ on $S_n \backslash\backslash \mathrm{PG}_{k-1}^*(q)^n$ appears, which is induced by the natural action of $\mathrm{GL}_k(q)$ on $\mathrm{PG}_{k-1}^*(q)$.

Proof: From the second part of Lehmann's Lemma we obtain

$$\varphi(\mathbb{F}_q^* \wr_n S_n(Af)) = S_n(\varphi(Af)).$$

Using Exercise 1.4.13 we deduce that $\varphi(Af) = A\varphi(f)$, since

$$\varphi(Af)(i) = \mathbb{F}_q^*(f(i) \cdot A^\top) = A\mathbb{F}_q^*(f(i)) = A\varphi(f)(i)$$

for all $i \in n$. Thus, $S_n(\varphi(Af)) = S_n(A\varphi(f))$ and this orbit equals $A(S_n(\varphi(f)))$, since A operates by matrix multiplication from the left, and π permutes the columns of (the matrix) f.   □

This way we have just replaced the action of $\mathrm{GL}_k(q) \times \mathbb{F}_q^* \wr_n S_n$ on $(\mathbb{F}_q^k \setminus \{0\})^n$ by the action of $\mathrm{GL}_k(q) \times S_n$ on $\mathrm{PG}_{k-1}^*(q)^n$, where this action is of the form 1.4.11. Therefore, $\mathrm{GL}_k(q)$ acts only on the range $\mathrm{PG}_{k-1}^*(q)$ and S_n acts only on the domain n. Instead of 6.1.10 we are finally dealing with

$$(\mathrm{GL}_k(q) \times S_n) \backslash\backslash \mathrm{PG}_{k-1}^*(q)^n.$$ **6.1.12**

This proves the following fundamental result:

Theorem *The linear isometry classes of linear, nonredundant (n,k)-codes over $\mathbb{F}_q$* **6.1.13**
are the orbits of $\mathrm{GL}_k(q) \times S_n$ on $\mathrm{PG}_{k-1}^*(q)^n$, the representatives of which are of rank k. They form a subset of

$$\mathrm{GL}_k(q) \backslash\backslash (S_n \backslash\backslash \mathrm{PG}_{k-1}^*(q)^n).$$

The inner orbit set $S_n \backslash\backslash \mathrm{PG}_{k-1}^*(q)^n$ can be represented by any complete system of mappings $f \colon n \to \mathrm{PG}_{k-1}^*(q)$ of pairwise different content

$$c(f) \colon \mathrm{PG}_{k-1}^*(q) \to \mathbb{N} \; : \; y \mapsto |f^{-1}(\{y\})|.$$

Hence, the set of all linear isometry classes of linear, nonredundant (n,k)-codes over $\mathbb{F}_q$ can be identified with the set of orbits of $\mathrm{GL}_k(q)$ on the set of mappings $f \in \mathrm{PG}_{k-1}^*(q)^n$ of pairwise different content which form $k \times n$-matrices of rank k. $\quad\Box$

Moreover, the class of bijective functions $f \colon n \to \mathrm{PG}_{k-1}^*(q)$ is the class of the simplex-codes. This fact demonstrates the particular role of simplex-codes and their dual codes, the Hamming-codes.

6.1.14   **Definition (projective codes and projective matrices)** A nonredundant (n,k)-code C is called *projective* if the columns of any generator matrix Γ of C are pairwise linearly independent. In this case, we call Γ a *projective matrix*. In other words, a $k \times n$ matrix Γ over $\mathbb{F}_q$ is called projective if no two columns are linearly dependent. If $n = 1$ we require that Γ is not the zero matrix. $\diamond$

Thus, projective codes have projective generator matrices and vice-versa. The columns of a projective generator matrix Γ are never zero and are representatives of pairwise distinct one-dimensional (punctured) subspaces of $\mathbb{F}_q^k$. Therefore, they give rise to an injective mapping

$$\overline{\Gamma} \colon n \to \mathrm{PG}_{k-1}(q) \quad \text{or} \quad \overline{\Gamma} \colon n \to \mathrm{PG}_{k-1}^*(q).$$

Here we prefer to use $\mathrm{PG}_{k-1}^*(q)$ since its elements are orbits under the action of $\mathbb{F}_q^*$. It is straightforward to verify that being projective is a property of the isometry class of a code. That is, for linearly isometric codes C_1 and C_2 the code C_2 is projective if and only if C_1 has this property.

Generalizing this definition, an arbitrary (n,k)-code C is called *injective* or *reduced* if the mapping

$$\overline{\Gamma} \colon n \to \mathrm{PG}_{k-1}^*(q) \cup \{0\},$$

corresponding to the columns of an arbitrary generator matrix Γ of C, is injective.

The numbers of linear isometry classes of linear codes will be obtained from a refinement of the metric classification 6.1.1. Besides the total number of linear isometry classes, we also evaluate the number of linear isometry classes of nonredundant codes as well as of projective codes.

The set of all k-dimensional *nonredundant* subspaces of $\mathbb{F}_q^n$ is indicated as

$$\mathcal{V}(n,k,q).$$

By $\overline{\mathcal{V}}(n,k,q)$ we denote the set of all projective $U \in \mathcal{V}(n,k,q)$, and we write $\overline{\mathcal{U}}(n,k,q)$ for the set of all injective $U \in \mathcal{U}(n,k,q)$. For the sets of linear isometry classes in $\mathcal{U}(n,k,q)$ and $\mathcal{V}(n,k,q)$ we use the symbols

$$\begin{aligned}
\mathcal{U}_{n,k,q} &:= M_n(q) \backslash\!\backslash \mathcal{U}(n,k,q), & \mathcal{V}_{n,k,q} &:= M_n(q) \backslash\!\backslash \mathcal{V}(n,k,q), \\
\overline{\mathcal{U}}_{n,k,q} &:= M_n(q) \backslash\!\backslash \overline{\mathcal{U}}(n,k,q), & \overline{\mathcal{V}}_{n,k,q} &:= M_n(q) \backslash\!\backslash \overline{\mathcal{V}}(n,k,q).
\end{aligned}$$

In addition, we introduce the following sets:

$$T_{n,k,q} := \bigcup_{l \leq k} V_{n,l,q} \quad (= V_{n,\leq k,q}),$$

$$\overline{T}_{n,k,q} := \bigcup_{l \leq k} \overline{V}_{n,l,q} \quad (= \overline{V}_{n,\leq k,q}),$$

comprising the classes of linear (n, l)-codes of dimension $l \leq k$. The cardinalities of these sets are denoted by

$$T_{nkq}, \overline{T}_{nkq}, V_{nkq}, \overline{V}_{nkq}, U_{nkq}, \overline{U}_{nkq}.$$

Of course, there is a close connection between these numbers. Using Exercise 6.1.6 we obtain the following basic results for the enumeration of linear isometry classes of linear codes (cf. Exercise 6.1.6):

Corollary   6.1.15

— *T_{nkq} is the number of orbits computed in 6.1.12,*

$$T_{nkq} = |(\mathrm{GL}_k(q) \times S_n) \backslash\!\backslash \mathrm{PG}^*_{k-1}(q)^n| = |\mathrm{GL}_k(q) \backslash\!\backslash (S_n \backslash\!\backslash \mathrm{PG}^*_{k-1}(q)^n)|.$$

If $k > 1$, then $T_{n,k-1,q}$ is also the number of $\mathrm{GL}_k(q) \times S_n$-orbits of mappings $f \in \mathrm{PG}^*_{k-1}(q)^n$ corresponding to matrices of rank not greater than $k - 1$.

— *$\overline{T}_{nkq}$ is the number of $\mathrm{GL}_k(q) \times S_n$-orbits on the set of injective functions in $\mathrm{PG}^*_{k-1}(q)^n$,*

$$\overline{T}_{nkq} = |(\mathrm{GL}_k(q) \times S_n) \backslash\!\backslash \mathrm{PG}^*_{k-1}(q)^n_{\mathrm{inj}}| = |\mathrm{GL}_k(q) \backslash\!\backslash (S_n \backslash\!\backslash \mathrm{PG}^*_{k-1}(q)^n_{\mathrm{inj}})|.$$

— *$V_{nkq} = T_{nkq} - T_{n,k-1,q}, \overline{V}_{nkq} = \overline{T}_{nkq} - \overline{T}_{n,k-1,q}$ for $1 < k \leq n$.*

— *$U_{nkq} = \sum_{i=k}^n V_{ikq}, \overline{U}_{kkq} = \overline{V}_{kkq}$, and $\overline{U}_{nkq} = \overline{V}_{n-1,k,q} + \overline{V}_{nkq}$ for $n > k$.*

The initial values for these recursions are $V_{n1q} = 1$ for $n \in \mathbb{N}^*$, $\overline{V}_{11q} = 1$ and $\overline{V}_{n1q} = 0$ for $n > 1$.   □

This way we have expressed $U_{nkq}, \overline{U}_{nkq}, V_{nkq}$, and $\overline{V}_{nkq}$ in terms of T_{nkq} and $\overline{T}_{nkq}$. The remaining problem is the evaluation of T_{nkq} and $\overline{T}_{nkq}$. In order to obtain these numbers we could, of course, use the Lemma of Cauchy–Frobenius 3.4.2 and compute the average number of fixed points. But it is our intention to get a more general result which gives a generating function for these numbers. It will turn out that the *weighted form* of the Lemma of Cauchy–Frobenius is more suitable for this purpose. For this reason we introduce *weight functions*. They are mappings defined on the set, on which the group acts, which are constant on each orbit. The range of these weight functions is usually a commutative ring (mostly a polynomial ring) which contains $\mathbb{Q}$ as a subring since we need to allow division by $|G|$. The following generalization of the Lemma of Cauchy–Frobenius allows us to count orbits with additional properties expressed by weights.

6.1.16 **The Lemma of Cauchy–Frobenius, weighted form** *Consider a finite action $_G X$ and suppose that $w : X \to R$ is a mapping from X into a commutative ring R which contains $\mathbb{Q}$ as a subring. If w is constant on the orbits of G on X, then, for each transversal T of the set of orbits we have*

$$\sum_{t \in T} w(t) = \frac{1}{|G|} \sum_{g \in G} \sum_{x \in X_g} w(x).$$

Proof: The following identities are obvious, possibly up to the final one which uses the fact that w is constant on the orbits:

$$\sum_{g \in G} \sum_{x \in X_g} w(x) = \sum_{x \in X} \sum_{g \in G_x} w(x) = \sum_{x \in X} |G_x| w(x)$$

$$= |G| \sum_{x \in X} |G(x)|^{-1} w(x) = |G| \sum_{t \in T} w(t). \qquad \square$$

If the values $w(f)$ of the weight function are monic monomials over $\mathbb{Q}$, then the values in $\{w(f) \mid f \in Y^X\}$ are linearly independent. Hence, the right hand side of the weighted form of the Lemma of Cauchy–Frobenius yields the number of orbits of any given weight.

For group actions on Y^X of the form 1.4.7 we introduce, for any given mapping $W : Y \to R$, the *multiplicative weight* $w : Y^X \to R$, by

6.1.17
$$w(f) := \prod_{x \in X} W(f(x)).$$

This mapping is clearly constant on the orbits of G on Y^X.

We recall from elementary theory of permutation groups that the permutation $\bar{g} : x \mapsto gx$ induced by $_G X$ possesses a decomposition into pairwise disjoint cycles. If this decomposition consists of $a_i(\bar{g})$ cycles of length i, for $i = 1, \ldots, |X|$, then the sequence

$$(a_1(\bar{g}), a_2(\bar{g}), \ldots, a_{|X|}(\bar{g}))$$

is called the *cycle type* of $\bar{g}$. In other words, $a_i(\bar{g})$ is the number of orbits of length i of the group $\langle \bar{g} \rangle$ on X, i.e.

$$a_i(\bar{g}) = \left| \{ \omega \in \langle \bar{g} \rangle \backslash\backslash X \mid |\omega| = i \} \right| = \left| \{ \omega \in \langle g \rangle \backslash\backslash X \mid |\omega| = i \} \right|.$$

The cycle type of $\bar{g}$ satisfies $\sum_{i=1}^{n} i a_i(\bar{g}) = |X|$, since X is the disjoint union of the cycles of $\langle \bar{g} \rangle$.

An application of the weighted form of the Lemma of Cauchy–Frobenius gives:

Pólya's Theorem *Let $_GX$ be a finite group action which induces according to 1.4.7* **6.1.18**
a group action on the finite set of mappings Y^X. Let R be a commutative ring which
contains $\mathbb{Q}$ as a subring. If T is a transversal of $G\backslash\backslash Y^X$, then for each $W: Y \to R$
and the corresponding multiplicative weight $w: Y^X \to R$ we have

$$\sum_{t\in T} w(t) = \frac{1}{|G|} \sum_{\bar g\in G} \prod_{i=1}^{|X|} \left(\sum_{y\in Y} W(y)^i\right)^{a_i(\bar g)} = \frac{1}{|G|} \sum_{\pi\in \overline{G}} \prod_{i=1}^{|X|} \left(\sum_{y\in Y} W(y)^i\right)^{a_i(\pi)},$$

where $a_i(\bar g)$ or $a_i(\pi)$ is the number of cyclic factors of length i of the permutation
$\bar g \in S_X$ *or* $\pi \in S_X$.   □

The most general multiplicative weight function is obtained by considering
the elements of Y as algebraically independent indeterminates in the polynomial ring $\mathbb{Q}[Y]$. The mapping $W: Y \to \mathbb{Q}[Y]$ which takes $y \in Y$ to itself gives
rise to the multiplicative weight

$$w: Y^X \to \mathbb{Q}[Y] \;:\; f \mapsto \prod_{x\in X} f(x) = \prod_{y\in Y} y^{|f^{-1}(\{y\})|}.$$

The image of f is a monic *monomial* in $\mathbb{Q}[Y]$, which uniquely describes the
content (cf. 6.1.13)

$$c(f): Y \to \mathbb{N} \;:\; y \mapsto \left|f^{-1}(\{y\})\right|$$

of f. The sum of weights of the elements in a transversal T of the orbits is

$$\sum_{t\in T} w(t) = \frac{1}{|G|} \sum_{\bar g\in G} \prod_{i=1}^{|X|} \left(\sum_{y\in Y} y^i\right)^{a_i(\bar g)}.$$

This result can be formulated – as it was already done by G. Pólya – in terms
of the *cycle index* polynomial corresponding to the action $_GX$.

Definition (cycle index polynomial) If G is a finite group acting on a finite set **6.1.19**
X, then the cycle index $C(G, X)$ of the action $_GX$ is the polynomial

$$C(G, X) := \frac{1}{|G|} \sum_{\bar g\in G} \prod_{i=1}^{|X|} z_i^{a_i(\bar g)} \in \mathbb{Q}[z_1, z_2, \dots, z_{|X|}],$$

where $(a_1(\bar g), \dots, a_{|X|}(\bar g))$ is the cycle type of $\bar g$.   ◇

Pólya's Theorem shows that the sum of the weights of the elements of a
transversal can be obtained from the cycle index by replacing the indeterminate z_i by the sum of the i-th powers of all weights, $\sum_{y\in Y} W(y)^i$, i.e.

$$\sum_{t\in T} w(t) = C(G, X)\big|_{z_i:=\sum_y W(y)^i}.$$

Our aim is to evaluate the *generating function* for T_{nkq}. For fixed k and q,
this is the formal power series whose coefficient of x^n is T_{nkq}. For this reason
we still give a short introduction to

6.1.20   **The ring of formal power series over a ring** Let R be an integral domain, then $R[\![x]\!]$, the ring of formal power series over R in the indeterminate x, is given by

$$R[\![x]\!] = \left\{ \sum_{n \geq 0} a_n x^n \;\middle|\; a_n \in R, \, n \in \mathbb{Z}, \, n \geq 0 \right\}.$$

Together with addition and multiplication

$$\sum_{n \geq 0} a_n x^n + \sum_{n \geq 0} b_n x^n := \sum_{n \geq 0} (a_n + b_n) x^n$$

$$\left(\sum_{n \geq 0} a_n x^n \right) \cdot \left(\sum_{n \geq 0} b_n x^n \right) := \sum_{n \geq 0} \left(\sum_{r=0}^{n} a_r b_{n-r} \right) x^n,$$

$R[\![x]\!]$ is an integral domain.

If f is a nonzero formal power series of the form $f = \sum_{n \geq N} a_n x^n \in R[\![x]\!]$ with $a_N \neq 0$, then N is called the *order* of f, for short

$$\mathrm{ord}(f) = N.$$

For technical reasons, we associate the zero series with the order $+\infty$.

A family $(f_j)_{j \in J}$ is called *summable* if for each $n \geq 0$ the cardinality of the index set

$$J_n := \{ j \in J \mid \mathrm{ord}(f_j) \leq n \}$$

is finite. In this case we set

$$\sum_{j \in J} f_j := \sum_{n \geq 0} s_n x^n,$$

where s_n is the coefficient of x^n in the (finite) sum

$$\sum_{j \in J_n} f_j.$$

Finally, if $(a_n)_{n \geq 0}$ is an arbitrary sequence of numbers, then the ordinary generating function for this sequence is given by

$$\sum_{n \geq 0} a_n x^n.$$

Using this, we can now prove the decisive result we need in order to enumerate linear codes:

6.1.21   **Theorem** *Let $_H Y$ be a finite group action. The generating function for the number of $(H \times S_n)$-orbits on Y^n is*

$$\sum_{n \in \mathbb{N}} |(H \times S_n) \backslash\backslash Y^n| \cdot x^n = C(H, Y)\big|_{z_i := \sum_{j=0}^{\infty} x^{i \cdot j}}.$$

For the subset Y^n_{inj} of the injective functions in Y^n we obtain

$$\sum_{n \in \mathbb{N}} |(H \times S_n) \backslash\backslash Y^n_{\mathrm{inj}}| \cdot x^n = C(H, Y)\big|_{z_i := 1 + x^i}.$$

Proof: 1. From Exercise 1.4.9 it follows that

$$(H \times S_n) \backslash\backslash Y^n = H \backslash\backslash (S_n \backslash\backslash Y^n).$$

Thus, according to Exercise 6.1.1, the orbit $(H \times S_n) \backslash\backslash Y^n$ corresponds to the set of H-orbits on the set of mappings $f \in Y^n$ of different content. The content $c(f)$ of $f \in Y^n$ maps $y \in Y$ to $c(f)(y) := |f^{-1}(\{y\})|$, the cardinality of the inverse image of y. It is a decomposition of n into $|Y|$ summands. Such a decomposition can be viewed as a mapping $\varphi \in \mathbb{N}^Y$ such that

$$\sum_{y \in Y} \varphi(y) = n.$$

2. If we now define a weight

$$W : \mathbb{N} \to \mathbb{Q}[x] \ : \ W(n) := x^n,$$

then we obtain the first assertion directly from the following generalization of 6.1.18.

Since the action of H on $\mathbb{N}^Y$ is *not a finite group action*, we need a generalization of Pólya's Theorem. For $\varphi \in \mathbb{N}^Y$ we define the weight $w(\varphi)$ by

$$w(\varphi) := \prod_{y \in Y} W(\varphi(y)) = x^{\sum_{y \in Y} \varphi(y)}.$$

Then φ is the content of some $f \in Y^n$ if and only if $w(\varphi) = x^n$. Thus, the set of S_n-orbits on Y^n is in bijection to the set

$$\mathbb{N}_n^Y := \left\{ \varphi \in \mathbb{N}^Y \ \middle| \ w(\varphi) = x^n \right\}.$$

Moreover, the $(H \times S_n)$-orbits on Y^n correspond to the H-orbits on $\mathbb{N}_n^Y$, where H acts on the domain Y as introduced in 1.4.7. The three families $(x^n)_{n \geq 0}$, $(|(H \times S_n) \backslash\backslash Y^n| \, x^n)_{n \geq 0}$, and $(|\mathbb{N}_n^Y| x^n)_{n \geq 0}$ are summable in $\mathbb{Q}[[x]]$, which is the ring of formal power series in the indeterminate x over $\mathbb{Q}$. Hence,

$$\sum_{n \in \mathbb{N}} x^n = \sum_{n \in \mathbb{N}} W(n), \quad \sum_{n \in \mathbb{N}} |(H \times S_n) \backslash\backslash Y^n| \, x^n, \quad \sum_{n \in \mathbb{N}} |\mathbb{N}_n^Y| x^n$$

exist as elements of $\mathbb{Q}[[x]]$. Since $\mathbb{N}^Y$ is the disjoint union of $\mathbb{N}_n^Y$ for $n \in \mathbb{N}$, the last sum is equal to $\sum_{\varphi \in \mathbb{N}^Y} w(\varphi)$. Moreover, all elements of an orbit $\omega = H(\varphi)$ have the same weight, which allows us to set $w(\omega) := w(\varphi)$. Consequently, we get

$$\sum_{n \in \mathbb{N}} |(H \times S_n) \backslash\backslash Y^n| \, x^n = \sum_{n \in \mathbb{N}} |H \backslash\backslash \mathbb{N}_n^Y| x^n = \sum_{\omega \in H \backslash\backslash \mathbb{N}^Y} w(\omega).$$

Since $(w(\varphi))_{\varphi \in \mathbb{N}^Y}$ is a summable family, also $(w(\varphi))_{\varphi \in (\mathbb{N}^Y)_h}$ is summable for $h \in H$, where $(\mathbb{N}^Y)_h$ is the set of fixed points of h. Moreover, $(|H_\varphi| w(\varphi))_{\varphi \in \mathbb{N}^Y}$

is summable, where H_φ is the stabilizer of φ. Following the ideas of the proof of Pólya's Theorem, we determine the sum of the weights of the fixed points of $h \in H$ in $\mathbb{N}^Y$ as

$$\sum_{\varphi \in (\mathbb{N}^Y)_h} w(\varphi) = \prod_{i=1}^{|Y|} \left(\sum_{n \in \mathbb{N}} W(n)^i \right)^{a_i(\bar{h})},$$

and finally

$$\sum_{\omega \in H \backslash\!\backslash \mathbb{N}^Y} w(\omega) = C(H, Y)\big|_{z_i = \sum_{n \in \mathbb{N}} W(n)^i} = C(H, Y)\big|_{z_i = \sum_{n \in \mathbb{N}} x^{i \cdot n}}.$$

3. The second assertion follows similarly, since the contents of injective functions are decompositions whose summands are either 0 or 1. Thus, instead of $\mathbb{N}^Y$ we consider $\{0, 1\}^Y$, and the weight $W \colon \{0, 1\} \to \mathbb{Q}[x]$ is defined by $W(0) := 1$ and $W(1) := x$. This gives the second assertion about the generating function for the number of orbits of injective functions.   □

We are now in a position to derive the generating functions for the numbers

$$T_{nkq} = |(\mathrm{GL}_k(q) \times S_n) \backslash\!\backslash \mathrm{PG}_{k-1}^*(q)^n|$$

and

$$\overline{T}_{nkq} = |(\mathrm{GL}_k(q) \times S_n) \backslash\!\backslash \mathrm{PG}_{k-1}^*(q)_{\mathrm{inj}}^n|$$

by an application of the last theorem. These numbers are numbers of orbits of the general linear group. As pointed out in Section 3.7, we can restrict our attention to the projective linear group

$$\mathrm{PGL}_k(q) := \mathrm{GL}_k(q) / \mathcal{Z}_k,$$

which is the factor group over the *center* $\mathcal{Z}_k$ of the general linear group. This reduction is possible, since the action of the general linear group is an action on mappings (to be exact, on orbits of mappings), the range of which is the projective space $\mathrm{PG}_{k-1}^*(q)$. It proves

6.1.22   **Corollary** *Since the general linear group $\mathrm{GL}_k(q)$ operates as the projective linear group $\mathrm{PGL}_k(q)$ on the projective space $\mathrm{PG}_{k-1}^*(q)$, we have*

$$T_{nkq} = \mathrm{GL}_k(q) \backslash\!\backslash (S_n \backslash\!\backslash \mathrm{PG}_{k-1}^*(q)^n) = \mathrm{PGL}_k(q) \backslash\!\backslash (S_n \backslash\!\backslash \mathrm{PG}_{k-1}^*(q)^n)$$

and

$$\overline{T}_{nkq} = \mathrm{GL}_k(q) \backslash\!\backslash (S_n \backslash\!\backslash \mathrm{PG}_{k-1}^*(q)_{\mathrm{inj}}^n) = \mathrm{PGL}_k(q) \backslash\!\backslash (S_n \backslash\!\backslash \mathrm{PG}_{k-1}^*(q)_{\mathrm{inj}}^n).$$ □

Using these identities we obtain, by an application of 6.1.21, the following result [61]:

Corollary *The generating functions for the numbers T_{nkq} and $\overline{T}_{nkq}$ can be obtained*   6.1.23
from the cycle index of the natural action of the projective linear group on the projective
space in the following way:

$$\sum_{n\in\mathbb{N}} T_{nkq}x^n = C(\mathrm{PGL}_k(q), \mathrm{PG}^*_{k-1}(q))\big|_{z_i := \sum_{j=0}^{\infty} x^{i\cdot j}},$$

and

$$\sum_{n\in\mathbb{N}} \overline{T}_{nkq}x^n = C(\mathrm{PGL}_k(q), \mathrm{PG}^*_{k-1}(q))\big|_{z_i := 1 + x^i}.$$ □

Example Let us consider isometry classes of binary linear codes. Since the   6.1.24
wreath product $\mathbb{F}_2^* \wr_n S_n$ is isomorphic to the symmetric group S_n, we are faced
with an action of $S_n \times \mathrm{GL}_k(2)$ on $(\mathbb{F}_2^k \setminus \{0\})^n$. In this situation the projective
linear group is simply the linear group, and from 6.1.23 we obtain that

$$\sum_{n=0}^{\infty} T_{nk2}x^n = C(\mathrm{GL}_k(2), \mathbb{F}_2^k \setminus \{0\})\big|_{z_i := \sum_{j=0}^{\infty} x^{i\cdot j}},$$

and

$$\sum_{n=0}^{\infty} \overline{T}_{nk2}x^n = C(\mathrm{GL}_k(2), \mathbb{F}_2^k \setminus \{0\})\big|_{z_i := 1 + x^i}.$$

These cycle indices are known for $q = 2$, see [50], [60], [82], [83], [184], and
programs for their evaluation are implemented in SYMMETRICA ([190]), so
that tables can be determined easily. Comparing Tables 6.2 and 6.1 shows that
the set of isometry classes of (n, k)-codes is much smaller than the set of of all
(n, k)-codes for given parameters n and k.   ◇

If the cycle indices $C(\mathrm{PGL}_k(q), \mathrm{PG}^*_{k-1}(q))$ are known for general q, it is possible to evaluate the numbers T_{nkq} and $\overline{T}_{nkq}$, from which we can deduce V_{nkq},
$\overline{V}_{nkq}$, U_{nkq}, and $\overline{U}_{nkq}$ for arbitrary fields $\mathbb{F}_q$. A method for computing these cycle indices is described in Section 6.3. Finally, in Section 6.4 we present several
tables of these numbers which were calculated using SYMMETRICA (cf. [59]).
They extend the results of D. Slepian on binary codes, see [184]. It is also possible to determine the number of linear isometry classes of linear (n, k)-codes
over $\mathbb{F}_q$ by using the software of the attached CD for moderate parameters n,
k and q.

For later applications to the construction of transversals of isometry classes
of projective codes in Chapter 9 we mention the following two facts: From
Exercise 6.1.2 it follows that

$$S_n \backslash\backslash \mathrm{PG}^*_{k-1}(q)^n_{\mathrm{inj}} = \binom{\mathrm{PG}^*_{k-1}(q)}{n},$$

the set of all n-subsets of $\mathrm{PG}^*_{k-1}(q)$. This implies

$$\mathcal{T}_{n,k,q} = \mathrm{PGL}_k(q) \backslash\backslash \binom{\mathrm{PG}^*_{k-1}(q)}{n}.$$

Exercises

Exercise Let the symmetric group S_n act on the set of mappings Y^n as described in 1.4.7. Show that two mappings $f_1, f_2 \in Y^n$ belong to the same orbit if and only if they are of the same content, i.e.

$$|f_1^{-1}(\{y\})| = |f_2^{-1}(\{y\})| \text{ for all } y \in Y.$$

Exercise Let Y^n_{inj} denote the set of mappings $f \in Y^n$ which are injective, i.e. with $|f^{-1}(\{y\})| \leq 1$, for all $y \in Y$. Show that the S_n-orbits on this set can be represented by n-subsets of Y.

Exercise Let x be an indeterminate over $\mathbb{R}$. Two nonnegative integers n and k define the rational function $\begin{bmatrix} n \\ k \end{bmatrix}$ by

$$\begin{bmatrix} n \\ k \end{bmatrix} := \begin{cases} \dfrac{[n]!}{[k]![n-k]!} & \text{if } k \leq n, \\ 0 & \text{otherwise,} \end{cases}$$

where

$$[0]! := 1, \qquad [n]! := [n][n-1] \cdots [1], \quad n \geq 1,$$

and $[n] = 1 + x + \ldots + x^{n-1}$ for $n \geq 1$. Prove that the number of subspaces of dimension k of $\mathbb{F}_q^n$ is the value of the *Gauss-polynomial* $\begin{bmatrix} n \\ k \end{bmatrix}$ at q:

$$|\mathcal{U}(n,k,q)| = \begin{bmatrix} n \\ k \end{bmatrix}(q) := \left. \frac{(x^n - 1) \cdots (x^{n-k+1} - 1)}{(x^k - 1) \cdots (x - 1)} \right|_{x=q}$$

$$= \frac{(q^n - 1) \cdots (q^{n-k+1} - 1)}{(q^k - 1) \cdots (q - 1)}.$$

The numbers $\begin{bmatrix} n \\ k \end{bmatrix}(q)$ are known as the *q-binomial numbers*. In the notation of Section 3.7, we have $\theta_{n-1}(q) = |\mathcal{U}(n,1,q)| = \begin{bmatrix} n \\ 1 \end{bmatrix}(q) = \frac{q^n-1}{q-1}$.

Exercise Show that the action of $M_n(q)$ on $\mathcal{U}(n,k,q)$ can be restricted to actions on $\mathcal{V}(n,k,q)$, on $\overline{\mathcal{U}}(n,k,q)$, and on $\overline{\mathcal{V}}(n,k,q)$, which means that these subsets of $\mathcal{U}(n,k,q)$ are unions of orbits of the linear isometry group.

Exercise Prove that for finite actions $_G X$, $_H Y$ and the corresponding canonical   E.6.1.5
actions on Y^X the following enumeration formulae hold true:

$$|G \backslash\backslash Y^X| = \frac{1}{|G|} \sum_{g \in G} |Y|^{c(\bar{g})} = C(G, X)\big|_{z_i := |Y|},$$

where $c(\bar{g}) := \sum_i a_i(\bar{g}) = |\langle g \rangle \backslash\backslash X|$ denotes the number of cycles in the cycle
decomposition of the permutation $\bar{g}$, while

$$|(H \times G) \backslash\backslash Y^X| = \frac{1}{|H||G|} \sum_{(h,g) \in H \times G} \prod_{i=1}^{|X|} |Y_{h^i}|^{a_i(\bar{g})} = \frac{1}{|H|} \sum_{h \in H} C(G, X)\big|_{z_i := |Y_{h^i}|}$$

and

$$|H \wr_x G \backslash\backslash Y^X| = C(G, X)\big|_{z_i := |H \backslash\backslash Y|}.$$

Exercise Prove the assertions in 6.1.15. Show that the rank of a matrix corre-   E.6.1.6
sponding to a mapping $\Gamma \in PG^*_{k-1}(q)^n$ does not depend on the choice of the
representatives of the elements $\Gamma(i)$ in $PG^*_{k-1}(q)$. Check that the matrices in
the orbit $(GL_k(q) \times S_n)(\Gamma)$ are all of the same rank.

For $\ell < k$, show that the mapping

$$GL_\ell(q) \to GL_k(q) : A \mapsto \left(\begin{array}{c|c} A & 0 \\ \hline 0 & I_{k-\ell} \end{array} \right),$$

where I_r is the unit matrix of rank r, is an embedding of $GL_\ell(q)$ into $GL_k(q)$.
Consider the natural embedding of $\mathbb{F}_q^\ell$ in $\mathbb{F}_q^k$ given by $v \mapsto (v \mid 0_{k-\ell})$.

If the function $\Gamma \in PG^*_{k-1}(q)^n$ describes a matrix of rank $\ell < k$, find a
function $\Gamma'' \in PG^*_{\ell-1}(q)^n$ which in a natural way can be identified with a suitable element of the orbit $(GL_k(q) \times S_n)(\Gamma)$. Show that all elements of the orbit $(GL_\ell(q) \times S_n)(\Gamma'')$ correspond in the same way to elements of the orbit
$(GL_k(q) \times S_n)(\Gamma)$. (Hint: For finding Γ'', determine by elementary row operations on Γ a matrix Γ' in which the last $k - \ell$ rows consist of zeros only. The
mapping Γ'' can be obtained from Γ' by omitting the last $k - \ell$ entries in each
column.)

In order to prove that for $k > 1$ the number of $GL_k(q) \times S_n$-orbits of mappings $\Gamma \in PG^*_{k-1}(q)^n$ corresponding to matrices of rank not greater than $k - 1$
is equal to $T_{n,k-1,q}$, show that all matrices in the orbit $(GL_k(q) \times S_n)(\Gamma)$ in
which the last row consists of zeros only, i.e. $GL_k(q)(\Gamma) \ni \Gamma' = \left(\dfrac{\Gamma''}{0_n} \right)$

with Γ'' corresponding to a mapping in $PG^*_{k-2}(q)^n$, are of the form

$$\left(\begin{array}{c|c} A & B \\ \hline 0_{k-1} & D \end{array} \right) \cdot \Gamma' \cdot M_\pi = \left(\frac{A \cdot \Gamma'' \cdot M_\pi}{0_n} \right)$$

for some $A \in \mathrm{GL}_{k-1}(q)$, $B \in \mathbb{F}_q^{(k-1) \times 1}$, $D \in \mathrm{GL}_1(q) = \mathbb{F}_q^*$ and a permutation matrix M_π for $\pi \in S_n$. This is the $(\mathrm{GL}_{k-1}(q) \times S_n)$-orbit of Γ''. Thus the $(\mathrm{GL}_k(q) \times S_n)$-orbits of matrices Γ of rank less than k and without zero columns correspond to the $(\mathrm{GL}_{k-1}(q) \times S_n)$-orbits on $\mathrm{PG}_{k-2}^*(q)^n$.

E.6.1.7   **Exercise** Let R be a ring. Consider the set S of all sequences $(r_n)_{n \geq 0}$ with $r_n \in R$ for $n \geq 0$. Prove that this set together with addition and multiplication

$$(r_n)_{n \geq 0} + (s_n)_{n \geq 0} = (r_n + s_n)_{n \geq 0}, \qquad (r_n)_{n \geq 0}, (s_n)_{n \geq 0} \in S,$$

$$(r_n)_{n \geq 0} \cdot (s_n)_{n \geq 0} = (t_n)_{n \geq 0}, \qquad t_n = \sum_{i=0}^{n} r_i s_{n-i}, \qquad (r_n)_{n \geq 0}, (s_n)_{n \geq 0} \in S,$$

is a ring. In addition, show that

— S is commutative if and only if R is commutative,

— S is a ring with 1 if and only if R is a ring with 1,

— S is an integral domain if and only if R is an integral domain.

Now assume that R is an integral domain. Let $s = (r_n)_{n \geq 0}$ be an element of S different from 0. Then $N = \min \{n \geq 0 \mid r_n \neq 0\}$ is called the order of s, in short $\mathrm{ord}(s)$. The order of 0 is defined to be $+\infty$. Show that the mapping

$$d \colon S \times S \to \mathbb{R} \ : \ d(s^{(1)}, s^{(2)}) := \begin{cases} 2^{-\mathrm{ord}(s^{(1)} - s^{(2)})} & \text{if } s^{(1)} \neq s^{(2)}, \\ 0 & \text{if } s^{(1)} = s^{(2)}, \end{cases}$$

is a metric on S. This metric induces a topology on S, the order topology. Prove that a topological basis of the system of neighborhoods of $s^{(0)} \in S$ is given by

$$U_n(s^{(0)}) = \left\{ s \in S \mid \mathrm{ord}(s - s^{(0)}) > n \right\}, \qquad n \in \mathbb{N}.$$

A family $(s^{(n)})_{n \geq 0}$ with $s^{(n)} \in S$ is called summable if the following limit exists with respect to the order topology:

$$\lim_{N \to \infty} \sum_{n=0}^{N} s^{(n)}.$$

Prove that $(s^{(n)})_{n \geq 0}$ is summable in S if and only if $\lim_{n \to \infty} \mathrm{ord}(s^{(n)}) = +\infty$, which is equivalent to $\lim_{n \to \infty} s^{(n)} = 0$.

If $(s^{(n)})_{n \geq 0}$ is a summable family, then we set

$$\sum_{n \geq 0} s^{(n)} = \lim_{N \to \infty} \left(\sum_{n=0}^{N} s^{(n)} \right).$$

We identify the elements r of R with the series $(r, 0, 0, \ldots)$ in S. Consider the particular element $x = (0, 1, 0 \ldots) \in S$. Show that any sequence $(r_n)_{n \geq 0} \in S$ can be written as

$$\sum_{n \geq 0} r_n x^n,$$

where x^n is the n-fold product of x introduced in Exercise 1.6.6. This representation as a sum makes sense, since the family $(r_n x^n)_{n \geq 0}$ is summable.

Finally, we identify S with the ring $R[\![x]\!]$ of formal power series over r in the indeterminate x.

Exercise Prove the following formulae for the order of formal series over an integral domain R. For $f, g \in R[\![x]\!]$ we have $\mathrm{ord}(f + g) \geq \min\{\mathrm{ord}(f), \mathrm{ord}(g)\}$ and $\mathrm{ord}(fg) = \mathrm{ord}(f) + \mathrm{ord}(g)$. We use the convention $+\infty > n$ for all $n \in \mathbb{N}$, $+\infty \leq +\infty$, and $(+\infty) + n = n + (+\infty) = (+\infty) + (+\infty) = +\infty$ for $n \in \mathbb{N}$.

E.6.1.8

6.2 Indecomposable Linear Codes

6.2

The enumerative formulae just derived and the corresponding tables of numbers give us a good idea about the multitude of linear isometry classes of linear codes without zero columns in their generator matrices. But we are mainly interested in the optimal codes, i.e. in the (n, k)-codes with maximal minimum distance d. Hence, we are in fact interested in a small fraction of the total variety of linear isometry classes which we have enumerated. To begin with, we mention that there exists a *Decomposition Theorem* for linear codes. D. Slepian has shown in [184], that every linear code can be decomposed in an essentially unique way into an outer direct sum of *indecomposable* codes, and we recall from Section 2.2, that the minimum distance of an outer direct sum is the least among the minimum distances of its components. This motivates enumeration and the construction of the linear isometry classes of *indecomposable linear codes*, the generator matrices of which do not contain zero columns and whose minimum distance is maximal, for given parameters n, k, q. In this section we restrict our investigations to nonredundant codes.

Definition (indecomposable codes) We call a code *decomposable*, if it is linearly isometric to a code with a generator matrix in the form of a block diagonal matrix

6.2.1

$$\Gamma = \left(\begin{array}{c|c} \Gamma_0 & 0 \\ \hline 0 & \Gamma_1 \end{array} \right) =: \Gamma_0 \dotplus \Gamma_1,$$

consisting of two generator matrices Γ_i of linear (n_i, k_i)-codes with $1 \leq k_i \leq n_i$ for $i \in 2$. Hence, it is linearly isometric to the outer direct sum of at least two codes. Correspondingly, we speak about a *decomposable generator matrix*. Otherwise, both the code and its generator matrix are said to be *indecomposable*.

◇

At first we prove a Decomposition Theorem for linear codes. For this purpose we recall some concepts and facts from Linear Algebra about independent families. We are dealing with finite families S of elements of $\mathbb{F}_q^k$. These are finite sequences $S = (v_i)_{i \in n}$ of vectors $v_i \in \mathbb{F}_q^k$ of length $n \geq 1$. The families $S_0 = (v_{0i})_{i \in n_0}, \ldots, S_{r-1} = (v_{r-1,i})_{i \in n_{r-1}}$ in $\mathbb{F}_q^k$ are called *independent* if an equation of the form

$$\sum_{i \in r} \sum_{j \in n_i} \alpha_{ij} v_{ij} = 0, \qquad \alpha_{ij} \in \mathbb{F}_q,$$

always implies that

$$\sum_{j \in n_i} \alpha_{ij} v_{ij} = 0 \ \text{ for } \ i \in r.$$

In other words, there are no linear relations between vectors of different independent families.

The proof of the next lemma is left to the reader.

6.2.2   **Lemma**   *If $S_0, \ldots, S_{r-1}$ are independent families in $\mathbb{F}_q^k$, and if R_i are nonempty subfamilies of S_i, then also $R_0, \ldots, R_{r-1}$ are independent families in $\mathbb{F}_q^k$.*   □

A family $S = (v_i)_{i \in I}$ in $\mathbb{F}_q^k$ is called *indecomposable*, if it cannot be expressed as the union of at least two (nonempty) independent subfamilies $(v_i)_{i \in I'}$ and $(v_i)_{i \in I''}$ where I is the disjoint union $I' \cup I''$. Otherwise, S is called *decomposable*. In the sequel, we want to prove that any decomposable sequence can be decomposed uniquely into the union of indecomposable subfamilies. For doing this, we need some notions about linear combinations.

Let $S = (v_i)_{i \in n}$ be a family in $\mathbb{F}_q^k$. A linear combination

$$\sum_{i \in n} \alpha_i v_i, \qquad \alpha_i \in \mathbb{F}_q,$$

is called *irreducible*, if there does not exist a proper partial sum (consisting of at least one and at most $n - 1$ summands) which yields zero. Otherwise, the linear combination is called *reducible*.

6.2.3   **Lemma**   *Let S be a family of vectors in $\mathbb{F}_q^k$. Any reducible linear combination of vectors of S which yields zero can be decomposed into a sum of irreducible linear combinations.*

Proof: If the linear combination

$$\sum_{i \in n} \alpha_i v_i = 0$$

is reducible, then there exist partial sums which also yield zero. Assume that

$$\alpha_{i_0} v_{i_0} + \ldots + \alpha_{i_{r-1}} v_{i_{r-1}} = 0$$

is such a partial sum of minimal length. Then this partial sum is irreducible. Moreover,

$$\sum_{i \in n} \alpha_i v_i - \sum_{j \in r} \alpha_{i_j} v_{i_j} = 0$$

is also a partial sum which yields zero. Either it is also irreducible, or we can repeat the procedure just described, in order to obtain, after a finite number of steps, the desired decomposition into irreducible linear combinations.   □

Using the sequence $S = (v_i)_{i \in n}$, we can form $q^n - 1$ different linear combinations such that not all coefficients α_i are equal to zero. Omitting all those linear combinations which do not yield the value zero and also those which are reducible, we end up with a finite list $\mathcal{L}$ of irreducible linear combinations which sum up to zero.

Two vectors v_i and v_j from S are called *directly connected*, if there exists a linear combination in $\mathcal{L}$ with coefficients $\alpha_i \neq 0 \neq \alpha_j$. A vector of S which does not occur in any of the linear combinations in $\mathcal{L}$ is called *directly connected with itself*. Two vectors v_i and v_j from S are called *connected*, if there exists an integer $m \geq 0$ and a sequence $(v_{i_0}, v_{i_1}, \ldots, v_{i_m})$ of vectors in S such that $i = i_0$, $j = i_m$ and v_{i_r} is directly connected with $v_{i_{r+1}}$ for $r \in m$. In order to indicate that v_i and v_j are connected we write $v_i \sim v_j$ and also $i \sim j$. (When v is directly connected with itself we also write $v \sim v$.)

Lemma *Let $S = (v_i)_{i \in n}$ be a family in $\mathbb{F}_q^k$.*   6.2.4

1. *The relation $\sim$, introduced above, is an equivalence relation on the set of vectors v_i for $i \in n$. The equivalence class of v_i corresponds to the subfamily $(v_j)_{j \sim i}$.*

2. *The family $(v_j)_{j \sim i}$ is indecomposable.*

3. *Let $\{v_i \mid i \in I'\}$ be a complete set of representatives with respect to $\sim$. Then the families $(v_j)_{j \sim i}$ for $i \in I'$ are independent.*

4. *If R is an indecomposable family in S, then there exists exactly one $i \in I'$ such that R is a subfamily of $(v_j)_{j \sim i}$.*

Proof: The proof of the first part is obvious. If we suppose that $(v_j)_{j \sim i}$ is decomposable, then there exist two nonempty, disjoint sets I' and I'' such that $I' \cup I'' = \{j \mid j \sim i\}$ and $(v_j)_{j \in I'}$ and $(v_j)_{j \in I''}$ are independent families. Choose $j_1 \in I'$ and $j_2 \in I''$. Since $v_{j_1} \sim v_{j_2}$, there exists a sequence $v_{j_1} = v_{i_0} \sim \ldots \sim v_{i_m} = v_{j_2}$ such that v_{i_r} is directly connected with $v_{i_{r+1}}$ for $r \in m$. From the special choice of j_1 and j_2 in I' and I'', respectively, we derive the existence of at least one index r such that i_r belongs to I' and i_{r+1} belongs to I''. Then v_{i_r} and $v_{i_{r+1}}$ are directly connected, which is a contradiction to the fact that they belong to two independent families. Consequently, $(v_j)_{j \sim i}$ is indecomposable.

In order to prove the third assertion, assume that

$$\sum_{i \in I'} \sum_{j \sim i} \alpha_{ij} v_j = 0$$

is a linear combination, which contains vectors from at least two different families $(v_j)_{j \sim i}$ with nonzero coefficients. Then this linear combination is not irreducible, since otherwise vectors of different equivalence classes would be directly connected. According to 6.2.3, this reducible linear combination can be written as a sum of irreducible linear combinations, each of which is zero. Since they are irreducible, none of these linear combinations contains vectors from different equivalence classes. Forming the sum of all irreducible linear combinations containing vectors from $(v_j)_{j \sim i}$ we get

$$\sum_{j \sim i} \alpha_{ij} v_j = 0$$

for each $i \in I'$.

Assume that $R = (v_i)_{i \in J}$ for $J \subseteq n$ is an indecomposable subfamily of S. For $i \in I'$ let $R_i = (v_j)_{j \in J, \, j \sim i}$. We have just proved that $(v_j)_{j \sim i}$ for $i \in I'$ are independent families. Then there is exactly one $i_0 \in I'$ such that R_{i_0} is not empty. If we suppose on the contrary that there are at least two nonempty families, then, according to 6.2.2, they are also independent families. Hence, R is the union of at least two independent families, which is a contradiction to the assumption that R is indecomposable. This finishes the proof of the last assertion. □

Based on these results we prove the next

6.2.5   Theorem *A finite family S of vectors in $\mathbb{F}_q^k$ can be written in a unique way as the union of independent, indecomposable sets.*

Proof: According to 6.2.4, we obtain a decomposition of S into independent, indecomposable families by determining the equivalence classes $(v_j)_{j \sim i}$ for $i \in I'$.

Conversely, consider a decomposition of S into independent, indecomposable families R_k for k in an index set K. From the last statement of 6.2.4 we deduce that for each $k \in K$ there exists exactly one $i \in I'$ such that R_k is a subfamily of $(v_j)_{j \sim i}$. Moreover, since the family $(v_j)_{j \sim i}$ is indecomposable, R_ℓ is not a subfamily of $(v_j)_{j \sim i}$ for $\ell \neq k$. Hence, the indecomposable families R_k correspond in a unique way to the independent, indecomposable families $(v_j)_{j \sim i}$ for $i \in I'$. $\square$

Remark Let S denote the family of columns of a generator matrix Γ of an (n, k)-code C. Then C is indecomposable if and only if S is indecomposable. This characterization is, first of all, independent of the choice of a generator matrix Γ of C, since the columns of $A \cdot \Gamma$, for $A \in \mathrm{GL}_k(q)$, satisfy the same dependency relations as the columns of Γ. Secondly, this characterization is independent of the choice of the representative C of its linear isometry class, since a linear isometry permutes the columns of Γ and multiplies them by nonzero elements of $\mathbb{F}_q^*$. (See Exercise 6.2.1.) $\diamond$

6.2.6

We are now in a position to prove Slepian's Theorem:

The Decomposition Theorem for Linear Codes *Any (n,k)-code C over $\mathbb{F}_q$ is linearly isometric to an outer direct sum of indecomposable codes C_i:*

6.2.7

$$C \simeq C_0 + \ldots + C_{r-1}.$$

This decomposition is unique in the following sense. If we are given another decomposition of C of the form

$$C \simeq C_0' + \ldots + C_{r'-1}'$$

with indecomposable codes C_i', then $r = r'$ and there exists a permutation $\sigma \in S_r$ so that C_i and $C_{\sigma(i)}'$ are linearly isometric.

Proof: We only have to prove the uniqueness of such a decomposition. Assume that C is linearly isometric to two decompositions, say,

$$C_0 + \ldots + C_{r-1}$$

and

$$C_0' + \ldots + C_{r'-1}'$$

with indecomposable (n_i, k_i)-codes C_i and indecomposable (n_i', k_i')-codes C_i' with generator matrices Γ_i and Γ_i', respectively. The parameters n_i, n_i', k_i, and k_i' satisfy the equations

$$\sum_{i \in r} n_i = n = \sum_{i \in r'} n_i' \quad \text{and} \quad \sum_{i \in r} k_i = k = \sum_{i \in r'} k_i'.$$

6.2.8

By assumption there exist matrices $A \in \mathrm{GL}_k(q)$ and $B \in M_n(q)$ such that

6.2.9
$$A \cdot \Gamma' \cdot B = \Gamma$$

such that
$$\Gamma := \Gamma_0 \dotplus \ldots \dotplus \Gamma_{r-1}$$
and
$$\Gamma' := \Gamma'_0 \dotplus \ldots \dotplus \Gamma'_{r'-1}.$$

The columns of Γ decompose into indecomposable families $S_0, \ldots, S_{r-1}$, where S_0 consists of the first n_0 columns of Γ, S_1 of the next n_1 columns, and so on. According to 6.2.9, the columns of $A \cdot \Gamma' \cdot B = \Gamma$ and $\Gamma' \cdot B$ satisfy the same dependency relations. Hence, the first n_0 columns of $\Gamma' \cdot B$ form an independent set $\tilde{S}_0$, the following n_1 an independent set $\tilde{S}_1$, and so on.

On the other hand, $\Gamma' \cdot B$ arises from Γ' by reordering the columns and multiplying the columns by elements of $\mathbb{F}_q^*$. Hence after some permutation, the columns of $\Gamma' \cdot B$ satisfy the same dependency relations as the columns of Γ'. But the columns of Γ' decompose into r' independent families which are given by the decomposition of Γ'. The first n'_0 columns form an independent set S'_0, the following n'_1 an independent set S'_1, and so on. From 6.2.5 we deduce that $r = r'$. Moreover, there exists a permutation $\sigma \in S_r$, such that for $i \in r$ the lengths $n'_{\sigma(i)}$ and n_i of the indecomposable families $S'_{\sigma(i)}$ and $\tilde{S}_i$ coincide, and the family $\tilde{S}_i$ consists – up to scalar multiples – of those columns of Γ' which contain the submatrix $\Gamma'_{\sigma(i)}$. Thus, $\Gamma' \cdot B$ can be written in the form $A' \cdot \Gamma''$, where A' is a suitable permutation matrix in $\mathrm{GL}_k(q)$ and Γ'' is given by

$$\Gamma'' = (\Gamma'_{\sigma(0)} \cdot B_0) \dotplus \ldots \dotplus (\Gamma'_{\sigma(r-1)} \cdot B_{r-1}),$$

for suitable matrices $B_i \in M_{n_i}(q)$. Finally, if we put $A'' := A \cdot A' \in \mathrm{GL}_k(q)$, then

6.2.10
$$A'' \cdot \Gamma'' = A \cdot \Gamma' \cdot B = \Gamma.$$

Let T_0 be the matrix consisting of the first n_0 columns of Γ, T_1 the matrix, consisting of the next n_1 columns, and so on. Analogously, we define the matrices T''_i as submatrices of Γ''. From this construction it follows immediately that T_i is a matrix of rank k_i and T''_i is of rank $k'_{\sigma(i)}$. Since $T_i = A'' \cdot T''_i$ and A'' is regular, we deduce that $k'_{\sigma(i)} \geq k_i$. This, together with 6.2.8, gives that $k'_{\sigma(i)}$ equals k_i. If we write the matrix A'' as block matrix $(A''_{ij})_{i,j \in r}$, consisting of blocks A''_{ij}, which are $k_i \times k_j$-matrices, from 6.2.10 we obtain

$$A''_{ii} \cdot \Gamma'_{\sigma(i)} \cdot B_i = \Gamma_i, \qquad i \in r.$$

Comparing the degrees we obtain that the diagonal blocks A''_{ii} are all regular. Hence, $\Gamma'_{\sigma(i)}$ and Γ_i are generator matrices of linearly isometric codes C_i and $C'_{\sigma(i)}$. $\square$

Using the notation introduced in Exercise 2.3.17, the last theorem can be restated for linear isometry classes of linear codes as:

Corollary *The linear isometry class $\hat{C}$ of any linear code C over $\mathbb{F}_q$ can be expressed*   **6.2.11**
as an outer direct sum of the linear isometry classes $\hat{C}_i$ of indecomposable codes C_i:

$$\hat{C} = \hat{C}_0 \dotplus \ldots \dotplus \hat{C}_{r-1}.$$

The indecomposable summands $\hat{C}_i$ are uniquely determined by $\hat{C}$ apart from their order. □

Another consequence is the following cancellation law:

Corollary   *Let $\hat{C}_0$, $\hat{C}_1$ and $\hat{C}_2$ be linear isometry classes of linear codes. From*   **6.2.12**
$\hat{C}_0 \dotplus \hat{C}_1 = \hat{C}_0 \dotplus \hat{C}_2$ *we obtain that* $\hat{C}_1 = \hat{C}_2$. □

For systematic linear codes there is an easy and obvious

Test on Indecomposability *A generator matrix $\Gamma = (I_k \mid A)$ of a linear (n,k)-code*   **6.2.13**
with $k < n$ is (together with the generated code) indecomposable if and only if there
exists a sequence $a_{ij}, a_{lm}, \ldots$ of nonzero entries in A such that each element (except
the first one, of course) lies in the same row or in the same column as its predecessor,
and so that each row is represented by at least one element of the sequence.

Proof: Because of the special form of Γ, the first k columns of Γ define k independent families. Each of these families consists of just that column. The remaining columns of Γ, i.e. the columns of A, can be represented as linear combinations of the first k columns. Moreover, the columns of Γ form an indecomposable family if and only if the first k columns are connected. This implies the statement. □

We can also represent the elements of A as the vertices of a graph $\mathcal{G}_A$. In this graph two vertices are connected by an edge, if they are both different from 0 and occur either in the same row or column of A. Then the code C is indecomposable, if and only if there is a walk in $\mathcal{G}_A$ which visits each of the k rows at least once.

In case $n = k$, this theorem does not apply. It is clear that (n,n)-codes are indecomposable if and only if $n = 1$.

If the codes do not have zero columns (as we assumed in this section), and if there exists a walk in $\mathcal{G}_A$ which visits each of the k rows of A at least once, then there exists a walk in $\mathcal{G}_A$ which visits all columns of A. With this characterization it is easy to prove

6.2.14   **Theorem** *A nonredundant linear code C is indecomposable if and only if its dual code $C^\perp$ is indecomposable.*   □

6.2.15   **Examples**

1. The code with generator matrix

$$
\Gamma = \begin{pmatrix} 1 & 0 & 0 & 1 & 1 & 1 \\ 0 & 1 & 0 & 0 & 0 & 1 \\ 0 & 0 & 1 & 0 & 0 & 1 \end{pmatrix}
$$

   is indecomposable, since the sequence $\gamma_{05}, \gamma_{15}, \gamma_{25}$ is a sequence of entries of the last $n - k = 3$ columns of Γ which has the required properties.

2. Any nonredundant $(n, 1)$-code is indecomposable.

3. Any (n, k)-MDS-code with $k < n$ is indecomposable.   ◇

Indecomposable codes are optimal in the following sense.

6.2.16   **Theorem** *Let C be an (n, k)-code with $k < n$ and with minimum distance d. Then there exists an indecomposable (n, k)-code C' such that $\mathrm{dist}(C') \geq d$.*

Proof: For $r \geq 2$, let $C \simeq C_0 + \ldots + C_{r-1}$ be a decomposable code, where C_i are (n_i, k_i, d_i)-codes. From the properties of the outer direct sum (cf. 2.2.11) it follows that $\mathrm{dist}(C) = \min\{d_i \mid i \in r\}$.

By induction on r we prove the assertion of the theorem: If $r = 2$, we consider the following generator matrix Γ of C:

$$
\Gamma = \left(\begin{array}{c|c|c|c} I_{k_0} & A_0 & 0 & 0 \\ \hline 0 & 0 & I_{k_1} & A_1 \end{array} \right)
$$

with $(n_i - k_i) \times k_i$-matrices A_i. Without restriction we suppose that $\mathrm{dist}(C) = \mathrm{dist}(C_0) \leq \mathrm{dist}(C_1)$. If $k_1 < n_1$, then 6.2.13 shows that the matrix

6.2.17 $$
\Gamma' := \left(\begin{array}{c|c|c|c} I_{k_0} & A_0 & 0 & B \\ \hline 0 & 0 & I_{k_1} & A_1 \end{array} \right) \quad \text{with } B := \begin{pmatrix} 1 & \cdots & 1 \\ 0 & \cdots & 0 \\ \vdots & & \vdots \\ 0 & \cdots & 0 \end{pmatrix}
$$

generates an indecomposable code C'. Let v denote a nontrivial linear combination of the rows of Γ'. Unless the first k_0 entries of v are all zero, we have $\mathrm{wt}(v) \geq \mathrm{dist}(C_0)$ since the first k_0 entries of v are a codeword in C_0. If the first k_0 entries of v are all zero then the second half of v is a nonzero codeword in C_1, whence $\mathrm{wt}(v) \geq \mathrm{dist}(C_1)$. Therefore, the minimum distance of C' is at least $\mathrm{dist}(C)$.

If $k_1 = n_1$, we have $n_1 = 1$, since C_1 was supposed to be indecomposable. Hence, $1 = \mathrm{dist}(C_1) \geq \mathrm{dist}(C) \geq 1$. But every indecomposable (n, k)-code has $d \geq 1$, and so the theorem is proved for the case $r = 2$.

Now we assume that $r > 2$. The induction assumption gives the existence of an indecomposable $(n - n_{r-1}, k - k_{r-1})$-code C' with

$$\mathrm{dist}(C') \geq \mathrm{dist}(C_0 + \ldots + C_{r-2}) = \min\{\mathrm{dist}(C_0), \ldots, \mathrm{dist}(C_{r-2})\}.$$

Moreover, it implies the existence of an indecomposable (n, k)-code C'' with

$$\mathrm{dist}(C'') \geq \mathrm{dist}(C' + C_{r-1}) = \min\{\mathrm{dist}(C'), \mathrm{dist}(C_{r-1})\}$$
$$\geq \min\{\min\{\mathrm{dist}(C_0), \ldots, \mathrm{dist}(C_{r-2})\}, \mathrm{dist}(C_{r-1})\} = \mathrm{dist}(C). \quad \square$$

Theorem *Any indecomposable code of length greater than 1 has minimum distance at least* 2. $\square$ **6.2.18**

Theorem *Up to linear isometry, for any field $\mathbb{F}_q$ and $n > 2$ there is a unique indecomposable $(n, n-1)$-code C over $\mathbb{F}_q$. It has a generator matrix of the form* **6.2.19**

$$\begin{pmatrix} 1 & & & 1 \\ & \ddots & & \vdots \\ & & 1 & 1 \end{pmatrix}.$$

Therefore, C is linearly isometric to the parity check code of $\mathbb{F}_q^{n-1}$. It is also linearly isometric to the dual of a one-dimensional code generated by the all-one vector. If $q = 2$, then C is the set of all vectors of $\mathbb{F}_2^n$ which have even weight.

Proof: Since C is indecomposable and of length greater than 1, by 6.2.18, its minimum distance is at least 2. By the Singleton-bound 2.1.1, it is at most 2, thus $\mathrm{dist}(C) = 2$. There exists a code linearly isometric to C with generator matrix $\Gamma = (I_{n-1} \mid A)$ where A is an $(n-1) \times 1$-matrix. Since the rows of Γ are codewords of weight not smaller than $\mathrm{dist}(C)$, each component of A is different from 0. By a suitable monomial transformation, there exists a code linearly isometric to C which has a generator matrix of the form $(I_{n-1} \mid A')$ where all components of A' are equal to 1. $\square$

We are now going to show how indecomposable codes can be enumerated. For this purpose, we introduce the following sets and symbols for their cardinalities:

— Let $\mathcal{R}_{nkq}$ denote the set of linear isometry classes of nonredundant, indecomposable (n,k)-codes over $\mathbb{F}_q$,

$$\mathcal{R}_{nkq} := \left\{ M_n(q)(C) \in \mathcal{V}_{nkq} \mid C \text{ is indecomposable} \right\}.$$

— $R_{nkq} := |\mathcal{R}_{nkq}|$ indicates the number of linear isometry classes of nonredundant, indecomposable (n,k)-codes over $\mathbb{F}_q$.

— The symbol $\overline{\mathcal{R}}_{nkq}$ denotes the set of linear isometry classes of (nonredundant), indecomposable, projective (n,k)-codes over $\mathbb{F}_q$, i.e.

$$\overline{\mathcal{R}}_{nkq} := \left\{ M_n(q)(C) \in \overline{\mathcal{V}}_{nkq} \mid C \text{ is indecomposable} \right\}.$$

— $\overline{R}_{nkq} := |\overline{\mathcal{R}}_{nkq}|$ indicates the number of linear isometry classes of (nonredundant), indecomposable, projective (n,k)-codes over $\mathbb{F}_q$.

From 6.2.19 it follows immediately that $R_{21q} = 1$, $\overline{R}_{21q} = 0$, and $R_{n,n-1,q} = 1 = \overline{R}_{n,n-1,q}$ for $n > 2$. Moreover, we already know $R_{11q} = 1 = \overline{R}_{11q}$, $R_{nnq} = 0 = \overline{R}_{nnq}$ for $n > 1$, $R_{n1q} = 1$ for $n \geq 1$, and $\overline{R}_{n1q} = 0$ for $n \geq 2$. The following theorem (cf. [61]) gives a recursive procedure for the evaluation of the numbers R_{nkq} and $\overline{R}_{nkq}$ from V_{nkq} and $\overline{V}_{nkq}$, respectively.

6.2.20   **Theorem** *For $n \geq 2$ we have*

$$R_{nkq} = V_{nkq} - \sum_a \sum_b \prod_{\substack{j=1 \\ a_j \neq 0}}^{n-1} \left(\sum_c U(c) \right),$$

where

6.2.21
$$U(c) = \prod_{i=1}^{j} C(S_{\nu(i,c)}, \nu(i,c))\Big|_{z_\ell = R_{jiq}}$$

is a product computed from substitutions into the cycle indices of symmetric groups of degree $\nu(i,c)$ for

$$\nu(i,c) = |\{\ell \in a_j \mid c_\ell = i\}|, \qquad 1 \leq i \leq j.$$

The first sum runs through the cycle types $a = (a_1, \ldots, a_{n-1})$ of n with at least two summands, i.e. $a_i \in \mathbb{N}$ and $\sum i a_i = n$, and with the additional property $\sum a_i \leq k$, whereas the second sum is taken over the $(n-1)$-tuples $b = (b_1, \ldots, b_{n-1}) \in \mathbb{N}^{n-1}$, for which $a_i \leq b_i \leq i a_i$, and $\sum b_i = k$. The third sum runs over all a_j-tuples $c = (c_0, \ldots, c_{a_j-1}) \in \mathbb{N}^{a_j}$ satisfying $j \geq c_0 \geq \ldots \geq c_{a_j-1} \geq 1$ and $\sum c_i = b_j$.

Analogously, $\overline{R}_{nkq}$ can be evaluated recursively from $\overline{V}_{nkq}$ and $\overline{R}_{jiq}$ with $j < n$.

We would like to remark that the numbers $U(c)$ in 6.2.21 are expressed solely in terms of cycle indices of symmetric groups in their natural action (see Exercise 6.3.3).

Proof: In order to obtain R_{nkq}, we have to subtract from V_{nkq} the number of all classes of nonredundant, decomposable (n,k)-codes over $\mathbb{F}_q$. In other words, we have to evaluate the number of isometry classes of (n,k)-codes which can be written as a direct sum of indecomposable (n_i,k_i)-codes where

$$\sum_{i\in r} n_i = n, \quad \sum_{i\in r} k_i = k, \quad 1 \le k_i \le n_i, \quad 2 \le r \le k. \qquad \textbf{6.2.22}$$

According to 6.2.7, the (n_i,k_i)-codes in a decomposition can be arranged so that $n_0 \ge n_1 \ge \ldots \ge n_{r-1}$ holds true, and, if successive n_i are equal, for example $n_i = n_{i+1}$, then we can assume, in addition, that the inequality $k_i \ge k_{i+1}$ is satisfied. In order to describe all decompositions, first we list all partitions of n into at least two but not more than k parts. Hence, we suppose that $n = n_0 + n_1 + \ldots + n_{r-1}$ is a partition with $n_0 \ge \ldots \ge n_{r-1} \ge 1$ and $2 \le r \le k$. Its type is of the form $(a_1, a_2, \ldots, a_{n-1})$ with $a_j := |\{i \mid i \in r, \, n_i = j\}|$. Decomposable codes corresponding to different types $(a_1, a_2, \ldots, a_{n-1})$ are not linearly isometric.

In a second step we calculate for each such partition of n all sequences $(k_0, \ldots, k_{r-1})$ satisfying 6.2.22. If we are given such a sequence $(k_0, \ldots, k_{r-1})$, we put

$$b_j := \sum_{i:n_i=j} k_i, \quad 1 \le j \le n-1.$$

Then

$$\sum_{j=1}^{n-1} b_j = \sum_{i\in r} k_i = k \quad \text{and} \quad a_j \le b_j \le j \cdot a_j. \qquad \textbf{6.2.23}$$

Decomposable codes corresponding to the same type $(a_1, a_2, \ldots, a_{n-1})$ which give rise to different vectors b are not linearly isometric. Conversely, we can start with any sequence $(b_1, \ldots, b_{n-1})$ satisfying 6.2.23 and evaluate all sequences $(k_0, \ldots, k_{r-1})$ with $b_j = \sum_{i:n_i=j} k_i$ which give linearly nonisometric codes with parameters (n_i, k_i) for $i \in r$. According to 6.2.7, for each j with $b_j \ne 0$ (which implies $a_j \ne 0$) we have to determine all partitions of b_j into exactly a_j parts of the following form:

$$b_j = \sum_{i\in a_j} c_i, \quad j \ge c_0 \ge \ldots \ge c_{a_j-1} \ge 1. \qquad \textbf{6.2.24}$$

These sequences c describe all possible ways of writing a $(j \cdot a_j, b_j)$-code as the outer direct sum of a_j codes of length j and dimension c_i for $i \in a_j$. Codes with different sequences are clearly not isometric.

In a final step we have to evaluate the number of linearly nonisometric decomposable $(j \cdot a_j, b_j)$-codes which are outer direct sums of a_j codes of length j. For each partition c of b_j with the properties 6.2.24 let $U(c)$ be the number of linearly nonisometric $(j \cdot a_j, b_j)$-codes which are the outer direct sum of indecomposable (j, c_i)-codes for $i \in a_j$.

We may assume that during the recursive procedure for the evaluation of the R_{nkq}, the numbers $R_{j,c_i,q}$ for $j < n$ have already been computed. If all components c_i of c are pairwise different, then the number $U(c)$ is equal to the product

6.2.25
$$\prod_{i \in a_j} R_{j,c_i,q},$$

which is a special case of 6.2.21. (See Exercise 6.2.8.)

Otherwise, there exist s, t with $s < t$ and $c_s = c_t$. Since $c_s = c_{s+1} = \ldots = c_t$, and according to 6.2.7, any permutation of the summands with the same parameters in a given direct decomposition into indecomposable codes leads to linearly isometric codes. Hence, for $1 \le i \le j$ let $v(i) := v(i,c)$ denote the cardinality of the set $\{\ell \in a_j \mid c_\ell = i\}$. Obviously, there is a bijection between the classes of codes which are outer direct sums of $v(i)$ indecomposable (j,i)-codes and the orbits of the symmetric group $S_{v(i)}$ acting on the set of all mappings from $v(i)$ into a set of R_{jiq} elements. In this case, the symmetric group acts canonically on the set of these mappings:

$$S_{v(i)} \times R_{jiq}^{v(i)} \to R_{jiq}^{v(i)} : (\pi, f) \mapsto f \circ \pi^{-1}.$$

A combination of Pólya's Theorem and the result of Exercise 6.1.5 completes the proof that $U(c)$ is given by 6.2.21.

Since $U(c)$ is the number of decomposable $(j \cdot a_j, b_j)$-codes which are an outer direct sum of indecomposable (j, c_i)-codes for $i \in a_j$, we can determine the number of all decomposable $(j \cdot a_j, b_j)$-codes which are the outer direct sum of a_j indecomposable codes of length j, by summing over all sequences c satisfying 6.2.24.

By summing these numbers over all cycle types $(a_1, \ldots, a_{n-1})$ of n with $\sum a_i = k$, and over all sequences b with the properties 6.2.23, we compute the number of all linearly nonisometric, nonredundant, decomposable (n,k)-codes over $\mathbb{F}_q$. It must be subtracted from V_{nkq} in order to obtain the number of all linearly nonisometric, nonredundant, indecomposable (n,k)-codes over $\mathbb{F}_q$. $\square$

In Section 6.4 we present tables of R_{nkq} and $\overline{R}_{nkq}$ which were computed by using SYMMETRICA. They can also be determined with the software included on the attached CD. In the case $q = 2$, these tables confirm (and in some parts also correct) the numbers given by D. Slepian in [184]. Moreover, these numbers lead to the conjecture that the sequences $(R_{nkq})_{1 \le k < n}$ are unimodal and symmetric for fixed n and q. The symmetry follows directly from the fact that the dual of an indecomposable code is again indecomposable (cf. Exercise 6.2.9). However, the unimodality has not yet been proved (see [61]).

For fixed n and q, the sequences R_{nkq} are symmetric, i.e.

$$R_{nkq} = R_{n,n-k,q}, \qquad 1 \le k \le \lfloor n/2 \rfloor.$$

Therefore, it is possible to use the formula from 6.2.20 in order to compute further values of V_{nkq}. Let n_0 be a positive integer and q the cardinality of a field. At first we compute the numbers V_{nkq} for $1 \leq n \leq n_0$ and $1 \leq k \leq \lfloor n_0/2 \rfloor$ as described in the previous section. This allows us to determine the numbers R_{nkq} for $1 \leq n \leq n_0$ and $1 \leq k \leq \lfloor n_0/2 \rfloor$. For $1 \leq n \leq n_0$ and $\lfloor n_0/2 \rfloor < k \leq n_0$ we determine the missing numbers R_{nkq} either by symmetry (for $k < n$) or by setting $R_{nkq} = 0$ for $k \geq n$. From 6.2.20 we immediately obtain the following formula

$$ V_{nkq} = R_{nkq} + \sum_a \sum_b \prod_{\substack{j=1 \\ a_j \neq 0}}^{n-1} \left(\sum_c U(c) \right), $$

which allows us to compute the missing values V_{nkq} for $1 \leq n \leq n_0$ and $\lfloor n_0/2 \rfloor < k \leq n_0$.

Example Let $n_0 = 12$ and $q = 2$. From Table 6.21 on page 508 we obtain the numbers R_{nk2} for $1 \leq n \leq 12$ and $1 \leq k \leq 6$. Now we determine the values R_{nk2} for $7 \leq k \leq 12$ as shown in Table 6.3 on the left hand side. This allows the computation of the values V_{nk2} for $1 \leq n \leq 12$ and $7 \leq k \leq 12$, shown in the right hand side of Table 6.3, without determining the cycle indices of $\mathrm{PGL}_k(2)$ for $7 \leq k \leq 12$. $\diamond$

6.2.26

Table 6.3 Extending tables by using the symmetry of R_{nk2}

			R_{nk2}							V_{nk2}			
$n\backslash k$	7	8	9	10	11	12	$n\backslash k$	7	8	9	10	11	12
1	0	0	0	0	0	0	1	0	0	0	0	0	0
2	0	0	0	0	0	0	2	0	0	0	0	0	0
3	0	0	0	0	0	0	3	0	0	0	0	0	0
4	0	0	0	0	0	0	4	0	0	0	0	0	0
5	0	0	0	0	0	0	5	0	0	0	0	0	0
6	0	0	0	0	0	0	6	0	0	0	0	0	0
7	0	0	0	0	0	0	7	1	0	0	0	0	0
8	1	0	0	0	0	0	8	7	1	0	0	0	0
9	7	1	0	0	0	0	9	35	8	1	0	0	0
10	51	8	1	0	0	0	10	170	47	9	1	0	0
11	361	79	10	1	0	0	11	847	277	61	10	1	0
12	2484	754	121	12	1	0	12	4408	1775	436	78	11	1

Exercises

E.6.2.1   **Exercise** Let Γ be a generator matrix of an (n,k)-code over $\mathbb{F}_q$ and let M be a monomial matrix in $M_n(q)$. Discuss the relations between the linear dependencies occurring between the columns of Γ and between the columns of $\Gamma \cdot M$.

E.6.2.2   **Exercise** Prove 6.2.11.

E.6.2.3   **Exercise** Find a proof of 6.2.12.

E.6.2.4   **Exercise** Use Exercise 1.3.9 in order to prove that 6.2.14 is true.

E.6.2.5   **Exercise** Prove that any (n,k)-MDS-code with $k < n$ is indecomposable.

E.6.2.6   **Exercise** Show that the code which is generated by the matrix in 6.2.17 is indecomposable.

E.6.2.7   **Exercise** Prove 6.2.18.

E.6.2.8   **Exercise** Prove that 6.2.25 is a special case of 6.2.21.

E.6.2.9   **Exercise** Prove that $R_{nkq} = R_{n,n-k,q}$ is true for $1 \leq k \leq \lfloor n/2 \rfloor$.

6.3 Cycle Indices of Projective Linear Groups

We have seen how the linear isometry classes of (n,k)-codes over $\mathbb{F}_q$ can be enumerated using cycle indices of projective linear groups $\mathrm{PGL}_k(q)$. It remains to discuss the evaluation of these multivariate polynomials. The formal definition

$$C(G, X) := \frac{1}{|G|} \sum_{g \in G} \prod_{i=1}^{|X|} z_i^{a_i(\bar{g})} \in \mathbb{Q}[z_1, z_2, \ldots, z_{|X|}]$$

of cycle indices, given in 6.1.19, shows that we must determine the cycle types

$$a(\bar{g}) = (a_1(\bar{g}), \ldots, a_{|X|}(\bar{g}))$$

of the homomorphic images $\bar{g}$ of the elements and the order of the acting group $\mathrm{GL}_k(q)$ or of its epimorphic image $\mathrm{PGL}_k(q)$. According to Exercise 6.3.1, the orders of these groups are

$$|GL_k(q)| = [q]_k := (q^k - 1)(q^k - q) \cdots (q^k - q^{k-1}) \qquad\qquad \textbf{6.3.1}$$

and

$$|PGL_k(q)| = [q]_k / (q - 1). \qquad\qquad \textbf{6.3.2}$$

These groups are quite big, and so it is not efficient to establish a complete catalog of all their elements, except for very small values of q and k. A much more economic way is to use the fact that the cycle types of conjugate elements (as well as of images of conjugate elements under homomorphisms) are the same (see Exercise 6.3.2). It reduces the problem to a characterization of the conjugacy classes and the evaluation of the cycle types of representatives of each of these classes. Using this fact we rewrite the cycle index of a group G which acts on a set X in the following form:

$$C(G, X) = \frac{1}{|G|} \sum_{\mathcal{C}} |\mathcal{C}| \prod_{i=1}^{|X|} z_i^{a_i(\bar{g}_\mathcal{C})}, \qquad\qquad \textbf{6.3.3}$$

where $g_\mathcal{C}$ is a representative of the conjugacy class $\mathcal{C}$, the summation is over all the conjugacy classes $\mathcal{C}$ of elements in G, and

$$a(\bar{g}_\mathcal{C}) = (a_1(\bar{g}_\mathcal{C}), \dots, a_{|X|}(\bar{g}_\mathcal{C}))$$

denotes the cycle type of $\bar{g}_\mathcal{C}$, the permutation induced by $g_\mathcal{C}$ on X. As we already know, the cycle type of $\bar{g}$ satisfies

$$\sum_{i=1}^{|X|} i a_i(\bar{g}) = |X|.$$

In general, we call a sequence $a = (a_1, \dots, a_n)$ of nonnegative integers a *cycle type of n*, if $\sum_{i=1}^{n} i \cdot a_i = n$ is satisfied. For short, we write $a \vdash n$, and we note in passing that each cycle type $a \vdash n$ occurs as type of a permutation of n.

Let us now concentrate on the evaluation of the cycle index of the natural action of $G := PGL_k(q)$ on $X := PG_{k-1}^*(q)$. The action 3.7.4 of $GL_k(q)$ on $PG_{k-1}^*(q)$ induces this action of the projective linear group. According to 3.7.6, it can be written as

$$(\mathbb{F}_q^*(A), \mathbb{F}_q^*(v)) \mapsto \mathbb{F}_q^*(v \cdot A^\top), \qquad A \in GL_k(q), \ v \in \mathbb{F}_q^k. \qquad\qquad \textbf{6.3.4}$$

Here in this section it is more convenient to represent vectors as column vectors, so 6.3.4 is written as

$$(\mathbb{F}_q^*(A), \mathbb{F}_q^*(v)) \mapsto \mathbb{F}_q^*(A \cdot v), \qquad A \in GL_k(q), \ v \in \mathbb{F}_q^k.$$

The notation $A \cdot v$ is similar to the notation of applying an endomorphism A of $\mathbb{F}_q^k$ to the vector v which we indicate just by Av.

In order to evaluate the cycle index of the projective linear group we proceed as follows. In a first step each conjugacy class of $\mathrm{GL}_k(q)$ will be described by a *normal form* which is a particular representative of the conjugacy class. Then we evaluate the cardinalities of the conjugacy classes and the cycle types of their representatives.

The announced normal forms of the elements in $\mathrm{GL}_k(q)$ are obtained by using a general approach known from linear algebra, and described in most of the standard lectures on this subject, e.g. in [155].

First we determine a normal form of an arbitrary endomorphism A of $\mathbb{F}^k$. Let x be an indeterminate over $\mathbb{F}$. Then the vector space $\mathbb{F}^k$ together with the outer composition

6.3.5
$$\mathbb{F}[x] \times \mathbb{F}^k \to \mathbb{F}^k \ : \ (f, v) \mapsto fv := \sum_{i=0}^{d} \kappa_i A^i v,$$

becomes an $\mathbb{F}[x]$-module, where f denotes the polynomial $f = \sum_{i=0}^{d} \kappa_i x^i$.

Let $\{e^{(0)}, \ldots, e^{(k-1)}\}$ be the canonical basis of $\mathbb{F}^k$ consisting of the unit vectors. Then
$$\mathbb{F}^k = \sum_{i \in k} \mathbb{F}\, e^{(i)} = \sum_{i \in k} \mathbb{F}[x] e^{(i)}.$$

Since $\mathbb{F}[x]e^{(i)}$ is a subset of $\mathbb{F}^k$, the cyclic $\mathbb{F}[x]$-module $\mathbb{F}[x]e^{(i)}$ is of finite dimension, and the canonical epimorphism from $\mathbb{F}[x]$ to $\mathbb{F}[x]e^{(i)}$ has a kernel different from 0. This kernel is an ideal in the principal ideal domain $\mathbb{F}[x]$, whence it is generated by a monic polynomial $g_i \in \mathbb{F}[x]$ of degree at least 1.

The polynomial $f \in \mathbb{F}[x]$ annihilates $v \in \mathbb{F}^k$ if $fv = 0$. The polynomial $f \in \mathbb{F}[x]$ annihilates $W \subseteq V$ if f annihilates each vector of W. The monic polynomial $f \in \mathbb{F}[x] \setminus \{0\}$ of smallest degree which annihilates v is called the *minimal polynomial* of v. The monic polynomial $f \in \mathbb{F}[x] \setminus \{0\}$ of smallest degree which annihilates $\mathbb{F}^k$ is called the *minimal polynomial* of A. It is usually indicated by M_A. The most important property of minimal polynomials is described in the next

6.3.6   **Lemma** *Let A be an endomorphism of $\mathbb{F}^k$. The polynomial $g \in \mathbb{F}[x]$ annihilates $v \in \mathbb{F}^k$ or $\mathbb{F}^k$ if and only if g is a multiple of the minimal polynomial of v or A, respectively.*   □

The proof is left to the reader.

From the Homomorphism Theorem (Exercise 3.2.3) we deduce that $\mathbb{F}[x]e^{(i)}$ is isomorphic to $\mathbb{F}[x]/I(g_i)$, and the polynomial g_i annihilates the module $\mathbb{F}[x]e^{(i)}$ completely, since $g_i e^{(i)} = 0$. If g denotes the least common multiple of

$g_0, \ldots, g_{k-1}$, then g annihilates the whole vector space $\mathbb{F}^k$. Consequently, g is the minimal polynomial of A, and $\mathbb{F}^k$ can also be seen as an $\overline{\mathbb{F}[x]} := \mathbb{F}[x]/I(g)$-module. Now we decompose g into its pairwise distinct monic, irreducible factors $f_i \in \mathbb{F}[x]$,

$$g = \prod_{i \in t} f_i^{c_i},$$

where t denotes the number of different factors, and $c_i \geq 1$ is the multiplicity of the i-th factor. For $i \in t$ the polynomials $h_i := \prod_{j \neq i} f_j^{c_j}$ are relatively prime by construction, i.e. $\gcd(h_0, \ldots, h_{t-1}) = 1$, and according to Bézout's Identity (cf. Exercise 3.1.6) there exist polynomials $H_i \in \mathbb{F}[x]$ such that 1 can be expressed as

$$1 = H_0 h_0 + \ldots + H_{t-1} h_{t-1}.$$

Putting $E_i := H_i h_i$, we obtain a decomposition of $\overline{1} \in \overline{\mathbb{F}[x]}$ into a sum of pairwise orthogonal and idempotent elements

$$\overline{1} = \overline{E}_0 + \ldots + \overline{E}_{t-1}.$$ 6.3.7

This decomposition of $\overline{1}$ yields, according to Exercise 4.5.2, a decomposition – the *primary decomposition* – of $\mathbb{F}^k$ as a direct sum of *primary components* of the form

$$\mathbb{F}^k = \overline{E}_0 \mathbb{F}^k \oplus \ldots \oplus \overline{E}_{t-1} \mathbb{F}^k.$$

The $\overline{\mathbb{F}[x]}$-module $\overline{E}_i \mathbb{F}^k$ and the $\mathbb{F}[x]$-module $E_i \mathbb{F}^k$ describe the same set, therefore the primary components are A-invariant, since

$$A(\overline{E}_i \mathbb{F}^k) = x E_i \mathbb{F}^k = E_i x \mathbb{F}^k \subseteq E_i \mathbb{F}^k = \overline{E}_i \mathbb{F}^k, \qquad i \in t.$$

Now we consider each of these components $E_i \mathbb{F}^k$ as an $\mathbb{F}[x]/I(f_i^{c_i})$-module. According to 4.7.11, the ring $\mathbb{F}[x]/I(f_i^{c_i})$ has exactly one composition series. Thus it follows from 4.7.12 that $E_i \mathbb{F}^k$ is a direct sum of submodules

$$E_i \mathbb{F}^k = U_{i0} \oplus \ldots \oplus U_{i,n_i-1}, \quad U_{ij} = \mathbb{F}[x] u_{ij} \simeq \mathbb{F}[x]/I(f_i^{t_{ij}}), \ 1 \leq t_{ij} \leq c_i,$$ 6.3.8

where U_{ij} is cyclic over the ring $\mathbb{F}[x]/I(f_i^{c_i})$. These submodules can be ordered in such a way that $1 \leq t_{i0} \leq t_{i1} \leq \ldots \leq t_{i,n_i-1} = c_i$ holds true. Also the submodules U_{ij} are A-invariant. Summarizing, the vector space $\mathbb{F}^k$ is the direct sum of cyclic subspaces

$$\mathbb{F}^k = \bigoplus_{i \in t} \bigoplus_{j \in n_i} U_{ij}, \quad U_{ij} = \mathbb{F}[x] u_{ij} \simeq \mathbb{F}[x]/I(f_i^{t_{ij}}), \ 1 \leq t_{ij} \leq c_i.$$ 6.3.9

Let $f := \sum_{i=0}^{d} \kappa_i x^i$, $\kappa_d = 1$, be a monic, irreducible polynomial of degree d. Assume that f is the minimal polynomial of $v \in \mathbb{F}^k$, whence $U = \mathbb{F}[x]v \simeq \mathbb{F}[x]/I(f)$ is a d-dimensional cyclic subspace of $\mathbb{F}^k$. Using the basis

$(v, Av, \ldots, A^{d-1}v)$ of U, the restriction of the endomorphism A to U is represented by the *companion matrix* $C(f)$ of f given by

$$
C(f) := \begin{pmatrix}
0 & 0 & \cdots & 0 & 0 & -\kappa_0 \\
1 & 0 & \cdots & 0 & 0 & -\kappa_1 \\
0 & 1 & \cdots & 0 & 0 & -\kappa_2 \\
\vdots & \vdots & \ddots & \vdots & \vdots & \vdots \\
0 & 0 & \cdots & 1 & 0 & -\kappa_{d-2} \\
0 & 0 & \cdots & 0 & 1 & -\kappa_{d-1}
\end{pmatrix}.
$$

Assume that f^n is the minimal polynomial of $v \in \mathbb{F}^k$, whence $U = \mathbb{F}[x]v \simeq \mathbb{F}[x]/\mathrm{I}(f^n)$ is an nd-dimensional cyclic subspace of $\mathbb{F}^k$. We choose a basis of U of the form $(v, Av, \ldots, A^{d-1}v, fv, Afv, \ldots, A^{d-1}fv, \ldots, f^{n-1}v, Af^{n-1}v, \ldots, A^{d-1}f^{n-1}v)$, so that the normal form of the restriction of A to U is the following square block-matrix

$$
H(f^n) := \left.\begin{pmatrix}
C(f) & 0 & 0 & \cdots & 0 & 0 \\
\hline
I'_d & C(f) & 0 & \cdots & 0 & 0 \\
0 & I'_d & C(f) & \cdots & 0 & 0 \\
\vdots & \vdots & \vdots & \ddots & \vdots & \vdots \\
\hline
0 & 0 & 0 & \cdots & C(f) & 0 \\
0 & 0 & 0 & \cdots & I'_d & C(f)
\end{pmatrix}\right\} \; n \text{ blocks,}
$$

where I'_d is the elementary matrix $B^{(2)}_{0,d-1,1}$ of dimension d (cf. Exercise 1.7.3), which is the identity matrix I_d with an additional 1 in the right upper corner. The matrix $H(f^n)$ is an $nd \times nd$-matrix and is called the *hyper companion matrix* of f^n. In the case $n = 1$ the matrices $H(f^1)$ and $C(f)$ coincide.

Now we introduce the following notion. Assume that $f_0, \ldots, f_{t-1}$ are pairwise distinct monic, irreducible polynomials over $\mathbb{F}$. If there exists a decomposition 6.3.9 of $\mathbb{F}^k$ with exactly $a_j^{(i)}$ cyclic subspaces isomorphic to $\mathbb{F}[x]/\mathrm{I}(f_i^j)$ for $1 \leq j \leq c_i$ and for $i \in t$, then the *Jacobi normal form* of A is a block-diagonal matrix of the form

6.3.10
$$
\mathrm{diag}\left(D(f_0, a^{(0)}), \ldots, D(f_{t-1}, a^{(t-1)})\right)
$$

where $a^{(i)}$ is a cycle type of $\sum_{j=1}^{c_i} ja_j^{(i)}$, for $i \in t$. The block-diagonal matrix $D(f, a)$, determined by a monic irreducible polynomial f and a cycle type a, is built from companion and hyper companion matrices of f in the following way:

$$
D(f, a) = \mathrm{diag}\left(\underbrace{C(f), \ldots, C(f)}_{a_1 \text{ times}}, \underbrace{H(f^2), \ldots, H(f^2)}_{a_2 \text{ times}}, \ldots\right).
$$

A different approach to normal forms can be found in [60].

The *characteristic polynomial* of an $(n \times n)$-matrix A over $\mathbb{F}$ is defined as $\chi_A(x) := \det(xI_n - A)$. Developing this determinant for $A = C(f)$ with respect to the top row, we get $\chi_{C(f)} = f$. Consequently, $\chi_{H(f^n)} = f^n$, and if $A = \operatorname{diag}\left(D(f_0, a^{(0)}), \ldots, D(f_{t-1}, a^{(t-1)})\right)$, then

$$\chi_A = \prod_{i \in t} f_i^{\gamma_i},$$

where $\gamma_i = \sum_j j a_j^{(i)}$. In other words, the sequence $a^{(i)}$ is a cycle type of γ_i.

By construction, the minimal polynomial of the companion matrix $C(f)$ is equal to f, and $M_{H(f^n)} = f^n$. Consequently, the minimal polynomial of $A = \operatorname{diag}\left(D(f_0, a^{(0)}), \ldots, D(f_{t-1}, a^{(t-1)})\right)$ is given by

$$M_A = \prod_{i \in t} f_i^{c_i},$$

where c_i is the maximal j such that $a_j^{(i)} \neq 0$. From this description it is obvious that M_A is a divisor of χ_A. This proves the

Cayley–Hamilton Theorem *If A is an endomorphism of $\mathbb{F}^k$, then $\chi_A(A) = 0$.*   □   **6.3.11**

Now we come back to our main situation $\mathbb{F} = \mathbb{F}_q$. As we have seen in the proof of 3.2.25, there exist exactly

$$m_q(d) = \frac{1}{d} \sum_{t \mid d} \mu(t) q^{\frac{d}{t}}$$

monic, irreducible polynomials of degree d over $\mathbb{F}_q$, where μ is the number theoretic Möbius function (cf. Exercise 3.2.15). Each of these polynomials of degree not greater than k, with exception of the polynomial $f(x) = x$, can occur as a divisor of the characteristic polynomial of a regular matrix $A \in \operatorname{GL}_k(q)$. We indicate these polynomials by $f_0, f_1, \ldots, f_{t_k-1}$, where

$$t_k := \left(\sum_{i=1}^{k} m_q(i)\right) - 1.$$

If, moreover, d_i indicates the degree of the polynomial f_i for $i \in t_k$, then we obtain the following description of the conjugacy classes in $\operatorname{GL}_k(q)$:

Theorem *For each conjugacy class in $\operatorname{GL}_k(q)$ there exists exactly one pair (γ, a),*   **6.3.12**
where $\gamma = (\gamma_0, \ldots, \gamma_{t_k-1}) \in \mathbb{N}^{t_k}$ is a solution of

$$\sum_{i \in t_k} \gamma_i d_i = k,$$ **6.3.13**

and $a = (a^{(0)}, \ldots, a^{(t_k-1)})$ *is a sequence of cycle types* $a^{(i)} \vdash \gamma_i$, *so that*

6.3.14
$$\mathrm{diag}\left(D(f_0, a^{(0)}), \ldots, D(f_{t_k-1}, a^{(t_k-1)}) \right)$$

is the normal form of this class. Conversely, to each such pair (γ, a) *there exists exactly one conjugacy class the normal form of which is the block-diagonal matrix 6.3.14.*   □

Our next task is the evaluation of the size of the conjugacy classes. Conjugation on $\mathrm{GL}_k(q)$ is a particular group action of $\mathrm{GL}_k(q)$ on itself (cf. Exercise 3.4.2). The centralizer of $A \in \mathrm{GL}_k(q)$ is the stabilizer of A with respect to this action.

6.3.15
Theorem (J.P.S. Kung [117]) *Let* $f \in \mathbb{F}_q[x]$ *be a monic, irreducible polynomial of degree* d, *and let* $a \vdash \gamma$ *be a cycle type of the positive integer* γ. *For* $i \in \{0, 1, \ldots, \gamma\}$ *determine* m_i *by*

$$m_i := \sum_{k=1}^{i} k a_k + \sum_{k=i+1}^{\gamma} i a_k.$$

Then the order of the centralizer of $D(f, a)$ *in* $\mathrm{GL}_{\gamma d}(q)$ *is*

6.3.16
$$b(d, a) := \prod_{i=1}^{\gamma} \prod_{j \in a_i} \left(q^{d m_i} - q^{d(m_i - j - 1)} \right).$$

Proof: Let $n = \gamma d$ be the dimension of a vector space V equipped with the basis $B = (e_0, \ldots, e_{n-1})$ so that the linear mapping $A : V \rightarrow V$ has a representation with respect to this basis in the form $D(f, a)$, where f is a monic, irreducible polynomial in $\mathbb{F}_q[x]$ of degree d and $a \vdash \gamma$. (The vectors e_i should not be mixed up with the unit vectors $e^{(i)}$.) We also consider V as an $\mathbb{F}_q[x]$-module. Determine c by

$$c := \max \{ i \mid 1 \leq i \leq \gamma, \ a_i \neq 0 \},$$

then $\ker f^c = V$, $m_c = \gamma$, and

$$\dim(\ker f^i) = d \left(\sum_{k=1}^{i} k a_k + \sum_{k=i+1}^{\gamma} i a_k \right) = d m_i \text{ for } 1 \leq i \leq c.$$

Consequently, the sets

$$U_i := \left\{ v \in V \mid f^i v = 0 \text{ and } f^{i-1} v \neq 0 \right\} = \ker f^i \setminus \ker f^{i-1}$$

contain $q^{d m_i} - q^{d m_{i-1}}$ elements for $1 \leq i \leq c$. Now we want to choose a particular series of elements of the given basis of V – called *canonical generators of* A – by taking exactly one element of B from each cyclic subspace in the decomposition 6.3.8 of V. For example, a list of canonical generators is given by

$$e_0, e_d, e_{2d}, \ldots, e_{(a_1-1)d},$$
$$e_{a_1 d}, e_{(a_1+2)d}, e_{(a_1+2\cdot2)d}, \ldots, e_{(a_1+2(a_2-1))d},$$
$$\ldots$$

$$e_{(a_1+\ldots+(c-1)a_{c-1})d}, e_{(a_1+\ldots+(c-1)a_{c-1}+c)d}, \ldots, e_{(a_1+\ldots+(c-1)a_{c-1}+c(a_c-1))d}.$$

Now we label these canonical generators consecutively as $\hat{e}_0, \hat{e}_1, \ldots$. To be more precise, for $j \in a_i, i \geq 1$ we have

$$\hat{e}_{a_1+\ldots+a_{i-1}+j} = e_{(a_1+2a_2+\ldots+(i-1)a_{i-1}+ij)d}.$$

In order to complete the proof, we still need to characterize the vector space automorphisms which commute with A.

Lemma *Let ψ be a vector space endomorphism which commutes with $A := D(f, a)$.* **6.3.17**
Then:

1. *ψ is uniquely determined on V by the values $\psi(\hat{e}_i)$ on the canonical generators.*

2. *If $v \in U_i$ is a canonical generator, then $\psi(v)$ belongs to $\ker f^i$ for $1 \leq i \leq c$.*

3. *ψ is a vector space automorphism if and only if there are no linear relations with coefficients in $\mathbb{F}_q[x]$ among the values $\psi(\hat{e}_i)$. In particular, any canonical generator $v \in U_i$ is mapped onto $\psi(v) \in U_i$.*

Proof: The proof of the first two assertions is left to the reader (cf. Exercise 6.3.8). As was shown in 6.3.8, assume that the vector space V has a decomposition into a direct sum of cyclic subspaces

$$V = \bigoplus_{\ell=0}^{a_1+\ldots+a_c-1} V_\ell \text{ with } V_\ell \simeq \mathbb{F}_q[x]/I(f^{j_\ell}) \text{ for } 1 \leq j_\ell \leq c.$$

Moreover, let $\hat{e}_\ell$ be the unique canonical generator of A which belongs to V_ℓ. Then

$$(\hat{e}_\ell, A \cdot \hat{e}_\ell, \ldots, A^{dj_\ell-1} \cdot \hat{e}_\ell)$$

is a basis of V_ℓ. Finally we assume that the monic polynomial f is of the form $f = \sum_{i=0}^d \alpha_i x^i$ with $\alpha_d = 1$.

If ψ is an automorphism of V, then

$$(\psi(\hat{e}_\ell), A \cdot \psi(\hat{e}_\ell), \ldots, A^{dj_\ell-1} \cdot \psi(\hat{e}_\ell))$$

is a basis of $\psi(V_\ell)$. In other words, $\psi(V_\ell)$ is also a dj_ℓ-dimensional cyclic subspace of V. And $\mathbb{F}_q[x]\psi(\hat{e}_\ell) = \psi(V_\ell)$, since $A^{dj_\ell} \cdot \psi(\hat{e}_\ell) = \psi(A^{dj_\ell} \cdot \hat{e}_\ell)$ and

$$\psi(A^{dj_\ell} \cdot \hat{e}_\ell) = \psi\left(\sum_{i\in d}(-\alpha_i)A^{ij_\ell} \cdot \hat{e}_\ell\right) = \sum_{i\in d}(-\alpha_i)A^{ij_\ell} \cdot \psi(\hat{e}_\ell) \in \psi(V_\ell).$$

Conversely, if ψ is an endomorphism which is not an automorphism of V, then the vectors $\psi(e_0), \ldots, \psi(e_{n-1})$ are linearly dependent. Thus, there exist $\alpha_i \in \mathbb{F}_q$, $i \in n$, not all equal to 0, such that

$$\sum_{i \in n} \alpha_i \psi(e_i) = 0.$$

This is a nontrivial linear combination of $\psi(e_i)$. Equipping each subspace V_ℓ with the basis $(\hat{e}_\ell, A \cdot \hat{e}_\ell, \ldots, A^{dj_\ell - 1} \cdot \hat{e}_\ell)$ described above, we derive

$$0 = \sum_{\ell=0}^{a_1 + \ldots + a_c - 1} \underbrace{\sum_{r \in dj_\ell} \alpha_{\ell r} A^r \cdot \psi(\hat{e}_\ell)}_{=:\phi_\ell(A)} = \sum_{\ell=0}^{a_1 + \ldots + a_c - 1} \phi_\ell(x) \psi(\hat{e}_\ell)$$

for suitable $\alpha_{\ell r} \in \mathbb{F}_q$. By construction, not all polynomials ϕ_ℓ are equal to zero, whence we have found a nontrivial linear relation between the vectors $\psi(\hat{e}_\ell)$ with coefficients in $\mathbb{F}_q[x]$. This contradicts our assumption. $\qquad\square$

6.3.17 shows that the image of a canonical generator $v \in U_i$ under an automorphism ψ is again an element of U_i. In the notation of 6.3.8, this means that ψ only permutes the subspaces U_{ij} of a submodule $E_i \mathbb{F}_q^k$ which are isomorphic to the *same* factor module $\mathbb{F}_q[x]/I(f^j)$.

In order to complete the proof of 6.3.15, we determine the number of all possible automorphisms ψ of V by an application of 6.3.17. Starting with the last canonical generator of A, the value $\psi(\hat{e}_{a_1 + \ldots + a_c - 1})$ must be chosen in U_c. There are $q^{dm_c} - q^{dm_c - 1}$ possibilities to do so. If $\hat{e}_{a_1 + \ldots + a_c - 2}$ also belongs to U_c, then there remain $q^{dm_c} - q^{dm_c - 1} q^d = q^{dm_c} - q^{d(m_c - 1 + 1)}$ possibilities to determine $\psi(\hat{e}_{a_1 + \ldots + a_c - 2})$ in U_c so that ψ is an automorphism. (This is just the overall number of vectors in V which do not belong to the $\mathbb{F}_q[x]$-submodule generated by $\ker f^{c-1}$ and $\psi(\hat{e}_{a_1 + \ldots + a_c - 1})$.) In a similar fashion, the values of the other canonical generators of A which also belong to U_c are determined. Altogether there are

$$\prod_{j \in a_c} \left(q^{dm_c} - q^{d(m_c - 1 + j)} \right) = \prod_{j \in a_c} \left(q^{dm_c} - q^{d(m_c - j - 1)} \right)$$

possibilities to determine an automorphism ψ on U_c.

Now assume that W_k, $0 \le k \le a_1 + \ldots + a_c - 1$, denotes the $\mathbb{F}_q[x]$-module generated by $\psi(\hat{e}_j)$ for $j \ge k$. Assume that the canonical generator $\hat{e}_k$ belongs to U_i and that the values $\psi(\hat{e}_j)$ are already determined for $j > k$. In order to determine an automorphism, the vector $\psi(\hat{e}_k)$ must be chosen from $\ker f^i$, but it may not belong to the $\mathbb{F}_q[x]$-module generated by $\ker f^{i-1}$ and $\ker f^i \cap W_{k+1}$. This shows that if ψ is an automorphism already determined on $U_{i+1}, \ldots, U_c$, then there are

$$\prod_{j \in a_i} \left(q^{dm_i} - q^{d(m_{i-1} + a_{i+1} + \ldots + a_c + j)} \right) = \prod_{j \in a_i} \left(q^{dm_i} - q^{d(m_i - j - 1)} \right)$$

possibilities to determine the values of ψ for the canonical generators belonging to U_i (these are the generators $\hat{e}_{a_1+\ldots+a_{i-1}}, \ldots, \hat{e}_{a_1+\ldots+a_i-1}$) such that ψ is also an automorphism of $\ker f^i$. Eventually, the product of these expressions for $i = 1, \ldots, c$ (or $i = 1, \ldots, \gamma$) yields $b(d, a)$. $\square$

As we have seen in the previous proof, the order $b(d, a)$ of the centralizer of $D(f, a)$ in $\mathrm{GL}_{\gamma d}(q)$, where f is an irreducible polynomial of degree d and $a \vdash \gamma$, depends only on the degree of f and on the cycle type a. It does not depend on the particular polynomial f itself. According to 3.4.1, the size of the conjugacy class of a normal form 6.3.14 is

$$\frac{[q]_k}{\prod_{i \in t_k} b(d_i, a^{(i)})}.$$

Before we compute the cycle type of the permutation representation of the natural action 6.3.4 of $\mathbb{F}_q^*(A) \in \mathrm{PGL}_k(q)$ on $\mathrm{PG}_{k-1}^*(q)$, we investigate once more the action of $\mathrm{GL}_k(q)$ on $\mathbb{F}_q^k$. From Exercise 1.4.13 it follows that this action can be reduced to an action on $\mathbb{F}_q^k \setminus \{0\}$. In the next step, we determine the subcycle index of the following action:

$$\mathrm{GL}_k(q) \times \mathbb{F}_q^k \setminus \{0\} \to \mathbb{F}_q^k \setminus \{0\} \ : \ (A, v) \mapsto A \cdot v,$$

from which we will later on determine the cycle index $C(\mathrm{PGL}_k(q), \mathrm{PG}_{k-1}^*(q))$. Recall that in the present section we write vectors as columns and not as rows.

We introduce subcycles and integral elements of vectors $v \in \mathbb{F}_q^k \setminus \{0\}$ in the following way: The vector v belongs to a *subcycle* of A *of length* s if and only if

$$s = \min \left\{ n \in \mathbb{N}^* \mid A^n \cdot v \in \mathbb{F}_q^*(v) \right\}.$$

The *integral element* of v is the element $\alpha_0 \in \mathbb{F}_q^*$ for which $A^s \cdot v = \alpha_0 v$. The set

$$\langle A \rangle (\mathbb{F}_q^*(v)) = \left\{ A^i \cdot \alpha v \mid i \in \mathbb{N}, \ \alpha \in \mathbb{F}_q^* \right\}$$

is the disjoint union of s subsets, each containing $q - 1$ elements, since

$$\langle A \rangle (\mathbb{F}_q^*(v)) = \bigcup_{i \in s} A^i \mathbb{F}_q^*(v) = \bigcup_{i \in s} \left\{ A^i \cdot \alpha v \mid \alpha \in \mathbb{F}_q^* \right\}$$

$$= \bigcup_{i \in s} \left\{ \alpha A^i \cdot v \mid \alpha \in \mathbb{F}_q^* \right\} = \bigcup_{i \in s} \mathbb{F}_q^*(A^i \cdot v).$$

These $s(q - 1)$ vectors in $\mathbb{F}_q^k \setminus \{0\}$ describe exactly s elements of the projective space $\mathrm{PG}_{k-1}^*(q)$, which are the elements of exactly one cycle of length s of $A \in \mathrm{GL}_k(q)$ or $\mathbb{F}_q^*(A) \in \mathrm{PGL}_k(q)$ on $\mathrm{PG}_{k-1}^*(q)$, namely

$$\left(\mathbb{F}_q^*(v), \ldots, \mathbb{F}_q^*(A^{s-1} \cdot v) \right).$$

Moreover, each vector $v' \in \langle A \rangle (\mathbb{F}_q^*(v))$ belongs to a subcycle of A of length s with integral element α_0. Using indeterminates z attached with two indices

– the first one giving the length s of a subcycle and the second one indicating the integral element α_0 corresponding to the subcycle – the operation of A on $\langle A \rangle (\mathbb{F}_q^*(v))$ is described by the subcycle expression $sc(A, v) := z_{s,\alpha_0}^{q-1}$. Since the set $\mathbb{F}_q^k \setminus \{0\}$ is the disjoint union of $\langle A \rangle (\mathbb{F}_q^*(v_i))$, $i \in I$, we define the *subcycle type* of A to be the product of the subcycle expressions $\prod_{i \in I} sc(A, v_i)$. A term of the form z_{s,α_0}^r in the subcycle type of A indicates that there exist $r \cdot s$ vectors $v \in \mathbb{F}_q^k \setminus \{0\}$ such that $s = \min\{n \in \mathbb{N}^* \mid A^n \cdot v \in \mathbb{F}_q^*(v)\}$ and $A^s \cdot v = \alpha_0 v$. Moreover, the exponent r is always a multiple of $q - 1$.

6.3.18   **Definition (subcycle index)** The *subcycle index* for the action of the general linear group $\mathrm{GL}_k(q)$ on $\mathbb{F}_q^k \setminus \{0\}$ is the sum of the subcycle types of $A \in \mathrm{GL}_k(q)$ divided by the order of $\mathrm{GL}_k(q)$, i.e.

$$SC(\mathrm{GL}_k(q), \mathbb{F}_q^k \setminus \{0\}) = \frac{1}{|\mathrm{GL}_k(q)|} \sum_{A \in \mathrm{GL}_k(q)} \prod_{\langle A \rangle (\mathbb{F}_q^*(v))} sc(A, v).$$

The last product must be computed over all $\langle A \rangle (\mathbb{F}_q^*(v))$ in the set

$$\left\{ \langle A \rangle (\mathbb{F}_q^*(v)) \mid v \in \mathbb{F}_q^k \setminus \{0\} \right\}. \qquad \diamond$$

6.3.19   **Remark (cycle index of $\mathrm{PGL}_k(q)$ on $\mathrm{PG}_{k-1}^*(q)$)** From the subcycle index of $\mathrm{GL}_k(q)$ on $\mathbb{F}_q^k \setminus \{0\}$ it is quite easy to obtain the cycle index of the action of $\mathrm{PGL}_k(q)$ on $\mathrm{PG}_{k-1}^*(q)$ by omitting the second index of each indeterminate and by dividing each exponent by $q - 1$. $\qquad \diamond$

Hence, as the next step we compute the subcycle index of $\mathrm{GL}_k(q)$ acting on $\mathbb{F}_q^k \setminus \{0\}$. Since the subcycle types of conjugate matrices in $\mathrm{GL}_k(q)$ are the same, it is enough to determine the subcycle types of the normal forms 6.3.14. First we determine them for hyper companion matrices, later we will deduce a method which allows us to compute the subcycle type of block-diagonal matrices.

The companion and hyper companion matrices depend on polynomials $f \in \mathbb{F}_q[x]$. The subcycle types of these matrices can be obtained from the subexponents of the corresponding polynomials. Therefore, next we introduce exponent and subexponent of a polynomial.

6.3.20   **Definition (exponent, order, period)** The *exponent, order,* or *period* of a polynomial $f \in \mathbb{F}_q[x]$ with $f(0) \neq 0$, is the smallest positive integer e, for which f is a divisor of $x^e - 1$ (cf. [131]). We indicate it as

$$\mathrm{Exp}(f) := \min \{e \in \mathbb{N}^* \mid f \text{ is a divisor of } x^e - 1\}. \qquad \diamond$$

Some properties of the exponent of a polynomial are collected in the next

Lemma *Let $f \in \mathbb{F}_q[x]$ be a monic, irreducible polynomial of degree d with $f(0) \neq 0$.*   **6.3.21**

1. *The exponent of f is equal to the order of an arbitrary root β of f in the multiplicative group $\mathbb{F}_{q^d}^*$. In other words, for any root β of f we have*

$$\mathrm{Exp}(f) = \min\{n \in \mathbb{N}^* \mid \beta^n = 1\} = \mathrm{ord}(\beta).$$

2. *$\mathrm{Exp}(f)$ is a divisor of $q^d - 1$, but it does not divide $q^r - 1$ for $1 \leq r < d$.*

3. *The set $E(d,q)$ of all positive integers, which occur as exponents of monic, irreducible polynomials of degree d over $\mathbb{F}_q$, is*

$$E(d,q) = \left\{ e \in \mathbb{N}^* \;\middle|\; e \mid (q^d - 1) \text{ and } e \nmid (q^r - 1) \text{ for } 1 \leq r < d \right\}.$$

4. *The number of all monic, irreducible polynomials f of degree d over $\mathbb{F}_q$ with $f(0) \neq 0$ and with exponent $e \in E(d,q)$ is $v(d,e) := \phi(e)/d$, where ϕ is the Euler function (cf. 3.4.15).*

5. *For $n \in \mathbb{N}^*$, the polynomial f is a divisor of $x^n - 1$ if and only if $\mathrm{Exp}(f)$ is a divisor of n. (This assertion holds true for arbitrary $f \in \mathbb{F}_q[x]$ with $f(0) \neq 0$.)*

6. *For $n \in \mathbb{N}^*$, the exponent $\mathrm{Exp}(f^n)$ is equal to $\mathrm{Exp}(f)p^t$, where p is the characteristic of $\mathbb{F}_q$, and t is given by $t := \min\{r \in \mathbb{N} \mid p^r \geq n\}$.*

Proof: 1. Let $\beta \in \mathbb{F}_{q^d}$ be a root of f. From 3.2.19 we know that f is the minimal polynomial of β. Moreover, $\beta, \beta^q, \ldots, \beta^{q^{d-1}}$ are all the roots of f, they all are simple and have the same order in $\mathbb{F}_{q^d}^*$. Consequently, β satisfies the equation $\beta^n = 1$ if and only if f is a divisor of $x^n - 1$. From the definitions of $\mathrm{ord}(\beta)$ and $\mathrm{Exp}(f)$ it is clear that $\mathrm{ord}(\beta) = \mathrm{Exp}(f)$.

2. Since β is an element of $\mathbb{F}_{q^d}^*$, its order is a divisor of $q^d - 1$, and moreover $d = \min\{n \in \mathbb{N}^* \mid \beta^{q^n} = \beta\}$, since β is a root of an irreducible polynomial over $\mathbb{F}_q$ of degree d. Hence, $d = \min\{n \in \mathbb{N}^* \mid \beta^{q^n-1} = 1\}$, and, therefore, $\mathrm{ord}(\beta)$ is not a divisor of $q^r - 1$ for $1 \leq r < d$.

3. Thus, $E(d,q)$ is a subset of

$$\left\{ e \in \mathbb{N}^* \;\middle|\; e \mid q^d - 1 \text{ and } e \nmid q^r - 1 \text{ for } 1 \leq r < d \right\}.$$

We still prove that for each positive integer e with $e \mid q^d - 1$ and $e \nmid q^r - 1$ for $1 \leq r < d$ there exists an irreducible polynomial f of degree d such that $\mathrm{Exp}(f) = e$. Assume that e is a divisor of $q^d - 1$ and $e \nmid q^r - 1$ for $1 \leq r < d$. Since $\mathbb{F}_{q^d}^*$ is cyclic, there exist $\phi(e)$ elements $\beta \in \mathbb{F}_{q^d}^*$, which are of order e. According to the particular choice of e, these β do not belong to a proper subfield

$\mathbb{F}_{q^r}$ of $\mathbb{F}_{q^d}$ for $r < d$. Thus, their minimal polynomials are of degree d and exponent e.

4. Each of these minimal polynomials has exactly d distinct roots in $\mathbb{F}_{q^d}$, which are all of the same order. Hence, there are $\phi(e)/d$ different monic, irreducible polynomials over $\mathbb{F}_q$ of degree d with exponent e.

5. Assume that $e = \mathrm{Exp}(f)$ is a divisor of n. Then

$$f \mid x^e - 1 \mid x^n - 1.$$

Conversely, let f be a divisor of $x^n - 1$. According to the division algorithm, there exist $m \in \mathbb{N}$ and $0 \le r < e$ such that $n = me + r$ and, therefore,

$$x^n - 1 = (x^{me} - 1)x^r + (x^r - 1).$$

Consequently, f is a divisor of $x^r - 1$. This is only possible for $r = 0$, which proves that e is a divisor of n.

6. Assume that $e = \mathrm{Exp}(f)$ and e_n denotes the exponent of f^n. From $f \mid f^n \mid x^{e_n} - 1$ and from the fifth assertion we deduce that $e \mid e_n$. As a consequence of $f \mid x^e - 1$, we derive

$$f^n \mid (x^e - 1)^n \mid (x^e - 1)^{p^t} = x^{ep^t} - 1,$$

whence $e_n \mid ep^t$. Hence, e_n is of the form $e_n = ep^r$ where, $0 \le r \le t$. Since e is a divisor of $q^d - 1$, the integers e and p are relatively prime, thus $x^e - 1$ has only simple roots. All roots of the polynomial $x^{ep^r} - 1 = (x^e - 1)^{p^r}$ occur with the multiplicity p^r, all roots of f^n, however, with the multiplicity n. Finally, f^n is a divisor of $x^{ep^r} - 1$, whence comparing the multiplicities of their roots we obtain that $n \le p^r$ and, consequently, $r = t$.   □

6.3.22   **Definition (subexponent)** The *subexponent* of a polynomial $f \in \mathbb{F}_q[x]$ with $f(0) \ne 0$ is defined as

$$\mathrm{Subexp}(f) := \min \left\{ n \in \mathbb{N}^* \mid \exists\, \alpha_0 \in \mathbb{F}_q^* \text{ such that } f \mid x^n - \alpha_0 \right\}.$$

If $f \mid x^n - \alpha_0$ with $\alpha_0 \in \mathbb{F}_q^*$ and $n = \mathrm{Subexp}(f)$, then α_0 is called the *integral element* of f (cf. [89]).   ◇

Using the notation from 6.3.21, some properties of the subexponent of a polynomial are collected in the next lemma, the proof of which is left as an exercise for the reader.

Lemma *Let $f \in \mathbb{F}_q[x]$ be a monic, irreducible polynomial of degree d with $f(0) \neq 0$.* **6.3.23**

1. *Any root $\beta \in \mathbb{F}_{q^d}$ of f satisfies*

$$\mathrm{Subexp}(f) = \min \left\{ n \in \mathbb{N}^* \mid \beta^n \in \mathbb{F}_q^* \right\}.$$

In other words, $\mathrm{Subexp}(f)$ is equal to the order of $\beta \mathbb{F}_q^*$ in the cyclic factor group $\mathbb{F}_{q^d}^* / \mathbb{F}_q^*$.

2. *$\mathrm{Subexp}(f)$ is a divisor of $(q^d - 1)/(q - 1)$.*

3. *For $n \in \mathbb{N}^*$, the subexponent $\mathrm{Subexp}(f^n)$ is equal to $\mathrm{Subexp}(f)p^t$, where p is the characteristic of $\mathbb{F}_q$ and t is given by $t := \min \{ r \in \mathbb{N} \mid p^r \geq n \}$. If α denotes the integral element of f, then α^{p^t} is the integral element of f^n.*

4. *$\mathrm{Subexp}(f)$ is a divisor of $\mathrm{Exp}(f)$ and the quotient*

$$h := \frac{\mathrm{Exp}(f)}{\mathrm{Subexp}(f)}$$

is a divisor of $q - 1$. Moreover, h is the multiplicative order of the integral element of f and $h = \gcd(q - 1, \mathrm{Exp}(f))$.

5. *The subexponent of f can be computed from its exponent by*

$$\mathrm{Subexp}(f) = \frac{\mathrm{Exp}(f)}{\gcd(q - 1, \mathrm{Exp}(f))}.$$

6. *Consider $e \in E(d, q)$ and let $h := \gcd(q - 1, e)$. For each $\alpha \in \mathbb{F}_q^*$ of multiplicative order h there exist exactly $\phi(e)/(d \cdot \phi(h))$ monic, irreducible polynomials $f \in \mathbb{F}_q[x]$ of degree d, exponent e, subexponent e/h, and with integral element α.*

7. *The number of all monic, irreducible polynomials over $\mathbb{F}_q$ of degree d and of subexponent s is*

$$\sum_e \frac{\phi(e)}{d},$$

where the sum is taken over all $e \in E(d, q)$ with $e/\gcd(e, q - 1) = s$.

8. *In the case $q = 2$ the subexponent and the exponent of f coincide.*

9. *Let $S(d, q)$ be the set of all pairs (s, α) such that there exists a monic, irreducible polynomial over $\mathbb{F}_q$ of degree d with subexponent s and integral element α. Then*

$$S(d, q) = \bigcup_{e \in E(d,q)} \left\{ (s, \alpha) \; \middle| \; s = \frac{e}{\gcd(e, q - 1)}, \; \mathrm{ord}(\alpha) = \gcd(e, q - 1) \right\}.$$

For each $(s, \alpha) \in S(d, q)$ there are exactly

$$m(d, s, \alpha) := \frac{\nu(d, s \, \mathrm{ord}(\alpha))}{\phi(\mathrm{ord}(\alpha))}$$

monic, irreducible polynomials over $\mathbb{F}_q$ of degree d with subexponent s and integral element α.   □

The connection between the subcycle type of a hyper companion matrix $H(f^r)$ and the subexponent and the integral element of f is described in

6.3.24   **Lemma** *Let $f \in \mathbb{F}_q[x]$ be a monic, irreducible polynomial of degree d with $f(0) \neq 0$, subexponent s, and integral element α. Then the subcycle type of $H(f^r)$ on $\mathbb{F}_q^{rd} \setminus \{0\}$ is equal to*

$$\prod_{i=1}^{r} z_{s_i,\alpha_i}^{(q^{id}-q^{(i-1)d})/s_i},$$

where $s_i = \mathrm{Subexp}(f^i)$ and $\alpha_i = \alpha^{s_i/s}$ is the integral element of f^i for $1 \leq i \leq r$.

Proof: For $1 \leq i \leq r$ let $U_i := \ker f^i \setminus \ker f^{i-1}$ be the set of those $v \in \mathbb{F}_q^{rd}$, which are annihilated by f^i, but not by f^{i-1}. Consider $v \in U_i$, $A = H(f^r)$, a positive integer n, and $\beta \in \mathbb{F}_q^*$. Since f^i is the minimal polynomial of v,

$$A^n \cdot v = \beta v \Longleftrightarrow A^n \cdot v - \beta v = 0 \Longleftrightarrow (x^n - \beta)v = 0 \Longleftrightarrow f^i \mid x^n - \beta.$$

Consequently, v belongs to a subcycle of $H(f^r)$ of length $s_i = \mathrm{Exp}(f^i)$ with integral element $\alpha_i = \alpha^{s_i/s}$, where α is the integral element of f. Since the set U_i contains $q^{id} - q^{(i-1)d}$ vectors, it contributes the term

$$z_{s_i,\alpha_i}^{(q^{id}-q^{(i-1)d})/s_i}$$

to the subcycle type of $H(f^r)$.   □

Next we describe the announced method for computing the subcycle type of a 2×2-block diagonal matrix from the known subcycle types of the two diagonal blocks. By induction, this allows us to compute the subcycle type of any matrix in normal form 6.3.14.

Assume that $A_1 \in \mathrm{GL}_{k_1}(q)$ and $A_2 \in \mathrm{GL}_{k_2}(q)$ are regular matrices. Then $\mathrm{diag}(A_1, A_2) \in \mathrm{GL}_{k_1+k_2}(q)$. The set $\mathbb{F}_q^{k_1+k_2} \setminus \{0\}$ can be decomposed in the following way

$$\left(\mathbb{F}_q^{k_1} \setminus \{0\} \times \{0\}^{k_2}\right) \dot\cup \left(\{0\}^{k_1} \times \mathbb{F}_q^{k_2} \setminus \{0\}\right) \dot\cup \left(\mathbb{F}_q^{k_1} \setminus \{0\} \times \mathbb{F}_q^{k_2} \setminus \{0\}\right).$$

In the sequel, let β denote a primitive element of $\mathbb{F}_q^*$.

6.3.25   **Lemma** *Assume that we have indeterminates $z_{n,\alpha}$ attached with two indices, where $n \in \mathbb{N}^*$ and $\alpha \in \mathbb{F}_q^*$. We define a multiplication $\circledast$ by*

$$z_{s_1,\beta^{r_1}}^{j_1} \circledast z_{s_2,\beta^{r_2}}^{j_2} := z_{s_3,\beta^{r_3}}^{j_3},$$

where

$$s_3 = \mathrm{lcm}(s_1, s_2) \frac{q-1}{\gcd\left(q-1, \mathrm{lcm}(s_1,s_2)r_1/s_1 - \mathrm{lcm}(s_1,s_2)r_2/s_2\right)},$$

$$r_3 \equiv \frac{r_1 s_3}{s_1} \equiv \frac{r_2 s_3}{s_2} \bmod q - 1,$$

and

$$j_3 = \frac{s_1 j_1 s_2 j_2}{s_3}.$$

Using this multiplication, we define a multiplication $\star$ of subcycle types by

$$\left(\prod_{i=1}^{v_1} z_{u_i,\alpha_i}^{t_i} \right) \star \left(\prod_{j=1}^{v_2} z_{v_j,\kappa_j}^{w_j} \right) := \left(\prod_{i=1}^{v_1} z_{u_i,\alpha_i}^{t_i} \right) \left(\prod_{j=1}^{v_2} z_{v_j,\kappa_j}^{w_j} \right) \prod_{i=1}^{v_1} \prod_{j=1}^{v_2} \left(z_{u_i,\alpha_i}^{t_i} \circledast z_{v_j,\kappa_j}^{w_j} \right).$$

The subcycle type of the matrix $\mathrm{diag}(A_1, A_2)$ is the $\star$-product of the subcycle types of A_1 and A_2. (The n-th power with respect to the multiplication $\star$ will be denoted by $(\ldots)^{\star n}$.) The operator $\star$ can be extended linearly to $\mathbb{Q}[\{z_{n,\alpha} \mid n \in \mathbb{N}^*, \alpha \in \mathbb{F}_q^*\}]$.

Proof: Assume that $v_1 \in \mathbb{F}_q^{k_1} \setminus \{0\}$ belongs to a subcycle of A_1 of length s_1 with integral element β^{r_1}. Then also $(v_1^\top \mid 0_{k_2}^\top)^\top$ belongs to a subcycle of $\mathrm{diag}(A_1, A_2)$ of length s_1 with integral element β^{r_1}. (In the present section we write vectors as columns, thus $(v_1^\top \mid 0_{k_2}^\top)^\top$ is a column of length $k_1 + k_2$.) Similarly, the subcycles of A_2 containing a vector $v_2 \in \mathbb{F}_q^{k_2} \setminus \{0\}$ correspond to the subcycles of $\mathrm{diag}(A_1, A_2)$ containing $(0_{k_1}^\top \mid v_2^\top)^\top$. Thus, we only have to investigate pairs $(v_1^\top \mid v_2^\top)^\top \in \mathbb{F}_q^{k_1} \times \mathbb{F}_q^{k_2}$ with $v_1 \neq 0$ and $v_2 \neq 0$. Moreover, we suppose that v_1 belongs to a subcycle of A_1 of length s_1 with integral element β^{r_1} and v_2 to a subcycle of A_2 of length s_2 with integral element β^{r_2}. Then $\mathrm{lcm}(s_1, s_2)$ is equal to

$$\min\left\{ n \in \mathbb{N}^* \mid \exists \alpha_1, \alpha_2 \in \mathbb{F}_q^* : \mathrm{diag}(A_1^n, A_2^n) \cdot (v_1^\top \mid v_2^\top)^\top = (\alpha_1 v_1^\top \mid \alpha_2 v_2^\top)^\top \right\}.$$

In particular, for $i = 1, 2$ we have

$$\alpha_i = (\beta^{r_i})^{\mathrm{lcm}(s_1,s_2)/s_i} = \beta^{r_i \mathrm{lcm}(s_1,s_2)/s_i}.$$

Now we determine the length s_3 and the integral element α of the subcycle containing $(v_1^\top \mid v_2^\top)^\top$. They satisfy the identity

$$s_3 = \min\left\{ n \in \mathbb{N}^* \mid \exists \alpha \in \mathbb{F}_q^* : \mathrm{diag}(A_1^n, A_2^n)(v_1^\top \mid v_2^\top)^\top = \alpha(v_1^\top \mid v_2^\top)^\top \right\}.$$

Thus, we have to determine the smallest positive integer n such that $\alpha_1^n = \alpha_2^n$. This number is the multiplicative order of $\alpha_1 \alpha_2^{-1}$ in $\mathbb{F}_q^*$, which can be computed by

$$\mathrm{ord}(\alpha_1 \alpha_2^{-1}) = \frac{\mathrm{ord}(\beta)}{\gcd(\mathrm{ord}(\beta), r_1 \mathrm{lcm}(s_1,s_2)/s_1 - r_2 \mathrm{lcm}(s_1,s_2)/s_2)}.$$

Hence, $s_3 = \mathrm{lcm}(s_1, s_2)\,\mathrm{ord}(\alpha_1\alpha_2^{-1})$ and the corresponding integral element is of the form

$$\beta^{r_3} = \alpha_i^{\mathrm{ord}(\alpha_1\alpha_2^{-1})} = \beta^{r_i\,\mathrm{lcm}(s_1,s_2)\,\mathrm{ord}(\alpha_1\alpha_2^{-1})/s_i} = \beta^{r_i s_3/s_i}.$$

If the subcycle type of A_i contains a term $z_{s_i,\beta^{r_i}}^{j_i}$, then, by construction, there are exactly $s_i j_i$ elements in $\mathbb{F}_q^{k_i} \setminus \{0\}$ in the subcycles of A_i of length s_i with integral element β^{r_i}, for $i = 1, 2$. Consequently, all pairs of these elements, these are $s_1 j_1 s_2 j_2$ vectors in $\mathbb{F}_q^{k_1} \times \mathbb{F}_q^{k_2}$, belong to subcycles of $\mathrm{diag}(A_1, A_2)$ of length s_3 with integral element β^{r_3}. Since all these subcycles are of length s_3, by this construction we get exactly $s_1 j_1 s_2 j_2 / s_3$ subcycles of length s_3 with integral element β^{r_3}. This yields the factor $z_{s_3,\beta^{r_3}}^{s_1 j_1 s_2 j_2/s_3} = z_{s_1,\beta^{r_1}}^{j_1} \circledast z_{s_2,\beta^{r_2}}^{j_2}$ in the subcycle type of $\mathrm{diag}(A_1, A_2)$. Therefore, the subcycle type of $\mathrm{diag}(A_1, A_2)$ is the product of expressions of the form

$$z_{s_1,\beta^{r_1}}^{j_1}, \quad z_{s_2,\beta^{r_2}}^{j_2}, \quad \left(z_{s_1,\beta^{r_1}}^{j_1} \circledast z_{s_2,\beta^{r_2}}^{j_2} \right)$$

which are due to the vectors of the form $(v_1^\top \mid 0_{k_2}^\top)^\top$, $(0_{k_1}^\top \mid v_2^\top)^\top$, and $(v_1^\top \mid v_2^\top)^\top$, where $v_i \in \mathbb{F}_q^{k_i} \setminus \{0\}$ is contained in a subcycle of A_i of length s_i with integral element β^{r_i}, for $i = 1, 2$. Finally considering all possible combinations $(v_1^\top \mid v_2^\top)^\top$ yields the desired subcycle type of $\mathrm{diag}(A_1, A_2)$.   □

The multiplication $\star$ is associative and commutative (cf. Exercise 6.3.11). Moreover the empty product is defined to be 1.

Collecting all the results of the present section, we have proved the following formula for the computation of the cycle index $C(\mathrm{PGL}_k(q), \mathrm{PG}_{k-1}^*(q))$.

6.3.26   **Theorem** *Assume that f_i for $i \in t_k$ are the monic, irreducible polynomials of degree $d_i \leq k$ over $\mathbb{F}_q$ which can occur as divisors of a characteristic polynomial of a regular matrix of rank k (thus $f_i \neq x$). For $n > 1$ we use 6.3.23.3 in order to compute both the subexponents $s_{i,n}$ of f_i^n and the corresponding integral elements $\alpha_{i,n}$ from $s_{i,1}$, the subexponent of f_i, and from $\alpha_{i,1}$, the integral element of f_i.*

The subcycle index $SC(\mathrm{GL}_k(q), \mathbb{F}_q^k \setminus \{0\})$ of the action of $\mathrm{GL}_k(q)$ on $\mathbb{F}_q^k \setminus \{0\}$ is

$$\frac{1}{[q]_k} \sum_\gamma \sum_a \frac{[q]_k}{\prod_{i \in t_k} b(d_i, a^{(i)})} \underset{i \in t_k}{\bigstar} \underset{j=1}{\overset{\gamma_i}{\bigstar}} \left(\prod_{\ell=1}^{j} z_{s_{i,\ell},\alpha_{i,\ell}}^{u_{i,\ell}} \right)^{\star a_j^{(i)}},$$

where $u_{i,\ell}$ is given by

$$u_{i,\ell} = \frac{q^{\ell d_i} - q^{(\ell-1)d_i}}{s_{i,\ell}}.$$

Moreover, $[q]_k$ denotes the order of $\mathrm{GL}_k(q)$, and $b(d_i, a^{(i)})$ is the order of the centralizer of $D(f_i, a^{(i)})$ as computed in 6.3.16. The first sum is taken over all solutions

$\gamma = (\gamma_0, \dots, \gamma_{t_k-1}) \in \mathbb{N}^{t_k}$ *of 6.3.13. For each solution γ and for each $i \in t_k$ we have to determine the set of all cycle types of γ_i*

$$CT(\gamma_i) := \{a \mid a \vdash \gamma_i\}.$$

The second sum is taken over all t_k-tuples

$$a = (a^{(0)}, \dots, a^{(t_k-1)}) \in \underset{i \in t_k}{\times} CT(\gamma_i).$$

As already mentioned before, by omitting the second index of each indeterminate and by dividing the exponent of each indeterminate (in the subcycle index of $\mathrm{GL}_k(q)$) by $q - 1$, we obtain the cycle index of the action of $\mathrm{PGL}_k(q)$ on $\mathrm{PG}_{k-1}^*(q)$.   □

Example In order to present a nontrivial example we determine the cycle index   **6.3.27**
of $\mathrm{PGL}_3(3)$ acting on $\mathrm{PG}_2^*(3)$. At first we need a list of all monic, irreducible polynomials different from $f = x$ of degree at most 3 over $\mathbb{F}_3$ together with their exponents, subexponents and integral elements (cf. Table 6.4).

Table 6.4 The irreducible polynomials of degree at most 3 over $\mathbb{F}_3$ different from $f = x$

i	f_i	d_i	e_i	s_i	α_i
0	$x + 1$	1	2	1	2
1	$x + 2$	1	1	1	1
2	$x^2 + 1$	2	4	2	2
3	$x^2 + x + 2$	2	8	4	2
4	$x^2 + 2x + 2$	2	8	4	2
5	$x^3 + 2x + 1$	3	26	13	2
6	$x^3 + 2x + 2$	3	13	13	1
7	$x^3 + x^2 + 2$	3	13	13	1
8	$x^3 + x^2 + x + 2$	3	13	13	1
9	$x^3 + x^2 + 2x + 1$	3	26	13	2
10	$x^3 + 2x^2 + 1$	3	26	13	2
11	$x^3 + 2x^2 + x + 1$	3	26	13	2
12	$x^3 + 2x^2 + 2x + 2$	3	13	13	1

With these polynomials we determine the following normal forms. In addition to each normal form we also indicate its subcycle type.

— The polynomials of degree 3 occur only in the form

$$D(f_i, (1,0,\dots)) \text{ for } i \geq 5.$$

They have subcycle types

$$z_{s_i,\alpha_i}^{26/s_i} = z_{13,\alpha_i}^2.$$

— In the normal forms of $GL_3(3)$ companion matrices of polynomials of degree 2 occur only in combination with polynomials of degree 1. These normal forms are described by

$$\operatorname{diag}(D(f_i, (1,0,\ldots)), D(f_j, (1,0,\ldots))) \text{ for } 0 \le i \le 1, \ 2 \le j \le 4.$$

They have subcycle types

$$z_{s_i,\alpha_i}^{2/s_i} \star z_{s_j,\alpha_j}^{8/s_j} = z_{1,\alpha_i}^{2} \star z_{s_j,2}^{8/s_j}.$$

— In all other normal forms just polynomials of degree 1 occur. For $0 \le i, j \le 1$ and $i \ne j$ they can be described as:

normal form	subcycle type
$D(f_i, (3,0,\ldots))$	$\left(z_{1,\alpha_i}^{2}\right)^{\star 3}$
$D(f_i, (1,1,0,\ldots))$	$z_{1,\alpha_i}^{2} \star \left(z_{1,\alpha_i}^{2} z_{3,\alpha_i}^{2}\right)$
$D(f_i, (0,0,1,0,\ldots))$	$z_{1,\alpha_i}^{2} z_{3,\alpha_i}^{8}$
$\operatorname{diag}(D(f_i, (2,0,\ldots)), D(f_j, (1,0,\ldots)))$	$\left(z_{1,\alpha_i}^{2}\right)^{\star 2} \star z_{1,\alpha_j}^{2}$
$\operatorname{diag}(D(f_i, (0,1,0,\ldots)), D(f_j, (1,0,\ldots)))$	$\left(z_{1,\alpha_i}^{2} z_{3,\alpha_i}^{2}\right) \star z_{1,\alpha_j}^{2}$

In order to derive the subcycle index of $GL_3(3)$ acting on $\mathbb{F}_3^3 \setminus \{0\}$, the subcycle type of every normal form must be multiplied by the cardinality of its conjugacy class and, finally, the sum of these subcycle types must be divided by the order of $GL_3(3)$.

$$SC(GL_3(3), \mathbb{F}_3^3 \setminus \{0\}) = 1/11232 \Big(1728 z_{13,1}^2 + 1728 z_{13,2}^2 + 702 z_{1,1}^2 z_{2,2}^4 z_{4,1}^4$$
$$+ 702 z_{1,2}^2 z_{2,2}^4 z_{4,1}^4 + 1404 z_{1,1}^2 z_{4,2}^2 z_{8,1}^2 + 1404 z_{1,2}^2 z_{4,2}^2 z_{8,1}^2 + z_{1,1}^{26} + z_{1,2}^{26}$$
$$+ 104 z_{1,1}^8 z_{3,1}^6 + 104 z_{1,2}^8 z_{3,2}^6 + 624 z_{1,1}^2 z_{3,1}^8 + 624 z_{1,2}^2 z_{3,2}^8 + 117 z_{1,1}^8 z_{1,2}^2 z_{2,1}^8$$
$$+ 117 z_{1,2}^8 z_{1,1}^2 z_{2,1}^8 + 936 z_{1,1}^2 z_{1,2}^2 z_{2,1}^2 z_{3,1}^2 z_{6,1}^2 + 936 z_{1,1}^2 z_{1,2}^2 z_{2,1}^2 z_{3,2}^2 z_{6,1}^2 \Big).$$

This yields the cycle index

$$C(PGL_3(3), PG_2^*(3)) = 1/5616 \Big(1728 z_{13} + 1404 z_1 z_4 z_8$$
$$+ 624 z_1 z_3^4 + 702 z_1 z_2^2 z_4^2 + 936 z_1^2 z_2 z_3 z_6 + 104 z_1^4 z_3^3 + 117 z_1^5 z_2^4 + z_1^{13} \Big). \quad \diamond$$

The computation of the subcycle index can still be simplified. Actually, it is not necessary to know all the different monic, irreducible polynomials over $\mathbb{F}_q$ of degree at most k. As we have seen in part 9 of 6.3.23, for each $(s, \alpha) \in S(d, q)$ it is possible to determine the exact number of monic, irreducible polynomials over $\mathbb{F}_q$ of degree d with subexponent s and integral element α. Since the subcycle type of $H(f^n)$ depends only on the three parameters (d, s, α) and on

n, of course, we need not determine the conjugacy classes of $GL_k(q)$ themselves. It suffices to know how many different monic, irreducible polynomials with parameters (d, s, α, n) occur in the normal forms. This approach motivates the following formula for the computation of the subcycle index of $GL_k(q)$ on $\mathbb{F}_q^k \setminus \{0\}$:

$$\sum_{c \vdash k} \star \sum_{d=1}^{k} \star \sum_{r} \star \sum_{(s,\alpha) \in S(d,q)} \sum_{t} \xi(m(d,s,\alpha),t) \stackrel{r(s,\alpha)}{\underset{j=1}{\star}} \left(\sum_{a \vdash j} \frac{1}{b(d,a)} z(d,s,\alpha,a) \right)^{\star t_j}$$

Here $z(d,s,\alpha,a)$ stands for the subcycle type of a matrix $D(f,a)$, where f is an arbitrary monic irreducible polynomial in $\mathbb{F}_q[x]$ of degree d with subexponent s and integral element α, and where $a \vdash j$ is a cycle type of j. This subcycle type can be computed by

$$z(d,s,\alpha,a) = \stackrel{j}{\underset{\ell=1}{\star}} \left(\prod_{n=1}^{\ell} z_{s_n,\alpha_n}^{u_n} \right)^{\star a_\ell},$$

where s_n stands for sp^t and α_n for α^{p^t}, where p is the characteristic of $\mathbb{F}_q$, and t is the smallest nonnegative integer such that $p^t \geq n$. The exponents u_n are computed via

$$u_n = \frac{q^{nd} - q^{(n-1)d}}{s_n}.$$

The first sum in the subcycle index of $GL_k(q)$ is taken over all cycle types $c \vdash k$. Here c is of the form $c = (c_1, \ldots, c_k)$ and the number c_d represents the number of monic, irreducible polynomials of degree d (counted with their multiplicities), which occur as factors of the characteristic polynomial of a normal form in $GL_k(q)$.

The second sum is taken over all functions r from $S(d,q)$ to $\mathbb{N}$ which satisfy

$$\sum_{(s,\alpha) \in S(d,q)} r(s,\alpha) = c_d.$$

If the characteristic polynomial has exactly c_d irreducible factors of degree d, then the value $r(s, \alpha)$ stands for the number of irreducible factors with parameters (d, s, α).

The third sum is taken over all cycle types $t \vdash r(s, \alpha)$ with the additional property that

$$\sum_j t_j \leq m(d,s,\alpha).$$

Such a cycle type t describes the type of a set-partition of a set of cardinality $r(s, \alpha)$ into at most $m(d, s, \alpha)$ subsets. For any $t \vdash r(s, \alpha)$ there are

$$\xi(m(d,s,\alpha),t) := \binom{m(d,s,\alpha)}{t_1, t_2 \ldots, m(d,s,\alpha) - \sum_j t_j}$$

possibilities to choose – among the $m(d,s,\alpha)$ different monic, irreducible polynomials with parameters (d,s,α) – for each j exactly t_j polynomials, which occur with the multiplicity j in the considered characteristic polynomial.

Finally the last sum is taken over all cycle types $a \vdash j$. These cycle types describe all possible normal forms whose characteristic polynomials are the j-th power of one monic irreducible polynomial. The reader should recall that the characteristic polynomial of $D(f,a)$ equals f^j in this situation.

6.3.28   **Example** We continue 6.3.27 by determining the sets $E(d,q)$, $S(d,q)$ for $q = 3$, $1 \le d \le 3$, and the numbers $v(d,e)$ and $m(d,s,\alpha)$ for $e \in E(d,q)$ and $(s,\alpha) \in S(d,q)$. This provides all the necessary information for computing the subcycle index of $GL_3(3)$. In fact, the information contained in Table 6.4 is not needed for this purpose.

$E(1,3) = \{1,2\}$	$v(1,1) = 1$	$v(1,2) = 1$
$E(2,3) = \{4,8\}$	$v(2,4) = 1$	$v(2,8) = 2$
$E(3,3) = \{13,26\}$	$v(3,13) = 4$	$v(3,26) = 4$
$S(1,3) = \{(1,1),(1,2)\}$	$m(1,1,1) = 1$	$m(1,1,2) = 1$
$S(2,3) = \{(2,2),(4,2)\}$	$m(2,2,2) = 1$	$m(2,4,2) = 2$
$S(3,3) = \{(13,1),(13,2)\}$	$m(3,13,1) = 4$	$m(3,13,2) = 4$

$\diamondsuit$

In [142], explicit formulae for the numbers T_{nkq}, $\overline{T}_{nkq}$, V_{nkq}, $\overline{V}_{nkq}$, R_{nkq}, and $\overline{R}_{nkq}$ are given for $k \le 3$. This is done by a careful analysis of the conjugacy classes of elements of $PGL_k(q)$. The formulae result from counting fixed points and applying the Lemma of Cauchy–Frobenius. Since in the general formula too many different cases must be considered, we present some of the resulting formulae for $n = 7$.

For example, for any field of characteristic $p = 2$ we obtain

$$\overline{T}_{73q} = \frac{q^6 + 7q^5 + 9q^4 + 183q^3 + 632q^2 - 364q + 1344}{5040} +$$

$$+ \left[\frac{q^2 + 18q + 20}{36}\right]_{3|q-1} + \left[\frac{16}{5}\right]_{5|q-1} + \left[\frac{6}{7}\right]_{7|q-1},$$

where

$$[x]_{a|b} := \begin{cases} x & \text{if } a \mid b, \\ 0 & \text{else.} \end{cases}$$

For characteristic $p > 2$ we get

$$\overline{T}_{73q} = \frac{q^6 + 7q^5 + 9q^4 + 183q^3 + 1157q^2 + 56q - 201}{5040} +$$

$$+ \left[\frac{q^2 + 10q - 15}{72}\right]_{3|q} + \left[\frac{q^2 + 18q + 77}{36}\right]_{3|q-1} + \left[\frac{4q + 13}{12}\right]_{4|q-1} +$$

$$+ \left[\frac{1}{3}\right]_{12|q-1} + \left[\frac{1}{6}\right]_{12|q-9} + \left[\frac{16}{5}\right]_{5|q-1} + \left[\frac{2}{5}\right]_{5|q} + \left[\frac{8}{7}\right]_{7|q-1} +$$

$$+ \left[\frac{6}{7}\right]_{7|q+1} + \left[\frac{2}{7}\right]_{7|q} + \left[\frac{2}{7}\right]_{7|q^2+q+1} .$$

Similar formulae can be found for $\overline{V}_{nkq}$ and $\overline{R}_{nkq}$. For $p = 2$ and $n = 7$ we obtain

$$\overline{V}_{73q} = \frac{q^6 + 7q^5 + 8q^4 + 197q^3 + 456q^2 + 420q + 384}{5040} +$$

$$+ \left[\frac{q^2 + 14q + 36}{36}\right]_{3|q-1} + \left[\frac{14}{5}\right]_{5|q-1} + \left[\frac{3}{7}\right]_{7|q-1} ,$$

and

$$\overline{R}_{73q} = \frac{q^6 + 7q^5 + 8q^4 + 190q^3 + 414q^2 + 588q + 272}{5040} +$$

$$+ \left[\frac{q^2 + 14q + 40}{36}\right]_{3|q-1} + [2]_{5|q-1} + \left[\frac{3}{7}\right]_{7|q-1} .$$

For $p > 2$ and $n = 7$ we have

$$\overline{V}_{73q} = \frac{q^6 + 7q^5 + 8q^4 + 197q^3 + 981q^2 + 1050q - 1896}{5040} +$$

$$+ \left[\frac{q^2 + 6q - 3}{72}\right]_{3|q} + \left[\frac{q^2 + 14q + 81}{36}\right]_{3|q-1} + \left[\frac{4q + 13}{12}\right]_{4|q-1} +$$

$$+ \left[\frac{1}{3}\right]_{12|q-1} + \left[\frac{1}{6}\right]_{12|q-9} + \left[\frac{14}{5}\right]_{5|q-1} + \left[\frac{2}{5}\right]_{5|q} + \left[\frac{5}{7}\right]_{7|q-1} +$$

$$+ \left[\frac{3}{7}\right]_{7|q+1} + \left[\frac{1}{7}\right]_{7|q} + \left[\frac{2}{7}\right]_{7|q^2+q+1}$$

and

$$\overline{R}_{73q} = \frac{q^6 + 7q^5 + 8q^4 + 190q^3 + 939q^2 + 903q - 2008}{5040} +$$

$$+ \left[\frac{q^2 + 6q + 13}{72}\right]_{3|q} + \left[\frac{q^2 + 14q + 85}{36}\right]_{3|q-1} + \left[\frac{2q + 5}{6}\right]_{4|q-1} +$$

$$+ \left[\frac{1}{3}\right]_{12|q-1} + \left[\frac{1}{6}\right]_{12|q-9} + [2]_{5|q-1} + \left[\frac{1}{5}\right]_{5|q} + \left[\frac{5}{7}\right]_{7|q-1} +$$

$$+ \left[\frac{3}{7}\right]_{7|q+1} + \left[\frac{1}{7}\right]_{7|q} + \left[\frac{2}{7}\right]_{7|q^2+q+1} .$$

The expressions for T_{nkq}, V_{nkq}, and R_{nkq} are even more complicated.

Exercises

E.6.3.1   **Exercise** Prove that the orders of the groups $\mathrm{GL}_k(q)$ and $\mathrm{PGL}_k(q)$ are given by

$$|\mathrm{GL}_k(q)| = (q^k - 1)(q^k - q)\ldots(q^k - q^{k-1}) =: [q]_k, \qquad |\mathrm{PGL}_k(q)| = \frac{[q]_k}{q-1}.$$

E.6.3.2   **Exercise** Let $_G X$ be a group action. Prove that conjugate elements $g_1, g_2 \in G$ induce permutations $\overline{g_1}, \overline{g_2}$ of X of the same cycle type. In other words, if $g_2 = g g_1 g^{-1}$ for some $g \in G$, then $a_i(\overline{g_1}) = a_i(\overline{g_2})$ for all i. Hint: Which relation holds between the cycles of π and $\rho\pi\rho^{-1}$ for $\pi, \rho \in S_X$?

E.6.3.3   **Exercise** Prove that the cycle index of the natural action of the symmetric group S_n on the set $n = \{0, 1, \ldots, n-1\}$ is given by

$$C(S_n, n) = \sum_{a \vdash n} \prod_{k=1}^{n} \frac{1}{a_k! k^{a_k}} z_k^{a_k}.$$

Hint: Prove first the following propositions:

1. The cycle type $a(\pi)$ of a permutation $\pi \in S_n$ characterizes the conjugacy class of π in S_n. Hence, elements in different conjugacy classes of S_n have different cycle types.

2. For each cycle type $a \vdash n$ there exist permutations $\pi \in S_n$ with $a(\pi) = a$.

3. The number of elements of S_n of cycle type $a \vdash n$ is

$$\frac{n!}{\prod_{k=1}^{n} a_k! k^{a_k}}.$$

E.6.3.4   **Exercise** Let A be an endomorphism of $\mathbb{F}^k$. Show that $\mathbb{F}^k$ together with the outer composition 6.3.5 is an $\mathbb{F}[x]$-module, that is, for all $f, f_1, f_2 \in \mathbb{F}[x]$ and all $v, v_1, v_2 \in \mathbb{F}^k$ we have $f_1(f_2 v) = (f_1 f_2)v$, $(f_1 + f_2)v = f_1 v + f_2 v$, $f(v_1 + v_2) = f v_1 + f v_2$ and $1_\mathbb{F} v = v$.

E.6.3.5   **Exercise** Prove 6.3.6.

E.6.3.6   **Exercise** Prove that 6.3.7 is a decomposition of 1 into pairwise orthogonal idempotents.

Exercise Let A be an endomorphism of $\mathbb{F}^k$. Prove that $\mathbb{F}^k$ is a cyclic $\mathbb{F}[x]$-module if and only if the characteristic polynomial χ_A and the minimal polynomial M_A of A coincide.   E.6.3.7

Exercise Prove 6.3.17.   E.6.3.8

Exercise Prove that conjugate matrices in $\mathrm{GL}_k(q)$ have the same subcycle type.   E.6.3.9

Exercise Prove 6.3.23.   E.6.3.10

Exercise Prove that the multiplication $\star$ of 6.3.25 is commutative and associative.   E.6.3.11

6.4 Numerical Data for Linear Isometry Classes

In Tables 6.7–6.12 we present the numbers of linear isometry classes of nonredundant linear codes and of projective linear codes for $q = 2, 3, 4$. For computing these numbers we had to determine the auxiliary data T_{nkq} and $\overline{T}_{nkq}$ given in Tables 6.13–6.18. The numbers of all linear isometry classes of linear codes are displayed in Tables 6.19–6.20. Some values for U_{nk2} were already presented in Table 6.2. Finally the numbers of indecomposable linear codes are presented in Tables 6.21–6.26. These numbers were computed with the computer algebra system SYMMETRICA ([190]). Due to restrictions of the page size in some tables the entries for $n = 13$ or $n = 14$ are omitted. It is also possible to determine tables of $\begin{bmatrix} n \\ k \end{bmatrix}(q)$, T_{nkq}, $\overline{T}_{nkq}$, V_{nkq}, $\overline{V}_{nkq}$, U_{nkq}, R_{nkq} and $\overline{R}_{nkq}$ with the software from the attached CD.

Table 6.5 Values of $\left[{n \atop k}\right](3)$

$n\backslash k$	1	2	3	4
1	1	0	0	0
2	4	1	0	0
3	13	13	1	0
4	40	130	40	1
5	121	1 210	1 210	121
6	364	11 011	33 880	11 011
7	1 093	99 463	925 771	925 771
8	3 280	896 260	25 095 280	75 913 222
9	9 841	8 069 620	678 468 820	6 174 066 262
10	29 524	72 636 421	18 326 727 760	500 777 836 042
11	88 573	653 757 313	494 894 285 941	40 581 331 447 162
12	265 720	5 883 904 390	13 362 799 477 720	3 287 582 741 506 063
13	797 161	52 955 405 230	360 801 469 802 830	266 307 564 861 468 823

Table 6.6 Values of $\left[{n \atop k}\right](4)$

$n\backslash k$	1	2	3	4
1	1	0	0	0
2	5	1	0	0
3	21	21	1	0
4	85	357	85	1
5	341	5 797	5 797	341
6	1 365	93 093	376 805	93 093
7	5 461	1 490 853	24 208 613	24 208 613
8	21 845	23 859 109	1 550 842 085	6 221 613 541
9	87 381	381 767 589	99 277 752 549	1 594 283 908 581
10	349 525	6 108 368 805	6 354 157 930 725	408 235 958 349 285
11	1 398 101	97 734 250 405	406 672 215 935 205	104 514 759 495 347 685

Table 6.7 Values of V_{nk2}

$n\backslash k$	1	2	3	4	5	6	7
1	1	0	0	0	0	0	0
2	1	1	0	0	0	0	0
3	1	2	1	0	0	0	0
4	1	3	3	1	0	0	0
5	1	4	6	4	1	0	0
6	1	6	12	11	5	1	0
7	1	7	21	27	17	6	1
8	1	9	34	63	54	25	7
9	1	11	54	134	163	99	35
10	1	13	82	276	465	385	170
11	1	15	120	544	1 283	1 472	847
12	1	18	174	1 048	3 480	5 676	4 408
13	1	20	244	1 956	9 256	22 101	24 297
14	1	23	337	3 577	24 282	87 404	143 270

Table 6.8 Values of V_{nk3}

$n\backslash k$	1	2	3	4	5	6	7
1	1	0	0	0	0	0	0
2	1	1	0	0	0	0	0
3	1	2	1	0	0	0	0
4	1	4	3	1	0	0	0
5	1	5	8	4	1	0	0
6	1	8	19	15	5	1	0
7	1	10	39	50	24	6	1
8	1	14	78	168	118	37	7
9	1	17	151	538	628	255	53
10	1	22	280	1 789	3 759	2 266	518
11	1	26	506	5 981	26 131	28 101	7 967
12	1	33	904	20 502	208 045	500 237	230 165
13	1	38	1 571	70 440	1 788 149	11 165 000	11 457 192
14	1	46	2 687	241 252	15 675 051	269 959 051	734 810 177

Table 6.9 Values of V_{nk4}

$n \backslash k$	1	2	3	4	5	6	7
1	1	0	0	0	0	0	0
2	1	1	0	0	0	0	0
3	1	2	1	0	0	0	0
4	1	4	3	1	0	0	0
5	1	6	9	4	1	0	0
6	1	9	24	17	5	1	0
7	1	12	55	70	28	6	1
8	1	17	131	323	189	44	7
9	1	22	318	1784	1976	490	65
10	1	30	772	12094	36477	13752	1240
11	1	37	1881	89437	923978	948361	102417
12	1	48	4568	668922	25124571	91149571	25983495
13	1	59	10857	4843901	665246650	9163203790	9229228790

Table 6.10 Values of $\overline{V}_{nk2}$

$n \backslash k$	1	2	3	4	5	6	7
1	1	0	0	0	0	0	0
2	0	1	0	0	0	0	0
3	0	1	1	0	0	0	0
4	0	0	2	1	0	0	0
5	0	0	1	3	1	0	0
6	0	0	1	4	4	1	0
7	0	0	1	5	8	5	1
8	0	0	0	6	15	14	6
9	0	0	0	5	29	38	22
10	0	0	0	4	46	105	80
11	0	0	0	3	64	273	312
12	0	0	0	2	89	700	1285
13	0	0	0	1	112	1794	5632
14	0	0	0	1	128	4579	26792

Table 6.11 Values of $\overline{V}_{nk3}$

$n\backslash k$	1	2	3	4	5	6	7
1	1	0	0	0	0	0	0
2	0	1	0	0	0	0	0
3	0	1	1	0	0	0	0
4	0	1	2	1	0	0	0
5	0	0	3	3	1	0	0
6	0	0	4	8	4	1	0
7	0	0	4	19	15	5	1
8	0	0	3	44	61	26	6
9	0	0	3	91	277	162	40
10	0	0	2	199	1 439	1 381	375
11	0	0	1	401	8 858	17 200	5 923
12	0	0	1	806	62 311	311 580	182 059
13	0	0	1	1 504	459 828	6 876 068	9 427 034
14	0	0	0	2 659	3 346 151	159 373 844	608 045 192

Table 6.12 Values of $\overline{V}_{nk4}$

$n\backslash k$	1	2	3	4	5	6	7
1	1	0	0	0	0	0	0
2	0	1	0	0	0	0	0
3	0	1	1	0	0	0	0
4	0	1	2	1	0	0	0
5	0	1	4	3	1	0	0
6	0	0	8	10	4	1	0
7	0	0	10	35	19	5	1
8	0	0	13	136	122	33	6
9	0	0	17	657	1 320	376	52
10	0	0	19	3 849	25 619	11 632	1 057
11	0	0	19	23 456	645 751	845 949	95 960
12	0	0	17	138 200	16 822 798	81 806 606	25 058 580
13	0	0	13	761 039	418 686 704	8 140 667 601	8 935 079 862

Table 6.13 Values of T_{nk2}

$n \backslash k$	1	2	3	4	5	6	7
1	1	1	1	1	1	1	1
2	1	2	2	2	2	2	2
3	1	3	4	4	4	4	4
4	1	4	7	8	8	8	8
5	1	5	11	15	16	16	16
6	1	7	19	30	35	36	36
7	1	8	29	56	73	79	80
8	1	10	44	107	161	186	193
9	1	12	66	200	363	462	497
10	1	14	96	372	837	1 222	1 392
11	1	16	136	680	1 963	3 435	4 282
12	1	19	193	1 241	4 721	10 397	14 805
13	1	21	265	2 221	11 477	33 578	57 875
14	1	24	361	3 938	28 220	115 624	258 894

Table 6.14 Values of T_{nk3}

$n \backslash k$	1	2	3	4	5	6	7
1	1	1	1	1	1	1	1
2	1	2	2	2	2	2	2
3	1	3	4	4	4	4	4
4	1	5	8	9	9	9	9
5	1	6	14	18	19	19	19
6	1	9	28	43	48	49	49
7	1	11	50	100	124	130	131
8	1	15	93	261	379	416	423
9	1	18	169	707	1 335	1 590	1 643
10	1	23	303	2 092	5 851	8 117	8 635
11	1	27	533	6 514	32 645	60 746	68 713
12	1	34	938	21 440	229 485	729 722	959 887
13	1	39	1 610	72 050	1 860 199	13 025 199	24 482 391
14	1	47	2 734	243 986	15 919 037	285 878 088	1 020 688 265

Table 6.15 Values of T_{nk4}

$n\backslash k$	1	2	3	4	5	6	7
1	1	1	1	1	1	1	1
2	1	2	2	2	2	2	2
3	1	3	4	4	4	4	4
4	1	5	8	9	9	9	9
5	1	7	16	20	21	21	21
6	1	10	34	51	56	57	57
7	1	13	68	138	166	172	173
8	1	18	149	472	661	705	712
9	1	23	341	2 125	4 101	4 591	4 656
10	1	31	803	12 897	49 374	63 126	64 366
11	1	38	1 919	91 356	1 015 334	1 963 695	2 066 112
12	1	49	4 617	673 539	25 798 110	116 947 681	142 931 176
13	1	60	10 917	4 854 818	670 101 468	9 833 305 258	19 062 534 048

Table 6.16 Values of $\overline{T}_{nk2}$

$n\backslash k$	1	2	3	4	5	6	7
1	1	1	1	1	1	1	1
2	0	1	1	1	1	1	1
3	0	1	2	2	2	2	2
4	0	0	2	3	3	3	3
5	0	0	1	4	5	5	5
6	0	0	1	5	9	10	10
7	0	0	1	6	14	19	20
8	0	0	0	6	21	35	41
9	0	0	0	5	34	72	94
10	0	0	0	4	50	155	235
11	0	0	0	3	67	340	652
12	0	0	0	2	91	791	2 076
13	0	0	0	1	113	1 907	7 539
14	0	0	0	1	129	4 708	31 500

Table 6.17 Values of $\overline{T}_{nk3}$

$n\backslash k$	1	2	3	4	5	6	7
1	1	1	1	1	1	1	1
2	0	1	1	1	1	1	1
3	0	1	2	2	2	2	2
4	0	1	3	4	4	4	4
5	0	0	3	6	7	7	7
6	0	0	4	12	16	17	17
7	0	0	4	23	38	43	44
8	0	0	3	47	108	134	140
9	0	0	3	94	371	533	573
10	0	0	2	201	1 640	3 021	3 396
11	0	0	1	402	9 260	26 460	32 383
12	0	0	1	807	63 118	374 698	556 757
13	0	0	1	1 505	461 333	7 337 401	16 764 435
14	0	0	0	2 659	3 348 810	162 722 654	770 767 846

Table 6.18 Values of $\overline{T}_{nk4}$

$n\backslash k$	1	2	3	4	5	6	7
1	1	1	1	1	1	1	1
2	0	1	1	1	1	1	1
3	0	1	2	2	2	2	2
4	0	1	3	4	4	4	4
5	0	1	5	8	9	9	9
6	0	0	8	18	22	23	23
7	0	0	10	45	64	69	70
8	0	0	13	149	271	304	310
9	0	0	17	674	1 994	2 370	2 422
10	0	0	19	3 868	29 487	41 119	42 176
11	0	0	19	23 475	669 226	1 515 175	1 611 135
12	0	0	17	138 217	16 961 015	98 767 621	123 826 201
13	0	0	13	761 052	419 447 756	8 560 115 357	17 495 195 219

Table 6.19 Values of U_{nk3}

$n \backslash k$	1	2	3	4	5	6	7
1	1	0	0	0	0	0	0
2	2	1	0	0	0	0	0
3	3	3	1	0	0	0	0
4	4	7	4	1	0	0	0
5	5	12	12	5	1	0	0
6	6	20	31	20	6	1	0
7	7	30	70	70	30	7	1
8	8	44	148	238	148	44	8
9	9	61	299	776	776	299	61
10	10	83	579	2565	4535	2565	579
11	11	109	1085	8546	30666	30666	8546
12	12	142	1989	29048	238711	530903	238711
13	13	180	3560	99488	2026860	11695903	11695903
14	14	226	6247	340740	17701911	281654954	746506080

Table 6.20 Values of U_{nk4}

$n \backslash k$	1	2	3	4	5	6	7
1	1	0	0	0	0	0	0
2	2	1	0	0	0	0	0
3	3	3	1	0	0	0	0
4	4	7	4	1	0	0	0
5	5	13	13	5	1	0	0
6	6	22	37	22	6	1	0
7	7	34	92	92	34	7	1
8	8	51	223	415	223	51	8
9	9	73	541	2199	2199	541	73
10	10	103	1313	14293	38676	14293	1313
11	11	140	3194	103730	962654	962654	103730
12	12	188	7762	772652	26087225	92112225	26087225

Table 6.21 Values of R_{nk2}

$n \backslash k$	1	2	3	4	5	6	7
1	1	0	0	0	0	0	0
2	1	0	0	0	0	0	0
3	1	1	0	0	0	0	0
4	1	1	1	0	0	0	0
5	1	2	2	1	0	0	0
6	1	3	5	3	1	0	0
7	1	4	10	10	4	1	0
8	1	5	18	28	18	5	1
9	1	7	31	71	71	31	7
10	1	8	51	165	250	165	51
11	1	10	79	361	809	809	361
12	1	12	121	754	2 484	3 759	2 484
13	1	14	177	1 503	7 240	16 749	16 749
14	1	16	254	2 893	20 341	72 828	113 662

Table 6.22 Values of R_{nk3}

$n \backslash k$	1	2	3	4	5	6	7
1	1	0	0	0	0	0	0
2	1	0	0	0	0	0	0
3	1	1	0	0	0	0	0
4	1	2	1	0	0	0	0
5	1	3	3	1	0	0	0
6	1	5	10	5	1	0	0
7	1	7	24	24	7	1	0
8	1	10	55	105	55	10	1
9	1	13	116	403	403	116	13
10	1	17	231	1 506	3 000	1 506	231
11	1	21	438	5 425	23 579	23 579	5 425
12	1	27	813	19 440	199 473	469 473	199 473
13	1	32	1 451	68 478	1 758 953	10 925 684	10 925 684
14	1	39	2 533	237 709	15 575 102	267 929 503	723 109 414

Table 6.23 Values of R_{nk4}

$n\backslash k$	1	2	3	4	5	6	7
1	1	0	0	0	0	0	0
2	1	0	0	0	0	0	0
3	1	1	0	0	0	0	0
4	1	2	1	0	0	0	0
5	1	4	4	1	0	0	0
6	1	6	14	6	1	0	0
7	1	9	38	38	9	1	0
8	1	13	104	238	104	13	1
9	1	18	276	1573	1573	276	18
10	1	25	711	11566	34288	11566	711
11	1	32	1793	88140	909664	909664	88140
12	1	42	4446	665736	25020688	90186547	25020688
13	1	53	10691	4836136	664473418	9137113963	9137113963

Table 6.24 Values of $\overline{R}_{nk2}$

$n\backslash k$	1	2	3	4	5	6	7
1	1	0	0	0	0	0	0
2	0	0	0	0	0	0	0
3	0	1	0	0	0	0	0
4	0	0	1	0	0	0	0
5	0	0	1	1	0	0	0
6	0	0	1	2	1	0	0
7	0	0	1	4	3	1	0
8	0	0	0	5	9	4	1
9	0	0	0	5	22	19	6
10	0	0	0	4	40	70	35
11	0	0	0	3	60	220	190
12	0	0	0	2	86	629	977
13	0	0	0	1	110	1700	4875
14	0	0	0	1	127	4463	24920

Table 6.25 Values of $\overline{R}_{nk3}$

$n \backslash k$	1	2	3	4	5	6	7
1	1	0	0	0	0	0	0
2	0	0	0	0	0	0	0
3	0	1	0	0	0	0	0
4	0	1	1	0	0	0	0
5	0	0	2	1	0	0	0
6	0	0	4	4	1	0	0
7	0	0	4	14	6	1	0
8	0	0	3	39	39	9	1
9	0	0	3	88	227	93	12
10	0	0	2	196	1 340	1 078	199
11	0	0	1	399	8 652	15 695	4 468
12	0	0	1	805	61 904	302 573	164 499
13	0	0	1	1 503	459 017	6 813 448	9 113 636
14	0	0	0	2 658	3 344 644	158 913 391	601 158 522

Table 6.26 Values of $\overline{R}_{nk4}$

$n \backslash k$	1	2	3	4	5	6	7
1	1	0	0	0	0	0	0
2	0	0	0	0	0	0	0
3	0	1	0	0	0	0	0
4	0	1	1	0	0	0	0
5	0	1	3	1	0	0	0
6	0	0	7	5	1	0	0
7	0	0	10	26	8	1	0
8	0	0	13	124	83	12	1
9	0	0	17	643	1 173	244	17
10	0	0	19	3 831	24 942	10 266	663
11	0	0	19	23 437	641 872	820 142	84 184
12	0	0	17	138 181	16 799 302	81 159 989	24 211 108
13	0	0	13	761 022	418 548 455	8 123 840 077	8 853 245 774

6.5 Critical Codes

According to 6.2.13, appending a nonzero column to an indecomposable code yields a code which is again indecomposable. This shows that there exists an infinite family of k-dimensional indecomposable linear codes over any field $\mathbb{F}_q$ and for any dimension k. On the other hand, the $(n-1,k)$-code obtained by deleting an arbitrary column of a generator matrix of an indecomposable (n,k)-code can be either decomposable or indecomposable. For this reason, we investigate a restricted class of indecomposable codes, the *critical, indecomposable* codes, for short *critical* codes, introduced in [6]. An indecomposable code is called critical if the removal of any column of a generator matrix results in a decomposable code. In this section we prove that for a given dimension there are only finitely many critical, indecomposable codes and that any indecomposable code is obtained from a critical code by appending columns to a generator matrix of the critical code. Similarly as in Section 6.2, we may always assume that the codes are nonredundant. The present section is mainly a summary of [6]. All theorems and examples are quoted or excerpted from [6]. However, the order of the material presented is changed slightly.

Given an arbitrary code C, we may consider the subcode which is generated by the codewords of weight 1. If such words exist, the subcode generated by them splits off as an outer direct summand. Therefore, C is not indecomposable. If $\text{dist}(C) > 1$, we investigate the subcode E of C which is generated by the vectors of weight 2. If such vectors exist, then E may or may not be an outer direct summand. The code E itself turns out to be an outer direct sum of codes, each summand being equivalent to a code which is the dual of a one-dimensional code generated by the all-one vector.

The support of a vector was defined in Section 1.6. The *support* of a vector space is the union of the supports of its elements. If the support of E is sufficiently large compared to the support of the code C and C is indecomposable, then we will prove that C is a critical code.

A particular class of vector space homomorphisms plays an important role for the following considerations.

Definition (code homomorphism) Let C and D be two linear codes over $\mathbb{F}_q$. A

code homomorphism is a vector space homomorphism $\varphi \colon C \to D$ such that

$$\text{wt}(\varphi(c)) \leq \text{wt}(c), \qquad c \in C. \qquad \diamond$$

In other words, code homomorphisms are linear mappings which are contractions with respect to the Hamming metric.

6.5.2   **Examples**

1. Let Y be a subset of $n = \{0, \ldots, n-1\}$ and $n \geq 1$. For $f \in \mathbb{F}_q^n$ let $f \downarrow Y$ be the restriction of f to Y. If C is a subspace of $\mathbb{F}_q^n$ and $D = \{f \downarrow Y \mid f \in C\}$, then the mapping $\varphi \colon C \to D$ defined by $\varphi(f) := f \downarrow Y$ is a code homomorphism. It is called a *projection* of C onto D. If $\dim(D) = \dim(C)$, then in coding theory we usually say that D is obtained from C by puncturing (cf. 2.2.8). We call D the projection of C onto Y.

2. If C' is a subspace of $C \subseteq \mathbb{F}_q^n$, then the natural injection of C' into C is a code homomorphism.

3. If C is a critical, indecomposable code of length $n > 1$, then the projections of C onto $n \setminus \{i\}$ are decomposable for $i \in n$.

4. A projection D of a decomposable code C is decomposable or indecomposable. If it is indecomposable, then $\dim(D) < \dim(C)$.   ◇

Based on code homomorphisms it is possible to introduce the category of linear codes. We will not do it here. For further details consult [6].

If $\varphi \colon C \to D$ is a vector space isomorphism so that both φ and its inverse $\varphi^{-1} \colon D \to C$ are code homomorphisms, then φ is called a *code isomorphism*. It follows, therefore, that a code isomorphism preserves weights. If $\varphi \colon C \to D$ is a code isomorphism and C and D are of the same length, then φ is a linear isometry in the sense of Section 1.4. Hence, if we do not restrict our attention to nonredundant codes, then the notion "up to isomorphism" is a generalization of the notion "up to linear isometry". Two codes which are the same up to isomorphism can have different block-lengths. Restricting ourselves to nonredundant codes the two notions mean the same. The projection $C \to D$ of a nonredundant code C is a code isomorphism if and only if $C = D$.

Even if the code homomorphism is a vector space isomorphism its inverse need not be a code homomorphism.

6.5.3   **Example** For $n > 1$ let C be the $(n, n-1)$-parity check code and D the puncturing of C in the first component. Then the projection $\varphi \colon C \to D$ is a vector space isomorphism. For each $c \in C$ whose first component is different from 0 we have $\mathrm{wt}(\varphi(c)) < \mathrm{wt}(c)$ and, consequently, $\mathrm{wt}(\varphi^{-1}(\varphi(c))) > \mathrm{wt}(\varphi(c))$. Thus, φ^{-1} is not a code homomorphism.   ◇

6.5.4   **Definition (critical code)** An indecomposable code C is called *critical, indecomposable* or just *critical* if, whenever $\varphi \colon C \to D$ is a projection which is not a code isomorphism, either $\dim(D) < \dim(C)$, or D is decomposable.   ◇

Examples 6.5.5

1. Up to isomorphism, there is exactly one critical code of dimension 1 over $\mathbb{F}_q$, namely $\mathbb{F}_q$.

2. In dimension 2 there is up to isomorphism one critical code, namely the $(3,2)$-code with generator matrix
$$\begin{pmatrix} 1 & 0 & 1 \\ 0 & 1 & 1 \end{pmatrix}.$$

3. For $n \geq 3$, the unique indecomposable $(n, n-1)$-code (cf. 6.2.19) is a critical code. It has a generator matrix of the form

$$\begin{pmatrix} 1 & & & 1 \\ & \ddots & & \vdots \\ & & 1 & 1 \end{pmatrix}$$

6.5.6

where all entries, which are not specified, are equal to 0.

4. There is no critical code of length 2.

5. Consider an indecomposable code C with a repeated column, the last column say. Projecting this code onto all but the last column yields a surjective code homomorphism, which is not an isomorphism. The image is indecomposable and has the same dimension as C, whence C is not critical. In particular, a critical, indecomposable code has no repeated columns. For example, according to Table 6.21 there exist two indecomposable binary $(5,2)$-codes. They are given by the generator matrices
$$\Gamma_1 = \begin{pmatrix} 1 & 0 & 0 & 0 & 1 \\ 0 & 1 & 1 & 1 & 1 \end{pmatrix} \quad \text{and} \quad \Gamma_2 = \begin{pmatrix} 1 & 0 & 1 & 0 & 1 \\ 0 & 1 & 0 & 1 & 1 \end{pmatrix}.$$
They both project onto the unique critical binary code of dimension 2.

6. According to Table 6.22, there exist exactly two indecomposable binary $(5,3)$-codes. They are given by the generator matrices
$$\Gamma_1 = \begin{pmatrix} 1 & 0 & 0 & 1 & 0 \\ 0 & 1 & 0 & 1 & 1 \\ 0 & 0 & 1 & 0 & 1 \end{pmatrix} \quad \text{and} \quad \Gamma_2 = \begin{pmatrix} 1 & 0 & 0 & 1 & 1 \\ 0 & 1 & 0 & 1 & 1 \\ 0 & 0 & 1 & 1 & 1 \end{pmatrix}.$$
Their weight distributions are
$$w_{C_1}(x) = 1 + 2x^2 + 4x^3 + x^4 \quad \text{and} \quad w_{C_2}(x) = 1 + 3x^2 + 3x^3 + x^5.$$
Deleting the second column of Γ_1 and the last column of Γ_2 shows that both codes project onto the same critical binary $(4,3)$-code with generator matrix
$$\begin{pmatrix} 1 & 0 & 0 & 1 \\ 0 & 1 & 0 & 1 \\ 0 & 0 & 1 & 1 \end{pmatrix}$$

◇

It is possible to generalize the last example.

6.5.7   **Theorem** *Over any field there is a unique critical code of dimension 3. Up to isomorphism, it is the* $(4,3)$*-code with generator matrix.*

$$\begin{pmatrix} 1 & 0 & 0 & 1 \\ 0 & 1 & 0 & 1 \\ 0 & 0 & 1 & 1 \end{pmatrix}.$$

Proof: Assume that C is a three-dimensional, critical code over $\mathbb{F}_q$. Then we can find a linearly isometric code with generator matrix $(I_3 \mid A)$.

If A has a column of weight three, then the columns of I_3 together with this additional column are the generator matrix of a projection of C. Moreover, this projection is a critical, three-dimensional, indecomposable code. Hence, it must be the generator matrix for the code linearly isometric to C. By changing the basis suitably, one can achieve that the column of weight 3 consists of three ones. A monomial transformation then gives the desired generator matrix.

If there is no column of weight 3 in A, then all columns of A have weight 2. There are no columns of weight 1, since they would be repeated columns, contradicting the fact that C is critical. Moreover, since C is indecomposable, there must be at least two columns in A whose zeros are in different rows. We want to prove that there is no critical $(5,3)$-code. Again, by a suitable monomial transformation we can assume that the generator matrix is of the form

$$\begin{pmatrix} 1 & 0 & 0 & 1 & 1 \\ 0 & 1 & 0 & 1 & 0 \\ 0 & 0 & 1 & 0 & 1 \end{pmatrix}.$$

Projecting C onto the last four coordinates gives a three-dimensional critical, indecomposable code which is easily seen to be linearly isometric to the $(4,3)$-code. This shows that no critical, three-dimensional, indecomposable code with block length greater than four exists.   □

The situation in dimension 4 is more interesting.

6.5.8   **Example** The binary $(5,4)$-parity check code is a critical code.

The projection of the binary $(7,4)$-Hamming-code onto any 6 coordinates is a critical $(6,4)$-code. A suitable generator matrix of this code is given by

$$\begin{pmatrix} 1 & 1 & 0 & 0 & 0 & 0 \\ 0 & 0 & 1 & 1 & 0 & 0 \\ 0 & 0 & 0 & 0 & 1 & 1 \\ 0 & 1 & 0 & 1 & 0 & 1 \end{pmatrix}.$$

This critical code belongs to an infinite class of critical binary $(2m, m + 1)$-codes, $m \geq 2$, with generator matrix

$$
\begin{pmatrix}
1 & 1 & 0 & 0 & \cdots & 0 & 0 \\
0 & 0 & 1 & 1 & \cdots & 0 & 0 \\
 & & & \ddots & & & \\
0 & 0 & 0 & 0 & \cdots & 1 & 1 \\
0 & 1 & 0 & 1 & \cdots & 0 & 1
\end{pmatrix}.
$$

For $m = 2$ we have the $(4, 3)$-parity check code, for $m = 3$ the code above.   ◇

If C is an indecomposable code which is not critical, then there exists a projection of C onto an indecomposable code of the same dimension but smaller length. This proves the next

Corollary *For any indecomposable code C, there exists a critical code D of the same dimension as C and a projection of C onto D.*   □   **6.5.9**

All indecomposable codes are given by adjoining columns to the generator matrix of a critical code. For example, all 2-dimensional binary indecomposable codes have generator matrices of the form

$$
\begin{pmatrix}
1 & \cdots & 1 & 0 & \cdots & 0 & 1 & \cdots & 1 \\
0 & \cdots & 0 & 1 & \cdots & 1 & 1 & \cdots & 1
\end{pmatrix}
$$

and project onto the unique critical $(3, 2)$-code over $\mathbb{F}_2$ by eliminating repeated columns.

By eliminating repeated columns we obtain the reduced code of C. By further deleting zero columns we obtain a projective code. The reduced code is indecomposable if and only if the original code had this property.

Definition (critical column) Let C be an indecomposable code of length $n > 1$.   **6.5.10**
The i-th column, $i \in n$, of C is *critical* if the projection of C onto $n \setminus \{i\}$ is a decomposable code. In other words, the i-th column is critical if the code which is obtained from C by puncturing the i-th coordinate is decomposable.

◇

Corollary *An indecomposable code C of length $n > 1$ is critical if and only if all its*   **6.5.11**
columns are critical.   □

Now we determine all critical $(n, n - 2)$-codes for $n > 2$.

Theorem *If C is a critical (nonredundant) $(n, n - 2)$-code over $\mathbb{F}_q$, then $n > 3$.*   **6.5.12**
It has minimum distance 2 and the subcode E of C generated by all codewords of

weight 2 *has support* n. *Moreover,* C *is linearly isometric to a code with a generator matrix of the form*

$$\Gamma = (I_{n-2} \mid A) \ \text{with} \ A = \begin{pmatrix} 1 & a_0 \\ \vdots & \vdots \\ 1 & a_{r-1} \\ 0 & 1 \\ \vdots & \vdots \\ 0 & 1 \end{pmatrix},$$

where the weight of the second column of A *is greater than* $n - 2 - r$ *but less than* $n - 2$. *Moreover, if* $a_i \neq 0$, *then there exists some* $j \in r \setminus \{i\}$ *with* $a_j = a_i$.

Conversely, any matrix A *as above yields a critical* $(n, n - 2)$-*code over* $\mathbb{F}_q$.

Proof: For $n = 3$ there is no critical, nonredundant $(3, 1)$-code. Hence, we assume that $n > 3$. The code C is linearly isometric to a code with generator matrix of the form $(I_{n-2} \mid A)$ where neither of the two columns of A can have weight $n - 2$, since otherwise C would not be critical. By a monomial transformation we can assure that the first column of A is a sequence of r ones, $r < n - 2$, followed by a sequence of zeros. If $a = (a_0, \ldots, a_{r-1}, a_r, \ldots, a_{n-3})^\top$ is the second column of A, then necessarily $a_i \neq 0$ for $i \geq r$, since C cannot have minimum weight 1. By a further monomial transformation, we can assume that these entries are equal to 1. Moreover, since C is indecomposable there must be some $i \in r$ so that $a_i \neq 0$. Thus, the weight of the second column of A is greater than $n - 2 - r$ but less than $n - 2$.

We next prove that for each $i \in r$ with $a_i \neq 0$ there exists some $j \in r$, $j \neq i$, such that $a_j = a_i$. If for i with $a_i \neq 0$ there were no j with $a_i = a_j$, then we can proceed as follows: Multiply the last column by a_i^{-1} so that there is a single 1 in the last column. (Then the elements in the last column are of the form $a_j a_i^{-1}$.) By elementary row operations it is possible to replace all nonzero entries different from 1 in the last column by 0. (For each j different from i we have to multiply the i-th row by $a_j a_i^{-1}$ and subtract the result from the j-th row.) After these row operations all entries of the last but one column are different from 0. Hence this matrix contains the $n - 2$ unit vectors, a column of weight $n - 2$ and a further column. Thus it is a generator matrix of a code which is not critical. This is a contradiction, since this code is linearly isometric to a critical code.

Lastly, we prove the assertion concerning the subcode E. The last $n - 2 - r$ rows of Γ belong to E, whence $\{i \in n \mid r \leq i \leq n - 3\} \cup \{n - 1\}$ is a subset of the support of E. Moreover, there exists $i \in r$ such that $a_i = 0$. Consequently, $\{n - 2\} \cup \{i \in r \mid a_i = 0\}$ is also contained in the support of E. Finally, consider some $i \in r$ with $a_i \neq 0$, then there is some $j \in r$ with $a_j = a_i$, and the sum of the i-th and j-th row is contained in E. Its support is $\{i,j\}$. Therefore, i and j also belong to the support of E. This proves that E has full support.   □

Now we want to describe the structure of critical codes. This way we find a "quasicanonical form" of critical and indecomposable codes.

First we need the following lemma describing the subspace generated by all codewords of weight 2.

Lemma *Let E be a code over $\mathbb{F}_q$ with minimum distance 2 which is generated by its vectors of weight 2. Then*

$$E = E_0 \dot{+} \ldots \dot{+} E_{r-1}$$

where each E_i is linearly isometric to an indecomposable $(n_i, n_i - 1)$-parity check code with $n_i \geq 2$.

6.5.13

Proof: We consider E as a code of length n with support $n = \{0, \ldots, n-1\}$. We introduce an equivalence relation on n by saying that i is in relation to j whenever there exists a codeword $c \in E$ of weight 2 so that $c_i \neq 0 \neq c_j$. Let $X_0, \ldots, X_{r-1}$ be the equivalence classes of this relation.

For $i \in r$ let E_i be the subspace of E which is generated by all vectors of weight 2 with support in X_i. Then $E_i \cap E_j = \{0\}$ for $i \neq j$ and $E_0 \dot{+} \ldots \dot{+} E_{r-1} = E$. By construction $|X_i| = n_i \geq 2$. Projecting E_i to its support X_i yields an indecomposable $(n_i, n_i - 1)$-code.   □

Corollary *Any code with minimum distance 2 which is generated by its vectors of weight 2 is linearly isometric to a code with generator matrix*

6.5.14

$$\left(\begin{array}{c|c|c|c} \Gamma_0 & 0 & \cdots & 0 \\ \hline 0 & \Gamma_1 & & 0 \\ \hline \vdots & & \ddots & \vdots \\ \hline 0 & 0 & \cdots & \Gamma_{r-1} \end{array} \right),$$

where Γ_i, $i \in r$, is an $(n_i - 1) \times n_i$-matrix, $n_i \geq 2$, of the form 6.5.6.   □

Note that this code is indecomposable if and only if $r = 1$.

From the test on indecomposability, 6.2.13, we obtain the following

---

Corollary *If C is a critical code with generator matrix of the form $(I_k \mid A)$, then any walk visiting all the k rows of the graph $\mathcal{G}_A$ defined on page 469 also visits every column of $\mathcal{G}_A$.*

6.5.15

Proof: If there were a walk visiting all rows but not all columns, then some columns of A could be eliminated and the resulting code would still be indecomposable. This is a contradiction to the assumption that C is critical.   □

This, however, is only a necessary, not a sufficient property for a code to be critical. We proceed with the following combinatorial lemma, which will later be applied to the supports of the columns of A.

6.5.16   **Lemma** *Let R be a finite set and C a collection of subsets of R satisfying the two conditions:*

1. *There is a sequence $R_0, R_1, \ldots, R_{m-1}$ of elements of C such that $R_i \cap R_{i+1} \neq \emptyset$ for $i \in m-1$ and $\bigcup_{i \in m} R_i = R$.*

2. *C is minimal with respect to the above property, i.e. no proper subset of C posses a sequence of elements satisfying this property.*

Then there exists some $r \in R$ and $i \in m$ such that $r \in R_i$ and $r \notin R_j$ for $j \neq i$. Moreover, if $|C| > 1$, then there exist at least two elements $r, r' \in R$, $r \neq r'$, and $i, i' \in m$, $i \neq i'$, with $r \in R_i$, $r \notin R_j$ for $j \neq i$ and $r' \in R_{i'}$, $r' \notin R_j$ for $j \neq i'$.

Proof: Any sequence from C having the required property must contain each element of C at least once by the minimality assumption. Choose a sequence $R_0, \ldots, R_{m-1}$ from C with the required property and with m minimal. If $m = 1$, then $C = \{R_0\}$ and the assertion is trivial. Otherwise, consider the shorter sequences $R_1, \ldots, R_{m-1}$ and $R_0, \ldots, R_{m-2}$. Since they both enjoy the intersection property, necessarily

$$\bigcup_{i=1}^{m-1} R_i \neq R \neq \bigcup_{i=0}^{m-2} R_i.$$

Due to the fact that m was minimal neither R_0 occurs among $R_2, \ldots, R_{m-1}$ nor R_{m-1} occurs among $R_0, \ldots, R_{m-2}$. Choose $r_0 \in R_0$, $r_0 \notin \bigcup_{i=1}^{m-1} R_i$ and $r_{m-1} \in R_{m-1}$, $r_{m-1} \notin \bigcup_{i=0}^{m-2} R_i$. Then $r_0 \neq r_{m-1}$ and we get the desired result.

□

6.5.17   **Theorem** *Every critical (n,k)-code with $n > 2$ has minimum distance 2. If $n \geq 4$, then the code contains at least two vectors of weight 2 with disjoint support.*

Let E be the subcode of C generated by all codewords of weight 2. Then either $C = E$ or there exists a subspace F of C of minimum distance greater than 2 so that $C = E + F$ and $E \cap F = \{0\}$. The subcode E can be expressed as $E = E_0 + \ldots + E_{r-1}$, $r \geq 1$, where each E_i is linearly isometric to an indecomposable $(n_i, n_i - 1)$-parity check code.

When $C = E + F$ with $F \neq \{0\}$, then F is an indecomposable code. Assume without loss of generality, that the support of E is equal to $s = \{0, \ldots, s-1\}$. If $s < n$, then the columns of F with column index in $\{s, \ldots, n-1\}$ are critical columns of F. The code F is also known as the auxiliary indecomposable code attached to C.

A generator matrix of a code linearly isometric to C is of the form

$$
\Gamma = \left(\begin{array}{cccc|c|cc}
\Gamma_0 & 0 & \cdots & & 0 & 0 \\
\hline
0 & \Gamma_1 & \cdots & & 0 & 0 \\
\vdots & & \ddots & & & \vdots \\
\hline
0 & 0 & \cdots & \Gamma_{r-1} & & 0 \\
\hline
\Lambda_0 & \Lambda_1 & \cdots & \Lambda_{r-1} & & \Lambda_r
\end{array} \right),
$$

where Γ_i, $i \in r$, is an $(n_i - 1) \times n_i$-matrix, $n_i \geq 2$, of the form 6.5.6. Each Λ_i, $i \in r$, is of the form

$$
\Lambda_i = \begin{pmatrix}
0 & \cdots & 0 & \ell_{0i} \\
\vdots & \ddots & \vdots & \vdots \\
0 & \cdots & 0 & \ell_{\delta-1,i}
\end{pmatrix}
$$

where $\delta = \dim(F) = k - \sum_i (n_i - 1)$, and $(\ell_{0i}, \dots, \ell_{\delta-1,i}) \in \mathbb{F}_q^\delta \setminus \{0\}$. Finally, all columns of the $\delta \times (n-s)$-matrix Λ_r are nonzero and critical. The matrix Γ is called a quasicanonical *form of C. The submatrix $(\Lambda_0 \mid \dots \mid \Lambda_r)$ is a generator matrix of F. The nonzero columns of this submatrix yield a generator matrix*

$$
\left(\begin{array}{ccc|c}
\ell_{00} & \cdots & \ell_{0,r-1} & \\
\vdots & \ddots & \vdots & \Lambda_r \\
\ell_{\delta-1,0} & \cdots & \ell_{\delta-1,r-1} &
\end{array} \right)
$$

of F projected onto its support which is a nonredundant, indecomposable code.

Proof: Assume that C is a nonredundant, critical (n,k)-code with $n > 2$ and systematic generator matrix $\Gamma = (I_k \mid A)$, where necessarily $k < n$. Let R be k, the set of all row-indices of Γ, and let $\mathcal{C}$ be the set of the supports of the columns of A. According to 6.5.15, $\mathcal{C}$ satisfies the assumptions of 6.5.16. Consequently, there exists some $i \in k$ such that i belongs to exactly one column of A, thus the i-th row of Γ is a codeword of weight 2. By 6.2.18, any indecomposable code of length greater than 1 has minimum distance at least 2. Hence, $\text{dist}(C) = 2$.

Assume that $n \geq 4$. If $n - k \geq 2$, then $|\mathcal{C}| > 1$, whence there exist $i, j \in k$, $i \neq j$, so that there is exactly one column of A the support of which contains i and there is exactly one column of A the support of which contains j. Consequently, the i-th and the j-th row of Γ are two codewords of weight 2 with disjoint support. If $n - k = 1$, then C is the $(n, n-1)$-parity check code which contains at least two codewords of weight 2 with disjoint support.

Let E be the subcode of C generated by the vectors of weight 2. By 6.5.13

$$
E = E_0 \dotplus \dots \dotplus E_{r-1},
$$

where each E_i is linearly isometric to a unique indecomposable $(n_i, n_i - 1)$-code, $n_i \geq 2$. If $n_i = 2$, then E_i is the repetition code, otherwise E_i is a critical code. It is possible that the support of E is a proper subset of n. In this case assume, without loss of generality, that the support of E is s. Moreover, we assume that the support X_0 of E_0 consists of the first n_0 columns, and the support X_i of E_i consists of the n_i columns following the support of E_{i-1}, for $1 \leq i < r$. Thus $X_0 = \{0, \ldots, n_0 - 1\} = n_0$, $X_1 = \{n_0, \ldots, n_0 + n_1 - 1\} = (n_0 + n_1) \setminus n_0$, and so on.

If $r = 1$ and $C = E$, we are done. Otherwise E is properly contained in C and $C = E + F$ where $F \cap E = \{0\}$. Since E contains all codewords of C of weight 2, the code F has minimum distance at least 3. By suitable row operations it is possible to find generators of F the support of which is contained in

$$S = \left\{ \sum_{j=0}^{i} n_j - 1 \;\middle|\; i \in r \right\} \cup \{s, s+1, \ldots, n-1\}.$$

Recall that $\sum_{j=0}^{i} n_j - 1$ belongs to the support of E_i, $i \in r$. Moreover, since C is indecomposable, S is the support of F.

The fact that C is indecomposable implies that also F is indecomposable. The fact that C is critical implies that all columns with index in $\{s, \ldots, n-1\}$ are critical. ☐

If $r > 1$ and $s = n$, then necessarily $r \geq 3$, since otherwise the weight of the generators of F would be less than 3, what is impossible since all codewords of weight 2 belong to E and there are no codewords of weight 1 in C.

This way we obtain only a quasicanonical form of critical codes since we have specified neither the order of the Γ_i nor the order of the nonzero columns of the matrices Λ_i. This description of the quasicanonical form yields a method for constructing critical codes and arbitrary indecomposable codes. Any indecomposable code is linearly isometric to a code with generator matrix

$$\Gamma = \left(\begin{array}{ccccc|c}
\Gamma_0 & 0 & \cdots & 0 & 0 & N_0 \\
0 & \Gamma_1 & \cdots & 0 & 0 & N_1 \\
\vdots & & \ddots & & \vdots & \vdots \\
0 & 0 & \cdots & \Gamma_{r-1} & 0 & N_{r-1} \\
\Lambda_0 & \Lambda_1 & \cdots & \Lambda_{r-1} & \Lambda_r & N_r
\end{array} \right),$$

with suitable matrices N_i, $0 \leq i \leq r$.

6.5.18 **Example** Consider the binary $(5, 2, 3)$-code F with generator matrix

$$\Gamma = \begin{pmatrix} 1 & 0 & 1 & 0 & 1 \\ 0 & 1 & 0 & 1 & 1 \end{pmatrix}$$

which is indecomposable and has one critical column, the last. Now we want to construct a nonredundant, critical $(9, 6)$-code with auxiliary code F. Since

$n_i \geq 2$, we have $r = 4$, $n_0 = n_1 = n_2 = n_3 = 2$ and $s = 8$. Therefore, a quasicanonical form of the critical $(9,6)$-code is

$$\begin{pmatrix} 1 & 1 & 0 & 0 & 0 & 0 & 0 & 0 & 0 \\ 0 & 0 & 1 & 1 & 0 & 0 & 0 & 0 & 0 \\ 0 & 0 & 0 & 0 & 1 & 1 & 0 & 0 & 0 \\ 0 & 0 & 0 & 0 & 0 & 0 & 1 & 1 & 0 \\ 0 & 1 & 0 & 0 & 0 & 1 & 0 & 0 & 1 \\ 0 & 0 & 0 & 1 & 0 & 0 & 0 & 1 & 1 \end{pmatrix}.$$

$\diamond$

Using these quasicanonical forms, we are able to classify the critical $(n, n-2)$-codes in more details.

Corollary *The quasicanonical generator matrix of a critical, indecomposable $(n, n-2)$-code over $\mathbb{F}_q$ with $n > 3$ is of the form*   **6.5.19**

$$\Gamma = \begin{pmatrix} \Gamma_0 & 0 & \cdots & 0 \\ 0 & \Gamma_1 & \cdots & 0 \\ \vdots & & \ddots & \\ 0 & 0 & \cdots & \Gamma_{r-1} \\ \Lambda_0 & \Lambda_1 & \cdots & \Lambda_{r-1} \end{pmatrix},$$

where $r \geq 3$, Γ_i is an $(n_i - 1) \times n_i$-matrix, $n_i \geq 2$, given by 6.5.6, $i \in r$. Moreover, $\Lambda_i = (0 \mid \ldots \mid 0 \mid e^{(i)})$ is an $(r-2) \times n_i$-matrix for $i \in r - 2$,

$$\Lambda_{r-2} = \begin{pmatrix} 0 & \cdots & 0 & 1 \\ \vdots & \ddots & \vdots & \vdots \\ 0 & \cdots & 0 & 1 \end{pmatrix} \quad \text{and} \quad \Lambda_{r-1} = \begin{pmatrix} 0 & \cdots & 0 & \ell_0 \\ \vdots & \ddots & \vdots & \vdots \\ 0 & \cdots & 0 & \ell_{r-3} \end{pmatrix}$$

with pairwise different, nonzero elements $\ell_0, \ldots, \ell_{r-3}$. Thus, we obtain the following estimates: $q - 1 \geq r - 2$ and $n \geq 6$.

Proof: The quasicanonical form of critical codes was described in 6.5.17. According to 6.5.12, the code E generated by all codewords of weight 2 has full support. Whence, $n - s = 0$ and the matrix Λ_r does not occur in this quasicanonical form. By construction $r = 1$ and $r = 2$ are impossible. If $r \geq 3$, then the auxiliary code F projected onto its nonzero columns is an $(r, r-2)$-code $\tilde{F}$ with minimum distance $d \geq 3$. Therefore, it is an MDS-code. According to 2.5.6, there exists a systematic generator matrix of a code linearly isometric to $\tilde{F}$ with generator matrix of the form

$$\begin{pmatrix} 1 & 0 & \cdots & 0 & 1 & \ell_0 \\ 0 & 1 & \cdots & 0 & 1 & \ell_1 \\ \vdots & & \ddots & & \vdots & \vdots \\ 0 & 0 & \cdots & 1 & 1 & \ell_{r-3} \end{pmatrix}.$$

$\square$

For the binary case we obtain even a canonical form of critical $(n, n-2)$-codes.

6.5.20   **Corollary** *The binary critical $(n, n-2)$-codes, $n \geq 6$, have the canonical form*

$$\Gamma = \begin{pmatrix} \Gamma_0 & 0 & 0 \\ \hline 0 & \Gamma_1 & 0 \\ \hline 0 & 0 & \Gamma_2 \\ \hline \Lambda_0 & \Lambda_1 & \Lambda_2 \end{pmatrix},$$

where Γ_i is an $(n_i - 1) \times n_i$-matrix given by 6.5.6, $i \in 3$, with $n_0 \geq n_1 \geq n_2 \geq 2$, and Λ_i is an $1 \times n_i$-matrix of the form

$$\Lambda_i = (0 \quad \dots \quad 0 \quad 1), \qquad i \in 3.$$

Proof: Since $\mathbb{F}_2$ contains exactly two elements, we obtain from $1 \geq r - 2$ that $r \leq 3$, thus $r = 3$. Another proof of this fact is based on 2.5.7, where we have shown that there exist only trivial binary MDS-codes. Hence, only for $r = 3$ there exist binary $(r, r-2, 3)$-codes.   □

6.5.21   **Corollary** *The number of linearly nonisometric critical binary $(n, n-2)$-codes with $n \geq 6$ is the same as the number of partitions of $n - 3$ into three parts.*

Proof: The matrices Λ_i in the last row of a canonical form of a critical binary $(n, n-2)$-code have exactly one row. Therefore,

$$\sum_{i=0}^{2}(n_i - 1) = n - 3$$

is the sum of the ranks of the matrices Γ_i for $i \in 3$. Since $n_0 \geq n_1 \geq n_2$ and $n_2 - 1 \geq 1$, the sequence $(n_0 - 1, n_1 - 1, n_2 - 1)$ is a partition of $n - 3$.   □

6.5.22   **Example** For $n = 6$ there is exactly one partition of 3 with three parts, namely $3 = 1 + 1 + 1$. We have met the corresponding critical $(6, 4)$-code in 6.5.8. For $n = 7$ there is the unique partition $4 = 2 + 1 + 1$ which yields the canonical form

$$\Gamma = \begin{pmatrix} 1 & 0 & 1 & 0 & 0 & 0 & 0 \\ 0 & 1 & 1 & 0 & 0 & 0 & 0 \\ 0 & 0 & 0 & 1 & 1 & 0 & 0 \\ 0 & 0 & 0 & 0 & 0 & 1 & 1 \\ 0 & 0 & 1 & 0 & 1 & 0 & 1 \end{pmatrix}.$$

◇

Theorem *A nonredundant, binary, critical $(n, n-2)$-code C contains the all-one **6.5.23**
vector if and only if it comes from a partition of $n-3$ into three parts all of the same
parity (this means, that all three parts are either odd or even).*

Proof: Assume that $n-3$ has a partition $k_0 + k_1 + k_2$ with $k_0 \geq k_1 \geq k_2 \geq 1$.
Using the canonical form 6.5.20 of C we have: If all three k_i are odd, then

$$(\underbrace{1,\ldots,1,}_{k_0}\underbrace{1,\ldots,1,}_{k_1}\underbrace{1,\ldots,1,}_{k_2}0)\cdot\Gamma=(1,\ldots,1).$$

If all three k_i are even, then

$$(\underbrace{1,\ldots,1,}_{k_0}\underbrace{1,\ldots,1,}_{k_1}\underbrace{1,\ldots,1,}_{k_2}1)\cdot\Gamma=(1,\ldots,1).$$

Conversely, assume that $c = (1,\ldots,1)$ is contained in C. Then there exists
some $v \in \mathbb{F}_2^{n-2}$ so that $v \cdot \Gamma = c$. Moreover, assume that Γ corresponds to
a partition $k_0 + k_1 + k_2 = n - 3$ with $k_i = n_i - 1$, $i \in 3$. If k_0 is odd, then
the first k_0 entries of v must be equal to 1. These entries guarantee that $c_0 = c_1 = \ldots = c_{k_0-1} = 1$. The first k_0 components of c are not influenced by the
remaining v_i, $k_0 \leq i < k$. Since $c_{k_0} = 1$, necessarily v_{n-3}, the last component
of v, must be 0. Therefore, k_1 and k_2 are also odd, since otherwise $c_{n_0+n_1-1} = 0$ or $c_{n_0+n_1+n_2-1} = 0$. If k_0 is even, then similar considerations show, that
necessarily $v_{n-3} = 1$, in order to have $c_{k_0} = 1$ and consequently, both k_1 and
k_2 must be even.   □

Now we investigate the dual of a critical code.

Examples   **6.5.24**

1. If C is the critical $(n, n-1)$-code, $n > 2$, over $\mathbb{F}_q$, then, according to Exercise 1.3.9 its dual code $C^\perp$ is generated by $(-1,\ldots,-1,1)$. Thus, it is linearly isometric to the code generated by the all-one vector and its reduced
   code is the $(1,1)$-code $\mathbb{F}_q$.

2. If C is a critical (n, k)-code over $\mathbb{F}_q$ different from the critical $(n, n-1)$-code,
   then C has an auxiliary code F. Let $\tilde{F}$ be the projection of F onto its support,
   then $\tilde{F}$ is a nonredundant, indecomposable code. We want to prove that
   the reduced code of $C^\perp$ is linearly isometric to the reduced code of $\tilde{F}^\perp$. By
   6.2.14 the dual of $\tilde{F}$, whence also the reduced code of $\tilde{F}$, is indecomposable.

Using the quasicanonical form described in 6.5.17, the code C is linearly isometric to a code C' with generator matrix

$$
\begin{pmatrix}
I_{k_0} & 0 & \cdots & 0 & 0 & A_0 \\
\hline
0 & I_{k_1} & \cdots & 0 & 0 & A_1 \\
\hline
\vdots & & \ddots & & \vdots & \vdots \\
\hline
0 & 0 & \cdots & I_{k_{r-1}} & 0 & A_{r-1} \\
\hline
0 & 0 & \cdots & 0 & I_\delta & A
\end{pmatrix}
$$

where $k_i = n_i - 1$, I_{k_i} is the unit matrix, $i \in r$, and $(I_\delta \mid A)$ is a systematic generator matrix of a code linearly isometric to $\tilde{F}$, where $\delta = n - \sum_i k_i$ and A is a $\delta \times (n-k)$-matrix. Moreover, the rows of the matrix A_i, $i \in r$, are copies of a nonzero multiple of a single row of A or they are unit vectors. The dual of C' has a generator matrix of the form

$$
(-A_0^\top \mid \cdots \mid -A_{r-1}^\top \mid -A^\top \mid I_{n-k}).
$$

All columns of $-A_i^\top$, $i \in r$, are nonzero multiples of columns of $-A^\top$ or they are unit vectors, therefore, the reduced code of $C^\perp$ is linearly isometric to the reduced code of $\tilde{F}^\perp$.   ◇

It seems natural to ask from which critical, indecomposable codes a given inde-composable code might arise by augmentation of their quasicanonical genera-tor matrices. Or, equivalently, given an indecomposable (n, k)-code C, what are the the critical, indecomposable (m, k)-codes which arise as projections from C?

6.5.25   **Definition (spectrum of a code)** The *spectrum* $\mathrm{spec}(C)$ of an indecomposable code C is the set of all linear isometry classes of critical, indecomposable codes D which satisfy

- $\dim(D) = \dim(C)$

- there exists a projection of C onto D.   ◇

6.5.26   **Theorem** *The spectrum of an (n, k)-MDS-code with $1 < k < n$ contains only the linear isometry class of the unique critical $(k+1, k)$-parity check code.*

Proof: In each systematic generator matrix $(I_k \mid A)$ of any code linearly iso-metric to C all columns of A have weight k. Thus, the only critical k-dimen-sional code obtained as a projection of C is linearly isometric to the unique critical, $(k+1, k)$-parity check code.   □

Corollary *The spectrum of the m-th order q-ary Hamming-code C contains only one* **6.5.27**
element.

If $m > 2$, projecting C onto all but one coordinate yields a critical code in which the code generated by all vectors of weight 2 is the sum of the unique indecomposable q-ary $(q, q - 1)$-code repeated $(q^{m-1} - 1)/(q - 1)$ times and the auxiliary code is the $(m - 1)$-th order q-ary Hamming-code.

Proof: The first assertion follows from 6.5.26. The proof of the second assertion is based on design theory. The reader should consult [7]. □

Example The second order ternary Hamming-code has a generator matrix **6.5.28**

$$\begin{pmatrix} 1 & 0 & 1 & 1 \\ 0 & 1 & 1 & -1 \end{pmatrix}.$$

Therefore, the quasicanonical form of the critical code in the spectrum of the third order ternary Hamming-code is

$$\begin{pmatrix}
1 & 0 & 1 & 0 & 0 & 0 & 0 & 0 & 0 & 0 & 0 & 0 \\
0 & 1 & 1 & 0 & 0 & 0 & 0 & 0 & 0 & 0 & 0 & 0 \\
0 & 0 & 0 & 1 & 0 & 1 & 0 & 0 & 0 & 0 & 0 & 0 \\
0 & 0 & 0 & 0 & 1 & 1 & 0 & 0 & 0 & 0 & 0 & 0 \\
0 & 0 & 0 & 0 & 0 & 0 & 1 & 0 & 1 & 0 & 0 & 0 \\
0 & 0 & 0 & 0 & 0 & 0 & 0 & 1 & 1 & 0 & 0 & 0 \\
0 & 0 & 0 & 0 & 0 & 0 & 0 & 0 & 0 & 1 & 0 & 1 \\
0 & 0 & 0 & 0 & 0 & 0 & 0 & 0 & 0 & 0 & 1 & 1 \\
0 & 0 & 1 & 0 & 0 & 0 & 0 & 0 & 1 & 0 & 0 & 1 \\
0 & 0 & 0 & 0 & 0 & 1 & 0 & 0 & 1 & 0 & 0 & -1
\end{pmatrix}.$$

No matter which nonzero column we append as the last column, we obtain an indecomposable $(13, 10)$-code. In order to obtain the Hamming-code, we must append a column so that the minimum distance of the new code is equal to 3. For this reason the nonzero entries in the first two rows must have opposite signs. Similar arguments hold for all but the last two rows. Using for instance $(1, -1, 1, -1, 1, -1, 1, -1, 0, 0)^\top$ as the last column we obtain a generator matrix of the Hamming-code. ◇

It is also possible that the spectrum contains more than one linear isometry class.

Example Consider the binary $(7, 4)$-code with the generator matrix **6.5.29**

$$\begin{pmatrix}
1 & 0 & 0 & 0 & 1 & 1 & 1 \\
0 & 1 & 0 & 0 & 1 & 1 & 1 \\
0 & 0 & 1 & 0 & 1 & 1 & 0 \\
0 & 0 & 0 & 1 & 1 & 0 & 1
\end{pmatrix}.$$

Projecting onto the first five columns gives the unique critical $(5,4)$-code while projecting onto all but the fifth column gives the critical $(6,4)$-code of 6.5.8. ◇

The proof of the following theorem is left to the reader.

6.5.30   **Theorem** *Let C be a nonredundant, binary, indecomposable (n,k)-code.*

1. *If C contains the all-one vector, then each code in its spectrum contains the all-one vector.*

2. *Let $k > 1$. If C contains the all-one vector, and $\mathrm{spec}(C)$ contains the critical $(k+1,k)$-parity check code, then k is odd.*

3. *Assume that k is even and C contains the all-one vector. Then the critical $(k+1,k)$-parity check code is not in $\mathrm{spec}(C)$. If a critical $(k+2,k)$-code is contained in $\mathrm{spec}(C)$, then it must come from a partition of $k-1$ into three odd parts.* □

Now we come back to binary Reed–Muller-codes.

6.5.31   **Theorem**

1. *The $(m-1)$-th order Reed–Muller-code $\mathrm{RM}^2_{m,m-1}$ of degree $m > 1$ is the unique critical $(2^m, 2^m - 1)$-code.*

2. *The spectrum of $\mathrm{RM}^2_{m,m-2}$, for $m \geq 2$, contains exactly one code. This is the critical code underlying the m-th order binary Hamming-code (cf. 6.5.27).*

3. *Assume that $m > 3$. The spectrum of $\mathrm{RM}^2_{m,1}$ consists of all critical codes of dimension $m+1$ containing the all-one vector. Thus, there is a difference between the spectra depending on the parity of m. If m is odd, then the critical $(m+2, m+1)$-code is not in the spectrum, whereas it is contained in the spectrum when m is even. The critical $(m+3, m+1)$-codes in the spectrum of $\mathrm{RM}^2_{m,1}$ are described in 6.5.23.*

Proof: 1. According to Exercise 2.4.3, the code $\mathrm{RM}^2_{m,m-1}$ contains all vectors of length 2^m of even weight, therefore, it is the unique critical $(2^m, 2^m - 1)$-code.

2. From 2.4.11 we obtain that $\mathrm{RM}^2_{m,m-2}$ is the parity extension of the m-th order binary Hamming-code. Projecting on all but 2 coordinates gives the underlying critical code. Any two coordinates yield the same critical code.

3. Every binary Reed–Muller-code contains the all-one vector. By the first part of 6.5.30, all codes in the spectrum of $\mathrm{RM}^2_{m,1}$ contain the all-one vector. Since $\dim(\mathrm{RM}^2_{m,1}) = m + 1$, (cf. 2.4.7) we obtain the assertion on the $(m+2, m+1)$-code from the second part of 6.5.30. □

The spectra of the ternary and binary Golay-codes are described in [6].

Exercises

Exercise Prove 6.5.30.

E.6.5.1

6.6 Random Generation of Linear Codes

6.6

In Sections 6.1–6.3 we have shown how to enumerate the linear isometry classes of linear codes, in Chapter 9 we will describe how to determine a (complete) set of representatives for given parameters n, k and q. From the tables of numbers of linear isometry classes we immediately realize that only for relatively small values of these parameters it will be possible to determine the sets of representatives completely. The order of the acting group increases, and the number of representatives quickly gets out of hand. In such situations, probabilistic methods may still allow the construction of linear codes which are distributed uniformly at random over all isometry classes.

The *Dixon–Wilf-algorithm* allows the generation of linear codes which are distributed uniformly at random over all linear isometry classes. Actually this algorithm was first developed for the random generation of unlabeled graphs (cf. [46]). It can always be applied for the random generation of objects, which are orbits of a finite group acting on a finite set.

Therefore, we present the algorithm for an arbitrary finite action of a group G on a set X. The algorithm describes a method how to choose elements x_0 of X at random such that the probability that x_0 belongs to a given orbit $\omega \in G\backslash\backslash X$ is $1/|G\backslash\backslash X|$ for each orbit ω. This allows us to sample elements of X which are uniformly distributed over the G-orbits on X.

Dixon–Wilf-algorithm *Let G be a finite group acting on a finite, nonempty set X. Choose a conjugacy class $\mathcal{C}$ of elements of G with the probability*

6.6.1

$$p(\mathcal{C}) := \frac{|\mathcal{C}| \cdot |X_g|}{|G| \cdot |G\backslash\backslash X|} \text{ for an arbitrary } g \in \mathcal{C}.$$

Pick any $g \in \mathcal{C}$ and determine at random a fixed point x of g. Then the probability that x lies in a given orbit $\omega \in G\backslash\backslash X$ is equal to $1/|G\backslash\backslash X|$.

Proof: Let $\mathcal{C}_0, \ldots, \mathcal{C}_{N-1}$ be the conjugacy classes of elements of G with representatives $g_i \in \mathcal{C}_i$. As a consequence of the Lemma of Cauchy–Frobenius 3.4.2, it follows

$$\sum_{i \in N} p(\mathcal{C}_i) = \frac{\sum_{i \in N} |\mathcal{C}_i| \, |X_{g_i}|}{\sum_{g \in G} |X_g|} = 1,$$

whence $p(.)$ is a probability distribution. Then for each $\omega \in G\backslash\backslash X$ we determine the probability that x belongs to ω as

$$
\begin{aligned}
p(x \in \omega) &= \sum_{i \in N} p(\mathcal{C}_i) p(x \in X_{g_i} \cap \omega) \\
&= \sum_{i \in N} p(\mathcal{C}_i) \frac{|X_{g_i} \cap \omega|}{|X_{g_i}|} = \sum_{i \in N} \frac{|\mathcal{C}_i| \, |X_{g_i}|}{|G| \, |G\backslash\backslash X|} \frac{|X_{g_i} \cap \omega|}{|X_{g_i}|} \\
&= \frac{1}{|G| \, |G\backslash\backslash X|} \sum_{i \in N} |\mathcal{C}_i| \, |X_{g_i} \cap \omega| = \frac{1}{|G| \, |G\backslash\backslash X|} \sum_{g \in G} |X_g \cap \omega|.
\end{aligned}
$$

The last sum is equal to $|G|$, since for $\omega = G(x)$ we have

$$
\sum_{g \in G} |X_g \cap \omega| = \sum_{g \in G} \sum_{x \in X_g \cap \omega} 1 = \sum_{x \in \omega} \sum_{g \in G_x} 1 = \sum_{x \in \omega} |G_x| = |G_x| \, |\omega| = |G|. \quad \square
$$

As we have seen in 6.1.15, the linear isometry classes of linear (n, l)-codes for $1 \le l \le k$ with $k \le n$ correspond to the $\mathrm{GL}_k(q) \times S_n$-orbits on the set of mappings from n to $\mathrm{PG}^*_{k-1}(q)$.

For this reason we formulate the Dixon–Wilf-algorithm for the canonical action of a direct product $H \times G$ on Y^X introduced in 1.4.11.

6.6.2

Corollary *Let* $_G X$ *and* $_H Y$ *be two finite group actions. Choose a conjugacy class* $\mathcal{C}$ *of elements of* $H \times G$ *with the probability*

$$
p(\mathcal{C}) := \frac{|\mathcal{C}| \, |Y^X_{(h,g)}|}{|G| \, |H| \, |(H \times G)\backslash\backslash Y^X|} \quad \text{for arbitrary } (h, g) \in \mathcal{C}.
$$

Pick any $(h, g) \in \mathcal{C}$ *and determine at random a function* $f \in Y^X$ *which is fixed under the action of* (h, g), *i.e.* $f(gx) = hf(x)$ *for all* $x \in X$. *Then the probability that* f *lies in a given orbit* $\omega \in (H \times G)\backslash\backslash Y^X$ *is equal to* $1/|(H \times G)\backslash\backslash Y^X|$. $\qquad\square$

According to Exercise 6.3.3, the conjugacy classes of $G := S_n$ are characterized by the cycle types $a \vdash n$. The conjugacy classes of $H := \mathrm{GL}_k(q)$ were described completely in 6.3.12. Hence, the conjugacy classes of $\mathrm{GL}_k(q) \times S_n$ are exactly the elements of the cartesian product $\mathcal{C}_1 \times \mathcal{C}_2$, where $\mathcal{C}_1$ is a conjugacy class of $\mathrm{GL}_k(q)$ and $\mathcal{C}_2$ is a conjugacy class of S_n. This shows how to obtain representatives of the conjugacy classes of $\mathrm{GL}_k(q) \times S_n$. In 6.3.14 the representatives of the conjugacy classes of $\mathrm{GL}_k(q)$ are described as block diagonal matrices of companion and hyper companion matrices of monic irreducible polynomials over $\mathbb{F}_q$. In order to list them all, it is necessary to know all these polynomials of degree up to k. As we have seen in Section 6.3, it was not necessary to know these polynomials explicitly as far as enumeration of linear isometry classes is concerned.

For certain values of k and q, tables of these polynomials exist. Recall from the beginning of Section 3.5 that all irreducible polynomials of a given degree n over $\mathbb{F}_q$ can be computed once a normal basis of $\mathbb{F}_{q^n}$ over $\mathbb{F}_q$ is known.

This motivates the following strategy. For $2 \le r \le k$ we generate monic polynomials of degree r over $\mathbb{F}_q$ at random. Using 3.5.20 we test these polynomials whether they are irreducible. We repeat this till for each r we have found an irreducible polynomial of degree r. With these polynomials we are able to determine a normal basis of $\mathbb{F}_{q^r}$ over $\mathbb{F}_q$ for each r. For more details see Section 6.9. Then we compute all Lyndon words of length r over an alphabet of q elements as described in 3.5.5. We consider these Lyndon words as the coefficient vectors of elements of $\mathbb{F}_{q^r}$ with respect to the normal basis of $\mathbb{F}_{q^r}$ over $\mathbb{F}_q$ just constructed. Using 3.5.2, we compute the minimal polynomials of these elements. These minimal polynomials provide a complete list of irreducible polynomials of degree r over $\mathbb{F}_q$.

The number of $GL_k(q) \times S_n$-orbits on $PG^*_{k-1}(q)^n$ was already computed as T_{nkq} in 6.1.23. The number of fixed points of $(A, \pi) \in GL_k(q) \times S_n$ in $PG^*_{k-1}(q)^n$ can be deduced from the next

---

Lemma *Assume that $_GX$ and $_HY$ are two finite group actions which induce natural*   **6.6.3**
actions of G, H and $H \times G$ on Y^X (as described in 1.4.7, 1.4.10, and 1.4.11).

— *The number of fixed points of $g \in G$ on Y^X is given by*

$$|Y|^{c(\overline{g})} \quad for \quad c(\overline{g}) := \sum_{i=1}^{|X|} a_i(\overline{g}),$$

where $(a_1(\overline{g}), \ldots, a_{|X|}(\overline{g}))$ is the cycle type of the induced permutation $\overline{g}$ on X.

— *The number of fixed points of $h \in H$ on Y^X is given by*

$$|Y_h|^{|X|},$$

where Y_h is the set of fixed points of h on Y.

— *The number of fixed points of $(h, g) \in H \times G$ on Y^X is given by*

$$\prod_{i=1}^{|X|} |Y_{h^i}|^{a_i(\overline{g})},$$

where $(a_1(\overline{g}), \ldots, a_{|X|}(\overline{g}))$ is the cycle type of the induced permutation $\overline{g}$ on X, and Y_h is the set of fixed points of h on Y. $\qquad\square$

A method for constructing the set of fixed points of (A, π) on $PG^*_{k-1}(q)^n$ is described in

6.6.4   **Lemma** *Consider the natural action of $H \times G$ on Y^X induced by two finite group actions $_G X$ and $_H Y$ as described in 1.4.11. The fixed points $f \in Y^X$ of $(h, g) \in H \times G$ have the following form. For each cycle Z of $\overline{g}$ on X, pick a representative $x_Z \in Z$. Then $f(x_Z) = y_0 \in Y$ with $|\langle h \rangle (y_0)|$ dividing $|Z|$ (that is, $y_0 \in Y_{h^{|Z|}}$, the set of fixed points of $h^{|Z|}$ on Y). The remaining values of f on Z are determined by*

$$f(g^i x_Z) := h^i y_0 \text{ for } 1 \le i < |Z|. \qquad \square$$

The proofs of the previous two lemmata are left to the reader as Exercise 6.6.1 and Exercise 6.6.2.

As mentioned above, applying the Dixon–Wilf-algorithm for the random generation of linear codes produces generator matrices of linear (n, l)-codes for $l \le k$. Therefore, after the generation the rank of each matrix must still be determined.

Some numerical results are presented in Table 6.27. For different parameters q, n and k, the table shows the distribution of ranks when 10 000 matrices were generated at random in each case.

For further illustration here are the numbers of conjugacy classes of $GL_k(2)$.

k	3	4	5	6	7	8	9	10
# of conjugacy classes	6	14	27	60	117	246	490	1002

The choice of a conjugacy class of S_n amounts to the choice of a cycle type (or partition) of n. The number of partitions of $n \in \mathbb{N}$ increases rapidly with n. Here are some of these numbers:

n	number of cycle types of n
10	42
15	176
20	627
25	1958
40	37 338
60	$\approx 10^6$
100	$\approx 2 \cdot 10^8$

For this reason we should try to avoid computing and storing the probabilities of all conjugacy classes of $GL_k(q) \times S_n$ before the generation process starts. For practical purposes we label the conjugacy classes by $\mathcal{C}_0, \ldots, \mathcal{C}_{N-1}$. Usually $\mathcal{C}_0$ is the conjugacy class of the identity element. The random choice of a conjugacy class $\mathcal{C}_{i_0}$ is done by first computing a random number $r \in [0, 1)$ and then determining the index $i_0 \in N$ so that

$$\sum_{j \in i_0} p(\mathcal{C}_j) \le r \text{ and } \sum_{j \in i_0 + 1} p(\mathcal{C}_j) > r.$$

Table 6.27 Distribution of ranks of 10 000 $k \times n$-matrices over $\mathbb{F}_q$ generated at random

q	n	k	rank distribution
2	15	3	(17, 534, 9449)
2	15	4	(1, 53, 677, 9269)
2	15	5	(0, 5, 68, 908, 9019)
2	15	6	(0, 0, 16, 142, 1488, 8354)
2	15	7	(0, 1, 5, 51, 492, 2672, 6779)
2	15	8	(0, 0, 1, 27, 272, 1523, 3970, 4207)
2	15	9	(0, 0, 1, 27, 246, 1374, 3289, 3507, 1556)
2	15	10	(0, 0, 2, 22, 228, 1179, 3279, 3434, 1531, 325)
2	20	3	(8, 218, 9774)
2	20	4	(0, 7, 185, 9808)
2	20	5	(0, 0, 3, 140, 9857)
2	20	6	(0, 0, 0, 2, 175, 9823)
2	20	7	(0, 0, 0, 0, 3, 225, 9772)
2	20	8	(0, 0, 0, 0, 0, 18, 529, 9453)
2	25	3	(3, 121, 9876)
2	25	4	(0, 2, 70, 9928)
2	25	5	(0, 0, 0, 30, 9970)
2	25	6	(0, 0, 0, 0, 10, 9990)
2	25	7	(0, 0, 0, 0, 0, 6, 9994)
2	25	8	(0, 0, 0, 0, 0, 0, 29, 9971)
3	15	3	(1, 122, 9877)
3	15	4	(0, 0, 50, 9950)
3	15	5	(0, 0, 0, 68, 9932)
3	25	5	(0, 0, 0, 0, 10000)

One can start the generation process immediately and evaluate probabilities of the conjugacy classes only if required. This means that we need to evaluate $p(C_i)$ only if the chosen random number exceeds $\sum_{j \in i} p(C_j)$. The efficiency of this revised method depends heavily on the numbering of the conjugacy classes. Clearly, the numbering should be chosen in such a way that $p(C_i) \geq p(C_{i+1})$.

We have applied the random generation of linear codes in order to describe the distribution of the minimum distance of linear codes with given parameters n, k and q. Two examples are presented in Table 6.28 and Table 6.29.

Table 6.28 Distribution of the minimum distances of 10 000 binary codes of length 20 and maximal dimension 8

$k\backslash d$	1	2	3	4	5	6
5	0	0	0	1	0	0
6	0	0	2	5	3	1
7	3	45	102	226	150	16
8	81	1 158	2 502	4 346	1 344	15

Table 6.29 Distribution of the minimum distances of 30 000 000 codes of length 12 and maximal dimension 5 over $\mathbb{F}_5$

$k\backslash d$	1	2	3	4	5	6	7	8
3	1	5	4	40	99	196	136	9
4	120	1060	5644	37440	137047	139665	5651	0
5	24017	243558	1486385	10048367	17047580	822975	0	0

Exercises

E.6.6.1   **Exercise** Prove 6.6.3.

E.6.6.2   **Exercise** Prove 6.6.4.

E.6.6.3   **Exercise** Use the enclosed software to obtain lower bounds for the minimum distance of linear (n, k)-codes over $\mathbb{F}_p$ for small parameters n, k and p. Compare these results with the list of best known linear codes [32].

6.7 Enumeration of Semilinear Isometry Classes

So far we were concerned only with the enumeration of linear isometry classes of codes. In this section we show how to generalize these methods in order to derive the number of semilinearly nonisometric codes.

In 1.5.10 we have described a semilinear isometry ι as $\iota = (\psi, (\alpha; \pi))$ where $\alpha \in \mathrm{Aut}(\mathbb{F}_q) = \mathrm{Gal}\,[\mathbb{F}_q : \mathbb{F}_p]$ and $(\psi; \pi)$ is a linear isometry. Thus $(\psi; \pi)$ belongs to the wreath product

$$\mathbb{F}_q^* \wr_n S_n = \left\{ (\psi; \pi) \mid \psi : n \to \mathbb{F}_q^*, \ \pi \in S_n \right\}.$$

Since $\mathrm{Gal}\,[\mathbb{F}_p : \mathbb{F}_p]$ contains just one element, we assume in this section that $q = p^r$ with $r > 1$. In the sequel we indicate the Galois group $\mathrm{Gal}\,[\mathbb{F}_q : \mathbb{F}_p]$, generated by the Frobenius automorphism $\tau(\kappa) = \kappa^q$, $\kappa \in \mathbb{F}_q$, by Gal. As we already know, it is a cyclic group of order r.

According to 1.5.11 two codes are called semilinearly isometric if there exists a semilinear isometry ι which maps one code onto the other code.

Our first aim is to show that the group of semilinear isometries is a generalized wreath product. Therefore, we apply the two semilinear isometries $\iota_2 = (\phi; (\beta, \rho))$ and $\iota_1 = (\psi; (\alpha, \pi))$ to the vector $v = (v_0, \dots, v_{n-1}) \in \mathbb{F}_q^n$ and indicate $\iota_1(v)$ by $v' = (v'_0, \dots, v'_{n-1})$. Then we obtain

$$
\begin{aligned}
\iota_2(\iota_1(v)) &= \iota_2(v') = \left(\phi(0)\beta(v'_{\rho^{-1}(0)}), \dots, \phi(n-1)\beta(v'_{\rho^{-1}(n-1)})\right) \\
&= \left(\dots, \phi(i)\beta\big(\psi(\rho^{-1}(i))\alpha(v_{\pi^{-1}(\rho^{-1}(i))})\big), \dots\right) \\
&= \left(\dots, \phi(i)\beta\big(\psi(\rho^{-1}(i))\big)(\beta \circ \alpha)(v_{(\rho\pi)^{-1}(i)}), \dots,\right).
\end{aligned}
$$

This formula motivates the following

Lemma *The group of all semilinear isometries of $\mathbb{F}_q^n$ is the semidirect product*   **6.7.1**

$$(\mathbb{F}_q^*)^n \rtimes (\mathrm{Gal} \times S_n),$$

with the normal subgroup on the left, where the multiplication is given by

$$(\phi; (\beta, \rho)) \cdot (\psi; (\alpha, \pi)) := (\phi\psi_{(\beta,\rho)}; (\beta\alpha, \rho \circ \pi)),$$

with

$$\psi_{(\beta,\rho)}(i) := \beta(\psi(\rho^{-1}(i))), \qquad i \in n,$$

and

$$\phi\psi(i) := \phi(i)\psi(i), \qquad i \in n. \qquad\qquad \square$$

Therefore, the identity element is $(1; (\tau^0, \mathrm{id}))$, where 1 is the mapping $i \mapsto 1$, $i \in n$. The inverse of $(\psi; (\alpha, \pi))$ is $(\psi_{(\alpha^{-1}, \pi^{-1})}^{-1}; (\alpha^{-1}, \pi^{-1}))$ where $\psi^{-1}(i) := (\psi(i))^{-1}, i \in n$, and $\psi_{(\alpha,\pi)}^{-1} := (\psi_{(\alpha,\pi)})^{-1} = (\psi^{-1})_{(\alpha,\pi)}$.

Representing the product of two semilinear isometries in this way, it is easy to realize certain similarities with the ordinary wreath product $H \wr_X G$. In 1.4.8 we had considered a group G acting on a set X and an arbitrary group H. For defining the multiplication in $H \wr_X G$ we used the canonically induced action of G on H^X given by 1.4.7.

Here in the situation of the group of semilinear isometries, we have $X = n$ and $H = \mathbb{F}_q^*$. The group $\mathrm{Gal} \times S_n$ does not act on n, but it operates already on $(\mathbb{F}_q^*)^n$ and we do not have to consider an induced action on $(\mathbb{F}_q^*)^n$. Therefore, we say that the group of semilinear isometries is the *generalized wreath product* of $\mathbb{F}_q^*$ and $\mathrm{Gal} \times S_n$ which we indicate by

$$\mathbb{F}_q^* \wr_n (\mathrm{Gal} \times S_n).$$

Its order is equal to $(q-1)^n \cdot r \cdot n!$, and the generalization of the natural action of a wreath product (cf. 1.4.9) to this generalized wreath product is

$$(\psi; (\alpha, \pi))(v) = \big(\psi(0)\alpha(v_{\pi^{-1}(0)}), \ldots, \psi(n-1)\alpha(v_{\pi^{-1}(n-1)})\big)$$

which is the action of the semilinear isometry $(\psi; (\alpha, \pi))$ on $\mathbb{F}_q^n$.

Similarly as in Section 6.1 we describe codes by their generator matrices, and obtain that the set of semilinear isometry classes of (n, k)-codes is equal to the set of orbits

$$\mathbb{F}_q^* \wr_n (\mathrm{Gal} \times S_n) \backslash\!\backslash \big(GL_k(q) \backslash\!\backslash \mathbb{F}_q^{k \times n, k} \big),$$

where the operation of $(\psi; (\alpha, \pi)) \in \mathbb{F}_q^* \wr_n (\mathrm{Gal} \times S_n)$ on the orbit $GL_k(q)(\Gamma)$ is given by

$$\big((\psi; (\alpha, \pi)), GL_k(q)(\Gamma)\big) \mapsto GL_k(q)(\hat{\Gamma}) \text{ where } \hat{\Gamma}(i) = \psi(i)\alpha(\Gamma(\pi^{-1}(i))).$$

Here again we identify the matrix Γ with the function $\Gamma: n \to \mathbb{F}_q^k$ where $\Gamma(i)^\top$ is the i-th column of Γ. When writing Af, we identify the function $f \in (\mathbb{F}_q^k)^n$ with the corresponding $k \times n$-matrix $(f(0)^\top \mid \ldots \mid f(n-1)^\top)$. Then $Af = (A \cdot f(0)^\top \mid \ldots \mid A \cdot f(n-1)^\top)$ and $Af(i) = (A \cdot f(i)^\top)^\top = f(i) \cdot A^\top$ for $A \in GL_k(q)$.

We want to prove that this operation is well-defined. For $A \in GL_k(q)$ and $\tilde{\Gamma}$ given by $\tilde{\Gamma}(i) := \psi(i)\alpha((A \cdot \Gamma)(\pi^{-1}(i)))$ we have $GL_k(q)(\tilde{\Gamma}) = GL_k(q)(\hat{\Gamma})$, since $\tilde{\Gamma}(i) = \psi(i)\alpha(A)\alpha(\Gamma(\pi^{-1}(i)))$ and $\alpha(A) \in GL_k(q)$. (In Exercise 3.7.5 we have mentioned that α induces a group automorphism of $GL_k(q)$ by applying α to all components of the matrices in $GL_k(q)$.)

In the situation of linear isometries the actions of the isometry group and of the linear group were commuting and we obtained an action of the direct product of these two groups on $\mathbb{F}_q^{k \times n, k}$ (cf. 6.1.3).

In general, the action of the semilinear isometry group does not commute with the action of $GL_k(q)$. For $A \in GL_k(q)$, $(\psi; (\alpha, \pi)) \in \mathbb{F}_q^* \wr_n (\mathrm{Gal} \times S_n)$ and $\Gamma \in \mathbb{F}_q^{k \times n, k}$ we have

$$A \cdot (\psi; (\alpha, \pi))\Gamma =$$
$$\big(\psi(0)A\alpha(\Gamma(\pi^{-1}(0))), \ldots, \psi(n-1)A\alpha(\Gamma(\pi^{-1}(n-1)))\big)$$

and

$$(\psi; (\alpha, \pi))A \cdot \Gamma =$$
$$\big(\psi(0)\alpha(A)\alpha(\Gamma(\pi^{-1}(0))), \ldots, \psi(n-1)\alpha(A)\alpha(\Gamma(\pi^{-1}(n-1)))\big).$$

Therefore, we do not get an action of the direct product as in 6.1.3.

Again, similarly as in Section 6.1 we eliminate the rank condition on the $k \times n$-matrices and consider the set of all $k \times n$-matrices over $\mathbb{F}_q$ which do not contain zero columns. Thus, our task is to determine the cardinality of

$$\mathbb{F}_q^* \wr_n (\mathrm{Gal} \times S_n) \backslash\!\backslash \big(GL_k(q) \backslash\!\backslash (\mathbb{F}_q^k \setminus \{0\})^n \big).$$

For this reason we describe a generalization of Lehmann's Lemma 6.1.8. We generalize it in two ways, since on the one hand we are dealing with an action of the generalized wreath product, and on the other hand this wreath product operates on $GL_k(q)$-orbits of functions and not just on a set of functions. However we do not formulate it for arbitrary group actions but for the situation of the present problem.

Generalization of Lehmann's Lemma *If the mapping*   6.7.2

$$\varphi: GL_k(q)\backslash\backslash(\mathbb{F}_q^k \setminus \{0\})^n \to GL_k(q)\backslash\backslash\left(\mathbb{F}_q^*\backslash\backslash(\mathbb{F}_q^k \setminus \{0\})\right)^n$$

is given by

$$GL_k(q)(\Gamma) \mapsto GL_k(q)(\overline{\Gamma}) \ \text{where} \ \overline{\Gamma}(i) = \mathbb{F}_q^*(\Gamma(i)),$$

then the mapping

$$\Phi: \left(\mathbb{F}_q^* \wr_n (\text{Gal} \times S_n)\right)\backslash\backslash(GL_k(q)\backslash\backslash(\mathbb{F}_q^k \setminus \{0\})^n) \to$$

$$(\text{Gal} \times S_n)\backslash\backslash\left(GL_k(q)\backslash\backslash(\mathbb{F}_q^*\backslash\backslash(\mathbb{F}_q^k \setminus \{0\}))^n\right)$$

defined by

$$\left(\mathbb{F}_q^* \wr_n (\text{Gal} \times S_n)\right)(GL_k(q)(\Gamma)) \mapsto (\text{Gal} \times S_n)(\varphi(GL_k(q)(\Gamma)))$$

is a bijection. On the right hand side we have an operation of $(\text{Gal} \times S_n)$ *on the set of orbits* $GL_k(q)\backslash\backslash(\mathbb{F}_q^*\backslash\backslash(\mathbb{F}_q^k \setminus \{0\}))^n$ *of the form*

$$(\alpha, \pi) GL_k(q)(\overline{\Gamma}) = GL_k(q)(\hat{\Gamma})$$

where $\hat{\Gamma}(i) = \alpha(\overline{\Gamma}(\pi^{-1}(i))) = \alpha(\mathbb{F}_q^*(\Gamma(\pi^{-1}(i)))) = \mathbb{F}_q^*(\alpha(\Gamma(\pi^{-1}(i)))), i \in n.$

Proof: As in the proof of 6.1.8 we see that for $f_1, f_2 \in Y^X$ the following facts are equivalent:

$\Phi(\mathbb{F}_q^* \wr_n (\text{Gal} \times S_n)(f_1)) = \Phi(\mathbb{F}_q^* \wr_n (\text{Gal} \times S_n)(f_2))$

$(\text{Gal} \times S_n)(\varphi(f_1)) = (\text{Gal} \times S_n)(\varphi(f_2))$

$\varphi(f_2) \in (\text{Gal} \times S_n)(\varphi(f_1))$

$\varphi(f_2) = \alpha \circ \varphi(f_1) \circ \pi \ \text{for some} \ \alpha \in \text{Gal and some} \ \pi \in S_n$

$\varphi(f_2)(x) = \alpha(\varphi(f_1)(\pi(x))) \ \text{for some} \ \alpha \in \text{Gal}, \pi \in S_n, \text{and all} \ x \in X$

$\varphi(f_2)(x) = \varphi(\alpha \circ f_1)(\pi(x)) \ \text{for some} \ \alpha \in \text{Gal}, \pi \in S_n, \text{and all} \ x \in X$

$\mathbb{F}_q^*(f_2(x)) = \mathbb{F}_q^*((\alpha \circ f_1)(\pi(x))) \ \text{for some} \ \alpha \in \text{Gal}, \pi \in S_n, \text{and all} \ x \in X$

$f_2(x) \in \mathbb{F}_q^*((\alpha \circ f_1)(\pi(x))) \ \text{for some} \ \alpha \in \text{Gal}, \pi \in S_n, \text{and all} \ x \in X$

$$f_2 = (\psi; (\alpha, \pi)) f_1 \text{ for some } (\psi; (\alpha, \pi)) \in \mathbb{F}_q^* \wr_n (\text{Gal} \times S_n)$$

$$f_2 \in \mathbb{F}_q^* \wr_n (\text{Gal} \times S_n)(f_1)$$

$$\mathbb{F}_q^* \wr_n (\text{Gal} \times S_n)(f_2) = \mathbb{F}_q^* \wr_n (\text{Gal} \times S_n)(f_1).$$

Reading these implications from bottom to top we deduce that Φ is well-defined. From top to bottom it follows that Φ is injective. In order to prove that Φ is surjective, we notice that φ is surjective. □

As an immediate consequence we obtain that

$$\left| \left(\mathbb{F}_q^* \wr_n (\text{Gal} \times S_n) \right) \backslash\backslash \left(\text{GL}_k(q) \backslash\backslash (\mathbb{F}_q^k \setminus \{0\})^n \right) \right| =$$

$$\left| (\text{Gal} \times S_n) \backslash\backslash \left(\text{GL}_k(q) \backslash\backslash \text{PG}_{k-1}^*(q)^n \right) \right|.$$

It is still possible to find a simpler expression for

$$(\text{Gal} \times S_n) \backslash\backslash \left(\text{GL}_k(q) \backslash\backslash \text{PG}_{k-1}^*(q)^n \right).$$

According to Exercise 1.4.9 we can split the action of the direct product obtaining

$$\text{Gal} \backslash\backslash \left(S_n \backslash\backslash \left(\text{GL}_k(q) \backslash\backslash \text{PG}_{k-1}^*(q)^n \right) \right)$$

what is the same as

$$\text{Gal} \backslash\backslash \left((\text{GL}_k(q) \times S_n) \backslash\backslash \text{PG}_{k-1}^*(q)^n \right)$$

since the actions of $\text{GL}_k(q)$ and S_n commute. An application of the automorphism α to the orbit $(\text{GL}_k(q) \times S_n)(\overline{\Gamma})$ yields the orbit $(\text{GL}_k(q) \times S_n)(\hat{\Gamma})$ where $\hat{\Gamma}(i) = \alpha(\overline{\Gamma}(i)) = \mathbb{F}_q^*(\alpha(\Gamma(i)))$. These orbits can be represented as the elements of

6.7.3
$$(\text{P}\Gamma\text{L}_k(q) \times S_n) \backslash\backslash \text{PG}_{k-1}^*(q)^n,$$

since $\text{P}\Gamma\text{L}_k(q) = (\text{GL}_k(q) \rtimes \text{Gal}) / \mathcal{Z}_k$.

The reader should carefully check the following

6.7.4   **Lemma** *Let C be a code and ι a semilinear isometry.*

— *C is nonredundant if and only if $\iota(C)$ is nonredundant.*

— *C is projective if and only if $\iota(C)$ is projective.*

— *C is injective if and only if $\iota(C)$ is injective.*

— *C is indecomposable if and only if $\iota(C)$ is indecomposable.* □

Analogously to Section 6.1 and Section 6.2 we introduce the notions

$$t_{nkq} := \left| (\text{P}\Gamma\text{L}_k(q) \times S_n) \backslash\backslash \text{PG}_{k-1}^*(q)^n \right|,$$

$$\bar{t}_{nkq} := \left| (P\Gamma L_k(q) \times S_n) \backslash\backslash PG^*_{k-1}(q)^n_{\text{inj}} \right|.$$

Moreover, let v_{nkq} denote the number of semilinear isometry classes of nonredundant (n, k)-codes over $\mathbb{F}_q$ and $\bar{v}_{nkq}$ the number of semilinear isometry classes of projective (n, k)-codes over $\mathbb{F}_q$. The symbols u_{nkq} and $\bar{u}_{nkq}$ indicate the number of semilinear isometry classes of all, respectively injective, (n, k)-codes which may contain columns of zeros. The number of semilinear isometry classes of nonredundant indecomposable (n, k)-codes over $\mathbb{F}_q$ is denoted by r_{nkq} and of projective indecomposable (n, k)-codes over $\mathbb{F}_q$ by $\bar{r}_{nkq}$. These symbols are the lowercase versions of the corresponding numbers of linear isometry classes. The relations corresponding to 6.1.15 and 6.2.20 are collected in

Corollary   6.7.5

— t_{nkq} is the number of semilinear isometry classes of linear codes of length n and dimension at most k. If $k > 1$, then $t_{n,k-1,q}$ is also the number of $P\Gamma L_k(q) \times S_n$-orbits of mappings $f \in PG^*_{k-1}(q)^n$ corresponding to matrices of rank not greater than $k - 1$.

— $\bar{t}_{nkq}$ is the number of semilinear isometry classes of injective linear codes of length n and dimension at most k.

— $v_{nkq} = t_{nkq} - t_{n,k-1,q}$, $\bar{v}_{nkq} = \bar{t}_{nkq} - \bar{t}_{n,k-1,q}$ for $1 < k \leq n$. The initial values for these recursions are $v_{n1q} = 1$ for $n \in \mathbb{N}^*$, $\bar{v}_{11q} = 1$ and $\bar{v}_{n1q} = 0$ for $n > 1$.

— $u_{nkq} = \sum_{i=k}^{n} v_{ikq}$, $\bar{u}_{kkq} = \bar{v}_{kkq}$, and $\bar{u}_{nkq} = \bar{v}_{n-1,k,q} + \bar{v}_{nkq}$ for $n > k$.

— For $n \geq 2$ we have

$$r_{nkq} = v_{nkq} - \sum_a \sum_b \prod_{\substack{j=1 \\ a_j \neq 0}}^{n-1} \left(\sum_c U(c) \right),$$

where

$$U(c) = \prod_{i=1}^{j} C(S_{v(i,c)}, v(i,c)) \big|_{z_\ell = r_{jiq}}$$

is a product computed from substitutions into the cycle indices of symmetric groups of degree $v(i, c)$ given by

$$v(i, c) = |\{\ell \in a_j \mid c_\ell = i\}|, \qquad 1 \leq i \leq j.$$

The first sum runs through the cycle types $a = (a_1, \dots, a_{n-1})$ of n with at least two summands, i.e. $a_i \in \mathbb{N}$, $\sum i a_i = n$ and $\sum a_i \leq k$, while the second sum is taken over the $(n-1)$-tuples $b = (b_1, \dots, b_{n-1}) \in \mathbb{N}^{n-1}$, for which $a_i \leq b_i \leq i a_i$, and $\sum b_i = k$. The third sum runs over all a_j-tuples $c = (c_0, \dots, c_{a_j-1}) \in \mathbb{N}^{a_j}$ with *the properties $j \geq c_0 \geq \ldots \geq c_{a_j-1} \geq 1$ and $\sum c_i = b_j$. Analogously, $\bar{r}_{nkq}$ can be recursively evaluated from $\bar{v}_{nkq}$ and $\bar{r}_{jiq}$ with $j < n$. The initial values for these recursions are $r_{11q} = 1 = \bar{r}_{11q}$.* $\square$

This way we have expressed all these numbers in terms of t_{nkq} and $\bar{t}_{nkq}$. The remaining problem is the evaluation of t_{nkq} and $\bar{t}_{nkq}$. In 6.7.3 we have the canonical action of a direct product on a set of functions. Since the group acting on the domain is the symmetric group it is possible to apply 6.1.21 in order to compute the generating function for the cardinalities of these orbit sets and we obtain the following

6.7.6   **Corollary**   *The generating functions for the numbers t_{nkq} and $\bar{t}_{nkq}$ can be obtained from the cycle index of the natural action of the projective semilinear group on the projective space in the following way:*

$$\sum_{n \in \mathbb{N}} t_{nkq} x^n = C(P\Gamma L_k(q), PG^*_{k-1}(q))\big|_{z_i := \sum_{j=0}^{\infty} x^{i \cdot j}},$$

and

$$\sum_{n \in \mathbb{N}} \bar{t}_{nkq} x^n = C(P\Gamma L_k(q), PG^*_{k-1}(q))\big|_{z_i := 1 + x^i}.$$ $\square$

Finally, it remains to determine the cycle index of the natural action of the projective semilinear group on the projective space. In order to obtain some numerical results we used the computer algebra system GAP [63] together with a particular extension for projective spaces [74]. Based on 6.3.3 we determined a complete system of representatives of the conjugacy classes of elements of $P\Gamma L_k(q)$. We computed the cardinality of each class, and for each representative we determined the cycle type of the natural action on $PG^*_{k-1}(q)$.

For $q = 4$ we obtain the Tables 6.30 to 6.35, which should be compared with the Tables 6.15 , 6.9, 6.23, 6.20, 6.12 and 6.26. (Differences between corresponding tables are marked by boldface numbers.) The next field where differences occur between linear and semilinear isometries is $\mathbb{F}_8$. On the pages 542–548 we present some tables comparing the numbers T_{nk8} and t_{nk8}, V_{nk8} and v_{nk8}, R_{nk8} and r_{nk8}, U_{nk8} and u_{nk8}, $\bar{T}_{nk8}$ and $\bar{t}_{nk8}$, $\bar{V}_{nk8}$ and $\bar{v}_{nk8}$, and $\bar{R}_{nk8}$ and $\bar{r}_{nk8}$. Extended tables can be found on the attached CD-ROM.

Exercises

E.6.7.1   **Exercise** Prove 6.7.1.

E.6.7.2   **Exercise** Prove 6.7.4.

Table 6.30 Values of t_{nk4}

$n\backslash k$	1	2	3	4	5	6
1	1	1	1	1	1	1
2	1	2	2	2	2	2
3	1	3	4	4	4	4
4	1	5	8	9	9	9
5	1	7	16	20	21	21
6	1	10	34	51	56	57
7	1	13	68	138	166	172
8	1	18	144	445	629	673
9	1	23	309	1728	3322	3775
10	1	30	670	8640	31045	40323
11	1	37	1468	52924	543062	1047635
12	1	47	3251	360473	13107137	59070798
13	1	57	7156	2503187	336291123	4922753104
14	1	70	15665	16976798	8362677597	452322657324

Table 6.31 Values of v_{nk4}

$n\backslash k$	1	2	3	4	5	6
1	1	0	0	0	0	0
2	1	1	0	0	0	0
3	1	2	1	0	0	0
4	1	4	3	1	0	0
5	1	6	9	4	1	0
6	1	9	24	17	5	1
7	1	12	55	70	28	6
8	1	17	126	301	184	44
9	1	22	286	1419	1594	453
10	1	29	640	7970	22405	9278
11	1	36	1431	51456	490138	504573
12	1	46	3204	357222	12746664	45963661
13	1	56	7099	2496031	333787936	4586461981
14	1	69	15595	16961133	8345700799	443959979727

Table 6.32 Values of r_{nk4}

$n\backslash k$	1	2	3	4	5	6
1	1	0	0	0	0	0
2	1	0	0	0	0	0
3	1	1	0	0	0	0
4	1	2	1	0	0	0
5	1	4	4	1	0	0
6	1	6	14	6	1	0
7	1	9	38	38	9	1
8	1	13	99	216	99	13
9	1	18	244	1213	1213	244
10	1	24	579	7479	20603	7479
11	1	31	1344	50328	480335	480335
12	1	40	3084	354655	12685278	45448958
13	1	50	6937	2490249	333368938	4573198774
14	1	62	15381	16948216	8342784710	443612918007

Table 6.33 Values of u_{nk4}

$n\backslash k$	1	2	3	4	5	6
1	1	0	0	0	0	0
2	2	1	0	0	0	0
3	3	3	1	0	0	0
4	4	7	4	1	0	0
5	5	13	13	5	1	0
6	6	22	37	22	6	1
7	7	34	92	92	34	7
8	8	51	218	393	218	51
9	9	73	504	1812	1812	504
10	10	102	1144	9782	24217	9782
11	11	138	2575	61238	514355	514355
12	12	184	5779	418460	13261019	46478016
13	13	240	12878	2914491	347048955	4632939997
14	14	309	28473	19875624	8692749754	448592919724

Table 6.34 Values of $\bar{v}_{nk4}$

$n\backslash k$	1	2	3	4	5	6
1	1	0	0	0	0	0
2	0	1	0	0	0	0
3	0	1	1	0	0	0
4	0	1	2	1	0	0
5	0	1	4	3	1	0
6	0	0	8	10	4	1
7	0	0	10	35	19	5
8	0	0	13	124	118	33
9	0	0	17	499	1018	342
10	0	0	18	2421	15076	7571
11	0	0	18	13113	336911	444690
12	0	0	17	72823	8495389	41172182
13	0	0	13	390069	209826910	4073567723
14	0	0	10	1963645	4881485820	387971461593

Table 6.35 Values of $\bar{r}_{nk4}$

$n\backslash k$	1	2	3	4	5	6
1	1	0	0	0	0	0
2	0	0	0	0	0	0
3	0	1	0	0	0	0
4	0	1	1	0	0	0
5	0	1	3	1	0	0
6	0	0	7	5	1	0
7	0	0	10	26	8	1
8	0	0	13	112	79	12
9	0	0	17	485	883	214
10	0	0	18	2403	14557	6507
11	0	0	18	13095	334460	429438
12	0	0	17	72805	8482236	40834575
13	0	0	13	390052	209754039	4065069206
14	0	0	10	1963632	4881095698	387761618484

Table 6.36 Values of T_{nk8}

$n \backslash k$	1	2	3	4
1	1	1	1	1
2	1	2	2	2
3	1	3	4	4
4	1	5	8	9
5	1	7	16	20
6	1	14	57	78
7	1	21	273	555
8	1	39	2 034	13 931
9	1	64	16 668	714 573
10	1	109	132 237	40 746 243
11	1	173	986 453	2 188 928 772
12	1	286	6 876 180	108 587 171 103
13	1	439	44 880 936	4 985 542 976 595
14	1	686	275 497 786	212 944 610 369 565

Table 6.37 Values of t_{nk8}

$n \backslash k$	1	2	3	4
1	1	1	1	1
2	1	2	2	2
3	1	3	4	4
4	1	5	8	9
5	1	7	16	20
6	1	12	43	62
7	1	17	143	289
8	1	27	792	4 979
9	1	40	5 806	239 355
10	1	61	44 619	13 586 393
11	1	89	329 959	729 659 322
12	1	136	2 294 446	36 195 786 755
13	1	197	14 965 218	1 661 847 901 869
14	1	292	91 842 474	70 981 537 714 473

Table 6.38 Values of V_{nk8}

$n\backslash k$	1	2	3	4
1	1	0	0	0
2	1	1	0	0
3	1	2	1	0
4	1	4	3	1
5	1	6	9	4
6	1	13	43	21
7	1	20	252	282
8	1	38	1995	11 897
9	1	63	16 604	697 905
10	1	108	132 128	40 614 006
11	1	172	986 280	2 187 942 319
12	1	285	6 875 894	108 580 294 923
13	1	438	44 880 497	4 985 498 095 659
14	1	685	275 497 100	212 944 334 871 779

Table 6.39 Values of v_{nk8}

$n\backslash k$	1	2	3	4
1	1	0	0	0
2	1	1	0	0
3	1	2	1	0
4	1	4	3	1
5	1	6	9	4
6	1	11	31	19
7	1	16	126	146
8	1	26	765	4 187
9	1	39	5 766	233 549
10	1	60	44 558	13 541 774
11	1	88	329 870	729 329 363
12	1	135	2 294 310	36 193 492 309
13	1	196	14 965 021	1 661 832 936 651
14	1	291	91 842 182	70 981 445 871 999

Table 6.40 Values of R_{nk8}

$n\backslash k$	1	2	3	4
1	1	0	0	0
2	1	0	0	0
3	1	1	0	0
4	1	2	1	0
5	1	4	4	1
6	1	10	33	10
7	1	17	231	231
8	1	34	1 956	11 596
9	1	59	16 529	695 614
10	1	103	131 993	40 595 108
11	1	167	986 040	2 187 791 284
12	1	279	6 875 485	108 579 157 553
13	1	432	44 879 807	4 985 490 082 276
14	1	678	275 495 976	212 944 281 977 581

Table 6.41 Values of r_{nk8}

$n\backslash k$	1	2	3	4
1	1	0	0	0
2	1	0	0	0
3	1	1	0	0
4	1	2	1	0
5	1	4	4	1
6	1	8	21	8
7	1	13	107	107
8	1	22	732	4 024
9	1	35	5 709	232 626
10	1	55	44 465	13 535 084
11	1	83	329 720	729 278 112
12	1	129	2 294 075	36 193 111 160
13	1	190	14 964 655	1 661 830 261 138
14	1	284	91 841 624	70 981 428 231 327

Table 6.42 Values of U_{nk8}

$n\backslash k$	1	2	3	4
1	1	0	0	0
2	2	1	0	0
3	3	3	1	0
4	4	7	4	1
5	5	13	13	5
6	6	26	56	26
7	7	46	308	308
8	8	84	2 303	12 205
9	9	147	18 907	710 110
10	10	255	151 035	41 324 116
11	11	427	1 137 315	2 229 266 435
12	12	712	8 013 209	110 809 561 358
13	13	1 150	52 893 706	5 096 307 657 017
14	14	1 835	328 390 806	218 040 642 528 796

Table 6.43 Values of u_{nk8}

$n\backslash k$	1	2	3	4
1	1	0	0	0
2	2	1	0	0
3	3	3	1	0
4	4	7	4	1
5	5	13	13	5
6	6	24	44	24
7	7	40	170	170
8	8	66	935	4 357
9	9	105	6 701	237 906
10	10	165	51 259	13 779 680
11	11	253	381 129	743 109 043
12	12	388	2 675 439	36 936 601 352
13	13	584	17 640 460	1 698 769 538 003
14	14	875	109 482 642	72 680 215 410 002

Table 6.44 Values of $\overline{T}_{nk8}$

$n\backslash k$	1	2	3	4
1	1	1	1	1
2	0	1	1	1
3	0	1	2	2
4	0	1	3	4
5	0	1	5	8
6	0	1	25	39
7	0	1	132	364
8	0	1	901	11 408
9	0	1	6 155	619 402
10	0	0	38 344	34 810 827
11	0	0	217 432	1 812 498 279
12	0	0	1 119 290	86 640 720 291
13	0	0	5 242 484	3 818 392 707 185
14	0	0	22 449 375	156 004 978 540 987

Table 6.45 Values of $\overline{t}_{nk8}$

$n\backslash k$	1	2	3	4
1	1	1	1	1
2	0	1	1	1
3	0	1	2	2
4	0	1	3	4
5	0	1	5	8
6	0	1	15	27
7	0	1	58	164
8	0	1	327	3 940
9	0	1	2 101	206 934
10	0	0	12 870	11 605 307
11	0	0	72 638	604 172 431
12	0	0	373 366	28 880 263 069
13	0	0	1 747 940	1 272 797 652 589
14	0	0	7 483 895	52 001 659 817 699

Table 6.46 Values of $\overline{V}_{nk8}$

$n\backslash k$	1	2	3	4
1	1	0	0	0
2	0	1	0	0
3	0	1	1	0
4	0	1	2	1
5	0	1	4	3
6	0	1	24	14
7	0	1	131	232
8	0	1	900	10 507
9	0	1	6 154	613 247
10	0	0	38 344	34 772 483
11	0	0	217 432	1 812 280 847
12	0	0	1 119 290	86 639 601 001
13	0	0	5 242 484	3 818 387 464 701
14	0	0	22 449 375	156 004 956 091 612

Table 6.47 Values of $\overline{v}_{nk8}$

$n\backslash k$	1	2	3	4
1	1	0	0	0
2	0	1	0	0
3	0	1	1	0
4	0	1	2	1
5	0	1	4	3
6	0	1	14	12
7	0	1	57	106
8	0	1	326	3 613
9	0	1	2 100	204 833
10	0	0	12 870	11 592 437
11	0	0	72 638	604 099 793
12	0	0	373 366	28 879 889 703
13	0	0	1 747 940	1 272 795 904 649
14	0	0	7 483 895	52 001 652 333 804

Table 6.48 Values of $\overline{R}_{nk8}$

$n\backslash k$	1	2	3	4
1	1	0	0	0
2	0	0	0	0
3	0	1	0	0
4	0	1	1	0
5	0	1	3	1
6	0	1	23	9
7	0	1	130	207
8	0	1	899	10 374
9	0	1	6 153	612 345
10	0	0	38 343	34 766 326
11	0	0	217 432	1 812 242 500
12	0	0	1 119 290	86 639 383 565
13	0	0	5 242 484	3 818 386 345 408
14	0	0	22 449 375	156 004 950 849 125

Table 6.49 Values of $\overline{r}_{nk8}$

$n\backslash k$	1	2	3	4
1	1	0	0	0
2	0	0	0	0
3	0	1	0	0
4	0	1	1	0
5	0	1	3	1
6	0	1	13	7
7	0	1	56	91
8	0	1	325	3 554
9	0	1	2 099	204 505
10	0	0	12 869	11 590 334
11	0	0	72 638	604 086 920
12	0	0	373 366	28 879 817 061
13	0	0	1 747 940	1 272 795 531 280
14	0	0	7 483 895	52 001 650 585 861

6.8 Local Isometries

Let C and C' be two (n,k)-codes over $\mathbb{F}_q$. A *local linear isometry* between these two codes is a vector space isomorphism $\iota\colon C \to C'$ which preserves the distances between all pairs of codewords, i.e. $d(c_1,c_2) = d(\iota(c_1),\iota(c_2))$ for all $c_1,c_2 \in C$. So far we have shown in Section 1.4 that the linear isometries of $\mathbb{F}_q^n$, the *global linear isometries*, are the elements of $M_n(q)$. From 1.4.12 we know that $M_n(q)$ is isomorphic to the wreath product $\mathbb{F}_q^* \wr_n S_n$.

A *local semilinear isometry* between the two codes C and C' is a semilinear bijection $\sigma\colon C \to C'$ which preserves the distances between all pairs of codewords, i.e. $d(c_1,c_2) = d(\sigma(c_1),\sigma(c_2))$ for all $c_1,c_2 \in C$. So far we have shown in Section 6.7 that the semilinear isometries of $\mathbb{F}_q^n$, the *global semilinear isometries*, are the elements of the generalized wreath product $\mathbb{F}_q^* \wr_n (\mathrm{Gal} \times S_n)$.

In general a *local isometry* is a local linear or semilinear isometry. We want to prove that every local isometry between two (n,k)-codes can be extended to a global isometry of $\mathbb{F}_q^n$. This means that the set of local linear isometries between two linear (n,k)-codes is the wreath product $\mathbb{F}_q^* \wr_n S_n$ (cf. also [84, second edition, Section 9.1]) and the set of local semilinear isometries between two linear (n,k)-codes is the generalized wreath product $\mathbb{F}_q^* \wr_n (\mathrm{Gal} \times S_n)$.

As a generalization of Exercise 1.2.6 we obtain

Theorem *If C is a linear code of length n over $\mathbb{F}_q$, then for any $i \in n$ either the i-th component of all codewords is equal to 0, or each element $\alpha \in \mathbb{F}_q$ occurs as the i-th component of exactly $|C|/q$ codewords.*   □

6.8.1

First we associate an (n,k)-code C over $\mathbb{F}_q$ with the $q^k \times n$-matrix

$$M(C) = \begin{pmatrix} c^{(0)} \\ \vdots \\ c^{(q^k-1)} \end{pmatrix},$$

where the rows of the matrix are the codewords of C in a fixed but arbitrary order. If ι is a local isometry between C and C', then we assume that

$$M(C') = M(\iota(C)) = \begin{pmatrix} \iota(c^{(0)}) \\ \vdots \\ \iota(c^{(q^k-1)}) \end{pmatrix},$$

where the ordering of the rows of $M(C')$ is determined by the ordering of the rows of C.

Moreover, let $d_i^\top$, $d_i \in \mathbb{F}_q^{q^k}$, $i \in n$, be the i-th column of the matrix

$$M(C) = \left(d_0^\top \mid \ldots \mid d_{n-1}^\top \right).$$

We introduce an equivalence relation on the columns of $M(C)$. Two columns $d_i^\top$ and $d_j^\top$ are considered to be equivalent if there exists some $\kappa \in \mathbb{F}_q^*$ such that $d_i = \kappa d_j$. We call them proportional. (In general, two vectors v, w over $\mathbb{F}_q$ are *proportional* if there exists some $\kappa \in \mathbb{F}_q^*$ such that $v = \kappa w$.) A column $d_i^\top$ is called a zero column if all the components of d_i are equal to 0. The equivalence class of a zero column consists of all zero columns of $M(C)$. If $d_i^\top$ is not a zero column, then the equivalence class of $d_i^\top$ consists of all columns of $M(C)$ which are proportional to $d_i^\top$.

6.8.2   **Lemma** *Two locally isometric linear (n, k)-codes C and C' have the same number of zero columns.*

Proof: Assume that $d_i^\top$ is not a zero column of $M(C)$. According to 6.8.1, each element $\kappa \in \mathbb{F}_q$ occurs exactly q^{k-1} times in d_i. If we assume that C and C' have r, respectively, r' zero columns, then we obtain

$$(n-r)q^{k-1}(q-1) = \sum_{c\in C} \mathrm{wt}(c) = \sum_{c\in C'} \mathrm{wt}(c) = (n-r')q^{k-1}(q-1).$$

Consequently, $r = r'$.   □

In the next step we want to describe the equivalence class of a nonzero column. The *cross section* of a code C is similarly defined as the shortening of C (cf. 2.2.17). Let i be the index of a column of $M(C)$ which is not a zero column, then the cross section of C at position i is the code

$$C_i := \{c = (c_0, \dots, c_{n-1}) \in C \mid c_i = 0\}.$$

Consequently, C_i is an $(n, k-1, \geq d, q)$-code. The shortening of C in position i is obtained from the cross section of C in position i by deleting the i-th column of C_i.

6.8.3   **Lemma** *Let C be a linear (n, k)-code over $\mathbb{F}_q$. Two columns $d_i^\top \neq 0 \neq d_j^\top$ of $M(C)$ are proportional if and only if the cross sections C_i and C_j coincide.*

Proof: Assume that $d_i^\top$ and $d_j^\top$ are proportional. Then for each $c \in C$ we have $c_i = 0$ if and only if $c_j = 0$. Hence, the cross sections C_i and C_j describe the same code.

Conversely, we assume that $C_i = C_j$. We choose any two codewords $c, \tilde{c}$ of C which do not belong to C_i, whence $c_i \neq 0$, $\tilde{c}_i \neq 0$, $c_j \neq 0$, and $\tilde{c}_j \neq 0$. Then $f := c_i^{-1}c - \tilde{c}_i^{-1}\tilde{c}$ belongs to C and $f_i = 0$. Thus $f \in C_i$ and, consequently, $f_j = 0$. Since $f_j = c_i^{-1}c_j - \tilde{c}_i^{-1}\tilde{c}_j$, we obtain $c_i^{-1}c_j = \tilde{c}_i^{-1}\tilde{c}_j = \alpha \in \mathbb{F}_q^*$, and thus $c_j = \alpha c_i$ and $\tilde{c}_j = \alpha\tilde{c}_i$. This fact holds true for fixed $c \in C \setminus C_i$ and for any $\tilde{c} \in C \setminus C_i$, whence the columns $d_i^\top$ and $d_j^\top$ are proportional.   □

Now we prove that if $\iota\colon C \to C'$ is a local linear isometry, then there exists a permutation $\pi \in S_n$ such that the i-th column of $M(C)$ is proportional to the $\pi(i)$-th column of $M(C')$ for $i \in n$. This fact shows then that ι can be described as an element $(\psi; \pi)$ of $\mathbb{F}_q^* \wr_n S_n$. Thus it is a linear isometry of $\mathbb{F}_q^n$.

Theorem   *Assume that $\iota\colon C \to C'$ is a local linear isometry between two linear*   **6.8.4**
(n,k)-codes over $\mathbb{F}_q$. Then there exists a permutation $\pi \in S_n$ such that the i-th column of $M(C)$ is proportional to the $\pi(i)$-th column of $M(C')$ for $i \in n$.

Proof: To begin with, we determine the equivalence classes of the columns of $M(C)$. From 6.8.2 we know that $M(C)$ and $M(C')$ have the same number of zero columns, which we indicate by s.

Let $d_i^\top$ be a nonzero column of $M(C)$, and let $i = i_0, \ldots, i_{r-1}$ indicate the indices of the columns of $M(C)$ proportional to $d_i^\top$. Then all the cross sections $C_i = C_{i_0}, C_{i_1}, \ldots, C_{i_{r-1}}$ determine the same $(n, k-1)$-code. The matrix $M(C_i)$ has $r + s$ zero columns, namely $d_{i_0}^\top, \ldots, d_{i_{r-1}}^\top$, which come from the construction as a cross section in these columns, and $d_{i_r}^\top, \ldots, d_{i_{r+s-1}}^\top$, which are the zero columns appearing already in $M(C)$.

Since ι is a local linear isometry between C and C', also the restriction $\iota|_{C_i}$ is a linear isometry between C_i and $\iota(C_i)$, whence by 6.8.2, $M(C_i)$ and $M(\iota(C_i))$ have the same number of zero columns. Let us assume that the indices of the zero columns of $M(\iota(C_i))$ are given by $j_0, \ldots, j_{r+s-1}$, and that $j_r, \ldots, j_{r+s-1}$ are the indices of the s zero columns of $M(C')$. From 6.8.1 we know that in any of the columns $d_{j_0}'^\top, \ldots, d_{j_{r-1}}'^\top$ of $M(C')$ each element of $\mathbb{F}_q$ occurs exactly q^{k-1} times. Hence, $\iota(C_i)$ is the cross section of C' in any of the components $j_0, \ldots, j_{r-1}$, for instance, $M(\iota(C_i)) = M(C_{j_0}')$. According to 6.8.3, the columns of $M(C')$ with indices $j_0, \ldots, j_{r-1}$ are proportional and form an equivalence class of columns of $M(C')$.

Next we claim that the columns $d_{i_0}^\top$ and $d_{j_0}'^\top$ are proportional, i.e. there exists an element $\lambda \in \mathbb{F}_q^*$ such that $d_{j_0}'^\top = \lambda d_{i_0}^\top$. Assume that $b = (b_0, \ldots, b_{n-1})$ with $b_{i_0} = 1$ belongs to $C \setminus C_{i_0}$. Then $\iota(b) \in \iota(C \setminus C_{i_0}) = \iota(C) \setminus \iota(C_{i_0}) = C' \setminus C_{j_0}'$, whence the j_0-th component of $\iota(b)$, which we indicate as $\iota(b)_{j_0}$, is different from zero. Now take an arbitrary $c \in C \setminus C_{i_0}$. Since C_{i_0} is a $(k-1)$-dimensional subspace of C, there exist uniquely determined $\tilde{c} \in C_{i_0}$ and $\kappa \in \mathbb{F}_q$ such that $c = \tilde{c} + \kappa b$. Consequently, $\kappa = c_{i_0} \neq 0$. Since $\iota(\tilde{c}) \in C_{j_0}'$, the j_0-th component of $\iota(c) = \iota(\tilde{c}) + \kappa\iota(b)$ is equal to $c_{i_0}\iota(b)_{j_0}$. This holds true for any $c \in C \setminus C_{i_0}$, whence the i_0-th column of $M(C)$ is proportional to the j_0-th column of $M(C')$ with the nonzero factor $\lambda = \iota(b)_{j_0}$.

Finally, this method allows us to determine a permutation $\pi \in S_n$ in the following way. From the previous discussion we already know that C and C'

have the same number of zero columns, and if $d_i^\top \neq 0$ belongs to an equivalence class of r columns of $M(C)$, then we can find an equivalence class containing exactly r columns of $M(C')$ which are all proportional to $d_i^\top$. Hence, it is possible to determine π so that π maps zero columns of $M(C)$ to zero columns of $M(C')$ and each nonzero column $d_i^\top$ of $M(C)$ to a proportional column of $M(C')$. $\square$

Thus for each $c \in C$ we have

$$\iota(c) = \left(\psi(0)c_{\pi^{-1}(0)}, \ldots, \psi(n-1)c_{\pi^{-1}(n-1)}\right),$$

for some $\psi(\mathbb{F}_q^*)^n$.

Now let $\sigma: C \to C'$ be a local semilinear isometry with $\sigma(\kappa c) = \alpha(\kappa)\sigma(c)$ for $c \in C$, $\kappa \in \mathbb{F}_q$, where $\alpha \in \mathrm{Gal} := \mathrm{Gal}\,[\,\mathbb{F}_q : \mathbb{F}_p\,]$. We want to show that there exists a permutation $\pi \in S_n$ such that the image of the i-th column of $M(C)$ under α is proportional to the $\pi(i)$-th column of $M(C')$ for $i \in n$. This fact shows then that σ can be described as an element $(\psi; (\alpha, \pi))$ of $\mathbb{F}_q^* \wr_n (\mathrm{Gal} \times S_n)$. Thus it is a semilinear isometry of $\mathbb{F}_q^n$. The proof is based on the fact that the image of a subspace under a semilinear mapping is again a subspace.

6.8.5 **Theorem** *Assume that $\sigma: C \to C'$ is a local semilinear isometry between two linear (n,k)-codes over $\mathbb{F}_q$ with $\sigma(\kappa c) = \alpha(\kappa)\sigma(c)$ for $c \in C$, $\kappa \in \mathbb{F}_q$, where $\alpha \in \mathrm{Gal}$. Let $d_i^\top$ and $d_j'^\top$ be the columns of $M(C)$, respectively $M(C')$. Then there exists a permutation $\pi \in S_n$ such that $\alpha(d_i^\top)$ is proportional to $d_{\pi(i)}'^\top$ for $i \in n$.*

Proof: Only a few arguments must be changed in order to adapt the previous proof to local semilinear isometries. From 6.8.2 we know that $M(C)$ and $M(C')$ have the same number of zero columns, which we indicate by s.

Let $d_i^\top$ be a nonzero column of $M(C)$, and let $i = i_0, \ldots, i_{r-1}$ indicate the indices of the columns of $M(C)$ proportional to $d_i^\top$. Then all the cross sections $C_i = C_{i_0}, C_{i_1}, \ldots, C_{i_{r-1}}$ determine the same $(n, k-1)$-code. The matrix $M(C_i)$ has $r + s$ zero columns, namely $d_{i_0}^\top, \ldots, d_{i_{r-1}}^\top$, which come from the construction as a cross section in these columns, and $d_{i_r}^\top, \ldots, d_{i_{r+s-1}}^\top$, which are the zero columns appearing already in $M(C)$.

Since σ is a local semilinear isometry between C and C', also the restriction $\sigma|_{C_i}$ is a semilinear isometry between C_i and $\sigma(C_i)$, whence by 6.8.2, $M(C_i)$ and $M(\sigma(C_i))$ have the same number of zero columns. Let us assume that the indices of the zero columns of $M(\sigma(C_i))$ are given by $j_0, \ldots, j_{r+s-1}$, and that $j_r, \ldots, j_{r+s-1}$ are the indices of the s zero columns of $M(C')$. As above, $\sigma(C_i)$ is the cross section of C' in any of the components $j_0, \ldots, j_{r-1}$, for instance, $M(\iota(C_i)) = M(C'_{j_0})$. According to 6.8.3, the columns of $M(C')$ with indices $j_0, \ldots, j_{r-1}$ are proportional and form an equivalence class of columns of $M(C')$.

Next we claim that the columns $\alpha(d_{i_0}^\top)$ and $d_{j_0}'^\top$ are proportional, i.e. there exists some $\lambda \in \mathbb{F}_q^*$ such that $d_{j_0}'^\top = \lambda \alpha(d_{i_0}^\top)$. Assume that $b = (b_0, \ldots, b_{n-1})$ with $b_{i_0} = 1$ belongs to $C \setminus C_{i_0}$. Then $\sigma(b) \in C' \setminus C_{j_0}'$, whence the j_0-th component of $\sigma(b)$, which we indicate as $\sigma(b)_{j_0}$, is different from zero. Now take an arbitrary $c \in C \setminus C_{i_0}$. Since C_{i_0} is a $(k-1)$-dimensional subspace of C, there exist uniquely determined $\tilde{c} \in C_{i_0}$ and $\kappa \in \mathbb{F}_q$ such that $c = \tilde{c} + \kappa b$. Consequently, $\kappa = c_{i_0} \neq 0$. Since $\sigma(\tilde{c}) \in C_{j_0}'$, the j_0-th component of $\sigma(c) = \sigma(\tilde{c}) + \alpha(\kappa)\sigma(b)$ is equal to $\alpha(c_{i_0})\sigma(b)_{j_0}$. This holds true for any $c \in C \setminus C_{i_0}$, whence $\alpha(d_{i_0}^\top)$, the image of the i_0-th column of $M(C)$ under α, is proportional to the j_0-th column of $M(C')$ with the nonzero factor $\lambda = \sigma(b)_{j_0}$.

Using the same ideas as in the previous proof, we determine a permutation $\pi \in S_n$ so that $\alpha(d_i^\top)$ and $d_{\pi(i)}'^\top$, $i \in n$, are proportional.   □

Thus for each $c \in C$ we have

$$\sigma(c) = \big(\psi(0)\alpha(c_{\pi^{-1}(0)}), \ldots, \psi(n-1)\alpha(c_{\pi^{-1}(n-1)})\big),$$

for some $\psi \in (\mathbb{F}_q^*)^n$.

Exercises

Exercise Prove 6.8.1.   E.6.8.1

6.9 Existence and Construction of Normal Bases
6.9

In Section 3.3 normal bases of a field extension were introduced. So far we have not shown that it is always possible to find a normal basis. Our proof is based on some notions from module theory, which were presented in the meantime. An interesting and detailed discussions of normal bases can be found in [62].

In Section 6.3 we have shown that for any endomorphism A of $\mathbb{F}_q^n$ the vector space $\mathbb{F}_q^n$ becomes an $\mathbb{F}_q[x]$-module by 6.3.5. Here we repeat the outer multiplication once again

$$\mathbb{F}_q[x] \times \mathbb{F}_q^n \to \mathbb{F}_q^n : (f, v) \mapsto fv := f(A)v := \sum_{i=0}^d \kappa_i A^i v,$$

where f is the polynomial $\sum_{i=0}^d \kappa_i x^i$. If A is represented by a matrix then $A^i v$ is the matrix multiplication $v \cdot (A^i)^\top$. The minimal polynomial M_A of A is the monic polynomial $f \in \mathbb{F}_q[x]$ of least degree so that $f(A) = 0$. If we have a matrix representation of the endomorphism A with respect to the basis B of $\mathbb{F}_q^n$ over $\mathbb{F}_q$, then the characteristic polynomial χ_A of A is defined as the determinant $\chi_A(x) := \det(xI_n - A) \in \mathbb{F}_q[x]$, where I_n is the $n \times n$-unit matrix. The characteristic polynomial is always a polynomial of degree n. It does not depend on the particular choice of the basis B. By the Cayley–Hamilton Theorem 6.3.11 it satisfies $\chi_A(A) = 0$, whence the minimal polynomial M_A is a divisor of the characteristic polynomial χ_A.

Considered as a linear $\mathbb{F}_q$-space, $\mathbb{F}_{q^n}$ is isomorphic to $\mathbb{F}_q^n$, thus it is also an $\mathbb{F}_q[x]$-module: For any endomorphism α of $\mathbb{F}_{q^n}$ we obtain a module structure

$$\mathbb{F}_q[x] \times \mathbb{F}_{q^n} \to \mathbb{F}_{q^n} \; : \; (f, \kappa) \mapsto f\kappa := f(\alpha)(\kappa) := \sum_{i=0}^d \kappa_i \alpha^i(\kappa),$$

where f is the polynomial $\sum_{i=0}^d \kappa_i x^i$. In the present section we always consider $\alpha = \tau$, the Frobenius automorphism of $\mathbb{F}_{q^n}$ over $\mathbb{F}_q$. In order to show that a normal basis exists for each extension field $\mathbb{F}_{q^n}$ over $\mathbb{F}_q$, we apply Dedekind's Independence Theorem 3.3.6 to the n distinct powers of the Frobenius automorphism τ.

6.9.1   **Lemma** *For $n \geq 1$ let $\tau \colon \mathbb{F}_{q^n} \to \mathbb{F}_{q^n}$ be the Frobenius automorphism $\tau(\beta) = \beta^q$. Then the vector space $\mathbb{F}_{q^n}$ is a cyclic $\mathbb{F}_q[x]$-module.*

Proof: Since τ^n is the identity on $\mathbb{F}_{q^n}$, the minimal polynomial of τ is a divisor of $x^n - 1$. The automorphisms $\tau^0, \tau^1, \ldots, \tau^{n-1}$ are pairwise distinct, whence by Dedekind's Independence Theorem they are linearly independent over $\mathbb{F}_q$. For this reason, the degree of the minimal polynomial of τ is at least n. Consequently, $x^n - 1$ is the minimal polynomial of τ.

Moreover, n is the dimension of the $\mathbb{F}_q$-vector space $\mathbb{F}_{q^n}$. Therefore, $x^n - 1$ is also the characteristic polynomial of τ. Thus, the minimal polynomial and the characteristic polynomial of τ coincide, and according to Exercise 6.3.7, the $\mathbb{F}_q[x]$-module $\mathbb{F}_{q^n}$ is cyclic.   □

This allows us to prove the existence of a normal basis.

6.9.2   **The Existence of normal bases** *Let n be a positive integer. For any finite field $\mathbb{F}_q$ and its extension $\mathbb{F}_{q^n}$ there exists $\kappa \in \mathbb{F}_{q^n}$ so that*

$$\left\{ \kappa, \tau(\kappa), \ldots, \tau^{n-1}(\kappa) \right\}$$

is a basis of $\mathbb{F}_{q^n}$ over $\mathbb{F}_q$.

Proof: Since $\mathbb{F}_{q^n}$ is a cyclic $\mathbb{F}_q[x]$-module, according to 6.9.1, there exists some $\kappa \in \mathbb{F}_{q^n}$ so that

$$\mathbb{F}_{q^n} = \mathbb{F}_q[x]\kappa = \{ f\kappa \mid f \in \mathbb{F}_q[x] \}.$$

Since the minimal polynomial of τ is of degree n, we can restrict ourselves to polynomials f of degree less than n, obtaining

$$\mathbb{F}_{q^n} = \{ f\kappa \mid f \in \mathbb{F}_q[x], \; \deg f < n \}.$$

Consequently, there exist n polynomials $f_0, \ldots, f_{n-1}$ with $\deg f_i < n$ for $i \in n$, so that $\{f_0\kappa, \ldots, f_{n-1}\kappa\}$ is a basis of $\mathbb{F}_{q^n}$. Since each f_i is a linear combination of x^j for $j \in n$, we finally deduce that $\{\kappa, \tau(\kappa), \ldots, \tau^{n-1}(\kappa)\}$ is also a basis of $\mathbb{F}_{q^n}$. (Here we use the polynomials $f_i(x) = x^i$.) This is a normal basis of $\mathbb{F}_{q^n}$ over $\mathbb{F}_q$.   □

It is even possible to show that for any finite field $\mathbb{F}_q$ and its extension $\mathbb{F}_{q^n}$, where n is a positive integer, there exists a primitive element $\kappa \in \mathbb{F}_{q^n}$ so that

$$\left\{ \kappa, \tau(\kappa), \ldots, \tau^{n-1}(\kappa) \right\}$$

is a basis of $\mathbb{F}_{q^n}$ over $\mathbb{F}_q$ (cf. [127]).

Now we describe how to construct a normal basis. There exist both probabilistic and deterministic algorithms for finding a normal basis of $\mathbb{F}_{q^n}$ over $\mathbb{F}_q$. We will present both approaches.

Definition (trace function) The *trace function* of $\mathbb{F}_{q^n}$ over $\mathbb{F}_q$ is defined by   6.9.3

$$\mathrm{Tr}: \mathbb{F}_{q^n} \to \mathbb{F}_q \ : \ \alpha \mapsto \mathrm{Tr}(\alpha) := \sum_{i \in n} \alpha^{q^i}.$$ ◇

It is easy to prove that the trace function is a homomorphism.

An element $\kappa \in \mathbb{F}_{q^n}$ is called *normal* over $\mathbb{F}_q$ if $\{\kappa, \tau(\kappa), \ldots, \tau^{n-1}(\kappa)\}$ is a normal basis of $\mathbb{F}_{q^n}$ over $\mathbb{F}_q$.

In order to characterize whether a given set of n elements forms a basis of $\mathbb{F}_{q^n}$ over $\mathbb{F}_q$ we introduce the *discriminant* $\Delta: \mathbb{F}_{q^n}^n \to \mathbb{F}_q$ defined by

$$\Delta(\alpha_0, \ldots, \alpha_{n-1}) := \det \begin{pmatrix} \mathrm{Tr}(\alpha_0 \alpha_0) & \cdots & \mathrm{Tr}(\alpha_0 \alpha_{n-1}) \\ \vdots & \ddots & \vdots \\ \mathrm{Tr}(\alpha_{n-1}\alpha_0) & \cdots & \mathrm{Tr}(\alpha_{n-1}\alpha_{n-1}) \end{pmatrix}.$$

Theorem *The set $\{\alpha_0, \ldots, \alpha_{n-1}\} \subseteq \mathbb{F}_{q^n}$ is a basis of $\mathbb{F}_{q^n}$ over $\mathbb{F}_q$ if and only if*   6.9.4
$\Delta(\alpha_0, \ldots, \alpha_{n-1}) \neq 0$.

Proof: Assume that $\{\alpha_0, \ldots, \alpha_{n-1}\}$ is a basis of $\mathbb{F}_{q^n}$ over $\mathbb{F}_q$. We show that the row vectors of the matrix used to define Δ are linearly independent over $\mathbb{F}_q$. Assume that for $c_0, \ldots, c_{n-1} \in \mathbb{F}_q$ we have

$$\sum_{i \in n} c_i \left(\mathrm{Tr}(\alpha_i \alpha_0), \ldots, \mathrm{Tr}(\alpha_i \alpha_{n-1}) \right) = 0,$$

then

$$\sum_{i \in n} c_i \, \mathrm{Tr}(\alpha_i \alpha_j) = 0, \qquad j \in n.$$

For $\beta := \sum_{i \in n} c_i \alpha_i$ we have

$$
\begin{aligned}
\text{Tr}(\beta \alpha_j) &= \sum_{k \in n} (\beta \alpha_j)^{q^k} = \sum_{k \in n} \left(\sum_{i \in n} c_i \alpha_i \alpha_j \right)^{q^k} \\
&= \sum_{k \in n} \sum_{i \in n} c_i \left(\alpha_i \alpha_j \right)^{q^k} = \sum_{i \in n} c_i \, \text{Tr}(\alpha_i \alpha_j) = 0, \qquad j \in n.
\end{aligned}
$$

Since the trace is a vector space homomorphism and $\{\alpha_0, \ldots, \alpha_{n-1}\}$ is a basis of $\mathbb{F}_{q^n}$, we have $\text{Tr}(\beta \alpha) = 0$ for all $\alpha \in \mathbb{F}_{q^n}$. This is only possible for $\beta = 0$, whence $\sum_{i \in n} c_i \alpha_i = 0$ and consequently $c_0 = \ldots = c_{n-1} = 0$.

Conversely, assume that $\Delta(\alpha_0, \ldots, \alpha_{n-1}) \neq 0$ and $\sum_{i \in n} c_i \alpha_i = 0$ for some $c_0, \ldots, c_{n-1} \in \mathbb{F}_q$. Then $\sum_{i \in n} c_i \alpha_i \alpha_j = 0$ for $j \in n$ and by applying the trace function

$$
0 = \text{Tr}(0) = \text{Tr}\left(\sum_{i \in n} c_i \alpha_i \alpha_j \right) = \sum_{i \in n} c_i \, \text{Tr}(\alpha_i \alpha_j), \qquad j \in n.
$$

By assumption the rows of the matrix in the definition of $\Delta(\alpha_0, \ldots, \alpha_{n-1})$ are linearly independent, whence $c_0 = \ldots = c_{n-1} = 0$ and, therefore, $\alpha_0, \ldots, \alpha_{n-1}$ are linearly independent over $\mathbb{F}_q$. □

6.9.5   **Corollary** *The set $\{\alpha_0, \ldots, \alpha_{n-1}\} \subseteq \mathbb{F}_{q^n}$ is a basis of $\mathbb{F}_{q^n}$ over $\mathbb{F}_q$ if and only if the matrix*

$$
A := \begin{pmatrix} \alpha_0 & \cdots & \alpha_{n-1} \\ \alpha_0^q & \cdots & \alpha_{n-1}^q \\ \vdots & \ddots & \vdots \\ \alpha_0^{q^{n-1}} & \cdots & \alpha_{n-1}^{q^{n-1}} \end{pmatrix}
$$

is regular.

Proof: $\{\alpha_0, \ldots, \alpha_{n-1}\}$ is a basis if and only if $\Delta(\alpha_0, \ldots, \alpha_{n-1}) \neq 0$. As a matter of fact, $\Delta(\alpha_0, \ldots, \alpha_{n-1}) = \det(A^\top \cdot A) = (\det A)^2$. □

The probabilistic algorithm for finding a normal basis is based upon

---

6.9.6   **Theorem (Artin [3])** *Consider an irreducible polynomial f of degree n over $\mathbb{F}_q$ and $\alpha \in \mathbb{F}_{q^n}$ a root of f. Let*

$$
g(x) := \frac{f(x)}{(x - \alpha) f'(\alpha)} \in \mathbb{F}_{q^n}[x].
$$

Then there exist at least $q - n(n-1)$ elements $\kappa \in \mathbb{F}_q$ so that $g(\kappa)$ is normal over $\mathbb{F}_q$.

Proof: For $i \in n$ let $\alpha_i := \tau^i(\alpha)$ and $g_i(x) := \tau^i(g(x))$, where τ is the Frobenius automorphism of $\mathbb{F}_{q^n}$ over $\mathbb{F}_q$. Then

$$g_i(x) = \frac{f(x)}{(x - \alpha_i)f'(\alpha_i)}$$

is a polynomial in $\mathbb{F}_{q^n}[x]$ of degree $n - 1$ with roots α_k for $k \neq i$ and $g_i(\alpha_i) = 1$. Hence,

$$g_i(x)g_k(x) \equiv 0 \bmod \mathrm{I}(f), \qquad i \neq k. \tag{6.9.7}$$

Moreover,

$$\sum_{i \in n} g_i(x) - 1 = 0, \tag{6.9.8}$$

since the left-hand side is a polynomial of degree at most $n - 1$ with n roots $\alpha_0, \ldots, \alpha_{n-1}$. Multiplying 6.9.8 by $g_i(x)$ and using 6.9.7 yields

$$g_i(x) \equiv (g_i(x))^2 \bmod \mathrm{I}(f). \tag{6.9.9}$$

Let D be the matrix

$$D := \begin{pmatrix} g_0(x) & g_1(x) & \cdots & g_{n-1}(x) \\ g_1(x) & g_2(x) & \cdots & g_0(x) \\ \cdots & \cdots & \cdots & \\ g_{n-1}(x) & g_0(x) & \cdots & g_{n-2}(x) \end{pmatrix},$$

then $D^\top = D$. Because of 6.9.9 and 6.9.8, the diagonal elements of $D^\top \cdot D$ are of the form

$$\sum_{i \in n} g_i(x)^2 \equiv \sum_{i \in n} g_i(x) = 1 \bmod \mathrm{I}(f).$$

All the other entries of $D^\top \cdot D$ are 0 because of 6.9.7. Let $D(x) := \det D$. We obtain $D(x)^2 \equiv 1 \bmod \mathrm{I}(f)$. This means that $D(x)$ is a nonzero polynomial. By construction its degree is at most $n(n - 1)$. Therefore, $D(x)$ has at most $n(n - 1)$ roots.

Consider some $u \in \mathbb{F}_q$ with $D(u) \neq 0$. Then the matrix

$$\begin{pmatrix} g_0(u) & g_1(u) & \cdots & g_{n-1}(u) \\ g_1(u) & g_2(u) & \cdots & g_0(u) \\ \cdots & \cdots & \cdots & \\ g_{n-1}(u) & g_0(u) & \cdots & g_{n-2}(u) \end{pmatrix} =$$

$$\begin{pmatrix} g(u) & \tau(g(u)) & \cdots & \tau^{n-1}(g(u)) \\ \tau(g(u)) & \tau^2(g(u)) & \cdots & g(u) \\ \cdots & \cdots & \cdots & \\ \tau^{n-1}(g(u)) & g(u) & \cdots & \tau^{n-2}(g(u)) \end{pmatrix}$$

is regular, whence by 6.9.5, $\{g(u), \tau(g(u)), \ldots, \tau^{n-1}(g(u))\}$ is a basis of $\mathbb{F}_{q^n}$ over $\mathbb{F}_q$. In fact, it is a normal basis. □

6.9.10   **Algorithm (Generate a normal element)**

Input: q, n, an irreducible polynomial $f \in \mathbb{F}_q[x]$ of degree n, and α a root of f.

Output: A normal element or an error message. If $q > n(n-1)$ the output β is a normal element of $\mathbb{F}_{q^n}$ over $\mathbb{F}_q$.

(1) If $q \leq n(n-1)$ terminate the algorithm and output an error message.

(2) Determine g as in 6.9.6.

(3) Choose $u \in \mathbb{F}_q$ at random.

(4) Let $\kappa = g(u)$.

(5) If κ is normal over $\mathbb{F}_q$ output κ. Otherwise goto (3).

If $q > 2n(n-1)$, then, by 6.9.6, κ is normal with probability at least $1/2$.   □

Finally, we present a deterministic algorithm, due to Lenstra (cf. [126]), for constructing a normal basis.

6.9.11   **Definition (τ-order)** Let τ be the Frobenius automorphism of $\mathbb{F}_{q^n}$ over $\mathbb{F}_q$. For $\kappa \in \mathbb{F}_{q^n} \setminus \{0\}$ determine the least positive integer k and $c_0, \ldots, c_{k-1} \in \mathbb{F}_q$ so that

$$\tau^k(\kappa) = \sum_{i \in k} c_i \tau^i(\kappa).$$

Then the polynomial

$$\mathrm{Ord}_\kappa(x) := x^k - \sum_{i \in k} c_i x^i \in \mathbb{F}_q[x]$$

is called the τ-*order* of κ.   ◇

The τ-order of $\kappa \neq 0$ is uniquely determined. Since $\tau^n(\kappa) = \kappa$, it is clear that $\mathrm{Ord}_\kappa(x)$ is a divisor of $x^n - 1$. Moreover, the element κ is normal over $\mathbb{F}_q$ if and only if $\mathrm{Ord}_\kappa(x) = x^n - 1$.

6.9.12   **Lemma** *Consider $\alpha \in \mathbb{F}_{q^n} \setminus \{0\}$ with $\mathrm{Ord}_\alpha(x) \neq x^n - 1$, and let*

$$g(x) := \frac{x^n - 1}{\mathrm{Ord}_\alpha(x)}.$$

Then there exists $\beta \in \mathbb{F}_{q^n}$ so that $g(x)\beta = \alpha$.

Proof: Let γ be a normal element of $\mathbb{F}_{q^n}$ over $\mathbb{F}_q$. Then there exists some $f \in \mathbb{F}_q[x]$ so that $f(x)\gamma = \alpha$. Since $\mathrm{Ord}_\alpha(x)\alpha = 0$, we have $(\mathrm{Ord}_\alpha(x)f(x))\gamma = 0$. So $\mathrm{Ord}_\gamma(x) = x^n - 1$ is a divisor of $\mathrm{Ord}_\alpha(x)f(x)$. Thus, $g(x)$ is a divisor of $f(x)$. Let $f(x) = g(x)h(x)$, then $\alpha = f(x)\gamma = g(x)(h(x)\gamma)$. This proves that $\beta := h(x)\gamma$ satisfies the assertion.   □

Lemma *Consider* $\alpha, \beta \in \mathbb{F}_{q^n} \setminus \{0\}$ *with* $\mathrm{Ord}_\alpha(x) \neq x^n - 1$, **6.9.13**

$$g(x) := \frac{x^n - 1}{\mathrm{Ord}_\alpha(x)},$$

and $\alpha = g(x)\beta$ *as in the previous lemma. If* $\deg \mathrm{Ord}_\beta(x) \leq \deg \mathrm{Ord}_\alpha(x)$, *then there exists a nonzero* $\eta \in \mathbb{F}_{q^n}$ *so that*

$$g(x)\eta = 0,$$ **6.9.14**

and

$$\deg \mathrm{Ord}_{\alpha+\eta}(x) > \deg \mathrm{Ord}_\alpha(x).$$ **6.9.15**

Proof: Let γ be a normal element of $\mathbb{F}_{q^n}$ over $\mathbb{F}_q$. Then $\eta := \mathrm{Ord}_\alpha(x)\gamma$ is different from 0 and satisfies

$$g(x)\eta = \frac{x^n - 1}{\mathrm{Ord}_\alpha(x)} \mathrm{Ord}_\alpha(x)\gamma = (x^n - 1)\gamma = 0.$$

Now we prove that each nonzero solution η of 6.9.14 satisfies 6.9.15. From $\mathrm{Ord}_\beta(x)\alpha = \mathrm{Ord}_\beta(x)g(x)\beta = 0$ we obtain that $\mathrm{Ord}_\alpha(x)$ divides $\mathrm{Ord}_\beta(x)$. From the assumption on the degrees of these two polynomials we deduce that $\mathrm{Ord}_\alpha(x) = \mathrm{Ord}_\beta(x)$. Thus, by Exercise 6.9.2 we have $\gcd(g(x), \mathrm{Ord}_\alpha(x)) = 1$. Since $\mathrm{Ord}_\eta(x)$ is a divisor of $g(x)$, also $\gcd(\mathrm{Ord}_\eta(x), \mathrm{Ord}_\alpha(x)) = 1$. An application of Exercise 6.9.3 yields that $\mathrm{Ord}_{\alpha+\eta}(x) = \mathrm{Ord}_\alpha(x)\,\mathrm{Ord}_\eta(x)$, whence $\deg \mathrm{Ord}_{\alpha+\eta}(x) > \deg \mathrm{Ord}_\alpha(x)$.   □

Algorithm (Construct a normal element)   **6.9.16**
  Input: q and n.
  Output: A normal element of $\mathbb{F}_{q^n}$ over $\mathbb{F}_q$.

(1) Choose $\alpha \in \mathbb{F}_q$ at random and determine $\mathrm{Ord}_\alpha(x)$.

(2) If $\mathrm{Ord}_\alpha(x) = x^n - 1$ then output α and terminate the algorithm.

(3) Calculate $g(x) := (x^n - 1)/\mathrm{Ord}_\alpha(x)$.

(4) Find $\beta \in \mathbb{F}_{q^n}$ so that $g(x)\beta = \alpha$ and determine $\mathrm{Ord}_\beta(x)$.

(5) If $\deg \mathrm{Ord}_\beta(x) > \deg \mathrm{Ord}_\alpha(x)$, replace α by β and goto (2).

(6) If $\deg \mathrm{Ord}_\beta(x) \leq \deg \mathrm{Ord}_\alpha(x)$, then find a nonzero element $\eta \in \mathbb{F}_{q^n}$ so that $g(x)\eta = 0$. Replace α by $\alpha + \eta$, determine $\mathrm{Ord}_\alpha(x)$ and goto (2).

This algorithm terminates after finitely many steps, because in (6) the degree of $\mathrm{Ord}_\alpha(x)$ increases at least by 1.   □

Exercises

Exercise Why is the τ-order of $\kappa \neq 0$ is uniquely determined?   **E.6.9.1**

E.6.9.2   **Exercise** For $\alpha \in \mathbb{F}_{q^n}$ and $g \in \mathbb{F}_q[x]$ show that if $g(x)\alpha \neq 0$, then the τ-order of $g(x)\alpha$ is equal to $\mathrm{Ord}_\alpha(x)/\gcd(\mathrm{Ord}_\alpha(x), g(x))$.

E.6.9.3   **Exercise** Consider $\alpha, \eta \in \mathbb{F}_{q^n} \setminus \{0\}$ such that $\mathrm{Ord}_\alpha(x)$ and $\mathrm{Ord}_\eta(x)$ are relatively prime. Show that

$$\mathrm{Ord}_{\alpha+\eta}(x) = \mathrm{Ord}_\alpha(x)\,\mathrm{Ord}_\eta(x).$$

E.6.9.4   **Exercise** Let $\alpha \in \mathbb{F}_{3^4}$ be a root of the irreducible polynomial $f(x) = x^4 + x^3 + 2 \in \mathbb{F}_3[x]$. Using α, compute a normal basis of $\mathbb{F}_{3^4}$ over $\mathbb{F}_3$ and determine by an application of 3.5.5 the list of all irreducible polynomials of degree 4 over $\mathbb{F}_3$.

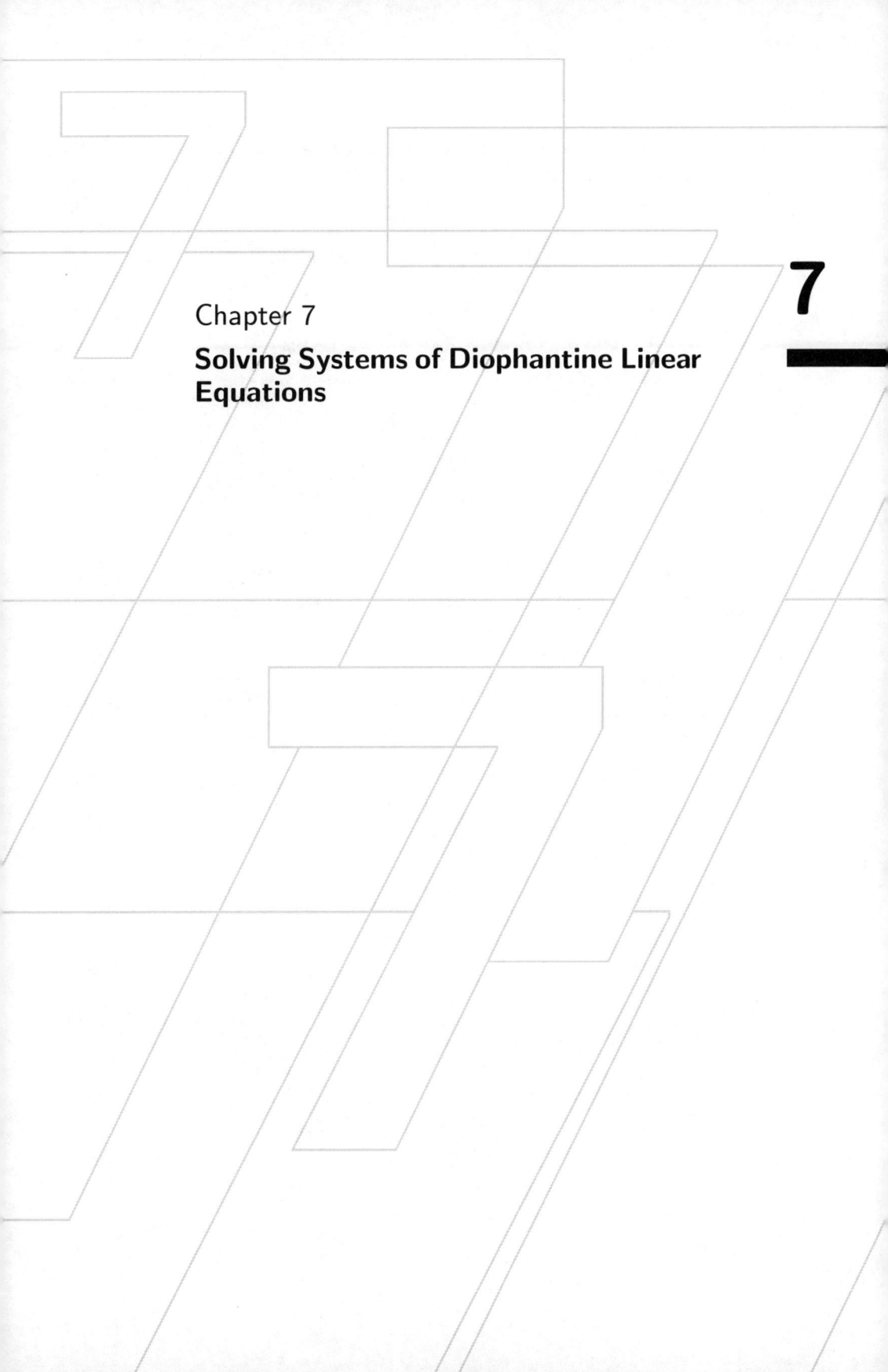

Chapter 7

**Solving Systems of Diophantine Linear
Equations**

7

7

7   **Solving Systems of Diophantine Linear Equations**

7 Solving Systems of Diophantine Linear Equations

In this chapter we consider systems of linear equations whose solutions are restricted to the integers. Linear equations of this form are called *Diophantine* linear equations. In Chapter 8 we will reduce the problem of finding linear codes with prescribed minimum distance to solving systems of Diophantine linear equations. If we try to solve these systems it is crucial to have fast methods at hand. Here, we study one possible approach based on so called lattice basis reduction. In Section 1.8 we saw an algorithm for determining the minimum distance of a linear code. Section 7.8 contains another minimum distance algorithm, also based on lattice basis reduction.

With Gaussian elimination we are able to solve linear systems $A \cdot x = d$ of equations for vectors $x \in \mathbb{R}^n$ easily[1]. The same algorithm works also if we restrict to $x \in \mathbb{Q}^n$. Then, since we can multiply the whole system with the least common multiple of all denominators, we can also solve these systems over $\mathbb{Z}$. Unfortunately, the size of the denominators can grow very rapidly. So, Gaussian Elimination does not longer run in polynomial time. But there exist algorithms to compute the Hermitian normal form (HNF) efficiently, i.e. in polynomial time, see for example [39]. Thus, with the help of the HNF we can solve systems of Diophantine linear equations easily.

The situation changes when we have to solve systems of linear *inequalities* over the integers or, equivalently, if we have to find nonnegative integral solutions of systems of linear equations. Equally hard problems arise if the variables x_i are restricted to integers from intervals $l_i \leq x_i \leq r_i$ for $i \in n$. The problem to decide if there is such a vector x is known to be NP-complete. At present, no algorithm is known which decides in a number of steps that is polynomial in the size of the input for this problem if there is a solution or not. Here, we restrict our attention to Diophantine linear equations of the following form.

$$A \cdot x = d, \quad l \leq x \leq r,$$

for given $A \in \mathbb{Z}^{m \times n}, d \in \mathbb{Z}^m, l, r \in \mathbb{Q}^n$, where $l \leq x \leq r$ means $l_i \leq x_i \leq r_i$, for each $i \in n$. We ask for solutions $x \in \mathbb{Z}^n$ with $l \leq x \leq r$.

Example In Chapter 8, Example 8.4.4, the following system of Diophantine linear equations occurs during the construction of linear codes with prescribed

7.0.1

---

[1]for technical reasons we use the column convention in the present chapter

minimum distance:

$$
\begin{pmatrix}
2 & 2 & 0 & -1 & 0 & 0 \\
1 & 1 & 2 & 0 & -1 & 0 \\
0 & 3 & 1 & 0 & 0 & -1 \\
3 & 6 & 4 & 0 & 0 & 0
\end{pmatrix}
\cdot
\begin{pmatrix}
x_0 \\ x_1 \\ x_2 \\ x_3 \\ x_4 \\ x_5
\end{pmatrix}
=
\begin{pmatrix}
0 \\ 0 \\ 0 \\ 14
\end{pmatrix},
$$

where $x_0 \in \{0,1,2,3,4\}$, $x_1 \in \{0,1,2\}$, $x_2 \in \{0,1,2,3\}$ and $x_i \in \{0,1,2,3,4,5\}$ for $i \in \{3,4,5\}$. It is easy to check that $x = (0,1,2,2,5,5)^\top$ is an integer solution of the system of equations which also satisfies the additional lower and upper bounds. ◇

Several equally hard variations of this problem exist. The knapsack problem and the subset sum problem are just two instances.

- The *knapsack problem:* Given nonnegative integers $c_i, w_i, i \in n$, and k, find a subset $S \subseteq \{0,1,\ldots,n-1\}$ such that $\sum_{j \in S} w_j \le k$ and $\sum_{j \in S} c_j$ is maximal.

- The *subset sum problem:* Given nonnegative numbers $w_i, i \in n$, and k, find a subset $S \subseteq \{0,1,\ldots,n-1\}$ such that $\sum_{j \in S} w_j = k$.

7.0.2   **Example** Let $w = (31,41,59,26,53,58,97,93,23,84,62)$, $k = 314$ and $c = (1,1,1,1,1,1,1,1,1,1,1)$.

- The subset sum problem asks for subsets $S \subseteq \{0,1,2,\ldots,10\}$ such that $\sum_{i \in S} w_i = 314$. There are three solutions: $\{3,4,5,7,9\}$, $\{0,3,4,5,9,10\}$, and $\{0,2,3,6,8,9\}$.

- In the knapsack problem, we ask for subsets $S \subseteq \{0,1,\ldots,10\}$ of maximal size subject to the condition that $\sum_{i \in S} w_i \le 314$. Here, the solution is $S = \{0,2,3,4,5,8,10\}$, $\sum_{i \in S} c_i = 7$ and $\sum_{i \in S} w_i = 312 \le 314$. ◇

As we will see in Section 7.8, the problem of computing the minimum distance of certain linear codes can be reduced to a problem of solving a Diophantine system of linear equations. Also, in Chapter 8, we will use systems of Diophantine linear equations to construct optimal codes. Many further objects from Discrete Mathematics can be constructed in a similar fashion. In fact, Combinatorial Designs, Steiner systems and covering codes have all been constructed in a similar way by means of Diophantine equations.

Many algorithms for solving these problems have been proposed. Some of them rely on relaxation techniques and use Linear Programming. Other approaches use backtracking. The approach used here is based on lattices[2] and

---

[2]in this chapter lattices are geometrical objects, different from the definition in 3.2.24

on a very important method invented by Lenstra, Lenstra and Lovász [125] – the celebrated LLL-algorithm. This method has been applied successfully to break certain cryptosystems based on the knapsack problem [119].

The first step is to transform the problem of finding the solutions of linear Diophantine equation systems into a problem involving lattices. A lattice is just the set of integer linear combinations of a given set of linearly independent vectors in a real vector space. In this setting, the problem can be reduced to the question of finding sufficiently short vectors in a suitable lattice. Here, short is usually meant in connection to a norm, like the ℓ_∞-norm or the Euclidean norm. To find these short vectors in polynomial time, we apply the LLL-algorithm. In a second step, we use exhaustive enumeration to find *all* vectors which are solutions of our original problem. This last step needs exponential time.

With this approach many finite incidence structures could be constructed, see [15], [16], [25], [28], [29], [199] and the literature cited there.

7.1 Lattices

Let us recall briefly the basic definitions and fundamental theorems of the theory of lattice. For a thorough introduction into the subject we refer the reader to [77], for instance.

— As usual, let $\mathbb{R}^n$ denote the real Euclidean n-dimensional space. Its elements $v \in \mathbb{R}^n$ are written as column vectors $v = (v_0, v_1, \ldots, v_{n-1})^\top$.

— For $q \in \mathbb{R}, q \geq 1$, we define the ℓ_q-*norm* by

$$\|-\|_q : \mathbb{R}^n \to \mathbb{R} : v \mapsto \left(\sum_{i \in n} |v_i|^q \right)^{1/q},$$

and the ℓ_∞-*norm* as follows:

$$\|-\|_\infty : \mathbb{R}^n \to \mathbb{R} : v \mapsto \max_{i \in n} |v_i|.$$

— For $m \in \mathbb{N}$, the vectors $b^{(0)}, b^{(1)}, \ldots, b^{(m-1)} \in \mathbb{R}^n$ span a subspace of $\mathbb{R}^n$ which we denote by

$$\langle b^{(0)}, b^{(1)}, \ldots, b^{(m-1)} \rangle := \left\{ \sum_{i \in m} x_i b^{(i)} \mid x_i \in \mathbb{R}, i \in m \right\}.$$

The notation for a subspace $\langle b^{(0)}, b^{(1)}, \ldots, b^{(m-1)} \rangle$ is not to be confused with the standard bilinear form

$$\langle v, w \rangle = \sum_{i \in n} v_i \cdot w_i$$

for $v, w \in \mathbb{R}^n$. But the meaning should be clear from the context.

The basic notions are the following ones:

7.1.1   **Definition (lattice)** Let $b^{(0)}, b^{(1)}, \ldots, b^{(m-1)}$ be m linearly independent vectors in $\mathbb{R}^n$.

— The set
$$L(b^{(0)}, b^{(1)}, \ldots, b^{(m-1)}) := \left\{ \sum_{i \in m} u_i \cdot b^{(i)} \mid u_i \in \mathbb{Z}, i \in m \right\} \subset \mathbb{R}^n$$
is called the *lattice (of vectors)* with *basis* $b^{(0)}, b^{(1)}, \ldots, b^{(m-1)}$.

— The *rank* m of a lattice L with basis $b^{(0)}, b^{(1)}, \ldots, b^{(m-1)}$ is the dimension of the $\mathbb{R}$-subspace $\langle b^{(0)}, b^{(1)}, \ldots, b^{(m-1)} \rangle$ which is spanned by the basis.

— We will write
$$B := \left(b^{(0)} \mid \ldots \mid b^{(m-1)} \right)$$
for the $n \times m$-matrix whose columns are the vectors $b^{(0)}, b^{(1)}, \ldots, b^{(m-1)}$. If $L = L(b^{(0)}, b^{(1)}, \ldots, b^{(m-1)})$, then B is called a *generator matrix* of L.   ◇

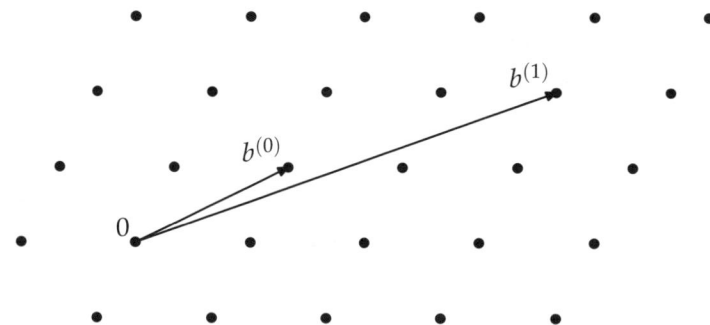

Fig. 7.1 A rank 2 lattice spanned by $b^{(0)}$ and $b^{(1)}$

It is well known [77, p. 18] that a lattice of vectors in $\mathbb{R}^n$ is a discrete additive subgroup of $\mathbb{R}^n$.

For a lattice $L \subset \mathbb{R}^n$, the most important (and difficult) algorithmic problems can be described as follows.

7.1.2   **Algorithmic problems for a given lattice L**

— The *shortest vector problem* (SVP): Find an ℓ_q-shortest vector in L, i.e. find an element w in L such that
$$\|w\|_q = \min\{\|w'\|_q \mid w' \in L \setminus \{0\}\}.$$

This question is most interesting for the Euclidean norm, the ℓ_1-norm, and the ℓ_∞-norm.

— The *closest vector problem* (CVP): Given a vector $v \in \mathbb{R}^n$ find a lattice vector w which is closest to v in the ℓ_q-norm, i.e. such that

$$\|v - w\|_q = \min\{\|v - w'\|_q \mid w' \in L\}.$$

— The *lattice basis reduction*: Given a basis $b^{(0)}, b^{(1)}, \ldots, b^{(m-1)}$ of the lattice L compute a new basis $b'^{(0)}, b'^{(1)}, \ldots, b'^{(m-1)}$ of L consisting of "shortest" vectors. Here, the meaning of short will have to be made precise.   ◇

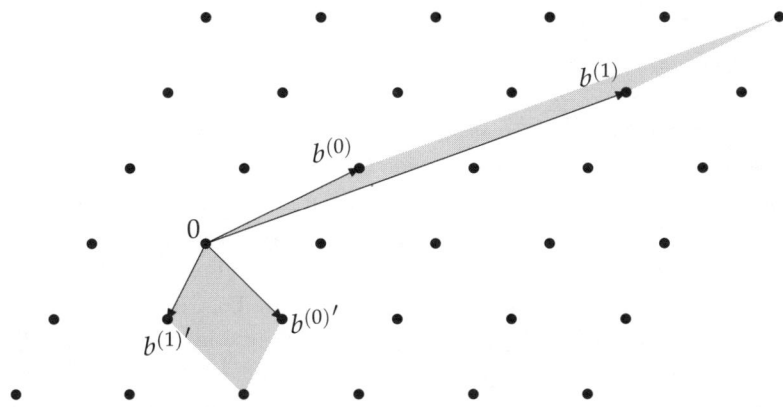

Fig. 7.2 Two different bases for $b^{(0)}, b^{(1)}$ and $b^{(0)'}, b^{(1)'}$ of the same lattice

For an overview on the algorithmic complexity of the above problems we refer to [147] and [199] and the literature cited there.

Concerning the last of the mentioned problems, we remark that the problem of finding a basis consisting of shortest vectors is not exactly defined provided the dimension is at least three. In fact, many different versions of the concept of a shortest basis exist. Two classical concepts are the reduced bases in the sense of Minkowski [150] and the reduced quadratic forms in the sense of Korkine and Zolotarev [113]. The latter aims at minimizing the orthogonality defect of a lattice basis, a concept which we will encounter in Section 7.5. Recently, one further variant has gained interest. In this, one finds a lattice basis minimizing the maximal length of any of its members, see [2], [23].

The above reduction concepts rely on the computation of shortest lattice vectors in sublattices and related lattices. Therefore, the problem of computing a reduced lattice basis in the sense of Minkowski or Korkine and Zolotarev is at least as hard as the shortest vector problem.

7.2 Diophantine Equations and Lattices

Subset sum problems. Lagarias and Odlyzko [119] have introduced the techniques of lattices and lattice basis reduction to the solution of subset sum problems. Recall that these problems can be written as finding all solutions $x \in \{0, 1\}^n$ of the system

$$A \cdot x = d$$

where A is a $1 \times n$ matrix over the integers and d is some integer. In fact, they introduced the lattice whose generator matrix is the $(m + n) \times (n + 1)$-matrix

$$B := \left(\begin{array}{c|c} N \cdot (-d) & N \cdot A \\ 0 & \\ \vdots & I_n \\ 0 & \end{array} \right)$$

where I_n denotes the identity matrix in $\mathbb{Z}^{n \times n}$ and N is a large integer constant. Let us call this lattice the Lagarias-Odlyzko lattice. It turns out that the solutions x of 7.2.1 are in bijection to certain short elements of the Lagarias-Odlyzko lattice. Namely, if $v = B \cdot w$ is an element of the lattice which is zero in the first m entries and where w is a $\{0, 1\}$-vector whose first component is equal to one, then $(w_1, \ldots, w_n)^\top$ is a solution of the Diophantine system 7.2.1 and vice-versa. Moreover, the first m components of v are zero, and hence no entry of v is a nonzero multiple of the large integer constant N. This means that v is short in the Lagarias-Odlyzko lattice. Therefore, we see that solutions of 7.2.1 are short vectors in the Lagarias-Odlyzko lattice. This means that it is useful to attack this kind of Diophantine problem with the method of finding short vectors in lattices. We illustrate this by an example.

Example Consider the subset sum problem of 7.0.2. Setting $N = 100$, the generator matrix B of 7.2.2 of the Lagarias-Odlyzko lattice is

−31400	3100	4100	5900	2600	5300	5800	9700	9300	2300	8400	6200
0	1	0	0	0	0	0	0	0	0	0	0
0	0	1	0	0	0	0	0	0	0	0	0
0	0	0	1	0	0	0	0	0	0	0	0
0	0	0	0	1	0	0	0	0	0	0	0
0	0	0	0	0	1	0	0	0	0	0	0
0	0	0	0	0	0	1	0	0	0	0	0
0	0	0	0	0	0	0	1	0	0	0	0
0	0	0	0	0	0	0	0	1	0	0	0
0	0	0	0	0	0	0	0	0	1	0	0
0	0	0	0	0	0	0	0	0	0	1	0
0	0	0	0	0	0	0	0	0	0	0	1

Multiplying this matrix by the vector $w = (1, 0, 0, 0, 1, 1, 1, 0, 1, 0, 1, 0)^\top$ gives the vector $v = (0, 0, 0, 0, 1, 1, 1, 0, 1, 0, 1, 0)^\top$. Therefore, $(w_1, w_2, \ldots, w_n)^\top$ is a solution to the subset sum problem: We have $w_i = 1$ if and only if $i \in S = \{4, 5, 6, 8, 10\}$. This set S solves the subset sum problem 7.0.2. Note that $\|v\|_2 = \sqrt{5}$ and $\|v\|_\infty = 1$.

Multiplying the matrix by the vector $w = (1, 1, 0, 1, 0, 2, 1, 1, -1, 1, 1, 0)^\top$ gives $v = (36500, 1, 0, 1, 0, 2, 1, 1, -1, 1, 1, 0)$. This means that $(w_1, w_2, \ldots, w_n)^\top$ does not solve the subset sum problem: The entries in $(w_1, w_2, \ldots, w_n)^\top$ are not all elements of $\{0, 1\}$. Furthermore, the linear equation $A \cdot (w_1, \ldots, w_n)^\top = d$ is violated. We note that $\|v\|_2 = \sqrt{1\,332\,250\,011}$ and $\|v\|_\infty = 36\,500$.   ◇

Below, we will employ the following strategy. We will start with the lattice basis which is given by the columns of the generator matrix B. We will then compute another basis for the same lattice which consists of short vectors. This transformation from one lattice basis to another is known as *lattice basis reduction*. A very important algorithm to achieve this transformation is the LLL-algorithm which we will discuss in Section 7.6.

In the context of the subset sum problem and the Lagarias-Odlyzko lattice, one hopes that through the process of lattice basis reduction one will eventually arrive at vectors $v \in \mathbb{Z}^{n+m}$ which are of the form $v_i = 0$ for $i \in m$ and either $v_i \in \{0, 1\}$ for $m \le i < m + n$ or, alternatively, $v_i \in \{0, -1\}$ for $m \le i < m + n$. It is proved in [119] that for a large class of subset sum problems a solution will correspond to the shortest vector of the lattice 7.2.2.

The Euclidean norm of such vectors is bounded above by $\sqrt{n}$. But not every short vector is a solution. Since the Euclidean distance of a vector does not distinguish between entries $+1$ and -1, it may happen that short vectors are computed whose entries are 0 or ± 1. In fact, the "mixed sign case" happens frequently among the vectors $v \in L$ with $\|v\|_\infty = 1$.

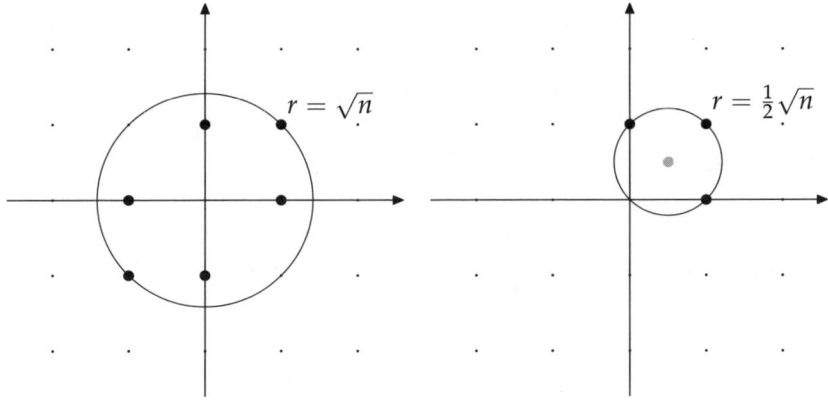

Fig. 7.3 Solution vectors for the lattice 7.2.2 (left) and for the lattice 7.2.4 (right) without the first component which is equal to zero

We can do better by appealing to the closest vector problem. The goal is to eliminate entries in the short lattice vectors which are -1. To do this, we introduce the vector $z = (0, \frac{1}{2}, \frac{1}{2}, \ldots, \frac{1}{2})^\top$, which is *not* contained in the Lagarias-Odlyzko lattice. We are now looking for the vectors closest to z in the Lagarias-Odlyzko lattice. In fact, since we are looking for $\{0, 1\}$-vectors, we may restrict our search to lattice vectors v at distance

$$\|v - z\| = \sqrt{\sum_{i \in n}(v_i - z_i)^2} = \sqrt{\sum_{i \in n} 1/4} = 1/2\sqrt{n}.$$

The situation for $n = 3$ is illustrated in Fig. 7.3. The first component of the lattice vectors is not shown as it is zero. The black dots indicate lattice points, leading to solutions, which are either short (left picture) or close to z (right picture). To solve the closest vector type problem, we use the augmented and embedded Lagarias-Odlyzko lattice, generated by the $(m + n + 1) \times (n + 1)$-matrix

$$B := \begin{pmatrix} -N \cdot d & N \cdot A \\ \hline -1/2 & \\ \vdots & I_n \\ -1/2 & \\ \hline 1 & 0 \quad \cdots \quad 0 \end{pmatrix}.$$

This means, we add a zero component to the original Lagarias-Odlyzko lattice and add the new basis vector $(-Nd, -\frac{1}{2}, \ldots, -\frac{1}{2}, 1)^\top$. The last component ensures that the columns of 7.2.4 are linearly independent. Also, it serves as a bookkeeping device. Namely, it keeps track of whether the new basis vector was used in the expression of a lattice element in terms of the new basis. As before, $N \in \mathbb{N}$ is a large integer constant. The solutions of the subset sum problem now correspond to elements $v = B \cdot w \in \mathbb{Z}^{m+n+1}$ of the new lattice 7.2.4 where $v_i = 0$ for $i \in m$, $v_i \in \{-1/2, 1/2\}$ for $m \le i < n + m$ and $|v_{n+m}| = |w_0| = 1$. For these vectors the maximum norm is equal to 1, and all lattice vectors which are not solutions of the subset sum problem have maximum norm greater than 1.

Example Consider again the subset sum problem from 7.0.2 and put $N = 100$. The extended Lagarias-Odlyzko lattice is generated by the matrix

$$
\left(
\begin{array}{c|ccccccccccc}
-31400 & 3100 & 4100 & 5900 & 2600 & 5300 & 5800 & 9700 & 9300 & 2300 & 8400 & 6200 \\
-1 & 2 & 0 & 0 & 0 & 0 & 0 & 0 & 0 & 0 & 0 & 0 \\
-1 & 0 & 2 & 0 & 0 & 0 & 0 & 0 & 0 & 0 & 0 & 0 \\
-1 & 0 & 0 & 2 & 0 & 0 & 0 & 0 & 0 & 0 & 0 & 0 \\
-1 & 0 & 0 & 0 & 2 & 0 & 0 & 0 & 0 & 0 & 0 & 0 \\
-1 & 0 & 0 & 0 & 0 & 2 & 0 & 0 & 0 & 0 & 0 & 0 \\
-1 & 0 & 0 & 0 & 0 & 0 & 2 & 0 & 0 & 0 & 0 & 0 \\
-1 & 0 & 0 & 0 & 0 & 0 & 0 & 2 & 0 & 0 & 0 & 0 \\
-1 & 0 & 0 & 0 & 0 & 0 & 0 & 0 & 2 & 0 & 0 & 0 \\
-1 & 0 & 0 & 0 & 0 & 0 & 0 & 0 & 0 & 2 & 0 & 0 \\
-1 & 0 & 0 & 0 & 0 & 0 & 0 & 0 & 0 & 0 & 2 & 0 \\
-1 & 0 & 0 & 0 & 0 & 0 & 0 & 0 & 0 & 0 & 0 & 2 \\
1 & 0 & 0 & 0 & 0 & 0 & 0 & 0 & 0 & 0 & 0 & 0 \\
\end{array}
\right).
$$

Note that we have scaled all rows except for the first and the last one by a factor of 2 to clear denominators.

Multiplying this matrix from the right by the same vector $w = (1,0,0,0,1, 1,1,0,1,0,1,0)^\top$ as in Example 7.2.3 produces the short lattice vector $v = (0,-1,-1,-1,1,1,1,-1,1,-1,1,-1,1)^\top$. Comparing this vector to the vector v of Example 7.2.3 shows that apart from the first entry we have replaced all zeros by -1s. Also, we have left in place the 1s and we have added a final entry. Furthermore, note that $\|v\|_2 = \sqrt{12}$ and $\|v\|_\infty = 1$. If the last component of v is equal to 1, then $v_i = -1$ corresponds to $i \notin S$, $v_i = 1$ corresponds to $i \in S$. If the last component of v is equal to -1 it is the other way round. ◇

As shown in [43], this improvement enlarges the class of subset sum problems whose solutions are shortest vectors in the original Lagarias-Odlyzko lattice 7.2.4 enormously.

Systems of Diophantine linear equations.   In order to solve the problem $A \cdot x = d$ for $A \in \mathbb{Q}^{m \times n}$ and $d \in \mathbb{Q}^m$ with $l \leq x \leq r$ for arbitrary bounds $l, r \in \mathbb{Q}^n$, our algorithm proceeds in two steps.

First, we compute a basis consisting of integer vectors $b^{(0)}, b^{(1)}, \ldots, b^{(n-m)}$ of the augmented system 7.2.6

$$
\underbrace{\left(\begin{array}{c|c} -d & A \end{array} \right)}_{=:\, A'} \cdot \begin{pmatrix} x_0 \\ \vdots \\ x_n \end{pmatrix} = 0 .
\qquad\qquad 7.2.6
$$

In this system, the negative of the right hand side has been added to the coefficient matrix A on the left, to form the extended coefficient matrix A'. Correspondingly, a component x_0 has been added to the vector x.

Since we can assume that the augmented matrix A' has full row-rank m, the kernel of the system 7.2.6 has dimension $n - m + 1$. Of course, only solutions of 7.2.6 with $x_0 = 1$ are interesting. Several polynomial-time algorithms are known to compute the integer basis of this kernel in $\mathbb{Z}^{n+1}$, as described in [39]. Since it is desirable for the second step of our algorithm to have a basis $b^{(0)}$, $b^{(1)}, \ldots, b^{(n-m)} \in \mathbb{Z}^{n+1}$ consisting of short vectors, algorithms based on lattice basis reduction are preferred [39], [198].

In order to handle the lower bounds, we reformulate the problem in such a way that the lower bounds on the variables are zero. Substituting $x := x - l$, $d := d - A \cdot l$ and $r := r - l$ yields the equivalent problem

$$A \cdot x = d \quad \text{and} \quad 0 \le x \le r.$$

Here, x is a vector in $\mathbb{Z}^n$ such that $0 \le x_i \le r_i$. This shows that we may assume that the lower bound l is zero.

Furthermore, we assume that $r_i > 0$ for $i \in n$. Otherwise, if there exists an $i \in n$ such that $r_i = 0$ or $r_i < 0$, it follows that $x_i = 0$ or $x_i < 0$, respectively. In the first case the variable x_i can be removed from the system of Diophantine linear equations, whereas in the second case we see immediately that the system has no solution.

For the above system 7.2.6 with lower bound 0 and arbitrary upper bounds $r \in \mathbb{Z}^n$ on the variables we introduce a modified version of the lattice 7.2.4. The basis of the new lattice consists of the columns of the following $(m + n + 1) \times (n + 1)$-matrix:

7.2.7

$$\left(\begin{array}{c|cccc} -N \cdot d & & N \cdot A & \\ \hline -r_{\max} & 2c_0 & 0 & \cdots & 0 \\ -r_{\max} & 0 & 2c_1 & \cdots & 0 \\ \vdots & \vdots & & \ddots & \vdots \\ -r_{\max} & 0 & \cdots & \cdots & 2c_{n-1} \\ r_{\max} & 0 & \cdots & \cdots & 0 \end{array} \right).$$

The entries $r_{\max}$ and c_i are defined by

$$r_{\max} = \operatorname{lcm}\{r_0, \ldots, r_{n-1}\} \quad \text{and} \quad c_i = \frac{r_{\max}}{r_i}, \quad i \in n,$$

and, as usual, $N \in \mathbb{N}$ is a large integer constant. In 7.6.17, we will compute a lower bound on the size of N.

After applying lattice basis reduction (see Section 7.6), the first $n - m + 1$ vectors of a reduced basis will have only zeros in the first m entries, provided N is large enough. These are relatively short vectors. The remaining m vectors contain at least one nonzero entry in the first m entries. Since entries in the first m rows are multiples of the large integer constant N, these vectors are long vectors. Thus, the new generator matrix of the lattice spanned by the columns of 7.2.7 has the following form:

$$
\left.\begin{array}{c}
m \text{ rows} \left\{ \\
\\
n+1 \text{ rows} \left\{
\end{array}
\left(
\begin{array}{ccc|ccc}
0 & \cdots & 0 & * & \cdots & * \\
\vdots & & \vdots & \vdots & & \vdots \\
0 & \cdots & 0 & * & \cdots & * \\
\hline
& * & & & * &
\end{array}
\right)
\right\}
\begin{array}{l}
\text{entries are} \\
\text{multiples of } N
\end{array}
$$

$$\underbrace{\qquad}_{n-m+1 \text{ columns}} \quad \underbrace{\qquad}_{m \text{ columns}}$$

7.2.8

The last m vectors cannot contribute to a solution of our original problem. Hence they can be removed from the basis. From the remaining $n - m + 1$ vectors we can delete the first m entries which are zero. This gives a basis $b^{(0)}$, $b^{(1)}, \ldots, b^{(n-m)} \in \mathbb{Z}^{n+1}$ of the kernel of 7.2.6.

Theorem *With the above definitions, let*   **7.2.9**

$$v = u_0 \cdot b^{(0)} + u_1 \cdot b^{(1)} + \ldots + u_{n-m} \cdot b^{(n-m)}$$ **7.2.10**

be an integer linear combination of the basis vectors with $v_n = r_{max}$. The vector v is a solution of the system $A \cdot x = d$, $0 \le x \le r$, if and only if

$$v \in \mathbb{Z}^{n+1} \quad where \quad -r_{max} \le v_i \le r_{max}, \ i \in n .$$ **7.2.11**

Proof: Let $v = u_0 \cdot b^{(0)} + u_1 \cdot b^{(1)} + \ldots + u_{n-m} \cdot b^{(n-m)}$ be an integer linear combination of the basis vectors with $v_n = r_{max}$. By looking at the initial basis 7.2.7 of the lattice we see that for every $i \in n$ there is an integer y_i such that $v_i = -r_{max} + 2y_i c_i$.

By using the definitions of r_{max} and c_i it is easy to verify that $-r_{max} \le v_i \le r_{max}$ is equivalent to $0 \le y_i \le r_i, i \in n$. $\square$

In a second step, the algorithm will search in the lattice of integer linear combinations of the basis vectors $b^{(0)}, b^{(1)}, \ldots, b^{(n-m)} \in \mathbb{Z}^{n+1}$. In this step, all lattice vectors which correspond to solutions of the original problem $A \cdot x = d$ are enumerated. Only solutions to 7.2.6 with $x_0 = 1$ are enumerated.

7.3   7.3 Basic Theory of Lattices

Let $L \subset \mathbb{R}^n$ be a lattice with basis $b^{(0)}, b^{(1)}, \ldots, b^{(m-1)}$. We want to estimate how short the vectors of a lattice basis for L can be. For this purpose, we introduce the determinant of a lattice. We will see that the determinant has a geometrical interpretation.

7.3.1   **Definition** If a set $S \subset \mathbb{R}^n$ is measurable in the sense of Lebesgue, then its Lebesgue measure is called the volume of S and denoted by Vol_S.   ◊

7.3.2   **Definition (fundamental parallelotope)** Let L be a lattice with basis $b^{(0)}, b^{(1)}, \ldots, b^{(m-1)}$. The set

$$F_B := \left\{ \sum_{i \in m} x_i b^{(i)} \mid 0 \le x_i < 1, i \in m \right\}$$

is the *fundamental parallelotope* of the lattice L with respect to the basis $b^{(0)}, b^{(1)}, \ldots, b^{(m-1)}$.   ◊

If $m = n$, i.e. if the lattice has full rank, the volume of the fundamental parallelotope F_B is equal to the absolute value of the determinant of the matrix $B = (b^{(0)} \mid \ldots \mid b^{(n-1)})$. If $m < n$, i.e. if the lattice is embedded in a space of higher dimension, the volume of the fundamental parallelotope in $\mathbb{R}^n$ is 0. Nevertheless, we will need the volume of the lattice L as a subset of the m-dimensional space $\langle b^{(0)}, b^{(1)}, \ldots, b^{(m-1)} \rangle \subset \mathbb{R}^n$. For this we introduce the *Gram matrix* $G(B)$ of the basis B.

7.3.3   **Definition (Gram matrix, determinant)** Let $B = (b^{(0)} \mid \ldots \mid b^{(m-1)})$ be a generator matrix of a lattice L with basis $b^{(0)}, \ldots, b^{(m-1)}$.

— The matrix

$$G(B) = \left(\langle b^{(i)}, b^{(j)} \rangle \right)_{i,j \in m} \in \mathbb{R}^{m \times m}$$

is called *Gram-matrix* $G(B)$ of the lattice basis.

— The *determinant* of the lattice L with respect to the generator matrix B is

$$\det(L) = \sqrt{\det(G(B))}.$$ ◊

It is easy to see that $\det(L)$ is well-defined and that it is equal to the volume of the fundamental parallelotope F_B in the space $\langle b^{(0)}, b^{(1)}, \ldots, b^{(m-1)} \rangle$.

It is well-known that the volume of the fundamental parallelotope of a lattice does not depend on the choice of the basis (cf. Fig. 7.2). Let $m \in \mathbb{Z}$, $m > 0$. A matrix $M \in \mathbb{Z}^{m \times m}$ with determinant ± 1 is called *unimodular*.

Lemma *The volume of the fundamental parallelotope of a lattice $L \subset \mathbb{R}^n$ of rank m is equal for all bases $b^{(0)}, b^{(1)}, \ldots, b^{(m-1)}$ of L.*   7.3.4

Proof: Let $A = \left(a^{(0)} \mid a^{(1)} \mid \ldots \mid a^{(m-1)}\right)$ and $B = \left(b^{(0)} \mid b^{(1)} \mid \ldots \mid b^{(m-1)}\right)$ be two generator matrices of the lattice L with fundamental parallelotopes F_A and F_B, respectively. Thus, we can express each basis vector in $\{b^{(0)}, b^{(1)}, \ldots, b^{(m-1)}\}$ as an integer linear combination of basis vectors in $\{a^{(0)}, a^{(1)}, \ldots, a^{(m-1)}\}$, and vice versa. That is, there exists a matrix $M \in \mathbb{R}^{m \times m}$ which describes the change from generator matrix A to generator matrix B with $B = A \cdot M$. The change from generator matrix B to A can then be expressed by $A = B \cdot M^{-1}$.

Since every lattice vector is an integer linear combination of basis vectors, the entries of the matrix M as well as the entries of the matrix M^{-1} are integers. Thus, also $\det(M)$ and $\det(M)^{-1} = \det(M^{-1})$ are integers. Since $\det(M) \neq 0$, the only possibility is that $\det(M) = \pm 1$. For the volume of the fundamental parallelotopes this gives

$$\mathrm{Vol}_{F_B} = \sqrt{\det(G(B))} = \sqrt{\det(M)^2 \cdot \det(G(A))} = \mathrm{Vol}_{F_A}. \qquad \square$$

Let $b^{(0)}, b^{(1)}, \ldots, b^{(m-1)}$ be a basis of a lattice $L \subset \mathbb{R}^n$ of rank m. From the above proof we see that the columns of the matrix $M \cdot \left(b^{(0)} \mid b^{(1)} \mid \ldots \mid b^{(m-1)}\right)$ form another basis of L provided that $M \in \mathbb{Z}^{m \times m}$ is a unimodular matrix. This means that there is a one-to-one correspondence between the unimodular matrices and the different bases of L.

A different kind of invariant of a lattice are the *successive minima* of Minkowski [150]. Again, this invariant does not depend on the choice of the basis.

Definition (successive minima) Let $L \subset \mathbb{R}^n$ be a lattice of rank m. For an   7.3.5
integer $i \in m$ let $\lambda_i(L)$ be the least positive real number for which there exist $i + 1$ linearly independent lattice vectors $v \in L \setminus \{0\}$ with $\|v\|_2 \leq \lambda_i(L)$. The numbers $\lambda_0(L), \lambda_1(L), \ldots, \lambda_{m-1}(L)$ are the *successive minima* of the lattice L. $\diamond$

From the definition it follows that

$$\lambda_0(L) \leq \lambda_1(L) \leq \ldots \leq \lambda_{m-1}(L).$$

Linearly independent vectors $v^{(i)} \in L$ with $\|v^{(i)}\| = \lambda_i(L)$ for $i \in m$ do not necessarily form a basis of the lattice. For example, the lattice

$$L = \left\{ u_0 e^{(0)} + u_1 e^{(1)} + \ldots + u_{n-1} e^{(n-1)} + u_n (\tfrac{1}{2}, \ldots, \tfrac{1}{2})^\top \mid u_0, u_1, \ldots, u_n \in \mathbb{Z} \right\}$$

in $\mathbb{Q}^n$ contains the vectors $e^{(0)}, e^{(1)}, \ldots, e^{(n-1)}$. Therefore, the successive minima of L are

$$\lambda_0(L) = \lambda_1(L) = \ldots = \lambda_{n-1}(L) = 1.$$

These successive minima are unique since the vectors $e^{(i)}$, $i \in n$, are the only vectors in L with Euclidean norm equal to one. But the vectors $e^{(0)}, e^{(1)}, \ldots, e^{(n-1)}$ do not form a basis of L.

The connection to quadratic forms. The arithmetic theory of lattices is closely related to the theory of positive definite quadratic forms whose long history dates back to Lagrange [120], Legendre [122], Gauss [66], Hermite [86] and Korkine and Zolotarev [112], [113].

Gauss [65] was first to notice the close connection between positive definite quadratic forms and lattices, i.e. the viewpoint of geometry. This geometric point of view was later developed systematically by Minkowski [150] and is now known as the "geometry of numbers".

7.3.6   **Definition (positive definite quadratic form)** A *positive definite quadratic form* is a map

$$f_A : \mathbb{Z}^m \to \mathbb{R} : x \mapsto x^\top \cdot A \cdot x,$$

where $A \in \mathbb{R}^{m \times m}$ is a symmetric positive definite matrix, i.e. $A^\top = A$ and $x^\top \cdot A \cdot x > 0$ for $x \in \mathbb{R}^n \setminus \{0\}$.   ◇

Let $B \in \mathbb{R}^{n \times m}$ be a matrix of rank m with $m \leq n$. Setting $A := B^\top \cdot B$, we note that $f_A(x) = x^\top \cdot (B^\top \cdot B) \cdot x = \|B \cdot x\|_2^2 \geq 0$ for $x \in \mathbb{Z}^m$. Since A has maximal rank m, $f_A(x) = 0$ is equivalent to $x = 0$. It follows that the matrix A is symmetric and positive definite. Therefore, the minimum value of $f_A(x)$ for all $x \in \mathbb{Z}^m \setminus \{0\}$ is equal to the square of the ℓ_2-shortest vector in the lattice with generator matrix B.

It is well-known that for any symmetric positive definite matrix $A \in \mathbb{R}^{m \times m}$ there exists a matrix $B \in \mathbb{R}^{m \times m}$ such that $A = B^\top \cdot B$. This is known as the Cholesky decomposition of A (see [39], for instance). This shows that for every positive definite quadratic form f_A there exists a lattice L, namely the lattice whose generator matrix is the matrix B with $A = B^\top \cdot B$.

Indeed many results in lattice theory were first formulated in the language of positive definite quadratic forms. An example is 7.5.4.

Exercises

E.7.3.1   **Exercise** Prove that the volume of a fundamental parallelotope of a lattice L is equal to $\det(L)$.

Exercise Let A be a symmetric, positive definite matrix $\in \mathbb{R}^{m \times m}$. Show that there exists a matrix $B \in \mathbb{R}^{m \times m}$ with $A = B^\top \cdot B$.

7.4 Gram–Schmidt Orthogonalization

Definition (orthogonal vectors) A set of vectors $v^{(0)}, \ldots, v^{(m-1)} \in \mathbb{R}^n \setminus \{0\}$ is called *orthogonal* if for $i, j \in m$

$$\langle v^{(i)}, v^{(j)} \rangle \begin{cases} \neq 0, & \text{if } i = j, \\ = 0, & \text{if } i \neq j. \end{cases}$$

◇

Lemma (Gram–Schmidt orthogonalization) *Let $b^{(0)}, b^{(1)}, \ldots, b^{(m-1)}$ be a set of linearly independent vectors $\in \mathbb{R}^n$. For $i = 0, 1, \ldots, m-1$, define vectors*

$$\hat{b}^{(i)} = b^{(i)} - \sum_{j=0}^{i-1} \mu_{ij} \cdot \hat{b}^{(j)},$$

where

$$\mu_{ij} = \frac{\langle b^{(i)}, \hat{b}^{(j)} \rangle}{\langle \hat{b}^{(j)}, \hat{b}^{(j)} \rangle}.$$

Then $\hat{b}^{(0)}, \hat{b}^{(1)}, \ldots, \hat{b}^{(m-1)}$ are orthogonal.

Proof: Let $b^{(0)}, b^{(1)}, \ldots, b^{(m-1)}$ be a set of linearly independent vectors $\in \mathbb{R}^n$. Then, $\hat{b}^{(0)} = b^{(0)}$ and $\hat{b}^{(1)} = b^{(1)} - \frac{\langle b^{(1)}, \hat{b}^{(0)} \rangle}{\langle \hat{b}^{(0)}, \hat{b}^{(0)} \rangle} \cdot \hat{b}^{(0)}$. Therefore,

$$\langle \hat{b}^{(1)}, \hat{b}^{(0)} \rangle = \langle b^{(1)}, \hat{b}^{(0)} \rangle - \frac{\langle b^{(1)}, \hat{b}^{(0)} \rangle}{\langle \hat{b}^{(0)}, \hat{b}^{(0)} \rangle} \cdot \langle \hat{b}^{(0)}, \hat{b}^{(0)} \rangle = 0.$$

By induction, it follows for $2 \leq k \leq m-1$ that

$$\begin{aligned} \langle \hat{b}^{(k)}, \hat{b}^{(j)} \rangle &= \langle b^{(k)}, \hat{b}^{(j)} \rangle - \sum_{i=0}^{k-1} \frac{\langle b^{(k)}, \hat{b}^{(i)} \rangle}{\langle \hat{b}^{(i)}, \hat{b}^{(i)} \rangle} \cdot \langle \hat{b}^{(i)}, \hat{b}^{(j)} \rangle \\ &= \langle b^{(k)}, \hat{b}^{(j)} \rangle - \frac{\langle b^{(k)}, \hat{b}^{(j)} \rangle}{\langle \hat{b}^{(j)}, \hat{b}^{(j)} \rangle} \cdot \langle \hat{b}^{(j)}, \hat{b}^{(j)} \rangle \\ &= 0, \end{aligned}$$

for $j = 0, 1, \ldots, k-1$. □

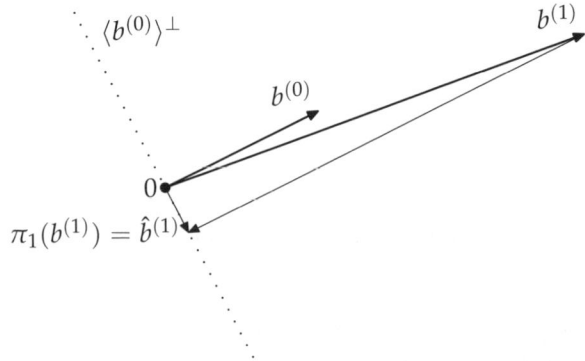

Fig. 7.4 Orthogonal projection π_1 of $b^{(1)}$ into $\langle b^{(0)} \rangle^\perp$

The procedure 7.4.2 is called *Gram–Schmidt orthogonalization*. The vectors $\hat{b}^{(i)}$, $i \in m$, are referred to as *Gram–Schmidt vectors* and the numbers μ_{ij}, $0 \le j \le i < m$, are called *Gram–Schmidt coefficients*. We note that in general the set of orthogonal vectors $\hat{b}^{(0)}, \hat{b}^{(1)}, \ldots, \hat{b}^{(m-1)}$ is not longer contained in L, since the Gram–Schmidt coefficients μ_{ij} are not necessarily integers.

For $i \in m$ we can think of $\hat{b}^{(i)}$ as the orthogonal projection of $b^{(i)}$ into the subspace $H_{i-1} := \langle b^{(0)}, b^{(1)}, \ldots, b^{(i-1)} \rangle^\perp$, which is the subspace of dimension $m - i$ orthogonal to $\langle b^{(0)}, b^{(1)}, \ldots, b^{(i-1)} \rangle$ in $\langle b^{(0)}, b^{(1)}, \ldots, b^{(m-1)} \rangle$.

7.4.3   **Definition (orthogonal projection)** With the above notation, for $t \in m$ the *orthogonal projection* $\pi_t(v)$ is defined by

$$\pi_t : \mathbb{R}^n \to \langle b^{(0)}, b^{(1)}, \ldots, b^{(t-1)} \rangle^\perp, \quad v \mapsto \sum_{j=t}^{m-1} \frac{\langle v, \hat{b}^{(j)} \rangle}{\langle \hat{b}^{(j)}, \hat{b}^{(j)} \rangle} \cdot \hat{b}^{(j)} . \qquad \diamond$$

We note that the orthogonal projection depends on the choice of the basis $b^{(0)}, b^{(1)}, \ldots, b^{(m-1)}$ of the lattice $L \subset \mathbb{R}^n$. Further, from the definition it can be seen that for $t \in m$ the orthogonal projection $\pi_t(v)$ of a vector $v \in \mathbb{R}^n$ is a linear combination of the Gram–Schmidt vectors $\hat{b}^{(t)}, \ldots, \hat{b}^{(m-1)}$. For any lattice basis and any vector $v \in \mathbb{R}^n$ we have $\pi_0(v) = v$.

The orthogonal projection π_t is a linear mapping. Therefore, the projection of the lattice $L(b^{(0)}, b^{(1)}, \ldots, b^{(m-1)})$ into $\langle b^{(0)}, b^{(1)}, \ldots, b^{(t-1)} \rangle^\perp$ is again a lattice

$$L_t\left(\pi_t(b^{(t)}), \ldots, \pi_t(b^{(m-1)})\right) := \left\{ \sum_{i \in m} u_i \pi_t(b^{(i)}) \mid u_i \in \mathbb{Z} \right\}$$

$$= \left\{ \sum_{i=t}^{m-1} u_i \pi_t(b^{(i)}) \mid u_i \in \mathbb{Z} \right\}$$

spanned by the basis $\pi_t(b^{(t)}), \pi_t(b^{(t+1)}), \ldots, \pi_t(b^{(m-1)})$ for $t \in m$. The rank of the lattice L_t is equal to $m - t$.

In matrix notation, the Gram–Schmidt orthogonalization can be written as

$$B = \hat{B} \cdot \mu^\top$$

with a lower triangular $m \times m$-matrix

$$\mu = \begin{pmatrix} \mu_{00} & & & \\ \mu_{10} & \mu_{11} & & \\ \vdots & \vdots & \ddots & \\ \mu_{m-1,0} & \mu_{m-1,1} & \cdots & \mu_{m-1,m-1} \end{pmatrix},$$

where $\mu_{ii} = 1$, $i \in m$, and $\mu_{ij} = 0$ for $0 \le i < j < m$. This shows that the Gram–Schmidt orthogonalization is a unimodular transformation. In particular, $\det(\mu) = 1$ and we see that we can compute the determinant $\det(L)$ from the Gram–Schmidt vectors $\hat{b}^{(0)}, \hat{b}^{(1)}, \ldots, \hat{b}^{(m-1)}$ of a lattice basis $b^{(0)}, b^{(1)}, \ldots, b^{(m-1)}$ via

$$\det(L) = |\det(\hat{B}) \cdot \det(\mu)| = \prod_{i \in m} \|\hat{b}^{(i)}\|_2 . \qquad \textbf{7.4.4}$$

We note that the orthogonal basis $\hat{b}^{(0)}, \hat{b}^{(1)}, \ldots, \hat{b}^{(m-1)}$ depends on the ordering of the basis $b^{(0)}, b^{(1)}, \ldots, b^{(m-1)}$.

Exercises

Exercise Prove that the orthogonal projection is a linear mapping.   **E.7.4.1**

Exercise Show that for vectors $b^{(0)}, b^{(1)}, \ldots, b^{(n-1)} \in \mathbb{R}^n$:   **E.7.4.2**

$$\sqrt{\det G(b^{(0)}, b^{(1)}, \ldots, b^{(n-1)})} = |\det(b^{(0)}, b^{(1)}, \ldots, b^{(n-1)})| .$$

Exercise Use elementary row and column transformations to bring the Gram   **E.7.4.3** matrix $G(b^{(0)}, b^{(1)}, \ldots, b^{(m-1)})$, of linearly independent $b^{(0)}, b^{(2)}, \ldots, b^{(m-1)} \in \mathbb{R}^n$, with $m \le n$, to the form $\left(\langle \hat{b}^{(j)}, \hat{b}^{(l)} \rangle \right)_{j,l \in m}$.

7.5 Bounds on Lattice Vectors   **7.5**

The Hadamard inequality is a well-known lower bound on the length of the vectors of a lattice basis.

7.5.1   **Lemma (Hadamard's Inequality)** *If* $b^{(0)}, b^{(1)}, \ldots, b^{(m-1)}$ *is a basis of a lattice* $L \subset \mathbb{R}^n$, *then*

$$\det(L) \leq \prod_{i \in m} \|b^{(i)}\|_2.$$

Proof: Using 7.4.2 together with the mutual orthogonality of the vectors $\hat{b}^{(j)}$ we have

$$\|b^{(i)}\|_2^2 = \|\hat{b}^{(i)}\|_2^2 + \sum_{j=0}^{i-1} \mu_{ij}^2 \|\hat{b}^{(j)}\|_2^2 \geq \|\hat{b}^{(i)}\|_2^2.$$

With 7.4.4, $\det(L) = \prod_{i \in m} \|\hat{b}^{(i)}\|_2$, the inequality follows.   □

7.5.2   **Remark** The inequality of Hadamard can be written as

$$1 \leq \prod_{i \in m} \frac{\|b^{(i)}\|_2}{\|\hat{b}^{(i)}\|_2} = \frac{1}{\det(L)} \cdot \prod_{i \in m} \|b^{(i)}\|_2$$

with equality if and only if the basis $b^{(0)}, b^{(1)}, \ldots, b^{(m-1)}$ is orthogonal. Therefore, the product $\prod_{i \in m} \|b^{(i)}\|_2 / \|\hat{b}^{(i)}\|_2$ is a measure of the "non-orthogonality" of a basis $b^{(0)}, b^{(1)}, \ldots, b^{(m-1)}$.

The inequality of Hadamard is trivially satisfied if the vectors $b^{(0)}, b^{(1)}, \ldots,$ $b^{(m-1)}$ are linearly dependent.   ◇

7.5.3   **Definition (orthogonality defect)** For a lattice basis $b^{(0)}, b^{(1)}, \ldots, b^{(m-1)}$

$$\prod_{i \in m} \frac{\|b^{(i)}\|_2}{\|\hat{b}^{(i)}\|_2}$$

is called *orthogonality defect* of $b^{(0)}, b^{(1)}, \ldots, b^{(m-1)}$.   ◇

Since $\det(L) = \prod_{i \in m} \|\hat{b}^{(i)}\|_2$ does not depend on the choice of the lattice basis, the orthogonality defect is a measure for the geometric mean of the Euclidean length of the basis vectors. Consequently, a basis consisting of short vectors has a small orthogonality defect.

A classical result due to Hermite [86] gives an upper bound for the ℓ_2-shortest vector of a lattice.

7.5.4   **Theorem (Hermite)** *Let* $L \subset \mathbb{Z}^n$ *be a lattice of rank* m. *Then* L *contains a nonzero vector* v *such that*

$$\|v\|^2 \leq (4/3)^{(m-1)/2} \cdot \det(L)^{2/m},$$

where $\| - \|$ *denotes the Euclidean norm.*

Proof: Using the first successive minimum, the claim of the theorem becomes

$$\lambda_0(L)^2 \le (4/3)^{(m-1)/2} \cdot \det(L)^{2/m}.$$

Let B be a generating matrix of the lattice L and $y \in \mathbb{Z}^m \setminus \{0\}$ be a vector for which $B \cdot y$ takes on its minimum value, i.e. $\|B \cdot y\| = \lambda_0(L)$. Then we know that $r = \gcd(y_0, y_1, \ldots, y_{m-1}) = 1$. For otherwise, since $\frac{1}{r} \cdot y$ is integral,

$$\|B \cdot (\frac{1}{r}y)\| = \frac{1}{r}\|B \cdot y\| = \frac{1}{r} \cdot \lambda_0(L),$$

which would contradict the minimality of $\lambda_0(L)$ in the case $r > 1$.

By induction it is easy to show that there exists a matrix $W \in \mathbb{Z}^{m \times m}$ such that $\det(W) = 1$ and the first column of W is equal to y (see Exercise 7.5.1). Such a matrix is unimodular and hence by the remark after 7.3.4, $B \cdot W$ is another generator matrix of L. Moreover,

$$\|B \cdot W \cdot (1, 0, \ldots, 0)^\top\| = \|B \cdot y\| = \lambda_0(L).$$

Therefore, we can assume that B is a generator matrix of the lattice L and that the successive minimum $\lambda_0(L)$ is attained for the first column of B, i.e. the first basis vector. Now, we search for a matrix $S \in \mathbb{R}^{m \times m}$ with $S^\top \cdot S = I_m$ and $a_0, a_1, \ldots, a_{m-1} \in \mathbb{R}$ such that

$$S \cdot B = \begin{pmatrix} a_0 & a_1 & \cdots & a_{m-1} \\ 0 & & & \\ \vdots & & B' & \\ 0 & & & \end{pmatrix}.$$

The matrix S can be constructed by taking as first row the entries of the first column of B divided by $\|b^{(0)}\|$. The remaining rows are filled with linearly independent vectors in $\mathbb{R}^m$. After that, the rows of B must be orthogonalized by the Gram–Schmidt process and scaled to have norm one. Then, for arbitrary $x \in \mathbb{R}^m$ we have that

$$\|B \cdot x\|^2 = \langle Bx, Bx \rangle$$
$$= x^\top B^\top S^\top S B x$$
$$= \langle SBx, SBx \rangle$$
$$= \left\| \begin{pmatrix} a_0 & a_1 & \cdots & a_{m-1} \\ 0 & & & \\ \vdots & & B' & \\ 0 & & & \end{pmatrix} \cdot x \right\|^2$$
$$= (\sum_{i \in m} a_i x_i)^2 + \|B' \cdot (x_1, \ldots, x_{m-1})^\top\|^2.$$

Since $a_0 = \frac{1}{\|b^{(0)}\|} b^{(0)^\top} \cdot b^{(0)} = \|b^{(0)}\|$ and $\|b^{(0)}\|$ is minimal, it follows that $a_0 = \lambda_0(L)$. For the lattice L' which is generated by B', the determinant is

$$\det(L)^2 = \det(G(B)) = \det(G(SB)) = a_0^2 \cdot \det(G(B')) = a_0^2 \cdot \det(L')^2.$$

Therefore,

$$\det(L') = \det(L) \cdot \frac{1}{\lambda_0(L)}.$$

We can now prove the initial claim by induction. The case $m = 1$ is clear. For $m > 1$, suppose that there exists a vector $x' = (x_1, \ldots, x_{m-1})^\top \in \mathbb{Z}^{m-1}$ with

$$\|B' \cdot x'\|^2 \leq (4/3)^{(m-2)/2} \cdot \det(L')^{2/(m-1)}$$
$$= (4/3)^{(m-2)/2} \cdot \det(L)^{2/(m-1)} \frac{1}{\lambda_0(L)^{2/(m-1)}}.$$

For the fixed values $x_1, \ldots, x_{m-1} \in \mathbb{Z}$ we can choose $x_0 \in \mathbb{Z}$ such that

$$\left| x_0 + \frac{a_1 x_1 + \ldots + a_{m-1} x_{m-1}}{a_0} \right| \leq \frac{1}{2}.$$

It follows that

$$\left(\sum_{i \in m} a_i x_i \right)^2 \leq \frac{1}{4} \lambda_0^2(L)$$

and therefore, since $\lambda_0^2(L) \leq \|B \cdot x\|^2$, we have

$$\lambda_0^2(L) \leq \frac{1}{4} \lambda_0^2(L) + (4/3)^{(m-2)/2} \cdot \det(L)^{2/(m-1)} \frac{1}{\lambda_0(L)^{2/(m-1)}}.$$

An easy calculation shows that

$$\lambda_0^2(L) \leq (4/3)^{(m-1)/2} \cdot \det(L)^{2/m},$$

which is the claim made at the beginning of the proof. Since this result is equivalent to the assertion, the proof is finished. $\qquad\square$

Minkowski started a systematic theory which is now known as the geometry of numbers. In [149], he proved the following fundamental theorem.

7.5.5   **Theorem (Minkowski)** *Let S be a convex set in $\mathbb{R}^n$ which is symmetric about the origin (i.e. $x \in S \Rightarrow -x \in S$). If the volume of S is greater than 2^n, then S contains a nonzero vector $v \in \mathbb{Z}^n$.*

For the proof of the theorem we use the following lemma by Blichfeldt [21].

7.5.6   **Lemma (Blichfeldt)** *Let S be a measurable set in $\mathbb{R}^n$. If $\mathrm{Vol}_S > 1$ or S is bounded and closed and $\mathrm{Vol}_S = 1$ then S contains two different vectors x and y such that $x - y$ is in $\mathbb{Z}^n \setminus \{0\}$.*

Proof: First we assume that $\mathrm{Vol}_S > 1$. Without loss of generality we suppose that S is bounded. The volume of the cube $Q := \{x \in \mathbb{R}^n \mid 0 \le x_i < 1,\ i \in n\}$ is equal to 1. Since S is bounded there are finitely many integral vectors $u^{(0)}, u^{(1)}, \dots, u^{(k-1)} \in \mathbb{Z}^n$ such that the intersection of S and $u^{(i)} + Q, i \in k$, is nonempty. Here, for $v \in \mathbb{R}^n$ the set $v + Q$ is defined as $v + Q = \{x \in \mathbb{R}^n \mid \exists q \in Q : x = v + q\}$.

For $i \in k$ set $S_i := S \cap (u^{(i)} + Q)$ and $S_i' := S_i - u^{(i)}$, compare Fig. 7.5. Then, for $i \in k$ the sets S_i' are contained in Q. On the other hand, we have

$$\sum_{i \in k} \mathrm{Vol}_{S_i'} = \sum_{i \in k} \mathrm{Vol}_{S_i} = \mathrm{Vol}_S > 1.$$

Therefore, these sets cannot be mutually disjoint and there are two sets S_j', S_l', $j, l \in k$, and a vector $z \in \mathbb{R}^n$ such that

$$z \in S_j' \cap S_l'.$$

It follows that both $x := u^{(j)} + z$ and $y := u^{(l)} + z$ are contained in S and $x - y = u^{(j)} - u^{(l)}$ is an integral vector.

Next, we assume that S is bounded and closed and $\mathrm{Vol}_S = 1$. Let $\theta_r > 1$ be a sequence of numbers with $\lim_{r \to \infty} \theta_r = 1$. For each r the set $\theta_r S$ has volume strictly greater than 1. From the previous result it follows that for each r there exist vectors $x^{(r)}, y^{(r)} \in \theta_r S$ such that $x^{(r)} - y^{(r)}$ is a nonzero integral vector. Since S is bounded and closed there exist subsequences $(x^{(r_t)})_{t \in \mathbb{N}}, (y^{(r_t)})_{t \in \mathbb{N}}$ converging to some vectors x and y in S, respectively. The difference $x - y$ must be a nonzero integral vector, which proofs the assertion.   □

Proof of Minkowski's theorem: Define $S/2 := \{x \in \mathbb{R}^n \mid 2x \in S\}$. Since the volume of S is greater than 2^n, $S/2$ has volume greater than 1. Using the lemma of Blichfeldt, we know that $S/2$ contains two different vectors x and y such that $x - y \in \mathbb{Z}^n \setminus \{0\}$. Therefore, $2x$ and $2y$ belong to S. Since S is symmetric about the origin, also $-2y$ belongs to S. The fact that S is convex implies that $\frac{1}{2}(2x) + \frac{1}{2}(-2y) = x - y \in S$. Since $x - y \in \mathbb{Z}^n$ this proves the theorem.   □

This result is sharp, as the n-dimensional cube $\{x \in \mathbb{R}^n \mid \|x\|_\infty < 1\}$ has volume 2^n and does not contain an integral vector $\ne 0$.

Remark As we can see from the proof, the theorem of Minkowski is also valid if the set S is bounded and closed and has $\mathrm{Vol}_S = 2^n$.   ◇   **7.5.7**

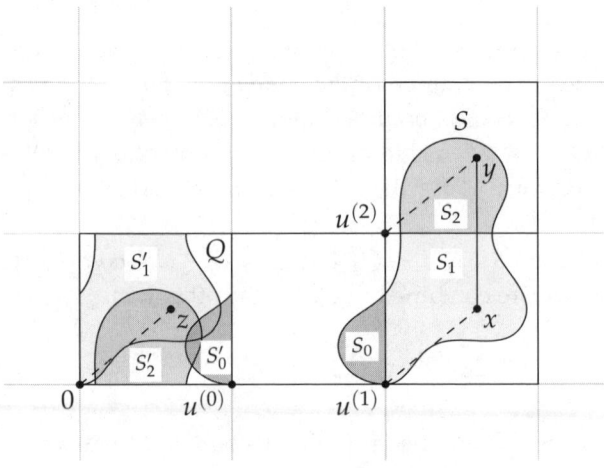

Fig. 7.5 $S_i = S \cap (u^{(i)} + Q)$ and $S_i' = S_i - u^{(i)}$, $i = 0, 1, 2$, in the Lemma of Blichfeldt

7.5.8 **Definition (Volume of the unit sphere)** We denote by ρ_n the volume of the unit sphere $S := \{x \in \mathbb{R}^n \mid \|x\|_2 \leq 1\}$ in $\mathbb{R}^n$:

$$\rho_n = \mathrm{Vol}_S = \frac{\pi^{n/2}}{\frac{n}{2}!},$$

where $\frac{n}{2}!$ is defined by $0! = 1$, $\frac{1}{2}! = \sqrt{\pi}/2$, and $\frac{n}{2}! = \frac{n}{2} \cdot (\frac{n}{2} - 1)!$ for $n \in \mathbb{Z}$, $n > 1$. ◇

As a direct consequence of 7.5.5, we have the following bound for the Euclidean length of an ℓ_2-shortest vector in a lattice L.

7.5.9 **Theorem (Minkowski)** *If $L \subset \mathbb{R}^n$ is a lattice of rank n, then there is a nonzero vector $v \in L$ with*

$$\|v\|_2 \leq 2 \left(\frac{\det(L)}{\rho_n} \right)^{1/n} = \frac{2}{\pi} \left(\frac{n}{2}! \cdot \det(L) \right)^{1/n}.$$

Proof: Let $s \in \mathbb{R}$ be a positive number and $b^{(0)}, b^{(1)}, \ldots, b^{(n-1)}$ be a basis of the lattice L. Consider the ellipsoid $K = \{x \in \mathbb{R}^n \mid \|B \cdot x\|_2^2 \leq s\}$ which is centered at 0 and whose volume is

$$\mathrm{Vol}(K) = \frac{\rho_n \cdot s^{n/2}}{\det(L)}.$$

Choose s such that

$$\frac{\rho_n \cdot s^{n/2}}{\det(L)} = 2^n.$$

By 7.5.5, there exists a nonzero lattice vector $v \in L$ with

$$\|v\|_2^2 \le 4 \left(\frac{\det(L)}{\rho_n} \right)^{2/n} ,$$

proving the theorem.   □

7.5.9 gives an upper bound for the ratio

$$\frac{\lambda_0(L)}{\det(L)^{1/n}} \le \frac{2}{\rho_n^{1/n}} .$$

Occasionally, the weaker estimate

$$\frac{\lambda_0(L)}{\det(L)^{1/n}} \le \sqrt{n}$$

of [107] suffices.

Definition (Hermite's constant) The supremum of the ratio   **7.5.10**

$$\frac{\lambda_0^2(L)}{\det(L)^{2/n}}$$

over all lattices in $\mathbb{R}^n$ of rank n is called *Hermite's constant* and is denoted as γ_n.

◇

Blichfeldt [22] provided the upper bound

$$\gamma_n \le \frac{n(1 + o(1))}{e\pi} .$$ **7.5.11**

Hermite's constant is known exactly for $n \le 8$. Meanwhile, the best known bounds for Hermite's constant are

$$\frac{n + \log(\pi \log n)}{2e\pi} + o(1) \le \gamma_n \le \frac{1.744n}{2e\pi} (1 + o(1)) .$$

The lower bound is from [148] and the upper bound is contained in [40].

Using the successive minima of a lattice, Minkowski [150] was able to sharpen the bounds of Theorems 7.5.5 and 7.5.9:

Theorem (Minkowski's Second Theorem)   **7.5.12**
If $L \subset \mathbb{R}^n$ is a lattice of rank n with successive minima $\lambda_0(L), \lambda_1(L), \ldots, \lambda_{n-1}(L)$, then

$$\lambda_0(L)\lambda_1(L) \cdots \lambda_{n-1}(L) \le 2^n \det(L) .$$

Proof: The proof can be found in [77, p. 59].   □

Exercises

E.7.5.1   **Exercise** Let $y_0, y_1, \ldots, y_{m-1}$ be integers with

$$\gcd(y_0, y_1, \ldots, y_{m-1}) = 1.$$

Show that there exists a matrix $W \in \mathbb{Z}^{m \times m}$ with $\det(W) = 1$ whose first column is equal to $(y_0, y_1, \ldots, y_{m-1})^\top$.

7.6 Lattice Basis Reduction

7.6

In this section we outline the classical concepts of lattice basis reduction as developed by Korkine and Zolotarev and later by Minkowski. Furthermore, we describe the celebrated LLL-algorithm which computes another type of reduced basis, the LLL-reduced or δ-reduced basis. We conclude this section by discussing some improvements and variations of the LLL-algorithm. Unless stated otherwise, by $\|-\|$ we always denote the Euclidean norm in this section.

Classical concepts of lattice basis reduction. Reduction methods for positive definite quadratic forms were first studied by Lagrange [120] for $n = 2$ and Gauss [66], [65] and Seeber [176] for $n = 3$. Hermite [87] was the first to propose a reduction method for positive quadratic forms for general values of n.

In his seminal work [150], Minkowski introduced the notion of a reduced basis of a lattice in dimension n for arbitrary positive integers n.

7.6.1   **Definition (Minkowski-reduced basis)** A basis $b^{(0)}, b^{(1)}, \ldots, b^{(n-1)}$ of the lattice $L \subset \mathbb{R}^n$ is *reduced in the sense of Minkowski*, if for $t = 0, 1, \ldots, n-1$

1. the vector $b^{(t)}$ is a shortest vector in L and

2. the set $\{b^{(0)}, b^{(1)}, \ldots, b^{(t)}\}$ can be extended to a basis of L.   ◊

In [151], Minkowski showed that

$$\frac{1}{\det(L)} \cdot \prod_{i \in n} \|b^{(i)}\| \le \frac{2^n}{\rho_n} \left(\frac{3}{2}\right)^{n(n-1)/2} = 2^{O(n^2)}$$

is an upper bound for the orthogonality defect of a Minkoswki-reduced basis. This bound is much larger than the bound which can be derived from Minkowski's Second Theorem 7.5.12. If there exists a basis $b^{(0)}, b^{(1)}, \ldots, b^{(n-1)}$

of the lattice L such that the lengths of the basis vectors equal the successive minima, then we can bound the orthogonality defect by

$$\frac{1}{\det(L)} \cdot \prod_{i \in n} \|b^{(i)}\| \leq 2^n.$$

Since the vector $b^{(0)}$ of a Minkowski-reduced basis $b^{(0)}, b^{(1)}, \ldots, b^{(n-1)}$ of a lattice L is an ℓ_2-shortest vector in L, the computation of a Minkowski-reduced basis of a lattice L is at least as hard as computing an ℓ_2-shortest vector in L.

From a computational point of view, the reduced bases of Korkine and Zolotarev [113] have turned out to be more useful.

Definition (Korkine–Zolotarev-reduced basis) A basis $b^{(0)}, b^{(1)}, \ldots, b^{(m-1)}$ of a lattice $L \subset \mathbb{R}^n$ is *reduced in the sense of Korkine and Zolotarev* [113], if

1. $b^{(0)}$ is an ℓ_2-shortest vector in L and

2. for all $t \in m$, $\hat{b}^{(t)}$ is an ℓ_2-shortest vector in the lattice $L_t(b^{(t)}, \ldots, b^{(m-1)})$. ◇

7.6.2

The upper bound on the orthogonality defect of a Korkine–Zolotarev-reduced basis is much better than that of a Minkowski-reduced basis. Lagarias, Lenstra and Schnorr [118] proved the following bounds.

Theorem Let $b^{(0)}, b^{(1)}, \ldots, b^{(n-1)}$ be a Korkine–Zolotarev-reduced basis of a lattice $L \subset \mathbb{Z}^n$. Then

7.6.3

$$\sqrt{\frac{4}{i+4}} \lambda_i(L) \leq \|b^{(i)}\| \leq \sqrt{\frac{i+4}{4}} \lambda_i(L) \quad \text{for } i \in n$$

and

$$\prod_{i \in n} \|b^{(i)}\| \leq \left(\gamma_n^n \cdot \prod_{i \in n} \frac{i+4}{4} \right)^{1/2} \cdot \det(L). \qquad \square$$

Let L be a lattice in $\mathbb{Z}^n$, and let $b^{(0)}, b^{(1)}, \ldots, b^{(n-1)}$ be a Korkine–Zolotarev-reduced basis for L. Using the asymptotic result 7.5.11 of Blichfeldt, the asymptotic upper bound for the orthogonality defect of a Korkine–Zolotarev-reduced basis $b^{(0)}, b^{(1)}, \ldots, b^{(n-1)}$ can be shown to be of order $O(n^n)$.

The vector $b^{(0)}$ of a Korkine–Zolotarev-reduced basis $b^{(0)}, b^{(1)}, \ldots, b^{(n-1)}$ of a lattice L is an ℓ_2-shortest vector in L. So, the computation of a Korkine–Zolotarev-reduced basis of an lattice L is at least as hard as computing an ℓ_2-shortest vector in L.

The LLL-algorithm. Summarizing, no fast algorithm to compute a Minkowski-reduced basis or a Korkine–Zolotarev-reduced basis is known. A major breakthrough was achieved by Lenstra, Lenstra, and Lovász in their seminal work

[125]. They compute a different type of reduced basis, which is now called an LLL-reduced basis. We only give a brief outline of the algorithm. For a detailed description, the reader is referred to the original paper [125] or to textbooks, like [39], for example.

Again, in this section the norm $\|-\|$ always denotes the Euclidean norm. For $r \in \mathbb{R}$, $\lfloor r \rceil$ denotes the nearest integer to r, i.e. $\lfloor r \rceil := \lfloor \frac{1}{2} + r \rfloor$.

A high-level description of the algorithm is as follows.

7.6.4   **Algorithm (LLL-algorithm [125])** The LLL (or L^3) algorithm computes an LLL-reduced basis. The input is a basis $b^{(0)}, \ldots, b^{(m-1)}$ of the lattice L of rank m.

(1) Let $\delta \in \mathbb{R}$ with $\frac{1}{4} < \delta < 1$.
(2) **Set** $k := 0$.
(3) **do**
(4)   1. **for** $j = 0, \ldots, k-1$
(5)   **replace** $b^{(k)}$ **by** $b^{(k)} - \lfloor \mu_{kj} \rceil b^{(j)}$,
(6)   where μ_{kj} is the Gram-Schmidt coefficient from 7.4.2.
(7)   2. **if** $\delta \|\pi_k(b^{(k)})\|^2 > \|\pi_k(b^{(k+1)})\|^2$ **then**
(8)   **interchange** $b^{(k+1)}$ and $b^{(k)}$
(9)   **update** $\hat{b}^{(k+1)}$, $\hat{b}^{(k)}$ and μ
(10)   **set** $k := \max(k-1, 0)$
(11)   **else**
(12)   **set** $k := k+1$
(13) **until** $k = m-1$.   □

Step 1 (line (4)) of the algorithm achieves that in each stage the basis vectors are "as orthogonal as possible". This means that the Gram–Schmidt orthogonalized vector is approximated by an integer linear combination of the basis vectors, compare Fig. 7.6. The hope is that for $0 \leq i \leq k$ the basis vectors $b^{(0)}, b^{(1)}, \ldots, b^{(i-1)}$ are close to being orthogonal. That is, they are good approximations of their Gram–Schmidt vectors $\hat{b}^{(0)}, \hat{b}^{(1)}, \ldots, \hat{b}^{(i-1)}$.

In Step 2 (line (7)) of the algorithm the Euclidean length of two vectors are compared:

7.6.5
$$\delta \|\pi_k(b^{(k)})\|^2 > \|\pi_k(b^{(k+1)})\|^2 .$$

The first vector

$$\pi_k(b^{(k)}) = \hat{b}^{(k)}$$

on the left hand side of 7.6.5 is the orthogonal projection of $b^{(k)}$ onto the subspace $\langle b^{(0)}, b^{(1)}, \ldots, b^{(k-1)} \rangle^{\perp}$. The second vector

$$\pi_k(b^{(k+1)}) = \sum_{i=k}^{m-1} \mu_{k+1,i} \hat{b}^{(i)} = \hat{b}^{(k+1)} + \mu_{k+1,k} \hat{b}^{(k)}$$

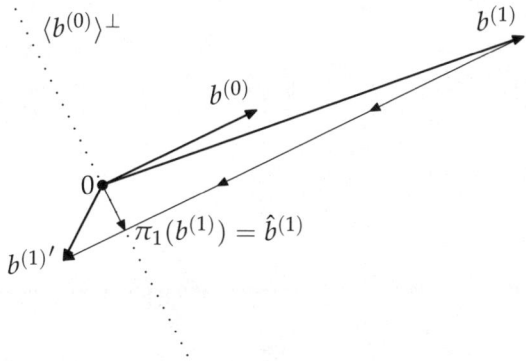

Fig. 7.6 $b^{(1)'}$ is the integer approximation of $\pi_1(b^{(1)})$

on the right hand side of 7.6.5 is the orthogonal projection of the vector $b^{(k+1)}$ into $\langle b^{(0)}, b^{(1)}, \ldots, b^{(k-1)} \rangle^{\perp}$. Depending on the length of their projected vectors onto $\langle b^{(0)}, b^{(1)}, \ldots, b^{(k-1)} \rangle^{\perp}$, we choose either $b^{(k)}$ or $b^{(k+1)}$ as the new vector $b^{(k)}$. In order to prove convergence of the algorithm, we only accept $b^{(k+1)}$ as the new basis vector $b^{(k)}$ if the length of the new orthogonal vector $\hat{b}^{(k)}$ is reduced significantly, i.e. if it is reduced by at least a factor of δ.

Example To illustrate the LLL-algorithm we consider the rank 2 lattice which is spanned by the vectors $b^{(0)} = \binom{4}{2}$ and $b^{(1)} = \binom{11}{4}$. Since $m = 2$, the variable k remains equal to zero throughout the algorithm. An LLL-reduced basis with $\delta = 1$ is computed by executing the following steps.

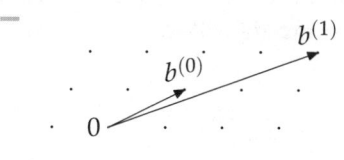

The input basis consisting of the vectors

$$b^{(0)} = \binom{4}{2} \text{ and } b^{(1)} = \binom{11}{4}.$$

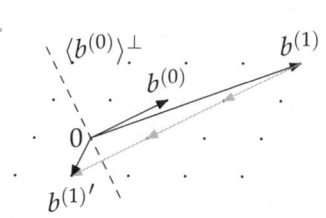

Since

$$\mu_{10} = \frac{\langle \binom{11}{4}, \binom{4}{2} \rangle}{\langle \binom{4}{2}, \binom{4}{2} \rangle} = \frac{13}{5} = 2.6,$$

we set $\lfloor \mu_{10} \rceil = 3$ in step 1 (line (5)). Then according to line (5), $b^{(1)}$ is replaced by

$$b^{(1)'} = \binom{11}{4} - 3 \cdot \binom{4}{2} = \binom{-1}{-2}.$$

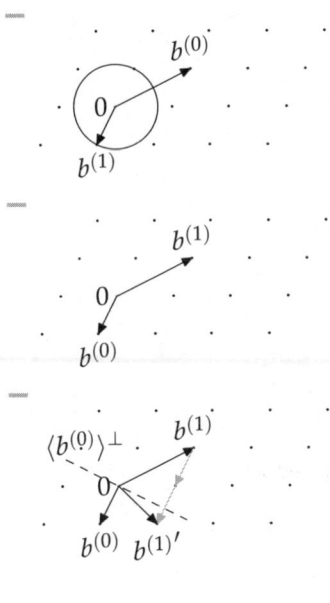

Step 2 (line (7)): $\|\pi_0(b^{(0)})\|^2 = 20$ and $\|\pi_0(b^{(1)})\|^2 = 5$ are compared.

Since

$$20 = \|\pi_0(b^{(0)})\|^2 > \|\pi_0(b^{(1)})\|^2 = 5,$$

the two vectors are swapped.

Again, in step 1 (line (5))

$$\mu_{10} = \frac{\langle \binom{4}{2}, \binom{-1}{-2} \rangle}{\langle \binom{-1}{-2}, \binom{-1}{-2} \rangle} = \frac{-8}{5} = -1.6.$$

Therefore, $\lfloor \mu_{10} \rceil = -2$ and the vector $b^{(1)}$ is replaced by

$$b^{(1)'} = \binom{4}{2} - (-2) \cdot \binom{-1}{-2} = \binom{2}{-2}.$$

Since

$$5 = \|\pi_0(b^{(0)})\|^2 < \|\pi_0(b^{(1)})\|^2 = 8,$$

the two vectors are not swapped in step 2 (line (7)) and the algorithm terminates. ◇

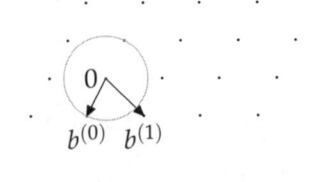

Algorithm 7.6.4 is only a very informal description of the algorithm. After swapping the two vectors $b^{(k)}$ and $b^{(k+1)}$, the Gram–Schmidt coefficients and the length of the Gram–Schmidt vectors require updating. The exact algorithm is given in [125].

The LLL-algorithm as described in Algorithm 7.6.4 works with rational numbers μ_{ij} and with rational vectors $\hat{b}^{(i)}$ for $0 \leq j \leq i < m$. Since numerator and denominator of rational numbers are integers which can be represented exactly on a computer, it was proposed in [125] that the LLL-algorithm might be formulated as an algorithm solely over the integers.

In order to represent all integers which appear in the course of the algorithm, the subdeterminants D_i of the Gram matrix can be used, i.e.

$$D_i := \det(\langle b^{(j)}, b^{(k)} \rangle)_{0 \leq j,k \leq i} = \prod_{k \in i} \|\hat{b}^{(k)}\| \quad \text{for } i \in m.$$

We know from [125] that

- $\|\hat{b}^{(i)}\|^2 = D_i/D_{i-1}$ for $i \in m$,

- $D_{i-1}\hat{b}^{(i)} \in L \subset \mathbb{Z}^n$ for $i \in m$ (and $D_{-1} := 1$),

- $D_j\mu_{ij} \in \mathbb{Z}$ for $0 \le j < i < m$.

Then, the LLL-algorithm can be modified to work with the integers $D_j\mu_{ij}$ and the integer vectors $D_{i-1}\hat{b}^{(i)}$ instead of μ_{ij} and $\hat{b}^{(i)}$, respectively. Thus the LLL-algorithm becomes an algorithm with exact arithmetic whose time complexity can then be estimated.

For the time complexity of the original LLL-algorithm, note that each interchange of the vectors $b^{(k)}$ and $b^{(k+1)}$ for $0 \le k < m - 1$ in step 2 of 7.6.4 reduces the value of $D := \prod_{i=0}^{m-2} D_i$ by at least a factor of δ. In step 1 of the algorithm, the Gram–Schmidt vectors and therefore also D remain unchanged. But, there exists a lower bound for D which is independent of the choice of the basis. This can be seen for example from 7.5.12: $D_i \ge \lambda_0(L)\lambda_1(L) \cdots \lambda_i(L)/2^i$ for $i \in m$. Therefore, Algorithm 7.6.4 terminates after a finite number of steps.

In [125] the following bounds on the time complexity of the LLL-algorithm are given.

Theorem Let $M \in \mathbb{R}$, $M \ge 2$, be such that $\|b^{(i)}\|^2 \le M$ for $i \in m$. The number of arithmetic operations needed by the LLL-algorithm is $O(m^4 \log M)$ and the integers on which these operations are performed each have binary length $O(m \log M)$. □   **7.6.7**

Definition (δ-reduced basis) The output $b^{(0)}, b^{(1)}, \dots, b^{(m-1)}$ of the LLL-algorithm with $\frac{1}{4} < \delta < 1$ is called δ-reduced basis of the lattice L. Sometimes, it is simply called an LLL-reduced basis of the lattice L.   ◇   **7.6.8**

In [125], the following bounds on the quality of the reduction are shown:

Theorem Let $b^{(0)}, b^{(1)}, \dots, b^{(m-1)}$ be a δ-reduced basis of the lattice $L \subset \mathbb{Q}^n$. Then   **7.6.9**

$$\|b^{(j)}\|^2 \le \left(\frac{4}{4\delta - 1}\right)^i \cdot \|\hat{b}^{(i)}\|^2 \text{ for } 0 \le j \le i < m.$$ **7.6.10**

$$\det(L) \le \prod_{i \in m} \|b^{(i)}\| \le \left(\frac{4}{4\delta - 1}\right)^{m(m-1)/4} \cdot \det(L).$$ **7.6.11**

$$\|b^{(0)}\| \le \left(\frac{4}{4\delta - 1}\right)^{(m-1)/4} \cdot \det(L)^{1/m}.$$ **7.6.12**

Proof: Since the basis $b^{(0)}, b^{(1)}, \ldots, b^{(m-1)}$ is δ-reduced, the condition of step 2 in the LLL-algorithm is not satisfied for all $0 \leq k < m - 1$. This together with the fact that the Gram–Schmidt vectors are mutually orthogonal implies that for $1 \leq i < m$

$$\|\hat{b}^{(i)}\|^2 + \mu_{i,i-1}^2 \|\hat{b}^{(i-1)}\|^2 \geq \delta \|\hat{b}^{(i-1)}\|^2.$$

With step 1 of the LLL-algorithm we can argue that $\mu_{i,i-1}^2 \leq \frac{1}{4}$, see [125]. This gives

$$\|\hat{b}^{(i)}\|^2 \geq \frac{4\delta - 1}{4} \cdot \|\hat{b}^{(i-1)}\|^2 \text{ for } 0 < i < m.$$

It follows for $0 \leq j \leq i < m$ that

$$\|\hat{b}^{(j)}\|^2 \leq \left(\frac{4}{4\delta - 1}\right)^{i-j} \|\hat{b}^{(i)}\|^2.$$

With

$$\|b^{(i)}\|^2 = \|\hat{b}^{(i)}\|^2 + \sum_{j \in i} \mu_{ij}^2 \|\hat{b}^{(j)}\|^2$$

and some elementary calculations (see Exercise 7.6.1) we get for $i \in m$:

$$\|b^{(i)}\|^2 \leq \|\hat{b}^{(i)}\|^2 + \sum_{j \in i} \frac{1}{4}\left(\frac{4}{4\delta - 1}\right)^{i-j} \|\hat{b}^{(i)}\|^2 \leq \left(\frac{4}{4\delta - 1}\right)^{i} \|\hat{b}^{(i)}\|^2.$$

Hence,

$$\|b^{(j)}\|^2 \leq \left(\frac{4}{4\delta - 1}\right)^{j} \cdot \|\hat{b}^{(j)}\|^2 \leq \left(\frac{4}{4\delta - 1}\right)^{i} \cdot \|\hat{b}^{(i)}\|^2.$$

Applying Hadamard's inequality 7.5.1 and 7.6.10 with $j = i$ we obtain

$$\det(L) \leq \prod_{i \in m} \|b^{(i)}\| \leq \prod_{i \in m} \left(\frac{4}{4\delta - 1}\right)^{i/2} \cdot \|\hat{b}^{(i)}\| = \left(\frac{4}{4\delta - 1}\right)^{\frac{m(m-1)}{4}} \cdot \det(L),$$

which is 7.6.11.

If we set $j := 0$ in 7.6.10 then the product of the right hand side of 7.6.10 for $i \in m$ gives $\|b^{(0)}\| \leq \left(\frac{4}{4\delta - 1}\right)^{(m-1)/4} \cdot \det(L)^{1/m}$.   □

It follows from 7.6.11 that the orthogonality defect of a δ-reduced basis can be bounded above by

$$\frac{1}{\det(L)} \cdot \prod_{i \in m} \|b^{(i)}\| \leq \left(\frac{4}{4\delta - 1}\right)^{m(m-1)/4},$$

which is $2^{O(m^2)}$ for $\delta = 3/4$. Thus, the orthogonality defect of an LLL-reduced basis has approximately the same size as the orthogonality defect of a basis which is reduced in the sense of Minkowski.

In [125], the authors provide upper bounds of the Euclidean lengths of the reduced basis vectors, compared to Euclidean lengths of a shortest lattice vector:

Theorem *Let $L \subset \mathbb{Q}^n$ be a lattice with δ-reduced basis $b^{(0)}, b^{(1)}, \ldots, b^{(m-1)}$. Then*   **7.6.13**

$$\|b^{(0)}\|^2 \leq \left(\frac{4}{4\delta - 1}\right)^{m-1} \cdot \lambda_0(L)^2 .$$ **7.6.14**

Proof: Let v be a vector in L such that $\|v\| = \lambda_0(L)$. Then we can write $v = \sum_{j \in m} r_j b^{(j)} = \sum_{j \in m} r'_j b^{(j)}$ with $r_j \in \mathbb{Z}$ and $r'_j \in \mathbb{Q}$ for $j \in m$. If t is the largest index such that $r_t \neq 0$ then we have $r_t = r'_t$ (cf. Exercise 7.6.2). Thus we deduce the inequality

$$\lambda_0^2(L) \geq r_t'^2 \|\hat{b}^{(t)}\|^2 \geq \|\hat{b}^{(t)}\|^2 .$$

With 7.6.10 we have the bound

$$\|b^{(0)}\|^2 \leq \left(\frac{4}{4\delta - 1}\right)^t \cdot \|\hat{b}^{(t)}\|^2 \leq \left(\frac{4}{4\delta - 1}\right)^{m-1} \cdot \|\hat{b}^{(t)}\|^2 .$$

Combining the two inequalities gives the required bound for $\|b^{(0)}\|^2$.   □

At first sight, the bound in 7.6.13 on the Euclidean length of the first basis vector of a LLL-reduced lattice basis does not look promising. However, there are situations where this theoretical bound is already good enough, i.e. where any nonzero vector in L which is not an ℓ_2-shortest vector has Euclidean length greater than $\left(\frac{4}{4\delta-1}\right)^{(m-1)/2} \cdot \lambda_0(L)$. Problems of this type can be solved by the LLL-algorithm in polynomial time. Examples are attacks on knapsack based cryptosystems with low-density [119], [43].

Secondly, in nearly all practical situations the LLL-algorithm behaves much better than the bound 7.6.14 indicates. It was already noted in [125] that the bound $\left(\frac{4}{4\delta-1}\right)^{m-1}$ in 7.6.13, which proved to be rather pessimistic in most instances, can be replaced by $\max\{\|b^{(i)}\|_2^2 / \|\hat{b}^{(j)}\|_2^2 \mid 0 \leq i \leq j < m\}$. If an LLL-reduced basis is available, then computing this bound is trivial. For $i = 0$ in many cases this bound turns out to be close to 1 and hence $b^{(0)}$ actually is an ℓ_2-shortest vector in the lattice.

On the other hand, Kannan [106] notes that there are lattices L of rank m for which the orthogonality defect of certain LLL-reduced bases reaches the bound 7.6.11 and the square of the norm of the first basis vector is larger than $\lambda_0(L)$ by a factor $2^{O(m^2)}$.

The following generalizations of 7.6.13 can be found in [125].

Theorem *Let $L \subset \mathbb{Q}^n$ be a lattice with LLL-reduced basis $b^{(0)}, b^{(1)}, \ldots, b^{(m-1)}$. For*   **7.6.15**
$t \in m$ let $v^{(0)}, v^{(1)}, \ldots, v^{(t-1)} \in L$ be t linearly independent lattice vectors. Then we have

$$\|b^{(t)}\|^2 \leq \left(\frac{4}{4\delta - 1}\right)^{m-1} \cdot \max\{\|v^{(i)}\|^2 \mid i \in t\} .$$

Proof: See [125, Prop. 1.12].   □

We note that if a lattice basis $b^{(0)}, b^{(1)}, \ldots, b^{(m-1)}$ of a lattice $L \subset \mathbb{Q}^n$ is δ-reduced with $\frac{1}{4} < \delta < 1$ then for $1 \leq t \leq m$ also the lattices $L_t(b^{(t)}, \ldots, b^{(m-1)})$ are δ-reduced. Moreover, the vector $\hat{b}^{(t)} = \pi_t(b^{(t)})$ is the first vector of the lattice basis $\pi_t(b^{(t)}), \ldots, \pi_t(b^{(m-1)})$ of the lattice $L_t(b^{(t)}, \ldots, b^{(m-1)})$. Applying 7.6.12 and 7.6.14 to the lattice $L_t(b^{(t)}, \ldots, b^{(m-1)})$ we get for $t \in m$:

7.6.16 **Corollary** *Let $L \subset \mathbb{Q}^n$ be a lattice with δ-reduced basis $b^{(0)}, b^{(1)}, \ldots, b^{(m-1)}$. Then, for $t \in m$:*

$$\|\hat{b}^{(t)}\| \leq \left(\frac{4}{4\delta - 1}\right)^{(m-1-t)/2} \cdot \lambda_0\left(L_t(b^{(t)}, \ldots, b^{(m-1)})\right),$$

$$\|\hat{b}^{(t)}\| \leq \left(\frac{4}{4\delta - 1}\right)^{(m-1-t)/4} \cdot \det\left(L_t(b^{(t)}, \ldots, b^{(m-1)})\right)^{1/(m-t)}. \qquad \square$$

The upper bounds on the basis vectors in 7.6.15 can be used to determine the size of the integer constant N in 7.2.7.

7.6.17 **Theorem** *Let $A \cdot x = d$ with $A \in \mathbb{Z}^{m \times n}$ and $d \in \mathbb{Z}^m$. There exists a constant N which depends only on the size of the entries of A and d such that the LLL-algorithm computes a basis of the form 7.2.8 if applied to the basis 7.2.7.*

Proof: Without loss of generality, we assume that the matrix A has rank m. Thus we can permute the columns of A such that the first m columns, i.e. $A^{(0)}, A^{(1)}, \ldots, A^{(m-1)}$, are linearly independent. Let $A' = \left(A^{(0)} \mid \ldots \mid A^{(m-1)}\right)$. Then, each of the $n - m + 1$ linear systems

$$A' \cdot x = -A^{(m+i)}, \quad 0 \leq i < n - m,$$

and

$$A' \cdot x = d,$$

possesses a unique solution in $\mathbb{Q}^m$.

For $0 \leq i < n - m$ let $v'^{(i)} \in \mathbb{Q}^m$ be the solution of the system of linear equations $A' \cdot x = -A^{(m+i)}$ and $v'^{(n-m)} \in \mathbb{Q}^m$ be the solution of the system $A' \cdot x = d$. Using Cramer's rule we can explicitly compute the solutions $v'^{(i)}$ for $0 \leq i < n - m$ via

$$v_k'^{(i)} = \frac{1}{\det(A')} \cdot \det\left((A^{(0)}, \ldots, A^{(k-1)}, -A^{(m+i)}, A^{(k+1)}, \ldots, A^{(m-1)})\right), \quad k \in m,$$

and

$$v_k'^{(n-m)} = \frac{1}{\det(A')} \cdot \det\left((A^{(0)}, \ldots, A^{(k-1)}, d, A^{(k+1)}, \ldots, A^{(m-1)})\right), \quad k \in m.$$

Setting $M := \max\{\|A^{(0)}\|, \|A^{(1)}\|, \dots, \|A^{(n-1)}\|, \|d\|\}$ and using Hadamard's inequality 7.5.1, we can bound the entries $|v_k'^{(i)}|$, $k \in m$, by

$$|v_k'^{(i)}| \leq \frac{1}{|\det(A')|} \cdot \|A^{(m+i)}\| \cdot \prod_{j \in m, j \neq k} \|A^{(j)}\| \leq \frac{1}{|\det(A')|} \cdot M^m$$

for $i \in n - m$ and

$$|v_k'^{(n-m)}| \leq \frac{1}{|\det(A')|} \cdot \|d\| \cdot \prod_{j \in m, j \neq k} \|A^{(j)}\| \leq \frac{1}{|\det(A')|} \cdot M^m .$$

Since all of the above determinants are integral, it follows that

$$\tilde{v}^{(i)} := \det(A') \cdot v'^{(i)}$$

are integer vectors for $0 \leq i \leq n - m$. Moreover, the vectors $\tilde{v}^{(i)}$, $0 \leq i \leq n - m$, are solutions of

$$A' \cdot x = -\det(A') \cdot A^{(m+i)} \quad \text{and} \quad A' \cdot x = \det(A') \cdot d,$$

respectively. We note that $\tilde{v}^{(i)}$ remains a solution of the linear system if we multiply the linear system with a nonzero constant N.

By filling in sufficiently many zeros, the solutions $\tilde{v}^{(i)}$ can be written as vectors in $\mathbb{Z}^{n+1}$, such that

$$(A' \mid A^{(m)} \mid \dots \mid A^{(n-1)} \mid -d) \cdot \underbrace{\begin{pmatrix} \tilde{v}_0^{(0)} & \tilde{v}_0^{(1)} & \cdots & \tilde{v}_0^{(n-m)} \\ \vdots & \vdots & & \vdots \\ \tilde{v}_{m-1}^{(0)} & \tilde{v}_{m-1}^{(1)} & \cdots & \tilde{v}_{m-1}^{(n-m)} \\ \det(A') & 0 & \cdots & 0 \\ 0 & \det(A') & & \\ \vdots & & \ddots & \\ 0 & & & \det(A') \end{pmatrix}}_{=: \tilde{V} \in \mathbb{Z}^{(n+1) \times (n-m+1)}} = 0.$$

The square of the Euclidean norm of the ith-column of $\tilde{V}$, $0 \leq i \leq n - m$, can be bounded by

$$\|\tilde{V}^{(i)}\|^2 \leq m \cdot M^{2m} + \det(A')^2 \leq (m+1) \cdot M^{2m} .$$

Multiplying $\tilde{V}$ by the lower part of the generator matrix 7.2.7, i.e. by the rows $m, \dots, m + n$, we get

$$V = \begin{pmatrix} -r_{\max} & 2c_0 & 0 & \cdots & 0 \\ -r_{\max} & 0 & 2c_1 & \cdots & 0 \\ \vdots & \vdots & & \ddots & \vdots \\ -r_{\max} & 0 & \cdots & \cdots & 2c_{n-1} \\ r_{\max} & 0 & \cdots & \cdots & 0 \end{pmatrix} \cdot \tilde{V} .$$

The resulting matrix V is the lower left part of a generator matrix of the lattice of the form 7.2.8. An elementary calculation shows that the norm of column $V^{(i)}$ of V, $0 \leq i \leq n - m$, can be bounded by

$$\|V^{(i)}\| \leq 2\sqrt{n+1} \cdot r_{\max} \cdot \|\tilde{V}^{(i)}\| \leq 2\sqrt{(n+1)(m+1)} \cdot r_{\max} \cdot M^m .$$

Now, let $b^{(0)}, b^{(1)}, \ldots, b^{(n)}$ be an LLL-reduced basis of a lattice generated by 7.2.7. Using 7.6.15, the length of the first $n - m + 1$ basis vectors $b^{(t)}$, $0 \leq t \leq n - m$, can be bounded by

$$\|b^{(t)}\|^2 \leq \left(\frac{4}{4\delta - 1}\right)^n \cdot \max\{\|V^{(i)}\|^2 \mid 0 \leq i \leq n - m\} .$$

If we choose N such that

$$N \geq \left(\frac{4}{4\delta - 1}\right)^{n/2} \cdot 2\sqrt{(m+1)(n+1)} \cdot r_{\max} \cdot M^m ,$$

then we have that

$$\|b^{(t)}\| < N$$

for $0 \leq t \leq n - m$. Thus, the LLL-algorithm will produce a basis whose first $n - m + 1$ vectors are all zero in the first m entries, for otherwise the Euclidean length of such a column would be greater than N. Therefore, the LLL-reduced basis has the form 7.2.8.   □

For practical purposes it is interesting to note that in almost all cases it suffices to choose N much smaller than the value of the previous bound.

Blockwise Korkine–Zolotarev reduction. As we have seen in Section 7.3, the bounds for the length of the vectors of a Korkine–Zolotarev-reduced basis are much better than the bounds 7.6.14 for an LLL-reduced basis of a lattice L. Unfortunately, no algorithm is known which computes a Korkine–Zolotarev-reduced basis in polynomial time.

In a sense, *Korkine–Zolotarev reduction* is a generalization of the LLL-algorithm. In Step 2 of the LLL-algorithm, we compare the Euclidean length of the projections of $b^{(k)}$ and $b^{(k+1)}$ onto the subspace $\langle b^{(0)}, b^{(1)}, \ldots, b^{(k-1)}\rangle^{\perp}$. In Korkine–Zolotarev reduction, we search for a nontrivial integer linear combination $u_k b^{(k)} + u_{k+1} b^{(k+1)} + \ldots + u_{m-1} b^{(m-1)}$ which minimizes the Euclidean length of

$$\pi_k\left(u_k b^{(k)} + u_{k+1} b^{(k+1)} + \ldots + u_{m-1} b^{(m-1)}\right) .$$

No algorithm is known which finds the integer linear combination of the shortest nontrivial projection in time which is polynomial in the number of vectors $(m - k)$. Therefore, Schnorr in [172] and [173] restricted the search to blocks

of β vectors at a time for some fixed integer constant β. A nontrivial integer
linear combination

$$u_k b^{(k)} + u_{k+1} b^{(k+1)} + \ldots + u_{k+\beta-1} b^{(k+\beta-1)}$$

minimizing the Euclidean length of

$$\pi_k(u_k b^{(k)} + u_{k+1} b^{(k+1)} + \ldots + u_{k+\beta-1} b^{(k+\beta-1)})$$

is then found by exhaustive enumeration. This algorithm is called *blockwise
Korkine–Zolotarev reduction*. For a further description of improved practical versions, we refer to [173] and [174]. In a blockwise Korkine–Zolotarev-reduced
basis of a lattice of rank m the factor $\left(\frac{4}{4\delta-1}\right)^{(m-1)/2}$ in 7.6.14 can be replaced
by $(1+\epsilon)^m$ for any fixed $\epsilon > 0$. Of course, the time complexity increases
exponentially as ϵ approaches 0.

To summarize the various reduction concepts, the Euclidean norm of the
vectors of a reduced basis can be bounded above as follows:

- The LLL-algorithm computes a basis $b^{(0)}, b^{(1)}, \ldots, b^{(m-1)}$ with

$$\|b^{(t)}\| \le \left(\frac{4}{4\delta-1}\right)^{(m-1)/2} \lambda_t(L) \text{ for } t \in m.$$

- If $b^{(0)}, b^{(1)}, \ldots, b^{(m-1)}$ is a Korkine–Zolotarev-reduced basis, then

$$\|b^{(t)}\| \le \left(\frac{t+4}{4}\right)^{1/2} \lambda_t(L) \text{ for } t \in m.$$

- If $b^{(0)}, b^{(1)}, \ldots, b^{(m-1)}$ is a blockwise Korkine–Zolotarev-reduced basis,
  then

$$\|b^{(0)}\| \le (1+\epsilon)^m \lambda_0(L).$$

Exercises

Exercise Let $\delta \in \mathbb{R}$ with $\frac{1}{4} < \delta < 1$. Show that for $i \in m$ E.7.6.1

$$\|\hat{b}^{(i)}\|^2 + \sum_{j \in i} \frac{1}{4}\left(\frac{4}{4\delta-1}\right)^{i-j} \|\hat{b}^{(i)}\|^2 \le \left(\frac{4}{4\delta-1}\right)^i \|\hat{b}^{(i)}\|^2.$$

Exercise Let $b^{(0)}, b^{(1)}, \ldots, b^{(m-1)}$ be a sequence of linearly independent vec-   E.7.6.2
tors in $\mathbb{R}^n$ and $\hat{b}^{(0)}, \hat{b}^{(1)}, \ldots, \hat{b}^{(m-1)}$ the associated Gram–Schmidt vectors. Any
vector $v \in \langle b^{(0)}, b^{(1)}, \ldots, b^{(m-1)} \rangle$ can then be written as $v = \sum_{j \in m} r_j b^{(j)} = \sum_{j \in m} r'_j \hat{b}^{(j)}$ with $r_j, r'_j \in \mathbb{R}$ for $j \in m$. Prove the following: If t is the largest
index with $r_t \ne 0$, then $r_t = r'_t$.

7.7 Lattice Point Enumeration

Let us again consider the problem of solving systems of Diophantine equations as described in 7.2.9. Usually, we are interested in finding all solutions to this problem, or to conclude that there are none. In terms of the associated lattice 7.2.7, this mean that we wish to enumerate all lattice points which are subject to a certain set of constraints. Such an approach has first been described by Ritter [170] for $\{0,1\}$ problems. Here we solve the general problem with arbitrary bounds on the variables.

A priori, a lattice $L = \{\sum_{i \in m} u_i b^{(i)} \mid u_i \in \mathbb{Z}\}$ of rank m contains infinitely many elements. It will turn out that there are bounds on the integers $|u_i|$, $i \in m$, which depend solely on the lattice basis $b^{(0)}, b^{(1)}, \ldots, b^{(m-1)}$. These bounds reduce the problem of finding vectors with the properties of 7.2.9 to a finite subset of the original lattice. Therefore we are left with the problem of enumerating all solution vectors of 7.2.9 in a finite subset of the lattice. In the following, we will describe this search process in more detail.

One possibility to compute the above mentioned bounds on the integers $|u_i|$ is by means of Linear Programming. This is not our approach here. Instead, we will use pruning tests to bound the integers $|u_i|$. Compared to the Linear Programming approach, these tests generally lead to a larger enumeration tree. Nevertheless, the pruning tests are very simple and easy to compute, therefore the overall enumeration time seems to be faster than the method based on Linear Programming. The pruning tests we use have quite a long history and are based on the work of [44], [45], [105], [107], [110], [170]. From 7.2.10 we see that a solution vector v, i.e. a vector v of the form 7.2.11, has the upper bounds

7.7.1	$$\|v\|_2^2 \leq (n+1) \cdot r_{\max}^2 \quad \text{and}$$
7.7.2	$$\|v\|_\infty \leq r_{\max}.$$

The exhaustive enumeration is arranged as a backtracking algorithm starting from $u_{n-m} \in \mathbb{Z}$, which successively chooses values $u_t \in \mathbb{Z}$ for $t = n - m, n - m - 1, \ldots, 1, 0$.

7.7.3 **Definition** In each level t of the backtracking algorithm $w^{(t)} = \pi_t(\sum_{j=t}^{n-m} u_j b^{(j)})$ is the projection of the linear combination of the already fixed variables u_t, $u_{t+1}, \ldots, u_{n-m}$ into the subspace of $\mathbb{R}^{n+1}$ which is orthogonal to the linear span $\langle b_0, \ldots, b_{t-1} \rangle$. ◇

Starting from $w^{(n-m+1)} = 0$, $w^{(t)}$ can be iteratively computed from $w^{(t+1)}$ by

7.7.4
$$w^{(t)} = \left(\sum_{i=t}^{n-m} u_i \mu_{it} \right) \hat{b}^{(t)} + w^{(t+1)},$$

with Gram-Schmidt coefficients μ_{it}. In each level t, $n - m \geq t \geq 0$, of the backtrack algorithm we test all possible integer values for the variable u_t. The following tests allow to restrict the possible values of u_t.

— **First pruning condition.** For all $j \leq t$ the vectors $\hat{b}^{(j)}$ are orthogonal to $w^{(t+1)}$ and therefore

$$\|w^{(t)}\|_2^2 = \left(\sum_{i=t}^{n-m} u_i \mu_{it} \right)^2 \|\hat{b}^{(t)}\|_2^2 + \|w^{(t+1)}\|_2^2 .$$

Further, we notice that $w^{(0)} = \sum_{j=0}^{n-m} u_j b^{(j)}$. Using $\|w^{(j)}\|_2 \geq \|w^{(t)}\|_2$ for $j \leq t$ and 7.7.1 we can backtrack as soon as

$$\|w^{(t)}\|_2^2 > c := (n+1) \cdot r_{\max}^2 .$$

For fixed $u_{t+1}, \ldots, u_{n-m}$, this gives a bound for u_t:

$$\left(u_t + \sum_{i=t+1}^{n-m} u_i \mu_{it} \right)^2 \leq \frac{c - \|w^{(t+1)}\|_2^2}{\|\hat{b}^{(t)}\|_2^2} .$$

This is the first pruning condition.

— **Second pruning condition.** Let $\overline{b}^{(i)}$, $i \in n - m + 1$, be a basis of the *dual lattice*, which is defined by the conditions $\langle \overline{b}^{(i)}, b^{(j)} \rangle = \delta_{ij}$ for $0 \leq i, j \leq n - m + 1$. If B is the matrix whose columns are the basis vectors $b^{(i)}$, $i \in n - m$, then it was observed in [45] that $u_i = \overline{b}^{(i)\top} \cdot B \cdot u$. Applying the inequality of Cauchy–Schwarz, i.e. $|\overline{b}^{(i)\top} \cdot (B \cdot u)| \leq \|\overline{b}^{(i)}\|_2 \cdot \|B \cdot u\|_2$, for $i \in n - m + 1$ then gives the bound

$$|u_i| \leq \|\overline{b}^{(i)}\|_2 \cdot \|B \cdot u\|_2 \leq \|\overline{b}^{(i)}\|_2 \cdot \sqrt{(n+1) \cdot r_{\max}^2}$$

and similarly

$$|u_i| \leq \|\overline{b}^{(i)}\|_1 \cdot r_{\max} .$$

Of course, the numbers $\|\overline{b}^{(i)}\|_1$, $\|\overline{b}^{(i)}\|_2$ can be precomputed before the enumeration.

— **Third pruning condition.** The third test is an adaption to the special situation that we are searching for an integer linear combination of the basis vectors which consists solely of components whose absolute value is bounded by $r_{\max}$. It is based on the following theorem, see [170].

7.7.5

Theorem *If the given sequence of integers $u_t, u_{t+1}, \ldots, u_{n-m} \in \mathbb{Z}$ can be extended to $u_0, \ldots, u_t, \ldots, u_{n-m} \in \mathbb{Z}$ such that $\sum_{i \in n-m+1} u_i b^{(i)}$ has the form 7.2.11, then for all $y_t, y_{t+1}, \ldots, y_{n-m} \in \mathbb{R}$:*

$$\left| \sum_{i=t}^{n-m} y_i \|w^{(i)}\|_2^2 \right| \leq r_{max} \cdot \left\| \sum_{i=t}^{n-m} y_i w^{(i)} \right\|_1 .$$

Proof: From 7.7.4 we see that $\langle w^{(l)}, w^{(i)} \rangle = \|w^{(i)}\|_2^2$ for $l < i$. If $w^{(0)}$ has the form 7.2.11 it follows from Hölder's inequality, see exercise 5.1.2, and 7.7.2 that

$$\left| \sum_{i=t}^{n-m} y_i \|w^{(i)}\|_2^2 \right| = \left| \langle w^{(0)}, \sum_{i=t}^{n-m} y_i w^{(i)} \rangle \right|$$

$$\leq \|w^{(0)}\|_\infty \cdot \left\| \sum_{i=t}^{n-m} y_i w^{(i)} \right\|_1$$

$$\leq r_{max} \cdot \left\| \sum_{i=t}^{n-m} y_i w^{(i)} \right\|_1 . \qquad \square$$

Of course the above inequality can be extended to arbitrary p-norms.

7.7.6

Remark We use this theorem in the enumeration algorithm in two ways.

— First, we take $(y_t, y_{t+1}, \ldots, y_{n-m}) = (1, 0, \ldots, 0)$, which results in the test

7.7.7

$$\|w^{(t)}\|_2^2 \leq r_{max} \|w^{(t)}\|_1 .$$

— Second, we will see that if the test 7.7.7 fails for some vector $w^{(t)} = x\hat{b}^{(t)} + w^{(t+1)}$, then it will also fail for all vectors $\tilde{w}^{(t)} = (x + r)\hat{b}^{(t)} + w^{(t+1)}$ with $r \in \mathbb{Z}$ and $xr > 0$. That means, we can stop the enumeration for these values of $r \in \mathbb{Z}$.

To show this, let $x \in \mathbb{R}$ and $r \in \mathbb{Z}$ such that $xr > 0$. For $w^{(t)} = x\hat{b}^{(t)} + w^{(t+1)}$ we define $\tilde{w}^{(t)} = (x + r)\hat{b}^{(t)} + w^{(t+1)}$ and we set $\eta := \frac{x}{x+r}$. Then, it is easy to see that

$$w^{(t)} = \eta \tilde{w}^{(t)} + (1 - \eta)w^{(t+1)} \quad \text{and} \quad 0 < \eta < 1.$$

If $\tilde{w}^{(t)}$ can lead to a solution, then we set $(y_t, y_{t+1}, \ldots, y_{n-m}) = (\eta, 1 - \eta, 0, \ldots, 0)$ and get with 7.7.5:

$$\eta \|\tilde{w}^{(t)}\|_2^2 + (1 - \eta)\|w^{(t+1)}\|_2^2 \leq r_{max} \|\eta \tilde{w}^{(t)} + (1 - \eta)w^{(t+1)}\|_1 .$$

Together, it follows

$$\|w^{(t)}\|_2^2 \leq \eta \|\tilde{w}^{(t)}\|_2^2 + (1 - \eta)\|w^{(t+1)}\|_2^2$$

$$\leq r_{max} \|\eta \tilde{w}^{(t)} + (1 - \eta)w^{(t+1)}\|_1$$

$$= r_{max} \|w^{(t)}\|_1 .$$

Therefore, if $\tilde{w}^{(t)}$ can lead to a solution, $w^{(t)}$ can also lead to a solution. On the contrary, if $w^{(t)}$ cannot lead to a solution, i.e. if $\|w^{(t)}\|_2^2 > r_{\max}\|w^{(t)}\|_1$, $\tilde{w}^{(t)}$ cannot lead to a solution for all $\tilde{w}^{(t)} = (x+r)\hat{b}^{(t)} + w^{(t+1)}$ with $r \in \mathbb{Z}$ and $xr > 0$.   ◇

Algorithm (Lattice point enumeration) Given the generator matrix 7.2.7 of the lattice $L \subset \mathbb{R}^{m+n+1}$ of rank $n+1$ from 7.2.7 all nonzero vectors $v \in L$ such that $\|v\|_\infty \le r_{\max}$ are determined.   7.7.8

— Compute an LLL-reduced basis $b^{(0)}, b^{(1)}, \ldots, b^{(n)}$ of the lattice L.

— Delete the unnecessary columns and rows of the generator matrix according to Section 7.2. The remaining basis $b^{(0)}, b^{(1)}, \ldots, b^{(n-m)} \subset \mathbb{R}^{n+1}$ has rank $n - m + 1$.

— Compute the Gram–Schmidt vectors $\hat{b}^{(0)}, \hat{b}^{(1)}, \ldots, \hat{b}^{(n-m)}$ together with the Gram–Schmidt coefficients μ_{ij}, see 7.4.2.

— Set $R := (n+1) \cdot r_{\max}^2$.

— The recursive backtracking algorithm enum() has two input parameters. The first parameter t is the search level, it runs from $n - m$ down to 0. The second parameter $w' \subset \mathbb{R}^{n+1}$ is the vector which has been computed in the level $t+1$. The enumeration is initiated with the call of enum$(n-m, 0)$.

```
(1)     function enum(t, w′)
(2)     begin
(3)         firstprune := false
(4)         yₜ := Σⁿ⁻ᵐᵢ₌ₜ₊₁ uᵢμᵢₜ
(5)         uₜ := ⌊−yₜ⌉
(6)         while true
(7)             w := (Σⁿ⁻ᵐᵢ₌ₜ uᵢμᵢₜ)b̂⁽ᵗ⁾ + w′
(8)             if ‖w‖² > R then return          /* step back */
(9)             if t > 0 then
(10)                if prune(uₜ) then
(11)                    if firstprune then return  /* step back */
(12)                    else
(13)                        next(uₜ)
(14)                        firstprune := true
(15)                        goto line (7)
(16)                    end if
(17)                else
(18)                    enum(t − 1, w)              /* step forward */
```

(19)   **else**   /* $t = 0 \rightarrow$ solution */
(20)   **if** w has the form 7.2.11 **then** print w
(21)   next(u_t)
(22)   **end while**
(23) **end**   □

The procedure next() in lines (13) and (21) determines the next possible integer value of the variable u_t. Initially, when entering a new level t, in line (5) u_t is set to be the closest integer value of $-y_t := -\sum_{i=t+1}^{n-m} u_i \mu_{it}$, say u_t^1. The next value u_t^2 of u_t is the second closest integer to $-y_t$ then follows u_t^3 and so forth. Therefore the values of u_t alternate around $-y_t$. If the function prune() returns true for w_t, then we do one more regular call of the procedure next() in line (13), i.e. u_t is set to be the next closest integer to $-y_t$. In Fig. 7.7 this happens while u_t^4 is determined.

After that, using 7.7.6, the enumeration proceeds only in this remaining direction. Compare the computation of u_t^5 in Fig. 7.7. Finally, the second time when the function prune() returns true, the algorithm steps back and increases the enumeration level, see line (11).

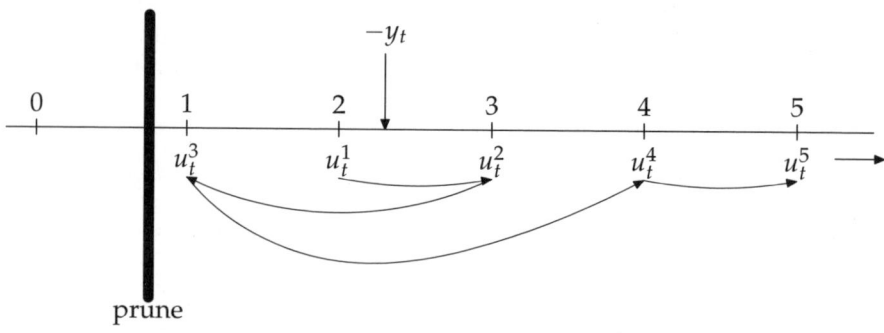

Fig. 7.7 Enumeration in level t and pruning after u_t^3

7.7.9   **Example** Suppose $y_t = -2.3$ for $0 \le t \le n - m$. Therefore, $-y_t = 2.3$ and according to line (5), $u_t = \text{round}(-y_t) = \lfloor 2.3 + \frac{1}{2} \rfloor = 2$. First, assume that the procedure prune() always return false. Therefore, in the subsequent calls of the procedure next() in line (21), the variable u_t takes the values $3, 1, 4, 0, 5, -1, 6, \ldots$.

Now assume, after testing the values $u_t = 2$ and 3, that the procedure prune() returns true for $u_t = 1$. Then the value of u_t is set to 4 in line (13).

After subsequent calls of next() in line (21), u_t takes the values $5, 6, 7$ and so forth until prune() returns true the next time.   ◇

The function prune for the third pruning test according to 7.7.7 can be implemented as follows.

Algorithm   7.7.10

    function prune(w_t)
(1)   **if** $\|w^{(t)}\|^2 \leq r_{max} \cdot \|w^{(t)}\|_1$
(2)   **return** false
(3)   **else**
(4)   **return** true
(5)   **end if**   □

The first pruning test from page 599 is done in line (8) in 7.7.8, the second pruning test from page 599 can be additionally added after line (6) in 7.7.8.
    If in the forward step of the algorithm, i.e. in line (18) a new level t is entered, then initially in the next call of enum() in line (5) u_t is set to the closest integer to $-\sum_{i=t+1}^{n-m} u_i \mu_{it}$. Since

$$\|w^{(t)}\|_2^2 = \left(u_t + \sum_{i=t+1}^{n-m} u_i \mu_{it} \right)^2 \|\hat{b}^{(t)}\|_2^2 + \|w^{(t+1)}\|_2^2,$$

this choice of u_t minimizes $\|w^{(t)}\|_2^2$.

Example We illustrate the algorithm by solving a system of Diophantine lin-   7.7.11
ear equations which occurs during the construction of linear codes with prescribed minimum distance in Chapter 8, Example 8.4.4. In order to find a
$(14, 3, 9)$-code over $\mathbb{F}_3$ the following system must be solved:

$$\begin{pmatrix} 2 & 2 & 0 & -1 & 0 & 0 \\ 1 & 1 & 2 & 0 & -1 & 0 \\ 0 & 3 & 1 & 0 & 0 & -1 \\ 3 & 6 & 4 & 0 & 0 & 0 \end{pmatrix} \cdot \begin{pmatrix} x_0 \\ x_1 \\ x_2 \\ x_3 \\ x_4 \\ x_5 \end{pmatrix} = \begin{pmatrix} 0 \\ 0 \\ 0 \\ 14 \end{pmatrix},$$

where $x_0 \in \{0, 1, 2, 3, 4\}$, $x_1 \in \{0, 1, 2\}$, $x_2 \in \{0, 1, 2, 3\}$ and $x_i \in \{0, 1, 2, 3, 4, 5\}$ for $i \in \{3, 4, 5\}$. According to 7.2.7, the lattice is generated by the matrix

$$
\left(
\begin{array}{c|cccccc}
0 & 2000 & 2000 & 0 & -1000 & 0 & 0 \\
0 & 1000 & 1000 & 2000 & 0 & -1000 & 0 \\
0 & 0 & 3000 & 1000 & 0 & 0 & -1000 \\
-14000 & 3000 & 6000 & 4000 & 0 & 0 & 0 \\
\hline
-60 & 30 & 0 & 0 & 0 & 0 & 0 \\
-60 & 0 & 60 & 0 & 0 & 0 & 0 \\
-60 & 0 & 0 & 40 & 0 & 0 & 0 \\
-60 & 0 & 0 & 0 & 24 & 0 & 0 \\
-60 & 0 & 0 & 0 & 0 & 24 & 0 \\
-60 & 0 & 0 & 0 & 0 & 0 & 24 \\
60 & 0 & 0 & 0 & 0 & 0 & 0
\end{array}
\right)
$$

where $r_{max} = \mathrm{lcm}(4,2,3,5) = 60$ and the constant N is set to $N = 1000$. In the first step of the algorithm, the LLL-reduction is applied to the columns of the above matrix. This results in the following new basis:

$$
\left(
\begin{array}{ccc|cccc}
0 & 0 & 0 & -1000 & 0 & 0 & 0 \\
0 & 0 & 0 & 0 & -1000 & 0 & 0 \\
0 & 0 & 0 & 0 & 0 & 0 & 1000 \\
0 & 0 & 0 & 0 & 0 & 1000 & 0 \\
\hline
-60 & -60 & 0 & 0 & 0 & -30 & 0 \\
60 & 0 & -60 & 0 & 0 & 0 & 0 \\
0 & 20 & 100 & 0 & 0 & 40 & 0 \\
-48 & -12 & -132 & 24 & 0 & -48 & 0 \\
-24 & 60 & 12 & 0 & 24 & 24 & 0 \\
72 & 60 & -60 & 0 & 0 & 24 & -24 \\
0 & 60 & -60 & 0 & 0 & 0 & 0
\end{array}
\right) .
$$

The first three columns correspond to solutions of the above system. But the first column corresponds to a solution where the right hand side of the above system is not included since the last entry is equal to zero. The second column corresponds to a solution, because all entries have absolute value at most 60. The solution $x = (x_1, x_2, \ldots, x_6)^{\top}$ of the original system of equations can now be obtained by solving

$$
\begin{pmatrix} -60 \\ 0 \\ 20 \\ -12 \\ 60 \\ 60 \\ 60 \end{pmatrix}
=
\begin{pmatrix}
-60 & 30 & 0 & 0 & 0 & 0 & 0 \\
-60 & 0 & 60 & 0 & 0 & 0 & 0 \\
-60 & 0 & 0 & 40 & 0 & 0 & 0 \\
-60 & 0 & 0 & 0 & 24 & 0 & 0 \\
-60 & 0 & 0 & 0 & 0 & 24 & 0 \\
-60 & 0 & 0 & 0 & 0 & 0 & 24 \\
60 & 0 & 0 & 0 & 0 & 0 & 0
\end{pmatrix}
\cdot
\begin{pmatrix} x_0 \\ x_1 \\ x_2 \\ x_3 \\ x_4 \\ x_5 \\ x_6 \end{pmatrix} ,
$$

which results in $x = (0,1,2,2,5,5)^\top$. For the exhaustive enumeration of all solutions we can remove the unnecessary rows and columns and use the lattice which is generated by the matrix

$$
\begin{pmatrix}
-60 & -60 & 0 \\
60 & 0 & -60 \\
0 & 20 & 100 \\
-48 & -12 & -132 \\
-24 & 60 & 12 \\
72 & 60 & -60 \\
0 & 60 & -60
\end{pmatrix}.
$$

Eventually, after 6 loops in the exhaustive enumeration step it is determined that there are no further solutions.   ◇

Exercises

Exercise Use the computer software on the CD-ROM of the book to compute   **E.7.7.1**
the solution of the system of Diophantine linear equations from 7.7.11.

7.8 Computing the Minimum Distance of Linear Codes   7.8

Let C be a binary or ternary linear code. It is possible to compute the minimum distance of such a code by using a variant of the lattice point enumeration algorithm from Section 7.7. For this purpose, we note that in the binary case we have $-1 \equiv 1 \bmod 2$ while in the ternary case $-1 \equiv 2 \bmod 3$. Thus, codewords of binary or ternary codes can be represented by vectors with integral entries in $\{0, 1, -1\}$.

Let $\mathbb{F}$ be a binary or ternary field, i.e. $q = 2$ or $q = 3$. Consider the lattice L_C which is spanned by the columns of the integral $(n + k) \times (k + n)$-matrix

$$
B_C = \left(\begin{array}{c|c}
N \cdot \Gamma^\top & N \cdot qI_n \\
\hline
I_k & 0
\end{array} \right),
$$

where Γ is a $k \times n$ generator matrix of the code C and N is a large integer constant. The matrix qI_n is used to reduce the integral linear combinations of the columns of $\Gamma^\top$ modulo q. Any lattice vector $v \in L_C$ with $v_i \in \{0, 1, -1\}$ for $i \in n$ corresponds to a codeword $v_C \in C$ and $\mathrm{wt}(v_C)$ is the number of nonzero entries in the first n coefficients of v. Thus, the minimum distance problem can be solved by finding a nonzero lattice vector with the least number (> 0) of nonzero entries in the first n rows.

If the constant N is large enough, the reduced lattice basis contains k vectors whose first n entries are all zero. These vectors can be removed. Further, the lower k components are no longer necessary and can be removed, too. To achieve an even better reduced basis, a useful strategy is to shuffle the remaining basis vectors randomly and apply lattice basis reduction to the reordered basis. This mixing and reduction step can be repeated several times. Finally, the resulting basis is enumerated with 7.7.8, as described below. Here is an example.

7.8.1   **Example** Since we write codewords as row vectors, we apply lattice basis reduction to rows in this example. So, the basis vectors are the rows of the generator matrix of the lattice L_C.

The goal is to determine the minimum distance of the ternary Golay code. It is also a quadratic-residue-code $C_Q(11,6)$ over $\mathbb{F}_3$, see Section 4.4. A generator matrix is

$$\Gamma = \begin{pmatrix} 2 & 2 & 1 & 2 & 0 & 1 & 0 & 0 & 0 & 0 & 0 \\ 0 & 2 & 2 & 1 & 2 & 0 & 1 & 0 & 0 & 0 & 0 \\ 0 & 0 & 2 & 2 & 1 & 2 & 0 & 1 & 0 & 0 & 0 \\ 0 & 0 & 0 & 2 & 2 & 1 & 2 & 0 & 1 & 0 & 0 \\ 0 & 0 & 0 & 0 & 2 & 2 & 1 & 2 & 0 & 1 & 0 \\ 0 & 0 & 0 & 0 & 0 & 2 & 2 & 1 & 2 & 0 & 1 \end{pmatrix}.$$

Using $N = 6$, the generator matrix $B_C^\top$ of the lattice L_C is

$$B_C^\top = \left(\begin{array}{ccccccccccc|cccccc} 12 & 12 & 6 & 12 & 0 & 6 & 0 & 0 & 0 & 0 & 0 & 1 & 0 & 0 & 0 & 0 & 0 \\ 0 & 12 & 12 & 6 & 12 & 0 & 6 & 0 & 0 & 0 & 0 & 0 & 1 & 0 & 0 & 0 & 0 \\ 0 & 0 & 12 & 12 & 6 & 12 & 0 & 6 & 0 & 0 & 0 & 0 & 0 & 1 & 0 & 0 & 0 \\ 0 & 0 & 0 & 12 & 12 & 6 & 12 & 0 & 6 & 0 & 0 & 0 & 0 & 0 & 1 & 0 & 0 \\ 0 & 0 & 0 & 0 & 12 & 12 & 6 & 12 & 0 & 6 & 0 & 0 & 0 & 0 & 0 & 1 & 0 \\ 0 & 0 & 0 & 0 & 0 & 12 & 12 & 6 & 12 & 0 & 6 & 0 & 0 & 0 & 0 & 0 & 1 \\ \hline 18 & 0 & 0 & 0 & 0 & 0 & 0 & 0 & 0 & 0 & 0 & 0 & 0 & 0 & 0 & 0 & 0 \\ 0 & 18 & 0 & 0 & 0 & 0 & 0 & 0 & 0 & 0 & 0 & 0 & 0 & 0 & 0 & 0 & 0 \\ 0 & 0 & 18 & 0 & 0 & 0 & 0 & 0 & 0 & 0 & 0 & 0 & 0 & 0 & 0 & 0 & 0 \\ 0 & 0 & 0 & 18 & 0 & 0 & 0 & 0 & 0 & 0 & 0 & 0 & 0 & 0 & 0 & 0 & 0 \\ 0 & 0 & 0 & 0 & 18 & 0 & 0 & 0 & 0 & 0 & 0 & 0 & 0 & 0 & 0 & 0 & 0 \\ 0 & 0 & 0 & 0 & 0 & 18 & 0 & 0 & 0 & 0 & 0 & 0 & 0 & 0 & 0 & 0 & 0 \\ 0 & 0 & 0 & 0 & 0 & 0 & 18 & 0 & 0 & 0 & 0 & 0 & 0 & 0 & 0 & 0 & 0 \\ 0 & 0 & 0 & 0 & 0 & 0 & 0 & 18 & 0 & 0 & 0 & 0 & 0 & 0 & 0 & 0 & 0 \\ 0 & 0 & 0 & 0 & 0 & 0 & 0 & 0 & 18 & 0 & 0 & 0 & 0 & 0 & 0 & 0 & 0 \\ 0 & 0 & 0 & 0 & 0 & 0 & 0 & 0 & 0 & 18 & 0 & 0 & 0 & 0 & 0 & 0 & 0 \\ 0 & 0 & 0 & 0 & 0 & 0 & 0 & 0 & 0 & 0 & 18 & 0 & 0 & 0 & 0 & 0 & 0 \end{array}\right).$$

Applying the LLL-algorithm to the *rows* of $B_C^\top$ gives

$$B_C = \left(\begin{array}{ccccccccccc|cccccc}
0 & 0 & 0 & 0 & 0 & 0 & 0 & 0 & 0 & 0 & 0 & 0 & 0 & 0 & 0 & 0 & -3 \\
0 & 0 & 0 & 0 & 0 & 0 & 0 & 0 & 0 & 0 & 0 & 0 & -3 & 0 & 0 & 0 & 0 \\
0 & 0 & 0 & 0 & 0 & 0 & 0 & 0 & 0 & 0 & 0 & 0 & 0 & 0 & -3 & 0 & 0 \\
0 & 0 & 0 & 0 & 0 & 0 & 0 & 0 & 0 & 0 & 0 & 0 & 0 & 3 & 0 & 0 & 0 \\
0 & 0 & 0 & 0 & 0 & 0 & 0 & 0 & 0 & 0 & 0 & 3 & 0 & 0 & 0 & 0 & 0 \\
0 & 0 & 0 & 0 & 0 & 0 & 0 & 0 & 0 & 0 & 0 & 0 & 0 & 0 & 0 & -3 & 0 \\
0 & -6 & -6 & -6 & 0 & -6 & -6 & 0 & -6 & 0 & 0 & 0 & 1 & 0 & -1 & 0 & 0 \\
0 & 0 & 0 & 0 & 0 & -6 & -6 & 6 & -6 & 0 & 6 & 0 & 0 & 0 & 0 & 0 & 1 \\
-6 & 0 & -6 & 0 & 0 & 0 & -6 & -6 & -6 & 0 & -6 & 1 & -1 & 0 & 1 & 0 & -1 \\
0 & -6 & -6 & 0 & 0 & -6 & 0 & 0 & 0 & 6 & 6 & 0 & 1 & 0 & 1 & 1 & 1 \\
0 & 0 & -6 & 0 & 6 & 0 & -6 & 0 & -6 & 6 & 0 & 0 & 0 & 1 & -1 & 1 & 0 \\
0 & -6 & 0 & 0 & -6 & 0 & -6 & -6 & -6 & 0 & 0 & 0 & 1 & -1 & -1 & 0 & 0 \\
0 & 0 & 6 & 0 & 6 & 6 & 6 & 0 & 0 & 0 & 6 & 0 & 0 & -1 & 1 & 0 & 1 \\
0 & 0 & 6 & 0 & 0 & 6 & 0 & 6 & 6 & 6 & 0 & 0 & 0 & -1 & 1 & 1 & 0 \\
-6 & -6 & 0 & 0 & 0 & -6 & 0 & 0 & -6 & 0 & -6 & 1 & 0 & 1 & 1 & 0 & -1 \\
-6 & 6 & 0 & 0 & -6 & 6 & 6 & 0 & 0 & 0 & 0 & 1 & 1 & 0 & 0 & 0 & 0 \\
-6 & -6 & -6 & 0 & 0 & 0 & -6 & 0 & 0 & -6 & 0 & 1 & 0 & -1 & 0 & -1 & 0
\end{array}\right).$$

We delete the unnecessary rows and columns, see 7.2.8. Then scaling and mixing the remaining rows gives

$$\left(\begin{array}{ccccccccccc}
0 & -1 & -1 & 0 & 0 & -1 & 0 & 0 & 0 & 1 & 1 \\
0 & 0 & 1 & 0 & 1 & 1 & 1 & 0 & 0 & 0 & 1 \\
0 & 0 & 1 & 0 & 0 & 1 & 0 & 1 & 1 & 1 & 0 \\
-1 & 0 & -1 & 0 & 0 & 0 & -1 & -1 & -1 & 0 & -1 \\
-1 & -1 & 0 & 0 & 0 & -1 & 0 & 0 & -1 & 0 & -1 \\
-1 & 1 & 0 & 0 & -1 & 1 & 1 & 0 & 0 & 0 & 0 \\
0 & 0 & 0 & 0 & 0 & -1 & -1 & 1 & -1 & 0 & 1 \\
0 & -1 & 0 & 0 & -1 & 0 & -1 & -1 & -1 & 0 & 0 \\
0 & -1 & -1 & -1 & 0 & -1 & -1 & 0 & -1 & 0 & 0 \\
0 & 0 & -1 & 0 & 1 & 0 & -1 & 0 & -1 & 1 & 0 \\
-1 & -1 & -1 & 0 & 0 & 0 & -1 & 0 & 0 & -1 & 0
\end{array}\right).$$

LLL-reduction of this lattice produces the following basis

$$\left(\begin{array}{ccccccccccc}
0 & -1 & -1 & 0 & 0 & -1 & 0 & 0 & 0 & 1 & 1 \\
-1 & 1 & 0 & 0 & -1 & 1 & 1 & 0 & 0 & 0 & 0 \\
0 & 0 & 0 & -1 & 0 & 0 & -1 & 0 & -1 & -1 & -1 \\
0 & 1 & 0 & -1 & 0 & 1 & 0 & 1 & -1 & 0 & 0 \\
0 & 0 & 0 & 0 & 0 & -1 & -1 & 1 & -1 & 0 & 1 \\
-1 & 0 & -1 & 1 & 0 & 0 & 0 & -1 & 0 & 1 & 0 \\
0 & -1 & 0 & -1 & 1 & 0 & 0 & 0 & -1 & 0 & 1 \\
0 & -1 & 0 & -1 & 0 & 0 & -1 & 1 & 0 & 1 & 0 \\
-1 & 0 & 1 & 1 & 0 & 0 & 1 & 0 & 0 & 0 & -1 \\
0 & 0 & 1 & 1 & -1 & 1 & 0 & -1 & 0 & 0 & 0 \\
-1 & 0 & 0 & 1 & 0 & 1 & 0 & 0 & 1 & -1 & 0
\end{array}\right).$$

The first row corresponds to a codeword of weight 5:

$$v = (0, -1, -1, 0, 0, -1, 0, 0, 0, 1, 1).$$

After 57 executions of the loop, algorithm 7.7.8 with the improvements described below determines that there exists no nonzero vector with weight ≤ 4. That means the minimum distance of the ternary Golay code is equal to $d = 5$.

If we use a systematic generator matrix of C of the form $\Gamma = (I_k \mid A)$, $A \in \mathbb{F}_q^{k \times n-k}$, we can do even better. We use the lattice L_C which is generated by

$$B_C = \left(\begin{array}{c|c} A^\top & qI_{n-k} \\ \hline I_k & 0 \end{array} \right),$$

It has the advantage that the constant N is not longer needed. In order to find a nonzero codeword of minimal weight, we have to find a nonzero lattice vector v in the rank n lattice $L_C \subseteq \mathbb{Z}^n$ with $\|v\|_\infty = 1$ which contains the minimal number of nonzero entries. Note that if $\text{wt}(v) = s$ and $\|v\|_\infty = 1$, then also $\|v\|_2^2 = s$.

The minimum distance of C can be computed by a variation of 7.7.8. Initially, in 7.7.8 we set $R = d - 1$, where d is an upper bound for the minimum distance of C. If no other bound is known, d is the weight of the shortest codeword in the generator matrix.

Then, the backtracking of the lattice point enumeration algorithm as described in 7.7.8 is started. If a lattice vector $v \in L_C$ with $\|v\|_\infty = 1$ and $\|v\|_2^2 \leq R$ is found during the enumeration then it is printed, after line (24) R is set to $R := \|v\|_2 - 1$, and the backtracking is continued. If it is known that the minimum distance of C is a multiple of some integer c – for example if C is a doubly even code – then we can even set $R := \|v\|_2 - c$ in this situation.

Further improvements in the enumeration can be obtained by modifying the lattice point enumeration of Section 7.7. For an integer $0 < t < n$ and a vector $v \in \mathbb{R}^n$, we define

$$\max_t(v)$$

to be the sum of the t largest absolute values of entries of v. For example, if $v = (-1, 2.5, -3, 0.5)^\top$, then $\max_2(v) = 3 + 2.5 = 5.5$.

Let $R = d - 1$, where $d > 1$ is an upper bound on the minimum distance of the code C and let $b^{(0)}, b^{(1)}, \ldots, b^{(n-1)}$ be a basis of the lattice L_C. With the notation of Section 7.7, 7.7.5 can be adapted to the computation of the minimum distance of a linear code in the following way.

7.8.2   **Theorem** Let $t \in n$. If for fixed $u_t, u_{t+1}, \ldots, u_{n-1} \in \mathbb{Z}$ there exist coefficients u_0, $u_1, \ldots, u_{t-1} \in \mathbb{Z}$ with $\|\sum_{i \in n} u_i b^{(i)}\|_\infty \leq 1$ and $\|\sum_{i \in n} u_i b^{(i)}\|_2^2 \leq R$, then for all y_t, $y_{t+1}, \ldots, y_{n-1} \in \mathbb{R}$:

7.8.3

$$\left| \sum_{i=t}^{n-1} y_i \|w^{(i)}\|_2^2 \right| \leq \max_R \left(\sum_{i=t}^{n-1} y_i w^{(i)} \right).$$

Proof: We have $\langle w^{(l)}, w^{(i)} \rangle = \langle w^{(i)}, w^{(i)} \rangle$ for $0 \le l < i < n$. If there exist $u_0, u_1, \ldots, u_{n-1} \in \mathbb{Z}$ such that for $w^{(0)} = \sum_{i \in n} u_i b^{(i)}$ simultaneously

$$\|w^{(0)}\|_\infty = 1 \text{ and } \|w^{(0)}\|_2^2 \le R,$$

then it is easy to see that for an arbitrary vector $v \in \mathbb{R}^n$ the inequality

$$|\langle w^{(0)}, v \rangle| \le \max_R(v)$$

holds. It follows that

$$\left| \sum_{i=t}^{n-1} y_i \langle w^{(i)}, w^{(i)} \rangle \right| = \left| \sum_{i=t}^{n-1} y_i \langle w^{(0)}, w^{(i)} \rangle \right|$$

$$= \left| \langle w^{(0)}, \sum_{i=t}^{n-1} y_i w^{(i)} \rangle \right|$$

$$\le \max_R \left(\sum_{i=t}^{n-1} y_i w^{(i)} \right). \qquad \square$$

Therefore, during the computation of the minimum distance of linear codes we can replace in the enumeration algorithm 7.7.8 the test in 7.7.5 by 7.8.3. Experiments show that 7.7.8 together with 7.8.3 can determine the minimum distance of quadratic-residue-codes for values of n at least up to 100.

Example For the quadratic-residue-code $C_Q(37, 19)$ over $\mathbb{F}_3$, whose generator   **7.8.4**
matrix is generated cyclically by the vector

$$(1,1,2,2,1,2,2,0,2,2,2,0,2,2,1,2,2,1,1,0,0,0,0,0,0,0,0,0,0,0,0,0,0,0,0,0,0),$$

the LLL-algorithm determines the vector

$$(0,0,0,1,0,0,0,1,0,1,1,0,1,0,0,0,0,0,1,2,0,0,1,1,0,0,0,0,0,0,1,0,0,0,0,0,0)$$

of weight 10. The enumeration 7.7.8 together with 7.8.3 needs 586 799 iterations to show that there is no codeword of lower weight. The parity extension of $C_Q(37, 19)$ has minimum weight 11. A vector with minimum weight is

$$(1,2,0,0,0,0,0,0,0,2,1,0,0,0,0,0,1,2,0,0,0,0,2,0,2,0,0,0,0,0,2,1,0,0,0,0,0,2). \qquad \diamond$$

Example The generator matrix of the quadratic-residue-code $C_Q(61, 31)$ over   **7.8.5**
$\mathbb{F}_3$ is generated by the vector

$$(1,0,2,1,2,2,0,0,0,1,0,2,1,1,2,1,2,1,1,2,0,1,0,0,0,2,2,1,2,0,$$
$$1,0,0,0,0,0,0,0,0,0,0,0,0,0,0,0,0,0,0,0,0,0,0,0,0,0,0,0,0,0,0).$$

The LLL-algorithm computes the vector

$$(0,0,0,0,0,0,0,0,0,0,0,0,0,1,0,0,0,0,1,0,0,0,0,0,0,0,0,2,1,0,$$
$$0,0,1,0,0,0,0,0,0,0,0,0,0,0,1,2,0,2,2,0,0,0,0,0,2,0,0,1,0,0,0)$$

which has weight 11. The exhaustive enumeration determines that there is no vector of lower weight. The parity extension of $C_Q(61,31)$ has minimum weight 12. A vector with minimum weight is

$$(2,0,1,0,0,0,0,0,0,0,0,0,0,0,0,0,0,0,2,0,1,0,0,2,0,0,0,0,0,0,$$
$$0,0,1,0,0,0,0,0,1,2,0,0,1,2,0,0,0,0,0,2,0,0,0,0,0,0,0,0,1,0,0,0).$$ ◇

7.8.6   **Example** The generator matrix of the quadratic-residue-code $C_Q(71,36)$ over $\mathbb{F}_3$ is generated by the vector

$$(2,2,2,2,2,2,2,0,0,2,2,1,1,2,2,2,0,2,2,2,0,1,1,0,2,0,2,0,1,0,0,0,0,0,0,1,$$
$$0,0,0,0,0,0,0,0,0,0,0,0,0,0,0,0,0,0,0,0,0,0,0,0,0,0,0,0,0,0,0,0,0,0,0,0).$$

The LLL-algorithm computes

$$(0,0,0,0,0,1,0,1,0,0,1,1,1,0,0,0,0,0,2,0,0,0,0,0,0,0,0,0,0,1,0,1,0,0,0,0,$$
$$2,2,0,0,0,0,0,0,0,0,0,2,0,0,0,0,2,1,0,0,0,0,2,0,2,0,2,0,0,2,0,0,0,0,0)$$

of weight 17. The parity extension of $C_Q(71,36)$ has 18 as upper bound for minimum distance. A vector attaining this bound is

$$(0,0,0,1,0,0,0,1,1,1,2,1,0,0,2,0,0,1,0,2,0,0,0,0,0,0,0,0,0,2,0,0,0,0,0,0,$$
$$1,0,0,0,2,0,0,0,0,1,0,0,1,0,0,0,1,0,0,2,0,0,1,0,0,0,0,0,1,0,0,0,0,0,0).$$ ◇

7.8.7   **Example** The generator matrix of the quadratic-residue-code $C_Q(83,42)$ over $\mathbb{F}_3$ is generated by

$$(2,0,1,2,1,2,0,0,1,2,2,0,0,2,1,2,1,1,0,0,1,1,1,1,0,0,2,2,1,1,2,1,0,2,0,1,0,0,1,2,1,$$
$$1,0,0,0,0,0,0,0,0,0,0,0,0,0,0,0,0,0,0,0,0,0,0,0,0,0,0,0,0,0,0,0,0,0,0,0,0,0,0,0,0).$$

The LLL-algorithm computes

$$(0,0,1,0,0,0,0,0,0,1,0,0,0,1,0,0,0,0,2,0,0,2,0,1,2,1,0,1,0,0,0,0,0,0,0,0,0,0,0,0,1,$$
$$0,2,0,0,0,0,0,0,2,0,0,1,0,0,0,2,0,0,0,0,0,0,0,0,0,0,0,2,0,2,0,1,0,2,0,0,1,1,0,0,0,0).$$

Therefore, an upper bound for the minimum distance is 20. It follows that the parity extension of $C_Q(83,42)$ has 21 as upper bound for the minimum distance.

The generator matrix of the quadratic-residue-code $C_Q(97,49)$ over $\mathbb{F}_3$ is generated by the vector

$$(1,1,1,0,0,1,1,0,2,1,0,0,0,1,0,2,1,2,0,0,2,2,2,0,1,0,2,2,2,0,0,2,1,2,0,1,0,0,0,1,2,0,1,1,0,0,1,1,$$
$$1,0,0,0,0,0,0,0,0,0,0,0,0,0,0,0,0,0,0,0,0,0,0,0,0,0,0,0,0,0,0,0,0,0,0,0,0,0,0,0,0,0,0,0,0,0,0,0,0).$$

The LLL-algorithm finds

$$(0,0,0,0,0,0,1,0,2,0,0,0,0,0,0,0,0,0,0,2,0,0,0,1,0,0,0,0,0,0,0,0,0,1,0,0,0,0,0,0,0,0,2,2,0,2,0,1,0,0,0,$$
$$0,0,2,0,2,1,1,0,0,0,0,1,0,0,0,2,0,1,1,1,0,0,2,0,0,0,0,0,0,0,0,1,0,0,0,0,0,0,0,0,0,0,2,0,0,2,0,1,0,0,0).$$

Therefore, an upper bound for the minimum distance is 23.   ◇

If the above examples are reproduced with the software from the enclosed CD-ROM the advantages and disadvantages of this algorithm can be seen. The LLL-algorithm is very good in computing codewords of small weight very fast. The second phase, which deterministically computes a codeword with minimum weight and proves that there are no codewords of smaller weight still needs exponential time.

Exercises

Exercise Use the computer software on the CD-ROM of the book to compute the minimum distance of the binary Golay code with generator matrix   **E.7.8.1**

$$G = \begin{pmatrix} 1&0&0&0&0&0&0&0&0&0&0&0&1&0&1&0&1&1&1&0&0&0&1&1 \\ 0&1&0&0&0&0&0&0&0&0&0&0&1&1&1&1&1&0&0&1&0&0&1&0 \\ 0&0&1&0&0&0&0&0&0&0&0&0&1&1&0&1&0&0&1&0&1&0&1&1 \\ 0&0&0&1&0&0&0&0&0&0&0&0&1&1&0&0&0&1&1&1&0&1&1&0 \\ 0&0&0&0&1&0&0&0&0&0&0&0&1&1&0&0&1&1&0&1&1&0&0&1 \\ 0&0&0&0&0&1&0&0&0&0&0&0&1&1&0&0&1&1&0&1&1&0&1 \\ 0&0&0&0&0&0&1&0&0&0&0&0&0&1&1&0&0&1&1&0&1&1&1 \\ 0&0&0&0&0&0&0&1&0&0&0&0&1&0&1&1&0&1&1&1&1&0&0&0 \\ 0&0&0&0&0&0&0&0&1&0&0&0&0&1&0&1&1&0&1&1&1&1&0&0 \\ 0&0&0&0&0&0&0&0&0&1&0&0&0&0&1&0&1&1&0&1&1&1&1&0 \\ 0&0&0&0&0&0&0&0&0&0&1&0&1&0&1&1&1&0&0&0&1&1&0&1 \\ 0&0&0&0&0&0&0&0&0&0&0&1&0&1&0&1&1&1&0&0&0&1&1&1 \end{pmatrix}.$$

Compute the minimum distance of this code over $\mathbb{F}_3$.

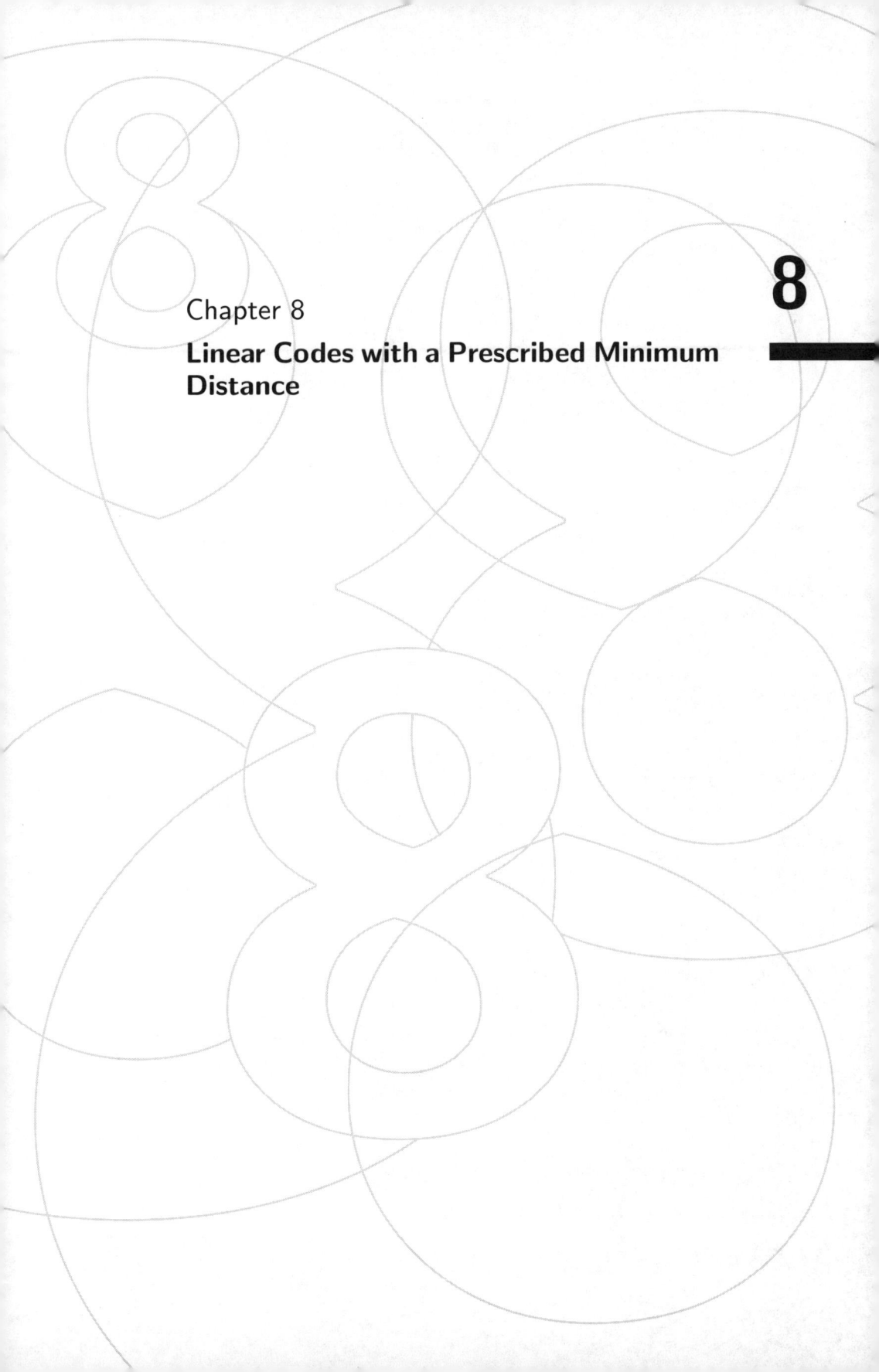

Chapter 8

Linear Codes with a Prescribed Minimum Distance

8

8

8 Linear Codes with a Prescribed Minimum Distance

After the *enumeration* of the isometry classes of codes in Chapter 6, we are now approaching the systematic *construction* of representatives of these classes. In this and the following chapter, we will present methods for constructing linear (n, k)-codes with *prescribed minimum distance d*. This means that for a given lower bound d on the minimum distance, we construct all $(n, k, \geq d)$-codes, i.e. codes whose minimum distance is at least as good as the lower bound we have chosen. We present essentially two different methods for solving this problem. Of course, both methods may fail to construct such codes, for instance if the lower bound d on the minimum distance was chosen too large. Nevertheless, in this case both methods provide *proof* that no code with the parameters under consideration exists. Needless to say that this construction problem is a very important and interesting one. In essence, all of coding theory is concerned with finding codes which allow one to transmit more data with fewer errors.

The above-mentioned construction of codes often leads to new and interesting codes either directly or indirectly by means of constructions and modifications in the sense of Section 2.2. In fact, A. Brouwer's helpful tables for the parameters of best known linear codes can sometimes be improved by such a search.

The construction that we have in mind in this chapter applies first of all to projective codes (cf. Exercise 1.3.21 and 6.1.14), i.e. codes whose columns can be taken as representatives of a *set* of points in projective space. The construction problem is then reduced to the problem of finding an equivalent structure in projective space called a *minihyper*. This is essentially a system of points in a suitable projective space with certain intersection properties with respect to hyperplanes. The necessary calculations amount to solving a system of Diophantine equations, similar to the techniques used for the construction of combinatorial designs.

Since for interesting parameter sets the coefficient matrix of the system often is too big to allow a direct solution, a well-known reduction is applied. Namely, we make an assumption about the presence of non-trivial automorphisms. This reduces the size of the coefficient matrix and thereby eases the problem to become more tractable. Of course, such a reduction is risky as it does not allow one to find solutions which do not satisfy the assumption on the presence of automorphisms. In this situation, the algorithm classifies codes with a given minimum distance which are invariant under the chosen group of automorphisms. In many cases this reduction indeed led to the discovery of new optimal codes (i.e. one could say that the end justifies the means in this case). Lastly, we also search for arbitrary nonredundant codes, i.e. codes which are not necessarily projective. The systematic construction of complete transversals of isometry classes of linear codes with a lower bound on their minimum distance is done in Chapter 9.

8.1 Minihypers

We begin with a closer examination of the use of generator matrices for encoding. For this purpose we introduce the notation $\gamma_{*,j}^{\top}$ for the j-th column of a generator matrix $\Gamma = (\gamma_{ij})$ and $\gamma_{i,*}$ for its i-th row. Using this notation, we can describe the generator matrix Γ of an (n, k)-code as

$$\Gamma = \left(\gamma_{*,0}^{\top} \mid \cdots \mid \gamma_{*,n-1}^{\top}\right) = \left(\begin{array}{c} \gamma_{0,*} \\ \hline \vdots \\ \hline \gamma_{k-1,*} \end{array}\right).$$

In terms of the standard bilinear form $\langle v, w \rangle = \sum_i v_i w_i$, we express a codeword $c := v \cdot \Gamma$ corresponding to a message $v \in \mathbb{F}_q^k$ as follows:

$$c = v \cdot \Gamma = \left(\langle v, \gamma_{*,0} \rangle, \ldots, \langle v, \gamma_{*,n-1} \rangle\right).$$

By definition, $n - \mathrm{wt}(v \cdot \Gamma)$ components of $v \cdot \Gamma$ are zero. In terms of the bilinear form, this means that $n - \mathrm{wt}(v \cdot \Gamma)$ columns of the generator matrix Γ are orthogonal to v, i.e., contained in the *hyperplane*

$$H(v) := P(v)^{\perp} = \left\{w \in \mathbb{F}_q^k \mid \langle v, w \rangle = 0\right\} \in \mathcal{U}(k, k-1, q).$$

This fact leads us to the following basic result:

Theorem *A $k \times n$-matrix over $\mathbb{F}_q$ generates an (n, k, d, q)-code C if and only if the columns of any generator matrix Γ of C satisfy the following two properties. Every hyperplane $H \in \mathcal{U}(k, k-1, q)$ contains at most $n - d$ columns of Γ and there is at least one hyperplane $H \in \mathcal{U}(k, k-1, q)$ which contains exactly $n - d$ columns of Γ. This property is independent of the choice of the generator matrix Γ of C, in that this property either holds for all generator matrices of C or none of the generator matrices of C has this property.*

Proof: 1. By 1.2.8, the minimum distance of a linear code equals the minimum weight of a nonzero codeword $c := v \cdot \Gamma$ for $v \in \mathbb{F}_q^k$, $v \neq 0$. The above argument shows that every hyperplane $H = H(v) \in \mathcal{U}(k, k-1, q)$ contains at most $n - d$ columns of Γ with equality if and only if the codeword $c = v \cdot \Gamma$ is of minimum weight d.

2. Conversely, if $\Gamma = (\gamma_{ij})$ is a $k \times n$-matrix over $\mathbb{F}_q$ satisfying

$$\max \left\{ |\{j \mid \gamma_{*,j} \in H\}| \mid H \in \mathcal{U}(k, k-1, q) \right\} = n - d,$$

then it is clear from the first part of the proof that its rows generate an (n, k', d)-code C over $\mathbb{F}_q$ of dimension $k' \leq k$. In order to show that $k' = k$ we have to check that the rows of Γ are linearly independent. Assume that Γ *does not* have full rank k. This means that the rows are linearly dependent, say

$$0 = c = v \cdot \Gamma,$$

for some $v \neq 0$. Since each $c_j = 0$, every column $\gamma_{*,j}$ is contained in the hyperplane $H(v)$, i.e. $|\{j \mid \gamma_{*,j} \in H(v)\}| = n$, contradicting the fact that

$$|\{j \mid \gamma_{*,j} \in H(v)\}| \leq n - d < n.$$

Thus Γ really generates an (n, k, d, q)-code. $\square$

We now recall from the metric classification of linear codes that permuting columns and/or multiplying columns of a generator matrix Γ with a nonzero element of $\mathbb{F}_q$ yields a generator matrix of a code which is linearly isometric. In fact, it is often simpler to deal not with the generator matrix Γ of a code but instead consider a certain map (or multiset), as described in the next remark. This applied to nonredundant codes only:

Remarks Let Γ denote a generator matrix of a nonredundant linear code C **8.1.2**
(which means that it does not contain a zero column). Then

— Γ can be identified with the mapping

$$\Gamma : n \to \mathbb{F}_q^k \backslash \{0\} \; : \; j \mapsto \gamma_{*,j}.$$

— Up to linear isometry, we may consider instead of column vectors the one-dimensional subspaces generated by the column vectors. They are the elements or *points* of the *projective geometry*

$$PG_{k-1}(q) = \left\{ P(v) \mid v \in \mathbb{F}_q^k \backslash \{0\} \right\}.$$

This means in fact that we can replace Γ by the mapping

$$\widetilde{\Gamma} : n \to PG_{k-1}(q) \; : \; j \mapsto P(\gamma_{*,j}).$$

The reason is that we easily obtain from $\widetilde{\Gamma}$ a matrix Γ' that generates a linear code C' linearly isometric to C, by simply taking from each value $P(\gamma_{*,j})$ of $\widetilde{\Gamma}$ a nonzero element and using it as the j-th column of Γ'.

— Moreover, because of isometry, it is possible to replace

$$\widetilde{\Gamma} = (P(\gamma_{*,0}), \ldots, P(\gamma_{*,n-1}))$$

by its orbit

$$S_n(\widetilde{\Gamma})$$

which consists of all the reorderings of this sequence $\widetilde{\Gamma}$. I.e. instead of the *sequence* of the points we consider the *multiset* of them. (In order to indicate a multiset we use the notation $\{\{\ldots\}\}$. In such a multiset, elements can occur several times, e.g. in $\{\{a, a, b, c, c, c\}\}$, a multiset of order 6, the element a occurs twice and c occurs three times.) This means that we replace Γ even by the multiset

$$\widetilde{\widetilde{\Gamma}} := \{\{P(\gamma_{*,0}), \ldots, P(\gamma_{*,n-1})\}\}$$

of cardinality n. It is clear that we can easily deduce from $\widetilde{\widetilde{\Gamma}}$ a matrix Γ'' that generates a code C'' linearly isometric to C.   ◇

For example the matrix

$$\Gamma = \begin{pmatrix} 1 & 0 & 0 & 1 & 1 & 1 & 1 & 1 \\ 0 & 1 & 0 & 0 & 2 & 2 & 2 & 2 \\ 0 & 0 & 1 & 1 & 1 & 1 & 2 & 2 \end{pmatrix}$$

generates an $(8, 3)$-code over $\mathbb{F}_3$. The corresponding mapping is

$$\widetilde{\Gamma} = (P(100), P(010), P(001), P(101), P(121), P(121), P(122), P(122))$$

and the resulting multiset is

$$\widetilde{\widetilde{\Gamma}} = \{\{P(010), P(001), P(101), P(121), P(121), P(122), P(122), P(100)\}\}.$$

8.1.3   **Corollary** *Both the mapping $\widetilde{\Gamma}$ and the multiset $\widetilde{\widetilde{\Gamma}}$ characterize the isometry class of the code C generated by Γ. Moreover, it is obvious how to obtain from $\widetilde{\Gamma}$ as well as from $\widetilde{\widetilde{\Gamma}}$ a generator matrix that generates a linear code linearly isometric to C.*   □

We are now in a position to rephrase 8.1.1 in terms of multisets. For this purpose we introduce the following kind of *restriction* of the multiset $\widetilde{\widetilde{\Gamma}}$ to a hyperplane H:

$$\widetilde{\widetilde{\Gamma}} \downarrow H := \{\{P \in \widetilde{\widetilde{\Gamma}} \mid P \subseteq H\}\}.$$

For example the restriction of the multiset $\widetilde{\widetilde{\Gamma}}$ defined by the above $(8,3)$-code to the hyperplane $H = H(110) = \{x \in \mathbb{F}_3^3 \mid x_0 + x_1 = 0\}$ is

$$\widetilde{\widetilde{\Gamma}} \downarrow H(110) = \{\!\!\{P(121), P(121), P(122), P(122)\}\!\!\}.$$

Its cardinality $|\widetilde{\widetilde{\Gamma}} \downarrow H(110)|$ is four. Using this notation we formulate the following corollary due to [88]:

Corollary *There is a nonredundant (n, k, d, q)-code if and only if there is a multiset $\mathcal{X}$ of order n, consisting of points of $\mathrm{PG}_{k-1}(q)$ such that*

$$\max\left\{|\mathcal{X} \downarrow H| \mid H \in \mathcal{U}(k, k-1, q)\right\} = n - d.\qquad\square$$

8.1.4

In fact, according to [88], we obtain even the weight distribution in this case:

Theorem *If $\mathcal{X}$ is a multiset of points in $\mathrm{PG}_{k-1}(q)$ with*

$$\max\left\{|\mathcal{X} \downarrow H| \mid H \in \mathcal{U}(k, k-1, q)\right\} = n - d,$$

8.1.5

then each matrix Γ whose columns are generators of the points of $\mathcal{X}$ generates an (n, k, d, q)-code C with weight distribution $W_C(x, y) = \sum_{i=0}^{n} A_i x^i y^{n-i}$, where $A_0 = 1$ and

$$A_i = (q-1) \cdot \left|\left\{H \in \mathcal{U}(k, k-1, q) \mid |\mathcal{X} \downarrow H| = n - i\right\}\right|, \quad \text{for } i > 0.$$

Proof: For each codeword $v \cdot \Gamma$ we have $|\mathcal{X} \downarrow H(v)| = n - \mathrm{wt}(v \cdot \Gamma)$. Since the generator matrix Γ has full rank k, a codeword $v \cdot \Gamma$ has weight 0 if and only if $v = 0$, and so $A_0 = 1$. The coefficients $A_i, i > 0$, are

$$
\begin{aligned}
A_i &= \left|\{c \in C \backslash \{0\} \mid \mathrm{wt}(c) = i\}\right| \\
&= \left|\left\{v \in \mathbb{F}_q^k \backslash \{0\} \mid \mathrm{wt}(v \cdot \Gamma) = i\right\}\right| \\
&= \left|\left\{v \in \mathbb{F}_q^k \backslash \{0\} \mid |\mathcal{X} \downarrow H(v)| = n - i\right\}\right| \\
&= (q-1) \cdot \left|\{H \in \mathcal{U}(k, k-1, q) \mid |\mathcal{X} \downarrow H| = n - i\}\right|,
\end{aligned}
$$

as stated. $\qquad\square$

Example (simplex-code) The k-th order q-ary simplex-code defined in 2.1.5 is an example of a nonredundant code. It is generated by any matrix Γ whose columns represent all $\theta_{k-1}(q) := (q^k - 1)/(q - 1)$ points of $\mathrm{PG}_{k-1}(q)$ (cf. 3.7.2). Using hyperplane intersections, we can easily deduce its parameters: Recall that every hyperplane contains $\theta_{k-2}(q)$ points, each of which is represented by

8.1.6

exactly one column of the generator matrix Γ. Therefore, the parameter of this code are $n = \theta_{k-1}(q)$ and $n - d = \theta_{k-2}(q)$, i.e.

$$(n, k, d) = \left((q^k - 1)/(q - 1), k, q^{k-1} \right).$$

The weight distribution is

$$1 + (q^k - 1)x^{q^{k-1}}.$$

Moreover, since

$$\frac{q^k - 1}{q - 1} = q^{k-1} + q^{k-2} + \ldots + q + 1 = \sum_{i \in k} \frac{q^{k-1}}{q^i} = \sum_{i \in k} \left\lceil \frac{d}{q^i} \right\rceil,$$

this code meets the Griesmer-bound, it is an optimal linear code.   ◇

Codes that are generated by a matrix Γ with pairwise linearly independent columns, so that $\widetilde{\Gamma}$ is a set in the strict sense, are called *projective* (cf. 6.1.14). For instance simplex-codes are projective. In other words the columns of generator matrices of projective linear (n, k)-codes correspond to pairwise distinct points. In order to emphasize this we shift from the calligraphic $\mathcal{X}$, that we used for multisets, to the notation X. Moreover we note that the restriction of sets to hyperplanes is the intersection. Projective codes are clearly nonredundant. As an immediate consequence we obtain

---

8.1.7   **Corollary** *There exists a projective linear (n, k, d)-code over $\mathbb{F}_q$ if and only if there exists a subset X of order n in $\mathrm{PG}_{k-1}(q)$ such that*

$$\max \left\{ |X \cap H| \mid H \in \mathcal{U}(k, k-1, q) \right\} = n - d.   \qquad \square$$

The complement of such a *set* X of points is called a minihyper. Minihypers are well-known objects in geometry. Several articles (cf. [25], [26], [54], [75], [79], or [143]) deal with minihypers and also with the connection between minihypers and linear codes. Hamada [78] discovered the relationship between Griesmer optimal linear codes and minihypers which we introduce now. We want to describe them in detail and we also give an algorithm for the construction of these objects.

8.1.8   **Definition (minihyper)** A (b, t)-*minihyper* in $\mathrm{PG}_{k-1}(q)$ is a set B of b points of $\mathrm{PG}_{k-1}(q)$ such that every hyperplane contains at least t points of B and at least one hyperplane contains exactly t points of B. Formally, a set $B \subseteq \mathcal{U}(k, 1, q)$ is a (b, t)-minihyper in $\mathrm{PG}_{k-1}(q)$ if and only if

$$|B| = b \quad \text{and} \quad \min \left\{ |B \cap H| \mid H \in \mathcal{U}(k, k-1, q) \right\} = t.   \qquad ◇$$

Using the concept of minihypers we reformulate the connection between projective codes and projective geometries.

Corollary *There is a projective (n, k, d)-code over $\mathbb{F}_q$ if and only if there is a (b, t)-minihyper in $\mathrm{PG}_{k-1}(q)$ where*

$$(b, t) = (\theta_{k-1}(q) - n, \theta_{k-2}(q) - n + d).$$

8.1.9

Proof: Let X be a set of n points with

$$\max\left\{|X \cap H| \mid H \in \mathcal{U}(k, k-1, q)\right\} = n - d.$$

Since every hyperplane contains $\theta_{k-2}(q)$ points, the set-theoretic complement $B := \mathcal{U}(k, 1, q) \setminus X$ satisfies the equation

$$\min\left\{|B \cap H| \mid H \in \mathcal{U}(k, k-1, q)\right\} = \theta_{k-2}(q) - (n - d).$$

Being the complement of X in $\mathrm{PG}_{k-1}(q) = \mathcal{U}(k, 1, q)$, the set B has $\theta_{k-1}(q) - n$ elements. Thus B is a

$$(\theta_{k-1}(q) - n, \theta_{k-2}(q) - n + d)$$

minihyper in $\mathrm{PG}_{k-1}(q)$. Since all arguments can be reversed, the existence of such a minihyper gives rise to a projective (n, k, d)-code.   □

Lemma *If $d \leq q^{k-1}$ and C is an (n, k, d, q)-code which attains the Griesmer-bound $n = \sum_{i \in k}\lceil d/q^i \rceil$, then C is a projective code.*   □

8.1.10

This lemma, the proof of which is left as Exercise 8.1.3, together with 8.1.9 implies the following corollary which is due to Hamada:

Corollary *Let $d \leq q^{k-1}$ and assume that $n = \sum_{i \in k}\lceil d/q^i \rceil$ which is taken from the Griesmer-bound. Then there exists a nonredundant linear (n, k, d)-code over $\mathbb{F}_q$ if and only if there exists a $(\theta_{k-1}(q) - n, \theta_{k-2}(q) - n + d)$-minihyper in $\mathrm{PG}_{k-1}(q)$.*   □

8.1.11

Example (Fano-plane) A well-known example is provided by the projective geometry $\mathrm{PG}_2(2)$, which is also known as the *Fano-plane*. It consists of the seven points and seven hyperplanes shown in the following table:

8.1.12

$$
\begin{array}{ll}
P_0 = P(100) = \{000, 100\} & H_0 = \{000, 100, 010, 110\} = P_0 \cup P_1 \cup P_3 \\
P_1 = P(010) = \{000, 010\} & H_1 = \{000, 010, 001, 011\} = P_1 \cup P_2 \cup P_4 \\
P_2 = P(001) = \{000, 001\} & H_2 = \{000, 100, 001, 101\} = P_0 \cup P_2 \cup P_5 \\
P_3 = P(110) = \{000, 110\} & H_3 = \{000, 100, 011, 111\} = P_0 \cup P_4 \cup P_6 \\
P_4 = P(011) = \{000, 011\} & H_4 = \{000, 010, 101, 111\} = P_1 \cup P_5 \cup P_6 \\
P_5 = P(101) = \{000, 101\} & H_5 = \{000, 001, 110, 111\} = P_2 \cup P_3 \cup P_6 \\
P_6 = P(111) = \{000, 111\} & H_6 = \{000, 110, 011, 101\} = P_3 \cup P_4 \cup P_5
\end{array}
$$

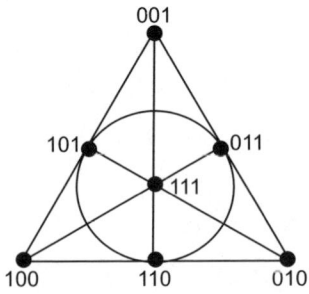

Fig. 8.1 The Fano-plane

The incidence relation between these points and hyperplanes is represented by the famous graph shown in Fig. 8.1. Each of the hyperplanes, which are the lines, together with the cycle, yields a $(3, 1)$-minihyper in $\mathrm{PG}_2(2)$. This property can easily be verified by looking at the figure. For example, take the line

$$B = \{P_3 = P(110), P_4 = P(011), P_5 = P(101)\}.$$

If we write the representatives of the four elements of the complement

$$X = \{P(100), P(010), P(001), P(111)\}$$

in a matrix column by column, we obtain the generator matrix

$$\Gamma = \begin{pmatrix} 1 & 0 & 0 & 1 \\ 0 & 1 & 0 & 1 \\ 0 & 0 & 1 & 1 \end{pmatrix}$$

of a binary $(4, 3, 2)$-code.   ◇

Now we are interested in a general approach to the construction of such a minihyper and, correspondingly, of codes with a prescribed minimum distance. For this purpose we introduce the following notion:

8.1.13   **Definition (blocking set)** A *t-blocking set* in $\mathrm{PG}_{k-1}(q)$ is a set B of points of $\mathrm{PG}_{k-1}(q)$ such that every hyperplane contains at least t points of B:

$$\min \{|B \cap H| \mid H \in \mathcal{U}(k, k-1, q)\} \geq t.$$ ◇

Hence, minihypers are t-blocking sets with additional properties. The largest possible size of an intersection of B and a hyperplane H is $\theta_{k-2}(q)$. Therefore B is a t-blocking set in $\mathrm{PG}_{k-1}(q)$ if

$$t \leq |B \cap H| \leq \theta_{k-2}(q)$$

for all hyperplanes $H \in \mathcal{U}(k, k-1, q)$. As t-blocking sets are suitable selections of points, they can be described by the *incidence matrix* $M_{k,q} = (m_{ij})$, the rows of which correspond to the hyperplanes $H_i \in \mathcal{U}(k, k-1, q)$, $i \in \theta_{k-1}(q)$, and the columns of which correspond to the points $P_j \in \mathcal{U}(k, 1, q)$, $j \in \theta_{k-1}(q)$. The entry m_{ij} of the i-th row and j-th column is defined as follows:

$$m_{ij} := \begin{cases} 1 & \text{if } P_j \subseteq H_i, \\ 0 & \text{otherwise.} \end{cases}$$

Hence a t-blocking set B is nothing but a selection of columns of the matrix $M_{k,q}$, or a 0-1-vector $x = (x_0, \dots, x_{\theta_{k-1}(q)-1})^\top$ which satisfies the condition

$$M_{k,q} \cdot x \in \{t, \dots, \theta_{k-2}(q)\}^{\theta_{k-1}(q)}.$$

This means that there is a vector $y = (y_0, \dots, y_{\theta_{k-1}(q)-1})^\top$ with components $y_i \in \{t, \dots, \theta_{k-2}(q)\}$ fulfilling the equation $M_{k,q} \cdot x = y$, which is equivalent to the equation

$$\left(M_{k,q} \mid -I \right) \cdot \begin{pmatrix} x \\ y \end{pmatrix} = 0,$$

where I is the identity matrix. Summarizing, we obtain the desired construction of blocking sets:

Corollary *There is a bijection between the set of all t-blocking sets in $\mathrm{PG}_{k-1}(q)$ and the set of vectors $\binom{x}{y}$ with $x_i \in \{0, 1\}$ and $y_i \in \{t, \dots, \theta_{k-2}(q)\}$ that solve the linear system of equations:*   **8.1.14**

$$\left(M_{k,q} \mid -I \right) \cdot \begin{pmatrix} x \\ y \end{pmatrix} = 0.$$

If $\binom{x}{y}$ denotes such a solution then the corresponding t-blocking set B in $\mathrm{PG}_{k-1}(q)$ is

$$B = \{ P_j \mid x_j = 1 \}. \qquad \qquad \square$$

Example (Fano-plane, cont.) We again consider the Fano-plane and construct   **8.1.15**
all 1-blocking sets in $\mathrm{PG}_2(2)$. There are seven hyperplanes and points as mentioned in Example 8.1.12. The corresponding incidence matrix is

$$M_{3,2} = \begin{pmatrix} 1 & 1 & 0 & 1 & 0 & 0 & 0 \\ 0 & 1 & 1 & 0 & 1 & 0 & 0 \\ 1 & 0 & 1 & 0 & 0 & 1 & 0 \\ 1 & 0 & 0 & 0 & 1 & 0 & 1 \\ 0 & 1 & 0 & 0 & 0 & 1 & 1 \\ 0 & 0 & 1 & 1 & 0 & 0 & 1 \\ 0 & 0 & 0 & 1 & 1 & 1 & 0 \end{pmatrix}.$$

Solving the corresponding linear system of equations from 8.1.14 we obtain 64 solutions $\binom{x}{y}$ with the required properties $x \in \{0,1\}^7$ and $y \in \{1,2,3\}^7$. Seven solutions correspond to the lines which are $(3,1)$-minihypers in $PG_2(2)$. The minihyper $B = \{P_3, P_4, P_5\}$ corresponds to the solution

$$\binom{x}{y} = (0,0,0,1,1,1,0;1,1,1,1,1,1,3)^\top.$$
◇

So far, we have constructed minihypers in the Fano-plane. In order to obtain new linear codes, we need to search for t-blocking sets or minihypers. The method proposed in 8.1.14 may not work because the incidence matrix $M_{k,q}$ can become too big for solving the system of Diophantine equations for interesting parameters. In these cases, the intention is to *reduce* the incidence matrix $M_{k,q}$ to a much smaller matrix so that it is possible to solve the corresponding system of Diophantine equations applying the lattice point enumeration algorithm described in the previous chapter. To achieve this goal we make an assumption about the presence of non-trivial automorphisms, similar to the methods that are used to construct combinatorial t-designs [18], [14], [15], [17]. In fact, such an assumption about the presence of non-trivial automorphisms leads to a very interesting area of Algebraic Combinatorics, the theory of groups acting on lattices. This is the topic of the following section.

Exercises

E.8.1.1   **Exercise** Show that the set of points in $PG_2(q)$ defined by the *conic*

$$\left\{ \langle (x,y,z) \rangle \mid x^2 = yz \right\}$$

corresponds to a q-ary $(q+1,3,q-1)$-code.

E.8.1.2   **Exercise** Verify that the set of points in $PG_3(q)$ defined by the *hyperbolic quadric*

$$\left\{ \langle (x,y,z,w) \rangle \mid zw = xy \right\}$$

corresponds to a q-ary $((q+1)^2,4,q^2)$-code.

E.8.1.3   **Exercise** Prove 8.1.10. Hint: Check that the generator matrix Γ of C does not contain zero columns. Assume that Γ has a repeated column. In this case, the matrix

$$\begin{pmatrix} 1 & 1 & \cdots \\ 0 & 0 & \\ \vdots & \vdots & \Gamma' \\ 0 & 0 & \end{pmatrix}$$

generates a code which is linearly isometric to C. Here, Γ' is a generator matrix of an $(n-2, k-1, d')$-code C' with $d' \geq d$. The inequality from the Griesmer-bound for C' then leads to a contradiction.

8.2 Group Actions on Lattices

In this section we investigate actions of subgroups of the general linear group $GL_k(q)$ on the set $\mathcal{U}(k, q) = PG(\mathbb{F}_q^k)$ of all subspaces of $\mathbb{F}_q^k$. This action is interesting because $\mathcal{U}(k, q)$ forms a lattice, the *linear lattice*, and since the action preserves the partial order, i.e. we have the implication

$$S \leq T \implies AS \leq AT$$

for all subspaces $S, T \in \mathcal{U}(k, q)$ and all $A \in GL_k(q)$. Hence let us introduce first the general concept of group actions on posets respectively lattices.

Definition (poset action) Let $(X, \leq)$ denote a poset on which a group G acts   8.2.1
from the left. Then we call the action $_G X$ a *poset action* if the implication

$$x \leq x' \implies gx \leq gx'$$

holds for all $x, x' \in X$ and $g \in G$. This will be abbreviated by

$$_G(X, \leq). \qquad \diamond$$

We note that we can in fact replace the implication by an equivalence since $gx \leq gx'$ also implies $x \leq x'$ if we apply g^{-1} from the left.

Analogously, we define a lattice action if the group elements commute with the infimum and supremum operator.

Definition (lattice action) Let $(X, \wedge, \vee)$ denote a lattice and let G be a group   8.2.2
acting on X. Then $_G X$ is called a *lattice action* if and only if

$$g(x \wedge x') = gx \wedge gx' \quad \text{and} \quad g(x \vee x') = gx \vee gx'$$

for all $x, x' \in X$ and $g \in G$. We indicate this situation as follows:

$$_G(X, \wedge, \vee). \qquad \diamond$$

Recall that a lattice $(X, \wedge, \vee)$ is always a poset, the corresponding order relation $\leq$ can be obtained by

$$x \leq x' : \iff x \wedge x' = x \iff x \vee x' = x'.$$

Using this equivalence we prove the following lemma.

8.2.3   **Lemma** *Let* $(X, \wedge, \vee)$ *be a lattice,* $(X, \leq)$ *the corresponding partial order and let G be a group acting on X. Then ${}_G X$ is a poset action if and only if ${}_G X$ is a lattice action.*

Proof: 1. Assume that ${}_G X$ is a poset action. We have $x \wedge x' \leq x$ and $x \wedge x' \leq x'$ for all $x, x' \in X$. Since G preserves the order relation we obtain $g(x \wedge x') \leq gx$ and $g(x \wedge x') \leq gx'$ for all $g \in G$ and hence $g(x \wedge x') \leq gx \wedge gx'$. If we assume that $g(x \wedge x') < gx \wedge gx'$ we obtain, after applying g^{-1} from the left, that

$$x \wedge x' = g^{-1}(g(x \wedge x')) < g^{-1}(gx \wedge gx') \leq g^{-1}(gx) \wedge g^{-1}(gx') = x \wedge x',$$

which yields the contradiction $x \wedge x' < x \wedge x'$. Thus we have $g(x \wedge x') = gx \wedge gx'$. The statement $g(x \vee x') = gx \vee gx'$ follows analogously.

2. Now we assume that ${}_G X$ is a lattice action. We have the following chain of equivalences:

$$x \leq x' \Leftrightarrow x = x \wedge x' \Leftrightarrow gx = g(x \wedge x') = gx \wedge gx' \Leftrightarrow gx \leq gx',$$

for all $x, x' \in X$ and $g \in G$. This completes the proof.   □

8.2.4   **Definition (poset automorphism)** Let $(X, \leq)$ denote a poset. Then a bijection $f : X \to X$ is called a *poset automorphism* if and only if

$$x \leq x' \implies f(x) \leq f(x')$$

for all elements $x, x' \in X$.   ◇

The set of all poset automorphisms of a poset $(X, \leq)$ forms a subgroup of the symmetric group S_X, the *automorphism group* of $(X, \leq)$, which will be abbreviated by $\mathrm{Aut}(X, \leq)$. A subgroup of this full automorphism group is called *a group of automorphisms* of $(X, \leq)$. Now recall the image $\overline{G} = \delta(G)$ of the permutation representation

$$\delta : G \to S_X : g \mapsto \overline{g} \text{ with } \overline{g} : x \mapsto gx$$

that obviously can be used to characterize a poset action: ${}_G X$ is a poset action if and only if

8.2.5 $$\overline{G} \leq \mathrm{Aut}(X, \leq).$$

For this reason we also say that G acts on a poset $(X, \leq)$ as a group of automorphisms in order to express that ${}_G X$ is a poset action. The most important properties of poset actions are the following ones:

Lemma *If $_G(X, \le)$ denotes a poset action with finite G, then it has the following*   **8.2.6**
properties:

1. *Any two elements in the same orbit are incomparable, i.e. the orbits are antichains.*

2. *If ω and ω' are orbits such that there exist $x \in \omega$ and $x' \in \omega'$ where $x < x'$, then we have, for any comparable pair of elements $y \in \omega$ and $y' \in \omega'$, that $y < y'$.*

3. *The partial order on X induces the following partial order on $G \backslash\backslash X$:*

$$\omega \le \omega' :\Longleftrightarrow \exists\, x \in \omega, x' \in \omega' : x \le x'.$$

4. *Consider an orbit $\omega \in G \backslash\backslash X$ and an arbitrary representative $x \in \omega$. For any orbit ω' the numbers*

$$\left|\{x' \in \omega' \mid x \le x'\}\right| \quad and \quad \left|\{x' \in \omega' \mid x \ge x'\}\right|$$

depend only on the orbit ω and not on the chosen representative $x \in \omega$.

5. *For any $x, x' \in X$, we have*

$$|G(x)| \cdot \left|\{z \in G(x') \mid x \le z\}\right| = |G(x')| \cdot \left|\{y \in G(x) \mid x' \ge y\}\right|.$$

Proof: 1. If $x \in X$ were comparable with $gx \ne x$, say (without restriction) $x < gx$, then we had $x < gx < g^2 x < \ldots < g^{-1}x < x$, which is a contradiction.

2. Suppose $x, y \in \omega$, $x', y' \in \omega'$, where $x < x'$ and y and y' are comparable. Then $y > y'$ would yield, for suitable $g, g' \in G$: $gx = y > y' = g'x'$, and hence also $x > g^{-1}g'x'$, which contradicts the first part that posets are antichains.

3. The reflexivity of $\le$ on $G \backslash\backslash X$ is obvious as well as the antisymmetry, and so it remains to prove the transitivity. Hence we assume that $\omega < \omega'$ and $\omega' < \omega''$, and consider elements $x \in \omega$, $x', y' \in \omega'$, $y'' \in \omega''$ which satisfy $x < x'$, $y' < y''$. There exists $g \in G$ with $gx' = y'$, and hence

$$\omega \ni gx < gx' = y' < y'' \in \omega'',$$

so that $\omega < \omega''$, as stated.

4. This follows from $x \le x' \iff gx \le gx'$.

5. Using 4., this follows from a trivial "double count" of the set

$$\{(y, z) \mid y \in G(x),\ z \in G(x'),\ y \le z\}.\qquad\qquad \square$$

As mentioned in Section 3.2 we represent a poset $(X, \leq)$ by its *zeta function* $\zeta : X \times X \rightarrow \{0, 1\}$ which is defined by

$$\zeta(x, x') := \begin{cases} 1 & \text{if } x \leq x', \\ 0 & \text{otherwise.} \end{cases}$$

If X is finite we can assume $X = \{x_0, \ldots, x_{m-1}\}$ to be topologically sorted, in the following sense:

8.2.7 **Definition (topological sorting)** A poset $(X, \leq)$ is *topologically sorted* if the elements of X are numbered in such a way that $x_i < x_j$ implies $i < j$ for all elements $x_i, x_j \in X$. ◇

It is not difficult to check (Exercise 8.2.1) that every finite poset $(X, \leq)$ can be sorted topologically. Therefore, in the following we always assume that the elements of the finite poset X in question have been numbered topologically as $\{x_0, \ldots, x_{m-1}\}$. In this case, the *zeta matrix*

$$Z(X, \leq) := (\zeta_{ij}), \text{ where } \zeta_{ij} := \zeta(x_i, x_j),$$

is upper triangular with ones along the main diagonal, and hence invertible over $\mathbb{Z}$. Its inverse

$$Z(X, \leq)^{-1} =: M(X, \leq) = (\mu_{ij}),$$

the Möbius matrix of the poset, defines the *Möbius function* of the finite poset: $\mu(x_i, x_j) := \mu_{ij} \in \mathbb{Z}$. In addition we remark that an action on a poset is a poset action if and only if

$$\zeta(x, x') = \zeta(gx, gx')$$

for all $g \in G$ and $x, x' \in X$. Here is our main example of a poset action:

8.2.8 **Example (the linear lattice)** As we have already mentioned at the beginning of this section, the set $\mathcal{U}(k, q)$ of subspaces of $\mathbb{F}_q^k$ forms a lattice with infimum $S \wedge T := S \cap T$ and supremum $S \vee T := \langle S \cup T \rangle$ (the subspace generated by the union of S and T). The general linear group $\mathrm{GL}_k(q)$ acts on this linear lattice in the following canonical way: For $M \in \mathrm{GL}_k(q)$ and $S \in \mathcal{U}(k, q)$ we have

$$MS := \left\{ v \cdot M^\top \mid v \in S \right\}.$$

This action is clearly a poset action, $\mathcal{U}(k, q)$ is partially ordered by inclusion, and it is obvious that the action respects inclusion:

$$S \leq T \Longrightarrow MS \leq MT.$$

Hence, by 8.2.3, this action is also a lattice action,

$$_{\mathrm{GL}_k(q)}\left(\mathcal{U}(k, q), \wedge, \vee\right),$$

and so the general linear group acts as a group of automorphisms on $\mathcal{U}(k, q)$. The zeta function of this linear lattice is

$$\zeta(S, T) = \begin{cases} 1 & \text{if } S \leq T, \\ 0 & \text{otherwise.} \end{cases}$$

Since this lattice is of great importance for the following, let us evaluate its Möbius function. To begin with, we claim that the sum of the values of the Möbius function over a full nontrivial interval is zero for each poset $(X, \leq)$, where all intervals are finite. Such posets are called *locally finite* (cf. 3.2.24 and Exercise 3.2.16).

$$\sum_{y: x \leq y \leq z} \mu(x, y) = \sum_{y: x \leq y \leq z} \mu(y, z) = \delta_{x,z} = \begin{cases} 0 & \text{if } x \neq z, \\ 1 & \text{if } x = z. \end{cases} \qquad \textbf{8.2.9}$$

In order to verify the first equation we use that the Möbius matrix is the inverse of the zeta matrix: $(\mu(x, y)) \cdot (\zeta(x, y)) = I$ gives

$$\sum_{y: x \leq y \leq z} \mu(x, y) = \sum_{y: x \leq y \leq z} \mu(x, y)\zeta(y, z) = (\mu * \zeta)(x, z) = \delta(x, z) = \delta_{x,z},$$

the second statement follows similarly. The next result is on the Möbius function of a finite lattice L with its elements

$$0 := \bigwedge_{\lambda \in L} \lambda \text{ and } 1 := \bigvee_{\lambda \in L} \lambda.$$

We state that

$$0 < \lambda \in L \Longrightarrow \sum_{\kappa: \kappa \vee \lambda = 1} \mu(0, \kappa) = 0. \qquad \textbf{8.2.10}$$

In order to prove this we consider the expression

$$\sigma(\lambda) := \sum_{\kappa, \nu} \mu(0, \kappa)\zeta(\kappa, \nu)\zeta(\lambda, \nu)\mu(\nu, 1) = \sum_{\kappa} \mu(0, \kappa) \sum_{\nu \geq \kappa \vee \lambda} \mu(\nu, 1).$$

Since, by 8.2.9, the inner sum $\sum_{\nu \geq \kappa \vee \lambda} \mu(\nu, 1)$ is zero, except for the case when $\kappa \vee \lambda = 1$, we find that

$$\sigma(\lambda) = \sum_{\kappa: \kappa \vee \lambda = 1} \mu(0, \kappa).$$

Hence it remains to prove that $\sigma(\lambda) = 0$. In order to do this we rewrite $\sigma(\lambda)$ in the following form:

$$\sigma(\lambda) = \sum_{\nu \geq \lambda} \mu(\nu, 1) \sum_{\kappa \leq \nu} \mu(0, \kappa).$$

The inner sum is zero, and hence $\sigma(\lambda) = 0$, which completes the proof.

We are now in a position to evaluate the Möbius function of the linear lattice. We claim that

$$\mu(S, T) = \begin{cases} (-1)^m q^{\binom{m}{2}} & \text{if } S \leq T, \\ 0 & \text{otherwise,} \end{cases} \qquad \textbf{8.2.11}$$

where $m := \dim(T) - \dim(S)$ and $\binom{0}{2} = \binom{1}{2} = 0$.

For its proof (by induction on m) we note first that if $S = T$, then $\mu(S, T) = 1$ and $m = 0$. If $S < T$, then $\mu(S, T) = \mu(0, \mathbb{F}_q^m)$, $m > 0$, since the lattice of subspaces between S and T is order isomorphic to the lattice of subspaces of $\mathbb{F}_q^m$. This is known from Linear Algebra (the Homomorphism Theorem). Hence it suffices to prove that

8.2.12
$$\mu(0, \mathbb{F}_q^m) = (-1)^m q^{\binom{m}{2}}, \qquad m > 0.$$

In order to check this we pick a one-dimensional subspace U and deduce from 8.2.10 that

$$\mu(0, \mathbb{F}_q^m) = - \sum_{S \vee U = \mathbb{F}_q^m, \, S \neq \mathbb{F}_q^m} \mu(0, S),$$

where the sum is taken over all proper subspaces S such that $S \vee U = \mathbb{F}_q^m$, i.e. over all the $(m-1)$-dimensional subspaces S of $\mathbb{F}_q^m$ that do not contain U. For all these S we have, by induction assumption, that

$$\mu(0, S) = (-1)^{m-1} q^{\binom{m-1}{2}}.$$

Moreover, the number of such subspaces is (Exercise 8.2.2)

8.2.13
$$\begin{bmatrix} m \\ m-1 \end{bmatrix}(q) - \begin{bmatrix} m-1 \\ m-2 \end{bmatrix}(q) = \begin{bmatrix} m \\ m-1 \end{bmatrix}(q) - \begin{bmatrix} m-1 \\ 1 \end{bmatrix}(q) = q^{m-1}.$$

Thus, we finally obtain that

$$\mu(0, \mathbb{F}_q^m) = (-1)^m q^{\binom{m}{2}},$$

which completes the proof of 8.2.12 on the values of the Möbius function of the linear lattice.   ◇

Our next step is a helpful *reduction process* that can be applied both to the zeta matrix and to the Möbius matrix of a poset or lattice, provided that we are given a poset or a lattice action. In this case, 8.2.6 implies the following

8.2.14   Corollary *Let $_GX$ be a poset action and let $\omega_0, \ldots, \omega_{l-1}$ be the orbits of G on the poset X. Then the values*

$$\sum_{x \in \omega_j} \zeta(x_i, x) \quad \text{and} \quad \sum_{x \in \omega_j} \zeta(x, x_i), \qquad i, j \in l,$$

are independent of the chosen representative $x_i \in \omega_i$.   □

This result enables us to introduce the *Plesken matrices* [162]

$$A^\wedge := A^\wedge(G) = (a_{ij}^\wedge), \text{ and } A^\vee := A^\vee(G) = (a_{ij}^\vee),$$

defined by

$$a_{ij}^{\wedge} := \sum_{x \in \omega_j} \zeta(x_i, x) = \left|\{x \in \omega_j \mid x_i = x_i \wedge x\}\right| = \left|\{x \in \omega_j \mid x_i \leq x\}\right|$$

and

$$a_{ij}^{\vee} := \sum_{x \in \omega_j} \zeta(x, x_i) = \left|\{x \in \omega_j \mid x_i = x_i \vee x\}\right|\right| = \left|\{x \in \omega_j \mid x \leq x_i\}\right|.$$

We note that these numbers are well-defined because of the 4th item of 8.2.6. In this language, the 5th item of 8.2.6 can be restated as

$$|\omega_i| \cdot a_{ij}^{\wedge} = |\omega_j| \cdot a_{ji}^{\vee}. \qquad\qquad\text{8.2.15}$$

Using topological sorting of the orbits we obtain

Corollary *For the Plesken matrices $A^{\wedge}(G)$ and $A^{\vee}(G)$ corresponding to a poset* **8.2.16** *action of a finite group G on a poset X the following is true:*

1. *If $D(G) := \mathrm{diag}\,(|\omega_0|,\ldots,|\omega_{l-1}|)$ denotes the diagonal matrix containing the lengths of the orbits of G on X on its diagonal, then*

$$A^{\vee}(G) = (D(G) \cdot A^{\wedge}(G) \cdot D(G)^{-1})^{\top}.$$

2. *The diagonal entries of the matrices $A^{\wedge}(G)$ and $A^{\vee}(G)$ are all one.*

3. *The orbits ω_i can be numbered such that $A^{\wedge}(G)$ is an upper triangular and $A^{\vee}(G)$ a lower triangular matrix.*   □

Example (the linear lattice cont.) The orbits ω_i of the general linear group on **8.2.17** the lattice $\mathcal{U}(k,q)$ are the sets of subspaces of the same dimension i, $0 \leq i \leq k$. Hence

$$A^{\vee}(\mathrm{GL}_k(q)) = \left(\begin{bmatrix} i \\ j \end{bmatrix}(q)\right)_{i,j \in k+1}.$$

Using 8.2.15 we obtain

$$A^{\wedge}(\mathrm{GL}_k(q)) = \left(\begin{bmatrix} k-i \\ j-i \end{bmatrix}(q)\right)_{i,j \in k+1}.$$

Of course, things become more complicated if we consider subgroups G of the general linear group $\mathrm{GL}_k(q)$. The reason is that the orbits of the general linear group, which are the sets of subspaces of same dimension, may split into several orbits. Nevertheless we consider this more general situation, since we find that certain submatrices of $A^{\wedge}(G)$ respectively $A^{\vee}(G)$ are crucial for the construction of minihypers respectively linear codes with prescribed minimum distance.

The dimension of a subspace is invariant under multiplication by an invertible matrix $M \in \mathrm{GL}_k(q)$. Thus

$$G \backslash\backslash \mathcal{U}(k, q) = \bigcup_{s=0}^{k} G \backslash\backslash \mathcal{U}(k, s, q).$$

This simple fact causes a block structure of the matrices $A^{\wedge} := A^{\wedge}(G)$ and $A^{\vee} := A^{\vee}(G)$. Namely, if

$$G \backslash\backslash \mathcal{U}(k, s, q) = \left\{ \omega_0^{(s)}, \ldots, \omega_{l_s-1}^{(s)} \right\}$$

is the set of orbits of G on the s-subspaces, we obtain, for $S_i \in \omega_i^{(s)}$, the matrix

$$A_{s,t}^{\wedge}(G) = \left(a_{ij}^{(\wedge, s, t)} \right)_{i \in l_s, \, j \in l_t}$$

with entries

$$a_{ij}^{(\wedge, s, t)} := \left| \left\{ T \in \omega_j^{(t)} \mid S_i = S_i \wedge T \right\} \right| = \left| \left\{ T \in \omega_j^{(t)} \mid S_i \leq T \right\} \right|.$$

In the same vein, we get

$$A_{s,t}^{\vee}(G) = \left(a_{ij}^{(\vee, s, t)} \right)_{i \in l_s, \, j \in l_t},$$

where

$$a_{ij}^{(\vee, s, t)} := \left| \left\{ T \in \omega_j^{(t)} \mid S_i = S_i \vee T \right\} \right| = \left| \left\{ T \in \omega_j^{(t)} \mid T \leq S_i \right\} \right|.$$

Recall once again that the entries of these matrices only depend on the respective orbit $\omega_i^{(s)}$, not on the chosen representative S_i.

These two matrices are exactly the submatrices of $A^{\wedge}$ respectively $A^{\vee}$ the rows of which belong to the orbits of G on the s-subspaces and whose columns belong to the orbits of G on the t-subspaces. If

$$\omega_0^{(0)}, \omega_0^{(1)}, \ldots, \omega_{l_1-1}^{(1)}, \ldots, \omega_0^{(k-1)}, \ldots, \omega_{l_{k-1}-1}^{(k-1)}, \omega_0^{(k)}$$

denotes the ordering of all orbits of G on $\mathcal{U}(k, q)$ we obtain the following block decomposition:

$$A^{\wedge}(G) = \left(A_{s,t}^{\wedge}(G) \right)_{s,t \in k+1} \quad \text{and} \quad A^{\vee}(G) = \left(A_{s,t}^{\vee}(G) \right)_{s,t \in k+1},$$

where $A_{s,t}^{\wedge}(G)$ and $A_{s,t}^{\vee}(G)$ are $l_s \times l_t$-matrices.   ◇

From 8.2.16 we deduce the relation between the different block matrices:

Corollary If $D_s(G) = \text{diag}(|\omega_0^{(s)}|,\ldots,|\omega_{l_s-1}^{(s)}|)$, $s \in k+1$, then the following is   **8.2.18**
true:

$$A_{s,t}^{\vee}(G) = \left(D_t(G) \cdot A_{t,s}^{\wedge}(G) \cdot D_s^{-1}(G)\right)^{\top}.$$ □

Here are several special cases of these matrices: For $t \in k+1$ we have

$$A_{t,t}^{\wedge}(G) = \begin{pmatrix} 1 & \cdots & 0 \\ \vdots & \ddots & \vdots \\ 0 & \cdots & 1 \end{pmatrix}, \quad A_{t,k}^{\wedge}(G) = \begin{pmatrix} 1 \\ \vdots \\ 1 \end{pmatrix}.$$

For all $s,t \in k+1$ with $s > t$:

$$A_{s,t}^{\wedge}(G) = \begin{pmatrix} 0 & \cdots & 0 \\ \vdots & & \vdots \\ 0 & \cdots & 0 \end{pmatrix}.$$

For all $t \in k+1$:

$$A_{0,t}^{\wedge}(G) = \left(|\omega_0^{(t)}|,\ldots,|\omega_{l_t-1}^{(t)}|\right).$$

The proofs are very easy. We continue with two numerical examples:

Example For the parameters $k := 8$, $q := 2$ and the general linear group $G :=$ **8.2.19**
$GL_k(q)$ we obtain the following Plesken matrices:

$$A^{\wedge} = \begin{pmatrix}
1 & 255 & 10795 & 97155 & 200787 & 97155 & 10795 & 255 & 1 \\
 & 1 & 127 & 2667 & 11811 & 11811 & 2667 & 127 & 1 \\
 & & 1 & 63 & 651 & 1395 & 651 & 63 & 1 \\
 & & & 1 & 31 & 155 & 155 & 31 & 1 \\
 & & & & 1 & 15 & 35 & 15 & 1 \\
 & & & & & 1 & 7 & 7 & 1 \\
 & & & & & & 1 & 3 & 1 \\
 & & & & & & & 1 & 1 \\
 & & & & & & & & 1
\end{pmatrix}$$

and

$$A^{\vee} = \begin{pmatrix}
1 & & & & & & & & \\
1 & 1 & & & & & & & \\
1 & 3 & 1 & & & & & & \\
1 & 7 & 7 & 1 & & & & & \\
1 & 15 & 35 & 15 & 1 & & & & \\
1 & 31 & 155 & 155 & 31 & 1 & & & \\
1 & 63 & 651 & 1395 & 651 & 63 & 1 & & \\
1 & 127 & 2667 & 11811 & 11811 & 2667 & 127 & 1 & \\
1 & 255 & 10795 & 97155 & 200787 & 97155 & 10795 & 255 & 1
\end{pmatrix}.$$

◇

8.2.20   **Example** Let $k := 3$ and $q := 2$. We consider the action of the complete monomial group $M_3(2)$ which is in fact isomorphic to the symmetric group S_3, acting by permuting the 3 coordinates:

$$M_3(2) = \left\{ \begin{pmatrix} 1 & 0 & 0 \\ 0 & 1 & 0 \\ 0 & 0 & 1 \end{pmatrix}, \begin{pmatrix} 0 & 1 & 0 \\ 0 & 0 & 1 \\ 1 & 0 & 0 \end{pmatrix}, \begin{pmatrix} 0 & 0 & 1 \\ 1 & 0 & 0 \\ 0 & 1 & 0 \end{pmatrix}, \right.$$

$$\left. \begin{pmatrix} 1 & 0 & 0 \\ 0 & 0 & 1 \\ 0 & 1 & 0 \end{pmatrix}, \begin{pmatrix} 0 & 0 & 1 \\ 0 & 1 & 0 \\ 1 & 0 & 0 \end{pmatrix}, \begin{pmatrix} 0 & 1 & 0 \\ 1 & 0 & 0 \\ 0 & 0 & 1 \end{pmatrix} \right\}.$$

Figrue 8.2 shows the Hasse diagram of this lattice. The vector space $\mathbb{F}_2^3$ corresponds to the vertex on top level. Subspaces in the same orbit are connected by a horizontal edge. The orbits, shown in the table next to the diagram, are arranged from the left to the right in each level.   ◇

Let us now restrict attention to the matrix $A^{\vee}_{k-1,1}(G)$, which can be used to construct minihypers with a prescribed group $G \leq \mathrm{GL}_k(q)$ of automorphisms, as we will see in the following section. Before that, let us mention an efficient way of computing this matrix.

8.2.21   **Lemma** For a matrix $M \in \mathrm{GL}_k(q)$ and a subspace S of $\mathbb{F}_q^k$ the following equation holds:

$$(MS)^{\perp} = (M^{\top})^{-1}S^{\perp}.$$

Proof: Since $(M^{\top})^{-1} = (M^{-1})^{\top}$, we have

$$\begin{aligned}
(MS)^{\perp} &= \left\{ v \in \mathbb{F}_q^k \mid \forall\, w \in MS : \langle v, w \rangle = 0 \right\} \\
&= \left\{ v \in \mathbb{F}_q^k \mid \forall\, w \in S : \langle v, w \cdot M^{\top} \rangle = 0 \right\} \\
&= \left\{ v \in \mathbb{F}_q^k \mid \forall\, w \in S : \langle v \cdot M, w \rangle = 0 \right\} \\
&= \left\{ v \cdot M^{-1} \in \mathbb{F}_q^k \mid v \in \mathbb{F}_q^k : \forall\, w \in S : \langle v, w \rangle = 0 \right\} \\
&= \left\{ v \cdot M^{-1} \in \mathbb{F}_q^k \mid v \in S^{\perp} \right\} \\
&= (M^{\top})^{-1}S^{\perp}.
\end{aligned}$$

□

8.2.22   **Definition (dual group)** Let G be a subgroup of $\mathrm{GL}_k(q)$, then we define G^* to be the set of all transposed matrices of G

$$G^* := \{ M^{\top} \mid M \in G \}$$

which is called the *dual group* of G.   ◇

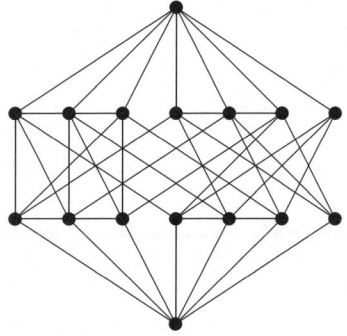

orbit	representative	orbit-length
ω_0	$\{(000)\}$	1
ω_1	$\langle(001)\rangle$	3
ω_2	$\langle(011)\rangle$	3
ω_3	$\langle(111)\rangle$	1
ω_4	$\langle(001),(010)\rangle$	3
ω_5	$\langle(111),(001)\rangle$	3
ω_6	$\langle(101),(011)\rangle$	1
ω_7	$\mathbb{F}_2^3$	1

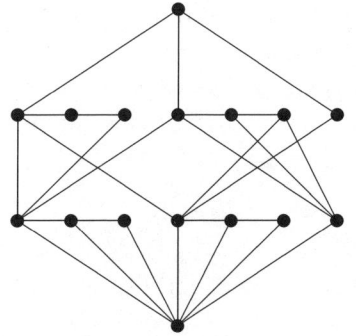

$$A^\wedge = \left(\begin{array}{c|ccc|ccc|c}
1 & 3 & 3 & 1 & 3 & 3 & 1 & 1 \\
 & 1 & 0 & 0 & 2 & 1 & 0 & 1 \\
 & & 1 & 0 & 1 & 1 & 1 & 1 \\
 & & & 1 & 0 & 3 & 0 & 1 \\\hline
 & & & & 1 & 0 & 0 & 1 \\
 & & & & & 1 & 0 & 1 \\
 & & & & & & 1 & 1 \\\hline
 & & & & & & & 1
\end{array}\right)$$

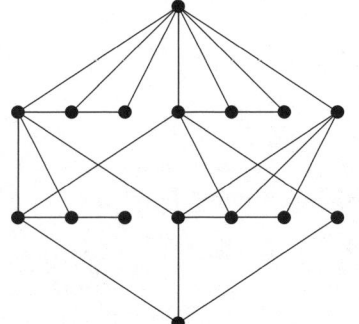

$$A^\vee = \left(\begin{array}{c|ccc|ccc|c}
1 & & & & & & & \\
1 & 1 & & & & & & \\
1 & 0 & 1 & & & & & \\
1 & 0 & 0 & 1 & & & & \\\hline
1 & 2 & 1 & 0 & 1 & & & \\
1 & 1 & 1 & 1 & 0 & 1 & & \\
1 & 0 & 3 & 0 & 0 & 0 & 1 & \\\hline
1 & 3 & 3 & 1 & 3 & 3 & 1 & 1
\end{array}\right)$$

Fig. 8.2 Lattice action of $M_3(2)$ on $\mathbb{F}_2^3$

The dual group G^* is isomorphic to G via the mapping

$$\iota : G \to G^* \ : \ M \mapsto (M^\top)^{-1}$$

since the equation

$$((M \cdot N)^\top)^{-1} = (M^\top)^{-1} \cdot (N^\top)^{-1}$$

holds for invertible matrices M and N.

8.2.23   **Corollary** *If $P(v)$ runs through a transversal of the orbits of G^* on the set of points $\mathcal{U}(k, 1, q)$, then $H(v)$ runs through a transversal of the orbits of G on the set of hyperplanes $\mathcal{U}(k, k-1, q)$. Furthermore, for the orbit of G on $H(v)$ we have*

$$G(H(v)) = \{ H(w) \mid P(w) \in G^*(P(v)) \} .$$ □

This corollary enables us to compute the orbits of a group $G \leq \mathrm{GL}_k(q)$ on the set of hyperplanes $\mathcal{U}(k, k-1, q)$. Instead of computing these orbits we construct orbit representatives of $G^* \backslash\backslash \mathcal{U}(k, 1, q)$ which is much easier since the representation of points needs one basis vector while hyperplanes are represented by $k - 1$ basis vectors.

Exercises

E.8.2.1   **Exercise** Prove that every finite poset can be sorted topologically.

E.8.2.2   **Exercise** Prove 8.2.13.

E.8.2.3   **Exercise** Prove that

$$\begin{bmatrix} r - s \\ r - t \end{bmatrix} (q) \cdot A^{\wedge}_{s,r} = A^{\wedge}_{s,t} \cdot A^{\wedge}_{t,r}.$$

for $s, t, r \in k + 1$ with $s \leq t \leq r$.

E.8.2.4   **Exercise** Show that we have, for the action of the monomial group,

$$A^{\wedge}_{t,k}(M_k(q)) = A^{\vee}_{n-t,n-k}(M_k(q)), \ \text{resp.} \ A^{\vee}_{t,k}(M_k(q)) = A^{\wedge}_{n-t,n-k}(M_k(q)).$$

8.3 Prescribing a Group of Automorphisms

As announced we are now going to use the prescription of a group of automorphisms for a construction of certain blocking sets respectively minihypers in projective geometries. Of course, such a prescription is risky since there may no exist a blocking set with this automorphism group. On the other hand, if there are such linear codes, then it will pay off since the number of columns of the incidence matrix $M_{k,q}$, which correspond to the points, will reduce to the number of orbits of the group on the set of 1-subspaces, and the same will happen to the rows which correspond to the hyperplanes. Quite often, this *data reduction* will bring the construction of linear codes within the reach of current computers.

Definition (automorphism of a blocking set) An element $M \in \mathrm{GL}_k(q)$ is called   8.3.1
an *automorphism* of a t-blocking set $B \subseteq \mathrm{PG}_{k-1}(q)$ if M permutes the points of B, i.e.

$$MB := \{MP \mid P \in B\} = B.$$

Recall from 8.2.8 that the action is $MP = MP(v) = P(v \cdot M^{\top})$.

A group consisting only of automorphisms of B is called *a* group of automorphisms of B, the maximal group with this property is called the *full* group of automorphisms and it is abbreviated by $\mathrm{Aut}(B)$. $\diamond$

The crucial facts for the construction of t-blocking sets with a prescribed group of automorphisms are the following ones.

Remarks Let G be a subgroup of $\mathrm{GL}_k(q)$.   8.3.2

— The group G is a group of automorphism of a t-blocking set B in $\mathrm{PG}_{k-1}(q)$ if and only if B is a union of G-orbits on $\mathcal{U}(k, 1, q)$.

— The incidence between points P and hyperplanes H is invariant under the action of G, i.e. if $P \subseteq H$ then $MP \subseteq MH$ for all $M \in G$.

— The number m of G-orbits on the set of points $\mathcal{U}(k, 1, q)$ is equal to the number of G-orbits on the set of hyperplanes $\mathcal{U}(k, k - 1, q)$.

— If $\{\omega_0, \ldots, \omega_{r-1}\}$ respectively $\{\Omega_0, \ldots, \Omega_{r-1}\}$ are the sets of G-orbits on $\mathcal{U}(k, 1, q)$ respectively $\mathcal{U}(k, k - 1, q)$ with representatives $P_i \in \omega_i$ respectively $H_i \in \Omega_i$, then the cardinality $|\omega_j \cap H_i| = |\{P \in \omega_j \mid P \subseteq H_i\}|$ is independent of the chosen representative H_i of the orbit Ω_i. $\diamond$

These facts embed the construction problem of blocking sets into the theory of group actions on lattices. They motivate the reduction of the incidence matrix $M_{k,q}$ between 1-subspaces and $(k - 1)$-subspaces to the incidence matrix

$M_{k,q}^G = (m_{ij}^G)$ between the *G-orbits* of 1-subspaces and the *G-orbits* of $(k-1)$-subspaces:

$$m_{ij}^G := |\omega_j \cap H_i|,$$

i.e. this matrix is a Plesken matrix:

8.3.3
$$M_{k,q}^G = A_{k-1,1}^\vee(G).$$

The following theorem describes the fundamental construction:

8.3.4   **Theorem**   *There is a bijection between the set of all t-blocking sets in* $\mathrm{PG}_{k-1}(q)$
with $G \le \mathrm{GL}_k(q)$ *as a group of automorphisms and the set of solutions* $\binom{x}{y}$, *with*
$x_i \in \{0,1\}$ *and* $y_i \in \{t,\ldots,\theta_{k-2}(q)\}$, *of the following system of linear equations:*

$$\left(M_{k,q}^G \mid -I\right) \cdot \binom{x}{y} = 0,$$

where $x = (x_0,\ldots,x_{r-1})^\top$, $y = (y_0,\ldots,y_{r-1})^\top$ *for* $r = |G\backslash\mathcal{U}(k,1,q)|$. *If* $\binom{x}{y}$
denotes a solution, then the corresponding t-blocking set B is

$$B = \bigcup_{j:x_j=1} \omega_j.$$

Proof: Let **B** be the set of all t-blocking sets in $\mathrm{PG}_{k-1}(q)$ having $G \le \mathrm{GL}_k(q)$
as a group of automorphisms and let **S** be the set of all solutions $\binom{x}{y}$ of the
linear system of equations $(M_{k,q}^G \mid -I) \cdot \binom{x}{y} = 0$ with $x_j \in \{0,1\}$ and $y_j \in \{t,\ldots,\theta_{k-2}(q)\}$. It is easy to see that the mappings

$$\varphi: \mathbf{S} \to \mathbf{B} : \binom{x}{y} \mapsto B \text{ with } B := \bigcup_{j:x_j=1} \omega_j$$

and

$$\psi: \mathbf{B} \to \mathbf{S} : B \mapsto \binom{x}{y} \text{ with } x_j := \begin{cases} 1 & \text{if } \omega_j \subseteq B, \\ 0 & \text{otherwise,} \end{cases} \text{ and } y_i := |B \cap H_i|,$$

are mutually inverse bijections.   □

If $\binom{x}{y}$ denotes an admissible solution of this linear system of equations and
$B = \bigcup_{j:x_j=1} \omega_j$ the corresponding t-blocking set in $\mathrm{PG}_{k-1}(q)$, then the cardinality of B is

$$b = \sum_{j:x_j=1} |\omega_j|.$$

If we add this equation as as further row to the linear system of equations we
obtain the corresponding construction of (b,t)-minihypers in $\mathrm{PG}_{k-1}(q)$ with a
prescribed group of automorphisms and hence projective codes.

Corollary *The set of all t-blocking sets in* $\text{PG}_{k-1}(q)$ *with cardinality b, having a subgroup* $G \leq \text{GL}_k(q)$ *as a group of automorphisms can be obtained from the set of vectors* $\binom{x}{y}$ *with* $x_i \in \{0,1\}$ *and* $y_i \in \{t,\ldots,\theta_{k-2}(q)\}$, $i \in r := |G\backslash\mathcal{U}(k,1,q)|$, *solving the linear system of equations:*

8.3.5

$$\left(\begin{array}{ccc|ccc} m_{0,0}^G & \cdots & m_{0,r-1}^G & -1 & & \\ \vdots & & \vdots & & \ddots & \\ m_{r-1,0}^G & \cdots & m_{r-1,r-1}^G & & & -1 \\ \hline |\omega_0| & \cdots & |\omega_{r-1}| & 0 & \cdots & 0 \end{array} \right) \cdot \left(\begin{array}{c} x_0 \\ \vdots \\ x_{r-1} \\ \hline y_0 \\ \vdots \\ y_{r-1} \end{array} \right) = \left(\begin{array}{c} 0 \\ \vdots \\ 0 \\ \hline b \end{array} \right)$$

If $\binom{x}{y}$ *denotes such a solution, then the corresponding t-blocking set B with* $|B| = b$ *is*

$$B = \bigcup_{j : x_j = 1} \omega_j.$$

This blocking set B is a (b,t)-*minihyper in* $\text{PG}_{k-1}(q)$ *if and only if the vector y contains a component which is exactly t, i.e. if and only if there is an index j with* $y_j = t$.

$\square$

Example We want to construct $(6,1)$-blocking sets in $\text{PG}_2(3)$, i.e. the parameters are $q = 3$, $k = 3$, $t = 1$, $b = 6$, $\theta_2(3) = (3^3 - 1)/(3 - 1) = 13$ and $\theta_1(3) = (3^2 - 1)/(3 - 1) = 4$. The projective geometry $\text{PG}_2(3)$ consists of the following 13 points

8.3.6

$$\begin{array}{lll} P_0 = P(001), & P_5 = P(101), & \\ P_1 = P(010), & P_6 = P(102), & P_{10} = P(120), \\ P_2 = P(011), & P_7 = P(110), & P_{11} = P(121), \\ P_3 = P(012), & P_8 = P(111), & P_{12} = P(122). \\ P_4 = P(100), & P_9 = P(112), & \end{array}$$

Hence the incidence matrix $M_{3,3}$ is of size 13×13. Now we prescribe the complete monomial group $G := M_3(3)$ which yields three orbits on the set of points $\mathcal{U}(3,1,3)$:

$$\begin{array}{l} \omega_0 = \{P_0, P_1, P_4\}, \\ \omega_1 = \{P_2, P_3, P_5, P_6, P_7, P_{10}\}, \\ \omega_2 = \{P_8, P_9, P_{11}, P_{12}\}. \end{array}$$

In addition we obtain the orbits on the set of hyperplanes $\mathcal{U}(3,2,3)$:

$$\begin{array}{l} \Omega_0 = \{P_0^\perp, P_1^\perp, P_4^\perp\}, \\ \Omega_1 = \{P_2^\perp, P_3^\perp, P_5^\perp, P_6^\perp, P_7^\perp, P_{10}^\perp\}, \\ \Omega_2 = \{P_8^\perp, P_9^\perp, P_{11}^\perp, P_{12}^\perp\} \end{array}$$

and the reduced matrix turns out to be of size 3×3:

$$M_{3,3}^G = A_{2,1}^{\vee}(M_3(3)) = \begin{pmatrix} 2 & 2 & 0 \\ 1 & 1 & 2 \\ 0 & 3 & 1 \end{pmatrix}.$$

This shows that we have obtained a data reduction by the *factor* $169/9$ which is nearly 20. The corresponding system of Diophantine equations is

$$\left(\begin{array}{ccc|ccc} 2 & 2 & 0 & -1 & 0 & 0 \\ 1 & 1 & 2 & 0 & -1 & 0 \\ 0 & 3 & 1 & 0 & 0 & -1 \\ \hline 3 & 6 & 4 & 0 & 0 & 0 \end{array} \right) \cdot \begin{pmatrix} x_0 \\ x_1 \\ x_2 \\ \hline y_0 \\ y_1 \\ y_2 \end{pmatrix} = \begin{pmatrix} 0 \\ 0 \\ 0 \\ 6 \end{pmatrix}$$

where $x_i \in \{0,1\}$ and $y_i \in \{1,2,3,4\}$. It is easy to see that $(0,1,0;2,1,3)^{\top}$ is the only solution of this system which corresponds to a $(6,1)$-blocking set

$$B = w_1 = \{P_2, P_3, P_5, P_6, P_7, P_{10}\}. \qquad \diamond$$

8.4 Linear Codes of Prescribed Type

We have seen that codes with minimum distance $d \leq q^{k-1}$ meeting the Griesmer-bound are always projective. If such a code is regarded as an n-set in $PG_{k-1}(q)$, then the complement of that n-set defines a minihyper and vice versa. The minihyper approach only works for projective codes. It does not work work general codes, since it is not clear how to define complements of multisets. In order to avoid such investigations we construct the n-multiset defining the linear code directly, using the same construction that we used for minihypers: We solve a linear system of Diophantine equations.

In Section 8.1, we have shown how to construct blocking sets with the aid of the incidence matrix $M_{k,q} = (m_{ij})$ by solving a system of Diophantine equations. The 0-1-vector x corresponds to a selection of points defining the blocking set B. The complement of the blocking set B then was a projective code. After changing some entries in the system of equations, this method allows us to construct the projective codes directly.

If $\mathcal{U}(k,1,q) = \{P_0, \ldots, P_{r-1}\}$ respectively $\mathcal{U}(k,k-1,q) = \{H_0, \ldots, H_{r-1}\}$, where $r := \theta_{k-1}(q)$ again denotes the number of points respectively hyperplanes, then the solutions $\binom{x}{y}$ with $x_j \in \{0,1\}$ and $y_j \in \{0,\ldots,n-d\}$ of the system

$$
\left(
\begin{array}{ccc|ccc}
m_{0,0} & \cdots & m_{0,r-1} & -1 & & \\
\vdots & & \vdots & & \ddots & \\
m_{r-1,0} & \cdots & m_{r-1,r-1} & & & -1 \\
\hline
1 & \cdots & 1 & 0 & \cdots & 0
\end{array}
\right)
\cdot
\begin{pmatrix}
x_0 \\
\vdots \\
x_{r-1} \\
y_0 \\
\vdots \\
y_{r-1}
\end{pmatrix}
=
\begin{pmatrix}
0 \\
\vdots \\
0 \\
n
\end{pmatrix}
$$

define the projective (n,k)-codes over $\mathbb{F}_q$ with minimum distance greater than or equal to d. The first part x of a solution $\binom{x}{y}$ defines a selection of points which determine the columns of a generator matrix: The point P_j is selected if and only if x_j is 1. Now if we permit values greater than 1 for the components x_j, then the vector x describes a multiset $\mathcal{X}$, containing the point P_j exactly x_j times.

Hence the solutions $\binom{x}{y}$ with $x_j \in \{0,\dots,n\}$ and $y_j \in \{0,\dots,n-d\}$ of the system of Diophantine equations correspond to the nonredundant (n,k)-codes over $\mathbb{F}_q$ with minimum distance greater than or equal to d.

Assume that $\mathcal{X}$ is such a multiset corresponding to a solution $\binom{x}{y}$ and consider $v_j \in \mathbb{F}_q^k \setminus \{0\}$ such that $H_j = H(v_j)$, then $y_j = |\mathcal{X} \downarrow H(v_j)|$. We obtain the weight distribution from 8.1.5:

$$
A_i = (q-1) \cdot \left| \{ j \in r \mid y_j = n - i \} \right|, \quad \text{for } i > 0.
$$

But as with the construction of blocking sets and projective codes the system of equations is still too big for an efficient computation of solutions. Therefore, again we reduce the dimension of the matrix of coefficients by prescribing a group of automorphisms. But first we have to make clear what a prescription of such a group means in the case of multisets and codes.

If $\Gamma = (\gamma_{ij})$ denotes a generator matrix of an (n,k)-code C and

$$
\mathcal{X}_\Gamma := \{\!\{ P(\gamma_{*,0}), \dots, P(\gamma_{*,n-1}) \}\!\}
$$

denotes the n-multiset of points in $\mathrm{PG}_{k-1}(q)$ defined by the columns of the generator matrix Γ, then the following holds true for each $M \in \mathrm{GL}_k(q)$:

$$
M\mathcal{X}_\Gamma := \{\!\{ MP(\gamma_{*,0}), \dots, MP(\gamma_{*,n-1}) \}\!\} = \mathcal{X}_{M\cdot\Gamma}.
$$

Definition (projective automorphism) A *projective automorphism* of a generator matrix Γ of a nonredundant (n,k)-code is an element $M \in \mathrm{GL}_k(q)$ which leaves the multiset $\mathcal{X}_\Gamma$ invariant:

$$
M\mathcal{X}_\Gamma = \mathcal{X}_\Gamma.
$$

8.4.1

A group consisting only of projective automorphisms of Γ is called a group of automorphisms of Γ. The largest group with this property is called the *full group of projective automorphisms* and it is abbreviated by $\mathrm{Aut}(\Gamma)$.   ◇

If Γ and Γ' denote two generator matrices of the same (n,k)-code C, then there is an element $N \in \mathrm{GL}_k(q)$ such that $\Gamma' = N \cdot \Gamma$. Now if M is a projective automorphism of Γ, i.e $M \mathcal{X}_\Gamma = \mathcal{X}_\Gamma$, then the conjugate element $N \cdot M \cdot N^{-1}$ defines a projective automorphism of Γ', since

$$N \cdot M \cdot N^{-1} \mathcal{X}_{N \cdot \Gamma} = N \cdot M \mathcal{X}_{N^{-1} \cdot N \cdot \Gamma} = N \cdot M \mathcal{X}_\Gamma = N \mathcal{X}_\Gamma = \mathcal{X}_{N \cdot \Gamma},$$

i.e. the conjugate group

$$NGN^{-1} := \left\{ N \cdot M \cdot N^{-1} \mid M \in G \right\}$$

is a group of projective automorphisms of $N \cdot \Gamma$. The set of matrices $N \cdot \Gamma$, where $N \in \mathrm{GL}_k(q)$, contains all the generator matrices of C, and thus all the conjugates NGN^{-1} of G are groups of projective automorphisms, so that we can introduce the following notion of type of a code:

8.4.2   **Definition (stabilizer type of a code)** Let G be a subgroup of $\mathrm{GL}_k(q)$. An (n,k)-code C over $\mathbb{F}_q$ has as *stabilizer type* the conjugacy class

$$\widetilde{G} := \left\{ NGN^{-1} \mid N \in \mathrm{GL}_k(q) \right\}$$

if there is a generator matrix Γ of C such that Γ has G as a group of projective automorphisms.   ◇

This concept allows us to formulate the following important consequence:

8.4.3   **Theorem** *Let G be a subgroup of $\mathrm{GL}_k(q)$ with orbits $\omega_0, \ldots, \omega_{r-1}$ on the set $\mathcal{U}(k,1,q)$ of points and orbits $\Omega_0, \ldots, \Omega_{r-1}$ on the set $\mathcal{U}(k, k-1, q)$ of hyperplanes. Consider representatives $H_i \in \Omega_i$ and put*

$$m_{ij}^G := \left| \{ P \in \omega_j \mid P \subseteq H_i \} \right|.$$

There is a bijection between the set of all linear (n,k)-codes over $\mathbb{F}_q$ with minimum distance at least d and type $\widetilde{G}$ and the set of vectors $\binom{x}{y}$ with $x_i \in \{0, \ldots, \lfloor n/|\omega_i| \rfloor\}$ and $y_i \in \{0, \ldots, n-d\}, i \in r := |G \backslash \backslash \mathcal{U}(k,1,q)|$, solving the linear system of equations:

$$
\left(
\begin{array}{ccc|ccc}
m_{0,0}^G & \cdots & m_{0,r-1}^G & -1 & & \\
\vdots & & \vdots & & \ddots & \\
m_{r-1,0}^G & \cdots & m_{r-1,r-1}^G & & & -1 \\
\hline
|\omega_0| & \cdots & |\omega_{r-1}| & 0 & \cdots & 0
\end{array}
\right)
\cdot
\left(
\begin{array}{c}
x_0 \\ \vdots \\ x_{r-1} \\ \hline y_0 \\ \vdots \\ y_{r-1}
\end{array}
\right)
=
\left(
\begin{array}{c}
0 \\ \vdots \\ 0 \\ \hline n
\end{array}
\right)
$$

If $\binom{x}{y}$ is a solution of this system, then the first part x defines an n-multiset $\mathcal{X}$ of points as follows

$$\mathcal{X} = \bigcup_{i:x_i>0} \bigcup_{j=1}^{x_i} \omega_j,$$

where $\cup$ means the union of multisets. Representatives of the points of $\mathcal{X}$, written column by column in a matrix, yield a generator matrix of an (n,k,d,q)-code C. Furthermore, the weight distribution $W_C(x,y) = y^n + \sum_{i=1}^n A_i x^i y^{n-i}$ is given by

$$A_i = (q-1) \sum_{j:y_j=n-i} |\Omega_j|. \qquad\qquad\qquad \square$$

Example Suppose we are now looking for a linear $(14,3,9)$-code over $\mathbb{F}_3$. Such a code is optimal. First note that a code with these parameters cannot be projective, since there are exactly 13 points in $\mathrm{PG}_2(3)$, i.e. at least one point has to occur twice in a generator matrix of such a code. The parameters $q = 3$ and $k = 3$ are the same as in example 8.3.6 and we also prescribe the group $M_3(3)$ as a group of automorphisms. The orbits on the points and hyperplanes are also shown in 8.3.6. The corresponding system of equations is

8.4.4

$$\begin{pmatrix} 2 & 2 & 0 & -1 & 0 & 0 \\ 1 & 1 & 2 & 0 & -1 & 0 \\ 0 & 3 & 1 & 0 & 0 & -1 \\ 3 & 6 & 4 & 0 & 0 & 0 \end{pmatrix} \cdot \begin{pmatrix} x_0 \\ x_1 \\ x_2 \\ \hline y_0 \\ y_1 \\ y_2 \end{pmatrix} = \begin{pmatrix} 0 \\ 0 \\ 0 \\ 14 \end{pmatrix},$$

where $x_0 \in \{0,1,2,3,4\}$, $x_1 \in \{0,1,2\}$, $x_2 \in \{0,1,2,3\}$ and $y_i \in \{0,1,2,3,4,5\}$, see also 7.7.11. A solution of this system is $(0,1,2;2,5,5)^\top$, which means that the orbit ω_1 occurs once in the corresponding multiset $\mathcal{X}$ and the orbit ω_2 occurs twice in $\mathcal{X}$. Thus we obtain the following multiset

$$\begin{aligned} \mathcal{X} &= \omega_1 \cup \omega_2 \cup \omega_2 \\ &= \{\!\{ P_2, P_3, P_5, P_6, P_7, P_{10}, P_8, P_9, P_{11}, P_{12}, P_8, P_9, P_{11}, P_{12} \}\!\} \end{aligned}$$

and finally the generator matrix

$$\Gamma = \begin{pmatrix} 0 & 0 & 1 & 1 & 1 & 1 & 1 & 1 & 1 & 1 & 1 & 1 & 1 & 1 \\ 1 & 1 & 0 & 0 & 1 & 2 & 1 & 1 & 2 & 2 & 1 & 1 & 2 & 2 \\ 1 & 2 & 1 & 2 & 0 & 0 & 1 & 2 & 1 & 2 & 1 & 2 & 1 & 2 \end{pmatrix}$$

of an optimal $(14,3,9)$-code. For the weight distribution we obtain:

$$W_C(x,y) = y^{14} + 20x^9 y^5 + 6x^{12}y^3,$$

since $A_9 = (3-1) \cdot (|\Omega_1| + |\Omega_2|) = 20$ and $A_{12} = (3-1) \cdot |\Omega_0| = 6.$ $\diamond$

8.5 Numerical Results

On the following pages we present new codes obtained with the proposed method. Applying the modifications to these codes described in the second chapter all in all we got more than 400 new codes (see [32]).

For each pair of values (q,k) we show a table with parameters n,d,G,r of 8.4.3. A row in such a table means that we have constructed a code over $\mathbb{F}_q$ with dimension k, with length n, minimum distance d and type $\widetilde{G}$. The number r is the number of orbits of the corresponding group G on the set of points $\mathcal{U}(k,1,q)$. A bold minimum distance d means, that the (n,k,d)-code is optimal.

For the finite field $\mathbb{F}_q = \mathbb{F}_{p^m}$ we use the additive representation, i.e. the elements of $\mathbb{F}_{p^m} = \mathbb{F}_p/I(f)$ which are cosets $k_0 + k_1 x + \ldots + k_{m-1} x^{m-1} + I(f)$ are coded as numbers $k_0 + k_1 p + \ldots + k_{m-1} p^{m-1}$. The corresponding irreducible polynomials can be found in Table 3.3.

Table 8.1 Linear codes for $q = 2$ and $k = 10$

n	d	G	r
177	84	$\left\langle \begin{array}{c} 1010100100 \\ 1100001000 \\ 1011000100 \\ 1011110100 \\ 1100000010 \\ 0010001111 \\ 1101010011 \\ 1011010110 \\ 0010110000 \\ 1010101100 \end{array} \right\rangle$	51

Table 8.2 Linear codes for $q = 3$ and $k = 6$

n	d	G	r
191	**126**	$\left\langle \begin{pmatrix} 100000 \\ 001000 \\ 000010 \\ 000001 \\ 000100 \\ 010000 \end{pmatrix}, \begin{pmatrix} 001000 \\ 000100 \\ 000001 \\ 100000 \\ 000010 \\ 010000 \end{pmatrix} \right\rangle$	20
202	132	$\left\langle \begin{pmatrix} 100000 \\ 001000 \\ 000010 \\ 000001 \\ 000100 \\ 010000 \end{pmatrix}, \begin{pmatrix} 001000 \\ 000100 \\ 000001 \\ 100000 \\ 000010 \\ 010000 \end{pmatrix} \right\rangle$	20
219	**144**	$\left\langle \begin{pmatrix} 000001 \\ 100002 \\ 010001 \\ 001002 \\ 000102 \\ 000012 \end{pmatrix} \right\rangle$	23

Table 8.3 Linear codes for $q = 3$ and $k = 7$

n	d	G	r
202	129	$\left\langle \begin{pmatrix} 2211010 \\ 1011220 \\ 1211220 \\ 0022000 \\ 0201010 \\ 0212120 \\ 0000001 \end{pmatrix} \right\rangle$	35
222	144	$\left\langle \begin{pmatrix} 0000010 \\ 1000010 \\ 0100000 \\ 0010010 \\ 0001000 \\ 0000120 \\ 0000001 \end{pmatrix} \right\rangle$	45

Table 8.4 Linear codes for $q = 3$ and $k = 8$

n	d	G	r
64	37	$\left\langle \begin{array}{l} 00000010 \\ 00010000 \\ 00000100 \\ 10000000 \\ 01000000 \\ 00001000 \\ 00100000 \\ 00000001 \end{array} , \begin{array}{l} 01000000 \\ 00001000 \\ 00000100 \\ 00000001 \\ 00010000 \\ 10000000 \\ 00000010 \\ 00100000 \end{array} \right\rangle$	72
224	141	$\left\langle \begin{array}{l} 00000100 \\ 10000000 \\ 01000200 \\ 00100200 \\ 00010100 \\ 00001000 \\ 00000010 \\ 00000001 \end{array} \right\rangle$	69
228	144	$\left\langle \begin{array}{l} 00000100 \\ 10000000 \\ 01000200 \\ 00100200 \\ 00010100 \\ 00001000 \\ 00000010 \\ 00000001 \end{array} \right\rangle$	69

Table 8.5 Linear codes for $q = 4$ and $k = 5$

n	d	G	r
56	**40**	$\left\langle \begin{pmatrix} 00001 \\ 00010 \\ 01000 \\ 10000 \\ 00100 \end{pmatrix} \right\rangle$	69
70	**50**	$\left\langle \begin{pmatrix} 01133 \\ 13012 \\ 03012 \\ 32022 \\ 13311 \end{pmatrix} \right\rangle$	33
99	**72**	$\left\langle \begin{pmatrix} 10000 \\ 00010 \\ 00001 \\ 01000 \\ 00100 \end{pmatrix}, \begin{pmatrix} 10000 \\ 02000 \\ 00100 \\ 00030 \\ 00002 \end{pmatrix} \right\rangle$	31
137	**100**	$\left\langle \begin{pmatrix} 22101 \\ 33213 \\ 31331 \\ 00020 \\ 13032 \end{pmatrix} \right\rangle$	33
163	**120**	$\left\langle \begin{pmatrix} 33221 \\ 21332 \\ 31212 \\ 21333 \\ 10013 \end{pmatrix} \right\rangle$	21
177	130	$\left\langle \begin{pmatrix} 00010 \\ 00100 \\ 10000 \\ 00001 \\ 01000 \end{pmatrix}, \begin{pmatrix} 00100 \\ 10000 \\ 00001 \\ 00010 \\ 01000 \end{pmatrix} \right\rangle$	25

Table 8.6 Linear codes for $q = 4$ and $k = 5$

n	d	G		r
182	134	$\left\langle \begin{pmatrix} 00010 \\ 00100 \\ 10000 \\ 00001 \\ 01000 \end{pmatrix} \right.$	$\left. , \begin{pmatrix} 00100 \\ 10000 \\ 00001 \\ 00010 \\ 01000 \end{pmatrix} \right\rangle$	25
189	**140**	$\left\langle \begin{pmatrix} 20200 \\ 31231 \\ 00022 \\ 30003 \\ 32102 \end{pmatrix} \right\rangle$		21
194	**144**	$\left\langle \begin{pmatrix} 10210 \\ 32020 \\ 30110 \\ 20000 \\ 00001 \end{pmatrix} \right\rangle$		21
226	**168**	$\left\langle \begin{pmatrix} 00010 \\ 00100 \\ 10000 \\ 00001 \\ 01000 \end{pmatrix} \right.$	$\left. , \begin{pmatrix} 00100 \\ 10000 \\ 00001 \\ 00010 \\ 01000 \end{pmatrix} \right\rangle$	25
236	**176**	$\left\langle \begin{pmatrix} 00010 \\ 00100 \\ 10000 \\ 00001 \\ 01000 \end{pmatrix} \right.$	$\left. , \begin{pmatrix} 00100 \\ 10000 \\ 00001 \\ 00010 \\ 01000 \end{pmatrix} \right\rangle$	25

Table 8.7 Linear codes for $q = 4$ and $k = 6$

n	d	G	r
102	72	$\left\langle \begin{pmatrix} 200000 \\ 002000 \\ 000200 \\ 000002 \\ 020000 \\ 000020 \end{pmatrix} , \begin{pmatrix} 002000 \\ 200000 \\ 020000 \\ 000002 \\ 000200 \\ 000020 \end{pmatrix} \right\rangle$	51
108	76	$\left\langle \begin{pmatrix} 000100 \\ 100000 \\ 000001 \\ 001000 \\ 000010 \\ 010000 \end{pmatrix} , \begin{pmatrix} 000001 \\ 000100 \\ 100000 \\ 000010 \\ 010000 \\ 001000 \end{pmatrix} \right\rangle$	51
134	96	$\left\langle \begin{pmatrix} 000100 \\ 100000 \\ 000001 \\ 001000 \\ 000010 \\ 010000 \end{pmatrix} , \begin{pmatrix} 000001 \\ 000100 \\ 100000 \\ 000010 \\ 010000 \\ 001000 \end{pmatrix} \right\rangle$	51
140	100	$\left\langle \begin{pmatrix} 010000 \\ 000010 \\ 000100 \\ 000001 \\ 100000 \\ 001000 \end{pmatrix} , \begin{pmatrix} 000100 \\ 000010 \\ 001000 \\ 100000 \\ 010000 \\ 000001 \end{pmatrix} \right\rangle$	51
146	104	$\left\langle \begin{pmatrix} 010000 \\ 000010 \\ 000100 \\ 000001 \\ 100000 \\ 001000 \end{pmatrix} , \begin{pmatrix} 000100 \\ 000010 \\ 001000 \\ 100000 \\ 010000 \\ 000001 \end{pmatrix} \right\rangle$	51

Table 8.8 Linear codes for $q = 4$ and $k = 6$

n	d	G	r
161	115	$\left\langle \begin{pmatrix} 000030 \\ 000300 \\ 300000 \\ 000003 \\ 003000 \\ 030000 \end{pmatrix}, \begin{pmatrix} 000100 \\ 000001 \\ 100000 \\ 010000 \\ 000010 \\ 001000 \end{pmatrix} \right\rangle$	51
165	118	$\left\langle \begin{pmatrix} 010000 \\ 000010 \\ 000100 \\ 000001 \\ 100000 \\ 001000 \end{pmatrix}, \begin{pmatrix} 000100 \\ 000010 \\ 001000 \\ 100000 \\ 010000 \\ 000001 \end{pmatrix} \right\rangle$	51
175	126	$\left\langle \begin{pmatrix} 001130 \\ 233003 \\ 322120 \\ 331110 \\ 103301 \\ 011332 \end{pmatrix} \right\rangle$	17
180	130	$\left\langle \begin{pmatrix} 010000 \\ 000010 \\ 000100 \\ 000001 \\ 100000 \\ 001000 \end{pmatrix}, \begin{pmatrix} 000100 \\ 000010 \\ 001000 \\ 100000 \\ 010000 \\ 000001 \end{pmatrix} \right\rangle$	51
185	134	$\left\langle \begin{pmatrix} 010000 \\ 000010 \\ 000100 \\ 000001 \\ 100000 \\ 001000 \end{pmatrix}, \begin{pmatrix} 000100 \\ 000010 \\ 001000 \\ 100000 \\ 010000 \\ 000001 \end{pmatrix} \right\rangle$	51

Table 8.9 Linear codes for $q = 4$ and $k = 6$

n	d	G		r
191	138	$\left\langle \begin{pmatrix} 010000 \\ 000010 \\ 000100 \\ 000001 \\ 100000 \\ 001000 \end{pmatrix} \right.$	$\left. \begin{pmatrix} 000100 \\ 000010 \\ 001000 \\ 100000 \\ 010000 \\ 000001 \end{pmatrix} \right\rangle$	51
195	141	$\left\langle \begin{pmatrix} 010000 \\ 000010 \\ 000100 \\ 000001 \\ 100000 \\ 001000 \end{pmatrix} \right.$	$\left. \begin{pmatrix} 000100 \\ 000010 \\ 001000 \\ 100000 \\ 010000 \\ 000001 \end{pmatrix} \right\rangle$	51
201	145	$\left\langle \begin{pmatrix} 000100 \\ 001000 \\ 000010 \\ 010000 \\ 100000 \\ 000001 \end{pmatrix} \right.$	$\left. \begin{pmatrix} 010000 \\ 001000 \\ 100000 \\ 000010 \\ 000001 \\ 000100 \end{pmatrix} \right\rangle$	51
205	148	$\left\langle \begin{pmatrix} 000100 \\ 001000 \\ 000010 \\ 010000 \\ 100000 \\ 000001 \end{pmatrix} \right.$	$\left. \begin{pmatrix} 010000 \\ 001000 \\ 100000 \\ 000010 \\ 000001 \\ 000100 \end{pmatrix} \right\rangle$	51
210	152	$\left\langle \begin{pmatrix} 000010 \\ 010000 \\ 100000 \\ 001000 \\ 000001 \\ 000100 \end{pmatrix} \right.$	$\left. \begin{pmatrix} 010000 \\ 000001 \\ 000010 \\ 001000 \\ 000100 \\ 100000 \end{pmatrix} \right\rangle$	51

Table 8.10 Linear codes for $q = 4$ and $k = 6$

n	d	G		r
220	160	$\left\langle \begin{matrix} 000010 \\ 010000 \\ 100000 \\ 001000 \\ 000001 \\ 000100 \end{matrix} \right.$, $\begin{matrix} 010000 \\ 000001 \\ 000010 \\ 001000 \\ 000100 \\ 100000 \end{matrix} \left. \right\rangle$		51
226	163	$\left\langle \begin{matrix} 000100 \\ 001000 \\ 000010 \\ 010000 \\ 100000 \\ 000001 \end{matrix} \right.$, $\begin{matrix} 010000 \\ 001000 \\ 100000 \\ 000010 \\ 000001 \\ 000100 \end{matrix} \left. \right\rangle$		51
232	168	$\left\langle \begin{matrix} 000010 \\ 010000 \\ 100000 \\ 001000 \\ 000001 \\ 000100 \end{matrix} \right.$, $\begin{matrix} 010000 \\ 000001 \\ 000010 \\ 001000 \\ 000100 \\ 100000 \end{matrix} \left. \right\rangle$		51
237	172	$\left\langle \begin{matrix} 000010 \\ 010000 \\ 100000 \\ 001000 \\ 000001 \\ 000100 \end{matrix} \right.$, $\begin{matrix} 010000 \\ 000001 \\ 000010 \\ 001000 \\ 000100 \\ 100000 \end{matrix} \left. \right\rangle$		51
242	176	$\left\langle \begin{matrix} 000010 \\ 010000 \\ 100000 \\ 001000 \\ 000001 \\ 000100 \end{matrix} \right.$, $\begin{matrix} 010000 \\ 000001 \\ 000010 \\ 001000 \\ 000100 \\ 100000 \end{matrix} \left. \right\rangle$		51

Table 8.11 Linear codes for $q = 4$ and $k = 7$

n	d	G	r
126	88	$\left\langle \begin{matrix} 3001233 \\ 2212232 \\ 0010311 \\ 2230310 \\ 1310312 \\ 1110332 \\ 2303023 \end{matrix} \right\rangle$	89
158	110	$\left\langle \begin{matrix} 1000010 \\ 0010001 \\ 0011000 \\ 1000000 \\ 1101000 \\ 1010111 \\ 0010100 \end{matrix} \right\rangle$	181
161	112	$\left\langle \begin{matrix} 1000010 \\ 0010001 \\ 0011000 \\ 1000000 \\ 1101000 \\ 1010111 \\ 0010100 \end{matrix} \right\rangle$	181
189	132	$\left\langle \begin{matrix} 1331321 \\ 1330022 \\ 2230231 \\ 0322221 \\ 2000032 \\ 3200131 \\ 2201032 \end{matrix} \right\rangle$	89

Table 8.12 Linear codes for $q = 5$ and $k = 5$

n	d	G	r
53	**40**	$\left\langle \begin{array}{c} 41233 \\ 22240 \\ 13413 \\ 32040 \\ 40040 \end{array} \right\rangle$	61
92	70	$\left\langle \begin{array}{c} 00002 \\ 10002 \\ 01001 \\ 00104 \\ 00010 \end{array} \right\rangle$	45
100	76	$\left\langle \begin{array}{c} 00002 \\ 10002 \\ 01001 \\ 00104 \\ 00010 \end{array} \right\rangle$	45
110	85	$\left\langle \begin{array}{c} 12344 \\ 31144 \\ 20332 \\ 34344 \\ 43110 \end{array} \right\rangle$	71

Table 8.13 Linear codes for $q = 5$ and $k = 6$

n	d	G	r
50	34	$\left\langle \begin{pmatrix} 212000 \\ 042000 \\ 023000 \\ 000130 \\ 000100 \\ 000001 \end{pmatrix} \right\rangle$	173
70	50	$\left\langle \begin{pmatrix} 010000 \\ 001000 \\ 100000 \\ 000010 \\ 000001 \\ 000100 \end{pmatrix}, \begin{pmatrix} 100000 \\ 010000 \\ 003000 \\ 000300 \\ 000010 \\ 000002 \end{pmatrix} \right\rangle$	35
73	52	$\left\langle \begin{pmatrix} 100000 \\ 000100 \\ 000010 \\ 000001 \\ 010000 \\ 001000 \end{pmatrix}, \begin{pmatrix} 001000 \\ 000100 \\ 000001 \\ 000010 \\ 010000 \\ 100000 \end{pmatrix} \right\rangle$	110

Table 8.14 Linear code for $q = 7$ and $k = 4$ with generator matrix

$$\Gamma := \begin{pmatrix} 0\ 0\ 0\ 1\ 1\ 1\ 1\ 1\ 1\ 1\ 1\ 1\ 1\ 1\ 1\ 1\ 1\ 1\ 1\ 1\ 1\ 1\ 1\ 1\ 1\ 1 \\ 0\ 1\ 1\ 2\ 3\ 3\ 6\ 0\ 0\ 2\ 4\ 4\ 6\ 0\ 2\ 4\ 5\ 5\ 6\ 1\ 1\ 2\ 5\ 5\ 6\ 2 \\ 0\ 3\ 6\ 5\ 0\ 6\ 0\ 1\ 2\ 2\ 1\ 5\ 6\ 4\ 4\ 6\ 1\ 5\ 2\ 4\ 5\ 3\ 3\ 6\ 4\ 6 \\ 1\ 6\ 1\ 5\ 6\ 2\ 1\ 4\ 1\ 2\ 4\ 4\ 5\ 4\ 6\ 1\ 1\ 5\ 1\ 2\ 6\ 6\ 4\ 1\ 4\ 0 \end{pmatrix}$$

n	d	G	r
26	**20**	$\left\langle \begin{pmatrix} 0010 \\ 1030 \\ 0110 \\ 0001 \end{pmatrix} \right\rangle$	74

Table 8.15 Linear codes for $q = 7$ and $k = 5$

n	d	G	r
28	20	$\left\langle \begin{pmatrix} 00100 \\ 00010 \\ 10000 \\ 01000 \\ 00001 \end{pmatrix}, \begin{pmatrix} 10000 \\ 03000 \\ 00300 \\ 00060 \\ 00002 \end{pmatrix} \right\rangle$	147
34	25	$\left\langle \begin{pmatrix} 53000 \\ 25000 \\ 00300 \\ 00001 \\ 00050 \end{pmatrix} \right\rangle$	189
48	36	$\left\langle \begin{pmatrix} 66240 \\ 44440 \\ 46200 \\ 10450 \\ 00001 \end{pmatrix} \right\rangle$	131

Table 8.16 Linear codes for $q = 8$ and $k = 4$

n	d	G	r
85	**72**	$\left\langle \begin{pmatrix} 0010 \\ 1070 \\ 0170 \\ 0001 \end{pmatrix} \right\rangle$	73
97	82	$\left\langle \begin{pmatrix} 0010 \\ 1000 \\ 0100 \\ 0001 \end{pmatrix}, \begin{pmatrix} 0001 \\ 1000 \\ 0010 \\ 0100 \end{pmatrix} \right\rangle$	57
103	**88**	$\left\langle \begin{pmatrix} 0010 \\ 1000 \\ 0100 \\ 0001 \end{pmatrix}, \begin{pmatrix} 0001 \\ 1000 \\ 0010 \\ 0100 \end{pmatrix} \right\rangle$	57
108	92	$\left\langle \begin{pmatrix} 0010 \\ 1000 \\ 0100 \\ 0001 \end{pmatrix}, \begin{pmatrix} 0001 \\ 1000 \\ 0010 \\ 0100 \end{pmatrix} \right\rangle$	57
117	**100**	$\left\langle \begin{pmatrix} 0571 \\ 3403 \\ 2247 \\ 1361 \end{pmatrix} \right\rangle$	45

Table 8.17 Linear codes for $q = 8$ and $k = 5$

n	d	G	r
79	63	$\left\langle \begin{array}{c} 00010 \\ 10010 \\ 01040 \\ 00160 \\ 00001 \end{array} \right\rangle$	121
98	80	$\left\langle \begin{array}{c} 10000 \\ 00100 \\ 01000 \\ 00001 \\ 00010 \end{array} , \begin{array}{c} 20000 \\ 03000 \\ 00400 \\ 00030 \\ 00002 \end{array} \right\rangle$	61
100	81	$\left\langle \begin{array}{c} 10000 \\ 00100 \\ 01000 \\ 00001 \\ 00010 \end{array} , \begin{array}{c} 20000 \\ 03000 \\ 00400 \\ 00030 \\ 00002 \end{array} \right\rangle$	61
103	84	$\left\langle \begin{array}{c} 10000 \\ 00100 \\ 01000 \\ 00001 \\ 00010 \end{array} , \begin{array}{c} 20000 \\ 03000 \\ 00400 \\ 00030 \\ 00002 \end{array} \right\rangle$	61
119	98	$\left\langle \begin{array}{c} 00010 \\ 10010 \\ 01040 \\ 00160 \\ 00001 \end{array} \right\rangle$	121
130	107	$\left\langle \begin{array}{c} 42000 \\ 56000 \\ 00120 \\ 00140 \\ 00007 \end{array} \right\rangle$	81

Table 8.18 Linear code for $q = 9$ and $k = 3$ with generator matrix

$$\Gamma := \begin{pmatrix} 0 & 1 & 0 & 1 & 1 & 1 & 1 & 1 & 1 & 1 & 1 & 1 & 1 & 1 & 1 & 1 & 1 \\ 1 & 0 & 1 & 0 & 1 & 1 & 1 & 2 & 2 & 5 & 7 & 5 & 7 & 5 & 7 & 6 & 8 \\ 1 & 1 & 6 & 6 & 5 & 7 & 8 & 5 & 7 & 0 & 0 & 1 & 7 & 8 & 6 & 1 & 8 \end{pmatrix}$$

n	d	G	r
17	**14**	$\left\langle \begin{pmatrix} 010 \\ 100 \\ 001 \end{pmatrix} \right\rangle$	51

Table 8.19 Linear codes for $q = 9$ and $k = 4$

n	d	G	r
41	34	$\left\langle \begin{pmatrix} 8418 \\ 5878 \\ 7312 \\ 0215 \end{pmatrix} \right\rangle$	20
102	88	$\left\langle \begin{pmatrix} 1600 \\ 8600 \\ 0006 \\ 0050 \end{pmatrix} \right\rangle$	46
123	106	$\left\langle \begin{pmatrix} 0001 \\ 1002 \\ 0100 \\ 0017 \end{pmatrix} \right\rangle$	20
130	112	$\left\langle \begin{pmatrix} 0600 \\ 3710 \\ 4130 \\ 0001 \end{pmatrix} \right\rangle$	50

Chapter 9

The General Case

9

9 The General Case

9

9 The General Case

After the construction of codes with prescribed automorphism group, we are now attacking the general case, i.e. we will no longer make any assumption on the presence of nontrivial automorphisms. The main goal is the evaluation of a transversal of the isometry classes, for any given parameter set. Of course, this daunting task can only be solved for small parameters. Also, since we are mostly interested in good codes, we shall restrict attention to codes with minimum distance at least 3, the reason is that this restriction makes things much easier.

The main point is that the construction of a transversal of codes and the classification by isometry classes are not two separate issues but rather go hand in hand. We will see that the classification is best done already during the construction of codes. In fact, the construction of codes is supported by the classification part in that not too much overhead is constructed which otherwise would have to be deleted later. The corresponding algorithmic principle is that of *orderly generation of discrete structures*. The order refers to an order which we impose on the objects, for instance the lexicographical order given by the columns of the generator matrices. This leads to a central problem in the systematic construction of transversals of orbits, e.g. of isometry classes: We have to introduce a *normal form*, following the request of D. Slepian, who wrote in 1960 ([184]):

> *"The task of analyzing group codes would be greatly simplified if a canonical form could be found for each equivalence class of Ω-matrices[1]. That is, for a given n and k, we should like to be able to write down one generator matrix from each equivalence class. This would provide a simple means of describing each of the essentially different (n, k)-codes."*

The plan of this chapter is as follows. We first show how to reduce the computation of transversals of isometry classes to a problem in finite projective geometry. This will give us control over the minimum distance. The remaining problem of computing orbits can be solved using methods from Computational Group Theory. We will give a very brief introduction to this area, focusing mainly on fundamental algorithms for permutation groups. After that, we describe the method of orderly generation, and we apply this to the construction of optimal linear codes. The major issue is that of computing orbits of a group on subsets. We treat the permutation representation of the projective linear group. Finally, we present numerical data which was computed. We

---

[1] a group code is a linear code, an Ω-matrix is a generator matrix Γ

classify the isometry classes of optimal linear codes for small parameters and over small fields.

9.1 The Problem

We are faced with the following problem: For a given length n, dimension k, minimum distance d and field $\mathbb{F}_q$, we would like to determine the isometry classes of linear (n, k, d, q)-codes. For instance, this could be done by listing generator matrices for each such code. Before we embark on this mission, let us recall what we have learned in the earlier chapters.

Remarks In order to evaluate a transversal of the isometry classes of linear (n, k, d, q)-codes, we can use the following facts:

— In Chapter 1 we saw that a linear code C can be described both by a generator matrix Γ and by a check matrix Δ. The check matrix is a generator matrix of the dual code $C^\perp$. Moreover, as $(C^\perp)^\perp = C$, the mapping

$$\perp \; : \; \mathcal{U}(n, k, q) \to \mathcal{U}(n, n - k, q), \; C \mapsto C^\perp$$

is a bijection from the set of (n, k)-codes to the set of $(n, n - k)$-codes over $\mathbb{F}_q$. In fact, as the map $\perp$ is compatible with the various types of isometries, this map descends to a bijection of the corresponding isometry classes. This fact holds true both for linear and for semilinear isometry classes. In the following, when we speak of isometry classes (unqualified) we mean that the result holds regardless of whether the isometry classes under consideration are linear or semilinear. It remains to investigate the map $\perp$ further.

— As we are interested mainly in good codes, we may ignore codes with minimum distance at most 2. Such codes cannot correct a single error, so this restriction does not exclude anything which would be interesting. So, from now on we consider only codes C with minimum distance at least 3, for short: linear $(n, k, \geq 3, q)$-codes.

— From 1.3.9 we know that check matrices of such codes have pairwise linearly independent columns. In the language of 6.1.14, this means that such codes have projective duals. Conversely, a code whose check matrix is projective has minimum distance at least 3 (see Exercise 1.3.21). Therefore, the duality map may be restricted to induce bijections between the following isometry classes of codes:

1. $(n, n - k)$ projective codes,

2. (n,k)-codes with minimum distance greater than or equal to three (or $(n,k,\geq 3)$-codes). ◇

Vector spaces are often difficult to handle with a computer. In part this results from the fact that there are usually many different bases for the same space. As far as Slepian's problem of computing transversals of linear codes is concerned, we have to consider orbits of the isometry group on vector spaces. This raises other issues, like how to represent these orbits on the Computer, when typically each orbit is very long and the orbit elements are vector spaces. In order to overcome these problems, we may look for different representations of codes. We take the approach indicated in the last item of the previous remark of looking at the projective dual code. We can build on ideas from Section 6.1. The fundamental result 6.1.13 identifies linear isometry classes of linear codes with certain orbits of groups on mappings into projective space. In 6.1.25, the result is specialized to injective functions, which can be identified with their image, since we have the symmetric group acting on the domain of the map. These results may be summarized and slightly generalized as

---

Theorem   9.1.2

1. *There is a one-to-one correspondence between the linear isometry classes of projective $(n, \leq k, q)$-codes and the set of orbits*

$$\mathrm{PGL}_k(q) \backslash\!\backslash \binom{\mathrm{PG}_{k-1}(q)}{n}.$$

2. *There is a one-to-one correspondence between the semilinear isometry classes of projective $(n, \leq k, q)$-codes and the set of orbits*

$$\mathrm{P\Gamma L}_k(q) \backslash\!\backslash \binom{\mathrm{PG}_{k-1}(q)}{n}.$$

In both cases, the isometry classes of projective (n, i, q)-codes correspond to the orbits on n-subsets of $\mathrm{PG}_{k-1}(q)$ with the property that the n points span a vector space of dimension i, for $0 \leq i \leq k$. □

In order to describe the underlying map between codes and orbits of points, we start with a generator matrix

$$\Gamma = (\gamma_{i,j}) \in \mathbb{F}_q^{k \times n}$$

of a projective (n,k,q)-code. Then

$$P(\Gamma) := \left\{ P(\gamma_{*,0}), P(\gamma_{*,1}), \ldots, P(\gamma_{*,n-1}) \right\} \subseteq \mathrm{PG}_{k-1}(q),$$ 9.1.3

is a set of n points in $\mathrm{PG}_{k-1}(q)$ with the property that these points span a vector space of dimension k. Here,

$$P(\gamma_{*,j}) = \langle \gamma_{*,j} \rangle$$

is the projective point whose homogeneous coordinates are listed in the j-th column of Γ.

From the definition of the map, it is clear that rearranging the columns of Γ does not change the set $P(\Gamma)$. The action of $\mathrm{GL}_k(q)$ on generator matrices is similar to the action of $\mathrm{PGL}_k(q)$ on n-sets of points in $\mathrm{PG}_{k-1}(q)$. This is because left-multiplying Γ by an invertible matrix A gives rise to the set $\{P(A \cdot \gamma_{*,j}) \mid j \in n\}$ which is the image of $P(\Gamma)$ under the projective transformation induced by A.

This shows that the map $\Gamma \mapsto P(\Gamma)$ descends to a map from the linear isometry classes of projective codes to the orbits of n-sets of points of $\mathrm{PG}_{k-1}(q)$ under the projective linear group $\mathrm{PGL}_k(q)$. This map is a one-to-one correspondence and preserves the dimension.

Furthermore, the action of $\Gamma\mathrm{L}_k(q)$ on generator matrices corresponds to the action of $\mathrm{P\Gamma L}_k(q)$ on n-sets of points in $\mathrm{PG}_{k-1}(q)$. Thus, the given map descends to a map from the semilinear isometry classes of projective codes to the orbits of n-sets of points of $\mathrm{PG}_{k-1}(q)$ under the projective semilinear group $\mathrm{P\Gamma L}_k(q)$. Again, the resulting map is a one-to-one correspondence and preserves the dimension.

In order to describe the inverse map, we introduce the following notation. To a set S of n different points $p_0, \ldots, p_{n-1}$ in $\mathrm{PG}_{k-1}(q)$ we associate the generator matrix

9.1.4
$$\Gamma(S) = (a_{i,j}) \in \mathbb{F}_q^{k \times n}$$

where $p_j = \langle a_{*,j} \rangle$ for $j \in n$. This construction is not unique for two reasons. At first, we are making a choice by ordering the points of the set. Furthermore, the vector $a_{*,j}$ with $p_j = \langle a_{*,j} \rangle$ is unique up to non-zero scalar multiples. Therefore, the matrix $\Gamma(S)$ is unique *up to order of its columns and multiplication of columns by nonzero scalars.* Changing to a different set

$$A \cdot S = \{P(A \cdot a_{*,j}) \mid j \in n\}$$

results in changing the generator matrix to $A \cdot \Gamma(S)$. Summarizing, the code generated by $\Gamma(S)$ is determined up to linear isometry.

Under the duality map, the previous result becomes

Corollary   9.1.5

1. *There is a one-to-one correspondence between the linear isometry classes of $(n, \geq k, \geq 3, q)$-codes and the set of orbits*

$$\mathrm{PGL}_{n-k}(q) \backslash\backslash \binom{\mathrm{PG}_{n-k-1}(q)}{n}.$$

2. *There is a one-to-one correspondence between the semilinear isometry classes of projective $(n, \geq k, \geq 3, q)$-codes and the set of orbits*

$$\mathrm{P\Gamma L}_{n-k}(q) \backslash\backslash \binom{\mathrm{PG}_{n-k-1}(q)}{n}.$$

In both cases, the isometry classes of $(n, k + i, \geq 3, q)$-codes correspond to the orbits on n-subsets of $\mathrm{PG}_{n-k-1}(q)$ with the property that the n points span a vector space of dimension $k - i$, for for some i with $0 \leq i \leq k$.   □

Here, if $S = \{p_0, \ldots, p_{n-1}\}$ is a set of n points in $\mathrm{PG}_{n-k-1}(q)$ we obtain a projective check matrix

$$\Delta(S) = (b_{i,j}) \in \mathbb{F}_q^{(n-k) \times n}$$ 9.1.6

where $p_j = \langle b_{*,j} \rangle$ for $j \in n$ (notice that this is a vector of length $n - k$). This matrix is well-defined up ordering of the columns and up to non-zero scalar multiples of the columns. Since we take this matrix as a representative of an isometry class of codes, this non-uniqueness does not bother us.

The last result is already very close to what we really want. Apart from codes with minimum distance 1 or 2, Slepian's problem of finding a transversal of all isometry classes of codes is solved (provided we can evaluate the orbits in question, this remains to be seen). But we can refine this approach a little, to better suit the application in coding theory. What if Slepian would have asked

> "For a given n and k and $d_{\min}$, we should like to write down one generator matrix from each equivalence class of (n, k)-codes whose minimum distance is at least $d_{\min}$."

That is, what if we are interested in codes with a given minimum distance. The point with codes is that we really are not interested all that much in the generality of *all* available codes. The focus is of course on "good" codes, i.e. codes whose minimum distance is high. That means, we wish to direct attention to finding only a subset of the set of all (n, k)-codes, namely those with minimum distance greater than or equal to $d_{\min}$, where $d_{\min}$ is some specified lower bound which we choose beforehand. Of course, in the spirit of Slepian we still want one generator matrix from each equivalence class, i.e. we still want to classify the codes exhaustively. In particular, if no such code exists, our construction procedure should prove this fact. As we will see shortly, it is possible to refine our approach and take into account the prescribed minimum distance $d_{\min}$ right from the start. Of course, this restriction will save us a lot of work since we can skip a whole lot of codes which do not meet the required minimum distance. In a sense, we are looking for the needle in the haystack.

Let us introduce the following terminology.

9.1.7   Definition In a projective space, a set of points $\langle v^{(0)} \rangle, \langle v^{(1)} \rangle, \ldots, \langle v^{(r-1)} \rangle$ is said to be in *in general position* (or *independent*) if they generate a vector space of dimension r. That is, the points are independent in projective space if and only if the representing vectors $v^{(0)}, v^{(1)}, \ldots, v^{(r-1)}$ are independent as vectors.   ◇

It is clear that this property does not depend on the choice of the representing non-zero vectors $v^{(i)}$ out of their respective subspace $\langle v^{(i)} \rangle$.

Using this language, we can rephrase 1.3.10 as follows. The generator matrices of linear codes over $\mathbb{F}_q$ of length n, dimension at least k and with minimum distance at least $d_{\min}$ for some integer $d_{\min} \geq 3$ correspond (up to ordering of the columns and multiplication of columns by non-zero scalars) to the n-subsets of $\mathrm{PG}_{n-k-1}(q)$ with the property that any $d_{\min} - 1$ points are in general position.

In fact, this correspondence descends to a correspondence between isometry classes of codes and orbits of projective groups on sets of points in projective space.

9.1.8   Theorem *For any given $d_{\min} \geq 3$, we have the following:*

1. *The linear isometry classes of linear $(n, \geq k, \geq d_{\min}, q)$-codes correspond one-to-one to the subset of*
$$\mathrm{PGL}_{n-k}(q) \backslash\!\backslash \binom{\mathrm{PG}_{n-k-1}(q)}{n},$$
consisting of the orbits of n-sets whose $d_{\min} - 1$-subsets are all in general position.
2. *Correspondingly, the semilinear isometry classes of linear $(n, \geq k, \geq d_{\min}, q)$-codes correspond one-to-one to the subset of*
$$\mathrm{P\Gamma L}_{n-k}(q) \backslash\!\backslash \binom{\mathrm{PG}_{n-k-1}(q)}{n},$$
consisting of the orbits of n-sets whose $(d_{\min} - 1)$-subsets are all in general position.

In both cases, the true minimum distance d of these codes is determined by the size of the smallest set of points which are dependent. Also, the true dimension of such a code *is determined as $n - r$, where r is the vector space dimension of the space spanned by the n points.*   □

The rest of this chapter is devoted to solving the problem of constructing and classifying codes algorithmically using Theorem 9.1.8. It involves techniques from Computational Group Theory. The major issue, namely that of computing orbits on sets is addressed in Sections 9.2 and 9.6. The following Section 9.2 handles the "base case", where the sets have size 1 and hence are in fact points. Section 9.6 treats the general case, building on the results of Section 9.2.

9.2 Computing with Permutation Groups

In this section we address the problem of explicit computations with permutation groups. Our main goal is to compute orbits of permutation groups on subsets. This is part of a rather new branch of mathematics called *Computational Group Theory*, or CGT for short. Our main references are the recent book by Holt, Eick and O'Brien [91], the book by Seress [177] and the one by Butler [35]. For more on combinatorial algorithms see the book by Kreher and Stinson [116]. Several computer algebra systems covering CGT are available. The two most prominent are GAP [63] and Magma [140].

Let G be a finite group acting on a finite set X. For technical reasons we prefer in this chapter actions from the right, i.e. mappings

$$X \times G \to X : (x, g) \mapsto xg,$$

such that $(xg)g' = x(gg')$ and $x1 = x$. But we still use the symbol $G(x)$ for the orbit of x and G_x for its stabilizer.

Let us assume that G acts faithfully, which means that only the identity element of G fixes every point in X. According to 1.4.5, G is isomorphic to the permutation group $\overline{G} = \delta(G)$ induced by G on X, a subgroup of the symmetric group S_X on X. Hence we can assume that G is a permutation group on X, i.e. that $G \leq S_X$. In this section, we are concerned with computational tasks like the following.

1. For $x \in X$, compute $G(x) = \{xg \mid g \in G\}$, the *orbit of x under G*.

2. For $x \in X$, compute $G_x = \{g \in G \mid xg = x\}$, the *stabilizer of x in G*.

3. For $x, y \in G$ with $y \in G(x)$, compute an element $g \in G$ with $xg = y$. We call such an element a *transporter element*.

A remark concerning the last problem is in order. The required element $g \in G$ with $xg = y$ may not be unique. In fact, by Lemma 3.4.1 the set of all elements $g \in G$ with this property forms a unique right coset of the stabilizer G_x, the stabilizer of x in G.

In order to get started, the group has to be specified in some concrete way. A very simple way is by a set of *generators*, i.e. a set S of elements of G which together generate G, i.e. $\langle S \rangle = G$. If G is finite, this means that each element $g \in G$ can be written as a word of finite length over the alphabet S (Exercise 9.2.1). This will suffice for the moment. A more sophisticated representation of a group will be presented in Section 9.7. So for now, let us always assume that G is given by a finite set of generators $S = \{s_0, \ldots, s_{r-1}\}$.

The first problem is that of computing the orbit of a point $x \in X$ under the group G. We start by introducing a graph which describes the action of G on the set X.

9.2.1   **Definition (action graph)** Let the group G act on the finite set X. Assume that G is generated by a set of generators $S = \{s_0, \ldots, s_{r-1}\}$. The *action-graph* of G on X with respect to the set S is the directed graph (digraph) $\mathcal{G} = (X, \mathcal{E})$. That is, the vertices of $\mathcal{G}$ are the elements of X. The edge set $\mathcal{E}$ consists of directed labeled edges. There is an edge from vertex x to vertex y labeled by s_j if

$$xs_j = y.$$

We write $x \rightarrow y$ to indicate that there is an edge from x to y. A *directed path* is a sequence of $x_0, x_1, \ldots, x_{u-1}$ of vertices which are pairwise distinct (except possibly for x_0 and x_{u-1} which may coincide) such that $x_0 \rightarrow x_1 \rightarrow \cdots \rightarrow x_{u-1}$. We write $x \rightsquigarrow y$ is there is a path from x to y. The length of a path is the number of edges used. We also define a cycle to be a path where the start and the endpoint coincide (i.e. with $x_0 = x_{u-1}$ in the above notation). A loop is a cycle of length 1, i.e. an edge from a vertex x to itself.   ◇

The action graph may have loops, i.e. edges of the form (x, x) for some vertex $x \in X$. Also, it may have several edges from vertex x to vertex y, namely if there are several elements $s \in S$ with $xs = y$.

9.2.2   **Lemma** *Let the group G act on the finite set X. Let $\mathcal{G} = (X, \mathcal{E})$ be the action graph with respect to the generating set S of G. Then the orbits of G on X correspond one-by-one to the connected components of $\mathcal{G}$. In particular, the connected components of $\mathcal{G}$ are well-defined and independent of the choice of the generating set S of G.*

Proof: Without loss of generality, we can replace G by the finite group G/K, where K is the kernel of the action of G on X, i.e. the pointwise stabilizer of the whole set X. The fact that G/K is finite follows from the fact that X is finite. Thus we may assume that G is a finite group. By Exercise 9.2.1, each $g \in G$ can be written as a word $s_{i_0} s_{i_1} \ldots s_{i_{u-1}}$ in the generators. Recall that we write $x \rightsquigarrow y$ if there is a directed path from x to y in $\mathcal{G}$. Such a path gives rise to a group element $g = s_{j_0} s_{j_1} \ldots s_{j_{u-1}}$ with $xg = y$. If $g^{-1} = s_{i_0} s_{i_1} \ldots s_{i_{v-1}}$, then there also is a path

$$y \to ys_{i_0} \to ys_{i_0}s_{i_1} \to \ldots \to yg^{-1} = x$$

in $\mathcal{G}$, i.e. $y \rightsquigarrow x$. This means that

$$x \rightsquigarrow y \iff y \rightsquigarrow x.$$

In other words, the relation "$\rightsquigarrow$" is undirected, and we can replace it by the symmetric $x \sim y$ (so that "$\sim$" really is an equivalence relation on X). We conclude that the concept of a connected component is well-defined in action graphs. Also, we have shown that $x \sim y$ if and only if x and y belong to the same G-orbit. This means that the connected components of $\mathcal{G}$ correspond bijectively to the G-orbits on X. It remains to show that the connected components in the action graph depend only on the group G, and not on the choice of the generating set S for G. To this end, let $T = \{t_0, \ldots, t_{s-1}\}$ be another generating set for G. Write $\mathcal{G}_S$ and $\mathcal{G}_T$ for the action graphs of G with respect to the generating sets S and T. We need to show that $x \rightsquigarrow y$ in $\mathcal{G}_S$ if and only if $x \rightsquigarrow y$ in $\mathcal{G}_T$. We note that $x \rightsquigarrow y$ in $\mathcal{G}_S$ implies that $xg = y$ for $g = s_{i_0} s_{i_1} \ldots s_{i_{u-1}}$. The element g has an expression in terms of the second generating set, say $g = t_{j_0} t_{j_1} \ldots t_{j_{v-1}}$. But then $x \rightsquigarrow y$ in $\mathcal{G}_T$. The converse follows by symmetry. $\square$

Remark In Computer Science, a subset U of vertices in a directed graph is called strongly connected if both $x \rightsquigarrow y$ and $y \rightsquigarrow x$ hold for all $x, y \in U$. The maximal strongly connected subsets of a graph are called strongly connected components and there are algorithms to compute these for a given graph (see [42]). It follows from 9.2.2 that the connected components of an action graph are strongly connected components. Nevertheless, there is a difference. The reason is that the strongly connected components in general digraphs may still have edges between them. The connected components in action graphs do not have this property. $\diamond$

9.2.3

Example Figure 9.1 shows action-graphs of S_6 with respect to two different generating systems. The left picture uses $s_0 = (0,1,2,3,4,5)$ and $s_1 = (0,1)$.

9.2.4

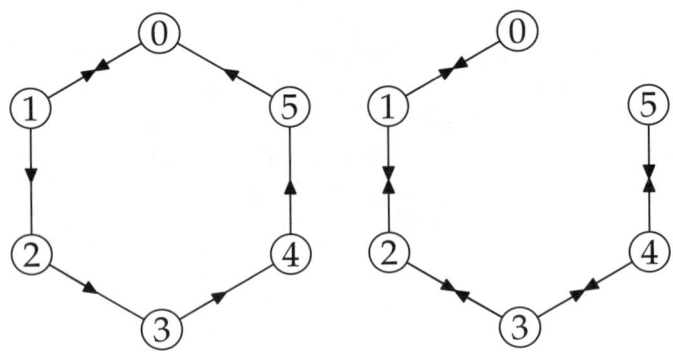

Fig. 9.1 Two action-graphs for S_6

The right picture is obtained by using the Coxeter generators $s_i = (i, i+1)$, where $i = 0, 1, \ldots, 4$. Edge labels and loops are not shown.   ◇

To compute the orbit $G(x)$ of a point $x \in X$, we compute a *spanning tree* of the connected component of $\mathcal{G}$ containing x. This spanning tree is a cycle-free connected subgraph of $\mathcal{G}$, rooted at x, whose vertices are the elements of $G(x)$. This means that there is a unique directed path from x to any element y in $G(x)$. This spanning tree can be described by the following data structure:

9.2.5   **Definition (Schreier-tree)** Let G be a group acting on a finite set X. Let G be given by generators $s_0, \ldots, s_{r-1}$. Let $\mathcal{G} = (X, \mathcal{E})$ be the action graph for G acting on X. Let x be an element of X. A *Schreier-tree* for the orbit of x is a spanning tree for the connected component of $\mathcal{G}$ containing x. The tree is rooted at x and all edges are pointing away from x.   ◇

We remark that a spanning tree for a connected component of a graph is in general not unique. For action graphs, this reflects the fact that there may be different ways to obtain a given element $y \in X$ as an image of x under group elements $g_1, g_2 \in G$. We will investigate these questions no further but we note, however, that the shape of the tree is important for performance considerations. For example, the average depth of a node should be small. There are special methods to build "shallow" Schreier-trees, see Seress [177]. The trick is to change the generating set S which is used for calculating the action graph beforehand.

The following basic orbit algorithm computes a Schreier-tree for the orbit of x under G. It uses a data structure called queue, which is similar to a waiting line. The new elements are appended to the end of the queue, and the elements are taken out in order. This means that the front-most element is processed

first, then the second element and so forth until all elements are processed and
the queue is empty.

Algorithm (orbits on points)   9.2.6

   Input: A permutation group G acting on a finite set $X = \{x_1, \ldots, x_n\}$, a
                   generating set $S = \{s_0, \ldots, s_{r-1}\}$ of G, a point $x \in X$.
   Output: A Schreier-tree $T = (\mathcal{O}, \mathcal{E})$ for the orbit $\mathcal{O} = G(x)$.

(1) let Q be a queue holding the element x
(2) let $\mathcal{O} := \{x\}, \mathcal{E} = \varnothing$, so that $T = (\{x\}, \varnothing)$ has only one node x
(3) **while** $Q \neq \varnothing$ **do**
(4)   let y be the first element of Q (remove y from Q)
(5)   **for** $i \in r$ **do**
(6) $z := y s_i$
(7)   **if** $z \notin \mathcal{O}$ **then**
(8)   append z to Q, add z to $\mathcal{O}$
(9)   add the edge (y, z) labeled by s_i to $\mathcal{E}$
(10)   **end if**
(11)   **end for**
(12) **end while**   □

Example Let G be the permutation group generated by   9.2.7

$$s_0 = (3,4)(9,14)(10,13)(11,12),$$
$$s_1 = (3,9)(4,14)(10,11)(12,13),$$
$$s_2 = (3,11)(4,12)(9,10)(13,14),$$
$$s_3 = (2,3)(6,9)(7,10)(8,11),$$
$$s_4 = (1,2)(5,6)(10,12)(11,13),$$
$$s_5 = (0,1)(6,7)(9,10)(13,14).$$

The action-graph and a spanning Schreier-tree are shown in Fig. 9.2. It can be
shown that $G \simeq \mathrm{PGL}_4(2)$. See also Examples 9.2.11, 9.3.11 and 9.8.12 below. $\diamond$

Let us now consider the problem of computing G_x, the stabilizer of x in G,
for $x \in X$. The following result, due to Schreier, provides a set of generators
for G_x, given generators for G.

Theorem (Schreier) *Let G be a finite group generated by a set of elements S. Let*   9.2.8
H be a subgroup of G and let $\mathcal{R}$ be a set of right coset representatives for H in G
containing 1. For $r \in \mathcal{R}$ and $s \in S$, let $\overline{rs}$ be the unique element in $\mathcal{R}$ with $rs \in H\overline{rs}$.
Then H is generated by all elements of the form $rs\overline{rs}^{-1}$, where $r \in \mathcal{R}$ and $s \in S$. Each
such element is called a Schreier-generator

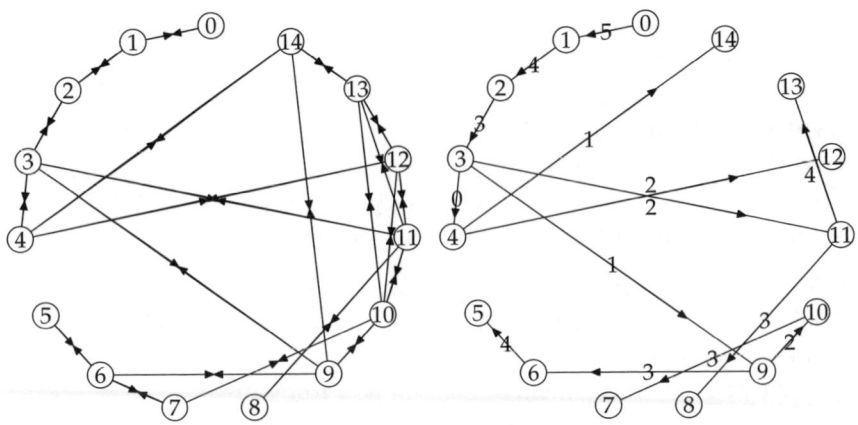

Fig. 9.2 Action-graph and Schreier-tree

Proof: The set $\mathcal{R}$ is a system of right coset representatives of H in G so that

$$G = \bigcup_{r \in \mathcal{R}} Hr.$$

We extend the function defined in the theorem to the whole group by letting $\bar{g}$, for $g \in G$, be the unique element in $\mathcal{R}$ with $g \in H\bar{g}$. Note that $\overline{hg} = \bar{g}$ if $h \in H$. Also, $\bar{\bar{g}} = \bar{g}$ for all $g \in G$. Finally, $\bar{g} = 1$ if and only if $g \in H$. Suppose $g = s_1 s_2 \cdots s_t \in G$ with each $s_i \in S$. Put

$$g_0 = 1, \ g_1 = s_1, \ g_2 = s_1 s_2, \ \ldots, \ g_t = s_1 s_2 \cdots s_t = g.$$

Write

$$u_0 = \bar{g}_0 = 1, \ u_1 = \bar{g}_1, \ \ldots, \ u_t = \bar{g}_t = \bar{g}.$$

Then

9.2.9
$$g u_t^{-1} = s_1 s_2 s_3 \cdots s_t u_t^{-1} = u_0 s_1 u_1^{-1} u_1 s_2 u_2^{-1} u_2 \cdots u_{t-1}^{-1} u_{t-1} s_t u_t^{-1},$$

which equals g if g is in H since then $u_t = \bar{g} = 1$. By definition of the function $g \mapsto \bar{g}$, we deduce from $\overline{g_{i-1}} = u_{i-1}$ that $g_{i-1} \in Hu_{i-1}$. Hence there exists an element $h \in H$ with $g_{i-1} = hu_{i-1}$. Therefore $g_{i-1}s_i = hu_{i-1}s_i$, which implies $\overline{g_{i-1}s_i} = \overline{u_{i-1}s_i}$. It follows that for $i \geq 1$

$$u_i = \bar{u}_i = \bar{\bar{g}}_i = \bar{g}_i = \overline{g_{i-1}s_i} = \overline{u_{i-1}s_i},$$

which is an element of the form $\overline{rs}$ with $r \in \mathcal{R}$ and $s \in S$. Now let $g \in H$ and hence $u_t = 1$. Then 9.2.9 becomes

$$g = g u_t^{-1} = \prod_{i=1}^{t} u_{i-1} s_i u_i^{-1} = \prod_{i=1}^{t} u_{i-1} s_i \overline{u_{i-1}s_i}^{-1},$$

i.e. g can be written as a product of elements of the form $r s \overline{r s}^{-1}$, with $r \in \mathcal{R}, s \in S$. This finishes the proof. □

One particular instance of this is the computation of point stabilizers in permutation groups.

Corollary *Let the group G act on the finite set X and let S be a set of generators for   **9.2.10** G. For $x \in X$, let $\mathcal{R} = \{r_1, r_2, \ldots, r_\ell\}$ with $r_1 = 1$ be a set of elements such that the following holds: For each $y \in G(x)$ there is one and only one element $r \in \mathcal{R}$ with $xr = y$ (and therefore $|G(x)| = \ell$). Then*

$$G_x = \langle r s \overline{r s}^{-1} \mid r \in \mathcal{R}, s \in S \rangle.$$ □

The last of the three problems is that of computing transporter elements g such that $yg = z$ (provided y and z are in the same G-orbit, say $G(x)$, of course). Such transporter elements can be computed from the Schreier-tree of the orbit. Let $T = (G(x), \mathcal{E})$, and let y and z be elements of the orbit. Following the edge labels along the path from x to y and from x to z, respectively, we obtain group elements u and v with $xu = y$ and $xv = z$. Then $yu^{-1}v = xv = z$, so that $u^{-1}v$ is a transporter element.

Example (continuation of Example 9.2.7) An element $g \in G$ mapping 6 to 13,   **9.2.11** for example, can be determined directly from the tree:

$$g = s_3^{-1} s_1^{-1} s_2 s_4 = (1, 2, 12, 14, 10, 7, 3)(4, 11, 8, 9, 5, 6, 13).$$

The reader should carefully note that this product of permutations has to be read from the left to the right, since in this chapter we prefer actions from the right. ◇

We need further notation concerning the solution of orbit type problems.

Definition (orbit data structure) Let G be a group which acts on the finite set   **9.2.12** X. The triple

$$\text{orbit}(G, X) = (T, \sigma, \varphi) := (T, \sigma, \varphi)$$

is the *orbit data* for G acting on X provided that

1. T is a transversal of the G-orbits on X,

2. $\sigma : X \to L(G) : x \mapsto G_x$,

3. $\varphi : X \to G : x \mapsto g$ with $xg \in T$.

Here, $L(G)$ denotes the lattice of subgroups of G as defined in 3.4.4. We call σ the *stabilizer map* and φ the *transporter map*. ◇

9.2.13   **Remarks**

1. It follows from 3.4.1 that the image of the map φ is unique only modulo elements of the stabilizer.
2. The orbit data structure is not a purely mathematical object. The point with the maps φ and σ is that we should be able to compute function values efficiently. When we say that the orbit data $(\mathcal{T}, \sigma, \varphi)$ for G on X is available, we mean that we have determined $\mathcal{T}$ and are able to evaluate the maps σ and φ with reasonably small effort.
3. It may be that $\sigma(y)$ is known only for elements of the transversal $\mathcal{T}$. If this is the case, and if in addition we are able to compute transporter elements, then for a given x in X we compute $g = \varphi(x)$, $y = xg$ and $\sigma(y) = G_y$. It follows from 3.4.3 that

$$\sigma(x) = G_x = g^{-1} G_y g = g^{-1} \sigma(y) g.$$ ◇

Exercises

E.9.2.1   **Exercise** Let G be a group, generated by a set $S = \{s_0, \ldots, s_{r-1}\}$ of generators. Then each $g \in G$ has an expression of the form $g = s_{i_0}^{\epsilon_0} s_{i_1}^{\epsilon_1} \ldots s_{i_{r-1}}^{\epsilon_{r-1}}$ with $i_k \in r$ and $\epsilon_k \in \{\pm 1\}$. Show that if G is finite, we can find an expression for g of this form with $\epsilon_k = 1$ for $k \in r$.

9.3 A Permutation Representation

To get back to codes, let us start by enumerating the points of finite projective spaces. This allows us to translate the action of the general linear group from that of a matrix group to that of a permutation group. We will do the case of $\mathbb{F}_q^k$ first, and then move on to the projective case.

Assume that $\kappa_0, \kappa_1, \kappa_2, \ldots, \kappa_{q-1}$ are the elements of the field $\mathbb{F}_q$, where we always require that $\kappa_0 = 0$ is the zero element and $\kappa_1 = 1$ is the unit element in the field.

We start by ranking the points in $\mathbb{F}_q^k = \{\sum_{i=0}^{k-1} v_i e^{(i)} \mid v_i \in \mathbb{F}_q\}$. Recall from 1.3.5 that for every integer $q \geq 2$ we have the base q expression of an integer $m \in q^k$

$$m = \sum_{i \in k} a_i q^i,$$

which we abbreviate as $m = (a_{k-1}, \ldots, a_0)_q$.

Lemma *Let q be a prime power. Let $m \in q^k$ be an integer with $m = (a_{k-1}, \ldots, a_0)_q$.*   **9.3.1**
The map

$$\mathrm{rk}_{k,q}^{-1} : q^k \rightarrow \mathbb{F}_q^k : m \mapsto (\kappa_{a_0}, \ldots, \kappa_{a_{k-1}}),$$ **9.3.2**

is a bijection, we call it the unrank *function for $\mathbb{F}_q^k$. Its inverse*

$$\mathrm{rk}_{k,q} : \mathbb{F}_q^k \rightarrow q^k : (\kappa_{a_0}, \ldots, \kappa_{a_{k-1}}) \mapsto m,$$ **9.3.3**

is the rank *function for $\mathbb{F}_q^k$.*   □

The proof is straightforward.

Example Let $\mathbb{F}_3 = \{\kappa_0, \kappa_1, \kappa_2\}$, with $\kappa_0 = \bar{0} = 0$, $\kappa_1 = \bar{1} = 1$, and $\kappa_2 = \bar{2} = 2$.   **9.3.4**
We obtain the following unrank function for $\mathbb{F}_3^2$.

$$\mathrm{rk}_{2,3}^{-1}(0) = (0,0), \ \mathrm{rk}_{2,3}^{-1}(1) = (1,0), \ \mathrm{rk}_{2,3}^{-1}(2) = (2,0),$$
$$\mathrm{rk}_{2,3}^{-1}(3) = (0,1), \ \mathrm{rk}_{2,3}^{-1}(4) = (1,1), \ \mathrm{rk}_{2,3}^{-1}(5) = (2,1),$$
$$\mathrm{rk}_{2,3}^{-1}(6) = (0,2), \ \mathrm{rk}_{2,3}^{-1}(7) = (1,2), \ \mathrm{rk}_{2,3}^{-1}(8) = (2,2).$$

Correspondingly

$$\mathrm{rk}_{2,3}((0,0)) = 0, \ \mathrm{rk}_{2,3}(e^{(0)}) = 1, \ \mathrm{rk}_{2,3}(e^{(1)}) = 3, \ldots \qquad \Diamond$$

Let us turn our attention to the projective space $\mathrm{PG}_d(q)$. We want to enumerate (i.e. label) the set of one-dimensional subspaces $\langle v \rangle$ of $\mathbb{F}_q^{d+1}$, where $v \neq 0$. Recall from Section 3.7 that we denote the number of points of $\mathrm{PG}_d(q)$ by

$$\theta_d(q) = \frac{q^{d+1} - 1}{q - 1} = |\mathrm{PG}_d(q)| = q^d + q^{d-1} + \ldots + q + 1.$$

In order to enumerate the points of a projective space $\mathrm{PG}_d(q)$, we are going to choose nonzero representatives out of each one-dimensional subspace of $V = \mathbb{F}_q^{d+1}$. Let $e^{(0)}, \ldots, e^{(d)}$ be the standard basis of V. We introduce the following notation. For

$$u = \langle u_0 e^{(0)} + \ldots + u_d e^{(d)} \rangle \in \mathrm{PG}_d(q),$$

let $\mathrm{lc}(u)$ be the largest index i for which $u_i \neq 0$ (and hence $u_{i+1} = \cdots = u_d = 0$). We call $\mathrm{lc}(u)$ the *leading coefficient* of u. Notice that this definition depends on the labeling of the basis vectors, which is intentional. To label the one-dimensional subspaces of V we need to pick one nonzero vector out of each such subspace. A simple way to do this is to take as representatives the vectors

$$u = (u_0, \ldots, u_d) \in \mathbb{F}_q^{d+1}$$

whose rightmost nonzero coordinate is one, i.e. with $u_k = 1, u_{k+1} = \cdots = u_d = 0$, where $k = \mathrm{lc}(u)$. Such vectors are called *standard*.

There is one more condition which we pose but which seems a little unmotivated at this point. We require that the unit vectors and the all-one vector get the smallest possible ranks, i.e. we ask that

$$\mathrm{rk}(\langle e^{(0)}\rangle) = 0,$$
$$\mathrm{rk}(\langle e^{(1)}\rangle) = 1,$$
$$\vdots$$
$$\mathrm{rk}(\langle e^{(d)}\rangle) = d,$$
$$\mathrm{rk}(\langle e^{(0)} + \ldots + e^{(d)}\rangle) = d + 1.$$

The reason for this requirement will become clear in Section 9.8, when we exhibit a special property of these vectors (namely, they form a "base" in the sense of Section 9.7).

The remaining vectors are of the form

$$u = (u_0, \ldots, u_{k-1}, 1, 0, \ldots, 0)$$

with $(u_0, \ldots, u_{k-1}) \in \mathbb{F}_q^k \setminus \{0\}$, where $k = \mathrm{lc}(u)$. If $k = d$ we also have that $(u_0, \ldots, u_{d-1}) \neq (1, \ldots, 1)$. We decide to order these vectors first according to the value of k (which can take any value from 1 to d). Among the vectors u for a given $k = \mathrm{lc}(u)$ we order according to the ranks of $(u_0, \ldots, u_{k-1})$ as points in $\mathbb{F}_q^k$ as given by 9.3.3. We skip the zero vector which cannot occur. If $k = d$ we also need to skip the all-one vector. This requires some additional effort. We will shift the rank before we apply 9.3.2 and conversely we will also shift the rank after application of 9.3.3. The all-one vector – as an element of $\mathbb{F}_q^d$ – has rank

$$1 + q + q^2 + \ldots + q^{d-1} = \frac{q^d - 1}{q - 1} = \theta_{d-1}(q).$$

Therefore, we need to increase all ranks which are greater than or equal to this number by one before calling 9.3.2. Conversely, if we are ranking a vector u with $\mathrm{lc}(u) = d$, we need to decrease all ranks of $(u_0, \ldots, u_{d-1}) \in \mathbb{F}_q^d$ by one if they happen to be greater than $\theta_{d-1}(q)$. To facilitate this we will introduce a shift function. Summarizing, we have the following unrank and rank functions for the points of $\mathrm{PG}_d(q)$. We remark that the if clauses are to be read in order, that is, the second and all following if clauses are to be understood as "otherwise if."

9.3.5

Lemma We define the unrank function $\mathrm{rk}_{d;q}^{-1} : \theta_d(q) \to \mathrm{PG}_d(q)$ by

9.3.6

$$\mathrm{rk}_{d;q}^{-1}(m) = \begin{cases} \langle e^{(m)} \rangle & \text{if } m \le d, \\ \langle \sum_{i=0}^{d} e^{(i)} \rangle & \text{if } m = d + 1, \\ \langle \mathrm{rk}_{d,1;q}^{-1}(m - d - 1) \rangle & \text{otherwise,} \end{cases}$$

where

$$
rk_{d,k;q}^{-1}(m) = \begin{cases} rk_{d,*;q}^{-1}(m) & if\ k=d \\ e^{(k)} + rk_{k,q}^{-1}(m) & if\ m < q^k \\ rk_{d,k+1;q}^{-1}(m - q^k + 1) & otherwise. \end{cases}
$$

9.3.7

Here,

$$
rk_{d,*;q}^{-1}(m) = e^{(d)} + rk_{d,q}^{-1}\left(shift_{\theta_{d-1}(q)}(m)\right)
$$

9.3.8

with

$$
shift_j(m) := \begin{cases} m & if\ m < j, \\ m+1 & otherwise. \end{cases}
$$

9.3.9

This map $rk_{d;q}^{-1}$ is a bijection. Its inverse is the rank function *for $PG_d(q)$, denoted as $rk_{d;q}$. For a point $\langle u \rangle$ with $u = (u_0, u_1, \ldots, u_d) \in \mathbb{F}_q^{d+1} \setminus \{0\}$ one has $rk_{d;q}(\langle u \rangle) =$*

$$
\begin{cases} k & if\ \langle u \rangle = \langle e^{(k)} \rangle \\ d+1 & if\ \langle u \rangle = \langle 1, \ldots, 1 \rangle \\ d+2-k+q\theta_{k-2}(q) + rk_{k,q}\left(\frac{u_0}{u_k}, \ldots, \frac{u_{k-1}}{u_k}\right) & if\ k = lc(u) < d \\ 2+q\theta_{d-2}(q) + shift_{\theta_{d-1}(q)}^{-1}\left(rk_{d,q}\left(\frac{u_0}{u_d}, \ldots, \frac{u_{d-1}}{u_d}\right)\right) & if\ lc(u) = d. \end{cases}
$$

9.3.10

$\square$

Example We have $\theta_2(2) = 2^2 + 2 + 1 = 7$, $\theta_2(3) = 3^2 + 3 + 1 = 13$ and $\theta_3(2) = 2^3 + 2^2 + 2 + 1 = 15$. Table 9.1 shows the labeling of points of $PG_2(2)$, $PG_2(3)$ and $PG_3(2)$. Let us see some specific examples. We have

9.3.11

$$
\begin{aligned}
rk_{3;2}^{-1}(4) &= \langle 1,1,1,1 \rangle & \text{by 9.3.6,} \\
rk_{3;2}^{-1}(5) &= \langle rk_{3,1;2}^{-1}(1) \rangle & \text{by 9.3.6} \\
&= \langle e^{(1)} + rk_{1,2}^{-1}(1) \rangle & \text{by 9.3.7} \\
&= \langle e^{(1)} + e^{(0)} \rangle = \langle 1,1,0,0 \rangle & \text{by 9.3.2,} \\
rk_{3;2}^{-1}(14) &= \langle rk_{3,1;2}^{-1}(10) \rangle & \text{by 9.3.6} \\
&= \langle rk_{3,2;2}^{-1}(9) \rangle & \text{by 9.3.7} \\
&= \langle rk_{3,3;2}^{-1}(6) \rangle & \text{by 9.3.7} \\
&= \langle rk_{3,*;2}^{-1}(6) \rangle & \text{by 9.3.7} \\
&= \langle e^{(3)} + rk_{3,2}^{-1}(shift_7(6)) \rangle & \text{by 9.3.7} \\
&= \langle e^{(3)} + rk_{3,2}^{-1}(6) \rangle & \text{by 9.3.9} \\
&= \langle e^{(3)} + e^{(2)} + e^{(1)} \rangle = \langle 0,1,1,1 \rangle & \text{by 9.3.2,} \\
rk_{2;3}^{-1}(12) &= \langle rk_{2,1;3}^{-1}(9) \rangle & \text{by 9.3.6} \\
&= \langle rk_{2,2;3}^{-1}(7) \rangle & \text{by 9.3.7} \\
&= \langle rk_{2,*;3}^{-1}(7) \rangle & \text{by 9.3.7}
\end{aligned}
$$

Table 9.1 Labeling $PG_2(2)$, $PG_2(3)$ and $PG_3(2)$

m	$rk_{2;2}^{-1}(m)$	$rk_{2;3}^{-1}(m)$	$rk_{3;2}^{-1}(m)$
0	$\langle 1,0,0 \rangle$	$\langle 1,0,0 \rangle$	$\langle 1,0,0,0 \rangle$
1	$\langle 0,1,0 \rangle$	$\langle 0,1,0 \rangle$	$\langle 0,1,0,0 \rangle$
2	$\langle 0,0,1 \rangle$	$\langle 0,0,1 \rangle$	$\langle 0,0,1,0 \rangle$
3	$\langle 1,1,1 \rangle$	$\langle 1,1,1 \rangle$	$\langle 0,0,0,1 \rangle$
4	$\langle 1,1,0 \rangle$	$\langle 1,1,0 \rangle$	$\langle 1,1,1,1 \rangle$
5	$\langle 1,0,1 \rangle$	$\langle 2,1,0 \rangle$	$\langle 1,1,0,0 \rangle$
6	$\langle 0,1,1 \rangle$	$\langle 1,0,1 \rangle$	$\langle 1,0,1,0 \rangle$
7		$\langle 2,0,1 \rangle$	$\langle 0,1,1,0 \rangle$
8		$\langle 0,1,1 \rangle$	$\langle 1,1,1,0 \rangle$
9		$\langle 2,1,1 \rangle$	$\langle 1,0,0,1 \rangle$
10		$\langle 0,2,1 \rangle$	$\langle 0,1,0,1 \rangle$
11		$\langle 1,2,1 \rangle$	$\langle 1,1,0,1 \rangle$
12		$\langle 2,2,1 \rangle$	$\langle 0,0,1,1 \rangle$
13			$\langle 1,0,1,1 \rangle$
14			$\langle 0,1,1,1 \rangle$

$$
\begin{aligned}
&= \langle e^{(2)} + rk_{2,3}^{-1}(\mathrm{shift}_4(7)) \rangle \quad \text{by 9.3.7} \\
&= \langle e^{(2)} + rk_{2,3}^{-1}(8) \rangle \quad \text{by 9.3.9} \\
&= \langle e^{(2)} + 2e^{(1)} + 2e^{(0)} \rangle = \langle 2,2,1 \rangle \quad \text{by 9.3.2.}
\end{aligned}
$$

Conversely, we have

$$
\begin{aligned}
rk_{3;2}(\langle 1,1,1,1 \rangle) &= 4 \quad \text{by 9.3.10,} \\
rk_{3;2}(\langle 1,1,0,0 \rangle) &= 3+2-1+\frac{0}{1}+rk_{1,2}((1)) \quad \text{by 9.3.10} \\
&= 4+1 = 5 \quad \text{by 9.3.3,} \\
rk_{3;2}(\langle 0,1,1,1 \rangle) &= 2+\frac{6}{1}+\mathrm{shift}_6^{-1}(rk_{3,2}((0,1,1))) \quad \text{by 9.3.10} \\
&= 8+\mathrm{shift}_7^{-1}(6) \quad \text{by 9.3.3,} \\
&= 8+6 = 14 \quad \text{by 9.3.9,} \\
rk_{2;3}(\langle 2,2,1 \rangle) &= 2+\frac{6}{2}+\mathrm{shift}_4^{-1}(rk_{2,3}((2,2))) \quad \text{by 9.3.10} \\
&= 5+\mathrm{shift}_4^{-1}(8) \quad \text{by 9.3.3} \\
&= 5+7 = 12 \quad \text{by 9.3.9.} \quad\quad\quad \diamond
\end{aligned}
$$

Example Using the ranks of the previous example, the permutations of Examples 9.2.7 and 9.2.11 can be written as matrices. Recall that we use row-vector convention, i.e. the images of a linear map are written in the rows of the corresponding matrix. For instance, the elements of the generating set S can be written as matrices as follows.

$$s_0 = \begin{pmatrix} 1\,0\,0\,0 \\ 0\,1\,0\,0 \\ 0\,0\,1\,0 \\ 1\,1\,1\,1 \end{pmatrix},$$

9.3.12

since $0s_0 = 0, 1s_0 = 1, 2s_0 = 2, 3s_0 = 4$ and $0, 1, 2, 4$ are the ranks of the projective points which are represented by the vectors in the rows of this matrix. Similarly, we obtain

$$s_1 = \begin{pmatrix} 1\,0\,0\,0 \\ 0\,1\,0\,0 \\ 0\,0\,1\,0 \\ 1\,0\,0\,1 \end{pmatrix}, \; s_2 = \begin{pmatrix} 1\,0\,0\,0 \\ 0\,1\,0\,0 \\ 0\,0\,1\,0 \\ 1\,1\,0\,1 \end{pmatrix}, \; s_3 = \begin{pmatrix} 1\,0\,0\,0 \\ 0\,1\,0\,0 \\ 0\,0\,0\,1 \\ 0\,0\,1\,0 \end{pmatrix},$$

$$s_4 = \begin{pmatrix} 1\,0\,0\,0 \\ 0\,0\,1\,0 \\ 0\,1\,0\,0 \\ 0\,0\,0\,1 \end{pmatrix}, \; s_5 = \begin{pmatrix} 0\,1\,0\,0 \\ 1\,0\,0\,0 \\ 0\,0\,1\,0 \\ 0\,0\,0\,1 \end{pmatrix}.$$

From this we see that the group of Example 9.2.7 really is a projective linear matrix group. That it is the full group $\mathrm{PGL}_4(2)$ will follow from a result in Section 9.8, where a special generating set ("strong generators") for this group is exhibited. The permutation

$$(1, 2, 12, 14, 10, 7, 3)(4, 11, 8, 9, 5, 6, 13)$$

of Example 9.2.11 is in fact the matrix

$$A = \begin{pmatrix} 1\,0\,0\,0 \\ 0\,0\,1\,0 \\ 0\,0\,1\,1 \\ 0\,1\,0\,0 \end{pmatrix},$$

and we have that

$$\mathrm{rk}_{3;2}^{-1}(6) \cdot A = \langle 1, 0, 1, 0 \rangle \cdot \begin{pmatrix} 1\,0\,0\,0 \\ 0\,0\,1\,0 \\ 0\,0\,1\,1 \\ 0\,1\,0\,0 \end{pmatrix}$$

$$= \langle 1, 0, 1, 1 \rangle = \mathrm{rk}_{3;2}^{-1}(13). \qquad \diamond$$

9.4 The Lexicographical Order

Let $(X, \leq)$ be a totally ordered set. In this section, we are concerned with the set of subsets of X, also known as the *power set* of X. In addition to our customary notation 2^X, we will introduce the notation $\mathcal{P}(X)$ for this set. That is,

$$\mathcal{P}(X) = \{A \mid A \subseteq X\}.$$

Clearly, the size of $\mathcal{P}(X)$ is $2^{|X|}$. Later on, we will also consider the set of subsets of size k of X, for some nonnegative integer $k \leq |X|$. We denote this set as

$$\mathcal{P}_k(X) = \{A \mid A \subseteq X, |A| = k\}.$$

We introduce the following notation. For a subset A of a totally ordered set X we write $A = \{a_0, a_1 \ldots, a_{m-1}\}_<$ to indicate that the elements of A are listed in order, i.e. that $a_0 < a_1 < \cdots < a_{m-1}$. The set $\mathcal{P}(X)$ can be ordered in a very natural way, using the ordering of elements of X. This is the *lexicographical order* which has already appeared in 3.4.20.

9.4.1   **Definition (the lexicographical order)** For subsets $A = \{a_0, a_1, \ldots, a_{m-1}\}_<$ and $B = \{b_0, b_1, \ldots, b_{n-1}\}_<$ of the totally ordered set X we put

$$A \preceq B \iff \begin{cases} \exists\, r < \min(m, n) : a_i = b_i \text{ for } i \in r \text{ and } a_r < b_r, \text{ or} \\ m \leq n \text{ and } a_i = b_i \text{ for } i \in m. \end{cases} \qquad \diamond$$

9.4.2   **Example** Let $X = \{a, b, \ldots, z\}$ be the Roman alphabet with the usual ordering of letters. Then $\mathcal{P}(X)$ is ordered lexicographically as follows (we leave out set brackets and commas for simplicity).

$$\emptyset \prec a \prec ab \prec abc \prec \cdots \prec abc \ldots wxyz \prec abc \ldots wxz$$
$$\prec abc \ldots wy \prec abc \ldots wyz \prec abc \ldots wz \prec \cdots$$
$$\prec b \prec bc \prec \cdots \prec bcd \ldots xyz \prec \cdots \prec y \prec yz \prec z. \qquad \diamond$$

Let $(X, \leq)$ be a totally ordered finite set. The lexicographical order on $\mathcal{P}(X)$ can be represented by a tree, the *order tree* $T_{(X, \preceq)}$ or simply $T_\preceq$. The nodes of $T_\preceq$ are the subsets of X, i.e. the elements of the power set $\mathcal{P}(X)$. The edges of $T_\preceq$ can be described as follows. For subsets A and B (of a totally ordered set X), we say that A is a *prefix* of B if $A \subseteq B$ and either $A = B$ or $\min(B \setminus A) > \max A$. In other words, the prefixes of a set $B = \{b_0, b_1, \ldots, b_{m-1}\}_<$ are just the sets $\{b_0, \ldots, b_i\}$ for $i \leq m - 1$. If A is a prefix of B then we say that B is a *descendant* (or *offspring*) of A or that A is an ancestor of B. We say that B is an *immediate descendant* of A if B is a descendant of A and $|B| = |A| + 1$, i.e. $B = A \cup \{\max B\}$. Two nodes are *siblings* if they are immediate descendants of

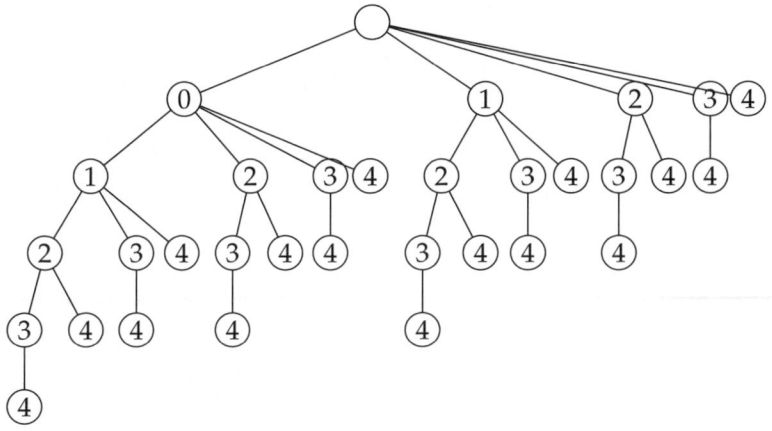

Fig. 9.3 Order tree of subsets of $\{0,1,2,3,4\}$

the same node. The edges of the tree $T_{\preceq}$ are between immediate descendants.
One can think of these edges as being directed, pointing from the smaller to
the larger set. In this sense, $T_{\preceq}$ is a rooted tree, with the empty set serving as
root. A *leaf* is a node without descendants. An *inner node* is a node which is not
a leaf. We say that a node has *distance i* from the root if the unique path from
the root to that node has length i. The i-th *level* of the tree is the set of nodes at
distance i from the root. A *common ancestor* of two sets A and B is an ancestor
of both A and B. The *immediate common ancestor* of A and B is the ancestor of A
and B which is largest in size. We arrange the siblings of a node according to
their largest element, using the original ordering of elements of X. The siblings
are drawn from left to right in increasing order. Thus $T_{\preceq}$ is an ordered tree. In
the Computer Science literature (cf. [42], for example) these trees are known as
binomial trees. We also mention that it is ubiquitous in the Computer Science
literature that trees grow top down. Thus the root node is the node on top of
the drawing, whereas the leaves are the nodes at the bottom.

Example Consider the five element set $X := \{0,1,2,3,4\}_{<}$. Figure 9.3 displays   **9.4.3**
the order tree $T_{(X,\preceq)}$. Here, we label the nodes by the largest element of the set
which they represent (the root node is represented as an empty node). Clearly,
the set corresponding to a node can be reconstructed by simply collecting all
labels of nodes along the unique path from the root to the node.

There are essentially two different ways of traversing the nodes of a tree.
The *depth first search* strategy (or *pre-order traversal*) is to move down the tree,
visiting a node and its offsprings recursively. This is done in such a way that

the leftmost offspring and its whole branch is visited first. Then on the way back one moves to the right, visiting possible siblings and their whole subtrees recursively. This way all the siblings are dealt with before the procedure returns to its ancestor. One may imagine this procedure as follows. Think of the tree as a fence in the plane. Walk around that fence, starting from the root node in the direction to the leftmost offspring, keeping the fence to your left. A different way to visit the nodes of a tree is the *breadth first search* order. Here, the nodes at any given level are visited in order, starting from the root and going down to deeper levels.

The ordering of subsets is encoded in the tree. Namely, we encounter the sets in lexicographical order if we visit the nodes of the tree in depth first strategy. In the above example, the depth first search arranges the subsets of $\{0, \ldots, 4\}$ in the following order, which is indeed lexicographical.

$$\varnothing \prec 0 \prec 01 \prec 012 \prec 0123 \prec 01234 \prec 0124 \prec 013 \prec 0134$$
$$\prec 014 \prec 02 \prec 023 \prec 0234 \prec 024 \prec 03 \prec 034 \prec 04$$
$$\prec 1 \prec 12 \prec 123 \prec 1234 \prec 124 \prec 13 \prec 134 \prec 14$$
$$\prec 2 \prec 23 \prec 234 \prec 24 \prec 3 \prec 34 \prec 4.$$

Using breadth first search, the nodes of the tree will be visited in the following order:

$$\varnothing,$$
$$0, 1, 2, 3, 4,$$
$$01, 02, 03, 04, 12, 13, 14, 23, 24, 34,$$
$$012, 013, 014, 023, 024, 034, 123, 124, 134, 234,$$
$$0123, 0124, 0134, 0234, 1234,$$
$$01234. \qquad \qquad \diamond$$

Let us collect fundamental properties of the order tree in the following lemma. The proofs are straightforward and therefore omitted.

9.4.4   **Lemma** Let $X = \{x_0, x_1, \ldots, x_{n-1}\}_<$ be a finite totally ordered set. Let $\preceq$ be the lexicographical order on $\mathcal{P}(X)$. Then the order tree $T_{\preceq}$ has the following properties.

1. For every node of the tree, the corresponding set is the union of the labels along the path leading to that node. Moreover, the labels are encountered in ascending order along this path.

2. The nodes at level i correspond to i-subsets of X, and hence there are $\binom{n}{i}$ of them.

3. For $A, B \subseteq X$, a common ancestor of A and B corresponds to a prefix of $A \cap B$ and vice-versa. The immediate common ancestor is the prefix of $A \cap B$ which is largest in size.

4. *The tree has 2^{n-1} leaves corresponding to the subsets of X which contain x_{n-1}. The tree has 2^{n-1} inner nodes corresponding to the subsets of X which do not contain x_{n-1}.*

5. *The subtree rooted at a set $A \subseteq X$ consists of the subsets $B \subseteq X$ for which A is a prefix of B. If $\max A = x_i$, there are 2^{n-1-i} such nodes. In particular, the subtree whose root is $\{x_i\}$ (i.e. the tree which is rooted at the i-th descendant of the global root), contains all subsets $A \subseteq X$ with $\min A = x_i$. There are 2^{n-1-i} such sets.*

6. *Two subtrees rooted at sets A and B (with $A, B \subseteq X$) are equal in shape and labeling of the nodes if and only if $\max A = \max B$.*

7. *If the order tree is traversed in depth first search, the subsets are encountered in lexicographic order. That is, a subset A precedes a subset B in the lexicographical order if and only if A is reached first when traversing the tree $T_{(X,\preceq)}$ in depth first search. In terms of the common ancestor C of A and B we can say that $A \preceq B$ if and only if either $C = A$ or $\min A \setminus C < \min B \setminus C$. That is, $A \preceq B$ if and only if either B is a descendant of A or the branch containing B is to the right of the branch containing A among the siblings of the immediate common ancestor of A and B.*

8. *Let A be a subset with $\max A = x_i$ (put $i = -1$ if $A = \emptyset$). The leftmost leaf in the subtree rooted at A is the set $A \cup \{x_{i+1}, \ldots, x_{n-1}\}$. The rightmost leaf in the subtree rooted at A is the set $A \cup \{x_{n-1}\}$.*

9. *Let A be a subset with $\max A = x_i$ (put $i = -1$ if $A = \emptyset$). The subtree rooted at A contains exactly $\binom{n-1-i}{k-|A|}$ sets of size k.*   □

It is of course useful to have rank and unrank functions for the set of subsets of a finite set.

Lemma *Let $X = \{x_0, x_1, \ldots, x_{n-1}\}_<$ be a totally ordered finite set of n elements.*   **9.4.5**
Define a function, the rank function, from $\mathcal{P}(X)$ to the set of integers 2^n as follows. For a set $A \subseteq X$ define

$$\mathrm{rk}_X : \mathcal{P}(X) \to 2^n : A \mapsto \begin{cases} 0 & \text{if } A = \emptyset, \\ |A| + \sum_{\substack{x_i \in X \setminus A \\ x_i < \max A}} 2^{n-1-i} & \text{otherwise.} \end{cases}$$

This function is one-to-one and onto. Its inverse is the unrank function, defined as

$$\mathrm{rk}_X^{-1}(r) := \mathrm{rk}_X^{-1}(r, 0),$$

where

$$\mathrm{rk}_X^{-1}(r, m) := \emptyset \quad \text{if } r = 0,$$

while for $0 < r < 2^{n-m}$ we have

$$\text{rk}_X^{-1}(r,m) := \begin{cases} \{x_m\} \cup \text{rk}_X^{-1}(r-1,m+1) & \text{if } 2^{n-1-m} \geq r, \\ \text{rk}_X^{-1}\left(r - 2^{n-1-m}, m+1\right) & \text{if } 2^{n-1-m} < r. \end{cases}$$

□

9.4.6   **Example** For $X = \{0,\dots,4\}$ as above, we have

$$\text{rk}_X(\{1,3,4\}) = 3 + 2^{5-1-0} + 2^{5-1-2} = 23,$$

which can be verified by counting the nodes in Fig. 9.3. The tree rooted at $\{0\}$ has 16 nodes, and the tree rooted at $\{1,2\}$ brings in another 4 nodes, so that the set $\{1,3,4\}$ has indeed rank 23. On the other hand, we have

$$
\begin{aligned}
\text{rk}_X^{-1}(23) &= \text{rk}_X^{-1}(23,0) \\
&\overset{16 \leq 23}{=} \text{rk}_X^{-1}(7,1) \\
&\overset{8 \geq 4}{=} \{1\} \cup \text{rk}_X^{-1}(6,2) \\
&\overset{4 \leq 6}{=} \{1\} \cup \text{rk}_X^{-1}(2,3) \\
&\overset{2 \geq 2}{=} \{1,3\} \cup \text{rk}_X^{-1}(1,4) \\
&\overset{1 \geq 1}{=} \{1,3,4\} \cup \text{rk}_X^{-1}(0,5) \\
&= \{1,3,4\},
\end{aligned}
$$

which is the original set again.   ◇

Sometimes we are only interested in the set $\mathcal{P}_k(X)$ of k-subsets of X, where k is some fixed integer with $0 \leq k \leq |X|$. The elements of $\mathcal{P}_k(X)$ can be ranked and unranked as well.

9.4.7   **Lemma** *Let $X = \{x_0, x_1, \dots, x_{n-1}\}_<$ be a totally ordered finite set of n elements. Let k be an integer with $0 \leq k \leq n$. Define a function, the rank function of $\mathcal{P}_k(X)$ to the set of integers $\binom{n}{k}$ as follows. For a k-subset $A = \{x_{a_0}, x_{a_1}, \dots, x_{a_{k-1}}\}_<$, put*

$$\text{rk}_{X,k} : \mathcal{P}_k(X) \rightarrow \binom{n}{k} : A \mapsto \sum_{i=0}^{k-1} \sum_{j=a_{i-1}+1}^{a_i-1} \binom{n-1-j}{k-1-i},$$

where $a_{-1} := -1$. The function $\text{rk}_{X,k}$ is one-to-one and onto. Its inverse is the function $\text{rk}_{X,k}^{-1}$, which is given by

$$\text{rk}_{X,k}^{-1}(r) = \text{rk}_{X,k}^{-1}(r,0),$$

where

$$\text{rk}_{X,k}^{-1}(r,m) := \emptyset \quad \text{if } k = 0,$$

whereas for k > 0

$$\text{rk}_{X,k}^{-1}(r,m) = \begin{cases} \{x_m\} \cup \text{rk}_{X,k-1}^{-1}(r, m+1) & \text{if } \binom{n-1-m}{k-1} > r, \\ \text{rk}_{X,k}^{-1}\left(r - \binom{n-1-m}{k-1}, m+1\right) & \text{if } \binom{n-1-m}{k-1} \leq r. \end{cases}$$

□

Example For $X = \{0,\dots,4\}$ as above, we have   9.4.8

$$\text{rk}_{X,3}(\{1,3,4\}) = \binom{5-1-0}{3-1-0} + \binom{5-1-2}{3-1-1} = \binom{4}{2} + \binom{2}{1} = 8,$$

which can of course be verified by counting nodes in Fig. 9.3. The tree rooted at $\{0\}$ contains six 3-subsets, and the tree rooted at $\{1,2\}$ brings in another two 3-subsets, so that the set $\{1,3,4\}$ has rank $6 + 2 = 8$. On the other hand, we have

$$\begin{aligned} \text{rk}_{X,3}^{-1}(8) &= \text{rk}_{X,3}^{-1}(8,0) \\ &\overset{\binom{5-1-0}{3-1}=6\leq 8}{=} \text{rk}_{X,3}^{-1}(2,1) \\ &\overset{\binom{5-1-1}{3-1}=3>2}{=} \{1\} \cup \text{rk}_{X,2}^{-1}(2,2) \\ &\overset{\binom{5-1-2}{2-1}=2\leq 2}{=} \{1\} \cup \text{rk}_{X,2}^{-1}(0,3) \\ &\overset{\binom{5-1-3}{2-1}=1>0}{=} \{1,3\} \cup \text{rk}_{X,1}^{-1}(0,4) \\ &\overset{\binom{5-1-4}{1-1}=1>0}{=} \{1,3,4\} \cup \text{rk}_{X,0}^{-1}(0,5) \\ &= \{1,3,4\}, \end{aligned}$$

which is the original set again.   ◇

Exercises

Exercise Compute the rank of $A = \{2,3,5,7\}$ as a subset of $\{0,\dots,7\}$.   E.9.4.1

Exercise Compute $\text{rk}_X^{-1}(99)$ where $X = \{0,\dots,7\}$.   E.9.4.2

Exercise Compute the rank of $A = \{2,3,5,7\}$ as a 4-subset of $\{0,\dots,7\}$.   E.9.4.3

Exercise Compute $\text{rk}_{X,4}^{-1}(66)$ where $X = \{0,\dots,7\}$.   E.9.4.4

Exercise If $X = \{apple, orange, pear, potato, banana, mango, lemon\}_<$, compute

1. $\mathrm{rk}_X(\{orange, potato, mango\})$,
2. $\mathrm{rk}_{X,3}(\{orange, potato, mango\})$,
3. $\mathrm{rk}_X^{-1}(79)$ and
4. $\mathrm{rk}_{X,3}^{-1}(27)$.

9.5 Orderly Generation of Codes

In order to construct linear codes, we need to direct attention to the technique of *orderly generation* of discrete structures. A discrete structure is simply a type of object which can be defined as an orbit of a group acting on a finite set. Examples in Combinatorics are graphs, codes, designs etc. When we speak of the construction of objects, we mean that we produce one object out of each isomorphism class. This object is called the representative, or the labeled object. In the 1970s, the technique of orderly generation has been invented independently by Read [165] and Faradžev [51, 52, 53] for the construction of graphs. The name comes from the fact that it generates representatives for the orbits in question in lexicographic order. A more refined version is described by McKay [146], who also presents an extensive literature list. McKay broadens the technique to general structures and introduces the concept of a canonical extension.

In the following, we will first discuss the technique of orderly generation in some detail and then come back to linear codes later. We start with an action of a group G on a finite set X, whose elements we call points. The group G also acts on subsets of X, via

$$\mathcal{P}(X) \times G \to \mathcal{P}(X) \colon (R, g) \mapsto Rg = \{xg \mid x \in R\}.$$

We call this the *induced action of G on $\mathcal{P}(X)$*. The *setwise stabilizer* of a set $R \subseteq X$ is the subgroup

$$G_R := \{g \in G \mid Rg = R\} = \{g \in G \mid \forall r \in R : rg \in R\}.$$

A related concept is the *pointwise stabilizer* of a set $R = \{r_0, \ldots, r_{s-1}\}$, which is the subgroup

$$G_{r_0,\ldots,r_{s-1}} := \{g \in G \mid r_i g = r_i \text{ for all } i \in s\} = \bigcap_{i \in S} G_{r_i}.$$

Occasionally, we will consider groups which are of mixed type. For instance, if we wish to stabilize the set R setwise, and in addition fix the point x, then we will write

$$G_{R,x} = G_R \cap G_x = \{g \in G \mid Rg = R, \text{ and } xg = x\}.$$

Here, the point x may or may not be a member of the set R. Another case is when the set R is enlarged by one further element x outside of R. The setwise stabilizer of $R \cup \{x\}$ is denoted as

$$G_{R \cup \{x\}} = \{g \in G \mid \forall r \in R : rg \in R \cup \{x\} \text{ and } xg \in R \cup \{x\}\}.$$

We would like to compute the orbits of G on the set of subsets of the finite set $X = \{x_0, \dots, x_{n-1}\}_<$. The following problems arise.

1. Compute a *transversal* $\mathcal{T}$ for the G-orbits on subsets of X, which is a set of subsets of X such that
   (a) each orbit of G on $\mathcal{P}(X)$ is represented by one subset in $\mathcal{T}$, and
   (b) no such orbit is represented twice.
   The elements of the transversal are called *orbit representatives*.

2. For $S \subseteq X$, compute $\sigma(S) = G_S = \{g \in G \mid Sg = S\}$, the setwise stabilizer of S in G.

3. For $S \subseteq X$, determine an element $\varphi(S) = g \in G$ which maps S to its orbit representative in $\mathcal{T}$, i.e. a transporter element (such an element might not be unique).

Of course, in many applications one is not interested in the totality of subsets. Instead, often one has restrictions coming from the particular problem one is interested. This means that we are only interested in a subset of $\mathcal{P}(X)$, or even subsets of

$$\mathcal{P}_i(X) = \{S \subseteq X \mid |S| = i\},$$

the set of subsets of size i. To formalize this idea, we may indicate this condition by a function

$$f : \mathcal{P}(X) \to \{0,1\}, \ S \mapsto f(S)$$

where $f(S)$ is one if and only if the set S is *admissible*, i.e. satisfies the condition. We require that the condition is invariant under the action of the group, i.e. that

$$f(S) = f(Sg) \quad \forall g \in G, \forall S \subseteq X. \tag{9.5.1}$$

Also, we require that the condition is *hereditary*, i.e. that

$$f(S) = 1 \implies f(T) = 1 \quad \forall T \subseteq S \subseteq X. \tag{9.5.2}$$

In the following, we will assume that such a function $f : \mathcal{P}(X) \to \{0,1\}$ has been defined. This is no restriction as one can always define $f(S) = 1$ for all $S \subseteq X$. If f is such a test-function, we may restrict the action of G to the set

$$\mathcal{P}^{(f)}(X) := \mathcal{P}(X) \cap f^{-1}(\{1\}) = \{S \in \mathcal{P}(X) \mid f(S) = 1\}$$

or to one of the sets

$$\mathcal{P}_i^{(f)}(X) := \mathcal{P}_i(X) \cap f^{-1}(\{1\}) = \{S \in \mathcal{P}_i(X) \mid f(S) = 1\}.$$

There are many different ways to choose a transversal. One particular is the *canonical transversal*. It consists of *canonical orbit representatives*, which are the sets $R \subseteq X$ with

$$R \preceq Rg \quad \text{for all } g \in G.$$

Each orbit $G(S)$, $S \subseteq X$ is represented in this transversal by its least element,

$$\overline{S} = \min_{R \in G(S)} R = \min_{g \in G} Sg,$$

where the minimum is taken with respect to the lexicographical order. The function which takes a set S to its canonical orbit representative $\overline{S}$ can be thought of as a projection map. It satisfies the property that $\overline{\overline{S}} = \overline{S}$. The image of this function is the canonical transversal

$$\mathcal{T} = \{\overline{S} \mid S \in \mathcal{P}(X)\}.$$

It consists of the canonical subsets.

Lemma *Let X be a totally ordered finite set, and let G be a group acting on X. Let A be a canonical subset of X in the sense of 9.5.3. Then every prefix B of A is also canonical.*

Proof: Let $A = \overline{A} = \{a_0, a_1, \ldots, a_{n-1}\}_<$ be a canonical subset of X. Let $B = \{a_0, \ldots, a_{m-1}\}$ with $m - 1 \leq n - 1$ be a prefix of A. Assume that B is not canonical. Thus there exists an element $g \in G$ with $Bg = C = \{c_0, \ldots, c_{m-1}\} \preceq B$ and $C \neq B$. Since $|C| = |B|$ it must be the case that there exists $r < m$ with $c_i = a_i$ for $i \in r$ and $c_r < a_r$. Also $Ag = C \cup \{a_m g, \ldots, a_{n-1} g\}$. In order to compare Ag with A in the lexicographical order, put

$$d = \min_{i=m}^{n-1} a_i g.$$

If $d > c_r$ then $Ag = \{c_0, c_1, \ldots, c_r, \ldots\}_<$ and therefore $Ag \preceq A$ but $Ag \neq A$, contradicting the fact that A is canonical. Otherwise, let s be the least index such that $d < c_s$. Then $Ag = \{c_0, c_1, \ldots, c_{s-1}, d, \ldots\}_<$ and because $c_i = a_i$ for $i \in s$ and $d < c_s = a_s$, again we have the contradiction that $Ag \preceq A$ but $Ag \neq A$. We conclude that the assumption was incorrect and thus B is canonical. □

The method of orderly generation looks at all extensions of the form

$$S \cup \{x\},$$

called *extension sets*. Here, S is a member of the transversal of i-subsets and x is in $X \setminus S$. In fact, one requires that x is the least element in its G_S-orbit and that $x > \max S$. Then, one employs a test for whether a given set $S \subseteq X$ is canonical. Such a test is not easy to provide, as it involves a systematic search over the whole group G, to test whether the set Sg is lexicographically less than S for any given $g \in G$. If no Sg precedes S in the lexicographic order, S is canonical and will be output by the algorithm. The automorphism group of S is just the set of all elements $g \in G$ for which $Sg = g$, so

$$G_S = \{g \in G \mid Sg = S\}$$

can be computed at the same time. Of course, this backtrack procedure can be refined. One would try to avoid looking at every group element $g \in G$. This can be done by taking into account the subgroup structure of G. In fact, the automorphism group will be constructed by successively extending the known part of the group with new automorphisms found during the search. We omit the details here. The algorithm orderly generation can be summarized as follows. We do not state this algorithm as a theorem since we do not prove its correctness. Nevertheless, we mention that correctness can be proved using 9.5.4. We define

$$\mathcal{T}_{\leq i} = \bigcup_{j=0}^{i} \mathcal{T}_j.$$

Algorithm (orderly generation)   9.5.5

   Input: A group G acting on a set $X_<$, a test-function f, an integer i

   Output: $\mathcal{T}_{\leq i}$, the canonical transversal for the G-orbits on admissible sets
                   of size $\leq i$.

(0) **if** $f(\emptyset) = 1$ **then** scan$(\emptyset, G)$ **end if**
(1) **end**

Where the function scan is defined as follows.

(2) scan(S, A)
(3) compute $\mathcal{T}_S$, the canonical transversal of the A-orbits on $X \setminus S$.
(4)   **for each** $x \in \mathcal{T}_S$ **do**
(5)   **if** $x > \max S$ **then**
(6)   **if** $f(S \cup \{x\}) = 1$ **then**

(7)   **if** $S \cup \{x\}$ is canonical **then**
(8)   print $S \cup \{x\}$
(9)   **if** $|S| + 1 < i$ **then**
(10)   scan$(S \cup \{x\}, G_{S \cup \{x\}})$
(11)   **end if**
(12)   **end if**
(13)   **end if**
(14)   **end if**
(15)   **end for**
(16) **end**   □

As already mentioned, testing whether a given set is the lexicographically least set among its G-orbit is a hard problem. It is actually easier to drop the requirement that the canonical element is the least among its orbit and replace it by some other kind of canonical form. This is McKay's variant. It relies on a function φ such that

9.5.6 $$R\varphi(R) = S\varphi(S) \quad \text{whenever} \quad R \sim_G S.$$

Such a function φ can be realized by a "partition backtrack" algorithm (cf. Leon's series of articles [128, 129, 130]). In addition, this algorithm computes the set-stabilizer of the set in question. If such a map φ is to be used for the orderly generation of orbits, the "scan" algorithm needs to change. This is because an extension $S \cup \{x\}$ where x is smallest among its G_S-orbit is not necessarily canonical with respect to the function φ. Also, the requirement that $x > \max S$ must be dropped. To make things work, one introduces another function

9.5.7 $$m : \mathcal{P}(X) \to X,$$

satisfying the two conditions

1. $m(R) \in R$, and
2. $m(Rg) \sim_{G_{Rg}} m(R)g$.

Such a function m is easily defined in terms of the map φ. For instance, one can take

$$m(R) = \left(\min R\varphi(R) \right) \varphi(R)^{-1}.$$

To see that this works, we argue as follows. It is clear that $m(R) \in R$. Since $\varphi(R)\varphi(Rg)^{-1}$ maps R to Rg, we deduce from 3.4.1 that there exists an element $h \in G_{Rg}$ such that

$$\varphi(R)\varphi(Rg)^{-1} = gh.$$

We conclude that

$$
\begin{aligned}
m(Rg) &= \Big(\min Rg\varphi(Rg) \Big)\varphi(Rg)^{-1} \\
&= \Big(\min R\varphi(R) \Big)\varphi(Rg)^{-1} \\
&= m(R)\varphi(R)\varphi(Rg)^{-1} \\
&= m(R)gh,
\end{aligned}
$$

which shows that $m(Rg)$ is in the same G_{Rg}-orbit as $m(R)g$.

We may summarize this algorithm as

Theorem (McKay [146]) *Let G act on the finite set $X_<$. Let $f : \mathcal{P}(X) \to \{0,1\}$*
be a test-function on X which is G-invariant and hereditary (in the sense of 9.5.1
and 9.5.2). Let φ and m be functions as in 9.5.6 and 9.5.7, respectively. Then for any
given integer $i \leq |X|$, Algorithm 9.5.9 computes a transversal $\mathcal{T}_{\leq i}$ of the orbits of G
on admissible subsets of X of size at most i together with the corresponding stabilizers
in G.

9.5.8

Algorithm (orderly generation by canonical augmentation)   9.5.9

   Input: A group G acting on a set $X_<$, a test-function f, an integer i,
                  functions φ and m as in 9.5.6 and 9.5.7, respectively.
   Output: $\mathcal{T}_{\leq i}$, a transversal for the G-orbits on admissible sets of size $\leq i$.

(0)   **if** $f(\varnothing) = 1$ **then** scan$(\varnothing, G)$ **end if**
(1)   **end**

Where the function scan is defined as follows.

(2) scan(S, A)
(3)   compute $\mathcal{T}_S$, a transversal of the A-orbits on $X \setminus S$.
(4)   **for each** $x \in \mathcal{T}_S$ **do**
(5)   **if** $f(S \cup \{x\}) = 1$ **then**
(6)   compute $y := m(S \cup \{x\})$ and $B := G_{S \cup \{x\}}$
(7)   **if** $x \sim_B y$ **then**
(8)   print $S \cup \{x\}$
(9)   **if** $|S| + 1 < i$ **then**
(10)   scan$(S \cup \{x\}, B)$
(11)   **end if**
(12)   **end if**
(13)   **end if**
(14)   **end if**

(15) **end for**

(16) **end**   □

Proof: We proceed by induction on j, the size of the subsets under considera-
tion. If $j = 0$, the algorithm outputs $\emptyset$ and $G_\emptyset = G$, provided that $f(\emptyset) = 1$.
Let us assume that $\mathcal{T}_i$, a transversal for the G-orbits on $\mathcal{P}_i^{(f)}(X)$ is computed
correctly (together with the corresponding stabilizers in G). We need to show
that each G-orbit on $(i+1)$-subsets is represented exactly once in the output
of the algorithm. We proceed in two steps.

At first, we claim that each G-orbit on admissible $(i+1)$-subsets is represented *at least once* in the output. To see this, let R be an admissible $(i+1)$-
subset of X. Since f is hereditary, the subset $R \setminus \{m(R)\}$ is again an admissible
i-subset. By induction hypothesis, there exists an element $g \in G$ such that

$$(R \setminus \{m(R)\})g = S \in \mathcal{T}_i.$$

We define

$$z := m(R)g \in X \setminus S.$$

Since $\mathcal{T}_S$ is a transversal of the G_S-orbits on $X \setminus S$, there exists an element
$h \in G_S$ such that

$$zh = x \in \mathcal{T}_S.$$

We conclude that

$$
\begin{aligned}
Rgh &= (R \setminus \{m(R)\})gh \cup \{m(R)gh\} \\
&= Sh \cup \{zh\} \\
&= S \cup \{x\}
\end{aligned}
$$

and $S \cup \{x\}$ is one of the extensions considered in lines (5)-(9). Since

$$y = m(S \cup \{x\}) = m(Rgh) \sim_{G_{S \cup \{x\}}} m(R)gh = zh = x,$$

the extension $S \cup \{x\}$ is accepted in line (7).

Secondly, we claim that each G-orbit on admissible $(i+1)$-subsets is represented *at most once* in the output. Assume the contrary. Let $R \sim_G S$ be two
admissible $(i+1)$-subsets computed by the algorithm. Then

$$R = U \cup \{x\}, \quad S = V \cup \{y\}$$

with $U, V \in \mathcal{T}_i, x \in \mathcal{T}_U, y \in \mathcal{T}_V$. In addition, we know that

$$x \sim_{G_R} m(R), \quad \text{and} \quad y \sim_{G_S} m(S),$$

since $U \cup \{x\}$ and $V \cup \{y\}$ must both have been accepted in line (7) of the algorithm. Since $R \sim_G S$, there exists an element $g \in G$ such that $Rg = S$. Thus $G_S = g^{-1}G_R g$. Since $x \sim_{G_R} m(R)$, there exists an element $r \in G_R$ such that $xr = m(R)$, and so

$$xg(g^{-1}rg) = m(R)g,$$

i.e.

$$xg \sim_{G_S} m(R)g,$$

since $g^{-1}rg \in g^{-1}G_R g = G_S$. Thus

$$y \sim_{G_S} m(S) = m(Rg) \sim_{G_S} m(R)g \sim_{G_S} xg \in S.$$

This means that there exists an element $h \in G_S$ with

$$y = xgh.$$

Thus

$$Ugh = (R \setminus \{x\})gh = (Sg^{-1} \setminus \{x\})gh = Sh \setminus \{xgh\} = S \setminus \{y\} = V,$$

i.e. $U \sim_G V$. But U and V are elements of the transversal $\mathcal{T}_i$, and by induction hypothesis, this transversal contains exactly one representative of each G-orbit. It follows that

$$U = V,$$

and therefore $x \neq y$ (since $R \neq S$ by assumption). Thus

$$Ugh = V = U,$$

i.e. $gh \in G_U$. From $xgh = y$ we conclude that $x \sim_{G_U} y$, which is a contradiction to the fact that the algorithm considers in line (4) only representatives $x \in \mathcal{T}_U$ of the U-orbits on $X \setminus U$. The assumption must be wrong and the claim is proved. This finishes the proof of the fact that algorithm 9.5.9 is correct.   □

Let us return to the problem of computing isometry classes of linear codes. Given a length n, a dimension k and a lower bound $d_{\min}$ for the minimum distance, let us now construct all (n, k)-codes over some finite field $\mathbb{F}_q$ whose minimum distance is at least $d_{\min}$ where $d_{\min} \geq 3$. From 9.1.5 we deduce the following. Depending on whether we want to compute linear or semilinear isometry classes, we are interested in the orbits of $G = \mathrm{PGL}_{n-k}(q)$ or $G = \mathrm{P\Gamma L}_{n-k}(q)$ on

$$\mathcal{P}_n(X), \quad X = \mathrm{PG}_{n-k-1}(q),$$

respectively.

It remains to take the prescribed minimum distance $d_{\min}$ into account. In order to apply 9.1.8, we need to check whether the n-subset of $\mathrm{PG}_{n-k-1}(q)$

under consideration has the property that any $d_{min} - 1$ points are independent. If $S \subseteq PG_{n-k-1}(q)$ is a set of size n, we put

$$f(S) = \begin{cases} 1 & \text{if any } d_{min} - 1 \text{ points of } S \text{ are independent,} \\ 0 & \text{otherwise.} \end{cases}$$

This function f is our test-function. We need to check if the function f satisfies the requirements listed in 9.5.1 and 9.5.2. Since both groups $PGL_{n-k}(q)$ and $P\Gamma L_{n-k}(q)$ preserve the linear structure of $\mathbb{F}_q^{n-k}$, the condition about linear independence of subsets is invariant under the action. It is clear that the condition is hereditary. Let us put

$$Y_{n,k,d_{min},q} = \mathcal{P}_n^{(f)}(PG_{n-k-1}(q)) = \left\{ S \subseteq PG_{n-k-1}(q) \,\middle|\, |S| = n, \ f(S) = 1 \right\}.$$

The next result shows a connection between canonical orbit representatives and systematic generator matrices.

9.5.10   **Lemma**   *Consider the action of $G \geq PGL_k(q)$ on n-subsets of points of $X = PG_{k-1}(q)$. Let X be totally ordered according to 9.3.5. Let*

$$A := \{ \langle u^{(0)} \rangle, \ldots, \langle u^{(n-1)} \rangle \}_<$$

be a canonical orbit representative. Let

$$\Gamma(A) = \left(u^{(0)\top} \,\middle|\, \cdots \,\middle|\, u^{(n-1)\top} \right)$$

be the generator matrix corresponding to A. Then the following conditions are equivalent.

1. *The rank of the matrix $\Gamma(A)$ is r.*
2. *$\langle u^{(i)} \rangle = \langle e^{(i)} \rangle$, for $i \in r$ and $u^{(j)} \in \langle e^{(0)}, \ldots, e^{(r-1)} \rangle$ for $j = r, \ldots, n - 1$.*
3. *$\langle u^{(i)} \rangle = \langle e^{(i)} \rangle$, for $i \in r$ and $\langle u^{(r)} \rangle \neq \langle e^{(r)} \rangle$.*

Proof:

1. $\Rightarrow$ 2.: Since G is transitive on r-dimensional subspaces, the rank condition implies that the orbit of A contains an element $B = \{ \langle e^{(0)} \rangle, \ldots, \langle e^{(r-1)} \rangle, \ldots \}$. But $\{ \langle e^{(0)} \rangle, \ldots, \langle e^{(r-1)} \rangle \}$ is the lexicographically least set of size r. Hence A contains this set, i.e.

$$A = \{ \langle e^{(0)} \rangle, \ldots, \langle e^{(r-1)} \rangle, \langle u^{(r)} \rangle, \ldots, \langle u^{(n-1)} \rangle \},$$

and – also by the rank condition – each column of $\Gamma(A)$ is in the span of these vectors:

$$u^{(j)} \in \langle e^{(0)}, \ldots, e^{(r-1)} \rangle, \ r \leq j \leq n - 1.$$

2. $\Rightarrow$ 3.: If $r = k$, there is nothing to show. Otherwise we have that $u^{(r)} \in \langle e^{(0)}, \ldots, e^{(r-1)} \rangle$ which implies $\langle u^{(r)} \rangle \neq \langle e^{(r)} \rangle$.

3. $\Rightarrow$ 1.: Since $\{ \langle u^{(0)} \rangle, \ldots, \langle u^{(r-1)} \rangle \} = \{ \langle e^{(0)} \rangle, \ldots, \langle e^{(r-1)} \rangle \}$, the rank of $\Gamma(A)$ is $\geq r$. Now assume that the rank of $\Gamma(A)$ is strictly greater than r. This implies that there is a column of $\Gamma(A)$ which is linearly independent from the first r columns. Let this be column $i \geq r$. Then $u^{(i)} \notin \langle e^{(0)}, \ldots, e^{(r-1)} \rangle$. Since G is transitive on subspaces of fixed dimension, there is an element $g \in G$ with

$$\{ \langle e^{(0)} \rangle, \ldots, \langle e^{(r-1)} \rangle, \langle u^{(i)} \rangle \} g = \{ \langle e^{(0)} \rangle, \ldots, \langle e^{(r)} \rangle \}.$$

But the latter set is the lexicographically least set of size $r + 1$. The prefix of length r of the canonical set A must therefore coincide with this set, i.e. we have shown that $\langle u^{(r)} \rangle = \langle e^{(r)} \rangle$, which contradicts 3. Thus the assumption that the rank of $\Gamma(A)$ is greater than r was incorrect. This means that the rank of $\Gamma(A)$ is indeed r.   □

Corollary   Let $A = \{ \langle u^{(0)} \rangle, \ldots, \langle u^{(n-1)} \rangle \}_<$ be a canonical representative for an orbit of $\mathrm{PGL}_k(q)$ acting on n-subsets of $\mathrm{PG}_{k-1}(q)$. If the vectors $u^{(i)}$ are standard, then the matrix   **9.5.11**

$$\Gamma(A) = \left(u^{(0)^\top} \mid \cdots \mid u^{(n-1)^\top} \right)$$

is systematic.   □

As described above, we compute the orbits of G on $Y_{(i,k,d_{\min},q)}$ where i goes from 0 to n. Here we choose $G = \mathrm{PGL}_{n-k}(q)$ or $G = \mathrm{P\Gamma L}_{n-k}(q)$, depending on whether we want to compute linear or semilinear isometry classes of codes. As described in 9.1.1, the sets in $Y_{(i,k,d_{\min},q)}$ give rise to $(i, \geq i - n + k, \geq d_{\min}, q)$-codes, which is sensible only for $i \geq n - k$. As pointed out before, we will have to go through all values $i \leq n$, since the orbits on $Y_{(i,k,d_{\min},q)}$ will be constructed inductively. At each step, the canonical transversal $\mathcal{T}_i$ for the orbits of G on the set $Y_{(i,k,d_{\min},q)}$ is computed, as well as some additional data. This additional data can be used to realize functions σ_i and φ_i with $(\mathcal{T}_i, \sigma_i, \varphi_i)$ as in 9.2.12. The union

$$\mathcal{T}_{\leq n} = \bigcup_{i=0}^{n} \mathcal{T}_i$$

of all canonical representatives is the *tree of canonical representatives*. The leaves at depth n comprise the isometry classes of codes. We follow the convention of labeling nodes by their largest element. We display the ranks of the projective points rather than the projective points themselves.

Given an orbit representative $A = \{ \langle u^{(0)} \rangle, \ldots, \langle u^{(s-1)} \rangle \}_<$, we construct the corresponding code as follows. As in 9.1.6, we form the check matrix $\Delta(A)$. At this point we come back to the initial remarks in Section 9.1 about the matrix $\Delta(A)$ being not uniquely defined. This non-uniqueness has two reasons. The first one lies in the fact that the elements of the set A may be rearranged freely. We have resolved that issue by requiring that the elements of A be ordered increasingly (according to their ranks as given by 9.3.5). The second problem lies in the choice of the representatives $u^{(i)}$ for the projective points. To this end, we simply require that $u^{(i)}$ is standard, i.e. that its rightmost nonzero coordinate is one. With these two conventions, the matrix $\Delta(A)$ becomes unique and we may take this matrix as a check matrix of a code. This $(n, \geq k, \geq d, q)$-code is a representative of an isometry class. More precisely, if the rank of $\Delta(A)$ is r, then we have found an $(n, n - r, \geq d, q)$-code. Here we have $n - r \geq k$ since $r \leq n - k$. In order to obtain a generator matrix, we proceed as follows.

By 9.5.10, $\Delta(A)$ is systematic provided that A is the lexicographically least element in its G-orbit. If r is determined as the index for which $\langle u^{(i)} \rangle = \langle e^{(i)} \rangle$ for $i = 0, \ldots, r - 1$ and $\langle u^{(r)} \rangle \neq \langle e^{(r)} \rangle$, then the rank of $\Delta(A)$ is r. Thus, we can write

$$\Delta(A) = \left(\begin{array}{c|c} I_r & M \\ \hline 0 & 0 \end{array} \right)$$

for some $r \times (n - r)$-matrix M. By 1.3.9, a generator matrix of the code is

$$\Gamma(A) = \left(-M^{\top} \mid I_{n-r} \right).$$

Let us consider an example.

9.5.12   **Example** We wish to construct and classify binary $(8, 4)$-codes with minimum distance at least $d_{\min} = 3$. For this we are looking for sets of 8 points in $PG_3(2)$. Since $d_{\min} - 1 = 2$, and since two distinct points of a projective space are always linearly independent, any subset is admissible. In order to construct the codes, we compute the orbits of $PGL_4(2)$ on $\mathcal{P}_{\leq 8}(PG_3(2))$ (here, $\mathcal{P}_{\leq i}(X) := \cup_{j=0}^{i} \mathcal{P}_i(X)$). The resulting tree of canonical orbit representatives is shown in Fig. 9.4 (see Example 9.3.11 for the ranks of points in $PG_3(2)$). We find 6 leaves at level 8, which comprise all essentially distinct $(8, 4, \geq 3)$-codes. Table 9.2 shows the corresponding check and generator matrices. The last code is equivalent to the extended $(7, 4)$-Hamming-code. Being the only code with distance 4, this is the optimal binary code with length 8 and dimension 4. Notice that the second to last code is decomposable. Its generator matrix contains a zero column. The code can thus be seen as the direct sum (in the sense of 2.2.11) of a $(6, 4)$-codes with a (rather trivial) $(1, 0)$-code. For a description of the algorithm to compute the orbits, we refer to the next section. In 9.6.12, we will pick this example up again and show more details of the actual computation. $\diamond$

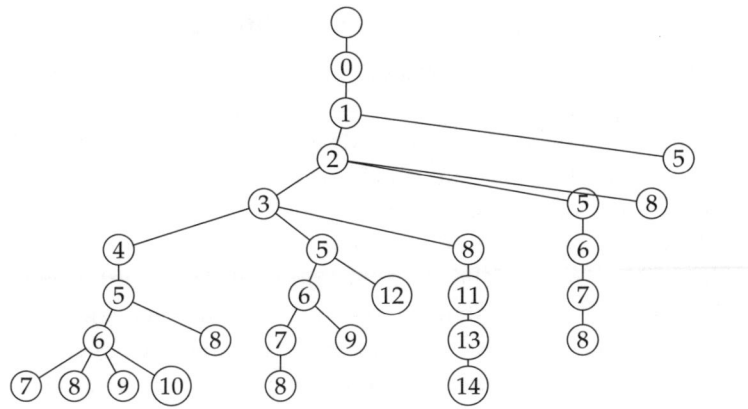

Fig. 9.4 Orbits of $\mathrm{PGL}_4(2)$ on $\mathcal{P}_{\leq 8}(\mathrm{PG}_3(2))$

Table 9.2 Binary $(8, 4, \geq 3)$-codes

A	$\Delta(A)$	$\Gamma(A)$	d
$\{0,1,2,3,4,5,6,7\}$	$\begin{pmatrix} 1\,0\,0\,0 & 1\,1\,1\,0 \\ 0\,1\,0\,0 & 1\,1\,0\,1 \\ 0\,0\,1\,0 & 1\,0\,1\,1 \\ 0\,0\,0\,1 & 1\,0\,0\,0 \end{pmatrix}$	$\begin{pmatrix} 1\,1\,1\,1 & 1\,0\,0\,0 \\ 1\,1\,0\,0 & 0\,1\,0\,0 \\ 1\,0\,1\,0 & 0\,0\,1\,0 \\ 0\,1\,1\,0 & 0\,0\,0\,1 \end{pmatrix}$	3
$\{0,1,2,3,4,5,6,8\}$	$\begin{pmatrix} 1\,0\,0\,0 & 1\,1\,1\,1 \\ 0\,1\,0\,0 & 1\,1\,0\,1 \\ 0\,0\,1\,0 & 1\,0\,1\,1 \\ 0\,0\,0\,1 & 1\,0\,0\,0 \end{pmatrix}$	$\begin{pmatrix} 1\,1\,1\,1 & 1\,0\,0\,0 \\ 1\,1\,0\,0 & 0\,1\,0\,0 \\ 1\,0\,1\,0 & 0\,0\,1\,0 \\ 1\,1\,1\,0 & 0\,0\,0\,1 \end{pmatrix}$	3
$\{0,1,2,3,4,5,6,9\}$	$\begin{pmatrix} 1\,0\,0\,0 & 1\,1\,1\,1 \\ 0\,1\,0\,0 & 1\,1\,0\,0 \\ 0\,0\,1\,0 & 1\,0\,1\,0 \\ 0\,0\,0\,1 & 1\,0\,0\,1 \end{pmatrix}$	$\begin{pmatrix} 1\,1\,1\,1 & 1\,0\,0\,0 \\ 1\,1\,0\,0 & 0\,1\,0\,0 \\ 1\,0\,1\,0 & 0\,0\,1\,0 \\ 1\,0\,0\,1 & 0\,0\,0\,1 \end{pmatrix}$	3
$\{0,1,2,3,4,5,6,10\}$	$\begin{pmatrix} 1\,0\,0\,0 & 1\,1\,1\,0 \\ 0\,1\,0\,0 & 1\,1\,0\,1 \\ 0\,0\,1\,0 & 1\,0\,1\,0 \\ 0\,0\,0\,1 & 1\,0\,0\,1 \end{pmatrix}$	$\begin{pmatrix} 1\,1\,1\,1 & 1\,0\,0\,0 \\ 1\,1\,0\,0 & 0\,1\,0\,0 \\ 1\,0\,1\,0 & 0\,0\,1\,0 \\ 0\,1\,0\,1 & 0\,0\,0\,1 \end{pmatrix}$	3
$\{0,1,2,3,5,6,7,8\}$	$\begin{pmatrix} 1\,0\,0\,0 & 1\,1\,0\,1 \\ 0\,1\,0\,0 & 1\,0\,1\,1 \\ 0\,0\,1\,0 & 0\,1\,1\,1 \\ 0\,0\,0\,1 & 0\,0\,0\,0 \end{pmatrix}$	$\begin{pmatrix} 1\,1\,0\,0 & 1\,0\,0\,0 \\ 1\,0\,1\,0 & 0\,1\,0\,0 \\ 0\,1\,1\,0 & 0\,0\,1\,0 \\ 1\,1\,1\,0 & 0\,0\,0\,1 \end{pmatrix}$	3
$\{0,1,2,3,8,11,13,14\}$	$\begin{pmatrix} 1\,0\,0\,0 & 1\,1\,1\,0 \\ 0\,1\,0\,0 & 1\,1\,0\,1 \\ 0\,0\,1\,0 & 1\,0\,1\,1 \\ 0\,0\,0\,1 & 0\,1\,1\,1 \end{pmatrix}$	$\begin{pmatrix} 1\,1\,1\,0 & 1\,0\,0\,0 \\ 1\,1\,0\,1 & 0\,1\,0\,0 \\ 1\,0\,1\,1 & 0\,0\,1\,0 \\ 0\,1\,1\,1 & 0\,0\,0\,1 \end{pmatrix}$	4

9.6 The Algorithm Snakes and Ladders

The two algorithms presented in the previous section rely on the fact that we are able to compute a canonical from of every subset. This is indeed a hard problem, and it can be tedious to provide a canonical form for a specific group action, since computing the canonical form depends very much of the nature of the group action under consideration. In this section, we will present an orbit algorithm which is general in the sense that it does not depend on the nature of the group in question. The algorithm proceeds in breadth first search, i.e. constructs the orbit representatives level by level. Also, it avoids backtracking as much as possible. The price one pays is that the amount of memory required correlates linearly to the number of orbits computed. In a sense, one trades computing time with memory. Of course, this is a limitation which may restrict the scope to which problems can be tackled. On the other hand, the speedup from the memory versus time tradeoff makes it realistic to tackle instances of hard problems, such as the computation of isometry classes of linear codes which is our main topic.

Essentially, this algorithm is due to Schmalz [171] ("Leiterspiel" loosely translated as "snakes and ladders"). Whereas Schmalz formulated his algorithm very much in the language of group theory, here we will stick to the concept of a group acting on a set. That is, we will describe the algorithm as computing orbits of a group G on subsets of a set X on which G acts. This is different from the approach taken by Schmalz, whose algorithm is formulated in the language of double cosets in finite groups. The name Leiterspiel ("ladder game") refers to the fact that the algorithm works along a sequence of subgroups which are alternately subgroups and overgroups of each other (what we will call the "down-and-up process").

We assume that orbits on points can be computed, for instance using the algorithms described in Section 9.2. The main goal is to provide a triple $(\mathcal{T}, \sigma, \varphi)$ which is a solution to the orbit problem for G acting on admissible subsets of X.

We will favor an inductive solution to the problem, namely by computing the orbits of G on $\mathcal{P}_i^{(f)}(X)$ for $i = 0, 1, \ldots$. In the search tree, this corresponds to a breadth-first search. Let

$$\mathrm{orbit}\big(G, \mathcal{P}_i^{(f)}(X)\big) = (\mathcal{T}_i, \sigma_i, \varphi_i),$$

be a solution to the orbit problem on i-subsets. Note that the case $i = 0$ is trivial, whereas $i = 1$ is the basic orbit problem on points, which we are able to solve using the methods provided in Section 9.2.

Assume that a transversal $\mathcal{T}_i$ of orbits of G on sets $\mathcal{P}_i^{(f)}(X)$, i.e. on admissible sets of size i has already been computed. Often this will be the canonical transversal, i.e. the transversal consisting of all sets which are canonical with respect to some ordering. For sake of simplicity, we will say that a set R is canonical if it belongs to one of the transversals $\mathcal{T}_i$ for some i.

In order to compute $\mathcal{T}_{i+1}$, we consider extensions of sets in $\mathcal{T}_i$. An *extension* is a set of the form

$$R \cup \{x\} \in \mathcal{P}_{i+1}^{(f)}(X),$$

where R is in $\mathcal{T}_i \subseteq \mathcal{P}_i^{(f)}(X)$ and x is in $X \setminus R$. There are four major tasks involved in computing the "next level" of orbits on $(i+1)$-sets:

Problem 1 Ensure that each G-orbit on admissible $(i+1)$-sets is reached.

Problem 2 Determine when two extensions $R \cup \{x\}$ and $S \cup \{y\}$ are isomorphic (i.e. belong to the same G-orbit). Note that here R and S are canonical, i.e. elements of the transversal $\mathcal{T}_i$.

Problem 3 Compute the stabilizer in G of an extension set $R \cup \{x\}$. Here we assume that the stabilizer of the canonical set R is known.

Problem 4 Provide a transporter map φ_{i+1} for $(i+1)$-sets. That is, given an $(i+1)$-subset $F \subseteq X$, compute an element $g \in G$ such that $Fg \in \mathcal{T}_{i+1}$.

Problem 1 is addressed easily. Let F be an admissible $(i+1)$-subset of X. Let $z := \max F$ and put $H := F \setminus \{z\}$, which is admissible since f is hereditary. Thus $H \in G(R)$ for some $R \in \mathcal{T}_i$ and $Hg = R$ for $g = \varphi(H)$. Let $x := zg \in X \setminus R$. This shows that $F \sim_G R \cup \{x\}$, which is one of the candidate sets which we considered. We note that later on, we will reduce the number of candidate sets further (see 9.6.2).

Problem 2 amounts to determining when two extensions $R \cup \{x\}$ and $S \cup \{y\}$ with $R, S \in \mathcal{T}_i$ belong to the same G-orbit. The following "exchange lemma" gives a necessary and sufficient condition for deciding that question (cf. Fig. 9.5).

Lemma *Assume that* $\mathrm{orbit}(G, \mathcal{P}_i(X)) = (\mathcal{T}_i, \sigma_i, \varphi_i)$. *For* $R, S \in \mathcal{T}_i$, $x \in X \setminus R$,   **9.6.1**
$y \in X \setminus S$, *we have* $R \cup \{x\} \sim_G S \cup \{y\}$ *if and only if one of the following two conditions holds*

1. *$R = S$ and $x \sim_{G_S} y$ or*

2. *there exists an $r \in R$ such that*

$$((R \setminus \{r\}) \cup \{x\})t = S \quad \text{and} \quad rt \sim_{G_S} y$$

where $t = \varphi_i((R \setminus \{r\}) \cup \{x\})$.

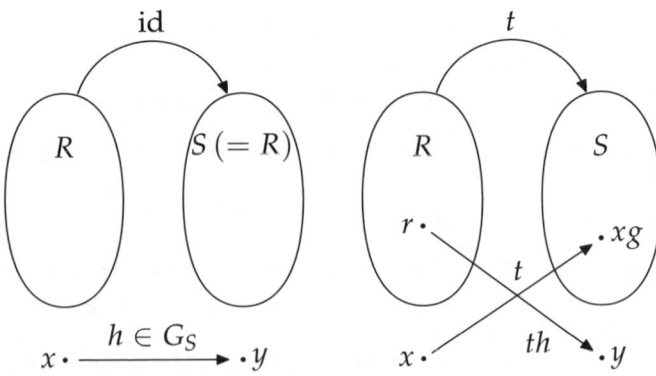

Fig. 9.5 The two cases of Lemma 9.6.1

Proof: Necessity: Assume that $R \cup \{x\} \sim_G S \cup \{y\}$. Then there exists an element $g \in G$ with

$$(R \cup \{x\})g = S \cup \{y\}.$$

We must show that 1. or 2. holds. Assume that $Rg = S$ and hence $xg = y$. Since R and S are both orbit representatives in $\mathcal{T}_i$, we must have $R = S$. But then $Sg = Rg = S$, i.e. $g \in G_S$ and hence $x \sim_{G_S} y$, which is 1. Otherwise we have $S \neq Rg \subseteq S \cup \{y\}$, and hence $xg \in S$. Let $r = yg^{-1} \in R$. Hence

$$((R \setminus \{r\} \cup \{x\})g = S.$$

But also

$$((R \setminus \{r\} \cup \{x\})t = S,$$

where $t = \varphi_i((R \setminus \{r\}) \cup \{x\})$ is the transporter element mapping $(R \setminus \{r\}) \cup \{x\}$ onto the canonical representative S. Thus g is contained in the left coset tG_S, i.e. $g = th$ for some $h \in G_S$. Now $rth = rg = y$, i.e. $rt \sim_{G_S} y$, which is 2. Sufficiency: If 1. is valid, and if $g \in G_S$ maps x to y, then clearly $(R \cup \{x\})g = Rg \cup \{xg\} = S \cup \{y\}$. If 2. holds with $h \in G_S$ mapping rt to y then

$$(R \cup \{x\})th = ((R \setminus \{r\}) \cup \{x\})th \cup \{r\}th = Sh \cup \{rth\} = S \cup \{y\}. \qquad \square$$

The first part of this result has an important implication for the candidate set of extensions (Problem 1):

9.6.2 **Corollary** *It suffices to consider only extensions of the form $R \cup \{x\}$ where R is a canonical i-set and $x \in X \setminus R$ is canonical under the stabilizer of R in G.* $\square$

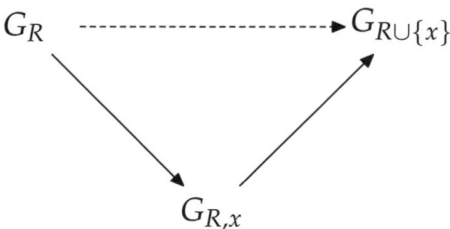

Fig. 9.6 The "down-and-up" process

From now on, we consider only extensions $R \cup \{x\}$ of the form described in the previous corollary.

Let us now describe the problem of computing the stabilizer of extension sets (Problem 3). If the extension is $R \cup \{x\}$, then this amounts to computing

$$G_{R \cup \{x\}},$$

the setwise stabilizer of $R \cup \{x\}$. We assume that G_R, the setwise stabilizer of the canonical set R is known. The difficulty is that there is no relationship between the groups G_R and $G_{R \cup \{x\}}$ (meaning that neither is a subgroup of the other in general). However, they share a common subgroup, namely the group $G_{R,x}$ which is the set of elements of G which stabilize R setwise and x pointwise. The idea is to first go down from G_R to $G_{R,x}$ (the "downstep"), which is relatively straightforward. Generators for $G_{R,x}$ can be computed from generators for G_R by means of 9.2.10. The difficulty is to compute the group $G_{R \cup \{x\}}$ from the subgroup $G_{R,x}$ (i.e. the "upstep"). In the following, we will address this problem first (we will call it the "down-and-up" process, cf. Fig. 9.6). Afterwards, we will present the algorithm to compute orbits on sets which combines all the ideas developed so far.

Recall that for a subgroup V of G we have

$$X_V = \{x \in X \mid \forall v \in V : xv = x\}.$$

The following result is a consequence of 3.4.1.

Lemma *Let the group H act on a set X. Let $V = H_x$ be the stabilizer of a point* $x \in X$. *In addition, let $\mathcal{R}$ be a set of elements of H such that for each $y \in H(x)$ there exists one and only one $g \in \mathcal{R}$ with $xg = y$. Then $\mathcal{R}$ is a set of right coset representatives of V in H.*   □

9.6.3

In the situation of the lemma, we call H the *extension of V w.r.t. the coset representatives* $\mathcal{R}$ and we write

$$H = \mathrm{Ext}(V, \mathcal{R}, x) = \bigcup_{r \in \mathcal{R}} Vr,$$

where the last union is over disjoint cosets. For $R \in \mathcal{T}_i$ and $x \in X \setminus R$, let us define

9.6.4
$$R^*(x) := \left\{ r \in R \;\middle|\; \begin{array}{l} (R \setminus \{r\}) \cup \{x\} \in G(R) \text{ and } rt \in G_R(x), \\ \text{where } t = \varphi_i((R \setminus \{r\}) \cup \{x\}) \end{array} \right\}$$

and put $R(x) = R^*(x) \cup \{x\}$. The next result describes the stabilizer of extension sets.

9.6.5
Lemma *Let G act on the finite set X. Assume that* $\mathrm{orbit}(G, \mathcal{P}_i(X)) = (\mathcal{T}_i, \sigma_i, \varphi_i)$. *For $R \in \mathcal{T}_i$ let* $\mathrm{orbit}(G_R, X \setminus R) = (\mathcal{T}_R, \sigma_R, \varphi_R)$. *Fix $x \in \mathcal{T}_R$. Then the orbit of x under $G_{R \cup \{x\}}$ is $R(x)$. In particular $G_{R \cup \{x\}} = \mathrm{Ext}(G_{R,x}, \mathcal{R}, x)$ where*

$$\mathcal{R} = \{1\} \;\cup\; \{ t \cdot \varphi_R(rt) \mid r \in R^*(x), \; t = \varphi_i((R \setminus \{r\}) \cup \{x\}) \}.$$

Here, $G_{R,x} = G_R \cap G_x = \{ g \in G \mid Rg = R, \; xg = x \}$ and $G_{R \cup \{x\}}$ is the setwise stabilizer of the set $R \cup \{x\}$.

Proof: Let $\mathcal{O}_x = G_{R \cup \{x\}}(x)$ be the orbit of x under $G_{R \cup \{x\}}$. We claim that $\mathcal{O}_x = R(x) = R^*(x) \cup \{x\}$.

Consider $r \in R$. Let $g \in G_{R \cup \{x\}}$ with $rg = x$. Since g maps $R \cup \{x\}$ onto itself, we have

$$R \cup \{x\} = (R \cup \{x\}) g^{-1} = Rg^{-1} \cup \{xg^{-1}\} = Rg^{-1} \cup \{r\}.$$

This implies that $Rg^{-1} = (R \setminus \{r\}) \cup \{x\}$ and therefore

$$((R \setminus \{r\}) \cup \{x\}) g = R \in \mathcal{T}_i,$$

i.e. $(R \setminus \{r\}) \cup \{x\} \in G(R)$. By definition, the group element $t = \varphi_i((R \setminus \{r\}) \cup \{x\})$ maps $(R \setminus \{r\}) \cup \{x\}$ to the canonical representative in $\mathcal{T}_i$ of its G-orbit, which must be the set R. Thus

$$((R \setminus \{r\}) \cup \{x\}) t = R = ((R \setminus \{r\}) \cup \{x\}) g.$$

We conclude that t and g belong to the same left coset of G_R, i.e. there is an element $h \in G_R$ such that $g = th$. Thus

$$((R \setminus \{r\}) \cup \{x\}) t = ((R \setminus \{r\}) \cup \{x\}) gh^{-1} = Rh^{-1} = R.$$

Also, $rt = rgh^{-1} = xh^{-1} \sim_{G_R} x$ and therefore $r \in R^*(x)$.

Since x is clearly an element of its own orbit under $G_{R \cup \{x\}}$, it remains to show that the elements $r \in R^*(x)$ lie in $\mathcal{O}_x$. If $r \in R^*(x)$, we have that $((R \setminus \{r\}) \cup \{x\})t = R$ and $rt \sim_{G_R} x \in \mathcal{T}_R$ where $t = \varphi_i((R \setminus \{r\}) \cup \{x\})$. The second condition implies that $rth = x$ where $h = \varphi_R(rt) \in G_R$. The equation

$$(R \cup \{x\})th = ((R \setminus \{r\}) \cup \{x\})th \cup \{r\}th = Rh \cup \{r\}th = R \cup \{x\}$$

shows that $g = th$ stabilizes $R \cup \{x\}$. In addition, since $xg^{-1} = r$ we have that $r \in \mathcal{O}_x$. This proves the claim.

We are now able to show that $G_{R \cup \{x\}} = \mathrm{Ext}(G_{R,x}, \mathcal{R}, x)$. It is clear that $G_{R,x}$ is a subgroup of $G_{R \cup \{x\}}$. Next, $G_{R \cup \{x\},x} = G_{R \cup \{x\}} \cap G_x = G_{R,x}$. Hence the cosets of $G_{R,x}$ in $G_{R \cup \{x\}}$ are in one-to-one correspondence with the distinct images of x under $G_{R \cup \{x\}}$. If we revisit the proof of the claim above, we notice that for $r \in R^*(x)$ the element $t\varphi_R(rt)$ where $t = \varphi_i((R \setminus \{r\}) \cup \{x\})$ is in $G_{R \cup \{x\}}$ and maps $r \in R^*(x) = \mathcal{O}_x \setminus \{x\}$ to x. Since the identity element (denoted as 1) is trivially contained in $G_{R \cup \{x\}}$ and maps x to x, the union $\mathcal{R}$ of 1 and all elements $t\varphi_R(rt)$ as above forms a transporter set for the distinct elements of $\mathcal{O}_x$ in $G_{R \cup \{x\}}$. By the standard argument alluded to above we have that $G_{R \cup \{x\}}$ is the union of the right cosets of $G_{R,x}$ with respect to the elements of $\mathcal{R}$. Therefore by 9.6.3, $G_{R \cup \{x\}}$ is the extension of $G_{R,x}$ with respect to the point x and the transversal $\mathcal{R}$.   □

The following result is helpful in computing the set $R^*(x)$. It may reduce the number of $r \in R$ which have to be tested.

Lemma *Let G act on the finite set X. Assume that* $\mathrm{orbit}(G, \mathcal{P}_i(X)) = (\mathcal{T}_i, \sigma_i, \varphi_i)$.   **9.6.6**
For $R \in \mathcal{T}_i$ let $\mathrm{orbit}(G_R, X \setminus R) = (\mathcal{T}_R, \sigma_R, \varphi_R)$. *Fix $x \in \mathcal{T}_R$ and $r \in R$. Then*

1. *If $r \in R^*(x)$ then $rs \in R^*(x)$ for all $s \in G_{R \cup \{x\}}$.*
2. *If $r \notin R^*(x)$ then $rs \notin R^*(x)$ for all $s \in G_{R \cup \{x\}}$.*

Proof: Let $r \in R^*(x)$ and $s \in G_{R \cup \{x\}}$. Then $(R \setminus \{r\}) \cup \{x\} \in G(R)$ and $rt_r \in G_R(x)$ for $t_r = \varphi_i((R \setminus \{r\}) \cup \{x\})$. The latter condition means that there is an element $h \in G_R$ such that

$$rt_r h = x.$$

We will now show that $rs \in R^*(x)$. Using the fact that $s \in G_{R \cup \{x\}}$ we obtain

$$
\begin{aligned}
((R \setminus \{rs\}) \cup \{x\})s^{-1}t_r &= ((R \cup \{x\}) \setminus \{rs\})s^{-1}t_r \\
&= (R \cup \{x\})s^{-1}t_r \setminus \{rs\}s^{-1}t_r \\
&= (R \cup \{x\})t_r \setminus \{r\}t_r \\
&= ((R \setminus \{r\}) \cup \{x\})t_r \\
&= R,
\end{aligned}
$$

i.e. $(R \setminus \{rs\}) \cup \{x\} \in G(R)$. Thus with $t_{rs} = \varphi_i((R \setminus \{rs\}) \cup \{x\})$ we have

$$((R \setminus \{rs\}) \cup \{x\})t_{rs} = R.$$

Putting the two equations together we see that $s^{-1}t_r$ and t_{rs} differ only by an element $u \in G_R$, i.e.

$$t_{rs} = s^{-1}t_r u.$$

Thus

$$rs \cdot t_{rs} = rss^{-1}t_r u = rt_r u = (rt_r h)h^{-1}u = xh^{-1}u \in G_R(x),$$

since h and u are both elements of G_R. This proves the first part. For the second part, assume that $r \notin R^*(x)$ but $rs \in R^*(x)$. Then by the first part we deduce that $r = rss^{-1} \in R^*(x)$, which is a contradiction. $\qquad\square$

9.6.7   Remarks

1. In the previous result, the group $G_{R \cup \{x\}}$ may be replaced by the subgroup $G_{R,x}$. The reason for doing this is that the group $G_{R \cup \{x\}}$ may not be known initially, whereas the smaller group $G_{R,x}$ may be known. In fact, we have

$$G_{R \cup \{x\}} = \mathrm{Ext}(G_{R,x}, \mathcal{R}, x)$$

where $\mathcal{R}$ is defined in terms of $R^*(x)$. Thus when testing elements r for membership in $R^*(x)$ we cannot use $G_{R \cup \{x\}}$. Since $G_{R,x}$ is simply a point stabilizer in the known group G_R, we may start with this group instead. Later on, when non-trivial elements

$$g^{(0)}, \ldots, g^{(i-1)}$$

in $\mathcal{R}$ have been found, we may form the overgroup

$$H^{(i)} = \langle G_{R,x}, g^{(0)}, \ldots, g^{(i-1)} \rangle \leq G_{R \cup \{x\}}$$

and apply 9.6.6 to $s \in H^{(i)}$.
2. To apply 9.6.6, one computes the orbits of $H = G_{R \cup \{x\}}$ (or any known subgroup thereof, see the previous remark) on the elements of R. For each orbit $H(r)$, only the representative r needs to be tested for membership in $R^*(x)$. If $r \in R^*(x)$ then $H(r) \subseteq R^*(x)$. Otherwise $H(r) \cap R^*(x) = \emptyset$. $\qquad\diamond$

Summarizing, we have seen in Lemma 9.6.1 how to decide whether or not two extensions $R \cup \{x\}$ and $S \cup \{y\}$ are in the same G-orbit, i.e. isomorphic. This is the main tool for reducing isomorphic copies. It is now time to take the lexicographical order into account. We use the following tie breaker. If two extensions are isomorphic then we always keep the lexicographically smaller one of the two and we discard the other one. So, if $R \preceq S$ then we keep $R \cup \{x\}$.

Or, if $R = S$ but $x < y$ then we keep $R \cup \{x\}$ and discard $R \cup \{y\}$. Essentially, we do a breadth first search in the tree of canonical orbit representatives. This step comprises the isomorph rejection.

We assume that all representatives of i-orbits are available, i.e. that

$$(\mathcal{T}_i, \sigma_i, \varphi_i)$$

has been computed. Next we examine the sets $R \in \mathcal{T}_i$ in lexicographically increasing order. For each such set R, we compute the orbits of its stabilizer $G_R = \sigma_i(R)$ on the remaining points $X \setminus R$. Let

$$(\mathcal{T}_R, \sigma_R, \varphi_R)$$

be the resulting orbit data. As usual, we assume that $\mathcal{T}_R$ is the canonical transversal. This means that the elements of $\mathcal{T}_R$ (which are just points) are the least among their respective G_R-orbit. Next, we consider the extensions of the form $R \cup \{x\}$ where $x \in \mathcal{T}_R$ (in increasing order). Since G_R is known by assumption, the stabilizer $G_{R,x}$ can be computed. Recall that

$$G_{R,x} = G_R \cap G_x$$

is the pointwise stabilizer of x in G_R. Actually,

$$G_{R,x} = \sigma_R(x)$$

is part of the orbit data which has been computed in the previous step. Next, we compute the set $R^*(x)$ of 9.6.4. For this, we try all $r \in R$ and see if the set $(R \setminus \{r\}) \cup \{x\}$ is contained in the G-orbit of R. This can be done by computing

$$t = \varphi_i((R \setminus \{r\}) \cup \{x\})$$

and testing whether

$$((R \setminus \{r\}) \cup \{x\})t = R.$$

If this is the case then we have to test the second condition, which requires that rt is in the same G_R-orbit as x. For this, we simply compute $h = \varphi_R(rt)$ and test if $rth = x$. If all these conditions hold then $r \in R^*(x)$, otherwise we proceed to test the next element in R.

Assume that $r \in R^*(x)$ has been found. Then

$$((R \setminus \{r\}) \cup \{x\})t = R, \quad \text{and} \quad rth = x$$

where t and h are as above. Thus

$$(R \cup \{x\})th = ((R \setminus \{r\}) \cup \{x\})th \cup \{r\}th = Rh \cup \{x\} = R \cup \{x\},$$

i.e. $a := a_r := th$ is an automorphism of the extension set $R \cup \{x\}$. This automorphism a has the property that $ra = x$, i.e. $xa^{-1} = r$. In other words, this automorphism is a coset representative for the subgroup $G_{R,x}$ in $G_{R\cup\{x\}}$. If $\mathcal{R}$ is the collection of all a_r for $r \in R^*(x)$ together with the identity, then $\mathcal{R}$ is a transversal of the cosets of $G_{R,x}$ in $G_{R\cup\{x\}}$. In other words,

$$G_{R\cup\{x\}} = \text{Ext}(G_{R,x}, \mathcal{R}, x).$$

As remarked above, once the first automorphism a_r has been found, we can immediately form the group $H^{(1)} := \langle G_{R,x}, a_r \rangle$, which is a subgroup of $G_{R\cup\{x\}}$. Later on, we may use $H^{(1)}$ to reduce the number of $r \in R$ which need to be tested for membership in $R^*(x)$. We proceed by induction on $i = 1, 2, \ldots$. Whenever another automorphism generator a_r has been found while testing an element $r \in R$, we define the group extension

$$H^{(i+1)} = \langle H^{(i)}, a_r \rangle.$$

Of course, once an element $r \in R$ has been proven to be outside of $R^*(x)$, we can eliminate the whole orbit $H^{(i)}(r) \subseteq R$ from the search. All this follows from 9.6.6.

What happens if $r \in R$ does not lie in $R^*(x)$? Then we have found a group element $g = th$ with $t = \varphi_i((R \setminus \{r\}) \cup \{x\})$ and $h \in G_S$ such that

$$((R \setminus \{r\}) \cup \{x\})t = S,$$

and $rth = y$. Thus

$$(R \cup \{x\})th = ((R \setminus \{r\}) \cup \{x\})th \cup \{r\}th = Sh \cup \{y\} = S \cup \{y\}.$$

This means that the extension $R \cup \{x\}$ is isomorphic to $S \cup \{y\}$, i.e.

$$R \cup \{x\} \sim_G S \cup \{y\}.$$

Here, we use the word isomorphic as a synonym for "being in the same G-orbit." In this language, we can say that the element th is an isomorphism between the two extensions. Note that $R = S$ is still possible (but then $x < y$). We claim that $R \cup \{x\}$ precedes $S \cup \{y\}$. To see this, recall that we proceed in a breadth first search fashion, i.e. we process the extensions at any given level in lexicographically increasing order. Hence, if $S \cup \{y\}$ were less than $R \cup \{x\}$ then we would have detected the fact that $R \cup \{x\} \sim_G S \cup \{y\}$ earlier, and we would have discarded $R \cup \{x\}$. So, at this point we decide to eliminate the extension $S \cup \{y\}$, since it is not canonical. However, we will not totally delete the extension from the search tree. Instead, we decide to save the isomorphism th which maps $R \cup \{x\}$ to $S \cup \{y\}$. Actually, we decide to store the inverse,

$$\psi_S(y) := (th)^{-1}.$$

and call this a *fusion element*. Also, we introduce a *fusion node* for the extension $S \cup \{y\}$. The fusion node serves as a means of recoding the information which we gained about isomorphic sets. If $S \cup \{y\}$ is a fusion node, we always have that

$$(S \cup \{y\})\psi_S(y) = R \cup \{x\} \text{ is canonical.}$$ **9.6.8**

The fusion nodes will help to speed up the algorithm when it comes to computing transporter elements, as we will see in the next paragraph. Summarizing, we have seen how to construct the canonical transversal $\mathcal{T}_{i+1}$ of orbits on sets of size $i + 1$ together with the respective stabilizers.

Let us now address the problem of defining the transporter map φ_{i+1} (since this map is needed for the induction). More specifically, given a set F of size $i + 1$, the question is to find the canonical representative

$$R \cup \{x\} \in \mathcal{T}_{i+1}$$

with $F \sim_G R \cup \{x\}$. In particular, we wish to determine an element $g \in G$ with $Fg = R \cup \{x\}$. This problem can be solved recursively. The set F is split into $z := \max F$ and $Z = F \setminus \{z\}$. By induction, we can compute an element $t := \varphi_i(Z)$. Then $S := Zt$ is a canonical orbit representative. Using the orbit data, we compute $h \in G_S$ such that $zth = y$ is canonical under G_S. If $S \cup \{y\}$ is canonical under G, we return th. Otherwise, if $S \cup \{y\}$ is a fusion node, then we have a fusion element $\psi_S(y)$ such that

$$(S \cup \{y\})\psi_S(y) = R \cup \{x\}$$

is canonical by 9.6.8 and we return $th\psi_S(y)$. This finishes the description of the algorithm. Let us present the algorithm as

---

Theorem *Let G act on the finite set X. Assume that we can compute stabilizers, group* **9.6.9**
extensions and orbits on points for subgroups of G. Furthermore, let $f : \mathcal{P}(X) \to \{0, 1\}$ be a test function which is G-invariant and hereditary (in the sense of 9.5.1 and 9.5.2). Then Algorithm 9.6.10 computes the orbits of G on $\mathcal{P}^{(f)}(X) = \mathcal{P}(X) \cap f^{-1}(\{1\})$, the set of admissible subsets of X. □

Algorithm (orbits on subsets) **9.6.10**

   Input: $\quad \operatorname{orbit}(G, \mathcal{P}_i^{(f)}(X)) = (\mathcal{T}_i, \sigma_i, \varphi_i)$

   Output: $\quad \operatorname{orbit}(G, \mathcal{P}_{i+1}^{(f)}(X)) = (\mathcal{T}_{i+1}, \sigma_{i+1}, \varphi_{i+1})$

(0) **for** $R \in \mathcal{T}_i$ **do**
(1)   compute $\operatorname{orbit}(G_R, X \setminus R) := (\mathcal{T}_R, \sigma_R, \varphi_R)$
(2) **end for**
(3) $\mathcal{T}_{i+1} := \emptyset$

(4)   **for** $R \in \mathcal{T}_i$ (in increasing order) **do**
(5)   **for** $x \in \mathcal{T}_R$ (in increasing order) with $f(R \cup \{x\}) = 1$
        and for which $\psi_R(x)$ has not yet been defined **do**
(6) $G_{R,x} := \sigma_R(x)$
(7) $H := G_{R,x}$
(8)   **for all** $r \in R$ which are least in their H-orbit **do**
(9) $t := \varphi_i((R \setminus \{r\}) \cup \{x\})$
(10) $S := ((R \setminus \{r\}) \cup \{x\})t$
(11) $h := \varphi_S(rt)$
(12) $y := rth$
            (now: $(R \cup \{x\})th = S \cup \{y\}$, $S \in \mathcal{T}_i$, $y \in \mathcal{T}_S$)
(13)   **if** $S = R$ **and** $y = x$ **then**   (case 1 of 9.6.1)
(14) $H := \langle H, th \rangle$
              (*th* is an automorphism of $R \cup \{x\}$)
(15)   **else**   (case 2 of 9.6.1)
(16) $\psi_S(y) := (th)^{-1}$
              (*th* is an isomorphism from $R \cup \{x\}$ to $S \cup \{y\}$)
(17)   **end if**
(18)   **end for**
(19)   append $R \cup \{x\}$ to $\mathcal{T}_{i+1}$
(20) $\sigma_{i+1}(R \cup \{x\}) := H \, (= G_{R \cup \{x\}})$
(21)   **end for**
(22) **end for**
(23) **return** $(\mathcal{T}_{i+1}, \sigma_{i+1}, \varphi_{i+1})$

Where the function φ_{i+1} is defined as follows.

(24) **function** $\varphi_{i+1}(F)$
(25) $z := \max F$, $Z := F \setminus \{z\}$ (a set of size i)
(26) $t := \varphi_i(Z)$
(27) $S := Zt$
(28) $h := \varphi_S(zt)$, $y := zth$
(29)   **if** $\psi_S(y)$ has been defined **then**
(30)   **return** $th\psi_S(y)$
(31)   **else**
(32)   **return** th
(33)   **end if**
(34) **end function** $\square$

Proof: The proof is by induction. The orbits of subsets of size 0 are trivially
known. The orbits of subsets of size 1 are known by assumption. Now assume

that orbit$(G, \mathcal{P}_i(X)) = (\mathcal{T}_i, \sigma_i, \varphi_i)$ has already been computed. In order to prove correctness of Algorithm 9.6.10, we verify that $\mathcal{T}_{i+1}$ is a transversal for the orbits of G on $\mathcal{P}_{i+1}(X)$, and that $\sigma_{i+1}(R) = G_R$ for $R \in \mathcal{T}_{i+1}$ and that $\varphi_{i+1}(S) = t$ such that $St \in \mathcal{T}_{i+1}$ for all $S \in \mathcal{P}_{i+1}(X)$.

First of all, each $(i+1)$-subset S can be written as $S = S' \cup \{y\}$ where S' is an i-subset and $y \in X \setminus S'$. Putting $g := \varphi_i(S')$ we get $S \sim_G Sg = S'g \cup \{yg\}$ where $S'g$ is an orbit representative in $\mathcal{T}_i$. Hence we get representatives of all G-orbits on $(i+1)$-sets from the extensions of the form $R \cup \{x\}$ where $R \in \mathcal{T}_i$ and $x \in X \setminus R$. In lines (0) and (4), (5) these extensions of orbit representatives are considered. In line (1), the orbits of G_R on $X \setminus R$ are computed for $R \in \mathcal{T}_i$. The result is $(\mathcal{T}_R, \sigma_R, \varphi_R)$, where

1. $\mathcal{T}_R$ is a transversal of the orbits of G_R on $X \setminus R$,

2. $\sigma_R : X \setminus R \to L(G)$ is such that $\sigma_R(x) = G_{R,x} = (G_R)_x$ is the stabilizer of x in G_R, and

3. $\varphi_R : X \setminus R \to G$ is a map with $\varphi_R(y) = g \in G$ such that $yg \in \mathcal{T}_R$.

The candidate set is the set of extensions $R \cup \{x\}$ where $R \in \mathcal{T}_i$ and $x \in \mathcal{T}_R$. In lines (4) and (5), the extensions $R \cup \{x\}$ are considered again.

We must now show that the extensions which are added to $\mathcal{T}_{i+1}$ in line (20) are pairwise not in the same G-orbit. Let $R \cup \{x\}$ and $S \cup \{y\}$ be two arbitrary distinct extension sets (with $R, S \in \mathcal{T}_i$ and $x \in \mathcal{T}_R$ and $y \in \mathcal{T}_S$). By 9.6.1, $R \cup \{x\} \sim_G S \cup \{y\}$ if and only if either $R = S$ and $x \sim_{G_R} y$, or for one $r \in R$ the equations

$$((R \setminus \{r\}) \cup \{x\})t = S \quad \text{and} \quad rt \sim_{G_S} y \qquad\qquad \text{9.6.11}$$

hold for $t = \varphi_i((R \setminus \{r\}) \cup \{x\})$. First, consider the case $R = S$ and $x \sim_{G_R} y$. Since $R \cup \{x\}$ is different from $S \cup \{y\}$, we must have $x \neq y$. But x and y are different elements of the transversal $\mathcal{T}_R$ of G_R orbits, which contradicts $x \sim_{G_R} y$. Hence we must be in the second case, i.e. 9.6.11 holds true for some $r \in R$. Without loss of generality, we assume that

$$R \cup \{x\} \preceq S \cup \{y\},$$

i.e. that $R \cup \{x\}$ has been considered before $S \cup \{y\}$ in lines (4) and (5). By 9.6.11, there is an element $r \in R$ for which $((R \setminus \{r\}) \cup \{x\})t = S$ with $t = \varphi_i((R \setminus \{r\}) \cup \{x\})$ and $rth = y \in \mathcal{T}_S$ with $h = \varphi_S(rt) \in G_S$. In this case, the fusion element $\psi_S(y) = (th)^{-1}$ will be defined in line (16) which prevents the extension $S \cup \{y\}$ from being considered in lines (4) and (5). This proves that the computed set $\mathcal{T}_{i+1}$ intersects each G-orbit at most once. From the above, we already know that $\mathcal{T}_{i+1}$ contains elements from every orbit, and hence $\mathcal{T}_{i+1}$ is a transversal of the $(i+1)$-orbits of G, as required. The fact that $G_{R \cup \{x\}} = \text{Ext}(G_{R,x}, \mathcal{R}, x)$ has been shown in 9.6.5. The transversal $\mathcal{R}$ is never explicitly computed. Instead, in line (7) the group H is initialized to be $H = G_{R,x} = \sigma_R(x)$. The if clause in line (13) evaluates to true if and only if $r \in R^*(x)$, which means that th is an element of $\mathcal{R}$. Therefore, the group H is extended by th in line (14). Line (8) reduces the number of $r \in R$ which have to be tested. According to 9.6.7, we require that $r \in R$ be minimal in its H-orbit. At the end of the loop, in line (18), the full group $G_{R \cup \{x\}}$ has been computed in H. In lines (19) and (20), the new canonical representative $R \cup \{x\}$ is added to $\mathcal{T}_{i+1}$ and the stabilizer $G_{R \cup \{x\}}$ is stored as $\sigma_{i+1}(R \cup \{x\})$. At the end of the for loops in lines (21) and (22), the transversal $\mathcal{T}_{i+1}$ is complete.

It remains to show that $\varphi_{i+1}(F)$ is an element $g \in G$ with $Fg \in \mathcal{T}_{i+1}$. In line (25), F is written as a union of an i-set Z and the element z. For $t = \varphi_i(Z)$ we then have $Zt = S \in \mathcal{T}_i$ in line (27). Hence the orbit data for the set S has been computed and we can evaluate $h = \varphi_S(zt)$ and define $y = zth$ in line (28). We now have

$$Fth\psi_S(y) = (Z \cup \{z\})th\psi_S(y) = S \cup \{y\}.$$

If $\psi_S(y)$ has not been defined then $S \cup \{y\}$ is canonical and we return th. Otherwise, there has been an extension $R \cup \{x\}$ and an element $r \in R \setminus R^*(x)$ such that $((R \setminus \{r\}) \cup \{x\})t' = S$, with $t' := \varphi((R \setminus \{r\}) \cup \{x\})$, and $y = xt'h'$ for $h' = \varphi_S(xt') \in G_{\{S\}}$. Since r is not in $R^*(x)$, the if clause in (13) did not hold and the element $(t'h')^{-1}$ has been stored as $\psi_S(y)$. Hence

$$\begin{aligned} Fth\psi_S(y) &= (Z \cup \{z\})th\psi_S(y) \\ &= (S \cup \{y\})\psi_S(y) \\ &= (S \cup \{y\})(t'h')^{-1} \\ &= (R \cup \{x\}) \in \mathcal{T}_{i+1}. \end{aligned}$$

This proves that in each case $\varphi_{i+1}(F)$ is an element that maps F to its canonical orbit representative, as required. This completes the proof that the algorithm computes the orbit data for G acting on subsets.   □

9.6.12   **Example (continuation of Example 9.5.12)** Let us consider the binary $(8,4)$-codes again. The generation tree is shown in Fig. 9.7. A node A is represented by a box, with the label max A indicated in a circle right above the box. The circled numbers immediately below the box are the possible extensions. Inside the box, information on the stabilizer is given. The first number is the order of the stabilizer. After that, the orbits of the stabilizer on points are indicated. Inside each orbit, the numbers are arranged in increasing order. Hence

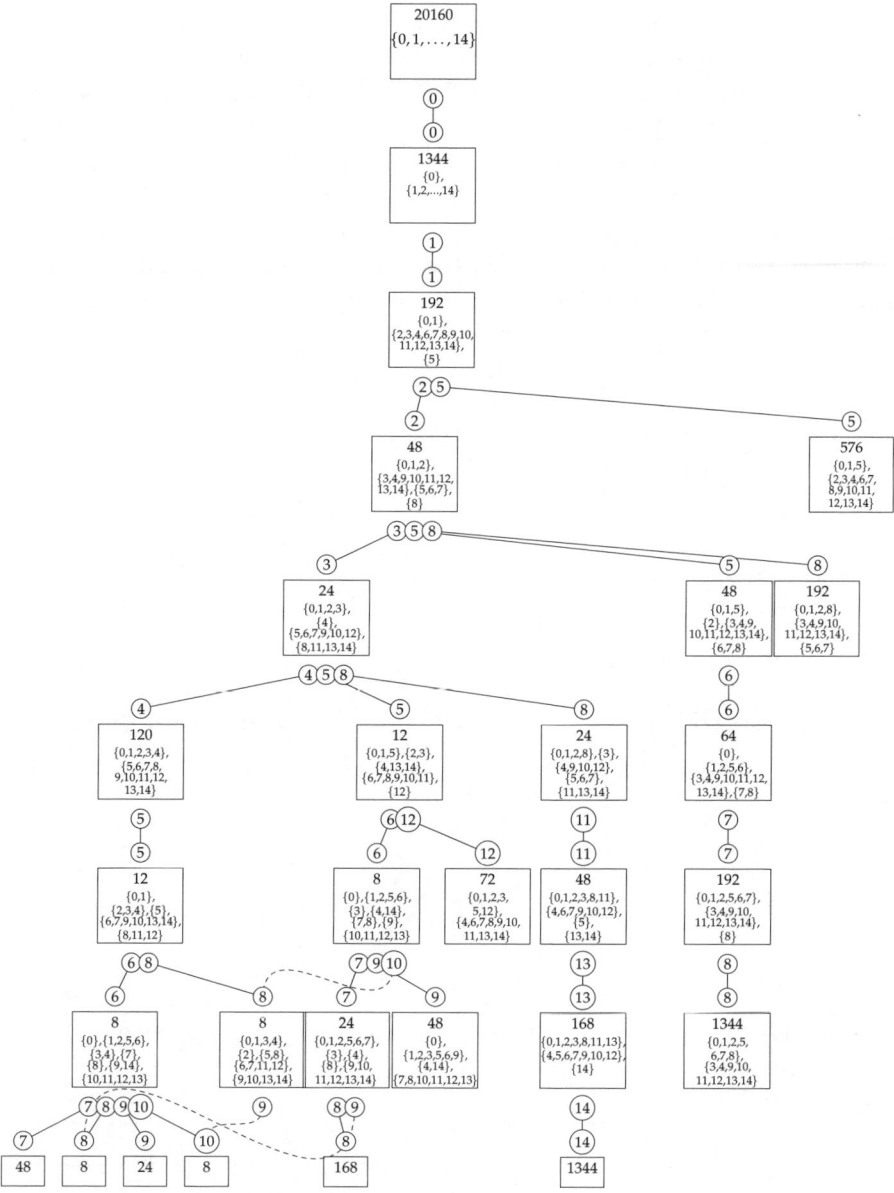

Fig. 9.7 Generation tree of $(8, 4, \geq 3, 2)$-codes

the first number is the least orbit representative. Not every orbit leads to an extension. The rank condition must be satisfied for possible extensions (since $d = 3$, the rank condition is always satisfied in this example). The solid lines stand for extensions leading to canonical sets, i.e. to new orbit representatives at depth one step further down the tree. These lines always connect circles with equal numbers. The three dashed and somewhat curly lines are related to fusion nodes. Recall that fusion nodes stand for extension sets which are not canonical. Each fusion node is connected by a curly line to the corresponding canonical node, which is to the left. Associated with every curly line is a fusion element. The three fusion nodes are

$$\{0,1,2,3,4,5,8,9\}, \{0,1,2,3,5,6,10\}, \text{ and } \{0,1,2,3,5,6,7,9\}.$$

The corresponding fusion elements are (in matrix form and as permutations of the points of $PG_3(2)$, respectively)

$$\psi_{\{0,1,2,3,4,5,8\}}(9) = \begin{pmatrix} 1 & 0 & 1 & 0 \\ 1 & 1 & 1 & 1 \\ 0 & 0 & 0 & 1 \\ 1 & 0 & 0 & 0 \end{pmatrix} = (0,6,13,12,9,2,3)(1,4,5,10,14,7,8),$$

$$\psi_{\{0,1,2,3,5,6\}}(10) = \begin{pmatrix} 1 & 1 & 1 & 0 \\ 1 & 1 & 0 & 0 \\ 0 & 0 & 0 & 1 \\ 0 & 1 & 0 & 0 \end{pmatrix} = (0,8,12,10)(1,5,2,3)(4,14,9,6)(7,11),$$

and

$$\psi_{\{0,1,2,3,5,6,7\}}(9) = \begin{pmatrix} 1 & 1 & 1 & 0 \\ 1 & 0 & 1 & 0 \\ 0 & 0 & 1 & 0 \\ 1 & 1 & 1 & 1 \end{pmatrix} = (0,8,7)(1,6,5)(3,4,9)(11,13,12).$$

For instance, the fusion node $\{0,1,2,3,5,6,10\}$ is connected to the canonical node $\{0,1,2,3,4,5,8\}$. This is because application of the fusion element maps one set onto the other:

$$\{0,1,2,3,5,6,10\}(0,8,12,10)(1,5,2,3)(4,14,9,6)(7,11) = \{8,5,3,1,2,4,0\}.$$

Let us trace the computation of the automorphism group of the extended Hamming code, for example. As pointed out in Example 9.5.12, the $(8,4)$ extended Hamming code is the rightmost leaf at level 8, i.e. the set

$$\{0,1,2,3,8,11,13,14\}.$$

Essentially, the computation consists of 8 repetitions of the "down-and-up" process described earlier in this section. For each prefix R of the set in question, we compute from the given group G_R the groups $G_{R,x}$ and $G_{R \cup \{x\}}$. We

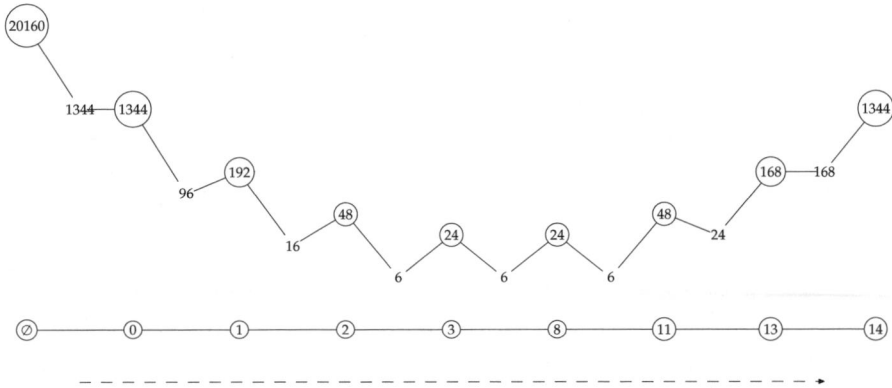

Fig. 9.8 Computing the automorphism group of the $(8, 4)$ extended Hamming code

proceed by induction on the size of the prefix R, i.e. we start with $R = \varnothing$, then consider $R = \{0\}$, after that $R = \{0, 1\}$ and so forth. This means that we are moving from the left to the right in Fig. 9.8, which shows the elements of R at the bottom. Above, the order of the groups G_R (circled) and $G_{R,x}$ is plotted on a logarithmic scale. The computation starts with the empty set, whose automorphism group is $G = \mathrm{PGL}(4, 2)$ of order 20160 (this is the root node in Fig. 9.7). Then the point 0 is chosen. Since 0 is in an orbit of length 15,

$$|G_0| = 20160/15 = 1344.$$

Next, we add the point 1 to the set. Since 1 is in an orbit of G_0 of length 14, we have

$$|G_{0,1}| = 1344/14 = 96.$$

The upstep results in an automorphism which interchanges 0 and 1, so that $G_{0,1}$ is of index 2 in the set stabilizer $G_{\{0,1\}}$, which must therefore be of order $2 \cdot 96 = 192$. Then the point 2 is added from an orbit of $G_{\{0,1\}}$ of length 12, so that $G_{\{0,1\},2}$ has order $192/12 = 16$ (recall that $G_{\{0,1\},2}$ denotes the intersection of the set stabilizer of $\{0, 1\}$ with the point stabilizer of 2. The following upstep detects that 2 is in an orbit of length 3 under $G_{\{0,1,2\}}$, so that the next set stabilizer is $G_{\{0,1,2\}}$ of order $16 \cdot 3 = 48$. The computation continues in this way. Eventually, the automorphism group of the extended Hamming code is computed to be the set stabilizer

$$G_{\{0,1,2,3,8,11,13,14\}}$$

of order 1344. $\diamond$

Fig. 9.9 The binary $(18, 9, 6)$-codes

Figure 9.9 shows the generation tree for the unique binary $(18, 9, 6)$-code. The sole purpose of this example is to give a rough idea of the nature of such trees. We suppress all labels and automorphism group order information.

9.7 Base and Strong Generating Sets

The orbit algorithm as described above depends on the availability of good algorithms to work with permutation groups. In particular, point stabilizer subgroups and extension overgroups need to be computed (as well as orbits on points). It turns out that our first attempt at these algorithms, based on sets of generators, does not perform well for large examples. The reason is that the number of generators produced by 9.2.10 may become too large, which in turn deteriorates the performance of the orbit algorithm on points.

In this section, we will overcome this bottleneck by introducing a better data structure for permutation groups. This data structure, introduced by Sims [181, 182], is called a stabilizer chain. It represents the group by means of a chain of subgroups terminating in the trivial group. Each group in the chain is the stabilizer of a point in the previous group.

To begin with, let G be a group acting faithfully on a set finite, totally ordered set $X = \{x_0, \ldots x_{n-1}\}_<$. A subset $B = \{b_0, \ldots, b_{r-1}\} \subset X$ is called *base* for G on X if the *pointwise stabilizer* $G_{b_0, \ldots, b_{r-1}} = 1$, i.e. if only the identity of G fixes all the points of B. An *ordered base* for G on X is a *sequence* $(b_0, \ldots, b_{r-1})$ such that the corresponding set $\{b_0, \ldots, b_{r-1}\}$ is a base for G. An ordered base B gives rise to a chain of subgroups

$$G = G^{(0)} \geq G^{(1)} \geq \cdots \geq G^{(r)} = \langle 1 \rangle,$$

9.7.1

where

$$G^{(i+1)} = G^{(i)}_{b_i}$$

9.7.2

is the stabilizer of b_i in $G^{(i)}$. This is called the *stabilizer chain* (or *Sims chain*) for G with respect to B. The base is called *irredundant* if no two (consecutive) terms of the sequence of subgroups coincide.

The images of the base points determine a permutation in the following sense.

Lemma *Let G be a permutation group with base $B = (b_0, \ldots, b_{r-1})$. Let g and h be two elements of G. Then $g = h$ if and only if $b_i g = b_i h$ for $i \in r$. In other words, knowing the images of all base points determines a permutation uniquely.*

9.7.3

Proof: The condition $b_i g = b_i h$ for $i \in r$ is equivalent to $b_i g h^{-1} = b_i$ for all $i \in r$, which in turn is equivalent to $g h^{-1} \in G^{(r)} = \langle 1 \rangle$, using the fact that B is a base. Thus $g h^{-1} = 1$, i.e. $g = h$. $\square$

By 3.4.1, the cosets of $G^{(i+1)}$ in $G^{(i)}$ correspond to the different elements in the orbit

$$\mathcal{O}^{(i)} = G^{(i)}(b_i),$$

which we call the *i-th basic orbit*. In particular, since $G^{(i+1)}$ is a point stabilizer in $G^{(i)}$ by 9.7.2, the index satisfies

$$\left| G^{(i)} \right| / \left| G^{(i+1)} \right| = |\mathcal{O}^{(i)}| =: \ell_i$$

and hence by 9.7.1

9.7.4
$$|G| = \prod_{i \in r} \left| G^{(i)} \right| / \left| G^{(i+1)} \right| = \prod_{i \in r} \ell_i.$$

For $i \in r$, we choose *coset representatives* $\sigma_{i,0}, \ldots, \sigma_{i,\ell_i-1}$ for $G^{(i+1)}$ in $G^{(i)}$, so that

9.7.5
$$G^{(i)} = \bigcup_{j \in \ell_i} G^{(i+1)} \sigma_{i,j}$$

is the decomposition of $G^{(i)}$ into cosets of $G^{(i+1)}$. We require that $\sigma_{i,0} = 1$, the identity element of G, for all $i \in r$. A *strong generating set* for G relative to B is a set S of elements of G with the property that

9.7.6
$$\langle S \cap G^{(i)} \rangle = G^{(i)} \quad \text{for } i \in r.$$

9.7.7   **Example** Consider the symmetric group $G = S_n$ acting on the set n. An ordered base for G is $B = (0, 1, \ldots, n-2)$. $G^{(i)}$ is isomorphic to S_{n-i} (acting on the set $\{i, \ldots, n-1\}$). The basic orbits are of length $\ell_i = n - i$. Coset representatives are $\sigma_{i,j} = (i, i+j)$ for $j \in \ell_i$ and $i \in n-1$. The sets

$$U = \{(0, 1, \ldots, n-1), (0, 1)\}$$

and

$$V = \{(0, 1), (1, 2), \ldots, (n-2, n-1)\}$$

both generate S_n. For $n \geq 3$, U is not a strong generating set (for example the group $G^{(1)}$, which is the symmetric group acting on $\{1, \ldots, n-1\}$ contains none of the generators). On the contrary, the generating set V is a strong generating set for all n. This is because $V \cap G^{(i)} = \{(i, i+1), \ldots, (n-2, n-1)\}$ generates S_{n-i} acting on the set $\{i, \ldots, n-1\}$. In fact, for each $n \geq 2$, the permutations $(i, i+1)$ for $i \in n-2$ form a strong generating set for S_n. $\diamond$

The point of knowing a strong generating set S for a permutation group G is that the basic orbits $G^{(i)}(b_i)$ can be computed easily. Namely, it is straightforward to compute the subsets

$$S^{(i)} := S \cap G^{(i)}, \quad i \in r,$$

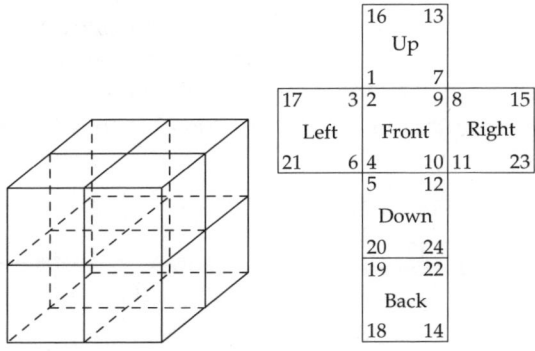

Fig. 9.10 Rubik's $2 \times 2 \times 2$ cube

which for fixed $i \in r$ contain those generators which fix the first i base points $b_0, \ldots, b_{i-1}$. Using the orbit algorithm of Section 9.2, one computes the corresponding basic orbit. From this orbit, coset representatives $\sigma_{i,0}, \ldots, \sigma_{i,\ell_i-1}$ can be determined (they are just the transporter elements of Section 9.2). The point is that the basic orbits and the corresponding Schreier trees can be constructed easily from the strong generating set. This is not the case for arbitrary generating sets, where one has to go through more complex algorithms, like the Schreier–Sims algorithm described in [91], for example. The difficulty lies in the fact that the basic orbits $\mathcal{O}^{(i)} = G^{(i)}(b_i)$ can only be computed when generators for $G^{(i)}$ are known. This explains why the set $S^{(i)}$ which generates $G^{(i)}$ is so valuable.

One further remark concerning the Schreier tree is in order. Recall that we require that $\sigma_{i,0} = 1$. This condition is automatically satisfied for Schreier trees, since the path from the root to itself corresponds to the empty word, which by definition is the identity element in the group. Let us consider another example.

***

Example Figure 9.10 shows Rubik's cube in the simplified version with sides   **9.7.8**
of length 2 instead of three. We label the faces with the integers in $\{1, \ldots, 24\}$ as indicated beneath. Here, we start labeling points from 1, since many current computer algebra systems have permutations act on $1, 2, 3, \ldots$ We will follow this convention throughout this example, for the sake of allowing the reader to verify the claims made by using a standard software package.

Consider the group G which is generated by the rotations of the sides. We follow the widely accepted notation due to Singmaster (cf. [183]), which denotes the quarter turns in clockwise direction of the left, right, front, back, up and down side of the cube by L, R, F, B, U and D, respectively. However, we stick to the notation $A^{-1}, A^{-2}, \ldots$ for the inverse, the square of the inverse etc. of the element A (as opposed to using A' for the inverse of A which is sometimes used). The permutations which correspond to these six generators are

$$
\begin{aligned}
R &= (7, 14, 24, 10)(8, 15, 23, 11)(9, 13, 22, 12) \\
B &= (13, 17, 20, 23)(14, 18, 19, 22)(16, 21, 24, 15) \\
D &= (4, 11, 22, 21)(5, 12, 24, 20)(6, 10, 23, 19), \\
L &= (1, 4, 20, 18)(2, 5, 19, 16)(3, 6, 21, 17), \\
F &= (1, 8, 12, 6)(2, 9, 10, 4)(3, 7, 11, 5), \\
U &= (1, 16, 13, 7)(2, 17, 14, 8)(3, 18, 15, 9).
\end{aligned}
$$

An ordered base for the group G is $(1, 4, 7, 10, 13, 16, 19)$. We get the following stabilizer chain, where we indicate the length of the fundamental orbit in parenthesis and where the grey area in the pictures indicates faces which have been stabilized.

$$G = G^{(0)}$$ (24)

$$\geq (G^{(1)} =\,) G_1$$ (21)

$$\geq (G^{(2)} =\,) G_{1,4}$$ (18)

$$\geq (G^{(3)} =\,) G_{1,4,7}$$ (15)

$$\geq (G^{(4)} =\,) G_{1,4,7,10}$$ (12)

$$\geq\ (G^{(5)} =)\, G_{1,4,7,10,13} \tag{9}$$

$$\geq\ (G^{(6)} =)\, G_{1,4,7,10,13,16} \tag{6}$$

$$\geq\ (G^{(7)} =)\, G_{1,4,7,10,13,16,19} = 1.$$

Hence by 9.7.4, the order of G (i.e. the number of positions) is

$$24 \times 21 \times 18 \times 15 \times 12 \times 9 \times 6 = 88\ 179\ 840.$$

Note that the generating set $\{L, R, F, B, U, D\}$ for G is *not* strong. A strong generating set can be found by considering moves which fix the grey part and permute the remaining faces among themselves. The point is that these moves may bring the grey part into disarray for a while. However, at the end of the move the grey faces are brought back into place. By computing Schreier trees it can be checked that

$$S = \{\nu, \tau, \delta, B, \omega, R, D, L\}$$

is a strong generating set, where

$$
\begin{aligned}
\tau &= (BLFRD)^3 = (19, 24)(20, 22)(21, 23), \\
\rho &= DFU^{-1}R^{-1}UFD^{-1}F^{-1} = (4, 5, 6)(7, 9, 8)(10, 12, 11)(19, 21, 20), \\
\nu &= \rho^2 B^{-1}\rho B = (19, 20, 21)(22, 24, 23), \\
\delta &= B\tau B^{-1} = (16, 20)(17, 19)(18, 21), \\
\omega &= DBD^{-1}B^{-1} = (10, 23, 11, 22, 12, 24)(16, 19, 18, 20, 17, 21).
\end{aligned}
$$

We find that

$$
\begin{aligned}
G^{(6)} &= G_{1,4,7,10,13,16} = \langle \nu, \tau \rangle, \\
G^{(5)} &= G_{1,4,7,10,13} = \langle \nu, \tau, \delta \rangle, \\
G^{(4)} &= G_{1,4,7,10} = \langle \nu, \tau, \delta, B \rangle, \\
G^{(3)} &= G_{1,4,7} = \langle \nu, \tau, \delta, B, \omega \rangle, \\
G^{(2)} &= G_{1,4} = \langle \nu, \tau, \delta, B, \omega, R \rangle, \\
G^{(1)} &= G_1 = \langle \nu, \tau, \delta, B, \omega, R, D \rangle, \\
G^{(0)} &= G = \langle \nu, \tau, \delta, B, \omega, R, D, L \rangle,
\end{aligned}
$$

which are groups of order 6, 54, 648, 9720, 174 960, 3 674 160 and 88 179 840, respectively. More details on the group of Rubik's cube (in particular, the version with sides of length 3) can be found in the books by Neumann, Stoy and Thompson [158] and in the above-mentioned book by Singmaster [183].   ◇

Our next goal is to identify group elements with integers, using a known stabilizer chain for the permutation group. This serves two purposes. Firstly, it is convenient, as integers are often easier to handle in computer programs. Secondly, this enables us to pick group elements uniformly at random, which is useful for randomized algorithms for permutation groups. To begin with, let us introduce the multibase representation of an integer.

9.7.9 **Lemma** *Let* $L = (\ell_0, \ldots, \ell_{r-1})$ *be a sequence of positive integers and define* $m = \prod_{i \in r} \ell_i$. *Any integer* n *in* $m = \{0, \ldots, m-1\}$ *has a unique representation of the form*

$$n = \sum_{i \in r} a_i \prod_{j \in i} \ell_j$$

with integers $a_i \in \ell_i$ *for* $i \in r$ *(here, an empty product is defined to be 1). We write*

$$n = (a_{r-1}, \ldots, a_0)_L$$

and call this the multibase representation *of* n *with respect to B.*

Proof: Put $m_i = \prod_{j \in i} \ell_j$ for $i \in r+1$, i.e. $m_r = m$.
Existence: If $r = 1$ we may put $a_0 = n$ and we are finished. Thus let us assume that $r \geq 2$. Given $n = n_{r-1}$ with $n \in m$, integral division yields unique integers $a_{r-1} \geq 0$ and n_{r-2} with

$$n = n_{r-1} = a_{r-1} m_{r-1} + n_{r-2} \quad \text{with } n_{r-2} \in m_{r-1}.$$

Here we have $a_{r-1} = \lfloor n_{r-1}/m_{r-1} \rfloor$, and since $n_{r-1} = n < m = m_{r-1}\ell_{r-1}$ we have $a_{r-1} \in \ell_{r-1}$. If $r \geq 3$, we may repeat this argument for n_{r-2} and obtain an equation of the form

$$n_{r-2} = a_{r-2} m_{r-2} + n_{r-3} \quad \text{with } n_{r-3} \in m_{r-2} \text{ and } a_{r-2} \geq 0.$$

Here we have $a_{r-2} = \lfloor n_{r-2}/m_{r-2} \rfloor$, and since $n_{r-2} < m_{r-1} = m_{r-2}\ell_{r-2}$ we have $a_{r-2} \in \ell_{r-2}$. If we proceed in this way, we define integers $a_i \in \ell_i$ and $n_{i-1} \in m_i$. Eventually we arrive at an equation of the form

$$n_1 = a_1 m_1 + n_0 \quad \text{with } n_0 \in m_1 \text{ and } a_1 \in \ell_1.$$

Note that by definition $m_1 = \ell_0$, so that we may simply put $a_0 = n_0 \in m_1 = \ell_0$. Thus, we have written n as

$$
\begin{aligned}
n &= n_{r-1} \\
&= a_{r-1} m_{r-1} + n_{r-2} \\
&= a_{r-1} m_{r-1} + a_{r-2} m_{r-2} + n_{r-3} \\
&\ \ \vdots \\
&= \sum_{i \in r} a_i m_i.
\end{aligned}
$$

Uniqueness: Let

$$(a_{r-1}, \ldots, a_0)_L = n = (b_{r-1}, \ldots, b_0)_L$$

be two expressions for n. Subtraction yields

$$0 = \sum_{i \in r}(b_i - a_i)m_i.$$

Put $\Delta_i := b_i - a_i$. Let j be such that $\Delta_j \neq 0$ (such an index j exists if we assume that the expressions are distinct). Therefore

$$\Delta_j m_j = -\sum_{\substack{i \in r \\ i \neq j}} \Delta_i m_i. \qquad\qquad \textbf{9.7.10}$$

Notice that

$$|\Delta_i| \leq b_i < \ell_i, \quad i \in r. \qquad\qquad \textbf{9.7.11}$$

If $j < r - 1$, we may consider 9.7.10 modulo m_{j+1} to get

$$\Delta_j m_j \equiv -\sum_{i \in j-1} \Delta_i m_i \bmod m_{j+1}. \qquad\qquad \textbf{9.7.12}$$

Using 9.7.11 we get that

$$|\Delta_i|m_i \leq (\ell_i - 1)m_i = \ell_i m_i - m_i = m_{i+1} - m_i.$$

Therefore, over the integers, the right hand side of 9.7.12 is bounded above by

$$\left| \sum_{i \in j-1} \Delta_i m_i \right| \leq \sum_{i \in j-1} |\Delta_i|m_i \leq \sum_{i \in j-1} (m_{i+1} - m_i) = m_j - m_0 = m_j - 1 < m_j.$$

But $\Delta_j \neq 0$, which means that 9.7.12 has no solution modulo m_{j+1}. Hence $\Delta_j \neq 0$ is impossible. If $j = r - 1$, 9.7.10 becomes

$$\Delta_{r-1} m_{r-1} = -\sum_{i \in r-1} \Delta_i m_i.$$

The same argument as before shows that the absolute value of the right hand side of this equation is bounded above by m_{r-1}, which contradicts the fact that Δ_{r-1} is nonzero. These contradictions show that the multibase representation is unique. □

We introduce some more notation. For a sequence $L = (\ell_0, \ldots, \ell_{r-1})$, let

$$\overleftarrow{L} = (\ell_{r-1}, \ldots, \ell_0)$$

be the *reversed sequence*. The following result enables us to identify group elements with integers.

9.7.13   **Lemma** *Let the group G be of order $|G|$ with base $B = (b_0, \ldots, b_{r-1})$ and basic orbits of lengths $|G^{(i)}(b_i)| = \ell_i$, $i \in r$. Furthermore, assume that coset representatives $\sigma_{i,j}$ for $j \in \ell_i$, $i \in r$ have been chosen. Put $L = (\ell_0, \ldots, \ell_{r-1})$. Define a map*

$$\mathrm{rk}^{-1} \colon |G| \to G \colon n \mapsto \sigma_{r-1,a_0}\sigma_{r-2,a_1} \cdots \sigma_{0,a_{r-1}},$$

where $(a_{r-1}, \ldots, a_0)_{\overleftarrow{L}}$ is the multibase representation of n with respect to $\overleftarrow{L}$. This map is bijective, we call it the unrank *function for G. Its inverse is the* rank *function for G.*

Proof: By 3.4.1, each element $g \in G^{(0)} = G$ can be written as

$$g = g^{(1)}\sigma_{0,a_{r-1}},$$

for a uniquely determined coset representative $\sigma_{0,a_{r-1}}$, $a_{r-1} in \ell_0$ and a unique element $g^{(1)} \in G^{(1)}$. Repeating this argument for $g^{(1)}$ yields a unique coset representative $\sigma_{1,a_{r-2}}$, $a_{r-2} \in \ell_1$, and a unique element $g^{(2)} \in G^{(2)}$ such that

$$g^{(1)} = g^{(2)}\sigma_{1,a_{r-2}}.$$

If $r > 2$, we may proceed in this fashion. In the i-th step we find an equation of the form

$$g^{(i)} = g^{(i+1)}\sigma_{i,a_{r-1-i}},$$

for some unique elements $g^{(i+1)}$ and $\sigma_{i,a_{r-1-i}}$, $a_{r-1-i} \in \ell_i$. This process terminates once we reach

$$g^{(r-1)} = g^{(r)}\sigma_{r-1,a_0},$$

with $a_0 \in \ell_{r-1}$, since then $g^{(r)} \in G^{(r)} = 1$, the trivial group, i.e. $g^{(r)} = 1$. Back-substituting the equations into each other gives

$$
\begin{aligned}
g &= g^{(1)}\sigma_{0,a_{r-1}} \\
  &= g^{(2)}\sigma_{1,a_{r-2}}\sigma_{0,a_{r-1}} \\
  &\;\;\vdots \\
  &= \sigma_{r-1,a_0}\sigma_{r-2,a_1} \cdots \sigma_{1,a_{r-2}}\sigma_{0,a_{r-1}},
\end{aligned}
$$

with $a_i \leq \ell_{r-1-i}$ for $i \in r$. This means that we are able to write the given group element g in a unique way as a product of coset representatives. In the literature, the indicated process is known as the *sift algorithm*. To turn this representation into a number, we simply consider the sequence $a_0, \ldots, a_{r-1}$ as multibase representation

$$(a_{r-1}, a_{r-2}, \ldots, a_0)_{\overleftarrow{L}} = n$$

of some integer $n \in |G| = \prod_{i \in r} \ell_i$. This process defines a rank function on the set of group elements. In fact, this function is bijective because different group elements give different factorizations and hence different multibase representations of numbers. The inverse process gives the unrank function.   □

Table 9.3 Unranking the elements of S_3

n	$(a_1, a_0)_{(2,3)}$	$\mathrm{rk}^{-1}(n) = \sigma_{1,a_0}\sigma_{0,a_1}$
0	$(0,0)$	$1 = 1 \cdot 1$
1	$(0,1)$	$(1,2) = (1,2) \cdot 1$
2	$(1,0)$	$(0,1) = 1 \cdot (0,1)$
3	$(1,1)$	$(0,1,2) = (1,2) \cdot (0,1)$
4	$(2,0)$	$(0,2) = 1 \cdot (0,2)$
5	$(2,1)$	$(0,2,1) = (1,2) \cdot (0,2)$

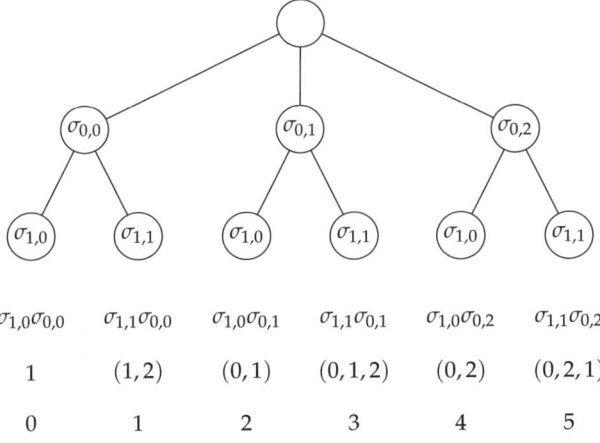

Fig. 9.11 The elements of S_3 by rank

We remark that the order of the terms in the function rk^{-1} of 9.7.13 matters, since we do not require the group to be abelian.

Example Consider the symmetric group S_3 acting on $\{0,1,2\}$ with base $B =$ **9.7.14**
$(0,1)$. The basic orbits are of length $\ell_0 = 3$ and $\ell_1 = 2$. Hence $\overleftarrow{L}=(3,2)=(2,3)$.
Coset representatives are

$$\sigma_{0,0} = 1, \ \sigma_{0,1} = (0,1), \ \sigma_{0,2} = (0,2), \ \sigma_{1,0} = 1, \ \sigma_{1,1} = (1,2).$$

The unrank function lists the elements in the order indicated in Table 9.3. The ordering may be visualized as in Fig. 9.11. The coset representatives are shown as the nodes of a tree. The leaves stand for elements of the group. The corresponding permutations and their ranks are shown at the bottom.   ◇

We are now in a position to define another important graph associated to a group. If G is a group and if S is a set of elements of G, the *Cayley-graph* of G

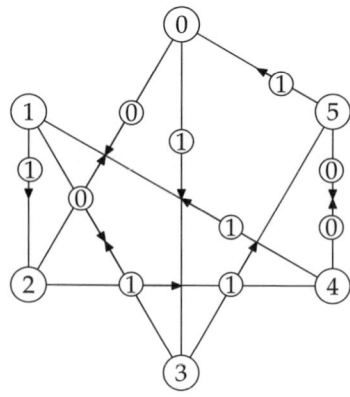

Fig. 9.12 The Cayley-graph of S_3

with respect to S is the action-graph whose vertices are the elements of G and whose edges are defined by the right-multiplication by elements $s \in S$. That is, the Cayley-graph of G with respect to S has an edge from x to y labeled by $s_i \in S$ if $xs_i = y$ holds in G. Figure 9.12 shows the Cayley graph of S_3 with respect to the generating set $S = \{s_0, s_1\}$ where $s_0 = (0,1,2)$ and $s_1 = (0,1)$. Cayley graphs are often used to investigate combinatorial problems theoretically, and they can also be useful for studying the concepts defined in this section.

9.7.15   **Example** Consider the Cayley graph of Rubik's cube. As noted above, for the $2 \times 2 \times 2$ cube, we may assume that one corner is fixed, for instance the front-top-left corner $1, 2, 3$. That leaves only the generators R, D and B as well as their inverses. We consider the Cayley graph of $G^{(1)} = G_1$ (of order $3\,674\,160$, see above) with respect to these 6 generators. Cayley graphs admit the defining group as vertex transitive automorphism group. Therefore, in order to compute the diameter of the graph it suffices to compute the distance of the vertex furthest away from a given vertex. If Γ_i is the set of vertices at distance i from the identity node, then Jianyi Yao, a student at Colorado State University, reports the following numbers:

i	$|\Gamma_i|$	i	$|\Gamma_i|$	i	$|\Gamma_i|$
0	1	5	2256	10	930588
1	6	6	8969	11	1350852
2	27	7	33058	12	782536
3	120	8	114149	13	90280
4	534	9	360508	14	276

In particular, this means that there are 276 "worst case" positions, i.e. positions which can be restored with no less than 14 quarter turns. This agrees with results obtained by Cooperman et al. [41], which report that the diameter of this graph is 14.   ◇

Summarizing, the concept of base and strong generating set defines a new data structure for permutation groups. This data structure is based on the stabilizer chain corresponding to the base. To represent that chain, one needs one Schreier tree for each basic orbit. At any particular level, one obtains coset representatives for the next subgroup in the chain from the Schreier tree. Using a fixed ordering of these representatives, 9.7.13 allows one to access group elements numerically. For further details on working with stabilizer chains, we refer to the above-mentioned books by Holt [91] and by Seress [177]. We only mention that randomization plays a key role in those algorithms.

Exercises

Exercise Let G be a finite group and let S be a subset of G. Show the following.   E.9.7.1

1. The Cayley-graph of G with respect to S is connected if and only if S generates G.

2. The Cayley-graph is undirected (i.e., (u, v) is an edge whenever (v, u) is an edge) if and only if S is closed under inverses (i.e., $s \in S \Leftrightarrow s^{-1} \in S$). In particular, this is the case if S consists of involutions, i.e. elements of order 2.

9.8 The Projective Linear Group   9.8

The goal of this section is to describe a stabilizer chain for $\mathrm{PGL}_k(q)$, the projective linear group of $\mathrm{PG}_{k-1}(q)$. We will find a base for this group, and we will list the coset representatives $\sigma_{i,j}$ explicitly. This leads us to determine a strong generating set. In the same vein, we will also treat the projective semilinear group $\mathrm{P\Gamma L}_k(q)$ in the following section.

For $-1 \leq s < d$, define the set
$$\mathrm{PG}_{d\backslash s}(q) = \{\langle u \rangle \in \mathrm{PG}_d(q) \mid \mathrm{lc}(u) > s\}.$$

We also put
$$\theta_{d\backslash s}(q) = |\mathrm{PG}_{d\backslash s}(q)| = \theta_d(q) - \theta_s(q) = \frac{q^{d+1} - q^{s+1}}{q - 1},$$
with $\theta_{-1}(q) = 0$. As usual, we rank and unrank the elements of this set.

9.8.1   **Lemma** *Let d, s and q be given, where q is a prime power and $-1 \leq s < d$. Define a map $\mathrm{rk}_{d\backslash s;q}^{-1}$ from $\theta_{d\backslash s}(q)$ to $\mathrm{PG}_{d\backslash s}(q)$ by*

$$\mathrm{rk}_{d\backslash s;q}^{-1}(n) = \begin{cases} \langle e^{(s+1+n)} \rangle & \text{if } n \leq d-s-1 \\ \langle \sum\limits_{i=0}^{d} e^{(i)} \rangle & \text{if } n = d-s \\ \langle \mathrm{rk}_{d,s+1;q}^{-1}(n-d+s) \rangle & \text{otherwise,} \end{cases}$$

where $\mathrm{rk}_{d,s+1;q}^{-1}$ is the function of 9.3.7. The map $\mathrm{rk}_{d\backslash s;q}^{-1}$ is a bijection, we call it the unrank function for $\mathrm{PG}_{d\backslash s}(q)$. Its inverse is the rank function for $\mathrm{PG}_{d\backslash s}(q)$, denoted as $\mathrm{rk}_{d\backslash s;q}$. For a point $\langle u \rangle \in \mathrm{PG}_{d\backslash s}(q)$, with $u = (u_0, u_1, \ldots, u_d) \in \mathbb{F}_q^{d+1} \backslash \{0\}$ one has $\mathrm{rk}_{d\backslash s;q}(\langle u \rangle) =$

9.8.2

$$\begin{cases} k & \text{if } \langle u \rangle = \langle e^{(s+1+k)} \rangle \\ d-s & \text{if } \langle u \rangle = \langle 1, \ldots, 1 \rangle \\ d+1-k+\frac{q^k-q^{s+1}}{q-1}+\mathrm{rk}_{k,q}\left(\frac{u_0}{u_k}, \ldots, \frac{u_{k-1}}{u_k}\right) & \text{if } k = \mathrm{lc}(u) < d \\ 1+\frac{q^d-q^{s+1}}{q-1}+\mathrm{shift}_{\theta_{d-1}(q)}^{-1}\left(\mathrm{rk}_{d,q}\left(\frac{u_0}{u_d}, \ldots, \frac{u_{d-1}}{u_d}\right)\right) & \text{if } \mathrm{lc}(u) = d. \end{cases}$$

For $s = -1$, we get the ordinary unrank function back, i.e.

$$\mathrm{rk}_{d\backslash -1;q} = \mathrm{rk}_{d;q} \quad \text{and} \quad \mathrm{rk}_{d\backslash -1;q}^{-1} = \mathrm{rk}_{d;q}^{-1}. \qquad \square$$

9.8.3   **Example** We have $\theta_{2\backslash -1}(3) = 13$, $\theta_{2\backslash 0}(3) = 13-1 = 12$, $\theta_{2\backslash 1}(3) = 13-4 = 9$. Table 9.4 shows the functions $\mathrm{rk}_{2\backslash s;3}(x)$ for $-1 \leq s \leq 1$. $\diamondsuit$

9.8.4   **Example** We have $\theta_{3\backslash -1}(2) = 15$, $\theta_{3\backslash 0}(2) = 14$, $\theta_{3\backslash 1}(2) = 12$ and $\theta_{3\backslash 2}(2) = 8$. Table 9.5 shows the functions $\mathrm{rk}_{3\backslash s;2}(x)$ for $-1 \leq s \leq 2$. $\diamondsuit$

Let us introduce some notation for special kinds of matrices. We denote by $F_{n,i}$ the $(n \times (n+1))$ matrix which is obtained from the identity matrix I_{n+1} by removing the i-th row. In other words, we put

$$F_{n,i} = \left(\begin{array}{c|c|c} I_i & 0_i^\top & 0 \\ \hline 0 & 0_{n-i}^\top & I_{n-i} \end{array} \right) = \left(\begin{array}{ccc|c|ccc} 1 & & & 0 & & & \\ & \ddots & & \vdots & & 0 & \\ & & 1 & 0 & & & \\ \hline & & & 0 & 1 & & \\ & 0 & & \vdots & & \ddots & \\ & & & 0 & & & 1 \end{array} \right),$$

a matrix whose i-th column is zero. In addition, let $E_{u,v}$ be the $k \times k$ matrix whose only nonzero entry is in the (u, v)-position, with value one. Formally

$$E_{u,v} = (\delta_{i,u}\delta_{v,j})_{i \in k, j \in k}.$$

Table 9.4 The functions $\mathrm{rk}_{2\backslash s;3}(\langle x\rangle)$ for $\langle x\rangle \in \mathrm{PG}_2(3)$

	$\mathrm{rk}_{2\backslash s;3}(\langle x\rangle)$		
$\langle x\rangle \in \mathrm{PG}_2(3)$	$s=-1$	$s=0$	$s=1$
$\langle 1,0,0\rangle$	0		
$\langle 0,1,0\rangle$	1	0	
$\langle 0,0,1\rangle$	2	1	0
$\langle 1,1,1\rangle$	3	2	1
$\langle 1,1,0\rangle$	4	3	
$\langle 2,1,0\rangle$	5	4	
$\langle 1,0,1\rangle$	6	5	2
$\langle 2,0,1\rangle$	7	6	3
$\langle 0,1,1\rangle$	8	7	4
$\langle 2,1,1\rangle$	9	8	5
$\langle 0,2,1\rangle$	10	9	6
$\langle 1,2,1\rangle$	11	10	7
$\langle 2,2,1\rangle$	12	11	8

Table 9.5 The functions $\mathrm{rk}_{3\backslash s;2}(\langle x\rangle)$ for $\langle x\rangle \in \mathrm{PG}_3(2)$

	$\mathrm{rk}_{3\backslash s;2}(\langle x\rangle)$			
$\langle x\rangle \in \mathrm{PG}_3(2)$	$s=-1$	$s=0$	$s=1$	$s=2$
$\langle 1,0,0,0\rangle$	0			
$\langle 0,1,0,0\rangle$	1	0		
$\langle 0,0,1,0\rangle$	2	1	0	
$\langle 0,0,0,1\rangle$	3	2	1	0
$\langle 1,1,1,1\rangle$	4	3	2	1
$\langle 1,1,0,0\rangle$	5	4		
$\langle 1,0,1,0\rangle$	6	5	3	
$\langle 0,1,1,0\rangle$	7	6	4	
$\langle 1,1,1,0\rangle$	8	7	5	
$\langle 1,0,0,1\rangle$	9	8	6	2
$\langle 0,1,0,1\rangle$	10	9	7	3
$\langle 1,1,0,1\rangle$	11	10	8	4
$\langle 0,0,1,1\rangle$	12	11	9	5
$\langle 1,0,1,1\rangle$	13	12	10	6
$\langle 0,1,1,1\rangle$	14	13	11	7

Lastly, we introduce the 2×2-matrix

$$P = \begin{pmatrix} 0 & 1 \\ 1 & 0 \end{pmatrix}.$$

The next result describes a base and strong generating set for the projective linear group $\mathrm{PGL}_k(q)$ in the standard action on $\mathrm{PG}_{k-1}(q)$. For sake of simplicity, we do not distinguish in our notation between the matrices and the induced permutations on the projective space. Also we let group elements be denoted either by matrices or by the corresponding permutations.

9.8.5   **Theorem (base and strong generating set for $\mathrm{PGL}_k(q)$)** *Let $q = p^h$ with p prime and h a positive integer. Let $\mathrm{PG}_{k-1}(q)$ be the one-dimensional subspaces of the vector space $V = \mathbb{F}_q^k$ with basis $e^{(0)}, \ldots, e^{(k-1)}$. Assume that $\kappa_0, \kappa_1, \ldots, \kappa_{q-1}$ are the elements of the field $\mathbb{F}_q$, ordered in such a way that $\kappa_0 = 0$ and $\kappa_1 = 1$.*

1. *For $i \in k + 1$, let*

$$b_i := \begin{cases} \langle e^{(i)} \rangle & \text{if } i < k, \\ \langle \sum_{i \in k} e^{(i)} \rangle & \text{if } i = k. \end{cases}$$

The sequence $B = (b_0, \ldots, b_k)$ is a base for $\mathrm{PGL}_k(q)$ acting on $\mathrm{PG}_{k-1}(q)$. The corresponding stabilizer chain has basic orbits of lengths

$$\ell_i = \begin{cases} \theta_{k-1 \setminus i-1}(q) & \text{for } i \in k, \\ (q-1)^{k-1} & \text{for } i = k. \end{cases}$$

2. *Coset representatives can be chosen as follows.*
   (a) For $i \in k$, and for $j \in \ell_i$, let

$$\sigma_{i,j} = \left(\begin{array}{c|c} I_i & 0 \\ \hline v & \\ \hline 0 & F_{k-i-1,s-i} \end{array} \right),$$

   where $\langle v \rangle = \mathrm{rk}^{-1}_{k-1 \setminus (i-1);q}(j)$ and $s = \mathrm{lc}(v) \geq i$.
   (b) For $j \in \ell_k$, define

$$\sigma_{k,j} = \mathrm{diag}(1, \kappa_{a_0+1}, \ldots, \kappa_{a_{k-2}+1}),$$

   where $j = (a_{k-2}, \ldots, a_0)_{q-1}$ is the base $(q-1)$ representation of j. If $q = 2$, the base point b_k is redundant.
3. *A strong generating set for $\mathrm{PGL}_k(q)$ is the set*

9.8.6
$$S = \Big\{ P_0, \ldots, P_{k-2}, \mathcal{E}_{r,j}, D_1, \ldots, D_{k-1} \ \Big| \ r \in h, \ j \in k-1 \Big\},$$

where

$$\mathcal{P}_i = \begin{pmatrix} I_i & 0 & 0 \\ 0 & P & 0 \\ 0 & 0 & I_{k-2-i} \end{pmatrix}, \qquad\qquad 9.8.7$$

$$\mathcal{E}_{r,j} = I_k + \beta_r E_{k-1,j}, \qquad\qquad 9.8.8$$

$$\mathcal{D}_i = I_k + (\alpha - 1)E_{i,i}. \qquad\qquad 9.8.9$$

Here, $(\beta_0, \ldots, \beta_{h-1})$ is an $\mathbb{F}_p$-basis for $\mathbb{F}_q$ (as vector space over $\mathbb{F}_p$) and α is a primitive element for $\mathbb{F}_q$, i.e. a generator of the multiplicative group $\mathbb{F}_q^*$. If $q = 2$, the elements $\mathcal{D}_i$ of 9.8.9 are all equal to I_k and may be omitted from the set S.

Proof: The pointwise stabilizer in $\mathrm{GL}_k(q)$ of the unit vectors $e^{(0)}, \ldots, e^{(k-1)}$ consists of the diagonal matrices with nonzero determinant. These are just the diagonal matrices whose diagonal entries are all nonzero. The stabilizer of the unit vectors and the vector $e^{(0)} + \ldots + e^{(k-1)}$ are the matrices of the center $\mathcal{Z}_k$, defined in 3.7.5, i.e. the matrices of the form λI_k where $\lambda \in \mathbb{F}_q^*$. Hence in the factor group $\mathrm{PGL}_k(q) = \mathrm{GL}_k(q)/\mathcal{Z}_k$, only the identity element stabilizes

$$b_0 = \langle e^{(0)} \rangle, \ldots, b_{k-1} = \langle e^{(k-1)} \rangle, \text{ and } b_k = \langle e^{(0)} + \ldots + e^{(k-1)} \rangle.$$

This shows that B is a base. The statement about the lengths of the basic orbits will follow once we have verified that the given coset representatives are a transversal for $G^{(i+1)}$ in $G^{(i)}$. For $i = 0$, we consider matrices of the form

$$\sigma_{0,j} = \begin{pmatrix} v \\ \hline F_{k-1,s} \end{pmatrix}, \quad j \in \ell_0,$$

where $\langle v \rangle = \mathrm{rk}_{k-1;q}^{-1}(j)$ and where $s = \mathrm{lc}(v)$. Developing the determinant of $\sigma_{0,j}$ along the nonzero entries of the matrix $F_{k-1,s}$ leaves a nonzero one by one matrix as last term. Thus $\sigma_{0,j}$ is an element of $\mathrm{PGL}_k(q)$. The fact that we can put any element $\langle v \rangle$ of $\mathrm{PG}_{k-1}(q)$ into the first row of the coset representative means that $\mathrm{PGL}_k(q)$ is transitive on the set of points of $\mathrm{PG}_{k-1}(q)$. Thus

$$\ell_0 = \theta_{k-1}(q) = \theta_{k-1 \setminus -1}(q) = \frac{q^k - 1}{q - 1}.$$

Next, consider the case where $0 < i < k$. Elements in $G^{(i)}$ stabilize pointwise the base points $b_0, \ldots, b_{i-1}$, which means that they fix the subspaces

$$\langle e^{(0)} \rangle, \ldots, \langle e^{(i-1)} \rangle$$

spanned by the first i unit vectors. Since the diagonal matrices are in $G^{(i+1)}$, we may choose these unit vectors themselves for the first i rows of $\sigma_{i,j}$, so that

$$\sigma_{i,j} = \left(\begin{array}{c|c} I_i & 0 \\ \hline * & * \end{array} \right).$$

The i-th row of $\sigma_{i,j}$ is the image $\langle v \rangle$ of $b_i = \langle e^{(i)} \rangle$ under $\sigma_{i,j}$. In order to make $\sigma_{i,j}$ invertible, v must not lie in the span of $e^{(0)}, \ldots, e^{(i-1)}$. Thus $\mathrm{lc}(v) \geq i$, i.e. $\langle v \rangle \in \mathrm{PG}_{k-1 \setminus i-1}(q)$. For $j \in \theta_{k-1 \setminus i-1}(q) = \ell_i$ we may take

$$\langle v \rangle = \mathrm{rk}_{k-1 \setminus i-1;q}^{-1}(j),$$

so that

$$\sigma_{i,j} = \left(\begin{array}{c|c} I_i & 0 \\ \hline & v \\ \hline 0 & F_{k-i-1,s-i} \end{array} \right).$$

By computing the determinant one verifies that this matrix $\sigma_{i,j}$ is invertible, provided that $s = \mathrm{lc}(v)$. This shows that the given set of matrices $\sigma_{i,j}$ form coset representatives for $G^{(i+1)}$ in $G^{(i)}$. Also, the lengths of the basic orbits are $\ell_i = \theta_{k-1 \setminus i-1}(q)$.

For $i = k$ we need coset representatives for $G^{(k+1)}$ in $G^{(k)}$. Recall that $G^{(k)}$ is the group of diagonal matrices (modulo scalars, i.e. modulo $\mathcal{Z}_k$) whereas $G^{(k+1)}$ is the identity modulo $\mathcal{Z}_k$. Thus coset representatives for $G^{(k+1)}$ in $G^{(k)}$ are diagonal matrices with nonzero elements on the diagonal. Modulo $\mathcal{Z}_k$, we may choose representatives of the form

$$\mathrm{diag}(1, \lambda_1, \ldots, \lambda_{k-1}),$$

where $\lambda_1, \ldots, \lambda_{k-1}$ are nonzero field elements which can be chosen independently. This shows that $\ell_k = (q-1)^{k-1}$. We consider the map which takes an integer $j \in (q-1)^{k-1}$ to the matrix

$$\mathrm{diag}(1, \kappa_{a_0+1}, \ldots, \kappa_{a_{k-2}+1}) \in \mathrm{PGL}_k(q),$$

where

$$j = (a_{k-2}, \ldots, a_0)_{q-1}$$

is the base $q-1$ representation of j. Since $\kappa_{a_i+1} \neq 0$ (recall that we require that $\kappa_0 = 0$ and $\kappa_u \neq 0$ for $u > 0$), this map is a bijection onto the mentioned set of coset representatives for $G^{(k+1)}$ in $G^{(k)}$. This finishes the proof of the first two parts of the theorem.

Let us now verify that the set given in 9.8.6 is a strong generating set for $\mathrm{PGL}_k(q)$. This is proved inductively, going from the small groups to the larger ones in the stabilizer chain, i.e. from the large indices to the smaller ones. Recall that we have set

$$S^{(i)} = S \cap G^{(i)}$$

for $i \in k+1$. Showing that the generating sets $S^{(i)}$ for $G^{(i)}$ are strong can be done by induction. We put

$$H^{(i)} = \langle S^{(i)} \rangle \le G^{(i)}, \quad i \in k+1,$$

and then show that $H^{(i)} = G^{(i)}$. In each step we need to show that

$$|H^{(i)}(b_i)| = \ell_i = |G^{(i)}(b_i)|,$$

since then by 3.4.1 and by induction hypothesis,

$$|H^{(i)}| = |H^{(i+1)}| \cdot \ell_i = |G^{(i+1)}| \cdot \ell_i = |G^{(i+1)}|$$

and therefore $H^{(i)} = G^{(i)}$.

The statement is clear for $i = k+1$, since $S^{(k+1)} = \emptyset$ and hence $H^{(k+1)} = G^{(k+1)} = 1$. For $i = k$,

$$S^{(k)} = S \cap G^{(k)} = \{\mathcal{D}_j \mid 1 \le j < k\}.$$

Modulo $\mathcal{Z}_k$, every diagonal matrix can be written as a product of (powers of) suitable $\mathcal{D}_j$. This shows that $G^{(k)} = H^{(k)} = \langle S^{(k)} \rangle$.

The set $S^{(k-1)} = S \cap G^{(k-1)}$ is

$$S^{(k-1)} = S^{(k)} \cup \{\mathcal{E}_{r,j} \mid r \in h, \ j \in k-1\},$$

with $\mathcal{E}_{r,j}$ as in 9.8.8. Written out, we have

$$\mathcal{E}_{r,j} = \left(\begin{array}{c|c} I_{k-1} & 0 \\ \hline v' & 1 \end{array} \right),$$

with

$$v' = \beta_r e^{(j)} \in \mathbb{F}_q^{k-1} \quad \text{for } r \in h, \ j \in k-1.$$

Now consider the basic orbit $G^{(k-1)}(b_{k-1})$. This is just the set

$$\mathrm{PG}_{k-1\backslash k-2}(q) = \{\langle v \rangle \in \mathrm{PG}_{k-1}(q) \mid \mathrm{lc}(v) = k-1\}.$$

Thus,

$$v = (v_0, \ldots, v_{k-2}, 1) = (v', 1)$$

with $v' = (v_0, \ldots, v_{k-2}) \in \mathbb{F}_q^{k-1}$ arbitrary. Notice that if $w = (w', 1)$ is another vector with $w' = (w_0, \ldots, w_{k-2}) \in \mathbb{F}_q^{k-1}$, then the corresponding coset representatives multiply as follows

$$\left(\begin{array}{c|c} I_{k-1} & 0 \\ \hline v' & 1 \end{array} \right) \cdot \left(\begin{array}{c|c} I_{k-1} & 0 \\ \hline w' & 1 \end{array} \right) = \left(\begin{array}{c|c} I_{k-1} & 0 \\ \hline v' + w' & 1 \end{array} \right).$$

This shows that in the factor group $G^{(k)}$ modulo $G^{(k+1)}$, multiplication of coset representatives

$$\sigma_{k,j} = \left(\begin{array}{c|c} I_{k-1} & 0 \\ \hline v & \end{array} \right) = \left(\begin{array}{c|c} I_{k-1} & 0 \\ \hline v' & 1 \end{array} \right)$$

results in addition of the first $k-1$ components of the vectors in the last rows. In particular, the coset representatives form a group by themselves (i.e. a "complement" of $G^{(k)}$ in $G^{(k-1)}$). It is clear that the first $k-1$ components form an additive group $\mathbb{F}_q^{k-1}$. Furthermore, since $\mathbb{F}_q \simeq \mathbb{F}_p^h$ (as additive groups), we have the isomorphism from the group of coset representatives onto $\mathbb{F}_q^{k-1} \simeq \mathbb{F}_p^{h(k-1)}$. Therefore, a basis for the group of coset representatives is given by the matrices $\mathcal{E}_{r,j}$, where $r \in h$ and $j \in k-1$. But these are exactly the elements of $S^{(k-1)} \setminus S^{(k)}$. This shows that the elements of $S^{(k-1)}$ generate the full basic orbit $G^{(k-1)}(b_{k-1})$, and hence by the remark that $\langle S^{(k-1)} \rangle = H^{(k-1)} = G^{(k-1)}$.

For $i \in k-1$, the only strong generator in $S^{(i)} \setminus S^{(i+1)}$ is the matrix $\mathcal{P}_i$ of 9.8.7. This matrix "swaps" the coefficients of the basis vectors $e^{(i)}$ and $e^{(i+1)}$. We claim that a Schreier-tree for the basic orbit $G^{(i)}(b_i)$ can be obtained from $S^{(i)} = S^{(i+1)} \cup \{P_i\}$. The points of $G^{(i)}(b_i)$ which are not in $G^{(i+1)}(b_{i+1})$ are the points of the set $\mathrm{PG}_{k-1 \setminus i-1}(q)$ which are not contained in $\mathrm{PG}_{k-1 \setminus i}(q)$. They are the elements of the form

$$\langle (v_0, \ldots, v_{i-1}, 1, 0, \ldots, 0) \rangle = \langle v_0 e^{(0)} + \ldots + v_{i-1} e^{(i-1)} + e^{(i)} \rangle$$

for arbitrary $v_0, \ldots, v_{i-1} \in \mathbb{F}_q$. Since

$$b_i P_i = \langle e^{(i)} \rangle P_i = \langle e^{(i+1)} \rangle = b_{i+1},$$

the points of $G^{(i+1)}(b_{i+1})$ can be reached from b_i using $\mathcal{P}_i$ and generators from $S^{(i+1)}$. The equation

$$\langle v_0 e^{(0)} + \ldots + v_{i-1} e^{(i-1)} + e^{(i+1)} \rangle P_i = \langle v_0 e^{(0)} + \ldots + v_{i-1} e^{(i-1)} + e^{(i)} \rangle$$

shows that all other points of $G^{(i)}(b_i) \setminus G^{(i+1)}(b_{i+1})$ can be reached as well. Hence $\langle S^{(i)} \rangle = H^{(i)} = G^{(i)}$. This finishes the proof of the theorem.   □

9.8.10   **Corollary** *The order of* $\mathrm{PGL}_k(q)$ *is*

$$(q-1)^{k-1} \prod_{i \in k} \theta_{k-1 \setminus i-1}(q) = \frac{1}{q-1} \prod_{i \in k} (q^k - q^i).$$ □

Example A stabilizer chain for $PGL_3(3)$ is obtained from the ordered base $(\langle e^{(0)}\rangle, \langle e^{(1)}\rangle, \langle e^{(2)}\rangle, \langle e^{(0)} + e^{(1)} + e^{(2)}\rangle)$. Coset representatives are $\sigma_{0,0} = I_3$,   **9.8.11**

$$\sigma_{0,1} = \begin{pmatrix} 010 \\ 100 \\ 001 \end{pmatrix}, \sigma_{0,2} = \begin{pmatrix} 001 \\ 100 \\ 010 \end{pmatrix}, \sigma_{0,3} = \begin{pmatrix} 111 \\ 100 \\ 010 \end{pmatrix}, \sigma_{0,4} = \begin{pmatrix} 110 \\ 100 \\ 001 \end{pmatrix}, \sigma_{0,5} = \begin{pmatrix} 210 \\ 100 \\ 001 \end{pmatrix},$$

$$\sigma_{0,6} = \begin{pmatrix} 101 \\ 100 \\ 010 \end{pmatrix}, \sigma_{0,7} = \begin{pmatrix} 201 \\ 100 \\ 010 \end{pmatrix}, \sigma_{0,8} = \begin{pmatrix} 011 \\ 100 \\ 010 \end{pmatrix}, \dots, \sigma_{0,12} = \begin{pmatrix} 221 \\ 100 \\ 010 \end{pmatrix}, \sigma_{1,0} =$$

$$I_3, \sigma_{1,1} = \begin{pmatrix} 100 \\ 001 \\ 010 \end{pmatrix}, \sigma_{1,2} = \begin{pmatrix} 100 \\ 111 \\ 010 \end{pmatrix}, \sigma_{1,3} = \begin{pmatrix} 100 \\ 110 \\ 001 \end{pmatrix}, \dots, \sigma_{1,11} = \begin{pmatrix} 100 \\ 221 \\ 010 \end{pmatrix},$$

$$\sigma_{2,0} = I_3, \sigma_{2,1} = \begin{pmatrix} 100 \\ 010 \\ 111 \end{pmatrix}, \sigma_{2,2} = \begin{pmatrix} 100 \\ 010 \\ 101 \end{pmatrix}, \dots, \sigma_{2,8} = \begin{pmatrix} 100 \\ 010 \\ 221 \end{pmatrix}, \sigma_{3,0} = I_3,$$

$\sigma_{3,1} = \mathrm{diag}(1,2,1), \sigma_{3,2} = \mathrm{diag}(1,1,2), \sigma_{3,3} = \mathrm{diag}(1,2,2)$.

Strong generators are $\mathcal{P}_1 = \sigma_{0,1}, \mathcal{P}_2 = \sigma_{1,1}, \mathcal{E}_{0,0} = \sigma_{2,2} = \begin{pmatrix} 100 \\ 010 \\ 101 \end{pmatrix}, \mathcal{E}_{0,1} =$

$\sigma_{2,4} = \begin{pmatrix} 100 \\ 010 \\ 011 \end{pmatrix}, \mathcal{D}_2 = \sigma_{3,1} = \mathrm{diag}(1,2,1), \mathcal{D}_3 = \sigma_{3,2} = \mathrm{diag}(1,1,2)$.   ◇

Example As pointed out in 9.3.12, the elements $s_0, \dots, s_5$ listed in 9.2.7 are   **9.8.12**
generators for $G = PGL_4(2)$. In fact, they are strong generators for G with
respect to the base (b_0, b_1, b_2, b_3), where $b_i = \mathrm{rk}_{3;2}^{-1}(i)$. In the following, to keep
the notation simple we will identify projective points with their ranks. Thus,
we would say that the base is $(0, 1, 2, 3)$. Let $G^{(i)} = G_{b_0, \dots, b_{i-1}} = G_{0, \dots, i-1}$ be the
stabilizer of the first i base points. Then

$$S^{(i)} = S \cap G^{(i)} = \begin{cases} \{s_0, s_1, s_2, s_3, s_4, s_5\} & \text{if } i = 0, \\ \{s_0, s_1, s_2, s_3, s_4\} & \text{if } i = 1, \\ \{s_0, s_1, s_2, s_3\} & \text{if } i = 2, \\ \{s_0, s_1, s_2\} & \text{if } i = 3. \end{cases}$$

The basic orbits $\mathcal{O}^{(i)}$ and the corresponding Schreier-trees are shown in
Fig. 9.13. From the Schreier-trees, coset representatives can be determined easily. For instance, an element of $G^{(2)}$ mapping $b_2 = 2$ to 10 (which is the 8-th
element in the orbit $\mathcal{O}^{(3)}$) is

$$\sigma_{2,7} = s_3 s_1 s_2$$

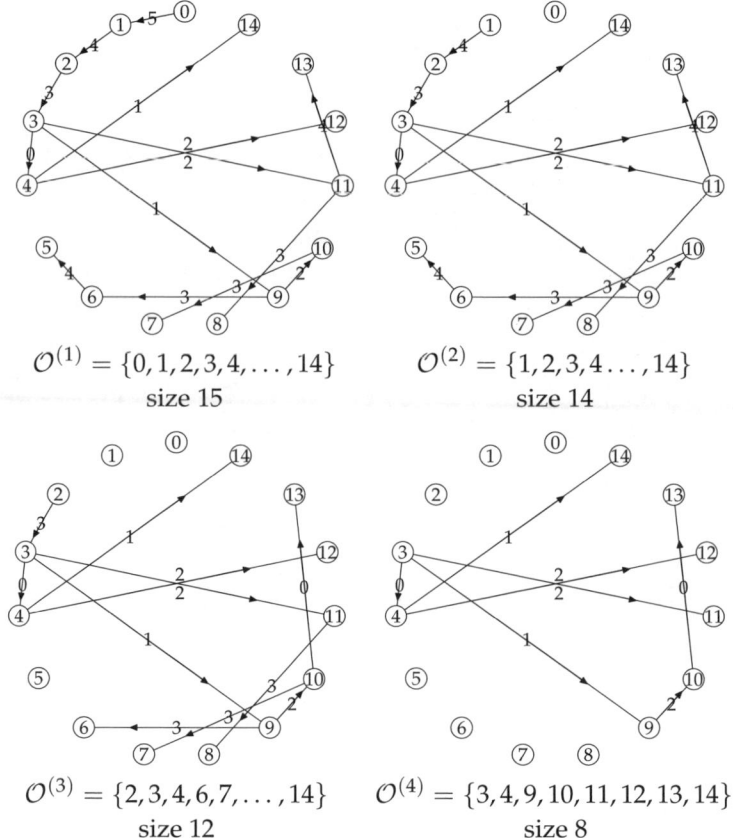

$$\mathcal{O}^{(1)} = \{0,1,2,3,4,\dots,14\}$$
size 15

$$\mathcal{O}^{(2)} = \{1,2,3,4\dots,14\}$$
size 14

$$\mathcal{O}^{(3)} = \{2,3,4,6,7,\dots,14\}$$
size 12

$$\mathcal{O}^{(4)} = \{3,4,9,10,11,12,13,14\}$$
size 8

Fig. 9.13 The basic orbits $\mathcal{O}^{(i)}$ for $\mathrm{PGL}_4(2)$

$$
\begin{aligned}
&= (2,3)(6,9)(7,10)(8,11)\\
&\cdot(3,9)(4,14)(10,11)(12,13)\\
&\cdot(3,11)(4,12)(9,10)(13,14)\\
&= (2,10,7,3)(4,13)(6,11,8,9)(12,14)
\end{aligned}
$$

Also, the group order is the product of the lengths of the basic orbits, which is $15 \cdot 14 \cdot 12 \cdot 8 = 20160$. It is now easy to access group elements numerically. For instance the group element 1777 (the birth year of Gauss) can be determined as follows. We write $1777 = ((14+4)12+6)8+1$, i.e. the multibase representation is $1777 = (1,4,6,1)_{8,12,14,15}$. Therefore we need coset representatives mapping $b_0,\dots,b_3$ to the second, 5-th, 7-th and second orbit element, respectively. That is, we need coset representatives $\sigma_{i,j}$ such that

$$\sigma_{0,1}(0) = 1, \ \sigma_{1,4}(1) = 5, \ \sigma_{2,6}(2) = 9, \ \sigma_{3,1}(3) = 4.$$

From the Schreier-trees, we obtain that

$$\sigma_{0,1} = s_5,$$
$$\sigma_{1,4} = s_4 s_3 s_1 s_3 s_4,$$
$$\sigma_{2,6} = s_3 s_1,$$
$$\sigma_{3,1} = s_0,$$

so that the group element 1777 is

$$\sigma_{3,1}\sigma_{2,6}\sigma_{1,4}\sigma_{0,1} = s_0 s_3 s_1 s_4 s_3 s_1 s_3 s_4 s_5$$
$$= (0,1,5)(2,10,12,6,3,4)(7,9,13,8,11,14).$$

It is also possible to compute the coset representatives $\sigma_{i,j}$ directly using 9.8.5 and the labeling of points as indicated in Table 9.5. This gives

$$\sigma_{3,1}\sigma_{2,6}\sigma_{1,4}\sigma_{0,1} = \begin{pmatrix} 1\ 0\ 0\ 0 \\ 0\ 1\ 0\ 0 \\ 0\ 0\ 1\ 0 \\ 1\ 1\ 1\ 1 \end{pmatrix} \begin{pmatrix} 1\ 0\ 0\ 0 \\ 0\ 1\ 0\ 0 \\ 1\ 0\ 0\ 1 \\ 0\ 0\ 1\ 0 \end{pmatrix} \begin{pmatrix} 1\ 0\ 0\ 0 \\ 1\ 1\ 0\ 0 \\ 0\ 0\ 1\ 0 \\ 0\ 0\ 0\ 1 \end{pmatrix} \begin{pmatrix} 0\ 1\ 0\ 0 \\ 1\ 0\ 0\ 0 \\ 0\ 0\ 1\ 0 \\ 0\ 0\ 0\ 1 \end{pmatrix}$$

$$= \begin{pmatrix} 0\ 1\ 0\ 0 \\ 1\ 1\ 0\ 0 \\ 0\ 1\ 0\ 1 \\ 1\ 1\ 1\ 1 \end{pmatrix}$$

This matrix sends the standard basis $0, 1, 2, 3$ to $1, 5, 10, 4$, respectively. Since group elements are the same whenever they have the same effect on all base points, this must be the same as the permutation

$$(0,1,5)(2,10,12,6,3,4)(7,9,13,8,11,14)$$

from above. Lastly, Fig. 9.14 depicts the coset representatives according to the 4 subgroups in the stabilizer chain of $PGL_4(2)$. The numbers shown are the actual elements in the basic orbits $\mathcal{O}^{(i)}$, each corresponding to one coset representative $\sigma_{i,j}$.   ◇

Exercises

Exercise Verify the statement of 9.8.1 that $rk_{d\setminus -1;q} = rk_{d;q}$ and that $rk_{d\setminus -1;q}^{-1} = rk_{d;q}^{-1}$.

E.9.8.1

Exercise It was noted after 9.2.5 that the Schreier-trees are not unique, for instance they depend on the choice of the generating set. On the other hand,

E.9.8.2

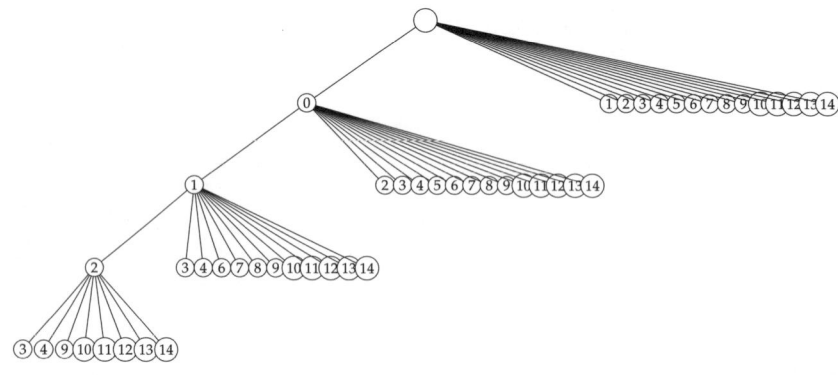

Fig. 9.14 The coset representatives for $\mathrm{PGL}_4(2)$

shortly before 9.7.9 it was noted that a stabilizer chain can be used to access group elements numerically. Convince yourself that the labeling of group elements using a stabilizer chain does not depend on the chosen generating set provided the elements of each of the fundamental orbits are ordered lexicographically. Therefore, a different choice of Schreier-trees in Example 9.8.12 would still yield the same group element with number 1777 as long as the elements of the basic orbits are listed in order.

E.9.8.3   **Exercise** Compute the position 999 999 of Rubik's $2 \times 2 \times 2$ cube, following the ideas developed in Exercise 9.8.2.

9.9   ## 9.9 The Projective Semilinear Group

The next result describes a base and strong generating set for $\mathrm{P\Gamma L}_k(q)$. The proof of this result follows easily from 3.7.11 and is omitted.

9.9.1   **Theorem (base and strong generating set for $\mathrm{P\Gamma L}_k(q)$)** *Let $q = p^h$ with p prime and h a positive integer. Let $\mathrm{PG}_{k-1}(q)$ be the one-dimensional subspaces of the vector space $V = \mathbb{F}_q^k$ with basis $e^{(0)}, \ldots, e^{(k-1)}$. If q is prime then $\mathrm{P\Gamma L}_k(q) \simeq \mathrm{PGL}_k(q)$ and 9.8.5 applies. Otherwise, if $q = p^h$ with $h > 1$, choose a primitive element α for $\mathbb{F}_q$. For $i \in k + 1$, let b_i be as in 9.8.5. Put $b_{k+1} = \langle \alpha e^{(0)} + e^{(1)} \rangle$.*

1. *The sequence $B = (b_0, \ldots, b_k, b_{k+1})$ is an ordered base for $\mathrm{P\Gamma L}_k(q)$ acting on $\mathrm{PG}_{k-1}(q)$. The corresponding stabilizer chain has basic orbits of lengths*

$$
\ell_i = \begin{cases} \theta_{k-1\backslash i-1}(q) & \text{for} \quad i \in k, \\ (q-1)^{k-1} & \text{for} \quad i = k, \\ h & \text{for} \quad i = k+1. \end{cases}
$$

2. *Coset representatives $\gamma_{i,j}$, $i \in k+2$, $j \in \ell_i$ can be chosen in the following way.*
   (a) *For $i \in k+1$, and for $j \in \ell_i$, let*

$$
\gamma_{i,j} = \left(\sigma_{i,j}, 0 \right)
$$

   with $\sigma_{i,j}$ as described in 9.8.5.

   (b) *For $j \in \ell_{k+1}$, let*

$$
\gamma_{k+1,j} = \left(I_k, j \right).
$$

3. *A strong generating set for $\mathrm{P\Gamma L}_k(q)$ is given by the elements*

$$
(\sigma, 0),
$$

where σ runs through all elements of a strong generating set of $\mathrm{PGL}_k(q)$ as described in 9.8.5, together with the element

$$
(I_k, 1). \qquad \square
$$

Corollary *The order of $\mathrm{P\Gamma L}_k(q)$ is* 9.9.2

$$
h(q-1)^{k-1} \prod_{i \in k} \theta_{k-1\backslash i-1}(q) = \frac{h}{q-1} \prod_{i \in k} (q^k - q^i). \qquad \square
$$

Example The field $\mathbb{F}_8$ is generated over $\mathbb{F}_2$ by a root α of the polynomial $X^3 +$ 9.9.3
$X^2 + 1$ (so that $\alpha^3 = 1 + \alpha^2$). In the additive labeling, the field elements are

$$
\begin{aligned}
\kappa_0 &= 0, \\
\kappa_1 &= 1, \\
\kappa_2 &= \alpha, \\
\kappa_3 &= \alpha + 1, \\
\kappa_4 &= \alpha^2, \\
\kappa_5 &= \alpha^2 + 1, \\
\kappa_6 &= \alpha^2 + \alpha, \\
\kappa_7 &= \alpha^2 + \alpha + 1.
\end{aligned}
$$

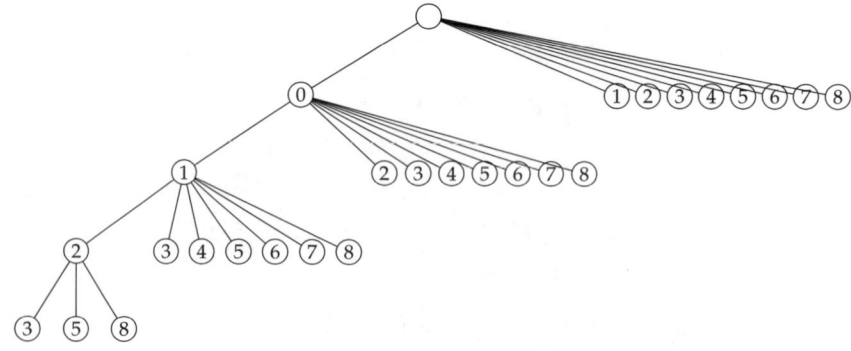

Fig. 9.15 The coset representatives for $P\Gamma L_2(8)$

Using the rank function of 9.3.5, the 9 points of the projective line $PG_1(8)$ are numbered as

$$
\begin{aligned}
0 &= \langle (1,0) \rangle, \\
1 &= \langle (0,1) \rangle, \\
2 &= \langle (1,1) \rangle, \\
3 &= \langle (\kappa_2,1) \rangle, \\
4 &= \langle (\kappa_3,1) \rangle, \\
5 &= \langle (\kappa_4,1) \rangle, \\
6 &= \langle (\kappa_5,1) \rangle, \\
7 &= \langle (\kappa_6,1) \rangle, \\
8 &= \langle (\kappa_7,1) \rangle.
\end{aligned}
$$

A base for $P\Gamma L_2(8)$ is $(0,1,2,3)$. Strong generators are

$$
s_0 = \left(\begin{pmatrix} 1 & 0 \\ 0 & 1 \end{pmatrix}, 1 \right) = (3,5,8)(4,6,7),
$$

$$
s_1 = \left(\begin{pmatrix} 1 & 0 \\ 0 & \kappa_2 \end{pmatrix}, 0 \right) = (2,7,4,8,6,5,3),
$$

$$
s_2 = \left(\begin{pmatrix} 1 & 0 \\ 1 & 1 \end{pmatrix}, 0 \right) = (1,2)(3,4)(5,6)(7,8),
$$

$$
s_3 = \left(\begin{pmatrix} 0 & 1 \\ 1 & 0 \end{pmatrix}, 0 \right) = (0,1)(3,7)(4,5)(6,8).
$$

The basic orbits have length $9, 8, 7$, and 3, respectively. We conclude that the group $P\Gamma L_2(8)$ has 1512 elements. Figure 9.15 depicts the coset representatives according to the 4 subgroups in the stabilizer chain of $P\Gamma L_2(8)$. The numbers shown are the elements in the basic orbits $\mathcal{O}^{(i)}$, each corresponding to one coset representative $\sigma_{i,j}$. $\diamondsuit$

Exercises

Exercise Compute a base and stabilizer chain for $P\Gamma L(3,4)$ using 9.9.1. List the coset representatives.

E.9.9.1

9.10 Numerical Data

9.10

Let us now present numerical data concerning the classification of isometry classes of linear indecomposable codes for small finite fields. In all cases, we classify the semilinear isometry classes over $\mathbb{F}_q$. If q is a prime, then of course the semilinear isometry classes are the same as the linear isometry classes. We present results for the fields $\mathbb{F}_q$ with $q \in \{2,3,4,5,8,9,16,25,27\}$ in Tables 9.6-9.24. For a given length n and dimension k, the corresponding entry in the table lists the number of semilinear isometry classes of (n,k)-codes with a given minimum distance. For instance, an entry of the form

$$d^x e^y f^z$$

indicates that there are x classes of codes with minimum distance d, y classes with minimum distance e and z classes with minimum distance f. The minimum distances are ordered decreasingly, and the first value, d, is the optimal minimum distance in that parameter case. Exponents whose value is 1 are omitted. Underlined entries indicate non-trivial MDS-codes.

Table 9.6 Optimal indecomposable $\mathbb{F}_2$ codes

$n \backslash k$	1	2	3	4	5	6	7
4	4						
5	5	3					
6	6	$4\,3$	3				
7	7	$4^2 3$	$4\,3^3$	3			
8	8	$5\,4^2$	$4^3 3^6$	$4\,3^4$			
9	9	$6\,5^2 4^2$	4^8	$4^4 3^{18}$	3^5		
10	10	$6^2 5^2 4^2$	$5^2 4^{18}$	4^{19}	$4^4 3^{36}$	3^4	
11	11	$7\,6^3 5^2$	$6\,5^8 4^{29}$	$5\,4^{66}$	4^{30}	$4^2 3^{58}$	3^3
12	12	$8\,7^2 6^3 5^2$	$6^6 5^{19}$	$6\,5^{12} 4^{201}$	4^{214}	4^{41}	$4^2 3^{84}$
13	13	$8^2 7^3 6^3$	$7\,6^{16} 5^{37}$	$6^6 5^{72}$	$5^{15} 4^{1159}$	4^{580}	4^{45}
14	14	$9\,8^3 7^3$	$8\,7^5 6^{37}$	$7\,6^{39} 5^{292}$	$6^6 5^{261}$	$5^{11} 4^{6704}$	4^{1488}
15	15	$10\,9^2 8^4 7^3$	$8^3 7^{17}$	$8\,7^5 6^{195}$	$7\,6^{91} 5^{2547}$	$6\,5^{995}$	$5^6 4^{41037}$
16	16	$10^2 9^3 8^4$	$8^{12} 7^{41}$	$8^4 7^{37}$	$8\,7^5 6^{1145}$	$6^{180} 5^{29826}$	$6^3 5^{4010}$
17	17	$11\,10^3 9^4 8^4$	$9^2 8^{32}$	$8^{18} 7^{241}$	$8^4 7^{84}$	7^3	6^{377}
18	18	$12\,11^2 10^4 9^4$	$10\,9^{11} 8^{71}$	8^{108}	$8^{34} 7^{1777}$	$8^2 7^{108}$	7^2
19		$12^2 11^3 10^5$	$10^6 9^{33}$	$9^7 8^{550}$	8^{411}	$8^{28} 7^{19021}$	$8\,7^{81}$
20			$11\,10^{21}$	$10^3 9^{81}$	$9^3 8^{6480}$	8^{1833}	8^{26}
21				10^{27}	$10^2 9^{178}$		
22					10^{37}	9^{248}	
23						10^{29}	9^{29}
24							10^6

Table 9.7 Optimal indecomposable $\mathbb{F}_2$ codes (cont.)

$n\backslash k$	8	9	10	11	12	13	14	15	16	17	18	19
12	3^2											
13	$4\,3^{109}$	3										
14	4^{48}	$4\,3^{126}$	3									
15	4^{3473}	4^{43}	$4\,3^{142}$	3								
16	$5\,4^{268258}$	4^{7456}	4^{47}	$4\,3^{143}$								
17	$6\,5^{13757}$	5	4^{14390}	4^{39}	3^{129}							
18	6^{918}	$6\,5^{29371}$		4^{25024}	4^{33}	3^{113}						
19	7	6^{1700}	5^{31237}		4^{39302}	4^{25}	3^{91}					
20	$8\,7^{33}$	7	6^{1682}	5^{14135}			4^{24}	3^{67}				
21	8^{12}	$8\,7^{20}$	7	6^{739}	5^{2373}			4^{16}	3^{50}			
22		8^9	$8\,7^{15}$	7	6^{128}	5^{128}			4^{15}	3^{34}		
23			8^8	$8\,7^{15}$	7	6^8	5			4^9	3^{21}	
24			8^9	$8\,7^{11}$			6				4^8	3^{14}
25					8^7							4^5

Table 9.8 Optimal indecomposable $\mathbb{F}_2$ codes (cont.)

$n\backslash k$	20	21	22	23	24	25	26
25	3^9						
26	4^4	3^5					
27		4^2	3^3				
28			4^2	3^2			
29				4	3		
30					4	3	
31						4	3
32							4

Table 9.9 Optimal indecomposable $\mathbb{F}_3$ codes

$n\backslash k$	1	2	3	4	5	6	7
3	3						
4	4	$\underline{3}$					
5	5	3^2					
6	6	$4^2 3^2$	3^4				
7	7	$5\,4^3 3^2$	$4^2 3^{12}$	3^4			
8	8	$6\,5^3 4^3$	$5\,4^{13} 3^{25}$	$4^3 3^{36}$	3^3		
9	9	$6^3 5^4$	$6\,5^8 4^{40}$	$5\,4^{41} 3^{185}$	$4\,3^{87}$	3^3	
10	10	$7^2 6^5$	$6^6 5^{39}$	$6\,5^{19} 4^{403}$	$5\,4^{134} 3^{1205}$	$4\,3^{195}$	3^2
11	11	$8\,7^4$	$7\,6^{35}$	$6^7 5^{452}$	$6\,5^{34} 4^{4840}$	$5\,4^{354} 3^{8297}$	3^{399}
12	12	$9\,8^4 7^6$	$8\,7^{15}$	6^{353}	$6^8 5^{8550}$	$6\,5^{36} 4^{73941}$	$4^{844} 3^{61060}$
13		$9^3 8^6$	$9\,8\,7^{107}$	7^{72}	6^{5037}	$6^9 5^{191851}$	5^6
14			$9^3 8^{72}$	$8^{14} 7^{5221}$	7^{236}	6^{47674}	6
15				9^3		7^{22}	
16					9		

Table 9.10 Optimal indecomposable $\mathbb{F}_3$ codes (cont.)

$n\backslash k$	8	9	10	11	12	13	14	15	16
11	3								
12	3^{805}	3							
13	$4^{1532} 3^{457485}$	3^{1503}	3						
14	5	4^{2020}	3^{2658}						
15			4^{1778}	3^{4304}					
16				4^{1019}	3^{6472}				
17					4^{337}	3^{8846}			
18						4^{90}	3^{11127}		
19							4^{20}	3^{12723}	
20								4^9	3^{13358}

Table 9.11 Optimal indecomposable $\mathbb{F}_3$ codes (cont.)

$n\backslash k$	17	18	19	20	21	22	23	24	25	26	27	28
21	3^{12723}											
22		3^{11127}										
23			3^{8846}									
24				3^{6472}								
25					3^{4304}							
26						3^{2659}						
27							3^{1505}					
28								3^{807}				
29									3^{402}			
30										3^{201}		
31											3^{94}	
32												3^{47}

Table 9.12 Optimal indecomposable $\mathbb{F}_3$ codes (cont.)

$n\backslash k$	29	30	31	32	33	34	35	36
33	3^{23}							
34		3^{12}						
35			3^6					
36				3^4				
37					3^2			
38						3		
39							3	
40								3

Table 9.13 Optimal indecomposable $\mathbb{F}_4$ codes

$n\backslash k$	1	2	3	4	5	6	7	8	9	10	11
3	3										
4	4	$\underline{3}$									
5	5	$\underline{4}3^2$	$\underline{3}$								
6	6	4^33^2	$\underline{4}3^6$								
7	7	5^24^4	4^73^{19}	3^{10}							
8	8	6^25^4	5^34^{38}	$4^{16}3^{96}$	3^{13}						
9	9	76^5	6^35^{39}	5^44^{326}	$4^{19}3^{466}$	3^{17}					
10	10	87^4	6^{45}	6^25^{642}	5^44^{4189}	$4^{23}3^{2380}$	3^{18}				
11		8^4	7^{25}	6^{841}	65^{19418}	54^{66475}	$4^{15}3^{13080}$	3^{18}			
12			8^{16}	7^{275}	6^{19181}			4^{13}	3^{17}		
13			8^{30}	7^{452}				4^4	3^{13}		
14				8^6	7^{14}				4^2	3^{10}	
15					8^3	7^4					4
16						8^2	7^3				
17							8^2	7^2			
18								8			

Table 9.14 Optimal indecomposable $\mathbb{F}_4$ codes (cont.)

$n\backslash k$	12	13	14	15	16	17	18
15	3^8						
16	4	3^5					
17		4	3^3				
18				3^2			
19					3		
20						3	
21							3

Table 9.15 Optimal indecomposable $\mathbb{F}_5$ codes

$n\backslash k$	1	2	3	4	5	6	7	8
3	3							
4	4	$\underline{3}$						
5	5	$\underline{4}3^2$	$\underline{3}$					
6	6	$\underline{5}4^43^2$	$\underline{4}3^9$	$\underline{3}$				
7	7	5^3	$4^{17}3^{29}$	3^{21}				
8	8	6^35^7	5^{16}	$4^{92}3^{344}$	3^{42}			
9		7^26^8	$6^{16}5^{248}$	5^{134}	$4^{387}3^{4570}$	3^{92}		
10			7^76^{486}	6^{93}	5^{558}	$4^{1568}3^{62846}$	3^{174}	
11					6^{60}	5^{503}	$4^{4089}3^{814405}$	3^{296}
12						6^{31}	5^{36}	4^{7062}

Table 9.16 Optimal indecomposable $\mathbb{F}_5$ codes (cont.)

$n\backslash k$	9	10	11	12	13	14	15	16	17	18	19	20	21	22
12	3^{476}													
13	4^{7258}	3^{669}												
14		4^{4678}	3^{832}											
15			4^{1810}	3^{948}										
16				4^{572}	3^{948}									
17					4^{183}	3^{832}								
18						4^{88}	3^{669}							
19							4^{36}	3^{476}						
20								4^{21}	3^{296}					
21									4^7	3^{174}				
22										4^4	3^{92}			
23											4	3^{42}		
24												4	3^{22}	
25													4	3^{12}

Table 9.17 Optimal indecomposable $\mathbb{F}_5$ codes (cont.)

$n \backslash k$	22	23	24	25	26	27	28
26	4	3^5					
27			3^3				
28				3^2			
29					3		
30						3	
31							3

Table 9.18 Optimal indecomposable $\mathbb{F}_8$ codes

$n \backslash k$	1	2	3	4	5	6	7	8	9
3	3								
4	4	$\underline{3}$							
5	5	$\underline{4}3^2$	$\underline{3}$						
6	6	$\underline{5}4^4$	$\underline{4}^33^{10}$	$\underline{3}$					
7	7	$\underline{6}$	$\underline{5}^24^{49}$	$\underline{4}^23^{54}$	$\underline{3}$				
8		$\underline{7}$	6^2	$\underline{5}4^{1700}$	$\underline{4}^23^{323}$	$\underline{3}$			
9			$\underline{7}^2$	$\underline{6}$	$\underline{5}4^{68877}$	$\underline{4}^23^{2097}$	$\underline{3}$		
10							$\underline{4}3^{12868}$		
11								3^{72638}	
12									3^{373366}

Table 9.19 Optimal indecomposable $\mathbb{F}_9$ codes

$n \backslash k$	1	2	3	4	5	6	7	8
3	3							
4	4	3^2						
5	5	$\underline{4}^2$	3^2					
6	6	$\underline{5}^2$	4^6	3^2				
7	7	$\underline{6}$	5^3	4^3	3			
8		$\underline{7}$	6^2	5^5	4^2	3		
9			$\underline{7}$	6^2	5^2	4	3	
10				$\underline{7}$	6^2	5	4	3

Table 9.20 Optimal indecomposable $\mathbb{F}_{16}$ codes

$n \backslash k$	1	2	3	4	5	6	7	8	9	10	11	12	13	14	15
3	3														
4	4	3^2													
5	5	$\underline{4}^3$	3^3												
6		5^4	$\underline{4}^{22}$	3^4											
7			5^{125}	$\underline{4}^{125}$	3^5										
8			5^{2981}	$\underline{4}^{685}$	3^6										
9				5^{6888}	$\underline{4}^{1534}$	3^6									
10					5^{356}	$\underline{4}^{1262}$	3^5								
11						5^{10}	4^{300}	3^4							
12							$\underline{5}^4$	4^{159}	3^3						
13								$\underline{5}^2$	4^{70}	3^2					
14									$\underline{5}$	4^{30}	3				
15										$\underline{5}$	4^9	3			
16											$\underline{5}$	4^5	3		
17												$\underline{5}$	4^3	3	
18															$\underline{4}^2$

Table 9.21 Optimal indecomposable $\mathbb{F}_{25}$ codes

$n\backslash k$	1	2	3	4	5	6	7	8	9	10	11	12	13	14
3	3													
4	4	$\underline{3^4}$												
5	5	$\underline{4^7}$	$\underline{3^7}$											
6		$\underline{5^{19}}$	$\underline{4^{205}}$	$\underline{3^{19}}$										
7				$\underline{4^{7163}}$	$\underline{3^{34}}$									
8						$\underline{3^{79}}$								
9							$\underline{3^{132}}$							
10								$\underline{3^{223}}$						
11									$\underline{3^{293}}$					
12										$\underline{3^{379}}$				
13											$\underline{3^{391}}$			
14												$\underline{3^{379}}$		
15													$\underline{3^{293}}$	
16														$\underline{3^{223}}$

Table 9.22 Optimal indecomposable $\mathbb{F}_{25}$ codes (cont.)

$n\backslash k$	15	16	17	18	19	20	21	22	23	24
17	$\underline{3^{132}}$									
18		$\underline{3^{79}}$								
19			$\underline{3^{34}}$							
20				$\underline{3^{19}}$						
21					$\underline{3^7}$					
22						$\underline{3^4}$				
23							$\underline{3}$			
24								$\underline{3}$		
25									$\underline{3}$	
26										$\underline{3}$

Table 9.23 Optimal indecomposable $\mathbb{F}_{27}$ codes

$n\backslash k$	1	2	3	4	5	6	7	8	9	10	11	12	13	14
3	3													
4	4	$\underline{3^3}$												
5	5	$\underline{4^4}$	$\underline{3^4}$											
6		$\underline{5^{14}}$	$\underline{4^{174}}$	$\underline{3^{14}}$										
7			$\underline{5^{8261}}$	$\underline{4^{8261}}$	$\underline{3^{29}}$									
8						$\underline{3^{72}}$								
9							$\underline{3^{134}}$							
10								$\underline{3^{257}}$						
11									$\underline{3^{390}}$					
12										$\underline{3^{565}}$				
13											$\underline{3^{670}}$			
14												$\underline{3^{738}}$		
15													$\underline{3^{670}}$	
16														$\underline{3^{565}}$

Table 9.24 Optimal indecomposable $\mathbb{F}_{27}$ codes (cont.)

$n\backslash k$	15	16	17	18	19	20	21	22	23	24	25	26
17	$3^{\underline{390}}$											
18		$3^{\underline{257}}$										
19			$3^{\underline{134}}$									
20				$3^{\underline{72}}$								
21					$3^{\underline{29}}$							
22						$3^{\underline{14}}$						
23							$3^{\underline{4}}$					
24								$3^{\underline{3}}$				
25									$\underline{3}$			
26										$\underline{3}$		
27											$\underline{3}$	
28												$\underline{3}$

Chapter A
Appendix: The Attached Compact Disc

A

A Appendix: The Attached Compact Disc

The enclosed compact disc contains both data and software. The data are tables of numbers of isometry classes or cycle index polynomials. There are also tables of all isometry classes of optimal linear codes for small parameters. The software enables the user to do research on linear codes. It permits to construct linear codes with prescribed minimum distance, to determine the minimum distance as well as weight enumerators of linear codes, and to determine cycle index polynomials for the natural actions of linear and projective linear groups. The software is powerful enough to improve international tables. For typographical reasons we denote the finite field of q elements by $GF(q)$. In the following sections we briefly describe how to use it.

Since both these tables and the software are in rapid progress, from time to time the interested user should consult the following address, where extensions or improvements and updates can be found:

http://linearcodes.uni-bayreuth.de

A.1 System Requirements

To use the programs you need a Windows system (any 32-bit version including 95/98/ME/NT/2000/XP) together with a modern browser (e.g. Firefox, Mozilla, Internet Explorer). There is also a version for a Linux environment (kernel version 2.2 and above) but you need some technical experience for installation. In both cases you need about 80MB of free space on the harddisc.

A.2 The Installation

On a *Windows* system proceed as follows:

— Copy the complete CD into a new directory on your PC.

— Start the application code.exe from the new directory.

In a window of your standard web-browser you will find a table showing links to various applications of programs (see below) and to precalculated and dynamic tables. If you use a browser different from the Microsoft Internet Explorer it may be necessary to deactivate the use of a proxy.

On a *Linux* system proceed as follows:

— Make sure that you have a running web-server on your PC.

— Copy the content of the directory htdocs from the CD into a sub-directory (say linearcodes), browsable by your web-browser (typically a sub-directory of /usr/local/httpd/htdocs). Make sure that all files in this directory are readable by your browser.

— Copy the contents of the directory cgi-bin from the CD into the directory executable by the web-browser (typically /usr/local/httpd/cgi-bin). The most important programs are: bsp_linux, mindist111, mindistter, mindistbin and solvediophant.
Make sure that the copied files in this directory are readable by your browser.

— Start the browser with the URL *http://...your..pc../linearcodes/index.html*

Usually, the maximum amount of time one process can use is restricted by the Linux web-server to 300 seconds. If you are using the web-server apache and you want to increase this value then the entry "Timeout" in the configuration file httpd.conf has to be changed. After that the web-server has to be restarted.

On both platforms (Linux and Windows) it may happen that you have to stop a running computation by hand. This is the case when you finish your browser, but the computation in the background was not stopped. For this you need to know the name of the executable program and knowing this you can stop it using the taskmanager (Windows) or the top program (Linux). The name of the Linux programs were given above, the name of the corresponding Windows programs are: bsp_windows, mindist111.exe, mindistter.exe, mindistbin.exe and solvediophant.exe.

A.3 The Programs

— *Minimum distance computations* For the computation of the minimum distance of binary of ternary linear codes from a generator matrix you can choose between two algorithms:

1. The straightforward algorithm 1.8.1 and

2. the algorithm of Section 7.8, based on lattice basis reduction.

— *Weight enumerator*

Using this program it is possible to compute the weight enumerator of a linear code from a generator matrix over an arbitrary field. This also gives the minimum distance in the cases we cannot use the above programs.

— *Construction of codes with given minimum distance*

This is an implementation of the algorithm described in Section 8.4. It was written using the SYMMETRICA library, which is public domain, see [190]. It chooses random subgroups of a corresponding linear group $\mathrm{GL}_k(q)$ and tries to find a code with the given parameters and the chosen group as a group of automorphisms. It terminates after 10 attempts to find such a code.

— *Random generation of linear codes*

This is an implementation of the Dixon–Wilf algorithm described in Section 6.6. It was written using the SYMMETRICA library, which is public domain, see [190]. The algorithm is only implemented for codes over prime fields $\mathbb{F}_p$. Additional information about the conjugacy classes and their probabilities is computed before the random generation starts to produce random codes. Therefore, depending on the input parameters n, k and q it takes some time till the first generator matrix is displayed. This information is also stored in files which are written to the hard disc. The next time you start the generator with the same parameter triple (n, k, q) these data are read and need not be computed again. This speeds up the generation of linear codes.

A.4 The Dynamic Tables

A.4

Using the methods described in Section 6.1 and Section 6.2, we have implemented routines to compute tables of numbers of the linear isometry classes of linear codes. This way it is possible to extend the tables of Section 6.4. The following numbers can be determined:

$\begin{bmatrix} n \\ k \end{bmatrix}(q)$:   the number of k-dimensional subspaces of $\mathbb{F}_q^n$

T_{nkq}:   the cardinality of $\mathrm{PGL}_k(q) \backslash\!\backslash \left(S_n \backslash\!\backslash \mathrm{PG}_{k-1}^*(q)^n \right)$

$\overline{T}_{nkq}$:   the cardinality of $\mathrm{PGL}_k(q) \backslash\!\backslash \left(S_n \backslash\!\backslash \mathrm{PG}_{k-1}^*(q)_{\mathrm{inj}}^n \right)$

V_{nkq}:   the number of linear isometry classes of nonredundant (n, k)-codes over $\mathbb{F}_q$

$\overline{V}_{nkq}$:   the number of linear isometry classes of projective (n, k)-codes over $\mathbb{F}_q$

U_{nkq}:   the number of linear isometry classes of (n,k)-codes over $\mathbb{F}_q$ that may contain columns of zeros

R_{nkq}:   the number of linear isometry classes of nonredundant indecomposable (n,k)-codes over $\mathbb{F}_q$

$\overline{R}_{nkq}$:   the number of linear isometry classes of projective indecomposable (n,k)-codes over $\mathbb{F}_q$

After the user has input three positive integers N, K and q, a table containing the corresponding numbers for $1 \leq n \leq N$ and $1 \leq k \leq K$ over $\mathbb{F}_q$ is computed.

According to 6.1.23, the numbers T_{nkq} and $\overline{T}_{nkq}$ are obtained by certain substitutions into the cycle index $C(\mathrm{PGL}_k(q), \mathrm{PG}_{k-1}^*(q))$. In Section 6.3 we have described a method for the computation of the cycle index for the natural action of the projective linear group $\mathrm{PGL}_k(q)$ on $\mathrm{PG}_{k-1}^*(q)$. A similar method is used in order to obtain the cycle index for the natural action of the general linear group $\mathrm{GL}_k(q)$ on $\mathbb{F}_q^k$ (cf. [60]). With the included software it is also possible to determine both these cycle index polynomials.

A.5 The Precomputed Tables: Enumerative Results

As was described in Section 6.7 we do not have routines for computing the cycle index for the natural action of the projective semilinear group $\mathrm{P\Gamma L}_k(q)$ on $\mathrm{PG}_{k-1}^*(q)$ for arbitrary values of k and q. In order to obtain some numbers of semilinear isometry classes we have determined these cycle index polynomials just for a few values of k for $q \in \{4,8\}$. These polynomials together with extensions of the tables of Section 6.7 can be found on the enclosed CD. For $q \in \{4,8\}$ tables of the following numbers are available:

t_{nkq}:   the cardinality of $\mathrm{P\Gamma L}_k(q) \backslash\backslash (S_n \backslash\backslash \mathrm{PG}_{k-1}^*(q)^n)$

$\overline{t}_{nkq}$:   the cardinality of $\mathrm{P\Gamma L}_k(q) \backslash\backslash (S_n \backslash\backslash \mathrm{PG}_{k-1}^*(q)_{\mathrm{inj}}^n)$

v_{nkq}:   the number of semilinear isometry classes of nonredundant (n,k)-codes over $\mathbb{F}_q$

$\overline{v}_{nkq}$:   the number of semilinear isometry classes of projective (n,k)-codes over $\mathbb{F}_q$

u_{nkq}:   the number of semilinear isometry classes of (n,k)-codes over $\mathbb{F}_q$ that may contain columns of zeros

r_{nkq}:   the number of semilinear isometry classes of nonredundant indecomposable (n,k)-codes over $\mathbb{F}_q$

$\bar{r}_{nkq}$:   the number of semilinear isometry classes of projective inde-
        composable (n,k)-codes over $\mathbb{F}_q$

A.6 The Precomputed Tables: Optimal Linear Codes

The CD contains tables of optimal linear codes for small parameters (n,k) and over small finite fields $\mathbb{F}_q$ (i.e. $q \leq 5$). For a given (n,k), the number of semi-linear isometry classes of indecomposable codes with minimum distance $\geq d_0$ are available. Of course, the largest $d \geq d_0$ for which codes exist is the optimal minimum distance. The value of d_0 varies, and depends on the given (n,k). However, for a given d_0 the classification of $(n,k, \geq d_0)$-codes is always complete.

The codes are listed by means of canonical generator matrices. In addition, information about the automorphism group (in $\mathrm{P\Gamma L}_k(q)$) is given. The information is stored in text-files, which are linked to the table entries. For each triple (n,k,d), a file containing representatives of all (n,k,d)-codes over the given field $\mathbb{F}_q$ is available. For example, in the table for binary codes, the entry for $(n,k,d) = (7,4,3)$ is 3^1, meaning that there is only one code (up to semilinear isometry). This code is of course the Hamming-code. Clicking on the link leads to a file which looks as follows:

```
1        the 1 isometry classes of irreducible [7,4,3]_2 codes are:
2
3        code no        1:
4        =================
5        1 1 1 1 0 0 0
6        1 1 0 0 1 0 0
7        1 0 1 0 0 1 0
8        0 1 1 0 0 0 1
9        the automorphism group has order 168
10       and is strongly generated by the following 7 elements:
11       (
12       1 0 0
13       0 1 0
14       1 1 1
15       ,
16       1 0 0
17       0 1 0
18       1 0 1
19       ,
20       1 0 0
21       1 1 0
22       1 0 1
23       ,
24       1 0 0
25       1 0 1
26       1 1 0
27       ,
```

```
28      1 0 0
29      0 1 1
30      1 1 0
31      ,
32      1 0 1
33      1 0 0
34      1 1 1
35      ,
36      0 1 1
37      1 0 0
38      0 0 1
39      )
40      acting on the columns of the generator matrix as follows (in
        order):
41      (3, 4)(6, 7),
42      (3, 6)(4, 7),
43      (2, 5)(3, 6),
44      (2, 6)(3, 5),
45      (2, 6, 7)(3, 4, 5),
46      (1, 2, 6)(3, 5, 4),
47      (1, 2, 6, 7)(4, 5)
48      orbits: { 1, 6, 7, 3, 2, 4, 5 }
49
```

At the beginning, the file indicates that there is just one isometry class of $(7,4,2,3)$-codes. After that, the code is listed as code 1. A generator matrix is given, as well as the order of the automorphism group. The generator matrix is in the form

$$\Gamma = (A \mid -I)$$

where A is a $k \times (n-k)$-matrix and $-I$ is the negative of the $k \times k$ identity matrix. It follows from Exercise 1.3.9 that

$$\Delta = (I \mid A^\top)$$

is the corresponding check matrix. The automorphism group itself is described in two different ways. At first, a list of strong generators in the form of matrices of size $(n-k) \times (n-k)$ is given. These are in fact the matrices which act on the dual code. More precisely, they act on the column vectors of Δ. We know that each automorphism induces a permutation of the columns. The permutations which are induced by the strong generators on the columns of the check matrix are listed next. For compatibility reasons with other Computer Algebra systems, we index the columns by the elements of the set $\{1,\ldots,n\}$ rather than $n = \{0,\ldots,n-1\}$. After the generators, the orbits of the automorphism group on the columns are listed. Let us show by example how the permutation of columns is obtained from the matrix. Consider the first strong generator

$$A = \begin{pmatrix} 1 & 0 & 1 \\ 0 & 1 & 1 \\ 1 & 1 & 1 \end{pmatrix}.$$

Since A acts on row-vectors from the right, and the columns of Δ are transposed row-vectors, we must consider

$$A^\top \cdot \Delta = \begin{pmatrix} 1 & 0 & 1 \\ 0 & 1 & 1 \\ 0 & 0 & 1 \end{pmatrix} \begin{pmatrix} 1 & 0 & 0 & 1 & 1 & 1 & 0 \\ 0 & 1 & 0 & 1 & 1 & 0 & 1 \\ 0 & 0 & 1 & 1 & 0 & 1 & 1 \end{pmatrix}$$

$$= \begin{pmatrix} 1 & 0 & 1 & 0 & 1 & 0 & 1 \\ 0 & 1 & 1 & 0 & 1 & 1 & 0 \\ 0 & 0 & 1 & 1 & 0 & 1 & 1 \end{pmatrix},$$

from which we see that we need to permute columns 3 and 4 and columns 6 and 7 of Δ to get to this matrix. In other words, the action of A induced the permutation $(3,4)(6,7)$. This is of course the first permutation in the list presented in the file.

Let us now describe how codes over fields $\mathbb{F}_q$ with q not prime are treated. We always write $q = p^h$ with p prime and $h > 1$ an integer. The elements of $\mathbb{F}_q$ are polynomials

$$a_{h-1}\alpha^{h-1} + \ldots + a_1\alpha + a_0$$

in α of degree at most $h - 1$ with coefficients $a_j \in \mathbb{F}_p$ for $j = 0, \ldots, h-1$. Here, α is root of an irreducible polynomial of degree h over $\mathbb{F}_p$. The field element as above is then represented by the integer whose base-p representation is $(a_{h-1}, a_{h-1}, \ldots, a_0)_p$. The polynomials used are in fact primitive, they are listed in Table 3.3.

The elements of $\mathrm{P\Gamma L}_k(q)$ are stored as pairs (A, i) where A is in $\mathrm{PGL}_k(q)$ and where i describes the power of the field automorphism which acts. Let us see an example. If $q = 4$, we have $\mathbb{F}_4 = \{0, 1, \alpha, 1+\alpha\}$ with α a root of $x^2 + x = 1$, i.e. with $\alpha^2 = 1 + \alpha$. We label the elements as follows: $0 \hat{=} 0, 1 \hat{=} 1, 2 \hat{=} \alpha, 3 \hat{=} \alpha + 1$. For $(n, k, d) = (6, 3, 4)$ we get the following file.

```
1        the 1 isometry classes of irreducible [6,3,4]_4 codes are:
2
3        code no       1:
4        =================
5        1 1 1 1 0 0
6        3 2 1 0 1 0
7        2 3 1 0 0 1
8        the automorphism group has order 720
9        and is strongly generated by the following 9 elements:
10       (
11       3 0 0
12       0 3 0
13       0 0 3
14       , 1
15       ;
16       2 0 0
17       0 3 0
18       0 0 1
```

```
19        , 1
20        ;
21        2 0 0
22        0 1 0
23        2 1 3
24        , 0
25        ;
26        3 0 0
27        0 2 0
28        3 2 1
29        , 1
30        ;
31        3 0 0
32        0 0 2
33        2 1 3
34        , 0
35        ;
36        0 0 3
37        3 0 0
38        0 3 0
39        , 0
40        ;
41        3 3 3
42        0 0 3
43        0 3 0
44        , 0
45        ;
46        1 3 2
47        0 1 0
48        3 0 0
49        , 0
50        ;
51        1 2 3
52        0 0 1
53        2 2 2
54        , 0
55        )
56        acting on the columns of the generator matrix as follows (in
          order):
57        (5, 6),
58        (4, 5),
59        (3, 4, 5),
60        (3, 4, 5, 6),
61        (2, 5, 3),
62        (1, 2, 3),
63        (1, 4)(2, 3),
64        (1, 3, 5),
65        (1, 4, 3, 2, 6)
66        orbits: { 1, 3, 4, 5, 6, 2 }
67
```

With the convention on labeling the elements of $\mathbb{F}_4$, the generator matrix can be written as

$$\Gamma = \begin{pmatrix} 1 & 1 & 1 & 1 & 0 & 0 \\ 1+\alpha & \alpha & 1 & 0 & 1 & 0 \\ \alpha & 1+\alpha & 1 & 0 & 0 & 1 \end{pmatrix}.$$

A.7 The Programs for Chapter 9

The programs for Chapter 9 can be found in the file ch9.tar.gz on the CD. On a unix environment, one can unpack the file by using tar -zxvf ch0.tar.gz. This will produce a directory CHAPTER9 with two subdirectories, called LIB and CODES. To compile the programs, a make / C++ environment is required. By default, the "GNU" C++ compiler is used to compile the programs, but this may be configured differently (by changing the makefiles). The source code is contained in LIB, which compiles 3 libraries. The directory CODES contains several programs related to the construction of isometry classes of codes as described in Chapter 9. All these programs are compiled by issuing make from within the CHAPTER9 directory. The main program in CODES is called codes. This program expects 4 parameters, which are n, k, q and d. Here, n is the length of the codes to be constructed, k is the dimension, q is the size of the field, and d is a lower bound on the minimum distance. In addition, the options -v, -vv, -vvv can be used to have the program produce more and more output (verbose mode). These options apply to all programs described here. If called

$$\text{codes } \langle n \rangle \ \langle k \rangle \ \langle q \rangle \ \langle d \rangle$$

the program will compute a transversal of the isometry classes of $(n, k, \geq d, q)$ codes, and store them in a file called

$$\text{codes\_}\langle n \rangle\text{\_}\langle k \rangle\text{\_}\langle q \rangle\text{\_}\langle d \rangle.$$

In fact, the program will compute all codes $(n - i, k - i, d, q)$ where $0 \leq i < k$. For instance, a call

$$\text{codes } 8 \ 4 \ 2 \ 3$$

would compute the $(5, 1, \geq 3, 2)$, $(6, 2, \geq 3, 2)$, $(7, 3, \geq 3, 2)$ and $(8, 4, \geq 3, 2)$ codes and therefore result in the creation of files

$$\text{codes\_5\_1\_2\_3, codes\_6\_2\_2\_3, codes\_7\_3\_2\_3, codes\_8\_4\_2\_3.}$$

The content of these files is as follows. The first row lists the values n, k, q and d. After that the codes are listed in a compact format, one row describing one code. The file is closed be a line starting in -1 and listing number of codes constructed and the number of codes considered internally during the construction process. Also, the computing time is given. For instance, the above-mentioned computation of $(8, 4, \geq 3, 2)$ codes produces the file

$$\text{codes\_8\_4\_2\_3.}$$

```
1    # 8 4 2 3
2    8 0 1 2 3 4 5 6 7 48 aaaaaaaeaaaaaaaeaaaaaaagaaaaaaaeaaaaaaab
     aaaaaaaccbpedbpfebicgdic
3    8 0 1 2 3 4 5 6 8 8 aaaaaaaeaaaaaaadaaaaaaabaaaaaaaeaaaaaaaba
     aaaaaaccbpeebicdbpf
4    8 0 1 2 3 4 5 6 9 24 aaaaaaaeaaaaaaafaaaaaaabaaaaaaagaaaaaaae
     aaaaaaabcbfjcbeiebicdbifjbde
5    8 0 1 2 3 4 5 6 10 8 aaaaaaaeaaaaaaadaaaaaaaeaaaaaaacaaaaaaab
     aaaaaaabfbpdckdpkfie
6    8 0 1 2 3 5 6 7 8 168 aaaaaaaeaaaaaaahaaaaaaahaaaaaaagaaaaaaa
     eaaaaaaabcbifcbihdbifebichbifgdicdhig
7    8 0 1 2 3 8 11 13 14 1344 aaaaaaaeaaaaaaaiaaaaaaaiaaaaaaahaaa
     aaaagaaaaaaaecbnecblecbhlcbeicbhoebnhilcelocn
8    -1 6 23 in 0:00
9
```

The format of the rows is as follows. The first integer gives the length, which is always n. The next n integers give the ranks of the columns of check matrix Δ of the code. After that, the order of the automorphism group is given. The following text string contains information on the automorphism group in coded form. Essentially, a strong generating set of the automorphism group is stored. To get access to the codes is a more human-readable version, the program codep is used. A call to

codep $\langle n \rangle$ $\langle k \rangle$ $\langle q \rangle$ $\langle d \rangle$

will process the file codes_$\langle n \rangle$_$\langle k \rangle$_$\langle q \rangle$_$\langle d \rangle$ and produce files

codes_$\langle n \rangle$_$\langle k \rangle$_$\langle q \rangle$_$\langle d' \rangle$.txt

where $d' \geq d$. There will be one file for each minimum distance occurring among the codes in the file which is processed. For instance, a call to

codep 8 4 2 3

will process the file codes_8_4_2_3 and produce files

codes_8_4_2_3.txt and codes_8_4_2_4.txt.

The first of these file will list the four $(8,4,3,2)$ codes, whereas the latter contains the unique $(8,4,4,2)$ code (i.e. the Hamming-code). This file lists the codes by generator matrices and with information in the automorphism group in the format described above. In this case, the file codes_8_4_2_3.txt starts as follows (cf. Tab. 9.2):

```
1    the 4 isometry classes of irreducible [8,4,3]_2 codes are:
2
3    code no        1:
4    =================
5    1 1 1 1 1 0 0 0
6    1 1 0 0 0 1 0 0
7    1 0 1 0 0 0 1 0
8    0 1 1 0 0 0 0 1
9    the automorphism group has order 48
```

```
10        and is strongly generated by the following 4 elements:
11        (
12        1 0 0 0
13        0 1 0 0
14        0 0 1 0
15        1 1 1 1
16        ;
17        1 0 0 0
18        1 1 0 0
19        1 0 1 0
20        1 1 1 1
21        ;
22        1 0 0 0
23        0 0 1 0
24        0 1 0 0
25        0 0 0 1
26        ;
27        1 1 0 0
28        0 1 1 0
29        0 1 0 0
30        0 0 0 1
31        )
32        acting on the columns of the generator matrix as follows (in
          order):
33        (4, 5),
34        (2, 6)(3, 7)(4, 5),
35        (2, 3)(6, 7),
36        (1, 7, 6)(2, 3, 8)
37        orbits: { 1, 6, 2, 7, 3, 8 }, { 4, 5 }
38
39        code no        2:
40        =================
41        1 1 1 1 1 0 0 0
42        1 1 0 0 0 1 0 0
43        1 0 1 0 0 0 1 0
44        1 1 1 0 0 0 0 1
45        the automorphism group has order 8
46
```

The third program codet is responsible for producing tables like 9.6–9.24 describing the number of isometries classes. A call to

   codet ⟨q⟩

produces such a table for the given field q. The optional parameter -c should be given if the program codep should be used to process the raw data files from the generator codes to the above-described text files. If the .txt files are already there (for instance from previous calls to codet), the option -c can be omitted.

The program make_BCH can be used to create generator polynomials for BCH-codes as described in Chapter 4. A call to

   make_BCH ⟨n⟩ ⟨q⟩ ⟨t⟩

produces a generator polynomial for the q-ary BCH-code of length n with designed distance t. For instance, a call to

```
make_BCH -v 15 2 7
```

will create the following output:

```
1      finite_field::init() GF(2) = GF(2^1):
2      field of order 2 initialized
3      GF(16) = GF(2^4) has 15-th roots of unity
4      this is a primitive BCH code
5      choosing the following irreducible and primitive polynomial:
6      X^{4} + X^{3} + 1
7      orbit of conjugate elements (in powers of \beta):
8      { 1 2 4 8 }
9      orbit of conjugate elements (in powers of \beta):
10     { 3 6 12 9 }
11     orbit of conjugate elements (in powers of \beta):
12     { 5 10 }
13     taking the minimum polynomials of { 1 3 5 }
14     minimal polynomial of \beta^1 is X^{4} + X^{3} + 1 of rank 25

15     minimal polynomial of \beta^3 is X^{4} + X^{3} + X^{2} + X +
       1 of rank 31
16     minimal polynomial of \beta^5 is X^{2} + X + 1 of rank 7
17     BCH(15,2,7) = X^{10} + X^{9} + X^{8} + X^{6} + X^{5} + X^{2}
       + 1 bose_distance = 7
18     BCH code with length n=15 designed distance t=7 over GF(2)
19     generated by
20     X^{10} + X^{9} + X^{8} + X^{6} + X^{5} + X^{2} + 1
21     $m_{1}m_{3}m_{5}$ where $m_{1}=25$, $m_{3}=31$, $m_{5}=7$
22     0:00
23
```

We can see that the program initializes the field $\mathbb{F}_{16}$, which has 15-th roots of unity. The field $\mathbb{F}_{16}$ is created using the polynomial $x^4 + x^3 + 1$. As this polynomial is primitive, the root $\alpha = \beta$ of this polynomial is a primitive element, i.e. an element of order 15. Since $t = 4$, we need the consecutive set $\beta, \beta^2, \ldots, \beta^5$ as roots. For this, the necessary 2-cyclotomic cosets modulo 15 are listed. We need $\{1, 2, 4, 8\}$, $\{3, 6, 12, 9\}$ and $\{5, 10\}$. The minimal polynomials are

$$\begin{aligned}
M_\beta &= x^4 + x^3 + 1, \\
M_{\beta^3} &= x^4 + x^3 + x^2 + x + 1, \quad \text{and} \\
M_{\beta^5} &= x^2 + x + 1.
\end{aligned}$$

Therefore, the code is generated by

$$g(x) = x^{10} + x^9 + x^8 + x^6 + x^5 + x^2 + 1,$$

which is the product of these minimal polynomials. Thus, we have found a $(15, 5)$ code with minimum distance at least 7.

The program compute_mindist allows us to compute the minimum distance of linear codes defined over arbitrary finite fields, using the algorithm of

Section 1.8. Assume we want to compute the minimum distance of the q-ary (n,k) code with generator matrix

$$\Gamma = \begin{pmatrix} a_{0,0} & a_{0,1} & \cdots & a_{0,n-1} \\ \vdots & & & \vdots \\ a_{k-1,0} & a_{k-1,2} & \cdots & a_{k-1,n-1} \end{pmatrix}.$$

In case that q is not prime, assume that $\mathbb{F}_q$ is defined by means of one of the minimal polynomials listed in Table 3.3. Using the additive representation of field elements, we assume that the entries a_{ij} of Γ are integers between 0 and $q - 1$. Then we would call the program `compute_mindist` with the following list of parameters:

$$n,\ k,\ q,\ a_{0,0},\ a_{0,1},\ \ldots,\ a_{k-1,n-1}.$$

For instance, if we want to compute the minimum distance of the second of the $(8, 4, \geq 3, 2)$ codes above, we would call

```
minimum_distance -v -vv -vvv 8 4 2 1 1 1 1 1 0 0 0 1
                 [ 1 0 0 0 1 0 0 1 0 1 0 0 0 1 0 1 1 1 0 0 0 0 1
```

and the program would output

```
1        computing minimum distance of the (8,4) code over GF(2)
2        which is generated by
3          1 1 1 1 1 0 0 0
4          1 1 0 0 0 1 0 0
5          1 0 1 0 0 0 1 0
6          1 1 1 0 0 0 0 1
7
8        finite field of order 2 initialized
9        multiplication table:
10         0 0
11         0 1
12       addition table:
13         0 1
14         1 0
15       the field: GF(2) = GF(2^1)
16       idx_zero = 0, idx_one = 1, idx_mone = 1
17       (8,4) code over GF(2), generated by
18         1 1 1 1 1 0 0 0
19         1 1 0 0 0 1 0 0
20         1 0 1 0 0 0 1 0
21         1 1 1 0 0 0 0 1
22
23       systematic generator matrix s[1]:
24         1 0 0 0 0 1 1 1
25         0 1 0 0 0 0 1 1
26         0 0 1 0 0 1 0 1
27         0 0 0 1 1 0 0 1
28
29       systematic generator matrix s[2]:
30         1 1 1 1 1 0 0 0
31         1 1 0 0 0 1 0 0
32         1 0 1 0 0 0 1 0
33         1 1 1 0 0 0 0 1
34       size of information subsets:
```

```
35      4 4
36      matrix 1 row 1 is 1 0 0 0 0 1 1 1   of weight 4 minimum is 4
37      matrix 1 row 2 is 0 1 0 0 0 0 1 1   of weight 3 minimum is 3
38      matrix 1 row 3 is 0 0 1 0 0 1 0 1   of weight 3 minimum is 3
39      matrix 1 row 4 is 0 0 0 1 1 0 0 1   of weight 3 minimum is 3
40      matrix 2 row 1 is 1 1 1 1 1 0 0 0   of weight 5 minimum is 3
41      matrix 2 row 2 is 1 1 0 0 0 1 0 0   of weight 3 minimum is 3
42      matrix 2 row 3 is 1 0 1 0 0 0 1 0   of weight 3 minimum is 3
43      matrix 2 row 4 is 1 1 1 0 0 0 0 1   of weight 4 minimum is 3
44      \bar{d}_1=3
45      mindist(C_{\le 1})=3
46      \underline{d}_1= +2-(4-4) +2-(4-4)=4
47      the minimum distance is 3
48      This was determined by looking at 8 codewords
49      (rather than 16 codewords)
50      The code has minimum distance 3
51
```

Thus, the code has minimum distance 3. To give an example for a minimum distance computation in the case when the field is not a prime field, consider one of the three $(6, 3, 4)$ MDS-codes over $\mathbb{F}_8$ (cf. Table 9.18). Using the additive representation of field elements, a generator matrix is

$$\Gamma = \begin{pmatrix} 1 & 1 & 1 & 1 & 0 & 0 \\ 3 & 2 & 1 & 0 & 1 & 0 \\ 2 & 3 & 1 & 0 & 0 & 1 \end{pmatrix}.$$

To compute the minimum distance of this code (assuming for the moment that we would not know that this code is MDS), we call

```
minimum_distance -v -vv -vvv 6 3 8 1 1 1 1 0 0 3 2
                                [1 0 1 0 2 3 1 0 0 1
```

and the program will output

```
1       computing minimum distance of the (6,3) code over GF(8)
2       which is generated by
3        1 1 1 1 0 0
4        3 2 1 0 1 0
5        2 3 1 0 0 1
6
7       finite field of order 8 initialized
8       multiplication table:
9        0 0 0 0 0 0 0 0
10       0 1 2 3 4 5 6 7
11       0 2 4 6 5 7 1 3
12       0 3 6 5 1 2 7 4
13       0 4 5 1 7 3 2 6
14       0 5 7 2 3 6 4 1
15       0 6 1 7 2 4 3 5
16       0 7 3 4 6 1 5 2
17      addition table:
18       0 1 2 3 4 5 6 7
19       1 0 3 2 5 4 7 6
20       2 3 0 1 6 7 4 5
21       3 2 1 0 7 6 5 4
22       4 5 6 7 0 1 2 3
23       5 4 7 6 1 0 3 2
```

```
24      6 7 4 5 2 3 0 1
25      7 6 5 4 3 2 1 0
26      the field: GF(8) = GF(2^3)
27      idx_zero = 0, idx_one = 1, idx_mone = 1
28      (6,3) code over GF(8), generated by
29      1 1 1 1 0 0
30      3 2 1 0 1 0
31      2 3 1 0 0 1
32
33      systematic generator matrix s[1]:
34      1 0 0 1 2 3
35      0 1 0 1 3 2
36      0 0 1 1 1 1
37
38      systematic generator matrix s[2]:
39      1 1 1 1 0 0
40      3 2 1 0 1 0
41      2 3 1 0 0 1
42      size of information subsets:
43      3 3
44      matrix 1 row 1 is 1 0 0 1 2 3  of weight 4 minimum is 4
45      matrix 1 row 2 is 0 1 0 1 3 2  of weight 4 minimum is 4
46      matrix 1 row 3 is 0 0 1 1 1 1  of weight 4 minimum is 4
47      matrix 2 row 1 is 1 1 1 1 0 0  of weight 4 minimum is 4
48      matrix 2 row 2 is 3 2 1 0 1 0  of weight 4 minimum is 4
49      matrix 2 row 3 is 2 3 1 0 0 1  of weight 4 minimum is 4
50      \bar{d}_1=4
51      mindist(C_{\le 1})=4
52      \underline{d}_1= +2-(3-3) +2-(3-3)=4
53      the minimum distance is 4
54      This was determined by looking at 6 codewords
55      (rather than 512 codewords)
56      The code has minimum distance 4
57
```

Thus the program computes the minimum distance 4 by looking at 6 rather than 512 codewords.

References

[1] A.V. Aho, J.E. Hopcroft, and J.D Ullman. *The design and analysis of computer algorithms*. Addison-Wesley, Reading, 1974.

[2] M. Ajtai. Generating hard instances of lattice problems. In *28th Ann. ACM Symp. on Theory of Computing*, pages 99–108, 1996.

[3] E. Artin. *Galois Theory*. Notre Dame Mathematical Lectures, no. 2. University of Notre Dame, Notre Dame, Ind., second edition, 1966.

[4] E. Artin. *Geometric Algebra*. Wiley Classics Library. John Wiley & Sons Inc., New York, 1988. Reprint of the 1957 original, A Wiley-Interscience Publication.

[5] E. F. Assmus, Jr. and J. D. Key. *Designs and their codes*, volume 103 of *Cambridge Tracts in Mathematics*. Cambridge University Press, Cambridge, 1992.

[6] E.F. Assmus Jr. The Category of Linear Codes. *IEEE Transactions on Information Theory*, 44(2):612–629, 1998.

[7] E.F. Assmus Jr. and H.F. Mattson Jr. On tactical configurations and error-correcting codes. *J. Comb. Theory*, 2:243–257, 1967.

[8] E.F. Assmus Jr. and H.F. Mattson Jr. New 5-designs. *J. Combinatorial Theory*, 6:122–151, 1969.

[9] S. Benedetto, D. Divsalar, G. Montorsi, and F. Pollara. Serial concatenation of interleaved codes: Performance analysis, design, and iterative decoding. Technical report, JPL TDA Progress Report, vol. 42-126, 1996. *http://tmo.jpl.nasa.gov/tmo/progress_report/42-126/126D.pdf*.

[10] E.R. Berlekamp. Goppa codes. *IEEE Trans. Information Theory*, IT-19:590–592, 1973.

[11] S.D. Berman. On the theory of group codes (in Russian). *Kibernetika*, 3:31–39, 1967. English translation: On the theory of group codes. *Cybernetics*, 3(1):25-31 (1969).

[12] C. Berrou, A. Glavieux, and P. Thitimajshima. Near Shannon limit error-correcting coding and decoding: turbo-codes. In *Proceedings of ICC'93*, pages 1064–1070, 1993. *http://courses.ece.cornell.edu/ece561/turbo_icc93.pdf*.

[13] A. Betten. *http://www.math.colostate.edu/~betten/index.html*.

[14] A. Betten. *Schnittzahlen von Designs. (Intersection numbers of designs)*. PhD thesis, University of Bayreuth, 2000. Bayreuther Mathematische Schriften 58.

[15] A. Betten, A. Kerber, A. Kohnert, R. Laue, and A. Wassermann. The discovery of simple 7-designs with automorphism group $P\Gamma L(2,32)$. In G. Cohen, M. Giusti, and T. Mora (eds.), *Applied Algebra, Algebraic Algorithms and Error-Correcting Codes, 11th International Symposium, AAECC-11, Paris, France, July 17–22, 1995.*, volume 948 of *Lecture Notes in Computer Science*, pages 131–145, Berlin, 1995. Springer-Verlag.

[16] A. Betten, A. Kerber, R. Laue, and A. Wassermann. Simple 8-designs with small parameters. *Designs, Codes and Cryptography*, 15:5–27, 1998.

[17] A. Betten, A. Kohnert, R. Laue, and A. Wassermann (eds.). *Algebraic combinatorics and applications. Proceedings of the Euroconference, ALCOMA, Gößweinstein, Germany, September 12–19, 1999.* Springer-Verlag, Berlin, 2001.

[18] A. Betten, R. Laue, and A. Wassermann. Some simple 7-designs. In J.W.P. Hirschfeld et al. (eds.), *Geometry, combinatorial designs and related structures. Proceedings of the first Pythagorean conference, Island of Spetses, Greece, June 1–7, 1996*, volume 245 of *Lond. Math. Soc. Lect. Note Ser.*, pages 15–25. Cambridge: Cambridge University Press, 1997.

[19] R.E. Blahut. *Theory and practice of error control codes.* Addison-Wesley Publishing Company, Reading, Massachusetts, 1983.

[20] I.F. Blake (ed.). *Algebraic coding theory: history and development.* Dowden Hutchinson & Ross Inc., Stroudsburg, Pa., 1973. Benchmark Papers in Electrical Engineering and Computer Science.

[21] H.F. Blichfeldt. A new principle in the geometry of numbers with some applications. *Trans. Amer. Math. Soc.*, 15:227–235, 1914.

[22] H.F. Blichfeldt. The minimum value of quadratic forms and the closest packing of spheres. *Math. Ann.*, 101:605–608, 1929.

[23] J. Blömer and J.-P. Seifert. On the complexity of computing short linearly independent vectors and short bases in a lattice. In *Annual ACM Symposium on Theory of Computing, Proceedings of the 31st Symposium (STOC '99) held in Atlanta, GA, May 1–4, 1999*, pages 711–720, 1999.

[24] R.C. Bose and D.K. Ray-Chaudhuri. On a class of error correcting binary group codes. *Inf. Control*, 3:68–79, 1960.

[25] M. Braun. Construction of linear codes with large minimum distance. *IEEE Transactions on Information Theory*, 50(8):1687–1691, 2004.

[26] M. Braun. *Konstruktion diskreter Strukturen unter Verwendung linearer Operationen auf dem linearen Verband.* PhD thesis, University of Bayreuth, 2004. Bayreuther Mathematische Schriften 69.

[27] M. Braun and A. Kohnert.
http://www.mathe2.uni-bayreuth.de/michael/codes/code_bounds.html.

[28]   M. Braun, A. Kohnert, and A. Wassermann. Construction of (n, r)-arcs in PG(2, q). *Innovations in Incidence Geometry*, 1:133–141, 2005.

[29]   M. Braun, A. Kohnert, and A. Wassermann. Optimal linear codes from matrix groups. *IEEE Transactions on Information Theory*, 2005. To appear.

[30]   R. Brigola. *Fourieranalysis, Distributionen und Anwendungen*. Vieweg, Braunschweig, 1997. ISBN 3-528-06619-9.

[31]   J. Brillhart, D.H. Lehmer, J.L. Selfridge, B. Tuckerman, and S.S. Wagstaff Jr. *Factorizations of $b^n \pm 1$, $b = 2, 3, 5, 6, 7, 10, 11, 12$ up to high powers*. Contemporary Mathematics, 22. Providence, RI: American Mathematical Society (AMS), second edition, 1988.

[32]   A.E. Brouwer. Linear code bounds.
http://www.win.tue.nl/˜aeb/voorlincod.html.

[33]   A.E. Brouwer. Bounds on the size of linear codes. In *Handbook of coding theory, Vol. I*, pages 295–461. North-Holland, Amsterdam, 1998.

[34]   H.O. Burton and E.J. Weldon. Cyclic product codes. *IEEE Trans. Information Theory*, IT-11:433–439, 1965.

[35]   G. Butler. *Fundamental algorithms for permutation groups*, volume 559 of *Lecture Notes in Computer Science*. Springer-Verlag, Berlin, 1991.

[36]   G. Castagnoli, J.L. Massey, Ph.A. Schoeller, and N. von Seemann. On repeated-root cyclic codes. *IEEE Trans. Inf. Theory*, 37(2):337–342, 1991.

[37]   P. Charpin. Une généralisation de la construction de Berman des codes de Reed et Muller p-aires. *Comm. Algebra*, 16(11):2231–2246, 1988.

[38]   M. Clausen and U. Baum. *Fast Fourier transforms*. B.I. Wissenschaftsverlag, Mannheim, 1993.

[39]   H. Cohen. *A course in computational algebraic number theory*, volume 138 of *Graduate Texts in Mathematics*. Springer-Verlag, Berlin, 1993.

[40]   J.H. Conway and N.J.A. Sloane. *Sphere Packings, Lattices and Groups*. Springer-Verlag, New York, third edition, 1999.

[41]   G. Cooperman, L. Finkelstein, and N. Sarawagi. Applications of Cayley graphs. In *Applied algebra, algebraic algorithms and error-correcting codes (Tokyo, 1990)*, volume 508 of *Lecture Notes in Comput. Sci.*, pages 367–378. Springer, Berlin, 1991.

[42]   T.H. Cormen, C.E. Leiserson, R.L. Rivest, and C. Stein. *Introduction to algorithms*. MIT Press, Cambridge, MA, second edition, 2001.

[43]   M.J. Coster, A. Joux, B.A. LaMacchia, A.M. Odlyzko, C.P. Schnorr, and J. Stern. Improved low-density subset sum algorithms. *Computational Complexity*, 2:111–128, 1992.

[44] R.R. Coveyou and R.D. MacPherson. Fourier analysis of uniform random number generators. *J. Assoc. Comp. Mach.*, 14:100–119, 1967.

[45] U. Dieter. How to calculate shortest vectors in a lattice. *Math. Comp.*, 29(131):827–833, 1975.

[46] J.D. Dixon and H.S. Wilf. The random selection of unlabeled graphs. *J. Algorithms*, 4:205–213, 1983.

[47] *http://www.log-1.com/Barres/Barcodes/en.*

[48] Ecma International, Geneva. *Standard ECMA-130, Data interchange on read-only 120 mm optical data disks (CD-ROM)*, second edition, 1966. *http://www.ecma-international.org/.*

[49] Ecma International, Geneva. *Standard ECMA-267, 120 mm DVD -Read-Only Disk*, third edition, 2001. *http://www.ecma-international.org/.*

[50] B. Elspas. The Theory of Autonomous Linear Sequential Networks. *IRE Transactions on Circuit Theory*, CT-6:45–60, 1959.

[51] I.A. Faradžev. Constructive enumeration of homogeneous graphs (in Russian). *Uspehi Mat. Nauk*, 31(1(187)):246, 1976.

[52] I.A. Faradžev. Constructive enumeration of combinatorial objects (in Russian). In *Algorithmic studies in combinatorics*, pages 3–11, 185. "Nauka", Moscow, 1978.

[53] I.A. Faradžev. Generation of nonisomorphic graphs with a given distribution of the degrees of vertices (in Russian). In *Algorithmic studies in combinatorics*, pages 11–19, 185. "Nauka", Moscow, 1978.

[54] S. Ferret and L. Storme. Minihypers and linear codes meeting the Griesmer bound: Improvements to results of Hamada, Helleseth and Maekawa. *Des. Codes Cryptography*, 25(2):143–162, 2002.

[55] G.D. Forney Jr. On decoding BCH codes. *IEEE Trans. Information Theory*, IT-11:549–557, 1965.

[56] G.D. Forney Jr. *Concatenated codes*. M.I.T. Research Monograph, No. 37. The M.I.T. Press, Cambridge, Mass., 1966.

[57] H. Fredricksen and J. Maiorana. Necklaces of beads in k colors and k-ary de Bruijn sequences. *Discrete Math.*, 23:207–210, 1978.

[58] H. Fripertinger. *http://www.mathe2.uni-bayreuth.de/frib/codes/tables.html.*

[59] H. Fripertinger. Enumeration of isometry-classes of linear (n,k)-codes over $GF(q)$ in SYMMETRICA. *Bayreuth. Math. Schr.*, 49:215–223, 1995.

[60] H. Fripertinger. Cycle indices of linear, affine, and projective groups. *Linear Algebra and its Applications*, 263:133–156, 1997.

[61] H. Fripertinger and A. Kerber. Isometry Classes of Indecomposable Linear Codes. In G. Cohen, M. Giusti, and T. Mora (eds.), *Applied Algebra, Algebraic Algorithms and Error-Correcting Codes, 11th International Symposium, AAECC-11, Paris, France, July 17–22, 1995*, volume 948 of *Lecture Notes in Computer Science*, pages 194–204, Berlin, 1995. Springer-Verlag.

[62] Shuhong Gao. *Normal Bases over Finite Fields*. PhD thesis, University of Waterloo, 1993.

[63] GAP – Groups, Algorithms, and Programming, Version 4.4. The GAP Group, Aachen, Germany and St. Andrews, Scotland, 2004.

[64] C. Gasquet and P. Witomski. *Fourier Analysis and Applications*, volume 30 of *Texts in Applied Mathematics*. Springer-Verlag, New York, 1998. ISBN 0-387-98485-2.

[65] C.F. Gauss. Besprechung des Buches von L.A. Seeber: Untersuchungen über die Eigenschaften der positiven ternären quadratischen Formen. *Göttingische gelehrte Anzeigen*, pages 188–196, 1831. Werke II.

[66] C.F. Gauss. *Disquisitiones Arithmeticae*. Chelsea Pub., New York, 1965. First published 1801 in Latin.

[67] E.N. Gilbert. A comparison of signaling alphabets. *Bell System Tech. J.*, 31:504–522, 1952. Also reprinted in [185] pp. 14–19 and [20] pp. 24–42.

[68] J. Gill. Handouts to EE 387, Error-Correcting Codes, 2003. *http://www.stanford.edu/class/ee387/*.

[69] D.G. Glynn. The nonclassical 10-arc of $PG(4,9)$. *Discrete Math.*, 59(1-2):43–51, 1986.

[70] M.J.E. Golay. Notes on digital coding. *Proc. IRE.*, 37, 1949.

[71] V.D. Goppa. A new class of linear correcting codes (in Russian). *Problemy Peredači Informacii*, 6(3):24–30, 1970.

[72] V.D. Goppa. Rational representation of codes and (L, g)-codes (in Russian). *Problemy Peredači Informacii*, 7(3):41–49, 1971.

[73] D.M. Gordon. Minimal permutation sets for decoding the binary Golay codes. *IEEE Trans. Inf. Theory*, 28:541–543, 1982.

[74] P. Govaerts and J. De Beule. pg, Projective Geometries, a share package for GAP 4. *http://cage.ugent.be/˜jdebeule/pg/*.

[75] P. Govaerts and L. Storme. On a particular class of minihypers and its applications. I: the result for general q. *Des. Codes Cryptography*, 28(1):51–63, 2003.

[76] J.H. Griesmer. A bound for error-correcting codes. *IBM J. Res. Develop.*, 4:532–542, 1960.

[77] P.M. Gruber and C.G. Lekkerkerker. *Geometry of Numbers*. North-Holland, Amsterdam, 1987.

[78] N. Hamada. A characterization of some $[n, k, d; q]$-codes meeting the Griesmer bound using a minihyper in a finite projective geometry. *Discrete Math.*, 116(1-3):229–268, 1993.

[79] N. Hamada and T. Helleseth. Arcs, blocking sets, and minihypers. *Comput. Math. Appl.*, 39(11):159–168, 2000.

[80] R.W. Hamming. Error detecting and error correcting codes. *Bell System Tech. J.*, 29:147–160, 1950.

[81] R.W. Hamming. *Coding and information theory*. Prentice-Hall Inc., Englewood Cliffs, N.J., 1986. ISBN 0-13-139072-4.

[82] M.A. Harrison. On the classification of boolean functions by the general linear and affine groups. *J. Soc. Ind. Appl. Math.*, 12:285–299, 1964.

[83] M.A. Harrison. Counting Theorems and their Applications to Switching Theory. In A. Mukhopadyay (ed.), *Recent Developments in Switching Functions*, chapter 4, pages 85–120. Academic Press, 1971.

[84] W. Heise and P. Quattrocchi. *Informations- und Codierungstheorie. Mathematische Grundlagen der Daten-Kompression und -Sicherung in diskreten Kommunikationssystemen*. Springer-Verlag, Berlin, Heidelberg, New York, Paris, Tokio, first, second and third edition, 1983, 1989 and 1995.

[85] H.J. Helgert and R.D. Stinaff. Minimum-distance bounds for binary linear codes. *IEEE Trans. Inf. Theory*, 19:344–356, 1973.

[86] Ch. Hermite. Extraits de lettres de M.Ch. Hermite à M. Jacobi sur différents objets de la théorie des nombres. *J. reine angew. Math.*, 40:279–290, 1850.

[87] Ch. Hermite. Première lettre à M. Jacobi. In *Ouvres*, volume I, pages 100–121. 1850.

[88] R. Hill and E. Kolev. A survey of recent results on optimal linear codes. In F.C. Holroyd et al. (eds.), *Combinatorial designs and their applications. Proceedings of the one-day conference, Milton Keynes, UK, 19 March 1997. London*, pages 127–152. Chapman & Hall/CRC. Chapman & Hall/CRC Res. Notes Math. 403, 1999.

[89] J.W.P. Hirschfeld. *Projective geometries over finite fields*. Oxford Mathematical Monographs. The Clarendon Press Oxford University Press, New York, second edition, 1998.

[90] A. Hocquenghem. Codes correcteurs d'erreurs. *Chiffres, Revue Assoc. franc. Calcul*, 2:147–156, 1959.

[91] D.F. Holt, B. Eick, and E. O'Brien. *Handbook of Computational Group Theory*. Chapmann and Hall / CRC, Boca Raton, Florida, 2005.

[92] W.C. Huffman. The automorphism groups of the generalized quadratic residue codes. *IEEE Trans. Inform. Theory*, 41(2):378–386, 1995.

[93] W.C. Huffman. Codes and groups. In *Handbook of coding theory, Vol. II*, pages 1345–1440. North-Holland, Amsterdam, 1998.

[94] W.C. Huffman and V. Pless. *Fundamentals of error-correcting codes*. Cambridge University Press, Cambridge, 2003.

[95] International Electrotechnical Commission, Geneva. *IEC 60908, Audio recording, Compact disc digital audio system*, second edition, 1999. *http://www.iec.ch/*.

[96] International Organization for Standardization (ISO), Geneva. *ISO/IEC 10149, Information technology – Data interchange on read-only 120 mm optical data disks (CD-ROM)*, 2001. *http://www.iso.org/*.

[97] K. Ireland and M. Rosen. *A classical introduction to modern number theory*, volume 84 of *Graduate Texts in Mathematics*. Springer-Verlag, New York, second edition, 1990.

[98] *http://www.isbn.org/*.

[99] *http://www.issn.org/*.

[100] N. Jacobson. *Basic algebra II*. W.H. Freeman and Company, San Francisco, 1980.

[101] N. Jacobson. *Basic algebra I*. W.H. Freeman and Company, New York, second edition, 1985.

[102] S.A. Jennings. The structure of the group ring of a p-group over a modular field. *Trans. Amer. Math. Soc.*, 50:175–185, 1941.

[103] A.J. Jerri. The Shannon sampling theorem. Its various extensions and applications: A tutorial review. *Proc. IEEE*, 65:1565–1596, 1977.

[104] D. Jungnickel. *Codierungstheorie*. Spektrum Akademischer Verlag GmbH, Heidelberg, 1995. ISBN 3-86025-432-4.

[105] M. Kaib and H. Ritter. Block reduction for arbitrary norms. Preprint, Universität Frankfurt, 1995.

[106] R. Kannan. Algorithmic geometry of numbers. *Annual Review of Computer Science*, 2:231–267, 1987.

[107] R. Kannan. Minkowski's convex body theorem and integer programming. *Math. Operations Research*, 12:415–440, 1987.

[108] A. Kerber. *Applied Finite Group Actions*, volume 19 of *Algorithms and Combinatorics*. Springer-Verlag, Berlin, Heidelberg, New York, 1999. ISBN 3-540-65941-2.

[109] A. Kerber and A. Kohnert. Modular irreducible representations of the symmetric group as linear codes. *European J. Comb.*, 25:1285–1299, 2004.

[110]   D.E. Knuth. *The Art of Computer Programming,* Vol. 2: *Seminumerical Algorithms.* Addison-Wesley, Reading, Mass., 1969.

[111]   D.E. Knuth. *The Art Of Computer Programming.* Vol. 4, Fascicle 2: *Generating all tuples and Permutations.* Addison-Wesley, Reading, Mass., 2005. ISBN 0-201-85393-0.

[112]   A. Korkine and G. Zolotareff. Sur les formes quadratiques positives ternaires. *Math. Ann.,* 5:581–583, 1872.

[113]   A. Korkine and G. Zolotareff. Sur les formes quadratiques. *Math. Ann.,* 6:366–389, 1873.

[114]   V.A. Kotel'nikov. *On the transmission capacity of 'ether' and wire in electrocommunications.* Izd. Red. Upr. Svyazi RKKA, 1933.

[115]   R. Kötter. A unified description of an error locating procedure for linear codes. In *Proc. ACCT-3, Voneshta Voda,* 1992.

[116]   D.L. Kreher and D.R. Stinson. *Combinatorial Algorithms.* CRC Press, Boca Raton, 1998.

[117]   J.P.S. Kung. The cycle structure of a linear transformation over a finite field. *Linear Algebra Appl.,* 36:141–155, 1981.

[118]   J.C. Lagarias, H.W. Lenstra Jr., and C.P. Schnorr. Korkin-Zolotarev bases and successive minima of a lattice and its reciprocal lattice. *Combinatorica,* 10:333–348, 1990.

[119]   J.C. Lagarias and A.M. Odlyzko. Solving low-density subset sum problems. *J. Assoc. Comp. Mach.,* 32:229–246, 1985. Appeared already in Proc. 24th IEEE Symp. Found. Comp. Sci. (1983), 1–10.

[120]   L.J. Lagrange. Recherches d'arithmétique. *Nouv. Mém. Acad. Roy. Sc. Beles Lettres, Berlin,* pages 265–312, 1773. Oeuvres 3, 693–758.

[121]   P. Landrock and O. Manz. Classical codes as ideals in group algebras. *Des. Codes Cryptogr.,* 2(3):273–285, 1992.

[122]   A.M. Legendre. *Essai sur la théorie des nombres.* Chez Duprat, Paris, 1798.

[123]   W. Lehmann. Das Abzähltheorem der Exponentialgruppe in gewichteter Form. *Mitt. math. Sem. Giessen,* 112:19–33, 1974.

[124]   W. Lehmann. *Ein vereinheitlichender Ansatz für die REDFIELD – PÓLYA – de BRUIJNSCHE Abzähltheorie.* PhD thesis, Universität Giessen, 1976.

[125]   A.K. Lenstra, H.W. Lenstra Jr., and L. Lovász. Factoring polynomials with rational coefficients. *Math. Ann.,* 261:515–534, 1982.

[126]   H.W. Lenstra Jr. Finding isomorphisms between finite fields. *Math. Comp.,* 56:329–347, 1991.

[127] H.W. Lenstra Jr. and R.J. Schoof. Primitive normal bases for finite fields. *Math. Comp.*, 48:217–231, 1987.

[128] J.S. Leon. Computing automorphism groups of combinatorial objects. In *Computational group theory (Durham, 1982)*, pages 321–335. Academic Press, London, 1984.

[129] J.S. Leon. Permutation group algorithms based on partitions. I. Theory and algorithms. *J. Symbolic Comput.*, 12(4-5):533–583, 1991. Computational group theory, Part 2.

[130] J.S. Leon. Partitions, refinements, and permutation group computation. In *Groups and computation, II (New Brunswick, NJ, 1995)*, volume 28 of *DIMACS Ser. Discrete Math. Theoret. Comput. Sci.*, pages 123–158. Amer. Math. Soc., Providence, RI, 1997.

[131] R. Lidl and H. Niederreiter. *Finite Fields*, volume 20 of *Encyclopedia of Mathematics and its Applications*. Addison-Wesley Publishing Company, London, Amsterdam, Don Mills – Ontario, Sydney, Tokyo, 1983. ISBN 0-201-13519-1.

[132] R.A. Liebler. On codes in the natural representations of the symmetric group. In T.V. Narayana et al. (eds.), *Combinatorics, representation theory and statistical methods in groups, Young Day Proc.*, volume 57 of *Lect. Notes pure appl. Math.*, pages 159–170. Marcel Dekker Inc., 1980.

[133] R.A. Liebler and K.-H. Zimmermann. Combinatorial S_n-modules as codes. *J. Algebr. Comb.*, 4(1):47–68, 1995.

[134] S. Lin and D.J. Costello. *Error Control Coding: Fundamentals and Applications*. Prentice Hall, Inc., Englewood Cliffs, N.J. 07632, 1983. ISBN 0-13-283796-X.

[135] S. Lin and E.J. Weldon. Further results on cyclic product codes. *IEEE Trans. Information Theory*, IT-16:453–459, 1970.

[136] H. Lüneburg. *On the rational normal form of endomorphisms. A primer to constructive algebra*. B.I.-Wissenschaftsverlag, Mannheim, Wien, Zürich, 1987.

[137] F.J. MacWilliams. A theorem on the distribution of weights in a systematic code. *Bell System Tech. J.*, 42:79–94, 1963. Also reprinted in [185] pp. 261–265 and [20] pp. 241–257.

[138] F.J. MacWilliams. Permutation decoding of systematic codes. *Bell System Tech. J.*, 43:485–505, 1964.

[139] F.J. MacWilliams and N.J.A. Sloane. *The theory of error-correcting codes*, volume 16 of *North-Holland Mathematical Library*. North-Holland Publishing Co., Amsterdam, 1977.

[140] Magma. The Computational Algebra Group within the School of Mathematics and Statistics of the University of Sydney, 2004.

[141] R.J. Marks II (ed.). *Advanced Topics in Shannon Sampling and Interpolation Theory*. Springer-Verlag, Berlin, 1993.

[142] R. Martí and E. Nart. Isometry classes of codes arising from sets of points in the projective plane. *European Journal of Combinatorics*, 15:1003–1023, 2004.

[143] T. Maruta. A characterization of some minihypers and its application to linear codes. *Geom. Dedicata*, 74(3):305–311, 1999.

[144] J.L. Massey. *Threshold Decoding*. M.I.T. Press, Cambridge, MA., 1963.

[145] H.F. Mattson and G. Solomon. A new treatment of Bose-Chaudhuri codes. *J. Soc. Indust. Appl. Math.*, 9:654–669, 1961.

[146] B.D. McKay. Isomorph-free exhaustive generation. *J. Algorithms*, 26(2):306–324, 1998.

[147] D. Micciancio and S. Goldwasser. *Complexity of Lattice Problems*. Kluwer Academic Publishers, 2002.

[148] J.W. Milnor and D. Husemoller. *Symmetric Bilinear Forms*. Springer-Verlag, Berlin, New York, 1973.

[149] H. Minkowski. Extrait d'une lettre addrésse à M. Hermite. *Bulletin des Sciences Mathématique (2)*, 17:24–29, 1893. Reprinted in [152], 266–270.

[150] H. Minkowski. *Geometrie der Zahlen*. Teubner, Leipzig, 1896.

[151] H. Minkowski. Diskontinuitätsbereich für arithmetische Äquivalenz. *J. reine angew. Math.*, 129:220–274, 1905. Reprinted in [152], 53–100.

[152] H. Minkowski. Extrait d'une lettre addrésse à M. Hermite. In D. Hilbert (ed.), *Gesammelte Abhandlungen*, volume I. Teubner, Leipzig, 1911. Reprinted Chelsea, New York, 1967.

[153] T.K. Moon. *Error correction coding : mathematical methods and algorithms*. Wiley-Interscience, Hoboken, N.J., 2005.

[154] D.E. Muller. Applications of boolean algebra to switching circuit design and to error correction. *IRE Trans. Elec. Comp.*, 3:6–12, 1954.

[155] W. Müller. *Lineare Algebra*, volume 42. Bayreuther Mathematische Schriften, second edition, 1992.

[156] H. Nakajima. *Digital Audio Technology*. TAB Books Inc., Blue Ridge Summit, Pa., 1983. ISBN 0-8306-1451-6.

[157] G. Nebe, E.M. Rains, and N.J.A. Sloane. *Self-Dual Codes and Invariant Theory*. Springer-Verlag, Berlin, 2006.

[158] P.M. Neumann, G.A. Stoy, and E.C. Thompson. *Groups and geometry*. Oxford Science Publications. The Clarendon Press Oxford University Press, New York, 1994.

[159] H.J. Nussbaumer. *Fast Fourier transform and convolution algorithms*, volume 2 of *Springer Series in Information Sciences*. Springer-Verlag, Berlin, Heidelberg, New York, 1981.

[160] H. Nyquist. Certain topics in telegraph transmission theory. *Transactions of the American Institute of Electrical Engineers*, 47(2):617–644, 1928.

[161] R. Pellikaan. On the efficient decoding of algebraic-geometric codes. In P. Camion et al. (eds.), *Eurocode 1992. International symposium on coding theory and applications, Udine, Italy, October 23 – 30, 1992.*, volume 339 of *CISM Courses Lect.*, pages 231–253, Wien, 1993. Springer-Verlag.

[162] W. Plesken. Counting with groups and rings. *J. Reine Angew. Math.*, 334:40–68, 1982.

[163] V.S. Pless, W.C. Huffman, and R.A. Brualdi (eds.). *Handbook of coding theory. Vol. I, II*. North-Holland, Amsterdam, 1998.

[164] K.C. Pohlmann. *The Compact Disc Handbook*. Oxford University Press, Oxford, New York, Toronto, second edition, 1992. ISBN 0-19-816327-4.

[165] R.C. Read. Every one a winner or how to avoid isomorphism search when cataloguing combinatorial configurations. *Ann. Discrete Math.*, 2:107–120, 1978. Algorithmic aspects of combinatorics (Conf., Vancouver Island, B.C., 1976).

[166] *http://searchstorage.techtarget.com/sDefinition/0,,sid5_gci503642,00.html*.

[167] I.S. Reed. A class of multiple-error correcting codes and the decoding scheme. *IEEE Trans. Inf. Theory*, 4:38–49, 1954.

[168] I.S. Reed and G. Solomon. Polynomial codes over certain finite fields. *J. Soc. Indust. Appl. Math.*, 8:300–304, 1960.

[169] S.H. Reiger. Codes for the correction of "clustered" errors. *Trans. IRE*, IT-6:16–21, 1960.

[170] H. Ritter. *Aufzählung von kurzen Gittervektoren in allgemeiner Norm*. PhD thesis, Universität Frankfurt, 1997.

[171] B. Schmalz. *t*-Designs zu vorgegebener Automorphismengruppe. *Bayreuth. Math. Schr.*, 41:1–164, 1992. Dissertation, Universität Bayreuth, Bayreuth, 1992.

[172] C.P. Schnorr. A hierachy of polynomial time lattice basis reduction algorithms. *Theoretical Computer Science*, 53:201–224, 1987.

[173] C.P. Schnorr and M. Euchner. Lattice basis reduction: Improved practical algorithms and solving subset sum problems. In *Proceedings of Fundamentals of Computation Theory '91, Lecture Notes in Computer Science 529*, pages 68–85, Heidelberg, 1991. Springer-Verlag.

[174] C.P. Schnorr and H.H. Hörner. Attacking the Chor-Rivest cryptosystem by improved lattice reduction. In *Advances in Cryptology – Eurocrypt '95, Lecture Notes in Computer Science 921*, pages 1–12, Heidelberg, 1995. Springer-Verlag.

[175] *http://searchnetworking.techtarget.com/*.

[176] L.A. Seeber. Untersuchungen über die Eigenschaften der positiven ternären quadratischen Formen. Freiburg, 1831.

[177] Á. Seress. *Permutation Group Algorithms*, volume 152 of *Cambridge Tracts in Mathematics*. Cambridge University Press, Cambridge, UK, 2003.

[178] C.E. Shannon. A mathematical theory of communication. *Bell System Tech. J.*, 27:379–423, 1948.

[179] C.E. Shannon. Communication in the presence of noise. *Proc. Institute of Radio Engineers*, 37(1):10–21, 1949.

[180] C.E. Shannon. *The mathematical theory of communication*. University of Illinois Press, Urbana, 1949. ISBN 0-252-72548-4.

[181] C.C. Sims. Computational methods in the study of permutation groups. In *Computational Problems in Abstract Algebra (Proc. Conf., Oxford, 1967)*, pages 169–183. Pergamon, Oxford, 1970.

[182] C.C. Sims. Computation with permutation groups. In *Proc. Second Symposium on Symbolic and Algebraic Manipulation*, pages 23–28. ACM Press, New York, 1971.

[183] D. Singmaster. *Notes on Rubik's magic cube*. Enslow Publishers, Hillside, N.J., fifth edition, 1981.

[184] D. Slepian. Some Further Theory of Group Codes. *Bell System Tech. J.*, 39:1219–1252, 1960. Also reprinted in [20] pp. 118–151.

[185] D. Slepian (ed.). *Key papers in the development of information theory*. IEEE Press [Institute of Electrical and Electronics Engineers, Inc.], New York, 1974. IEEE Press Selected Reprint Series.

[186] G. Solomon and J.J. Stiffler. Algebraically punctured cyclic codes. *Information and Control*, 8:170–179, 1965.

[187] H. Stichtenoth. *Algebraic function fields and codes*. Universitext. Springer-Verlag, Berlin, 1993.

[188] Y. Sugiyama, M. Kasahara, S. Hirasawa, and T. Namekawa. Correction to "An erasures-and -errors decoding algorithm for Goppa codes" (IEEE Trans. Information Theory **it-22** (1976), no. 2, 238–241). *IEEE Trans. Information Theory*, IT-22(6):765, 1976.

[189] Y. Sugiyama, M. Kasahara, S. Hirasawa, and T. Namekawa. An erasures-and-errors decoding algorithm for Goppa codes. *IEEE Trans. Information Theory*, IT-22(2):238–241, 1976.

[190] SYMMETRICA. A program system devoted to representation theory, invariant theory and combinatorics of finite symmetric groups and related classes of groups. Lehrstuhl II für Mathematik, Universität Bayreuth, 95440 Bayreuth, *http://www.symmetrica.de*.

[191] A. Tietävänien. On the nonexistence of perfect codes over finite fields. *SIAM J. appl. Math.*, 24:88–96, 1973.

[192] J.H. van Lint. *Introduction to coding theory*, volume 86 of *Graduate Texts in Mathematics*. Springer-Verlag, Berlin, third edition, 1999.

[193] J.H. van Lint and G. van der Geer. *Introduction to coding theory and algebraic geometry*, volume 12 of *DMV Seminar*. Birkhäuser Verlag, Basel, 1988.

[194] R.R. Varshamov. Estimate of the number of signals in error correcting codes. *Dokl. Akad. Nauk SSSR*, 117:739–741, 1957. English translation in [20] pp. 68–71.

[195] T. Verhoeff. An updated table of minimum-distance bounds for binary linear codes. *IEEE Trans. Inf. Theory*, 33:665–680, 1987.

[196] J.L. Walker. *Codes and curves*, volume 7 of *Student Mathematical Library*. American Mathematical Society, Providence, RI, 2000. IAS/Park City Mathematical Subseries.

[197] H.N. Ward. Visible codes. *Arch. Math. (Basel)*, 54(3):307–312, 1990.

[198] A. Wassermann. Finding simple t-designs with enumeration techniques. *J. Combinatorial Designs*, 6:79–90, 1998.

[199] A. Wassermann. Lattice point enumeration and applications. *Bayreuther Math. Schr.*, 73, 2006.

[200] E.T. Whittaker. On the functions which are represented by the expansion of interpolating theory. *Proc. Roy. Soc. Edinburgh*, 35:181–194, 1915.

[201] J.M. Whittaker. *Interpolatory Function Theory*, volume 33 of *Cambridge Tracts in Mathematics and Mathematics Physics*. Cambridge Univ. Press, Cambridge, 1935.

[202] *http://en.wikipedia.org/*.

[203] M. Wirtz. *Konstruktion und Tabellen linearer Codes*. PhD thesis, Universität Münster, 1991.

[204] E. Witt. Die 5-fach transitiven Gruppen von Mathieu. *Abh. Math. Semin. Univ. Hamburg*, 12:256–264, 1938.

[205] K.-H. Zimmermann. Beiträge zur algebraischen Codierungstheorie mittels modularer Darstellungstheorie. (Contributions to algebraic coding theory using modular representation theory). *Bayreuther Math. Schr.*, 48, 1994.

[206] K.-H. Zimmermann. On weight spaces of polynomial representations of the general linear group as linear codes. *J. Comb. Theory, Ser. A*, 67(1):1–22, 1994.

[207] V.A. Zinovjev and V.K. Leontjev. On perfect codes. *Problems of Info. Trans.*, 8:17–24, 1972.

Index

Printing: Krips bv, Meppel
Binding: Stürtz, Würzburg